Figures available in three downloadable sizes (resolutions)

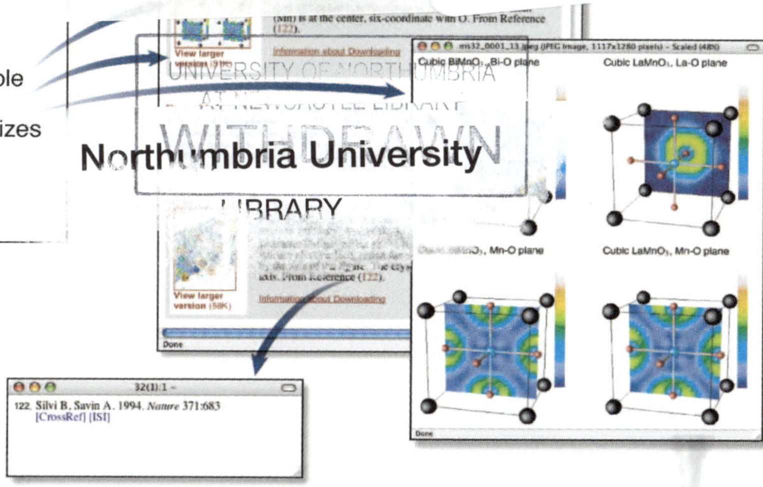

Citations in text link to references in bibliography

References in Annual Reviews chapter bibliography link out to sources of cited articles online

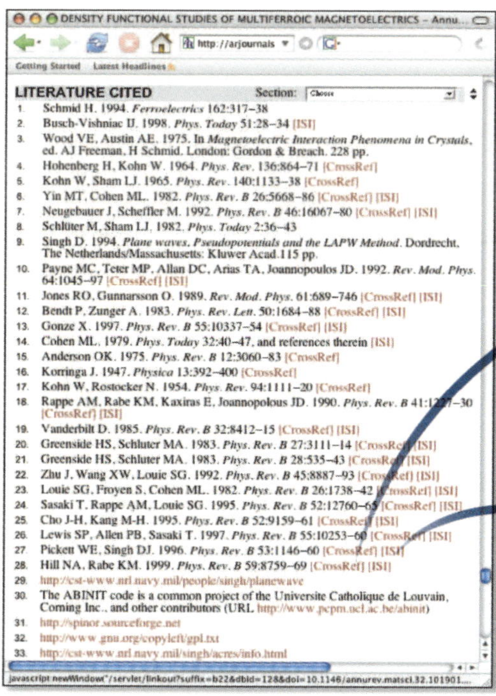

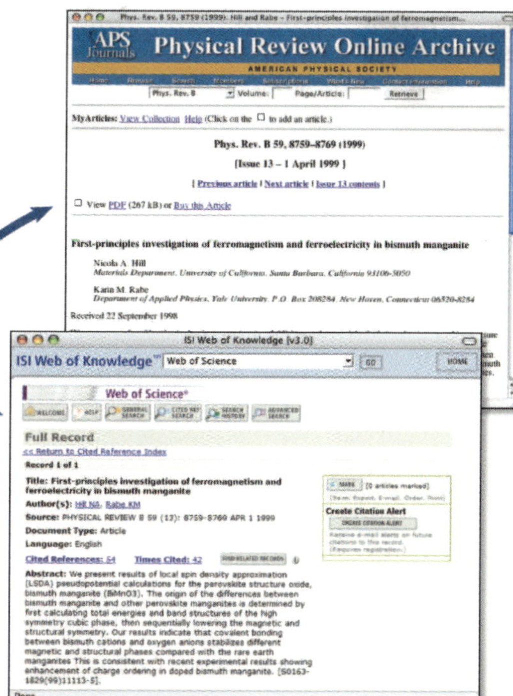

Annual Review of Materials Research

Production Editor: SANDRA H. COOPERMAN
Bibliographic Quality Control: MARY A. GLASS
Electronic Content Coordinator: SUZANNE K. MOSES
Subject Indexer: BRUCE TRACY

ANNUAL REVIEW OF MATERIALS RESEARCH

VOLUME 35, 2005

SIR JOHN M. THOMAS, *Guest Editor*
The Royal Institute of Great Britain

PRATIBHA L. GAI, *Guest Editor*
DuPont and the University of Delaware

DAVID R. CLARKE, *Co-Editor*
University of California, Santa Barbara

MANFRED RÜHLE, *Co-Editor*
Max-Planck Institute, Stuttgart

JOHN C. BRAVMAN, *Associate Editor*
Stanford University

www.annualreviews.org science@annualreviews.org 650-493-4400

ANNUAL REVIEWS
4139 El Camino Way • P.O. Box 10139 • Palo Alto, California 94303-0139

ANNUAL REVIEWS
Palo Alto, California, USA

International Standard Serial Number: 1531-7331
International Standard Book Number: 0-8243-1735-1
Library of Congress Catalog Card Number: 72-172108

∞ The paper used in this publication meets the minimum requirements of American National Standards for Information Sciences—Permanence of Paper for Printed Library Materials, ANSI Z39.48-1992.

TYPESET BY TECHBOOKS, FAIRFAX, VA
PRINTED AND BOUND BY MALLOY INCORPORATED, ANN ARBOR, MI

PREFACE

This volume of the *Annual Reviews of Materials Research* once again illustrates the exceptional breadth of materials research today. In addition to a collection of reviews on various materials aspects of the keynote topic, Catalysis, we have a series of general reviews on topics ranging from composites for windpower turbines, to modern methods of analytical electron microscopy, to molecular approaches to adhesion of metal/polymer interfaces.

The collection of topical reviews on the subject of Catalysis has been organized by our guest editors, Sir John M. Thomas and Pratibha L. Gai. Both are recognized authorities, as well as leading proponents, of the new scientific and technology opportunities offered by new catalysts, and they have assembled a notable collection of reviews that represent many of the exciting advances being made today. We are grateful to all the authors for accepting the challenge of presenting their perspectives of the topics they have reviewed.

Three of the general reviews present recent advances facilitated by developments in microscopy: analytical electron microscopy by Sigle, and the imaging with chromatic aberration corrected scanning transmission electron microscopy by Pennycook and co-workers. Together they illustrate how advances in materials are being paced by developments in instrumentation. A radically new method of establishing diffusional paths in multiple component solid systems and extracting physical data as a function of alloy composition is described by Zhao. This innovative approach enables phase diagram information to be obtained in a "parallel" rather than the traditional "serial" experimentation manner. Combined with recent developments in several measurement techniques, such as nano-indentation, this promises more rapid structural materials characterization.

The power of computational methods is graphically illustrated in the review by Finnis and colleagues, who have studied the early stages of oxidation of NiAl, a subject beyond the scope of first principles calculations until very recently. Similarly, substantial progress is being made in understanding the molecular basis of adhesion and failure of polymer/metal interfaces as described by Grundmeier and Stratmann. New insights are also being gained by combining theory and experiments in two more traditional areas of materials: quasicrystals as described by Gratias and the phenomenon of grain growth in polycrystalline materials as described by Rohrer. In many areas of materials today, technological need drives materials research. This is rarely better illustrated than in the development of windpower turbines in a review by Brøndsted. Although composite materials are now a familiar part of everyday life, the new windpower turbines on the drawing board require composite materials and structures of a size that was unimaginable only a few years ago—that is, the size of a football field!

The passage from a set of suggestions of possible topics and the names of prospective authors proposed at an Editorial Committee meeting all the way to a bound volume of reviews is lengthy and can be tortuous. The appearance of this volume has been possible only because of the care and constant attention of Sandra Cooperman, our production editor. This is the last volume of our series that Sandra has been shepherding, and its appearance is once again a lasting testament to her commitment to the series. We wish her well in her "retirement" and in the new activities she is embarking upon. Thank you, Sandra.

DAVID R. CLARKE
MANFRED RÜHLE
Co-Editors

Annual Review of Materials Research
Volume 35, 2005

Contents

Indexes

Errata

An online log of corrections to *Annual Review of Materials Research* chapters may be found at http://matsci.annualreviews.org/errata.shtml

Related Articles

From the ***Annual Review of Biomedical Engineering***, Volume 6 (2004)

Biomaterials: Where We Have Been and Where We Are Going, Buddy D. Ratner and Stephanie J. Bryant

Tissue Engineering of Ligaments, G. Vunjak-Novakovic, Gregory Altman, Rebecca Horan, and David L. Kaplan

Advances in High-Field Magnetic Resonance Imaging, Xiaoping Hu and David G. Norris

Micro-Computed Tomography—Current Status and Developments, Erik L. Ritman

Ocular Biomechanics and Biotransport, C. Ross Ethier, Mark Johnson, and Jeff Ruberti

From the ***Annual Review of Physical Chemistry***, Volume 56 (2005)

Ultrafast Chemistry: Using Time-Resolved Vibrational Spectroscopy for Interrogation of Structural Dynamics, Erik T.J. Nibbering, Henk Fidder, and Ehud Pines

Microfluidic Tools for Studying the Specific Binding, Adsorption, and Displacement of Proteins at Interfaces, Matthew A. Holden and Paul S. Cremer

Fourier Transform Infrared Vibrational Spectroscopic Imaging: Integrating Microscopy and Molecular Recognition, Ira W. Levin and Rohit Bhargava

Transport Spectroscopy of Chemical Nanostructures: The Case of Metallic Single-Walled Carbon Nanotubes, Wenjie Liang, Marc Bockrath, and Hongkun Park

Metal to Insulator Transitions in Clusters, Bernd von Issendorff and Ori Cheshnovsky

Ultrafast Electron Transfer at the Molecule-Semiconductor Nanoparticle Interface, Neil A. Anderson and Tianquan Lian

ANNUAL REVIEWS is a nonprofit scientific publisher established to promote the advancement of the sciences. Beginning in 1932 with the *Annual Review of Biochemistry*, the Company has pursued as its principal function the publication of high-quality, reasonably priced *Annual Review* volumes. The volumes are organized by Editors and Editorial Committees who invite qualified authors to contribute critical articles reviewing significant developments within each major discipline. The Editor-in-Chief invites those interested in serving as future Editorial Committee members to communicate directly with him. Annual Reviews is administered by a Board of Directors, whose members serve without compensation.

Annual Reviews of

Anthropology
Astronomy and Astrophysics
Biochemistry
Biomedical Engineering
Biophysics and Biomolecular Structure
Cell and Developmental Biology
Clinical Psychology
Earth and Planetary Sciences
Ecology, Evolution, and Systematics
Entomology
Environment and Resources
Fluid Mechanics
Genetics
Genomics and Human Genetics
Immunology
Law and Social Science
Materials Research
Medicine
Microbiology
Neuroscience
Nuclear and Particle Science
Nutrition
Pathology: Mechanisms of Disease
Pharmacology and Toxicology
Physical Chemistry
Physiology
Phytopathology
Plant Biology
Political Science
Psychology
Public Health
Sociology

SPECIAL PUBLICATIONS
Excitement and Fascination of Science, Vols. 1, 2, 3, and 4

The Materials Design and Chemistry of Environmentally Acceptable Catalysts

Long before the use of terms such as nanoscience and nanotechnology became pervasive in the vocabulary of pure and applied scientists, the science and technology of catalysts had involved the preparation, characterization, and utilization of materials that had reached the highest degrees of subdivision—from the atomic to the molecular to the supramolecular. An important, industrially relevant example is the cracking catalyst used extensively for the past 40 years in the manufacture of gasoline and a range of other petrochemical products. Here, metals such as platinum are believed to be dispersed atomically or as 2 to 10 atom clusters (often bimetallic) within the pores of thermally and hydrothermally stable zeolite (aluminosilicate) supports. Degrees of dispersion of this magnitude called for novel, powerful methods of characterization so that techniques such as electron microscopy and X-ray absorption spectra, which figure eminently in this volume, were invoked early on by the community of materials scientists engaged in the development of new, and the improvement of existing, solid catalysts.

Nowadays, with the phenomenon of catalysis looming ever larger in the development of new technologies required for the economic production of energy and fuel, as well as in the crusade to achieve sustainability (while complying with the environmentally desirable features of green chemistry and clean technology), materials aspects of solid and other catalysts are especially important. The choice of contributors to this keynote topic reflects these facts.

On the preparation front, we have a synoptic account by Schüth on engineered porous materials. The chapters by Thomas and Raja and by Hutchings place special emphasis on so-called mesoporous solids (pore diameters ranging from 15 to 500 Å) that have great potential as the basis of new families of catalysts. Hutchings highlights the various strategies, using both mesoporous and microporous solids (with diameter of pores from 3.5 to 15 Å), for effective asymmetric conversions, a topic of enormous commercial significance at present, particularly in the pharmaceutical and agrochemical industries.

Microporous aluminophosphates—as described here by Pastore, Coluccia, and Marchese—are also promising materials for the design of new solid-acid catalysts, a topic of great importance in the quest to achieve environmentally responsible new industrial processes. (There will doubtless come a time when ecologically unacceptable acids such as hydrogen fluoride will be banned by legislators and rejected by industrial practitioners.)

The preparation of skeletal catalysts, described here by Smith and Trimm, is of particular value when monometallic catalysts (such as nickel and copper) are employed. Photocatalysis lies at the heart of the work of Anpo and his colleagues, and they show how it becomes possible to confer unusual additional properties such as superhydrophilicity to a desirable photocatalyst such as TiO_2, which possesses admirable optical and unique visible light-responsive properties if it is prepared by advanced methods of ion-implantation.

Over and above considerations of novel selectivity and activity, catalysts, in order to be of real practical value, must also possess longevity. This is particularly true for finely divided metals supported on appropriate oxidic materials (for example, in the car exhaust system or in industrially significant hydrogenation and dehydrogenation reactors). Spencer and Twigg have focused on this area. Such systems lend themselves to some of the techniques described elsewhere in this volume, especially the powerful atomic resolution in situ electron microscopy, which directly probes catalysis at the atomic level, and electron beam–induced analytical spectroscopy techniques employed by Gai and Calvino to investigate catalytic processes. The methods of these workers are valid for hydrocarbon oxidation technology as well as for environmental catalytic technologies, and are of crucial importance in the quest for new sources of energy and new methods of environmental protection.

Of late, more has been heard about ionic liquids as catalysts than ever before. Their potential is outlined here by Hardacre. In particular, he elaborates the application of extended X-ray absorption fine structure (EXAFS) to describe the structure of these relatively new types of catalytic materials. EXAFS is one of the relatively few techniques for characterization that may be used under genuinely realistic in situ reactor conditions; this is well described in several other reviews presented here, especially in that of Tada and Iwasawa. In conjunction with X-ray diffractometry, which may be recorded in parallel with EXAFS data, the techniques yield both long-range and short-range atomic order.

SIR JOHN M. THOMAS
Guest Editor

PRATIBHA L. GAI
Guest Editor

Annu. Rev. Mater. Res. 2005. 35:1–27
doi: 10.1146/annurev.matsci.35.100303.121340

The Preparation and Characterization of Highly Efficient Titanium Oxide–Based Photofunctional Materials

Masakazu Anpo, Satoru Dohshi, Masaaki Kitano, Yun Hu, Masato Takeuchi, and Masaya Matsuoka
Department of Applied Chemistry, Graduate School of Engineering, Osaka Prefecture University, Sakai, Osaka 599-8531, Japan; email: anpo@ok.chem.osakafu-u.ac.jp; dohshi@ok.chem.osakafu-u.ac.jp; kitano@ok.chem.osakafu-u.ac.jp; huyun@ok.chem.osakafu-u.ac.jp; masato-t@ok.chem.osakafu-u.ac.jp; matsumac@ok.chem.osakafu-u.ac.jp

Key Words visible light-responsive TiO_2, metal-ion-implantation method, titanium oxide–based binary catalysts, RF magnetron sputtering, Ti-MCM-41

■ **Abstract** Recent research trends of the preparation and characterization of highly efficient titanium oxide–based photocatalysts are reviewed on the basis of studies done in our laboratory. Special attention is focused on the preparation and characterization of active sites of highly dispersed titanium oxide species within mesoporous matrices and on titanium oxide–based binary catalysts such as Ti/Si and Ti/B oxides and unique visible light-responsive second-generation TiO_2 photocatalysts prepared by an advanced metal-ion-implantation and RF magnetron sputtering deposition method. Highly dispersed titanium oxide species within mesoporous frameworks and titanium oxide–based binary catalysts exhibit high and unique photocatalytic activities for various reactions, such as the direct decomposition of NO_x. These species also have photo-induced superhydrophiric properties. It is of significance that visible light-responsive TiO_2 photocatalysts can be applied to the decomposition of NO_x in air and the degradation of pollutants in water, as well as to the splitting of water under UV and visible light irradiation.

INTRODUCTION

Environmental pollution and damage on a global scale, as well as a shortage of clean and natural energy sources, have drawn much attention to the vital need for ecologically clean chemical technologies, materials, and processes—one of the most urgent challenges facing chemical scientists today. In 1972 Fujishima & Honda (1) reported the photosensitized decomposition of water into H_2 and O_2 using an electrochemical cell consisting of a Pt electrode and a TiO_2 semiconductor electrode. Since that time, photocatalysis, through the use of various powdered semiconductors, has received much attention for its potential in the conversion of light energy into useful chemical energy (2–4). Semiconductor photocatalysis, with its early focus on TiO_2 as an effective photocatalyst, has been applied to a variety of reactions to address the reduction and/or elimination of environmental pollutants in water and air. Photocatalysis has also been shown to be useful for the decomposition of micro-organisms such as bacteria (5) and viruses (6), the deactivation of cancer cells (7, 8), the degradation and elimination of offensive odors (9), the photo-splitting of water to produce H_2 gas (10–15), the fixation of nitrogen (16–19), and the clean up of oil spills (20–22).

Semiconductors such as TiO_2, ZnO, and Fe_2O_3 can act as sensitizers for light-induced redox processes owing to their electronic structure, which is characterized by a filled valence band and an empty conduction band (23). As shown in Figure 1 (see color insert), when the energy of a photon corresponds to or exceeds the band gap energy of the semiconductor, an electron is promoted into the conduction band, leaving a hole in the valence band. The electrons in the conduction band and holes in the valence band can recombine and dissipate the input energy as heat and become trapped in metastable surface states, or they can react with electron acceptors and electron donors adsorbed on the semiconductor surface or within the surrounding electrical double layer of the charged particles.

In the absence of suitable electron and hole scavengers, the stored electrons and holes are dissipated within a few nanoseconds by their recombination (24). If a suitable scavenger or surface defect state is available to trap the electron and hole, recombination is prevented and subsequent redox reactions may occur because of the remaining free electrons and holes. The valence band holes act as powerful oxidants, whereas the conduction band electrons are good reductants (25). Most organic photodegradation reactions can be attributed to the high oxidizing power of the holes either directly or indirectly. However, to prevent a buildup of charge, one must also provide a reducible species such as oxygen to react with the electrons.

Applications of illuminated semiconductors in the remediation of contaminants have been successful for a wide variety of compounds (26–33). In many cases, complete mineralization of the organic compounds has been reported.

Recently, the discovery of photo-induced superhydrophilicity and the self-cleaning effect of TiO_2 thin films has led to even wider applications for TiO_2 photocatalysts (34).

In this review, research on such photofunctional catalysts and photocatalytic systems are introduced: (*a*) design and development of second-generation TiO_2 photocatalysts that can operate under visible light irradiation; (*b*) photocatalytic performance of titanium oxide–based thin film catalysts; (*c*) photocatalytic decomposition of water into H_2 and O_2 using TiO_2 thin film photocatalysts that can operate under visible light irradiation prepared by the radio frequency (RF) magnetron sputtering deposition method; and (*d*) characterizations of the local structures of the active sites of Ti-MCM-41 catalysts and their photocatalytic reactivity for the decomposition of NO into N_2 and O_2.

DESIGN AND DEVELOPMENT OF SECOND-GENERATION TiO_2 PHOTOCATALYSTS

A highly advanced metal-ion-implantation method was applied to modify the electronic properties of bulk TiO_2 photocatalysts by bombarding them with high-energy metal ions. It was discovered that such implantation with various transition metal ions, including V, Cr, Mn, Fe, and Ni, accelerated by high voltage, enabled a large shift in the absorption band of the TiO_2 catalysts toward the visible light region with differing levels of effectiveness. However, Ar, Mg, or Ti ion-implanted TiO_2 exhibited no shift in the absorption spectra, indicating that such a shift is not caused by the high-energy implantation process itself but by some interaction between the transition metal ions and the TiO_2 catalyst. The absorption band of the Cr ion-implanted TiO_2 shifts smoothly toward the visible light region; the extent of the red shift depending on the number of Cr ions, with the absorption maximum and minimum values always remaining constant, as can be seen in Figure *2b–d* (see color insert). The order of effectiveness in the red shift is V > Cr > Mn > Fe > Ni ions. Such a shift allows the metal-ion-implanted titanium oxide to use solar irradiation more effectively with efficiencies in the range of 20–30%.

Furthermore, such red shifts in the absorption band of the metal-ion-implanted TiO_2 catalysts were observed for all types of titanium oxide except the amorphous types, and the extent of the shift differed from sample to sample. It was also found that such shifts in the absorption band could be observed only after calcination of the metal-ion-implanted TiO_2 samples in O_2 at around 723–823 K. Therefore, it is clear that calcination under O_2 atmosphere in combination with metal-ion-implantation is instrumental in the red shift of the absorption spectrum toward visible light regions. These results show that such shifts in the absorption band of the TiO_2 catalysts by metal-ion-implantation are a general phenomenon and not a special feature of certain types of bulk TiO_2 catalysts.

Figure 3 (see color insert) shows the absorption spectra of the TiO_2 catalysts impregnated or chemically doped with large amounts of Cr ions compared with the aborption spectra of the Cr ion-implanted samples. The Cr ion-doped catalysts showed no shift in the absorption edge of TiO_2; however, a new absorption band, formed by the impurity energy level within the band gap, appeared as a shoulder peak at around 420 nm, its intensity increasing with the number of chemically doped Cr ions. These results indicate that the method of doping causes the electronic properties of the TiO_2 catalyst to be modified in completely different ways, thus confirming that only metal-ion-implanted TiO_2 catalysts show such shifts in the absorption band toward visible light regions, even with many fewer ions, compared with those for the chemically doped systems.

With the unimplanted original TiO_2 or with the chemically doped TiO_2 catalysts, the photocatalytic reaction does not proceed under visible light irradiation ($\lambda > 450$ nm). However, visible light irradiation of such advanced metal-ion-implanted TiO_2 catalysts enabled them to initiate various photocatalytic reactions. Visible light irradiation ($\lambda > 450$ nm) of the Cr ion-implanted TiO_2 catalysts in the presence of NO at 275 K led to the decomposition of NO into N_2, O_2, and N_2O with a good linearity against the light irradiation time. Under the same conditions of visible light irradiation, the unimplanted TiO_2 catalyst did not exhibit any photocatalytic reactivity. Moreover, the action spectrum for this reaction on the metal-ion-implanted TiO_2 was in good agreement with the absorption spectrum of the catalyst shown in Figure 2, indicating that only metal-ion-implanted TiO_2 catalysts were effective for the photocatalytic decomposition reaction of NO under visible light irradiation. Thus it was found that metal-ion-implanted TiO_2 catalysts could absorb visible light up to a wavelength of 400–600 nm and could operate effectively as photocatalysts; hence the name second-generation TiO_2 photocatalysts (2, 34–46).

It should be emphasized that the photocatalytic reactivity of the metal-ion-implanted TiO_2 catalysts retained the same photocatalytic efficiency as that of the unimplanted original TiO_2 catalyst under UV light irradiation ($\lambda < 380$ nm). When metal ions were incorporated into the TiO_2 catalyst, the photocatalytic efficiency decreased dramatically, even under UV irradiation, because of the rapid recombination of the photo-formed electrons and holes resulting from the impurity energy levels formed by the doped metal ions within the band gap of the catalyst (Figure 3). These results show that physically implanted metal ions do not work as electron and hole recombination centers but act only to modify the electronic property of the catalyst.

Various fieldwork experiments were conducted to test the photocatalytic reactivity of the newly developed TiO_2 catalysts under solar beam irradiation. Under outdoor solar insolation at ordinary temperatures, the Cr and V ion-implanted TiO_2 catalysts showed photocatalytic reactivity for the decomposition of NO several times higher than that of the unimplanted TiO_2 catalysts. It was also found the V ion-implanted TiO_2 catalysts showed photocatalytic reactivity for the hydrogenation of C_3H_4 with H_2O several times higher than that of the original TiO_2. These results, together with the results shown in Figure 2, indicate that by using metal-ion implanted titanium oxide photocatalysts, the efficient use of visible and solar light energy by photocatalysis is now possible.

The relationship between the depth profiles of equal amounts of V or Cr implanted within TiO_2, and their photocatalytic efficiency under visible light irradiation was investigated. For example, when the same amounts of metal ions were implanted into the deep bulk of the TiO_2 catalyst using a relatively high acceleration energy, the catalyst exhibited high photocatalytic efficiency under visible light irradiation. When low acceleration energies were used, the catalyst exhibited low photocatalytic efficiency under the same conditions of visible light irradiation.

Increasing the number of metal ions implanted into the deep bulk of the TiO_2 catalyst caused the photocatalytic efficiency to increase under visible light irradiation, passing through a maximum at around 6×10^{16} ions/cm^2, then decreasing with a further increase in the number of metal ions implanted. The presence of ions at the near surfaces could be observed by electron spectroscopy chemical analysis (ESCA) measurements only on samples implanted with a large number of metal ions. These results show that there are optimal conditions in the depth and number of metal ions implanted to achieve high photocatalytic reactivity under visible light irradiation.

The electron spin resonance (ESR) spectra of the V ion-implanted TiO_2 catalysts were measured before and after calcination of the samples in O_2 at around 723–823 K, respectively. Distinct and characteristic reticular V^{4+} ions were detected only after calcination at around 723–823 K. When a shift in the absorption band toward visible light regions by the V ion-implantation was observed, the presence of the reticular V^{4+} ions could be detected by ESR measurements. No such V ions having the same local structure or shifts in the absorption band could be observed with TiO_2 catalysts chemically doped with V ions.

The X-ray absorption near-edge structure (XANES) and Fourier transforms of the extended X-ray absorption fine structure (FT-EXAFS) oscillations for the TiO_2 catalysts chemically doped and physically implanted with Cr ions were measured to clarify the local structures of these Cr ions. Results obtained from the analyses of the X-ray absorption fine structure (XAFS) spectra showed that for the TiO_2 catalysts chemically doped with Cr ions by an impregnation or sol-gel method, the ions were present as aggregated Cr oxides in an octahedral coordination similar to Cr_2O_3, and in a tetrahedral coordination similar to CrO_3, respectively. In the catalysts physically implanted with Cr, the ions were present in a highly dispersed and isolated state in an octahedral coordination, suggesting that the Cr ions are incorporated in the lattice positions of the TiO_2 catalyst in place of the Ti ions (2, 34–46).

These findings show that the success in modifying the electronic state of TiO_2 by metal-ion-implantation, which enables the absorption of visible light wavelengths longer than 550 nm, is closely associated with the strong and long-distance interaction that arises between the TiO_2 and the metal ions implanted. The formation of impurity energy levels within the band gap of the catalysts that resulted from the formation of aggregated metal oxide clusters, which often are observed in the chemical doping of metal ions and/or oxides, had no effect (Figure 3).

These results obtained in the photocatalytic reactions and by various spectroscopic measurements of the catalysts indicate that the implanted metal ions are highly dispersed within the deep bulk of the TiO_2 catalysts and work to modify the electronic nature of the catalysts without any changes in the chemical properties of the surfaces. These modifications are closely associated with an improvement in the reactivity and sensitivity of the TiO_2 photocatalyst, thus enabling the TiO_2 catalyst to absorb and operate effectively not only under UV but also under visible light irradiation. As a result, under outdoor solar light irradiation at ordinary temperatures, transition metal-ion-implanted TiO_2 catalysts showed several times higher photocatalytic efficiency than the unimplanted TiO_2 catalysts.

PHOTOCATALYTIC PERFORMANCE OF TITANIUM OXIDE–BASED THIN FILM CATALYSTS

TiO_2 thin films have been widely investigated not only for their high photocatalytic reactivity but also for their unique photo-induced superhydrophilic properties (34, 47–54). Some of the recent main preparation techniques reported for TiO_2 thin films are the sol-gel method and metal oxide chemical vapor deposition (MOCVD) method. However, we also have recently shown that an ionized cluster beam (ICB) deposition method can be applied to produce highly homogeneous and efficient TiO_2 thin film photocatalysts without the need for special treatment or calcination at high temperatures for the crystallization process after deposition (37, 38, 55, 56). In addition, we have also reported that binary oxide thin films of different compositions can be easily prepared by the ICB method using multi-ion sources and controlling each deposition rate (57–59).

Elsewhere, we have reported that the photocatalytic reactivity of Ti/Si binary oxides prepared by a coprecipitation method is strongly affected by changes in the composition of the catalyst, i.e., by changing the Ti/Si ratios (60). Also, Imamura et al. reported that Ti/Si binary oxides prepared by the sol-gel method exhibit a high catalytic activity for the selective epoxidation of alkenes (61). On the other hand, to improve the photo-induced superhydrophilic properties of TiO_2 thin films, a combination or mixing of the Ti-oxide moiety with other oxides, such as SiO_2 or B_2O_3, has been reported. Significantly, the addition of SiO_2 fine particles or the deposition of the SiO_2 thin layer onto the surface of the TiO_2 thin films is shown to be effective in maintaining superhydrophilic properties, even under dark conditions and for long periods after light irradiation has ceased (62).

Ti/Si and Ti/B binary oxide thin films were also prepared by the ICB method using multi-ion sources. Characterization studies were carried out by spectroscopic measurements, such as XRD, XAFS, and UV-Vis, in order to elucidate the local structure of the Ti-oxide species of these thin films. The photocatalytic reactivity was also evaluated for significant reactions such as the decomposition of NO to produce N_2 and O_2, as well as N_2O, a minor product under UV light irradiation. Moreover, the photo-induced superhydrophilic properties were evaluated by measuring the changes in the contact angle of water droplets under UV irradiation and also under dark conditions at 298 K.

The XAFS spectra (XANES and FT-EXAFS) of the Ti/Si and Ti/B binary oxide thin films were measured in order to elucidate the local structure of the Ti-oxide species within these thin films. Figure 4 shows the XAFS spectra of the Ti/Si binary oxide thin films. The XANES spectra of these thin films with low TiO_2 compositions (left in Figure 4) exhibit an intense, single-pre-edge peak that can be attributed to the tetrahedrally coordinated TiO_2 species. Moreover, the FT-EXAFS spectra of the Ti/Si thin films (right in Figure 4) exhibit a strong peak at around 1.6 Å, which can be attributed to the Ti-O peak.

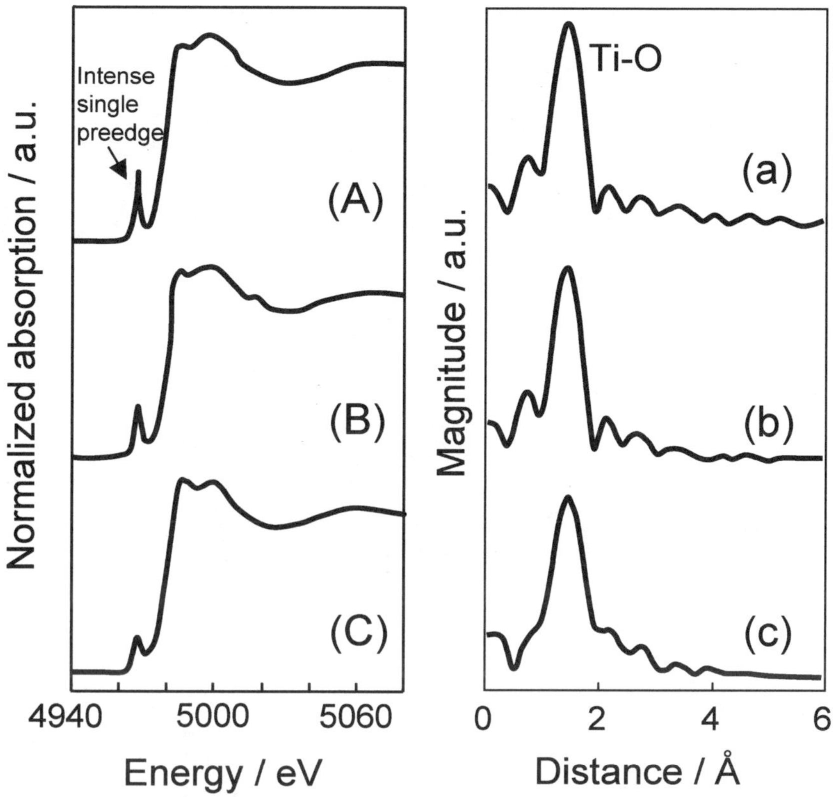

Figure 4 XAFS (XANES and FT-EXAFS) spectra of Ti/Si binary oxide thin films. TiO_2 content (%): (*a*, *A*) 6.6; (*b*, *B*) 9.5; (*c*, *C*) 50.1.

On the other hand, for the Ti/B binary oxide thin films with low TiO_2 compositions, the XANES spectra (left in Figure 5) exhibit three small, characteristic pre-edge peaks that can be attributed to the TiO_2 with an anatase crystalline structure. In the FT-EXAFS spectra for the Ti/B thin films (right in Figure 5), not only the Ti–O peak but also the Ti–O–Ti peak can be seen.

The results of the curve-fitting analysis of the FT-EXAFS spectra to obtain the coordination number and the bond distance between the Ti and O atoms of the Ti-oxide species in the Ti/Si and Ti/B binary oxides are summarized in Table 1. It is clear that the coordination numbers of the Ti/Si binary oxides with lower TiO_2 compositions are close to 4.0, indicating that the major Ti-oxide species are located in a tetrahedral coordination, whereas the coordination numbers for the Ti/B binary oxides are always close to 6.0 even for the Ti/B with low TiO_2 content, showing that the major Ti-oxide moiety is present in an octahedral coordination. These results indicate that for Ti/Si binary oxide thin films with low TiO_2 composition, the tetrahedrally coordinated Ti-oxide species are highly dispersed in

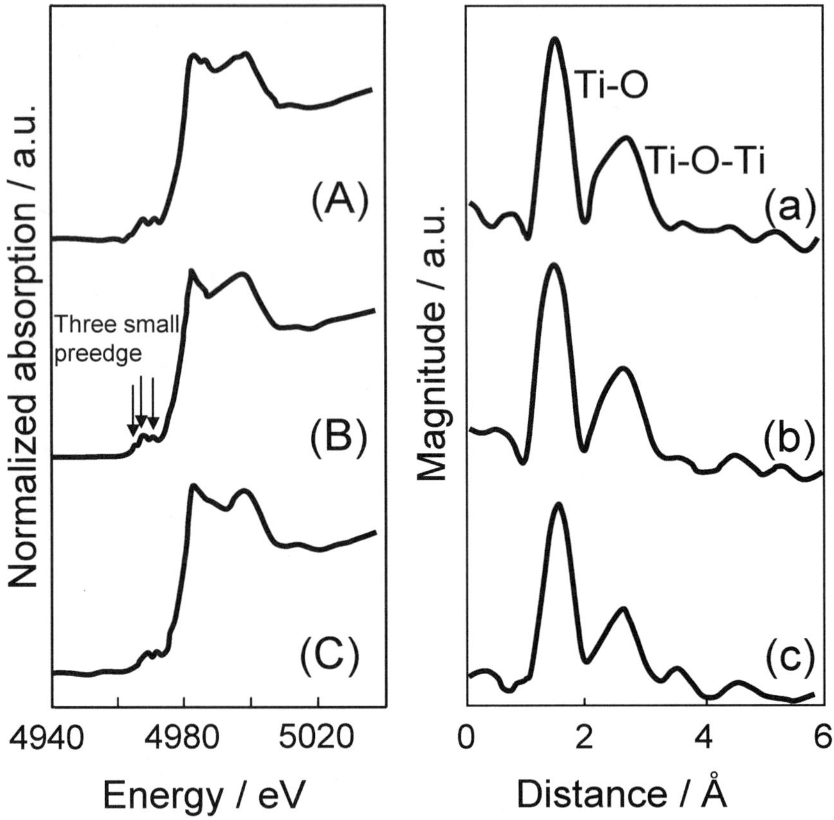

Figure 5 XAFS (XANES and FT-EXAFS) spectra of Ti/B binary oxide thin films. TiO_2 content (%): (*a*, *A*) 5.0; (*b*, *B*) 10.0; (*c*, *C*) 50.0.

the host SiO_2 matrices, whereas for the Ti/B binary oxide thin films with low TiO_2 composition, the octahedrally coordinated ultrafine TiO_2 nanoparticles are formed in the host B_2O_3 matrices. Furthermore, the bond distance between the Ti and O atoms of the Ti-oxide species in the Ti/Si binary oxide thin films varies from 1.81 Å to 1.85 Å, with an increase in the TiO_2 compositions from 6.6 to 50.1 wt%, reflecting the changes in the structures of the Ti-oxides from highly dispersed isolated tetrahedrally coordinated Ti-oxide species to TiO_2 anatase nanoparticles dispersed within the SiO_2 matrices. The bond distance between the Ti and O atoms of the Ti-oxide species in the Ti/B binary oxide thin films scarcely changes and is almost constant at around 1.90 Å, in a good agreement with the fact that Ti-oxides in Ti/B are always present as TiO_2 nanoparticles within the B_2O_3 matrices.

The UV-Vis absorption spectra using a transmittance method for the Ti/Si and Ti/B binary oxide thin films with lower TiO_2 content are shown in Figure 6.

TABLE 1 The results of the curve fitting analysis of the FT-EXAFS data for TiO2 thin films as well as Ti/Si and Ti/B binary oxide thin films

Catalyst	Shell	Bond distance (Å)	Coordination number
Pure TiO	Ti-O	1.91	6.0
Ti/Si (6.6/93.4)	Ti-O	1.81	4.3
Ti/Si (9.5/90.5)	Ti-O	1/82	4.4
Ti/Si (50.1/49.9)	Ti-O	1/85	4.9
Ti/B (5/95)	Ti-O	1.91	5.98
Ti/B (10/90)	Ti-O	1/90	5.97
Ti/B (50/50)	Ti-O	1.90	5.97

For both the Ti/Si and Ti/B binary oxide thin films, when the TiO_2 composition is low, the absorption band of TiO_2 shifts toward shorter wavelength regions. This can be attributed to the presence of the tetrahedrally coordinated Ti-oxide species in the SiO_2 matrices and the quantum size effect owing to the formation of extremely small octahedrally coordinated TiO_2 nanoparticles in the B_2O_3

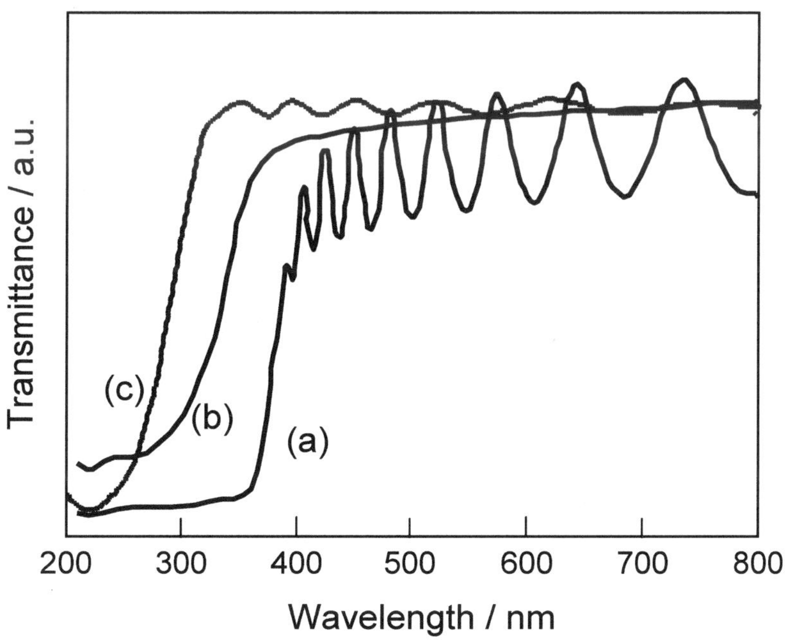

Figure 6 UV-Vis absorption spectra by transmittance method of Ti/Si and Ti/B binary oxide thin films: (*a*) TiO_2, (*b*) Ti/B (5/95), (*c*) Ti/Si (6.6/93.4).

matrices, respectively. It is also found that these binary oxide thin films are highly transparent and colorless and have a high transmittance in the visible light region owing to their highly homogeneous crystallinity and surface morphology. It should also be noted that the interference fringes, which could be remarkably observed for the pure TiO_2 thin film, almost disappear for the Ti/Si and Ti/B binary oxide thin films; in particular for Ti/B compared with TiO_2 thin films with the same film thickness through combination or by mixing with SiO_2 and B_2O_3. This is a great advantage for the application of these thin films on various substrates, significantly, in that they do not lose their original transparency and colorlessness.

UV light irradiation of Ti/Si and Ti/B binary oxide thin films in the presence of NO led to the photocatalytic decomposition of NO into N_2 and O_2, as well as N_2O at 298 K. Figure 7 shows the reaction time profile of the photocatalytic decomposition of NO on the Ti/Si (Ti/Si ratio = 6.6/93.4) binary oxide thin films. The reaction proceeds with a good linearity against the irradiation time, demonstrating that the decomposition reaction proceeds photocatalytically. We found the effect of the TiO_2 composition on the selectivity for the formation of N_2 as

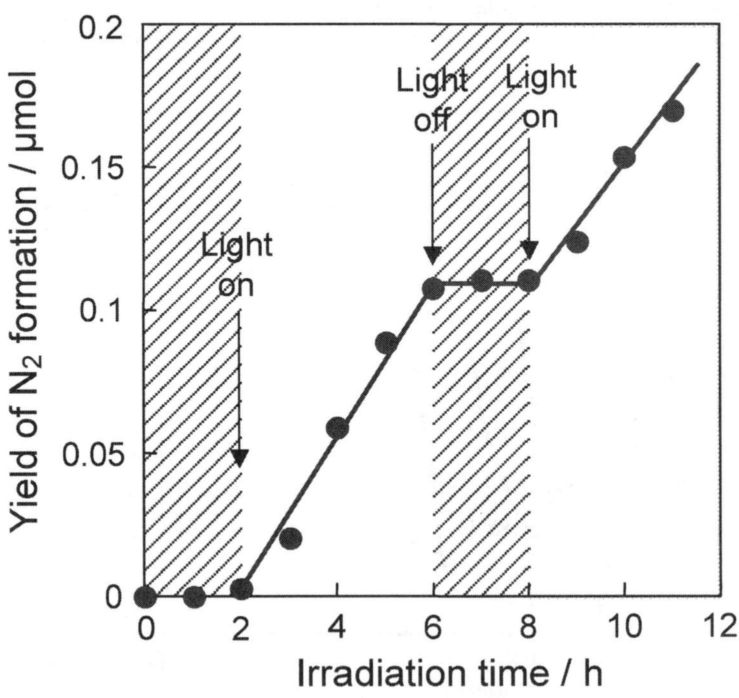

Figure 7 The reaction time profile of the photocatalytic decomposition of NO on Ti/Si binary oxide thin films (TiO_2 content; 6.6%) under UV light irradiation at 275 K.

well as the yields of N_2 and N_2O in the photocatalytic decomposition of NO on Ti/Si. As mentioned above, when the TiO_2 composition is low, the tetrahedrally coordinated Ti-oxide species are formed as the major Ti-oxide species. Upon the excitation of these Ti-oxide species by UV light, the charge transfer excited state of these species, $(Ti^{3+}–O^{-})^*$, is formed. These excited $(Ti^{3+}–O^{-})^*$ species have a high and unique photocatalytic activity for various reactions (63), interacting with the NO molecules to decompose into N_2 and O_2 with high efficiency and selectivity at 298 K. However, when the TiO_2 content is increased, the nanoclusters and even nanoparticles of TiO_2 with octahedral coordination are formed. Their photocatalytic behavior is the same as that of pure TiO_2 semiconducting photocatalysts where UV irradiation induces the formation of electrons in the conduction band and holes in the valence band, respectively, and these species react with NO to lead to the formation of N_2O and NO_2. Thus when the TiO_2 content was increased, the yields and the selectivity for the formation of N_2 decreased. These results are in good agreement with our previous study in which it was found that only highly dispersed tetrahedrally coordinated Ti-oxide species decompose NO into N_2 and O_2, whereas octahedrally aggregated Ti-oxides or nanoparticles decompose NO into N_2O and NO_2 (39, 64, 65).

On the other hand, for the Ti/B binary oxide thin films of low TiO_2 composition, ultrafine TiO_2 nanoparticles of octahedral coordination are formed in the host B_2O_3. Moreover, in contrast to the characteristics for Ti/Si binary oxide systems, UV irradiation of the Ti/B binary oxide thin films with low TiO_2 content in the presence of NO leads to the formation of N_2 and O_2 with a high selectivity (57, 58). Thus in the host B_2O_3, it was shown that highly dispersed octahedrally coordinated Ti-oxide nanoclusters can decompose NO molecules into N_2 and O_2 with high efficiency and selectivity at 298 K (66). Studies are now underway on the true nature and mechanisms behind the photocatalytic activity of highly dispersed octahedrally coordinated Ti-oxide nanoclusters formed in the host B_2O_3.

The contact angle of water droplets for the TiO_2 thin films dramatically decreased under UV light irradiation and finally reach 0°, i.e., highly efficient superhydrophilicity could be achieved. However, for the Ti/Si binary oxide thin films, the contact angle of the water droplets did not change dramatically, whereas for the Ti/B binary oxide thin films (see Figure 8), all of the samples except the pure B_2O_3 achieved superhydrophilicity, reaching a water contact angle of 0° under UV light irradiation. The inserted figure shows the initial decrease rate for the photo-induced hydrophilicity of these binary oxide thin films. The decrease rate in the contact angle for Ti/B was found to be much larger than that for the pure TiO_2 thin films. In particular, for the binary oxide thin films containing more than 90% B_2O_3, the rate was several times larger than for pure TiO_2. Moreover, as soon as UV light irradiation was discontinued, the contact angle of the water droplets for the TiO_2 thin films immediately increased and returned to its initial value, showing that it is a light-induced phenomenon. However, for the Ti/B binary oxide thin films, only half the value could be recovered.

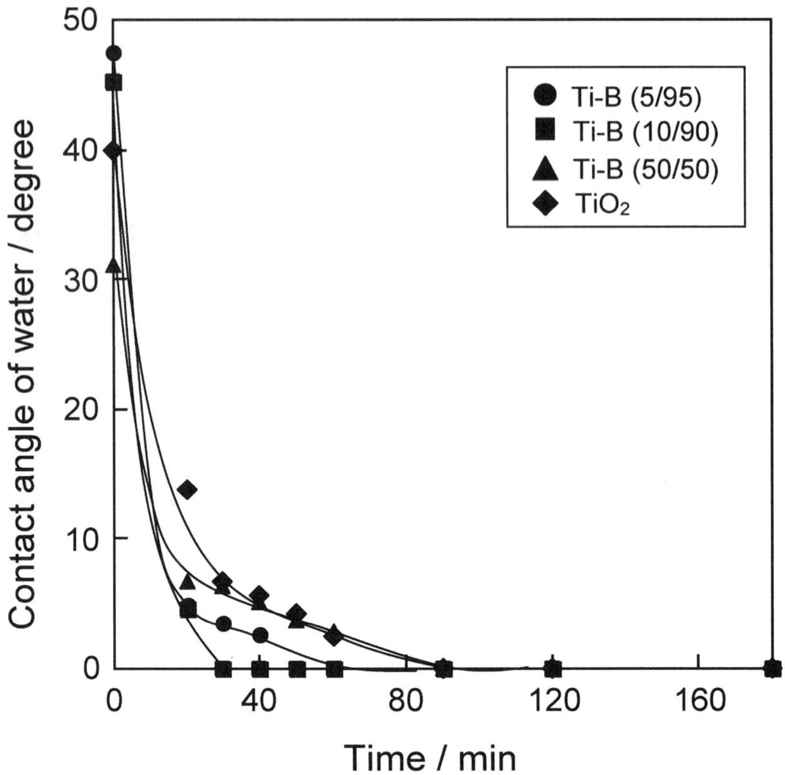

Figure 8 The changes in the contact angle of water droplets under UV light irradiation on Ti/B binary oxide thin films.

PHOTOCATALYTIC DECOMPOSITION OF WATER INTO H_2 AND O_2 USING TiO_2 THIN FILM PHOTOCATALYSTS

Recently, the production of H_2 and O_2 through the photocatalytic decomposition of water using solar energy has received a great deal of attention in studies for the design and development of artificial photosynthesis systems because it would enable the conversion of solar energy into useful chemical energy. Intensive studies of photocatalysis on semiconducting materials were initiated by the pioneering work of Fujishima & Honda (1), which reports on the first photo-assisted production of H_2 and O_2 from water by utilizing a photoelectrochemical cell consisting of a Pt and TiO_2 electrode under a small electric charge. It was then observed that TiO_2 semiconducting nanopowdered photocatalysts loaded with small amounts of Pt could work to decompose H_2O into H_2 and O_2 as a mixture gas under UV light irradiation at wavelengths shorter than 380 nm (67–72). It was also observed that in such powdered photocatalytic systems, the back reaction of the photo-formed

H_2 and O_2 to form the original H_2O could proceed on these Pt catalysts, resulting in low reaction yields (73). In addition to the decomposition reaction of H_2O into H_2 and O_2, various other significant properties of TiO_2 photocatalysts are being exploited. For example, such properties as their photo-induced superhydrophilicity (47, 57) are being investigated for the purification of toxic compounds in polluted water and air (74, 75). In addition to their abundance, TiO_2 photocatalytic materials are also attractive for their nontoxic, clean, safe properties, and thermal stability. Until recently, they have worked only under UV light irradiation of wavelengths shorter than 380 nm. However, new types of TiO_2 photocatalysts, which are able to absorb and operate effectively and efficiently even under visible or solar light, are being developed (40–42, 76, 77).

We were able to modify the electronic properties of TiO_2 photocatalysts by applying an advanced metal-ion-implantation method in which ion-implanted transition metal cations, such as V and Cr, were located at the lattice position to replace the Ti^{4+} of the TiO_2 (43, 44, 75, 78–80). Although this method is both theoretically interesting and successful in actual experiments, it is not practical for mass production owing to the high cost of using such an advanced, high-velocity ion-implantation apparatus to develop these TiO_2 photocatalysts.

RF magnetron sputtering is a novel and effective way to develop visible light-responsive TiO_2 thin film photocatalysts (45, 75). This method has great potential for the practical production of new types of thin film photocatalysts because it enables the high-speed deposition of uniform thin films with large areas and the precise control of the morphologies and chemical composition of the thin films.

Figure 9 shows the UV-Vis absorption spectra of the TiO_2 thin film photocatalysts prepared on quartz substrates with a TiO_2 thickness of 1.2 μm under differing substrate temperatures. As can be seen, the thin films prepared at lower substrate temperatures ($T = 473$ K) are colorless and transparent in the visible light regions, enabling the absorption of UV light shorter than only 380 nm (hereafter, referred to as UV-TiO_2 thin films). X-ray diffraction (XRD) analysis of the UV-TiO_2 thin film showed that it is made up of a homogeneous anatase crystalline structure. TiO_2 thin films prepared on substrates at higher temperatures ($T = 873$ K) are seen as yellow, which enables the absorption of visible light, the extent of which strongly depends on the variations in the substrate temperature (hereafter, referred to as Vis-TiO_2 thin films). The thin film photocatalyst prepared at a substrate temperature of 873 K enables the absorption of visible light up to 600 nm (Figure 9). XRD analysis of these Vis-TiO_2 thin films showed that they consist of a mixture of the anatase and rutile crystalline structural phases, with different composition ratios for anatase and rutile. Thus it is possible that visible light-responsive TiO_2 thin film photocatalysts can be prepared on various substrates through the control of the substrate temperature during the TiO_2 deposition process (45, 75).

The photocatalytic decomposition of water using these UV-TiO_2 and Vis-TiO_2 thin film photocatalysts onto which small amounts of Pt were loaded has been successful. Both the UV-TiO_2 and Vis-TiO_2 thin film photocatalysts can split water into H_2 and O_2 under UV light irradiation. It was found that the Pt-loaded Vis-TiO_2 thin film photocatalysts show much higher photocatalytic reactivity than

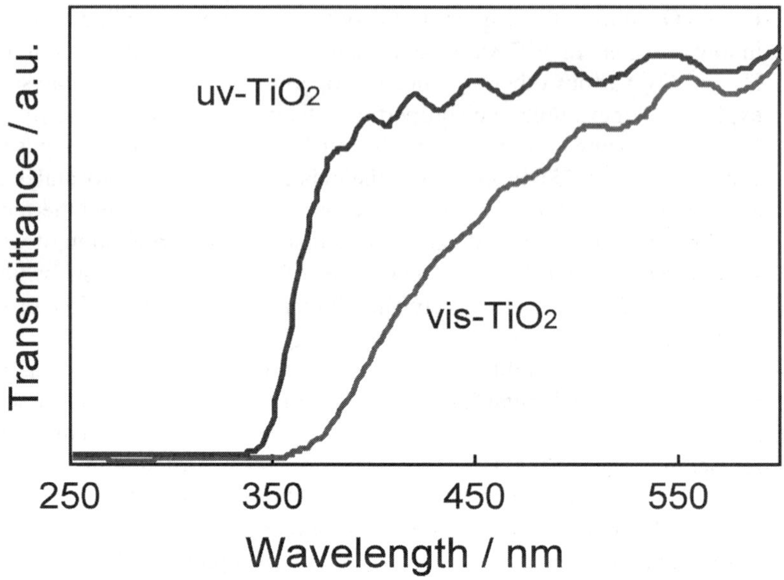

Figure 9 UV-Vis absorption spectra (in transmittance) of TiO_2 thin films prepared on a quartz substrate by a RF magnetron sputtering deposition method at different substrate temperatures. Substrate temperatures in K: UV-TiO_2, 473 K; Vis-TiO_2, 873 K.

the Pt-loaded UV-TiO_2 thin film photocatalysts under UV light irradiation ($\lambda < 380$ nm), although the surface areas of Vis-TiO_2 (22.7 m^2/g) were much smaller than those of UV-TiO_2 (50.9 m^2/g).

The wavelength effect of the irradiation light (action spectra) on the decomposition of H_2O using these types of TiO_2 thin film photocatalysts was investigated for the evolution of H_2 in water involving methanol as the sacrificial reagent. H_2 evolution could be seen when the Pt-loaded UV-TiO_2 thin films were irradiated under UV light of $\lambda < 380$ nm, but was not seen under visible light of $\lambda > 400$ nm, which was quite similar to the reactions observed on the Pt-loaded TiO_2 thin films prepared by a sol-gel method. On the other hand, Pt-loaded Vis-TiO_2 exhibited reactivity for the evolution of H_2 even under the visible light irradiation of $\lambda \geq 420$ nm. Moreover, the photocatalytic activity of Pt-loaded Vis-TiO_2 was also higher than that of Pt-loaded UV-TiO_2 under UV light irradiation.

The photocatalytic reactivity of these two types of TiO_2 thin film photocatalysts for the evolution of O_2 from H_2O was also examined in an aqueous silver nitrate solution in which the following stoichiometric reaction takes place:

$$4Ag^+ + 2H_2O \rightarrow 4Ag^0 + O_2 + 4H^+.$$

Under UV light irradiation, both types of Pt-loaded TiO_2 evolved O_2; however, the Pt-loaded Vis-TiO_2 exhibited much higher reactivity than did the Pt-loaded UV-TiO_2. Conversely, under visible light irradiation ($\lambda \geq 420$ nm), only the

Pt-loaded Vis-TiO_2 evolved O_2, its yield increasing linearly against the visible light irradiation time, whereas no O_2 evolution could be observed with the Pt-loaded UV-TiO_2 photocatalysts.

These results suggest not only that the band gap value of the Vis-TiO_2 is much smaller than that of UV-TiO_2 but also that the conduction band edge or valence band edge of Vis-TiO_2 locates at a position of sufficient potential for the decomposition of H_2O into H_2 or O_2. In fact, Pt-loaded Vis-TiO_2 could decompose water into H_2 and O_2 under light irradiation of longer than 390 nm.

In order to clarify the origin or cause for the absorption of visible light up to 600 nm by Vis-TiO_2, various spectroscopic investigations using XRD, ESCA, SIMS, TEM, and STM were undertaken. Analyses of these studies showed that Vis-TiO_2 exhibits columnar structures that stand perpendicularly on the substrate, whereas the depth profiles for the concentration of the Ti^{4+} and O^{2-} ions, which constitute Vis-TiO_2, gradually change from the top surface to the deep inside bulk, i.e., a O^{2-}/Ti^{4+} ratio starting from 2.00 at the top surface to 1.933 in the inside bulk can be observed. For UV-TiO_2, a smooth wall-like structure standing on the substrate could be observed with the O^{2-}/Ti^{4+} ratio scarcely changing and keeping a constant of 2.00 from the top surface to the deep inside bulk, providing a remarkable contrast to the Vis-TiO_2 thin film photocatalysts. With Vis-TiO_2, we believe that such a decline in the values in the Ti^{4+} and O^{2-} composition of the structure causes a profound perturbation in the electronic structure of TiO_2, thereby enabling absorption and operation under visible light. The results obtained by SIMS analyses are shown together with a cross-sectional TEM image of the Vis-TiO_2 thin films in Figure 10.

Recently, many studies have been devoted to the production of H_2 and O_2 by water splitting using various types of powdered photocatalysts under UV light irradiation (81–83), and, in some cases, it has been achieved even under visible light irradiation (84). However, powdered photocatalytic systems always yield a mixture of H_2 and O_2 as a result of the water splitting reaction. The separate evolution of H_2 and O_2 from water (85, 86) is strongly desired to obtain pure H_2 gas, which can be easily and safely utilized as fuel on a large scale. For this purpose, an H-type cell was constructed consisting of two water phases separated by a proton-exchange membrane and the Vis-TiO_2/Ti/Pt photocatalyst. The Vis-TiO_2/Ti/Pt photocatalyst was prepared by depositing Vis-TiO_2 on one side of a Ti metal foil (1.2 μm) and Pt on the other side of the foil, using RF magnetron sputtering.

Irradiation of Vis-TiO_2 by a 500-W Xe arc lamp led to the stoichiometrical evolution of H_2 and O_2 from the Pt and Vis-TiO_2 sides, respectively (Figure 11). The stoichiometric and separate evolution of H_2 and O_2 was thus successively carried out by the development of a Vis-TiO_2/Ti/Pt photocatalyst used in an H-type cell. The yields of the evolved H_2 and O_2 were found to increase as the pH difference between the two aqueous phases increased. This suggests that the pH difference improves the charge separation of the photo-formed electrons and holes, i.e., the electron transfer from the photo-irradiated Vis-TiO_2 to the Pt deposited on the other side of the Ti metal foil. We have shown that this system can be used for

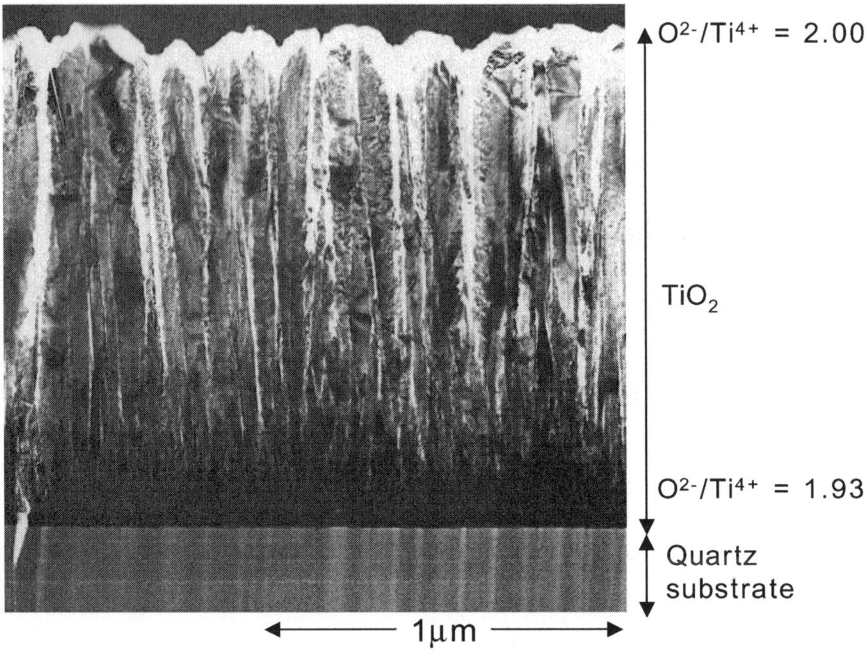

Figure 10 Cross-sectional TEM image of the Vis-TiO_2 thin film photocatalyst prepared on a quartz substrate and the changes in the ratio of Ti^{4+} and O^{2-} composition from the top surface to the inside bulk.

the separate evolution of H_2 and O_2 under visible light irradiation of wavelengths longer than 450 nm and even under solar irradiation.

The Vis-TiO_2 thin film photocatalysts, prepared by controlling the substrate temperature and the Ar pressure during the RF magnetron sputtering, have a unique columnar structure with a declining O/Ti ratio from the surface to the deep bulk near the substrate. The band gap of Vis-TiO_2 is smaller than that of UV-TiO_2, whereas the conduction band edge and valence band edge of Vis-TiO_2 has sufficient potential to decompose water into H_2 or O_2, respectively. We have also succeeded in the stoichiometric and separate evolution of H_2 and O_2 by utilizing the unique H-type photocatalytic cell combined with a Vis-TiO_2/Ti/Pt photocatalyst and proton-exchange membrane.

LOCAL STRUCTURES OF THE ACTIVE SURFACE SITES OF Ti-MCM-41 CATALYSTS AND THEIR PHOTOCATALYTIC REACTIVITY

NO_x is an especially harmful atmospheric pollutant and the main cause of acid rain and photochemical smog. The removal of nitrogen oxides (NO_x: NO, N_2O, NO_2), i.e., the direct decomposition of NO_x into N_2 and O_2, has been a great challenge

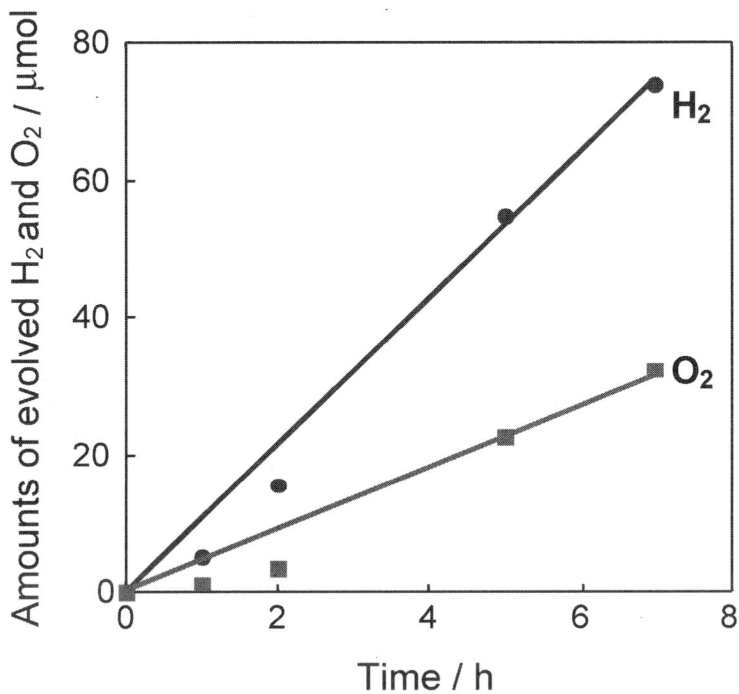

Figure 11 The photocatalytic decomposition of H_2O and the separate evolution of H_2 and O_2 under light irradiation with a 500-W Xe arc lamp on the Vis-TiO_2/Ti/Pt thin film photocatalyst having an H-type cell with two aqueous phases of different pH values, at 298 K.

for many researchers (39, 87–89). As a promising way to address such concerns, mesoporous materials incorporating transition metal ions (TMI) such as Ti^{4+} and V^{5+} have attracted much attention not only as active catalysts for the partial oxidation of alkenes but also as effective photocatalysts for the decomposition of NO_x into N_2 and O_2 or the reduction of CO_2 with H_2O to produce CH_4 and CH_3OH. (4, 41, 90–93). Such systems exhibit the advantage of having a high dispersion of TMI owing to their high internal surface area and nanoscaled pore reaction fields. In fact, these mesoporous materials incorporated with TMI are found to exhibit high photocatalytic activity for various reactions. However, the relationship between the local structure of TMI and their photocatalytic reactivity has yet to be clarified. We have investigated the relationship between the local structure of the Ti^{4+} centers in MCM-41 type materials and their photophysical and photocatalytic characteristics by applying various spectroscopic investigative techniques.

In this study, MCM-41 and Ti-MCM-41(x) (content of Ti as wt%; $x = 0.15$, 0.60, 0.85, 2.00) were synthesized in accordance with previous literature (94), using tetraethyl orthosilicate and tetraisopropyl orthotitanate as the starting materials and cetyltrimethylammonium bromide as the template. After the as-synthesized products were recovered by filtration, washed thoroughly with deionized water, and dried at 373 K for 12 h, calcination of the samples was performed in air at 823 K for 6 h. The titanium content in these materials was determined by atomic absorption analysis.

The XRD patterns showed that the Ti-MCM-41 catalysts have a MCM-41 mesoporous structure. The XAFS experiments indicated that in all the Ti-MCM-41 catalysts the majority of the Ti centers exist in a tetrahedral coordination. In particular for Ti-MCM-41, which has a Ti content of less than 2.00 wt%, the Ti^{4+} centers are highly dispersed and located in a tetrahedral coordination with the 4 oxygen atoms having a Ti–O bond length of 1.81 Å. Nevertheless, as can be seen in Figure 12, their diffuse reflectance UV-Vis spectra exhibited different degrees of dependence on the Ti content. In fact, in addition to a progressive increase in intensity, the increase in the Ti content resulted in a progressive shift of the absorption maximum from 205–208 nm [Ti-MCM-41(0.15) and Ti-MCM-41(0.60)] (Figure 12 lines *a*,*b*) to 215 nm [Ti-MCM-41(2.00)] (Figure 12 line *d*), whereas a shoulder at ~230 nm became evident in the spectrum of Ti-MCM-41(0.85) (Figure 12 line *c*), which largely contributed to the spectral profile of the Ti-MCM-41(2.00) sample. All the absorptions

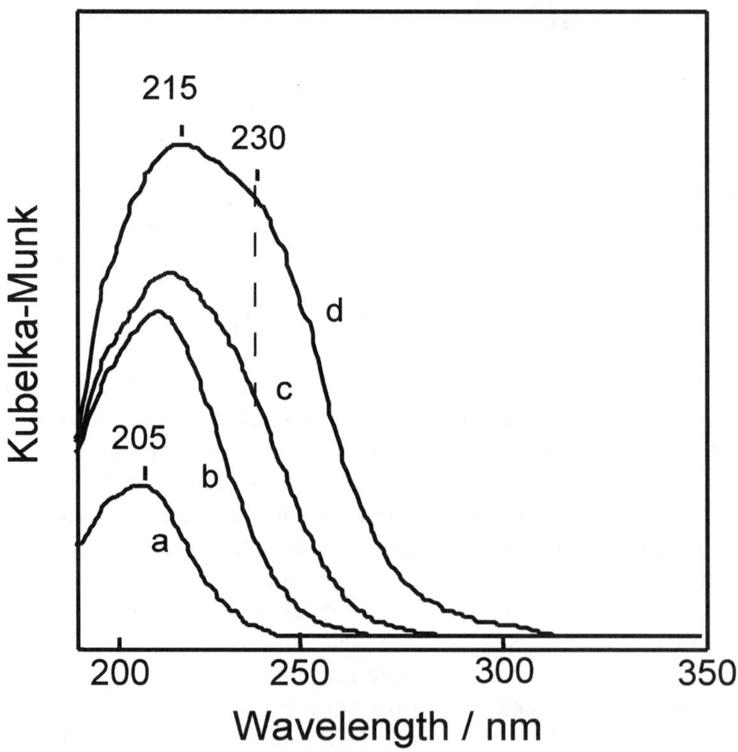

Figure 12 Diffuse reflectance UV-Vis spectra of Ti-MCM-41 with different Ti content: (*a*) 0.15, (*b*) 0.60, (*c*) 0.85, (*d*) 2.00 wt%.

cited above are in the ranges where the ligand-to-metal charge transfer involving an electron transfer from O^{2-} to the Ti^{4+} of the tetrahedrally coordinated Ti-oxide species forms a charge transfer excited state, $(Ti^{3+}–O^-)^*$. The bands at 205–215 nm can be attributed to the isolated tetrahedrally coordinated Ti-oxide species, whereas the band at 230–250 nm may be due to the dimeric or oligomeric tetrahedrally coordinated Ti-oxide species, the formation of which seems to be favored at higher Ti loadings. Moreover, the tail at $\lambda > 250$ nm observed for Ti-MCM-41 (2.00) can be attributed to a minor moiety of the Ti(IV) centers in penta- and/or octahedral coordination (95).

In order to monitor the relative number of the Ti-oxide species exposed at the surface walls of the various Ti-MCM-41 materials, the IR spectra of the ammonia irreversibly adsorbed at room temperature were recorded. No bands from the adsorbed species were observed for pure MCM-41. Conversely, the spectra of the Ti-MCM-41 samples in contact with ammonia exhibited, in the 1800–1350 cm^{-1} range, a characteristic band at 1608 cm^{-1} (δ_{asym} NH_{3ads}) owing to the NH_3 molecules irreversibly adsorbed on the Ti(IV) centers of the tetrahedrally coordinated Ti-oxide species (96, 97). The increase in the Ti content led to an increase in the intensity of this band, indicating that the number of tetra-coordinated Ti(IV) sites exposed at the surface walls of the channels of the Ti-MCM-41 materials increased with the Ti content. The bands at 1708, 1660, and 1453 cm^{-1} can be assigned to the vibrational bending modes of NH_4^+, formed by a protonation of the ammonia molecules by the acidic surface hydroxyl groups. The band at 1553 cm^{-1} is attributed to the Ti-NH_2 or Si-NH_2 bending mode formed by the irreversible reaction of NH_3 with the Si-O-Ti bridges or distorted surface Si-O-Si bridges formed by the incorporation of Ti into the Si-O-Si networks (97).

Nevertheless, the photoluminescence spectra exhibited different degrees of dependence on the Ti content. As shown in Figure 13, Ti-MCM-41 exhibits a typical photoluminescence spectrum at around 450–550 nm upon excitation of its charge transfer band at around 220–260 nm at 295 K, which coincides well with those previously observed for isolated tetrahedrally coordinated Ti-oxides species highly dispersed in silica matrices (44, 65, 95). The photoluminescence spectra are attributed to the reverse radiative decay process from the charge transfer excited state to the ground state of the isolated Ti-oxides in tetrahedral coordination (46). As can be seen in Figure 13, the intensity of this photoluminescence increases with an increase in the Ti content up to 0.60 wt%, and then sharply decreases with a higher Ti content. Furthermore, it was found that an increase in the Ti content from 0.60 wt% to 0.85 wt% leads to a decrease in the phosphorescence lifetime from 0.1 ms to 0.025 ms. On the basis of these data, when the Ti-MCM-41 materials are photo-excited by UV light, only the isolated tetrahedrally coordinated Ti-oxide species, which are more abundant for Ti content up to 0.60 wt%, can induce an excited state with a lifetime long enough to allow the appearance of a photoluminescence as a radiative decay to the ground state. Moreover, the dimeric and/or oligomeric tetrahedrally coordinated Ti-oxide species, which are most likely the dominant species present at higher Ti contents, decay quickly, apparently because

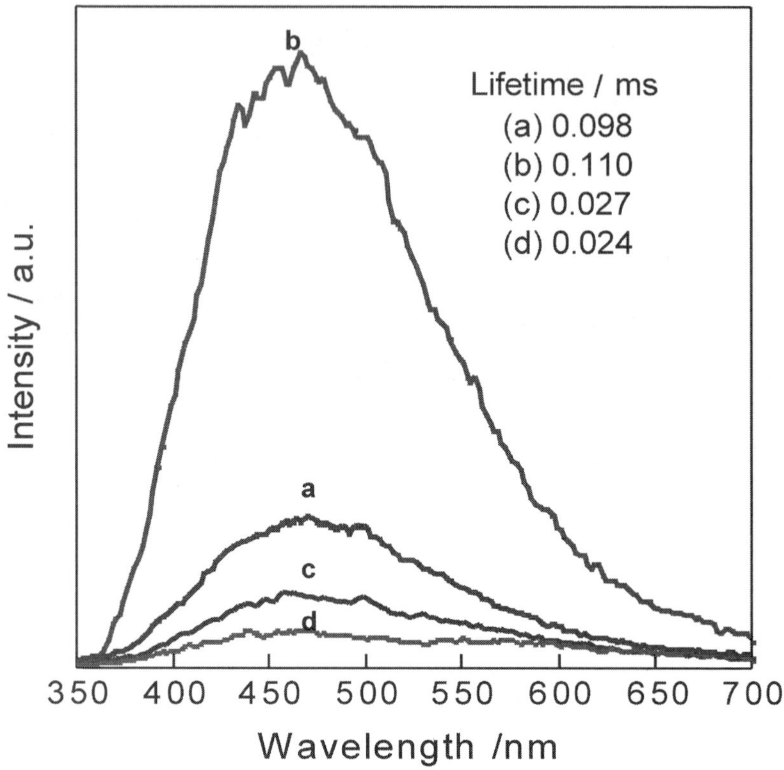

Figure 13 The photoluminescence spectra of Ti-MCM-41 with different Ti content measured at 295 K: (*a*) 0.15, (*b*) 0.60, (*c*) 0.85, (*d*) 2.00 wt%.

of their clustered structure, to the ground state through nonradiative vibrational processes.

It was found that UV irradiation of Ti-MCM-41 in the presence of NO leads to the formation of N_2, O_2, and N_2O, their yields increasing in proportion to the irradiation time, whereas under dark conditions, no products could be detected. These results indicate that the reaction proceeds photocatalytically on Ti-MCM-41 at 295 K. Figure 14 (see color insert) shows the effect of the Ti content on the reactivity of the photocatalytic decomposition of NO. The yield and selectivity of N_2 in the decomposition of NO is highest in the case of Ti-MCM-41 (0.60 Ti wt%), and an increase in Ti content led to a decrease in the reactivity, corresponding well with the intensities of the photoluminescence spectra from the isolated tetrahedrally coordinated Ti-oxide species (Figure 14). Thus the number of isolated tetrahedrally coordinated Ti-oxide species, not the total number of the Ti-oxide species exposed at the surface walls of the Ti-MCM-41 channels, as monitored by the intensity of the $\delta_{asym}NH_3$ band, appears to play an important role in this reaction. These results also indicate that only the highly dispersed isolated tetrahedrally coordinated

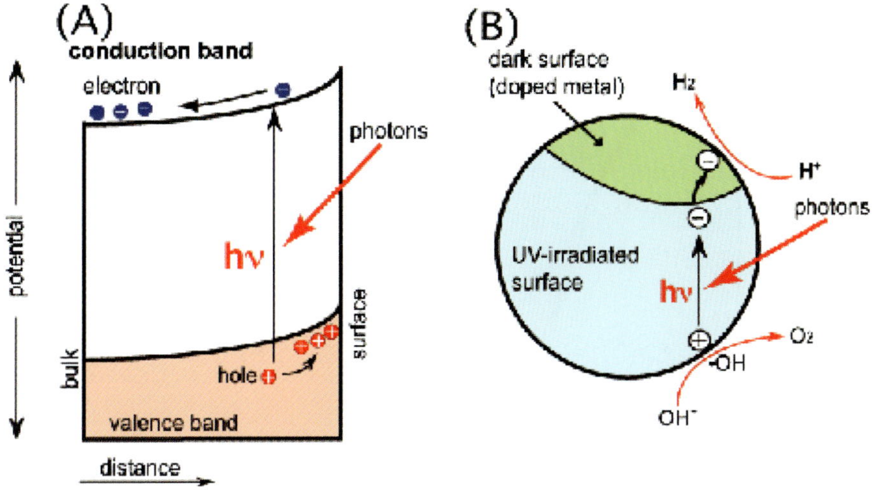

Figure 1 Photo-production of electrons and holes in the n-type semiconductor TiO_2 particles (*left*), and photocatalytic production of H_2 and O_2 on TiO_2 particles suspended in water (*right*).

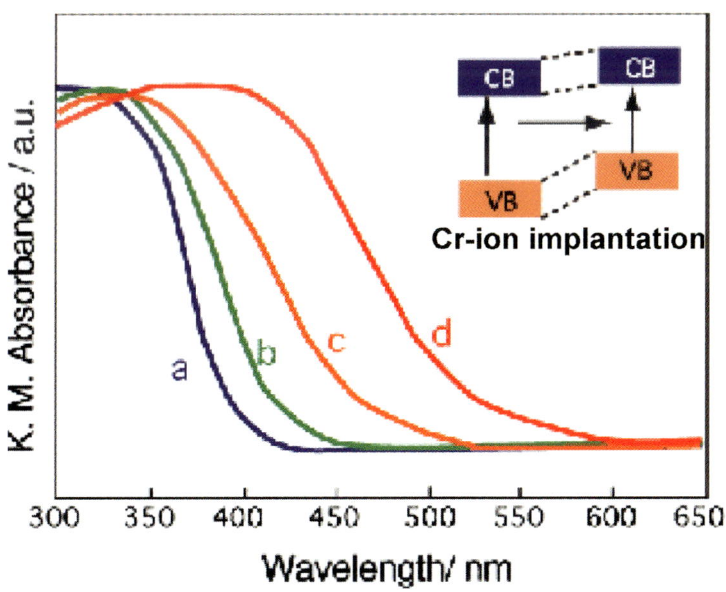

Figure 2 The UV-Vis absorption spectra of the TiO_2 (*a*) and Cr ion-implanted TiO_2 photocatalysts (*b–d*). The number of implanted Cr ions (μ mol/g): (*a*) 0, (*b*) 0.22, (*c*) 0.66, (*d*) 1.3.

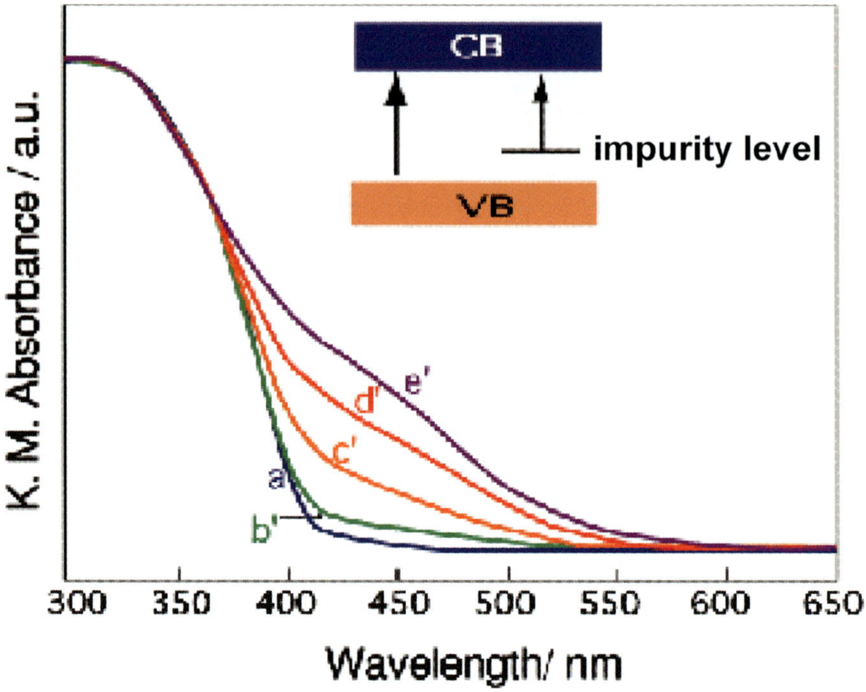

Figure 3 The UV-Vis absorption spectra of TiO_2 (*a*) and Cr ion-doped TiO_2 (*b′–d′*) photocatalysts prepared by an impregnation method. The number of doped Cr ions (wt%): (*a*) 0, (*b′*) 0.01, (*c′*) 0.1, (*d′*) 0.5, (*e′*) 1.0 (0.1 wt% equal to 4.9 μmol/g TiO_2).

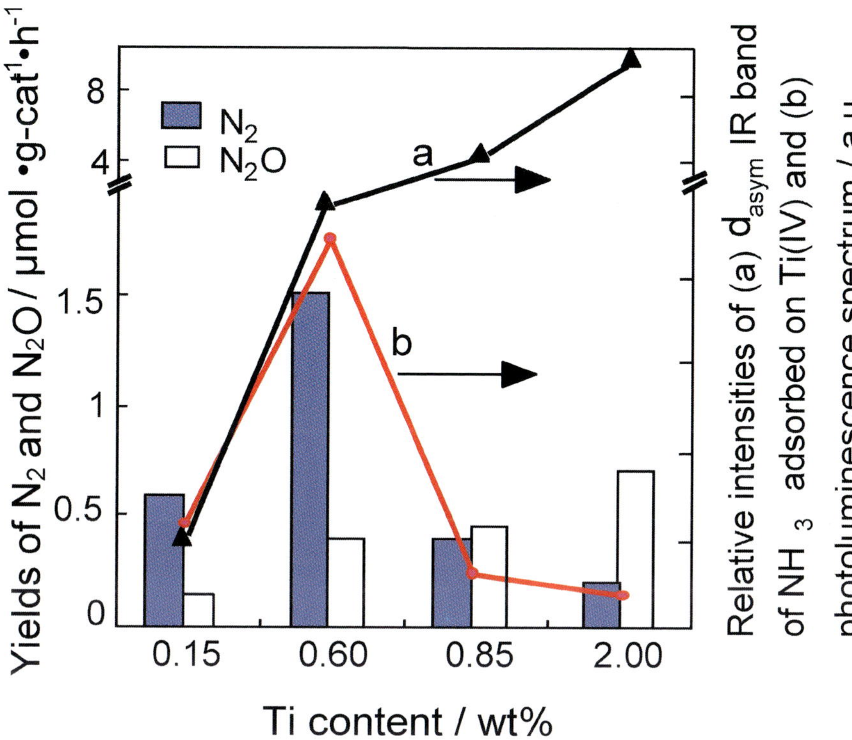

Figure 14 Relationship of the yields of N_2, N_2O, the intensities of the IR band of $\delta_{asym}NH_3$ adsorbed on Ti(IV) (*a*) and the photoluminescence spectra (*b*) of Ti-MCM-41 having differing Ti content.

Ti-oxide species act as active sites in the photocatalytic decomposition of NO into N_2 and O_2.

Mesoporous Ti-MCM-41 catalysts prepared at ambient temperatures with Ti contents of up to 0.60 wt% involve isolated tetrahedrally coordinated Ti-oxide species as the major species, whereas dimeric or oligomeric Ti-oxide species with Ti(IV) in tetrahedral coordination become overwhelming within the Ti-MCM-41 with high Ti contents. The photocatalytic reactivity of these Ti-MCM-41 for the decomposition of NO was found to depend strongly on the local structure of the Ti-oxide species. Furthermore, the yield and selectivity of N_2 corresponds with the yield of the photoluminescence of the isolated tetrahedrally coordinated Ti-oxides species, indicating that only these species are responsible for such photocatalytic reactivity.

SUMMARY

On the basis of studies in our laboratory, we have summarized recent research on the photocatalytic reactivity and photo-induced superhydrophilic properties of titanium oxide–based catalysts. Special attention was focused on the preparation of nanosized TiO_2 particles, highly dispersed titanium oxide species within zeolite cavities, titanium oxide–based binary catalysts, and second-generation TiO_2 photocatalysts that can operate under visible light irradiation fabricated using a metal-ion-implantation method. Furthermore, a new, low-cost RF magnetron sputtering method was presented, and detailed characterizations of these visible light-responsive TiO_2 thin film photocatalysts were examined at the molecular level along with investigations into the various significant photocatalytic reactions that could be initiated.

A combination of conventional chemical preparation methods and physical ion-engineering techniques will provide new approaches in the utilization of solar energy as the most abundant and safe energy source. These advances should successfully address environmental pollution on a large scale and enable the production of H_2 in the photocatalytic splitting of H_2O using visible light-responsive Ti-oxide photocatalysts and solar irradiation.

LITERATURE CITED

1. Fujishima A, Honda K. 1972. Electrochemical photolysis of water at a semiconductor electrode. *Nature* 238:37–38
2. Anpo M. 2000. Application of titanium oxide photocatalysts to improve our environment. In *Green Chemistry: Challenging Perspectives*, ed. P Tundo, P Anastas, pp. 1–19. New York: Oxford Univ. Press. 269 pp.
3. Anpo M, Takeuchi M, Yamashita H, Kishiguchi S. 2003. Design and development of new titanium dioxide semiconductor

photocatalysts. In *Molecular and Supramolecular Photochemistry: Semiconductor Photochemistry and Photophysics*, ed. V Ramamurthy, KS Schanze, 10:283–99. New York: Dekker. 361 pp.
4. Anpo M, Takeuchi M, Ikeue K, Dohshi S. 2002. Design and development of titanium oxide photocatalysts operating under visible and UV light irradiation: the applications of metal ion-implantation techniques to semiconducting TiO_2 and Ti/zeolite catalysts. *Curr. Opin. Solid State Mater. Sci.* 6:381–88
5. Ireland JC, Klostermann P, Rice EW, Clark RM. 1993. Inactivation of *Escherichia coli* by titanium-dioxide photocatalytic oxidation. *Appl. Environ. Microbiol.* 59:1668–70
6. Sjogren JC, Sierka RA. 1994. Inactivation of phage MS2 by iron-aided titanium-dioxide photocatalysis. *Appl. Environ. Microbiol.* 60:344–47
7. Cai RX, Kubota Y, Shuin T, Sakai H, Hashimoto K, Fujishima A. 1992. Induction of cytotoxicity by photoexcited TiO_2 particles. *Cancer Res.* 52:2346–48
8. Cai RX, Hashimoto K, Kubota Y, Fujishima A. 1992. Increment of photocatalytic killing of cancer-cells using TiO_2 with the aid of superoxide-dismutase. *Chem. Lett.* pp. 427–30
9. Suzuki K. 1993. In *Photocatalytic Purification and Treatment of Water and Air*, ed. DF Ollis, H Al-Ekabi. Tokyo/Amsterdam: Elsevier. 820 pp.
10. Karakitsou KE, Verykios XE. 1993. Effects of altervalent cation doping of TiO_2 on its performance as a photocatalyst for water cleavage. *J. Phys. Chem.* 97:1184–89
11. Gratzel M. 1981. Artificial photosynthesis: water cleavage into hydrogen and oxgen by visible light. *Acc. Chem. Res.* 14:376–84
12. Kalyanasundaram K, Borgarello E, Duonghong D, Gratzel M. 1981. Cleavage of water by visible light irradiation of colloidal CdS solutions. Inhibitation of photocorrosion by RuO_2. *Angew. Chem. Int. Ed. Engl.* 20:987–88
13. Duonghong D, Borgarello E, Gratzel M. 1981. Dynamics of light-induced water cleavage in colloidal systems. *J. Am. Chem. Soc.* 103:4685–90
14. Borgarello E, Kiwi J, Pelizzetti E, Visca M, Gratzel M. 1981. Photochemical cleavage of water by photocatalysis. *Nature* 289:158
15. Wold A. 1993. Photocatalytic properties of TiO_2. *Chem. Mater.* 5:280–83
16. Khan MMT, Rao NN. 1991. Stepwise reduction of coordinated dinitrogen to ammonia via diazinido and hydrazido intermediates on a visible-light irradiated $Pt/CdS/Ag_2S/RuO_2$ particulate system suspended in an aqueous-solution of $K[Ru(EDTA\text{-}H)Cl]_2H_2O$. *J. Photochem. Photobiol. A* 56:101–11
17. Schiavello M. 1993. Some working principles of heterogeneous photocatalysis by semiconductors. *Electrochim. Acta* 38:11–14
18. Khan MMT, Chatterjee D, Krishnaratnam M, Bala M. 1992. Photosensitized reduction of N_2 by $Ru(II)(bipy)_3{}^{2+}$ adsorbed on the surface of $Pt/TiO_2/RuO_2$ semiconductor particulate system containing Ru(III)-EDTA complex and L-ascorbic-acid. *J. Mol. Catal.* 72:13–18
19. Khan MMT, Chatterjee D, Bala M. 1992. Photocatalytic reduction of N_2 to NH_3 sensitized by the [Ru(III)-ethylendiaminetetraacetate-2,2′-bipyridyl] complex in a Pt-TiO_2 semiconductor particulate system. *J. Photochem. Photobiol. A* 67: 349–52
20. Gerischer H, Heller A. 1992. Photocatalytic oxidation of organic-molecules at TiO_2 particles by sunlight in aerated water. *J. Electrochem. Soc.* 139:113–18
21. Jackson NB, Wang CM, Luo Z, Schwitzgebel J, Ekerdt JG, et al. 1991. Attachment of TiO_2 powders to hollow glass microbeads—activity of the TiO_2-coated beads in the photoassisted oxidation of ethanol to acetaldehyde. *J. Electrochem. Soc.* 138:3660–64
22. Nair M, Luo ZH, Heller A. 1993. Rates of photocatalytic oxidation of crude-oil on salt-water on buoyant, cenosphere-attached

titanium-dioxide. *Ind. Eng. Chem. Res.* 32: 2318–23

23. Boer KW. 1990. Electron and other particles in bulk semiconductors. In *Survey of Semiconductor Physics*, p. 249. New York: Van Nostrand Reinhold. 1423 pp.
24. Rothenberger G, Moser J, Gratzel M, Serpone N, Sharma DK. 1985. Charge carrier trapping and recombination dynamics in small semiconductor particles. *J. Am. Chem. Soc.* 107:8054–59
25. Gratzel M. 1989. *Heterogeneous Photochemical Electron Transfer*. Boca Raton, FL: CRC Press. 159 pp.
26. Mills G, Hoffmann MR. 1993. Photocatalytic degradation of pentachlorophenol on TiO_2 particles—identification of intermediates and mechanism of reaction. *Environ. Sci. Technol.* 27:1681–89
27. Kormann C, Bahnemann DW, Hoffmann MR. 1991. Photolysis of chloroform and other organic molecules in aqueous TiO_2 suspensions. *Environ. Sci. Technol.* 25: 494–500
28. Carraway ER, Hoffmann AJ, Hoffmann MR. 1994. Photocatalytic oxidation of organic acids on quantum-sized semiconductor colloids. *Environ. Sci. Technol.* 28:786–93
29. Chemseddine A, Boehm HP. 1990. A study of the primary step in the photochemical degradation of acetic acid and chloroacetic acids on a TiO_2 photocatalyst. *J. Mol. Catal.* 60:295–311
30. D'Oliveira JC, Minero C, Pelizzetti E, Pichat P. 1993. Photodegradation of dichlorophenols and trichlorophenols in TiO_2 aqueous suspension—kinetic effects of the positions of the Cl atoms and identification of the intermediates. *J. Photochem. Photobiol. A* 72:261–67
31. D'Oliveira JC, Al-Sayyed G, Pichat P. 1990. Photodegradation of 2-chlorophenol and 3-chlorophenol in TiO_2 aqueous suspensions. *Environ. Sci. Technol.* 24:990–96
32. Hidaka H, Zhao J, Pelizzetti E, Serpone N. 1992. Photodegradation of surfactants. Comparison of photocatalytic processes between anionic sodium dodecylbenzenesulfonate and cationic benzyldodecyldimetylammonium chloride on the TiO_2 surface. *J. Phys. Chem.* 96:2226–30
33. Pelizzetti E, Minero C, Piccinini P, Vincenti M. 1993. Phototransformations of nitrogen-containing organic compounds over irradiated semiconductor-metal oxides—nitrobenzene and atrazine over TiO_2 and ZnO. *Coord. Chem. Rev.* 125:183–93
34. Anpo M, Takeuchi M, Yamashita H, Hirao T, Itoh N, Iwamoto N. 1999. The design and development of TiO_2 thin film photocatalysts which work under visible light irradiation by the application of ion engineering techniques. *Proc. 4th Int. Conf. ECOMATERIAL Gifu*. pp. 333–36. Gifu City, Japan: Sci. Technol. Agency
35. Anpo M, Takeuchi M, Kishiguchi S, Yamashita H. 1999. Preparation of unique titanium oxide photocatalysts capable of operating under visible light irradiation by an advanced metal ion-implantation method. *Surf. Sci. Jpn.* 20:60–65
36. Yamashita H, Harada M, Misaka J, Takeuchi M, Ikeue K, Anpo M. 2002. Photocatalytic degradation of propanol diluted in water under visible light irradiation using metal ion-implanted titanium dioxide photocatalysts. *J. Photochem. Photobiol. A* 148:257–61
37. Takeuchi M, Yamashita H, Matsuoka M, Anpo M, Hirao T, et al. 2000. Photocatalytic decomposition of NO on titanium oxide thin film photocatalysts prepared by an ionized cluster beam technique. *Catal. Lett.* 66:185–87
38. Yamashita H, Harada M, Misaka J, Takeuchi M, Ichihashi Y, et al. 2001. Application of ion beam techniques for preparation of metal ion-implanted TiO_2 thin film photocatalyst available under visible light irradiation: metal ion-implantation and ionized cluster beam method. *J. Synchrotron Rad.* 8:569–71
39. Anpo M. 1997. Photocatalysis on titanium oxide catalysts: approaches in achieving

highly efficient reactions and realizing the use of visible light. *Catal. Survey Jpn.* 1: 169–79

40. Anpo M. 2000. Utilization of TiO_2 photocatalysts in green chemistry. *Pure Appl. Chem.* 72:1265–70
41. Anpo M. 2000. Applications of titanium oxide photocatalysts and unique second-generation TiO_2 photocatalysts able to operate under visible light irradiation for the reduction of environmental toxins on a global scale. *Stud. Surf. Sci. Catal.* 130: 157–66
42. Anpo M, Takeuchi M. 2001. Design and development of second-generation titanium oxide photocatalysts to better our environment-approaches in realizing the use of visible light. *Int. J. Photoenergy* 3: 1–6
43. Anpo M, Ichihashi Y, Takeuchi M, Yamashita H. 1998. Design of unique titanium oxide photocatalysts by an advanced metal ion-implantation method and photocatalytic reactions under visible light irradiation. *Res. Chem. Intermed.* 24:143–49
44. Takeuchi M, Yamashita H, Matsuoka M, Anpo M, Hirao T, et al. 2000. Photocatalytic decomposition of NO under visible light irradiation on the Cr-ion-implanted TiO_2 thin film photocatalyst. *Catal. Lett.* 67:135–37
45. Takeuchi M, Anpo M, Hirao T, Itoh N, Iwamoto N. 2001. Preparation of TiO_2 thin film photocatalysts working under visible light irradiation by applying a RF magnetron sputtering deposition method. *Surf. Sci. Jpn.* 22:561–65
46. Anpo M, Che M. 2000. Applications of photoluminescence techniques to the characterization of solid surfaces in relation to adsorption, catalysis and photocatalysis. *Adv. Catal.* 44:119–257
47. Wang R, Hashimoto K, Fujishima A, Chikuni M, Kojima E, et al. 1997. Light-induced amphiphilic surfaces. *Nature* 388: 431–32
48. Wang R, Sakai N, Fujishima A, Watanabe T, Hashimoto K. 1999. Studies of surface wettability conversion on TiO_2 single-crystal surfaces. *J. Phys. Chem. B* 103: 2188–94
49. Miyauchi M, Nakajima A, Fujishima A, Hashimoto K, Watanabe T. 2000. Photoinduced surface reactions on TiO_2 and $SrTiO_3$ films: photocatalytic oxidation and photoinduced hydrophilicity. *Chem. Mater.* 12: 3
50. Sun RD, Nakajima A, Fujishima A, Watanabe T, Hashimoto K. 2001. Photoinduced surface wettability conversion of ZnO and TiO_2 thin films. *J. Phys. Chem. B* 105: 1984–90
51. Yu JC, Yu J, Tang HY, Zhang L. 2002. Effect of surface microstructure on the photoinduced hydrophilicity of porous TiO_2 thin films. *J. Mater. Chem.* 12:81–85
52. Miyauchi M, Nakajima A, Watanabe T, Hashimoto K. 2002. Photocatalysis and photoinduced hydrophilicity of various metal oxide thin films. *Chem. Mater.* 14: 2812–16
53. Nakamura M, Sirghi L, Aoki T, Hatanaka Y. 2002. Study on hydrophilic property of hydro-oxygenated amorphous TiO_x:OH thin films. *Surf. Sci.* 507:778–82
54. Nakamura M, Kato S, Aoki T, Sirghi L, Hatanaka Y. 2001. Role of terminal OH groups on the electrical and hydrophilic properties of hydro-oxygenated amorphous TiO_x:OH thin films. *J. Appl. Phys.* 90:3391–95
55. Yamashita H, Harada M, Tanii A, Honda M, Takeuchi M, et al. 2000. Preparation of efficient titanium oxide photocatalysts by an ionized cluster beam (ICB) method and their photocatalytic reactivities for the purification of water. *Catal. Today* 63:63–69
56. Harada M, Tanii A, Yamashita H, Anpo M. 1999. Preparation of titanium oxide photocatalysts loaded on activated carbon by an ionized cluster beam method and their photocatalytic reactivities for the degradation of 2-propanol diluted in water. *Z. Phys. Chem.* 213:59–65
57. Dohshi S, Takeuchi M, Anpo M. 2001. Photoinduced superhydrophilic properties of

Ti-B binary oxide thin films and their photocatalytic reactivity for the decomposition of NO. *J. Nanosci. Nanotechnol.* 1:337–42
58. Anpo M, Dohshi S, Takeuchi M. 2002. Preparation of Ti/B binary oxide thin films by the ionized cluster beam (ICB) method: their photocatalytic reactivity and photoinduced superhydrophilic properties. *J. Ceramic Process. Res.* 3:258–60
59. Takeuchi M, Matsuoka M, Yamashita H, Anpo M. 2001. Preparation of Ti-Si binary oxide thin film photocatalysts by the application of an ionized cluster beam method. *J. Synchrotron Rad.* 8:643–44
60. Anpo M, Nakaya H, Kodama S, Kubokawa Y, Domen K, Onishi T. 1986. Photocatalysis over binary metal-oxides: enhancement of the photocatalytic activity of TiO_2 in titanium-silicon oxides. *J. Phys. Chem.* 90:1633–36
61. Imamura S, Nakai T, Kanai H, Itoh T. 1995. Effect of tetrahedral Ti in titania-silica mixed oxides on epoxidation activity and lewis acidity. *J. Chem. Soc. Faraday Trans.* 91:1261–66
62. Hattori A, Kawahara T, Uemoto T, Suzuki F, Tada H, Ito S. 2000. Ultrathin SiO_x film coating effect on the wettability change of TiO_2 surfaces in the presence and absence of UV light illumination. *J. Colloid Interface Sci.* 232:410–13
63. Notari B. 1996. Microporous crystalline titanium silicates. *Adv. Catal.* 41:253–334
64. Anpo M, Aikawa N, Kubokawa Y. 1984. Photocatalytic hydrogenation of alkynes and alkenes with water over TiO_2-Pt loading effect on the primary processes. *J. Phys. Chem.* 88:3998–4000
65. Anpo M, Aikawa N, Kubokawa Y, Che M, Louis C, Giamello E. 1985. Photoluminescence and photocatalytic activity of highly dispersed titanium oxide anchored onto porous vycor glass. *J. Phys. Chem.* 89: 5017–21
66. Anpo M, Shima T, Kodama S, Kubokawa Y. 1987. Photocatalytic hydrogenation of CH_3CCH with H_2O on small-parrticle TiO_2: size quantization effects and reaction intermediates. *J. Phys. Chem.* 91:4305–10
67. Kiwi J, Grätzel M. 1984. Optimization of conditions for photochemical water cleavage. Aqueous Pt/TiO_2 (anatase) dispersions under ultraviolet light. *J. Phys. Chem.* 88: 1302–7
68. Yamaguchi K, Sato S. 1985. Photocatalysis of water over metallized powdered titanium-oxide. *J. Chem. Soc. Faraday Trans.* 81:1237–46
69. Sayama K, Arakawa H. 1997. Effect of carbonate salt addition on the photocatalytic decomposition of liquid water over Pt-TiO_2 catalyst. *J. Chem. Soc. Farady Trans.* 93:1647–54
70. Moon SC, Mametsuka H, Suzuki E, Anpo M. 1998. Stoichiometric decomposition of pure water over Pt-loaded Ti/B binary oxide under UV-irradiation. *Chem. Lett.* pp.117–18
71. Yoshida Y, Matsuoka M, Moon SC, Mametsuka H, Suzuki E, Anpo M. 2000. Photocatalytic decomposition of liquid-water on the Pt-loaded TiO_2 catalysts: effects of the oxidation states of Pt species on the photocatalytic reactivity and the rate of the back reaction. *Res. Chem. Intermed.* 26:567–74
72. Moon SC, Matsumura Y, Kitano M, Matsuoka M, Anpo M. 2003. Hydrogen production using semiconducting oxide photocatalysts. *Res. Chem. Intermed.* 29:233–56
73. Tabata S, Nishida H, Masaki Y, Tabata K. 1995. Stoichiometric photocatalytic decomposition of pure water in Pt/TiO_2 aqueous suspension system. *Catal. Lett.* 34: 245–49
74. Hoffmann MR, Martin ST, Choi W, Bahnemann DW. 1995. Environmental applications of semiconductor photocatalysis. *Chem. Rev.* 95:69–96
75. Anpo M, Takeuchi M. 2003. The design and development of highly reactive titanium oxide photocatalysts operating under visible light irradiation. *J. Catal.* 216:505–16

76. Choi W, Termin A, Hoffmann MR. 1994. The role of metal-ion dopants in quantum-sized TiO_2-correlation between photoreactivity and charge-carrier recombination dynamics. *J. Phys. Chem.* 98:13669–79
77. Asahi R, Morikawa T, Ohwaki T, Aoki K, Taga Y. 2001. Visible-light photocatalysis in nitrogen-doped titanium oxides. *Science* 293:269–71
78. Yamashita H, Ichihashi Y, Takeuchi M, Kishiguchi S, Anpo M. 1999. Characterization of metal ion-implanted titanium oxide photocatalysts operating under visible light irradiation. *J. Synchrotron. Rad.* 6:451–52
79. Anpo M. 2000. Use of visible light. Second-generation titanium oxide photocatalysts prepared by the application of an advanced metal ion-implantation method. *Pure Appl. Chem.* 72:1787–92
80. Yamashita H, Harada M, Misaka J, Takeuchi M, Ikeue K, Anpo M. 2002. Degradation of propanol diluted in water under visible light irradiation using metal ion-implanted titanium dioxide photocatalysts. *J. Photochem. Photobiol. A* 148:257–61
81. Ishihara T, Nishiguchi H, Fukamachi K, Takita Y. 1999. Effects of acceptor doping to $KTaO_3$ on photocatalytic decomposition of pure H_2O. *J. Phys. Chem.* 103:1–3
82. Domen K, Kondo JN, Hara M, Takata T. 2000. Photo- and mechano-catalytic overall water splitting reactions to form hydrogen and oxygen on heterogeneous catalysts. *Bull. Chem. Soc. Jpn.* 73:1307–31
83. Kato H, Kudo A. 2001. Water splitting into H_2 and O_2 on alkali tantalate photocatalysts $ATaO_3$ (A = Li, Na, and K). *J. Phys. Chem. B* 105:4285–92
84. Zou Z, Ye J, Sayama K, Arakawa H. 2001. Direct splitting of water under visible light irradiation with an oxide semiconductor photocatalyst. *Nature* 414:625–27
85. Fujishima A, Kohayakawa K, Honda K. 1975. Hydrogen production under sunlight with an electrochemical photocell. *J. Electrochem. Soc.* 122:1487–89
86. Fukamachi J, Ohnishi Y, Miyamoto C, Izumi I. 1996. Water decomposition using a p-n type photochemical diode in a dual electrolyte cell. *J. Technol. Educ.* 5:113–20
87. Zhang J, Minagawa M, Ayusawa T, Natarajan S, Yamashita H, et al. 2000. In situ investigation of the photocatalytic decomposition of NO on the Ti-HMS under flow and closed reaction systems. *J. Phys. Chem. B* 104:11501–5
88. Zhang J, Minagawa M, Matsuoka M, Yamashita H, Anpo M. 2000. Photocatalytic decomposition of NO on Ti-HMS mesoporous zeolite catalysts. *Catal. Lett.* 66: 241–43
89. Zhang J, Hu Y, Matsuoka M, Yamashita H, Minagawa M, et al. 2001. Relationship between the local structures of titanium oxide photocatalysts and their reactivities in the decomposition of NO. *J. Phys. Chem. B* 105:8395–98
90. Anpo M, Higashimoto S. 2001. Characteristics in the photocatalytic reactivity of the tetrahedrally coordinated Ti-oxide species designed within various types of zeolites and on support surfaces. *Stud. Surf. Sci. Catal.* 140:27–35
91. Matsuoka M, Anpo M. 2003. Local structures, excited states, and photocatalytic reactivities of highly dispersed catalysts constructed within zeolites. *J. Photochem. Photobiol. C* 3:225–52
92. Ikeue K, Yamashita H, Takewaki T, Anpo M. 2001. Photocatalytic reduction of CO_2 with H_2O on Ti-β zeolite photocatalysts: effect of the hydrophobic and hydrophilic properties. *J. Phys. Chem. B* 105:8350–55
93. Ikeue K, Nozaki S, Ogawa M, Anpo M. 2002. Characterization of the self-standing Ti-containing porous silica thin films and their reactivity for the photocatalytic reduction of CO_2 with H_2O. *Catal. Today* 74: 241–48
94. Zhang W, Froba M, Wang J, Tanev PT, Wong J, Pinnavaia TJ. 1996. Mesoporous titanosilicate molecular sieves prepared at ambient temperature by electrostatic

(S+I−, S(+)(−)I(+)) and neutral (S degrees I degrees) assembly pathways: a comparison of physical properties and catalytic activity for peroxide oxidations. *J. Am. Chem. Soc.* 118:9164–71
95. Marchese L, Gianotti E, Dellarocca V, Maschmeyer T, Rey F, et al. 1999. Structure-functionality relationships of grafted Ti-MCM41 silicas. Spectroscopic and catalytic studies. *Phys. Chem. Chem. Phys.* 1:585–92
96. Marchese L, Gianotti E, Maschmeyer T, Martra G, Coluccia S, Thomas JM. 1997. Spectroscopic tools for probing the isolated titanium centers in MCM41 mesoporous catalysts. *Nuovo Cimento* 19:1707–18
97. Raimondi ME, Gianotti E, Marchese L, Martra G, Maschmeyer T, et al. 2000. A spectroscopic study of group IV transition metal incorporated direct templated mesoporous catalysts Part 1: a comparison between materials synthesized using hydrophobic and hydrophilic Ti precursors. *J. Phys. Chem. B* 104:7102–9

Annu. Rev. Mater. Res. 2005. 35:29–49
doi: 10.1146/annurev.matsci.35.100303.121832

First published online as a Review in Advance on January 20, 2005

APPLICATION OF EXAFS TO MOLTEN SALTS AND IONIC LIQUID TECHNOLOGY

Christopher Hardacre
School of Chemistry and CenTACat, Queen's University, Belfast BT9 5AG, Northern Ireland, United Kingdom; email: c.hardacre@qub.ac.uk

Key Words RTIL, structure, speciation, XAFS, absorption, X ray

■ **Abstract** This paper describes the use of extended X-ray absorption fine structure spectroscopy (EXAFS) to examine the structure of molten salts and ionic liquids and species dissolved in them. The EXAFS theory is briefly described as are the methods by which EXAFS of these systems can be studied. A range of applications have used EXAFS to investigate the structure of metallic species in ionic liquids from extraction studies to catalysts. The area of structural investigations of ionic liquids is still being developed, although growing rapidly, whereas the structure of molten salts has been studied using EXAFS in more detail.

INTRODUCTION

The structure of liquids has been studied for many years. These investigations have, in general, focused on the arrangements in molecular solvents, in particular water. The structure of molten salts has been much less studied, and within this field the study of the structure of room-temperature ionic liquids is in its infancy. For molten salts/ionic liquids, the main experimental techniques that have been applied are neutron (1, 1a–e) and X-ray diffraction (2, 2a–c). In contrast, the use of extended X-ray absorption fine structure (EXAFS) has been much less researched. This review gives an introduction to molten salts and compares them with room-temperature ionic liquids and to the EXAFS technique. We illustrate how EXAFS has been used to elucidate the structure of molten salts and provide mechanistic information on processes performed in both room-temperature ionic liquids and molten salts.

MOLTEN SALTS AND IONIC LIQUIDS

Molten salts and ionic liquids are substances that contain only ions; however, they are distinguished by the difference in their melting points. Ionic liquids have been defined as salts with melting points below 100°C and preferably below room temperature (3, 3a). The melting points found in ionic liquids and molten salts

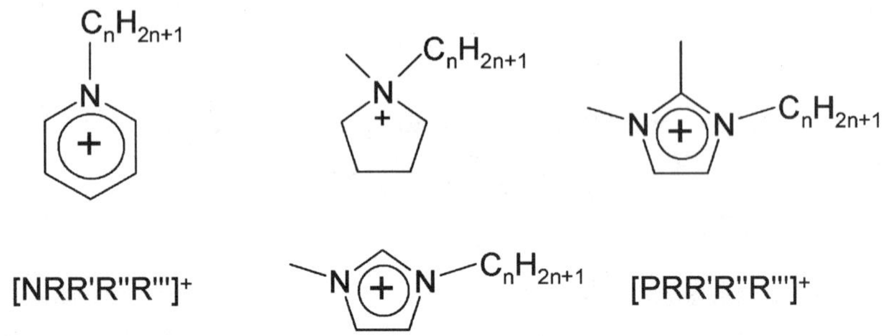

$[PF_6]^-$ $[BF_4]^-$ $[EtSO_4]^-$ $[CF_3COO]^-$

$[SO_3CF_3)]^-$ = $[OTf]^-$ $[N(SO_2CF_3)]^-$ = $[NTf_2]^-$

Figure 1 Common anions and cations used in ionic liquids.

differ because of the structure of the anions and cations. Molten salts generally consist of simple anions and cations that pack efficiently and, therefore, have a high lattice energy. By decreasing the symmetry of the cation and increasing the charge delocalization in both the cation and anion, the lattice energy is decreased and hence the melting point is lowered. Figure 1 shows the structures of some of the common anions and cations used to form ionic liquids. The most frequently used cation is based on 1-alkyl, 3-methyl imidazolium and throughout the text this is referred to as $[C_n\text{mim}]^+$, where n denotes the alkyl chain length, e.g., 1-ethyl, 3-methyl imidazolium is denoted by $[C_2\text{mim}]^+$.

Historically, ionic liquids were first investigated as electrochemical media, in particular for aluminum electrodeposition (4, 4a). These first-generation ionic liquids used combinations of aluminum(III) chloride with organic halides, for example 1-ethyl,3-methyl imidazolium chloride, $[C_2\text{mim}]Cl$, which generated molten salts that were liquid at room temperature. Subsequently, these materials were investigated as electrolytes for use in high-power batteries, where the room-temperature ionic liquid replaced high-temperature molten salt technology (5, 5a). Although these materials show good potential windows, they suffer from the drawback that they are highly moisture sensitive and corrosive. To alleviate this problem, Wilkes & Zaworotko developed second-generation ionic liquids by replacing the chloroaluminate system with a range of anions, including $[BF_4]^-$, $[PF_6]^-$, and $[NO_3]^-$ (6). The latter anions are much less corrosive than either molten salts or the first generation ionic liquids and have led to the massive upsurge of interest in ionic liquids. A more comprehensive review of the historical development of ionic liquids has recently been reported by Wilkes (7).

Both first- and second-generation ionic liquids have been used in a wide variety of applications and have the potential to revolutionize the process industry (8, 8a). By changing the cation and anion, the properties of these liquids may be altered systematically. In addition, owing to their immeasurably low vapor pressure, these ionic liquids are ideal engineering solvents. The applications applied to ionic liquid technology include, for example, clean synthesis (9, 9a,b), electrochemistry (10, 10a–c) and liquid crystals (11, 11a,b); all of which have recently been reviewed at length (12).

EXTENDED X-RAY ABSORPTION FINE STRUCTURE SPECTROSCOPY

Theory

Extended X-ray absorption fine structure spectroscopy measures the X-ray absorption as a function of energy and allows the local arrangement of atoms to be elucidated. The absorption results from the excitation of a core electron in an atom. Conventional EXAFS is usually associated with hard X rays, >3–4 keV, to allow measurements to be made outside a vacuum, and requires synchrotron radiation to provide the intensity at the energies involved. At these energies, the core electrons ejected correspond to the 1s (K-edge), 2s (L_I-edge), $2p_{1/2}$(L_{II}-edge), and $2p_{3/2}$(L_{III}-edge) states. As the photon energy is increased past the absorption edge, an oscillatory structure is found, which is described as the X-ray fine structure. The X-ray fine structure starts at approximately 30 eV past the edge and extends to a range of 1000 eV.

The EXAFS is observed as a modulating change in the absorption coefficient caused by the ejected electron wave back scattering from the surrounding atoms, which results in interference between ejected and back-scattered waves. It is defined as

$$\chi(k) = \frac{\mu(k) - \mu_0(k)}{\Delta\mu_0},$$

where $\chi(k)$ is the EXAFS as a function of the wavenumber of the photoelectron, k; $\mu(k)$ is the measured absorption above the absorption edge; $\mu_0(k)$ is the absorption spectrum without the EXAFS oscillations, i.e., the background; and $\Delta\mu_0$ is a normalization factor.

The wavenumber is defined at a photon energy, E, above the absorption edge energy, E_0, with respect to the mass of the electron, m_e.

$$k = \sqrt{\frac{2m_e}{\hbar^2}(E - E_0)}.$$

The EXAFS is related to the wavenumber by

$$\chi(k) \approx \sum_i \frac{N_i f_i(k)}{kr_i^2} e^{-2\sigma_i^2 k^2} e^{-2r_i/\lambda} \sin[2kr_i + \alpha_i(k)],$$

where $\chi(k)$ is the sum over N_i back-scattering atoms i, f_i is the scattering amplitude term characteristic of the atom, σ_i is the Debye-Waller factor associated with the vibration of the atoms, r_i is the distance from the absorbing atom, λ is the mean free path of the photoelectron, and α_i is the phase shift of the spherical wave as it scatters from the back-scattering atoms. By taking the Fourier transform of the amplitude of the fine structure, $\chi(k)$, a real-space radial distribution function of the back-scattering atoms around the absorbing atom is produced. It should be noted that this representation describes only the single-scattering approximation. Multiple scattering is important in many systems, as described in detail below.

On analysis of the EXAFS data, the local environment around a given absorbing atom can be obtained, i.e., the type, number, and distance of the back-scattering atoms. It should be noted that it is not necessary for the surrounding atoms to be formally bonded to the absorbing atom. Typically, the distance has an uncertainty of $\pm 1\%$ within a radius of approximately 6 Å; however, the error in the coordination number is strongly dependent on the system studied and can be high. In this regard, comparison with standard materials and the use of EXAFS in conjunction with other techniques to ensure a realistic interpretation of the data are vital.

Because the fine structure observed is associated with only the particular absorption edge being studied, and the energy of the absorption edge is dependent on the element and its oxidation state, the EXAFS examines the local structure around one particular element, and in some cases, an element in a given oxidation state. Therefore by studying more than one absorbing element in the sample, a fuller picture can be obtained.

Two recent reviews treat the theory more rigorously and summarize much of the data on general liquid systems (13, 14).

Experimental Liquid Set-Ups

Two main methods are employed in studying liquid samples (see Figure 2). These methods consist of supporting the liquid in an inert low-atomic-weight matrix such as graphite or boron nitride or sandwiching thin films between low-atomic-weight plates. The choice of the matrix materials used is a balance between its chemical inertness toward the liquid being studied and at the same time being thermally stable and transparent to the X rays at the absorption energy. The latter becomes less problematic as the energy of the absorption edge increases. Figure 3 shows two experimental cells that have been used to measure the EXAFS of low-temperature ionic liquid samples and high-temperature studies using self-supporting discs and thin films of liquid, respectively (15–17).

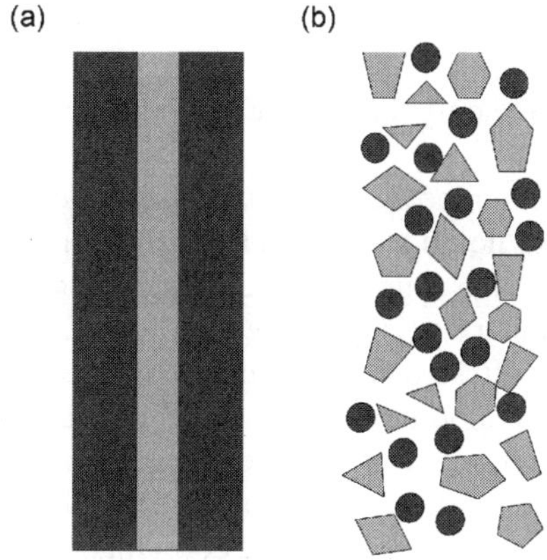

Figure 2 Schematic of the sample preparation methods used to study liquid EXAFS: (*a*) thin liquid film sandwiched between low-atomic-weight plates and (*b*) the liquid (*circles*) dispersed in a low-atomic-number matrix (*polyhedrons*). The figure has been redrawn from Reference 13 with permission.

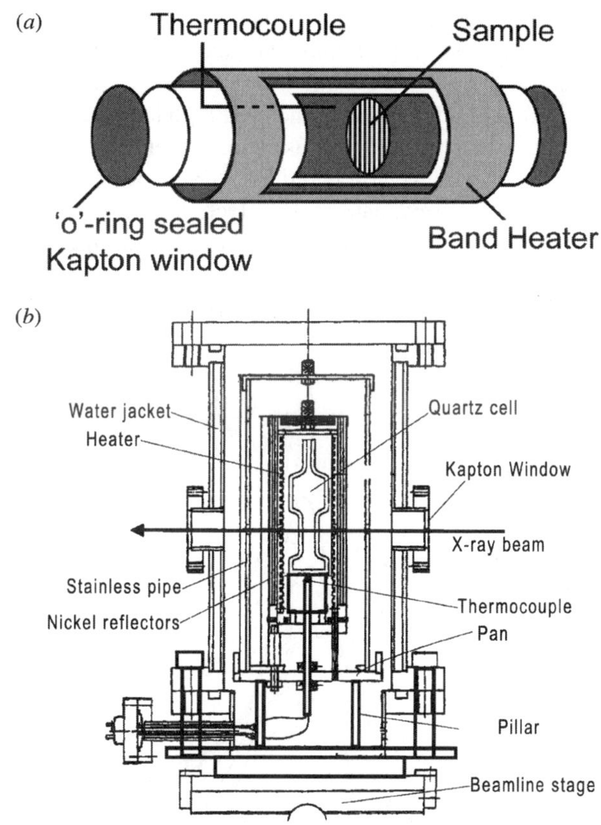

Figure 3 Schematic of (*a*) a simple combined transmission/fluorescence EXAFS cell used for low-temperature ionic liquids studies and (*b*) a transmission EXAFS cell used for high-temperature molten salts studies. Reproduced from Reference 17 with permission.

MOLTEN SALTS AND IONIC LIQUID STRUCTURE

To date, EXAFS has been used to examine only the structure of high-temperature molten salts in detail.

One of the earliest uses of EXAFS to examine the liquid structure of molten salts was reported by Crozier et al. (18). The liquid structure of $[Bu_4N][MnBr_3]$ and $[Bu_4N]_2[MnBr_4]$ was examined at temperatures up to 400 K. Of particular interest in this study was the extent to which anharmonicity in the two-body interaction potential used to analyze the EXAFS data affected the parameters extracted. At high temperatures the anharmonicity becomes important. After taking into account the asymmetry in the atomic distributions of bromine around the manganese and vice-versa, Mn-Br bond distances of 2.46 and 2.50 Å were found for $[Bu_4N][MnBr_3]$ and $[Bu_4N]_2[MnBr_4]$, respectively. From this data, a bromine coordination number of 3.14 around the manganese was determined in the liquid state of $[Bu_4N][MnBr_3]$ compared with 6 for the solid-state structure. This trigonal structure had not been previously considered in the liquid state and demonstrated the power of EXAFS analysis.

The analysis and effect of asymmetry at high temperatures in molten salts have been the subjects of a number of publications recently (see below). In particular, Di Cicco and coworkers have compared EXAFS, neutron diffraction and simulation data in detail in order to evaluate analysis techniques. An example of the asymmetry and how it changes with temperature is illustrated in Figure 4 for solid AgBr (19).

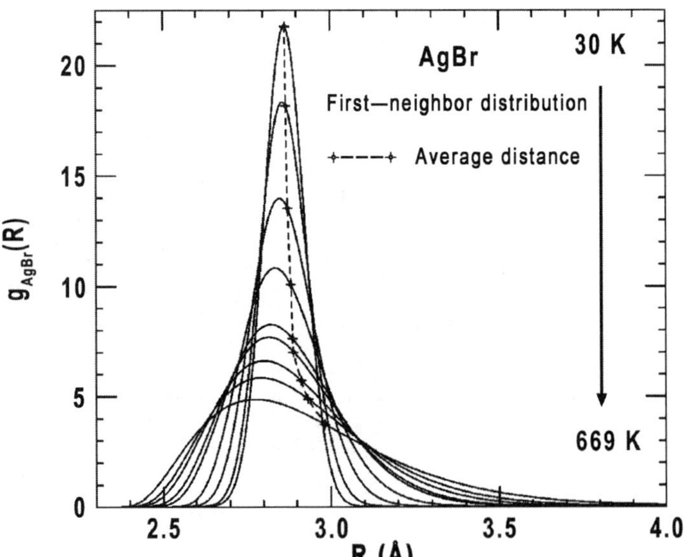

Figure 4 First neighbor AgBr distribution as a function of temperature. Reproduced from Reference 19 with permission.

Di Cicco and co-workers (20–22) have examined the structure of molten CuBr using Cu and Br K-edge EXAFS. From the EXAFS data, the Cu–Br bond length distribution was found to be shorter than that derived from neutron data. In addition, there was little contribution to the Cu data of Cu–Cu distances and only a small contribution to the Br EXAFS data of Br–Br distances. The data indicate that the melt has more covalent character than previously reported. The nearly total covalent bonding in CuBr is similar in the solid and liquid phases, and on melting a quasicontinuous phase change process occurs with a significant decrease in the Cu–Br distance in the liquid compared with that in the solid. With this methodology, the liquid structure of CuI (23) has also been determined, and similar inadequate descriptions in modeling, for example, have been shown. In this case, the existence of short Cu–Cu interactions are shown to be present.

Similarly, by analyzing the liquid structure of AgBr and AgI using multiple edge EXAFS, Di Cicco et al. have shown that, again, the molecular dynamics used to date is not sufficiently robust to predict the EXAFS data obtained (19, 24). For example, in Figure 5 the molecular dynamics simulations and neutron diffraction are compared using reverse Monte-Carlo methods with the EXAFS data for molten AgBr. The combined analysis of the Ag and Br K-edge data enabled the short-range structure of the liquid to be elucidated more accurately. From this analysis it has been shown that the most probable Ag–Br distance decreases with increasing temperature; however, the radial distribution function becomes increasingly asymmetric toward longer distances indicating that the average bond distance is expanding, as expected from thermal expansion (Figures 4 and 5). As found for CuBr, there was an almost complete absence of the Ag–Ag distance in the EXAFS.

In these studies, Di Cicco et al. used multiple-edge EXAFS and analyzed the data using the GNXAS method incorporating multiple scattering where the many body interactions contributing to the EXAFS were calculated using ab initio methods (19–23). Clearly, for these complex systems, simple Gaussian data analysis is not sufficient. However, despite the problems many studies have used simpler data analysis methods for the study of molten salts.

Gaussian analysis was used by Inui et al. for determination of the liquid structure of silver halides. They reported much shorter bond distances for Ag-X (X = Br or I) in both the molten and solid states compared with those from crystalline data and the neutron diffraction data. The discrepancy was of the order 0.15 Å. Despite this disparity, the authors also investigated mixtures of AgBr and AgI in the molten state. From the EXAFS, they concluded that as the concentration of iodide increased the first shell distance decreased slightly, and this was attributed to the increased covalency (25, 25a).

For comparison, Takeda et al. also examined the EXAFS through the melting point of AgCl-AgI mixtures (26). In this study, the effect of anharmonicity was taken into account using an asymmetric distribution model and a curve-fitting method. For $AgCl_{0.43}I_{0.57}$, the first shell distance is observed to decrease with increasing temperature from 2.76 Å at 360°C to 2.71 Å at 800°C. In addition, the

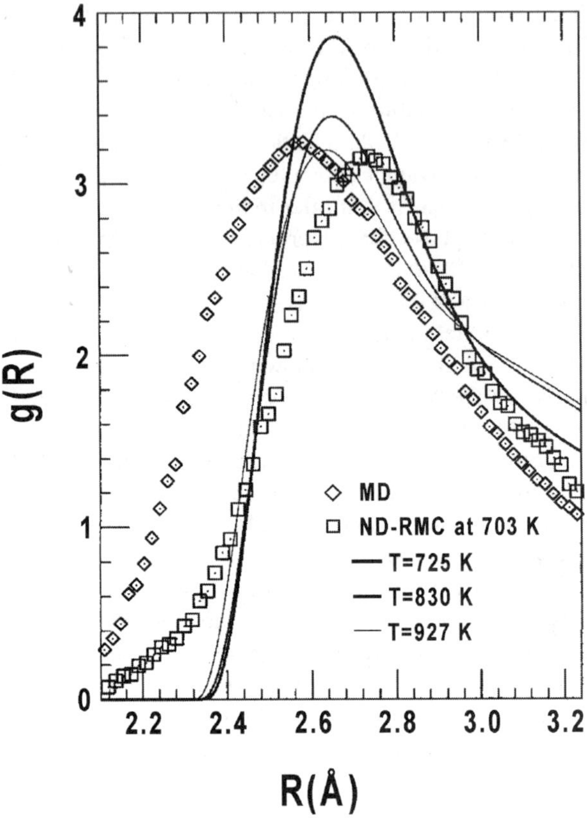

Figure 5 Comparison of the $g_{AgBr}(r)$ determined by neutron diffraction-reverse Monte Carlo (ND-RMC), molecular dynamics (MD) and EXAFS refinement as a function of temperature. Reproduced from Reference 19 with permission.

first shell coordination number decreases significantly from 4.1 to 3.3 over the same temperature range. By examining the partial radial distribution functions from a curve-fitting procedure using an asymmetric model distribution model, below the melting point, coordination numbers of 2 for both Cl and I were observed around the central silver cation. This was reported as indicative of a tetrahedral arrangement in the solid phase. As the temperature is increased the radial distribution function associated with the iodine atoms was invariant, whereas the chlorine atom radial distribution function amplitude decreased significantly. The overall decrease in the total coordination number on melting is therefore thought to occur because the chlorine atoms are removed from the first coordination shell as the bonding strength between the Ag and I increases.

In contrast, multiple-edge EXAFS studies on KBr and RbBr are in good agreement with the established theory (20, 22, 27, 28). A detailed study of KBr using multiple scattering, i.e., the analogous approach as used for CuX and AgX (see below), has been reported by Di Cicco et al. (27, 28). Although some changes to the shape of the radial distribution functions were observed in the liquid phase, the effect of the high degree of ionicity in the salt reduced the discrepancy between the theory and the experiment. The presence of three-body and second-neighbor interactions are significant only if there is a high degree of covalency present in the system. As found for CuBr and AgBr, it is noticeable that compared with the solid, the K-Br distance is shifted to slightly shorter distances.

Zn and Rb K-edge EXAFS have also been used to examine the melting of Rb_2ZnCl_4 compared with that of the liquid structure of $ZnCl_2$ and RbCl (29, 29a). On melting $ZnCl_2$, zinc is found to have tetrahedrally coordinated chlorine linked by corner-sharing chlorines in a weak extended network. In RbCl, significant disorder in the chloride shell around the rubidium is evident and indicates significant movement of the Rb^+ and Cl^- in the molten state. In the crystal structure of Rb_2ZnCl_4, the chlorine coordination number around the Rb is between 8 and 9, whereas the Zn is found in isolated $ZnCl_4$ units. In the molten state, the EXAFS also indicates isolated $[ZnCl_4]^{2-}$ units with a chlorine coordination of 7.6 around the Rb. This may be compared with a chlorine coordination of 4.8 in liquid RbCl. The EXAFS clearly shows that the solid and liquid structures of Rb_2ZnCl_4 are similar and that the melt does not rearrange into a simple combination of the component parts.

Similar EXAFS spectra have been recorded for $ZnBr_2$ by Okamoto et al. (30). It is clear from this study that multiple-scattering contributions are necessary to fit the data arising from the anharmonic effects. This was achieved by using a molecular dynamics simulation to act as the input to the EXAFS fitting program. From multiple edge analysis from the Zn and Br K-edge data, a Zn^{2+}-Br^- distance of 2.45 Å was found and a coordination number of 3.9 of bromides around the zinc cation. From the ratio of the Br^--Br^- distance to Zn^{2+}-Br^-, the structure of the $[ZnBr_4]^{2-}$ was indicated to be tetrahedral with most of the bromides shared between tetrahedra. The data from $ZnCl_2$ showed an analogous structure despite the difference in the ionic character in both melts.

The liquid structure of lanthanide halides has also been examined recently using EXAFS (31, 32). Inui et al. (25, 25a) used a cumulative expansion method to allow for the anharmonic vibration effect. On melting, the La^{3+}-Cl^- distance is observed to decrease from 2.94 to 2.89 Å and is in agreement with neutron diffraction studies. In addition a first shell coordination number of 7.4 was found in the EXAFS, again in agreement with both molecular dynamics calculations and neutron diffraction data. The EXAFS has been interpreted in this case as a mixture of $[LaCl_7]^{4-}$ and $[LaCl_8]^{5-}$. Second and third shell lanthanum distances were also observed in the molten state showing the extended order found in the liquid. Liquid yttrium(III) chloride has been compared with the $LaCl_3$ liquid. Much shorter M-Cl distances are observed for yttrium compared with those of lanthanum by 0.17 Å. Similarly, a large decrease in the first shell cation was observed, and the La^{3+}-La^{3+} distance was found to be 4.90 Å, whereas in YCl_3, the Y^{3+}-Y^{3+} distance was 3.85 Å. This large difference has been attributed to two different linkages found in the two salts. In $LaCl_3$, corner-sharing sixfold complexes are thought to be present, whereas in YCl_3, edge-sharing complexes dominate. In addition, because the distances are not simply linked to the La-Cl distance, the corner-sharing octahedra are likely to be distorted. In comparison, the structure of molten YBr_3 and $LaBr_3$ is similar to that observed for the chloride analogues (32, 33).

Okamoto et al. extended this study by examining the structure of YCl_3 and YBr_3 in eutectic LiCl-KCl and LiBr, respectively (32). On mixing the yttrium halides with the alkali halides, the first coordination shell becomes sharper, and a small decrease occurs in the Y^{3+}-halide bond distance with increasing dilution with the alkali salt. This narrowing of the peak comes from stabilization of the octahedra in the liquid. However, this is also accompanied by a decrease in the coordination number of Y^{3+} around the halide. This signifies a reduction in the number of edge-sharing octahedra, i.e., the liquid network becomes less extensive as the concentration of yttrium halide decreases in the melt.

As part of this ongoing study, Okamoto et al. (34) recently reported the structure of molten $SrCl_2$. Again the data analysis was performed using a cumulant expansion technique, and the effect of the anharmonic vibration effect observed at high temperature is shown clearly. Using the higher cumulants allows for the effect, and clearly much better fits are obtained using this procedure for the high-temperature samples. For the room-temperature samples, sufficiently good fits were obtained without the need to use higher cumulants. For comparison, the Sr^{2+}-Cl^- distance was 2.79 Å with a chloride coordination number of 7.0 without the anharmonic effects, whereas by using the third and fourth cumulants, the distance increased to 2.99 Å and the coordination number decreased to 6.6. From this analysis, it was concluded that the ion arrangement found in the molten state of $SrCl_2$ is simple compared with that of $ZnCl_2$ and YCl_3, for example, where edge/corner-shared tetrahedral and octahedral were observed.

Molten $PbCl_2$ has been examined using laboratory EXAFS by Ablanov et al. (35). A significant decrease was observed in the chlorine coordination number around the lead in the solid compared with the melt, from nine in the solid to six in the melt, at a distance of 2.83 Å. No model is proposed for the structure in the liquid; however, the authors conclude that the changes found in the liquid compared with the solid structure may be from Pb-Cl bond breaking, which results in a distortion in the local structure.

In the studies described above, the samples were supported in low-atomic-weight matrices, melted in situ and measured in transmission mode. Similarly, second-generation ionic liquids have been studied. Carmichael et al. (36) showed that it was possible to support and melt $[C_2mim]_2[NiCl_4]$ and $[C_{14}mim]_2[NiCl_4]$ in inert matrices such as boron nitride, graphite, and lithium fluoride without the EXAFS being affected by the sample matrix used. In these samples, the Ni K-edge EXAFS was investigated between room temperature and 131°C, and even in LiF, where halide exchange was possible, little difference was found in the Ni coordination on melting.

SPECIES DISSOLVED IN MOLTEN SALTS AND IONIC LIQUIDS

In contrast to the liquid structure of the melts, the structures found on dissolved species have been dominated by studies using ionic liquids. Both first- and second-generation ionic liquids have been examined in detail.

Dent et al. (37) studied the dissolution of $[C_2mim][MCl_4]$ in $[C_2mim]Cl\text{-}AlCl_3$ binary mixtures, for M = Mn, Co, and Ni, at $AlCl_3$ mol fractions (X_{AlCl_3}) of 0.35 and 0.60 using the M K-edges. Due to the problems associated with the high concentration of chloride, it was not possible to perform transmission experiments and thus fluorescence measurements were used. In this case, self-absorption problems were overcome by using a thin film of liquid pressed between two sheets of polythene sealed in a glove box. The coordination of Ni, Co, and Mn was found to change from $[MCl_4]^{2-}$ to $[M(AlCl_3)_4]^-$ as the mol fraction of $AlCl_3$ increased. Figure 6 shows the EXAFS and pseudoradial distribution functions for M = Co in both the acidic and basic chloroaluminate ionic liquids. Aluminum coordination was observed only in acidic mixtures, $X_{AlCl_3} > 0.5$. The latter was surprising given that at $X_{AlCl_3} = 0.60$ $AlCl_3$, almost half the anion species are in the form $[Al_2Cl_7]^-$, yet no coordinating $[Al_2Cl_7]^-$ was observed.

Dent et al. (38) also investigated the V K-edge EXAFS for the dissolution of $[C_2mim][VOCl_4]$ and $[NEt_4][VO_2Cl_2]$ in a basic $[C_2mim]Cl\text{-}AlCl_3$ ionic liquid ($X_{AlCl_3} < 0.5$) and compared the data with that from solid samples. In both cases, the dissolved and solid samples showed similar EXAFS and no coordination of the chloroaluminate species was found to the vanadyl oxygen, for example.

McMath recently studied the catalytically active species formed during the dimerization of but-1-ene in a range of chloroaluminate ionic liquids (16). As with the other chloroaluminate studies, fluorescence measurements were used; however, the EXAFS was performed in the presence of the gas in order to follow the active species in situ. Ni K-edge data was taken for $NiCl_2(PiPr_3)_2$ and $[Ni(MeCN)_6][BF_4]_2$ in basic and acidic chloroaluminate ionic liquids using mixtures of $AlCl_3$ and $[C_2mim]Cl$ in the absence and presence of ethylaluminium chloride. Unsurprisingly, under basic conditions ($X_{AlCl_3} < 0.5$) the $[Ni(MeCN)_6][BF_4]_2$ catalysts were found to be inactive and showed the presence of $[NiCl_4]^{2-}$ in solution. Surprisingly, melts with a small excess of $AlCl_3$ ($X_{AlCl_3} = 0.55$) also showed formation of only $[NiCl_4]^{2-}$ with this catalyst. Although, it was observed that this system had a small amount of dimerization activity, the reaction kinetics indicated that the catalyst underwent rapid deactivation over a 30-min period, which illustrates the limitation of the EXAFS technique. The EXAFS had to be taken over 4 h in order to obtain sufficient signal-to-noise, and in this case the data are not representative of the active system. On addition of $EtAlCl_2$, an active system with $[Ni(MeCN)_6][BF_4]_2$ was formed (mol ratio $[C_4mim]Cl{:}AlCl_3{:}EtAlCl_2$ 1:1.2:0.1). In this system the EXAFS data clearly indicated three bidentate aluminum chloride species coordinated through a pair of chlorines on each ligand, i.e., $[Ni[AlCl_4]_3]^-$, i.e., the analogous species reported by Dent et al. (37) (see above). This species was

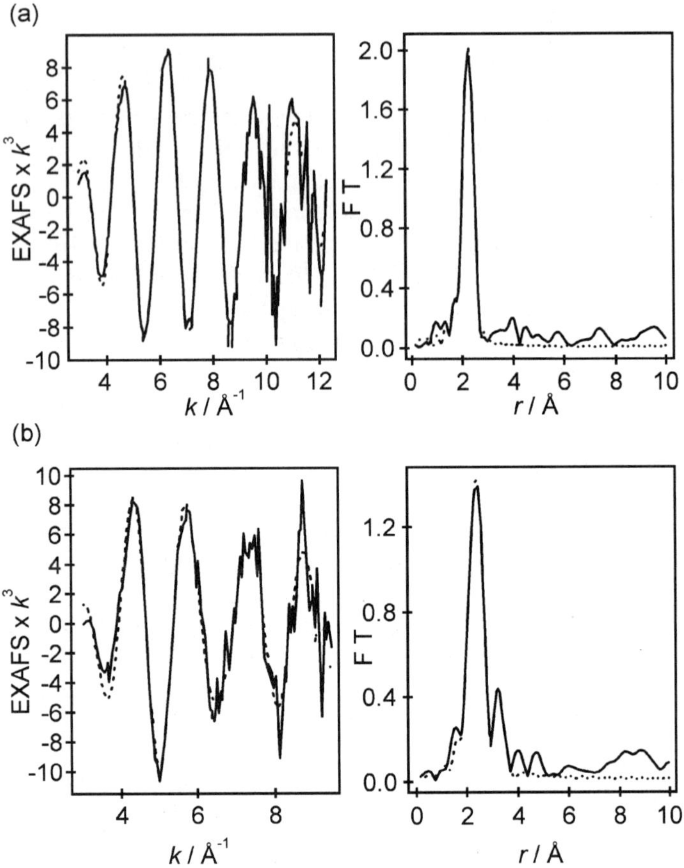

Figure 6 The EXAFS data and pseudoradial distribution functions of Co(II) in a (*a*) basic and (*b*) acidic chloroaluminate ionic liquid. Reproduced from Reference 37 with permission.

found to be present in all the solutions that showed high activity over extended reaction times. McMath postulated that because some activity was observed in the slightly acidic melts ($X_{AlCl_3} = 0.55$) under reaction conditions, it is possible that $[Ni[AlCl_4]_3]^-$ formed initially but converted to the inactive tetrachloronickel species over time. Although addition of trialkylphosphine increased the reaction rate, little change in the EXAFS was observed using $NiCl_2(PiPr_3)_2$ with the species identified in each case as $[Ni[AlCl_4]_3]^-$ in the active melts. All chloroaluminates described above contained $AlCl_3$. Some activity was also observed in melts synthesized from only $EtAlCl_3$ and [C$_4$mim]Cl using $NiCl_2(PiPr_3)_2$. In this case, the EXAFS results were fitted to the formation of a carbene complex shown in Figure 7, which is thought to be formed by the $EtAlCl_2$ acting as a proton scavenger that removes C(2) hydrogen thus allowing reaction with the Ni active center

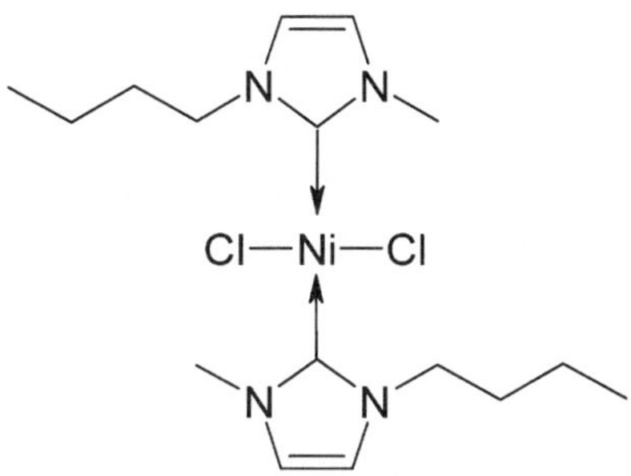

Figure 7 Ni carbene complex formed during the dimerization of but-1-ene in the ionic liquid containining 55 mol% $EtAlCl_3$ and 45 mol% $[C_4mim]Cl$ using $NiCl_2(PiPr_3)_2$ as the catalyst.

to form the carbene. For the same ionic liquid using $[Ni(MeCN)_6][BF_4]_2$ no activity was observed and the solution rapidly became dark. In this case, the EXAFS indicated the presence of metallic nickel.

Reactions have also been investigated in high-temperature melts, specifically, LiCl–KCl-, NaCl–CsC-l-, and NaCl–KCl-based melts, which are used in the reprocessing of spent nuclear fuels. Volkovich et al. reacted Re, ReO_2, and TcO_2 with Cl_2 and HCl in these melts at high temperature in order to chlorinate the metals before the possibility of electrochemically purifying the metals (39). Understanding the speciation of the elements in the melts is important and EXAFS provides a method of elucidating the structures. In the case of rhenium metal and rhenium dioxide, only octahedral $[ReCl_6]^{2-}$ was observed with an average Re-Cl distance of 2.36–2.38 Å, which was independent of the particular melt used, or whether Cl_2 or HCl was used as the chlorinating agent. In contrast, TcO_2 showed formation of Tc with 6 coordinating chlorine atoms at 2.36 Å on reaction with HCl, whereas on reaction with Cl_2, 4 oxygen atoms are found in the first shell at 1.72 Å. These correspond to the following reactions:

$$2MCl + TcO_2 + 4HCl \rightarrow [TcCl_6]^{2-} + 2M^+ + H_2O$$

$$2TcO_2 + 5/2Cl_2 + MCl \rightarrow [TcO_4]^- + TcCl_6 + M^+$$

$$3TcO_2 + 4Cl_2 + MCl \rightarrow TcCl_6 + 2TcO_3Cl.$$

This study shows the importance of performing the reactions under reaction conditions. Quenched melts were investigated in this study and cannot explain the measured electronic absorption spectra. Two clear anomalies are seen: In the reactions with rhenium, only one species is observed in the quenched solid, whereas at 650–700°C, it is clear that a range of species are present that depend on which binary molten is used. Furthermore, only $[TcO_4]^-$ is found on reacting TcO_2 with Cl_2, whereas chlorinated species must be present from the reactions shown above. The disappearance of the chlorinated moieties is explained by sublimation of the volatile TcO_xCl_y on quenching.

As noted above, the studies examining high chloride content melts had to be performed in fluorescence. In contrast, owing to the decrease in average atomic weight of the medium compared with that of the chloroaluminate systems, second-generation ionic liquids may be investigated in transmission mode. For example, Carmichael et al. (36) have shown that solutions of $[C_2mim]_2[NiCl_4]$ in $[C_4mim][PF_6]$ may be studied by supporting the liquid between two boron nitride discs. The resulting Ni K-edge EXAFS showed a local structure similar to that of the molten $[C_2mim]_2[NiCl_4]$ described above.

Baston et al. (40) studied samples of ionic liquid after the anodization of uranium metal in $[C_2mim]Cl$ using the U L_{III}-edge EXAFS to establish both the oxidation state and the speciation of uranium in the ionic liquid. This was part of an ongoing study to replace high-temperature melts such as LiCl:KCl (41) with ionic liquids. Although it was expected that, when anodized, the uranium would be in the +3 oxidation state, electrochemistry showed that the uranium is actually in a mixture of oxidation states. The EXAFS of the solution showed an edge jump at 17,166.6 eV, indicating a mixture of uranium(IV) and uranium(VI). Two peaks were fitted to the data corresponding to a 1:1 mixture of $[UCl_6]^{2-}$ and $[UO_2Cl_4]^{2-}$, in agreement with the position of the edge. Oxidation to uranium(VI) was surprising in this system and may have arisen from the presence of water owing to the highly hygroscopic nature of $[C_2mim]Cl$.

Nitrate-based second-generation ionic liquids have also been studied for the reprocessing of uranium (42). In particular, it has been shown that dinuclear dioxouranium(VI) salt containing a bridging oxalate ligand, 1-butyl-3-methylimidazolium μ_4-(*O*,*O*,*O*′,*O*′-ethane-1,2-dioato)-bis{bis(nitrato-*O*′,*O*)dioxouranate(VI)} ($[C_4mim]_2[\{UO_2(NO_3)_2\}_2(\mu\text{-}C_2O_4)]$), is formed on the oxidative dissolution of uranium(IV) oxide in $[C_4mim][NO_3]$ using concentrated nitric acid as the oxidizing agent. However, it was not clear how the oxalate species was formed. Ex-situ EXAFS was used to examine the speciation, in particular the reaction in the presence and absence of acetone. Figure 8 shows the pseudoradial distribution functions obtained from solutions of $UO_2(NO_3)_2.6H_2O$ and $[C_4mim]_2[\{UO_2(NO_3)_2\}_2(\mu_4\text{-}C_2O_4)]$ dissolved in $[C_4mim][NO_3]$ at room temperature and the reaction solution following the oxidative dissolution of UO_2 in $[C_4mim][NO_3]$, without added acetone and in the presence of 0.2 M acetone. On dissolution of $[C_4mim]_2[\{UO_2(NO_3)_2\}_2(\mu_4\text{-}C_2O_4)]$ in $[C_4mim][NO_3]$, the EXAFS indicated that the oxalate "dinuclear" species remains intact in solution and the hydrated dioxouranium(VI) cation is not formed. The analysis also shows that the η^2-NO_3 are not displaced by water. A comparison of the EXAFS data for the pure dinuclear oxalate and mononuclear nitrate complexes dissolved in the ionic liquid with the solutions obtained after dissolution of UO_2 in $[C_4mim][NO_3]$ indicates that a mixture of species is present. Furthermore, the addition of acetone had little effect on the EXAFS data. For both solutions, following dissolution of UO_2, the best fit was obtained with 15:85 mononuclear nitrate:dinuclear oxalate dimer mol ratio present in solution. It was not possible to determine whether the dimer is still present or whether a mononuclear species has been formed. For example, the fact that Debye-Waller factors for the U-C distances are so high suggests that the oxalate species is μ_2-bonded.

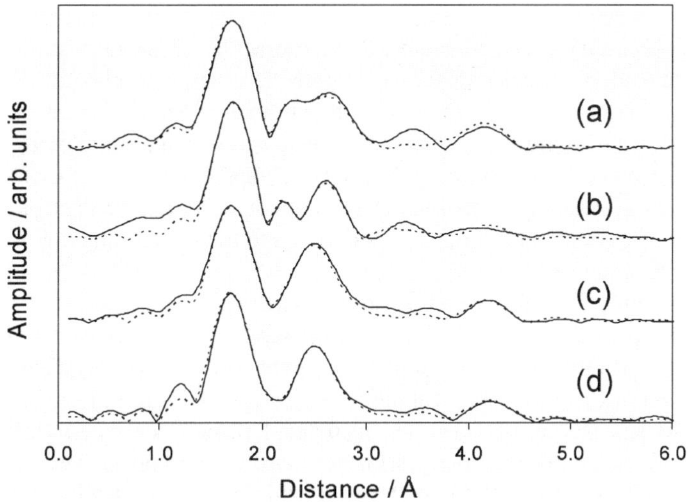

Figure 8 Comparison of the experimental (*solid line*) and fitted (*dashed line*) pseudoradial distribution (*lower*) functions from (*a*) $UO_2(NO_3)_2.6H_2O$ and (*b*) $[C_4mim]_2[\{UO_2(NO_3)_2\}_2(\mu_4\text{-}C_2O_4)]$ dissolved in $[C_4mim][NO_3]$, and following UO_2 dissolution in $[C_4mim][NO_3]$; (*c*) in the absence of acetone and (*d*) in the presence of 0.2 M acetone. Reproduced from Reference 42 with permission.

Although important information is generated from the EXAFS, this study does illustrate the limitations of the technique (42). The EXAFS is consistent with the hydrated dioxouranium(VI) nitrate species; however, there are a number of other possible structural arrangements on dissolution of $UO_2(NO_3)_2.6H_2O$ in the nitrate ionic liquid involving monodentate nitrate coordination, namely $[UO_2(\eta^2\text{-}NO_3)_2(\eta^1\text{-}NO_3)(H_2O)]^-$ and $[UO_2(\eta^2\text{-}NO_3)_2(\eta^1\text{-}NO_3)_2]^{2-}$. Similar fit factors were reported for each of these species, although slightly different χ^2 values are found. It is likely that a range of neutral and anionic species is present in each case and that the solution contains an equilibrium mixture that cannot be determined using EXAFS owing to its average nature. However, it is clear from the EXAFS, in combination with NMR data, that the formation of the oxalate is probably caused by organic impurities in either the nitric acid or ionic liquid that are oxidized to the oxalate ligand, which is trapped between two dioxouranium(VI) nitrate units.

The use of ionic liquids for the separation of uranium from nitric acid solutions has also been investigated as an alternative to dodecane in the TRUEX process (43). Ionic liquids provide an environment that can increase the concentration of actinides, and EXAFS has been used to understand the coordination of uranium. With use of carbamoylphosphine oxide (CMPO) and tri(n-butyl)phosphate, the EXAFS indicated significant changes in the inner sphere complex on changing the extracting solvent between dodecane and $[C_4mim][PF_6]$/$[C_8mim][NTf_2]$ but not between the ionic liquids. In dodecane, the uranium is surrounded by 2 uranyl oxygens, with 2 monodentate CMPO molecules (P = O bound) and 2 bidentate nitrate anions, i.e., 6 equatorial oxygens. In contrast, in the ionic liquid, although the 2 uranyl oxygens were found, the number of equatorial oxygens decreased to 4–4.5. The difference is attributed to water molecules complexing to the uranium and also the formation of $[UO_2(NO_3)(CMPO)]^+$, which causes the ionic liquid to cation exchange with ions in the aqueous phase. The exact number of water molecules could not be ascertained owing to the uncertainty in the denticity of the CMPO and nitrate species.

A similar mechanism for cation extraction from aqueous species was also reported by Jensen et al. for Sr^{2+}(aq) studied using EXAFS (44). With use of a representative neutral complexing agent, *cis*-syn-*cis*-dicyclohexyl-18-crown-6 (DCH18C6), and various counterions ($[NO_3]^-$, Cl^-, $[SO_4]^{2-}$), the speciation was studied in $[C_5mim][NTf_2]$ and in octan-1-ol and compared with dissolution of $Sr(NO_3)_2(18C6)$. No changes in the coordination sphere were observed on dissolving $Sr(NO_3)_2(18C6)$ in the ionic liquid compared with those in the solid state structure with a oxygen coordination number of 10.4. However, following extraction of Sr^{2+} in the presence of DCH18C6 from an aqueous solution of $Sr(NO_3)_2$ into the ionic liquid, the oxygen coordination number decreases to 8.5. This indicated that although the crown ether was retained, little nitrate coordination to the strontium cation was observed. In contrast, extraction using octan-1-ol showed a structure similar to that found in $Sr(NO_3)_2(18C6)$, i.e., with bidentate nitrate coordination and an oxygen coordination number of 10.2. The difference between the solvents was thought to be because the water displaced the nitrate anion in the case of the ionic liquid. The proposed structures are shown in Figure 9. In addition, if the nitrate salt is replaced by chloride or sulfate, the EXAFS also indicates no coordination of either anion, and a similar oxygen coordination number in each case. The absence of anion in the inner shell of the strontium cation shows that anion extraction is poor when the ionic liquids are used in contrast to use with the organic solvent. In fact, as described by Visser et al. for uranium extraction (43) (see above), significant solubilization of the

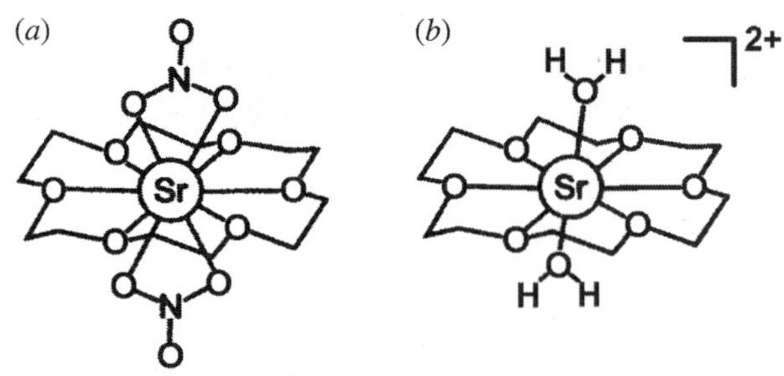

Figure 9 Schematic showing the coordination environments of (*a*) $Sr(NO_3)_2$ (DCH18C6) in octan-1-ol and (*b*) the $[Sr(DCH18C6)(H_2O)_2]^{2+}$ cation found in $[C_5mim][NTf2]$. Reproduced from Reference 44 with permission.

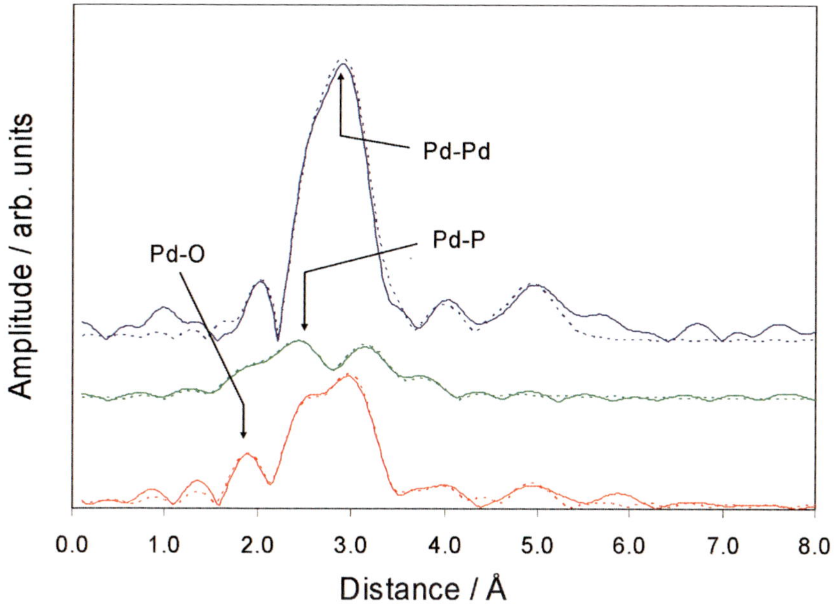

Figure 10 Comparison of the experimental (*solid line*) and fitted (*dashed line*) pseudoradial distribution functions from palladium acetate in $[C_4mim][PF_6]$ without (*blue*) and with (*green*) triphenylphosphine at 80°C and in the presence of triphenylphosphine and reagents at 50°C for 20 min (*red*). Reproduced from Reference 15 with permission.

ionic liquid into the aqueous phase occurs, despite the hydrophobicity of the ionic liquid used.

The large difference between the structure of complexes in ionic liquids and organic solvents has also been illustrated using EXAFS of europium(III) complexed to 1-(2-thienyl)-4,4,4-trifluoro-1,3-butanedione (Htta) (45). Following dissolution of Htta and Eu^{3+} in $[C_4mim][NTf_2]$, the EXAFS showed strong scattering due to approximately eight oxygen atoms in the inner shell at 2.39 Å; however, there was little evidence for second shell contributions from carbon, from the carbonyl groups, or from sulfur from coordination of $[NTf_2]^-$ anions. The EXAFS is consistent with the formation of $[Eu(tta)_4]^-$ complex in the ionic liquid, whereas in organic solvents neutral species are formed, e.g., $[Eu(tta)_3(Htta)]$ or $[Eu(tta)_3(H_2O)]$.

In all these studies EXAFS has illustrated some of the problems associated with using ionic liquids as extraction media (43–45). In each case, the anions formed caused the ionic liquid to become more hydrophilic by forming a binary ionic liquid, and the ionic liquid was leached into the aqueous phase.

Recently, EXAFS has been used to examine the structure of lead complexes formed during the preparation of acetate-based ionic liquids from lead(II) salts (46, 47). The synthesis is not clean, and a significant amount of lead was found to remain in the ionic liquid. Similar structures were found from EXAFS analysis of the solution and of single-crystal X-ray data obtained from precipitating the lead salt from the ionic liquid. These studies indicate that the salt contained $[Pb(O_2CMe)_4]^{2-}$ with the acetate groups is associated with the lone pair of the lead.

Hamill et al. (15) have investigated the Heck reaction in room-temperature ionic liquids catalyzed by palladium salts and complexes. On dissolution of palladium ethanoate in $[C_4mim]^+$ and *N*-butylpyridinium ($[C_4py]^+$) hexafluorophosphate and tetrafluoroborate ionic liquids, and triethyl-hexyl ammonium *bis*(trifluorome thanesulfonyl)imide, a gradual change from ethanoate coordination to the formation of palladium metal was observed in the Pd K-edge EXAFS, as shown in Figure 10, see color insert. In pyridinium chloride ionic liquids and in 1,2-dimethyl-3-hexyl imidazolium chloride ($[C_6mmim]Cl$), which is where the C(2) position is protected by a methyl group, only $[PdCl_4]^{2-}$ was observed. In $[C_6mim]Cl$, the EXAFS showed the formation of a *bis*-carbene complex similar to that found for the nickel-catalyzed dimerization of but-1-ene shown in Figure 7 (see above). In the presence of triphenylphosphine, Pd-P coordination was observed in all ionic liquids except where the carbene complex was formed. During the Heck reaction, the formation of palladium was found to be quicker than in the absence of reagents. Overall, the EXAFS showed the presence of small, approximately 1 nm diameter, palladium clusters formed in solution.

CONCLUSIONS

EXAFS has been used to investigate both the structure of ionic liquids/molten salts as well as species dissolved in the ionic liquids. Valuable information has been obtained in both areas that complement other data such as from neutron diffraction and UV-vis/IR spectroscopy. However, EXAFS interpretation can be difficult, as demonstrated by the large differences found in the structure of copper and silver halides depending on which type of analysis was used. In addition, for ionic liquids and molten salts, which are based on halides, problems associated with self-absorption occur, requiring thin films to be investigated. EXAFS has been used to study a wide range of applications in ionic liquids, in particular. However, EXAFS is an average technique and therefore systems that contain a high degree of speciation are often complex to analyze. Therefore, it is important to understand the limitations of the technique and use other spectroscopic and chemical information to analyze the system under investigation.

ACKNOWLEDGMENTS

C.H. thanks Mark Niewenhuyzen (Queen's University, Belfast) for useful discussions and comments in the preparation of this manuscript.

LITERATURE CITED

1. Edwards FG, Enderby JE, Howe RA, Page DI. 1975. The structure of molten sodium chloride. *J. Phys. C* 8:3483–90
1a. Biggin S, Enderby JE. 1982. Comments on the structure of molten-salts. *J. Phys. C* 15:L305–9
1b. Allen DA, Howe RA. 1992. A measurement of the structure of the superionic conductor copper(I) bromide in its liquid-phase by neutron-diffraction. *J. Phys. Condens. Matter* 4:6029–38
1c. Blander M, Bierwagen E, Calkins KG, Curtiss LA, Price DL, Saboungi M-L. 1992. Structure of acidic haloaluminate melts—neutron-diffraction and quantum chemical calculations. *J. Chem. Phys.* 97:2733–41
1d. Lee Y-C, Price DL, Curtiss LA, Ratner MA, Shriver DF. 2001. Structure of the ambient temperature alkali metal molten salt $AlCl_3$/LiSCN. *J. Chem. Phys.* 114:4591–94
1e. Trouw FR, Price DL. 1999. Chemical applications of neutron scattering. *Annu. Rev. Phys. Chem.* 50:571–601
2. Takahashi S, Maruoka K, Koura N, Ohno H. 1986. X-ray-diffraction analysis of molten $AlCl_3$-NaCl system. *J. Chem. Phys.* 84:408–15
2a. Takahashi S, Muneta TN, Koura N, Ohno H. 1985. Structural-analysis of the molten-salt 50 mol-percent $AlCl_3$-50 mol-percent NaCl by X-ray-diffraction. *J. Chem. Soc. Faraday Trans. II* 81:1107–15
2b. Igarashi K, Okamoto Y, Mochinaga J, Ohno H. 1988. X-ray-diffraction study of molten eutectic LiF-NaF-KF mixture. *J. Chem. Soc. Faraday Trans. I* 84:4407–15
2c. Takahashi S, Koura N, Murase M, Ohno H. 1986. Structural-analysis of $AlCl_3$-*N*-butylpyridinium chloride electrolytes by X-ray-diffraction. *J. Chem. Soc. Faraday Trans. II* 82:49–60
3. Gordon CM. 2002. New developments in catalysis using ionic liquids. *Appl. Catal.* A222:101–17
3a. Seddon KR. 1997. Ionic liquids for clean technology. *J. Chem. Biotechnol.* 68:351–56
4. Hurley FH, Wier TP. 1951. Electrodeposition of metals from fused quaternary ammonium salts. *J. Electrochem. Soc.* 98:203–6

4a. Hurley FH, Wier TP. 1951. The electrodeposition of aluminum from nonaqueous solutions at room temperature. *J. Electrochem. Soc.* 98:207–12
5. Gale RJ, Gilbert B, Osteryoung RA. 1978. Raman spectra of molten aluminum chloride: 1-butylpyridinium chloride systems at ambient temperatures. *Inorg. Chem.* 17:2728–29
5a. Nardi JC, Hussey CL, King LA. 1978. *U.S. Patent No.* 4,122,245
6. Wilkes JS, Zaworotko MJ. 1992. Air and water stable 1-ethyl-3-methylimidazolium based ionic liquids. *Chem. Commun.* 965–67
7. Wilkes JS. 2002. A short history of ionic liquids—from molten salts to neoteric solvents. *Green Chem.* 4:73–80
8. Rogers RD, Seddon KR. 2003. Ionic liquids—solvents of the future? *Science* 302:792–93
8a. Freemantle M. 1998. Designer solvents—ionic liquids may boost clean technology development. *Chem. Eng. News* 76:32–37
9. Earle MJ, Seddon KR. 2000. Ionic liquids. Green solvents for the future. *Pure Appl. Chem.* 72:1391–98
9a. Sheldon R. 2001. Catalytic reactions in ionic liquids. *J. Chem. Soc. Chem. Commun.* 2399–407
9b. Dupont J, de Souza RF, Suarez PAZ. 2002. Ionic liquid (molten salt) phase organometallic catalysis. *Chem. Rev.* 102: 3667–91
10. Lin M-C, Chen P-Y, Sun I-W. 2001. Electrodeposition of zinc telluride from a zinc chloride-1-ethyl-3-methylimidazolium chloride molten salt. *J. Electrochem. Soc.* 148:C653–58
10a. Chen P-Y, Lin Y-F, Sun I-W. 1999. Electrochemistry of gallium in the Lewis acidic aluminum chloride-1-methyl-3-ethylimidazolium chloride room-temperature molten salt. *J. Electrochem. Soc.* 146:3290–94
10b. Endres F, El Abedin SZ. 2002. Electrodeposition of stable and narrowly dispersed germanium nanoclusters from an ionic liquid. *J. Chem. Soc. Chem. Commun.* 892–93
10c. Nanjundiah C, McDevitt F, Koch VR. 1997. Differential capacitance measurements in solvent-free ionic liquids at Hg and C interfaces. *J. Electrochem. Soc.* 144:3392–97
11. Abdallah DJ, Robertson A, Hsu H-F, Weiss RG. 2000. Smectic liquid-crystalline phases of quaternary group VA (especially phosphonium) salts with three equivalent long *n*-alkyl chains. How do layered assemblies form in liquid-crystalline and crystalline phases? *J. Am. Chem. Soc.* 122:3053–62
11a. Bradley AE, Hardacre C, Holbrey JD, Johnston S, McMath SEJ, Nieuwenhuyzen M. 2002. Small-angle X-ray scattering studies of liquid crystalline 1-alkyl-3-methylimidazolium salts. *Chem. Mater.* 14:629–35
11b. Bowlas CJ, Bruce DW, Seddon KR. 1996. Liquid-crystalline ionic liquids. *J. Chem. Soc. Chem. Commun.* 1625–26
12. Wasserscheid P, Welton T, eds. 2003. *Ionic Liquids in Synthesis*. Weinheim: Wiley-VCH
13. Filipponi A. 2001. EXAFS for liquids. *J. Phys. Condens. Matter* 13:R23–60
14. Rehr JJ, Albers RC. 2000. Theoretical approaches to X-ray absorption fine structure. *Rev. Mod. Phys.* 72:621–54
15. Hamill NA, Hardacre C, McMath SEJ. 2002. In situ XAFS investigation of palladium species present during the Heck reaction in room temperature ionic liquids. *Green Chem.* 4:139–42
16. McMath SEJ. 2003. *Structure-reactivity relationships in ionic liquids*. PhD thesis. Queen's University Belfast. 273 pp.
17. Okamoto Y, Akabori M, Motohashi H, Itoh A, Ogawa T. 2002. High-temperature XAFS measurement of molten salt systems. *Nucl. Instrum. Methods Phys. Res. A* 487:605–11
18. Crozier ED, Alberding N, Sundheim BR. 1983. EXAFS study of bromomangante

ions in molten salts. *J. Chem. Phys.* 79:939–43

19. Di Cicco A, Taglienti M, Minicucci M, Filipponi A. 2000. Short-range structure of solid and liquid AgBr determined by multiple-edge X-ray absorption spectroscopy. *Phys. Rev. B* 62:12001–13
20. Di Cicco A, Minicucci M, Filipponi A. 1997. New advances in the study of local structure of molten binary salts. *Phys. Rev. Lett.* 78:460–63
21. Minicucci M, Di Cicco A. 1997. Short-range structure in solid and liquid CuBr probed by multiple-edge X-ray-absorption spectroscopy. *Phys. Rev. B* 56:11456–64
22. Di Cicco A. 1996. Local structure in binary liquids probed by EXAFS. *J. Phys. Condens. Matter* 8:9341–45
23. Trapananti A, Di Cicco A, Minicucci M. 2002. Structural disorder in liquid and solid CuI at high temperature probed by X-ray absorption spectroscopy. *Phys. Rev. B* 66:014202(1–11)
24. Di Cicco A, Minicucci M. 1999. Solid and liquid short-range structure determined by EXAFS multiple-scattering data analysis. *J. Synchrotron Rad.* 6:255–57
25. Inui M, Takeda S, Maruyama K, Shirakawa Y, Tamaki S. 1995. XAFS measurements on molten silver-halides. *J. Non-Cryst. Solids* 192–193:351–54

25a. Inui M, Maruyama K, Shirakawa Y, Takeda S, Tamaki S. 1996. EXAFS measurements on molten AgBr-AgI mixtures. *J. Non-Cryst. Solids* 205–207:155–58

26. Takeda S, Kawakita Y, Inui M, Maruyama K. 1999. Local structure of molten $Ag(Cl_{1-x}I_x)$ mixtures. *J. Non-Cryst. Solids* 250–52:410–14
27. Di Cicco A, Rosolen MJ, Tossici R, Marassi R, Rybicki J. 1997. Radial distribution function in solid and liquid KBr probed by EXAFS. *J. Phys. IV* 7:C245–46
28. Di Cicco A, Rosolen MJ, Marassi R, Tossici R, Filipponi A, Rybicki J. 1996. Short-range order in solid and liquid KBr probed by EXAFS. *J. Phys. Condens. Matter* 8:10779–97
29. Li HF, Lu KQ, Wu ZH, Dong J. 1994. EXAFS studies of molten $ZnCl_2$, RbCl and Rb_2ZnCl_4. *J. Phys. Condens. Matter* 6: 3629–40

29a. Lu KQ, Li HF, Wu ZH, Dong J, Cheng ZN. 1995. Structural studies of some chloride melts with EXAFS. *Physica B* 208&209:339–43

30. Okamoto Y, Fukushima K, Iwadate Y. 2002. XAFS study of molten zinc dibromide. *J. Non-Cryst. Solids* 312–14:450–53
31. Okamoto Y, Shiwaku H, Yaita T, Narita H, Tanida H. 2002. Local structure of molten $LaCl_3$ by K-absorption edge XAFS. *J. Mol. Struct.* 641:71–76
32. Okamoto Y, Akabori M, Motohashi H, Shiwaku H, Ogawa T. 2001. X-ray absorption study of molten yttrium trihalides. *J. Synchrotron Rad.* 8:1191–99
33. Okamoto Y, Motohashi H, Akabori M, Ogawa T. 1999. Structural analysis of molten rare earth halides by EXAFS. In *Molten Salts XII*, ed. PC Trulove, HC DeLong, GR Stafford, S. Deki, 41:240–46. Pennington, NJ: Electrochem. Soc.
34. Okamoto Y, Yaita T, Minato K. 2004. High-temperature XAFS study of solid and molten $SrCl_2$. *J. Non-Cryst. Solids* 333:182–86
35. Ablanov M, Matsuura H, Takagi R, 1999. Extended X-ray absorption fine structure spectroscopic study of $PbCl_2$ molten salt. *Denki Kagaku* 67:839–42
36. Carmichael AJ, Hardacre C, Holbrey JD, Nieuwenhuyzen M, Seddon KR. 1999. A method for studying the structure of low-temperature ionic liquids by XAFS. *Anal. Chem.* 71:4572–74
37. Dent AJ, Seddon KR, Welton T. 1990. The structure of halogenometallate complexes dissolved in both basic and acidic room-temperature halogenoaluminate(III) ionic liquids, as determined by EXAFS. *J. Chem. Soc. Chem. Commun.* 315–16

38. Dent AJ, Lees A, Lewis RJ, Welton T. 1996. Vanadium chloride and chloride oxide complexes in an ambient-temperature ionic liquid. The first use of bis(trichloromethyl) carbonate as a substitute for phosgene in an inorganic system. *J. Chem. Soc. Dalton Trans.* 2787–92
39. Volkovich VA, May I, Charnock JM, Lewin B. 2002. Reactions and speciation of technetium and rhenium in chloride melts: a spectroscopy study. *Phys. Chem. Chem. Phys.* 4:5753–60
40. Baston GMN, Bradley AE, Gorman T, Hamblett I, Hardacre C, et al. 2002. Ionic liquids for the nuclear industry: A radiochemical, structural, and electrochemical investigation. *ACS Symp. Industrial Applications of Ionic Liquids* 818:162–77
41. Laidler JJ, Battles JE, Miller WE, Ackerman JP, Carls EL. 1997. Development of pyroprocessing technology. *Progress Nucl. Energy* 31:131–40
42. Bradley AE, Hardacre C, Nieuwenhuyzen M, Pitner WR, Sanders D, et al. 2004. A structural and electrochemical investigation of 1-alkyl-3-methylimidazolium salts of the nitratodioxouranate(VI) anions, $[\{UO_2(NO_3)_2\}_2(\mu_4\text{-}C_2O_4)]^{2-}$, $[UO_2(NO_3)_3]^-$, and $[UO_2(NO_3)_4]^{2-}$. *Inorg. Chem.* 43:2503–14
43. Visser AE, Jensen MP, Laszak I, Nash KL, Choppin GR, Rogers RD. 2003. Uranyl coordination environment in hydrophobic ionic liquids: an in situ investigation. *Inorg. Chem.* 42:2197–99
44. Jensen MP, Dzielawa JA, Rickert P, Dietz ML. 2002. EXAFS investigations of the mechanism of facilitated ion transfer into a room-temperature ionic liquid. *J. Am. Chem. Soc.* 124:10664–65
45. Jensen MP, Neuefeind J, Beitz JV, Skanthakumar S, Soderholm L. 2003. Mechanisms of metal ion transfer into room-temperature ionic liquids: the role of anion exchange. *J. Am. Chem. Soc.* 125:15466–73
46. Ellis B. 1996. *Intl. Patent WO 96/18459*
47. Hamill JT, Hardacre C, Nieuwenhuyzen M, Seddon KR, Thompson SA, Ellis B. 2000. Comment on the preparation of the ionic liquid 1-ethyl-3-methylimidazolium ethanoate: a unique monomeric, homoleptic pentacoordinate lead ethanoate complex. *J. Chem. Soc. Chem. Commun.* 1929–30

Annu. Rev. Mater. Res. 2005. 35:51–73
doi: 10.1146/annurev.matsci.35.100303.111314

First published online as a Review in Advance on January 21, 2005

THE DIFFUSION-MULTIPLE APPROACH TO DESIGNING ALLOYS

Ji-Cheng (J.-C.) Zhao
GE Global Research, General Electric Company, Niskayuna, New York 12309; email: zhaojc@research.ge.com

Key Words phase diagrams, nano-indentation, thermal conductivity, combinatorial materials science, high throughput screening

■ **Abstract** The diffusion-multiple approach—the creation of composition gradients and intermetallic phases by long-term annealing of junctions of three or more phases/alloys—enables the generation of a large number of phases and compositions for efficient mapping of phase diagrams, phase properties, and kinetics. With an efficiency much higher than that of a conventional approach, many critical materials data, which otherwise would be too time-consuming and expensive to acquire, can be obtained and employed to accelerate alloy design. The critical data for structural materials design include phase diagrams, diffusion coefficients, precipitation kinetics, solution-strengthening effects, and precipitation-strengthening effects. All these data can be obtained from diffusion multiples. The combination of the diffusion-multiple approach with the CALPHAD approach can impact computational design of structural materials. The diffusion-multiple approach can also be applied to combinatorial screening of functional materials having unique physical, chemical, or other properties when micro-scale measurement techniques are available for these properties. Examples are shown to illustrate the progress made to date on applying the diffusion-multiple approach to accelerated design/discovery of materials.

INTRODUCTION

Combinatorial approaches based on thin-film libraries have now been extensively used to screen functional materials such as phosphors, magnetostrictive materials, and dielectrics (1–5). Unlike functional materials that most often are used for one specific physical or chemical property, structural materials often need to satisfy simultaneously several different property requirements such as strength, ductility, modulus, fracture toughness, and environmental resistance. It is currently not possible to combinatorially screen all these properties. Instead, high-efficiency approaches can be used to map critical properties and data in order to accelerate alloy design.

Compositional design of alloys involves several aspects: introduction of desired phases; exclusion of deleterious phases; solution alloying of individual phases (terminal solid solutions and intermetallics) to optimize properties; and control of size and distribution of second phases. The diffusion-multiple approach can assist with all but the last aspect.

The strength of structural alloys is controlled by dislocation mobility, and it comes predominantly from three sources: solution strengthening, precipitation/dispersion strengthening, and microstructural strengthening. The first case involves atoms of different sizes distorting a crystal lattice, thus making dislocations more difficult to move. Alloy designers would like to know which elements are more effective than others in solution-strengthening and by what amount. Currently, predictions from first principles are still not up to the challenge. Experimental mapping of solution-strengthening effects would be very useful for alloy design, especially for new alloy systems. The second case involves in situ formation (precipitation) or intentional addition (dispersion) of second-phase particles in a crystalline matrix to slow/block dislocation movements. In this case, it is critical to know the stability as well as the nucleation and growth kinetics of the particles. To predict these properties requires phase diagrams and diffusion coefficients. Phase diagrams are also very useful to guide alloy designers to induce phases that have been identified as beneficial precipitates and to avoid detrimental ones. The third case involves microstructural imperfections such as grain boundaries and dislocation tangles/pile-ups induced by alloy processing (such as cold rolling). The grain size effect (the Hall-Petch effect) is an example of this type of strengthening. Microstructural strengthening is accomplished by processing and is often not part of the early alloy composition design. (Some alloy compositions are tailored so that they can be processed in a certain way, and some of them can only be processed in a specific way.) Therefore, phase diagrams, solution-strengthening effects and diffusion coefficients are among the most basic information required for alloy design.

This review summarizes the diffusion-multiple approach (6–10) for mapping of phase diagrams, solution-hardening effects, diffusion coefficients, and modulus variation with composition and phases. Unlike thin-film combinatorial libraries, the continuous composition variations in solid solutions and intermetallic compounds can be created in bulk diffusion multiples by thermal interdiffusion to allow rapid mapping of these properties/data for accelerated alloy design. Although it is especially useful for designing structural materials, the diffusion-multiple approach can also be applied to discover new functional materials and to measure fundamental physical and chemical properties as a function of composition and crystal structure.

The diffusion-multiple approach is a simple yet very useful concept grown out of the traditional diffusion couple/triple methods that have been used for studying diffusion coefficients and phase diagrams (11–13). A diffusion multiple consists of two or multiple pieces of metals or alloys in close contact to allow thermal interdiffusion to create composition gradients and intermetallic compounds. Previous diffusion couple/triple work has been focused on measuring phase diagrams by taking advantage of local equilibrium at the phase interfaces. With micro-scale measurement tools for properties having become available, it is now possible to take advantage of the composition variations (libraries) and compound formation in diffusion multiples to map composition-phase-property relations to design or discover new materials. The area where the diffusion-multiple approach can have the most impact for structural materials is in rapid mapping of phase diagrams, solution-hardening effects, binary diffusivity matrices, and modulus dependency on composition and phases to provide critical data for computational design of materials.

PHASE DIAGRAM DETERMINATION

As one example, in order to explore phase relations and properties of the Ni-Fe-Mo system and the effects of these elements on modifying the properties of a commercial Ni-base superalloy Inconel 706 (IN706), a diffusion multiple of Ni, Fe, Mo and IN706 was made by placing quarter-pie-shaped components into a pure Ni tube, as shown in Figure 1, see color insert (6, 7). Two Ni caps were placed on top and bottom of the Ni tube and welded in vacuum using an electron beam. The welded assembly was treated by hot isostatic pressing for 4 hours at 1100°C under a 200 MPa pressure to provide good contact among the pieces. The assembly was then annealed for 1500 hours at 1100°C to cause extensive interdiffusion. At the Ni-Mo-Fe tri-junction where these three elements met, extensive interdiffusion took place. All the intermetallic compounds, both binary and ternary, NiMo (δ), Fe_7Mo_6 (μ) and $Fe_{11}Ni_{36}Mo_{53}$ (P-phase), formed by diffusion reactions (see Figure 1). By performing electron probe microanalysis (EPMA) across the interfaces and at the three-phase junctions, both tie-lines and tie-triangles were obtained to construct the entire isothermal section of the ternary phase diagram. This is based on local equilibrium at the phase interfaces. For instance, the left-hand edge of the tri-junction (points 0 to 5 in Figure 1*c*) was a binary Ni-Mo diffusion couple because Fe could not reach there by diffusion. EPMA analysis along points 0 to 5 gave tie-lines 1–2 and 3–4 in Figure 1*e*. Similarly, points 5 to 12 at the top edge of Figure 1*b* gave binary Mo-Fe equilibria and points 12 to 14 at the right-hand edge gave the binary Fe-Ni information (Fe and Ni are completely soluble at 1100°C). Closer to the center of the tri-junction, more interdiffusion of the three elements was achieved, and ternary tie-lines were obtained. At three-phase junctions such as locations 15 and 16 in Figure 1*c*, the respective tie-triangles $\gamma - \delta - \mu$ and $\gamma - \alpha\text{Fe} - \mu$ were obtained. Therefore, only one diffusion multiple was needed to survey all the phases and equilibria in the whole isothermal section. The diffusion-multiple method was much more efficient in mapping of phase diagrams than traditional equilibrated binary and ternary alloys and diffusion couple methods. For example, the same 1100°C Ni-Mo-Fe isothermal section of Figure 1*e* was determined by van Loo et al. (14) using 23 alloys and 7 diffusion couples. The isothermal section of Figure 1*e* measured from the single diffusion triple looks identical to that obtained by van Loo et al.

An important feature of a diffusion multiple is its ability to generate full libraries of all the single-phase compositions of a ternary system by the thermal interdiffusion process. It is clear that points 0-1-15-16-11-12-13-14-0 in Figure 1*c* correspond to those same points in Figure 1*e* in composition, and they include the entire composition region of the γ-phase. Similarly, all the other single-phase compositions (including the intermetallic compounds) for the Ni-Mo-Fe ternary system are also generated by thermal interdiffusion. This was not fully appreciated and had not been taken advantage of before. By using nano-indentation and other techniques, researchers can now map hardness, elastic modulus, and other properties as a function of composition for each phase (6, 7). This is discussed further below.

The CALculation of PHAse Diagram (CALPHAD) approach (15, 16) is a major part of computational design of structural materials. It provides multicomponent phase equilibrium data, the driving force for precipitation, the segregation and latent heat data for solidification modeling, and thermodynamic factors for constructing the diffusion coefficient matrices for multicomponent alloys. Since most binary phase diagram data already exist, the critical data required for building a reliable multicomponent thermodynamic database are experimental ternary phase diagrams. The diffusion-multiple approach for mapping of phase diagrams will have a high impact on the establishment of reliable thermodynamic databases for computational design of materials.

The state-of-the-art thermodynamic database for Ni-base superalloys can make good predictions of phase stability when the alloy composition is well inside the $\gamma + \gamma'$ (fcc + $L1_2$) two-phase region (17). However, this kind of superalloy has less than optimum strength. Much effort in designing advanced superalloys is placed on strengthening the superalloys with refractory elements while avoiding the formation of the topologically close-packed (TCP) phases [sigma (σ), mu (μ), P, and Laves], i.e., pushing the alloy composition to the very limit of the $\gamma + \gamma'$ phase region. Because the thermodynamic stability of the TCP phases is not well assessed in the current Ni database, the limit of the $\gamma + \gamma'$ boundary cannot be well defined. In other words, currently it is not possible to accurately predict how much of the refractory elements can be added to the alloy while still avoiding the TCP formation. Accurate predictions of TCP phase stability from thermodynamic calculations would make a big impact in superalloy design. This will require better thermodynamic assessments of ternary phase diagrams related to TCP phase formation, but many of the critical ternary phase diagrams are unavailable experimentally. To provide some of these data, four diffusion multiples were made to obtain 16 ternary isothermal sections related to TCP phases (Figure 2, see color insert). These isothermal sections will serve as input to CALPHAD assessments of TCP phase stability, thus improving the reliability of the thermodynamic databases of Ni-base alloys.

More complicated diffusion multiples can be designed to map many ternary isothermal sections from a single sample. An example is shown for a Pd-Pt-Rh-Ru-Cr diffusion multiple, where 10 ternary phase diagrams were mapped from a single sample (7, 18). The efficiency of diffusion multiples in mapping phase diagrams is now so high that the time-limiting step in building thermodynamic databases will no longer be the experimental determination of phase diagrams; instead it will be the thermodynamic assessment (optimization) process (from which the thermodynamic parameters are optimized from experimental phase diagrams and measured thermodynamic quantities such as heat of formation and activity).

EXTRACTION OF DIFFUSION COEFFICIENTS AND CONSTRUCTION OF DIFFUSIVITY DATABASE

Diffusion coefficients are essential data required to perform simulation of many metallurgical processes for materials design and application. Because many binary diffusion couples are created in diffusion multiples, binary diffusion coefficients can be readily extracted from the binary diffusion profiles using a Boltzmann-Matano analysis or other techniques (19, 20). Examples are shown in Figure 3. The Co-Ni diffusion profiles in Figure 3*b* were used to extract the diffusion coefficients shown in Figure 3*c*. Diffusion multiples made up of multicomponent superalloys were used to generate multicomponent diffusion profiles (such as the ones shown in Figure 3*e*,*f*) to test the diffusivity database predictions. A diffusivity database (19, 20) is usually constructed from binary and limited ternary diffusion coefficients together with the thermodynamic factors calculated from the CALPHAD-type thermodynamic databases. The solid lines in Figure 3*e*,*f* are calculation results from a diffusivity database based on binary and a few ternary diffusivity data (19, 20). The results show good agreement with experimental data for Figure 3*e* and somewhat less for Figure 3*f*. In cases when disagreement between experimental and simulated profiles is observed, some particular binary or ternary diffusion coefficients may be identified for further experimental study to improve upon. The validated diffusivity database can then be used to simulate many processes such as precipitation kinetics, coarsening kinetics, and grain growth.

MAPPING OF SOLUTION-HARDENING EFFECT AND ELASTIC MODULUS

As discussed above with respect to Figure 1, an important feature of a diffusion multiple is its ability to generate full libraries of all the single-phase compositions of a ternary system. Instrumented nano-indentation allows one to take advantage of the composition variations and phase formation in diffusion multiples to simultaneously evaluate hardness and elastic modulus at a micron or sub-micron scale (21, 22). Nano-indentation scans were made for the Pd-Pt-Rh ternary system for which all three elements are completely soluble. An optical image taken after the nano-indentation scans at the Pd-Pt-Rh tri-junction of a diffusion multiple is shown in Figure 4*a*. EMPA analysis was performed after nano-indentation in order to correlate the composition to the locations of the indentation (18).

The elastic moduli for the entire Pd-Rh-Pt ternary system were obtained. A negative deviation from a linear rule-of-mixtures modulus (Figure 4*b*, dashed line)

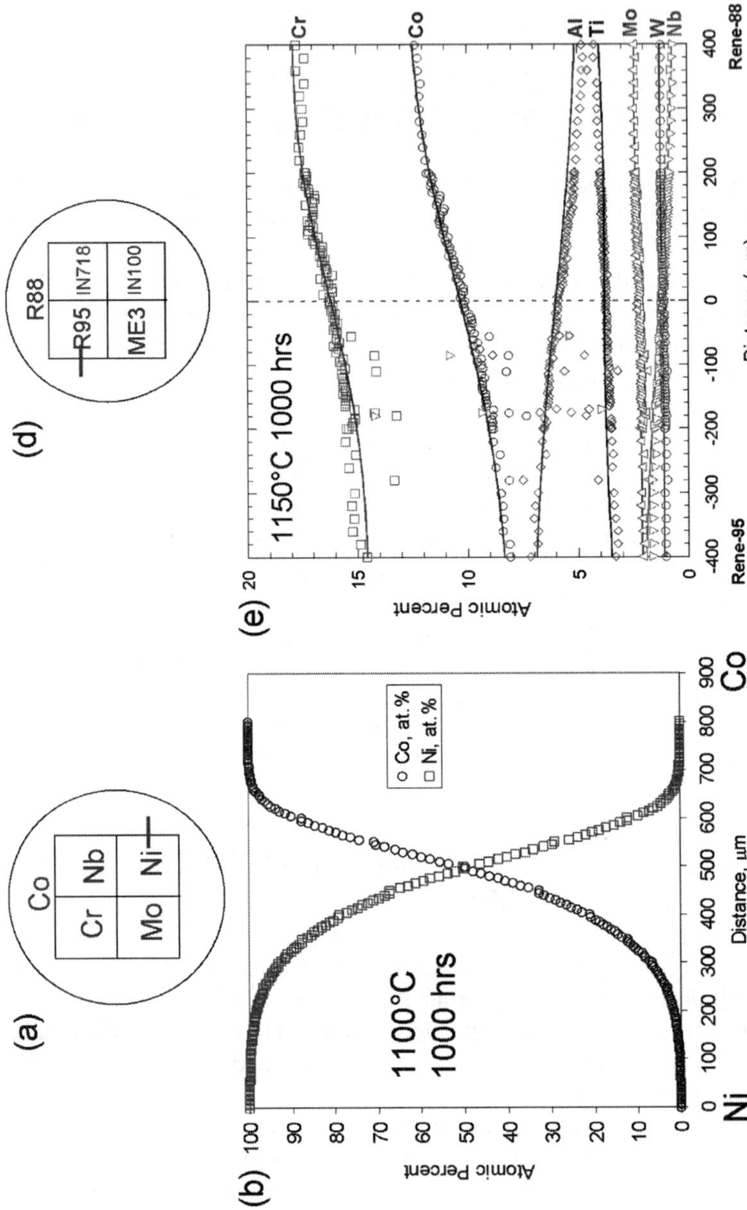
(a)
Co
Cr
Nb
Mo
Ni
(b)
1100°C
1000 hrs
Co, at.%
Ni, at.%
Atomic Percent
Distance, µm
Ni
Co
(d)
R88
R95
IN718
ME3
IN100
(e)
1150°C 1000 hrs
Atomic Percent
Distance (µm)
Rene-95
Rene-88
Cr
Co
Al
Ti
Mo
W
Nb

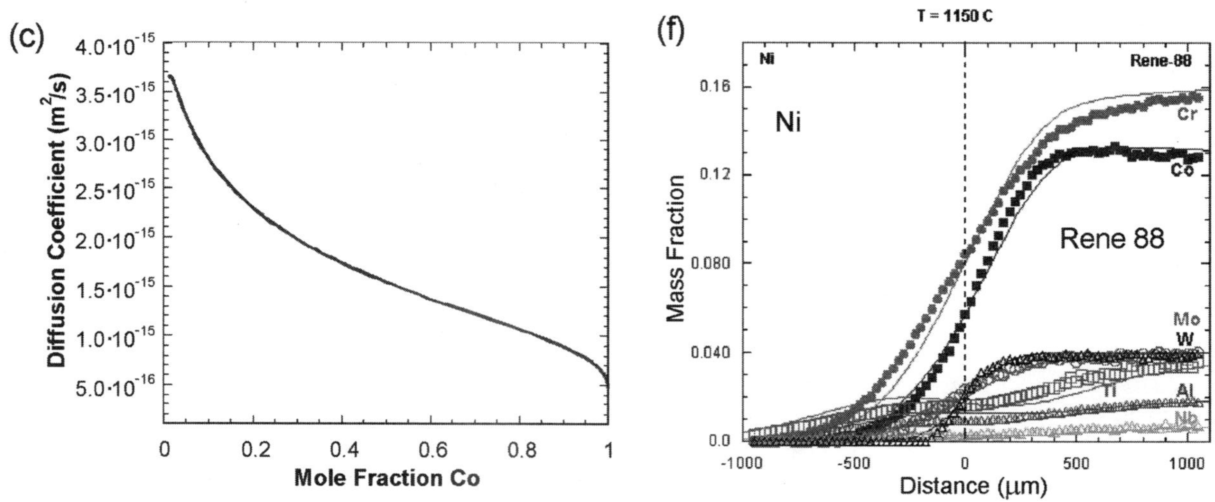

Figure 3 Evaluation of diffusion coefficients from composition profiles obtained from diffusion multiples: (*a*) a diffusion multiple made up of Co, Cr, Mo, Nb, and Ni; (*b*) a pair of Co-Ni diffusion profiles; (*c*) the interdiffusion coefficient obtained from the profiles shown in (*b*); (*d*) a diffusion multiple made up of five multicomponent superalloys to produce multicomponent diffusion profiles to test the reliability of a diffusivity database; (*e*) the diffusion profiles between Ni-base superalloys René 88 and René 95 after annealing at 1150°C for 1000 h; (*f*) the diffusion profiles between pure Ni and superalloy René 88 after annealing at 1150°C for 1000 h. The solid lines in (*e*) and (*f*) are predictions from a diffusivity database (9, 20).

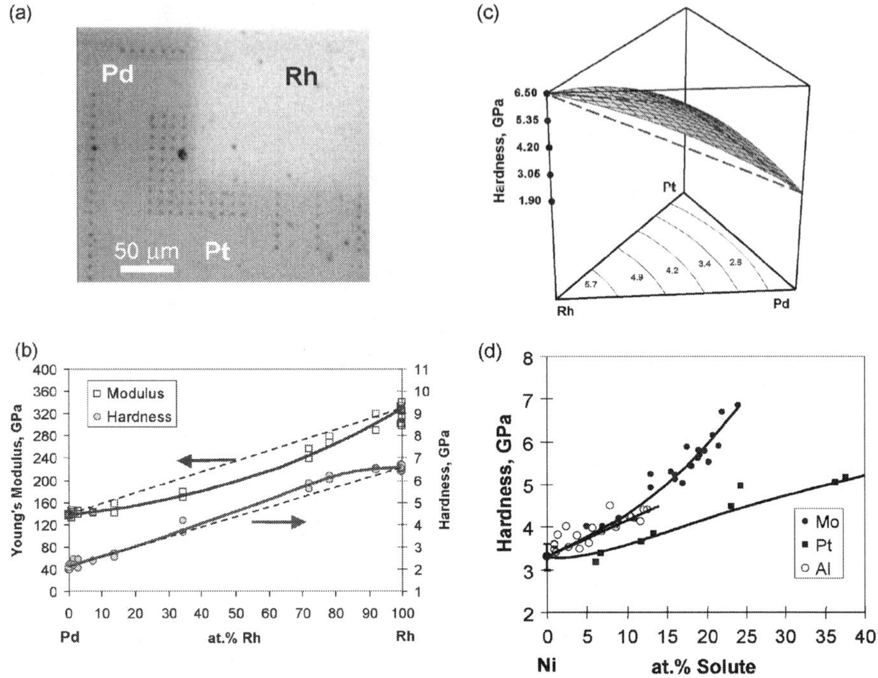

Figure 4 Efficient construction of solution-hardening databases using nano-indendation on samples with composition variations created in diffusion multiples: (*a*) Optical image of nano-indentation grid made at the Pd-Pt-Rh tri-junction region of a diffusion multiple. (*b*) Hardness and Young's modulus variation with composition for the Pd-Rh binary system that shows negative deviation from the linear behavior for the modulus and slight positive deviation for hardness (solution hardening). (*c*) Hardness data for the entire Pd-Pt-Rh ternary system obtained from the diffusion multiple. (*d*) Hardness data obtained from nano-indentation tests of diffusion multiples can be used to quickly construct a solution-hardening database as shown here for Ni-base systems (7, 18).

can be seen for the Pd-Rh binary system. For Pt-Rh, the modulus deviates slightly positively from linearity, while for Pd-Pt, the modulus essentially follows the linearity for Pd-rich compositions and deviates positively for Pt-rich compositions. Thus the three-dimensional surface representing the compositional dependence of modulus in the Pd-Rh-Pt system shows a more complex behavior than do simple linear rule-of-mixtures. It is interesting to note that the results are consistent with first-principles calculations of modulus for the binaries: negative deviation for Pd-Rh and positive deviations for both Pd-Pt and Pt-Rh (P.E.A. Turchi, private communication). The utility of the diffusion-multiple approach to generate data needed to verify first-principles models should have an impact on the validation of fundamental models of alloying. To my knowledge, the results for the Pd-Rh-Pt system represent the first complete mapping of Young's modulus for a ternary system.

Results for the hardness survey across the entire Pd-Rh-Pt ternary system are shown in Figure 4*c*. All three binaries and the ternary system show positive deviation from a simple linear rule-of-mixtures hardening everywhere in the alloy space. Such data can be used to construct solution-hardening effect databases that will be useful for alloy design, because solution hardening correlates directly to solution strengthening.

For instance, by combining the solution-hardening effect from the Ni-NiAl and Ni-Pt diffusion couples and the Ni-Mo-Fe-IN706 diffusion multiple, the solution-hardening potency of various elements in the γ-phase can be compared and assessed (7), as shown in Figure 4*d*. As more data are accumulated, a database of solution-hardening effects can be built. Such a database, which is not readily available and would be very time-consuming and expensive to obtain using the conventional one-alloy-at-a-time approach, will be very valuable for materials design by predicting the strengthening effect from complex compositions and by providing guidance for elemental selection, especially for new alloy systems where a selection of solution-strengthening alloying elements needs to be made.

It should be noted that the evaluation of the solution-hardening effect is not confined to terminal solid solutions but also applies to intermetallic compounds. Many intermetallic compounds have substitutional solubility of one element for another to form so-called line compounds. For instance, Ni has a high solubility ($\sim$40 at.%) substituting for Fe in the Fe_7Mo_6-based μ phase (Figure 1*e*). Nano-indentation tests on the Ni-Mo-Fe tri-junction region (Figure 1*c*) can quickly determine whether the Ni substitution of Fe in the μ phase changes its modulus and hardness. In this case, no significant change was observed because Ni and Fe are very similar in atomic size (7). Because of the brittle nature of most intermetallics, synthesis can be challenging, especially for those with very high melting points. The hardness and elastic modulus properties of intermetallic compounds can now be evaluated effectively without going through the time-consuming synthesis processes for individual compounds.

The electron backscatter diffraction (EBSD) technique (23) is useful for identifying the crystal structures of the intermetallic compounds in a diffusion multiple without going through the time-consuming sample making process for transmission electron microscopy (TEM). The intermetallic compounds formed in diffusion multiples can usually be identified by combining the compositional information from EPMA and crystal structure information from EBSD. In some cases when EBSD cannot index the phases having complex structures or completely new phases, it may be necessary to use micro X-ray diffraction or TEM electron diffraction for phase identification. In the latter case, TEM samples can be cut from specific intermetallic compounds in diffusion multiples using focused ion beam (FIB).

EVALUATION OF PRECIPITATION KINETICS

Precipitation hardening is the most important strengthening mechanism for most structural materials and precipitation kinetics is critical for controlling the microstructure and thus the properties of alloys. Diffusion multiples allow assessment of precipitation kinetics simultaneously for a range of compositions (7, 9), in contrast to the traditional approach where a single alloy (composition) is studied. The results of a binary Ni-NiAl diffusion couple are used as an example to discuss the methodology. The diffusion couple was heat-treated at 1204°C for 100 h in a vacuum furnace and then furnace-cooled to ambient temperature. The composition profiles based on EPMA measurements are shown in Figure 5*b*. Significant interdiffusion took place in the diffusion couple and γ' (Ni_3Al) phase was formed. It can be seen from the composition profiles in Figure 5*b* that a full composition variation (library) of all the single-phase regions between Ni and NiAl was obtained, e.g., for the γ-phase, 0 to 18 at.% Al (points 0 to 1). By performing nano-indentation tests on the diffusion couple, researchers obtained the hardness and modulus variation with distance (as shown in Figure 5*d*) that combined the results of several parallel nano-indentation scans. The hardness and modulus profiles (Figure 5*d*) can be converted into composition-phase-property data (Figure 5*c*) because the composition of each location is known from Figure 5*b*. The hardness of the γ-phase increased from ~3.3 to 4.5 GPa as the Al concentration increased from 0 to ~13 at.%, which was the result of solution-hardening. The further significant increase of hardness in the Al-rich compositions (13 to 18 at.%) of the γ-phase (highlighted in Figure 5*c*,*d*) came from precipitation hardening, where fine γ' particles formed in the γ matrix (when the diffusion couple was cooled from 1204°C to room temperature). Such precipitation hardening provides most of the strength for advanced Ni-based superalloys, and it is an important mechanism for structural materials design. Precipitation hardening requires supersaturation—a condition arising from decreasing solubility with lowering temperature (e.g., 18 and 13 at.% Al in γ at 1204 and 800°C, respectively; Figure 5*a*). Both phase diagrams and precipitation kinetics information are critical for designing precipitation-hardened alloys.

By cooling the 1204°C-exposed diffusion couple at different cooling rates and examining the hardness and microstructures at different locations (compositions) in the diffusion couple, the continuous cooling transformation kinetics of γ'

---→

Figure 5 Efficient study of phase precipitation kinetics using diffusion multiples: (*a*) Ni-rich side of Ni-Al phase diagram; (*b*) composition profiles of an Ni-NiAl diffusion couple after annealing at 1204°C for 100 h; (*c*) hardness variation with compositions and phases; (*d*) hardness and modulus profiles obtained from nano-indentation tests; (*e*) optical image of the sample after etching showing martensite formation and γ' precipitation in the Ni-NiAl diffusion couple. The circles show the location and composition of regions undergoing martensitic transformation during cooling from 1204°C to room temperature (7, 9).

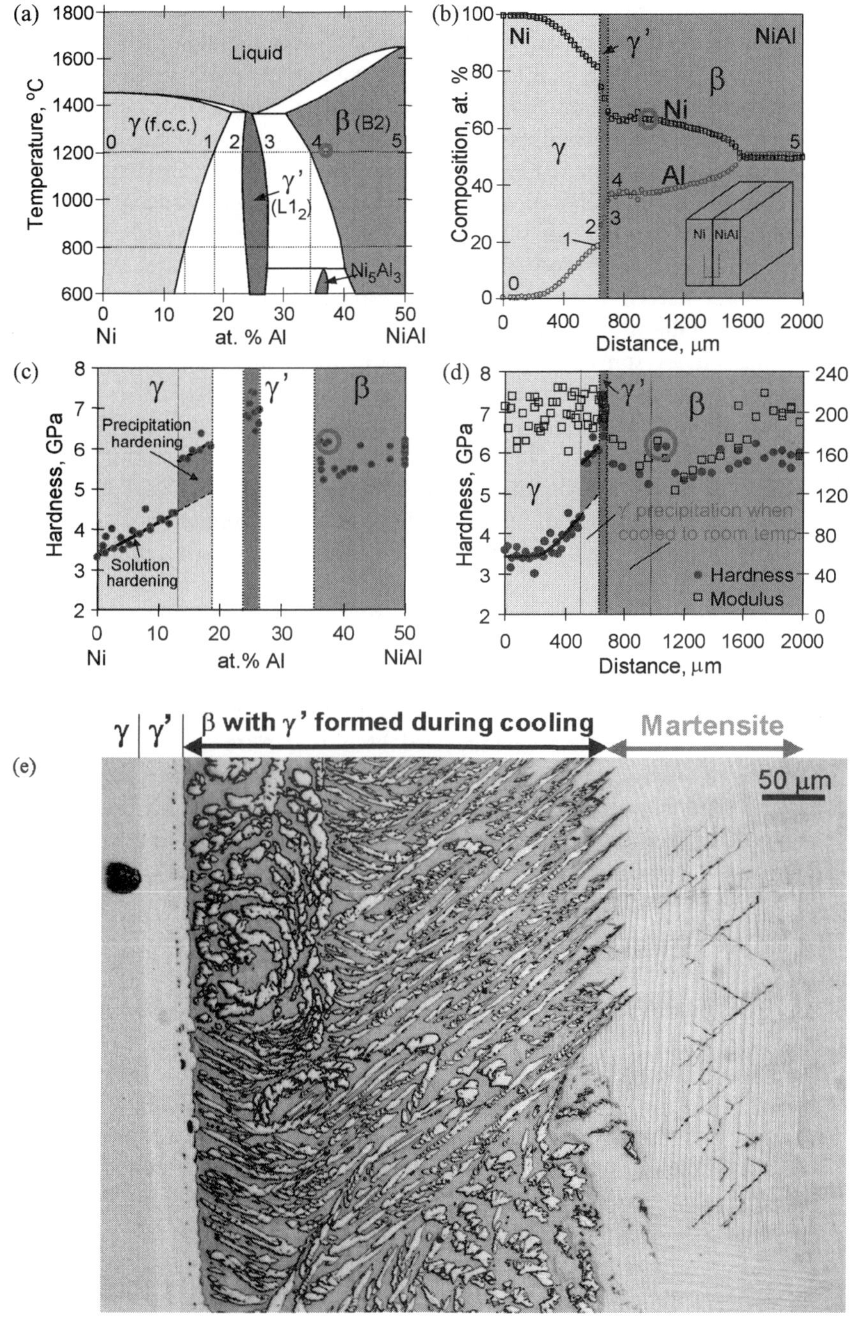
(a)
Liquid
γ (f.c.c.)
β (B2)
γ'
(L1$_2$)
Ni_5Al_3
Temperature, °C
Ni
at. % Al
NiAl
(b)
Ni
NiAl
γ'
β
γ
Ni
Al
Composition, at. %
Distance, μm
(c)
γ
γ'
β
Precipitation hardening
Solution hardening
Hardness, GPa
Ni
at.% Al
NiAl
(d)
γ'
β
γ
γ' precipitation when cooled to room temp.
Hardness
Modulus
Hardness, GPa
Modulus, GPa
Distance, μm
γ
γ'
β with γ' formed during cooling
Martensite
(e)
50 μm

precipitation can be evaluated simultaneously for a range of γ compositions. Alternatively, by quenching the 1204°C-exposed diffusion couple to ambient temperature (to retain the high temperature γ-phase), subsequently exposing the couple at lower temperatures (e.g., 800°C) for different time periods, and then examining the microstructure and hardness at different locations (compositions), the isothermal precipitation kinetics of the γ'-phase can be evaluated simultaneously for a library of γ compositions.

The solubility of Ni in the β-phase also decreases significantly with decreasing temperature, resulting in supersaturation at low temperatures. When the diffusion multiple was cooled from 1204°C to room temperature, the compositions near point 4 were supersaturated with respect to Ni solubility. Precipitation of γ' phase from the β-phase occurred, as shown in Figure 5*e*. Adjacent to this region, the martensitic transformation took place. The composition and location of the martensitic transformation region are shown in Figure 5 by circles. The regions of γ' precipitation and martensitic transformation will change with different cooling rates. In this way, the precipitation kinetics can be studied simultaneously for a range of β-phase compositions, as illustrated here for the Ni-NiAl diffusion couple, or similarly for a range of ternary and higher compositions.

The diffusion-multiple approach can also be applied to study precipitation-hardening effects with the help of nano-indentation, scanning electron microscopy (SEM) and/or TEM. For instance, the precipitation-hardening effect in Al-rich compositions (13 to 18 at.% Al) of the Ni-base fcc solid-solution was revealed by nano-indentation as shown in Figure 5*c,d*. The size of the γ' precipitates in this case was small and could not be revealed by SEM without special etching. Nano-indentation was able to reveal the hardness increase from precipitation hardening. When the precipitates are large, as in the case of γ' precipitates and martensite precipitated from the β-phase, they can be clearly seen in SEM images. Thus it is possible to identify compositions that have desired precipitates to strengthening the alloys. It is also possible to find the conditions to form the desired size of the precipitates by studying the precipitation kinetics.

MICRO-COMPRESSION TESTS ON MICRO-PEDESTALS

It is highly desirable to perform localized (micro-scale) measurements of yield strength and ductility from the compositions and phases created in diffusion multiples to discover new interesting alloy and intermetallic compositions. Efforts have been made to extract yield strength from the load-displacement curves by analyzing the elasto-plastic behavior during nano-indentation (24, 25). The procedures have not been widely accepted. A potentially more practical way to obtain compressive strength and perhaps even some rough indication of ductility is to make micro-pedestals and then perform micro-compression tests with a flat-tip nano-indenter, as elegantly demonstrated by Uchic et al. (26, 27). The micro-pedestals are such that their bottom ends are still attached to the sample. These micro-pedestals can be made by FIB, micro-EDM (electro-discharge machining) (Figure 6*b*, see color insert), or femtosecond laser machining (Figure 6*c*) (28; Q. Feng & T. Pollock, unpublished data). The latter two methods take only minutes to hours to make a pedestal, but the pedestal surface quality is not as good as that made by FIB even though the surface damage layer thickness is still minimal. It may be necessary to use FIB for final surface finish/cleaning for micro-pedestals made with micro-EDM or femtosecond laser.

The data analysis from the micro-compression tests is not as straightforward as one might think. The major problem is associated with the size dependence of mechanical properties (26, 27). As the dislocation movement becomes confined by small micro-pedestal diameters, the deformation behavior also changes, as does the measured apparent strength. One needs to test several diameter sizes to evaluate the minimum diameter required to obtain equivalent bulk property values.

Even with this size-dependence complication, the micro-pedestal and micro-compression tests may be an important tool for combinatorial screening of interesting compositions for structural alloys and intermetallics. One can envision making micro-pedestals on various intermetallic compounds such as those in the diffusion multiple (as shown in Figure 6*f*) and evaluating the compressive strength of the phases, and perhaps even getting some indication of their ductility. Such data would take a large amount of time to obtain using traditional methods by casting individual compositions of the compounds, machining them into test samples, and testing them (each of these steps is challenging for high melting point and brittle compounds). One could also directly screen compositions for strength using diffusion multiples made up of multicomponent Ni-base alloys, as shown in Figure 6*g*. The wide interdiffusion zone formed between/among multicomponent alloys represents new localized multicomponent alloys waiting to be tested for properties. Again, one needs to be careful with the selection of pedestal diameter and the interpretation of the experimental data. The pedestals may contain multiple phases and the dimensions and interfaces/boundaries may contribute differently to the strength and ductility.

MAPPING OF THERMAL CONDUCTIVITY AND ITS IMPLICATION

A tool has been developed to map thermal conductivity at a 3.5-μm spatial resolution, 90% accuracy, and a high mapping speed (e.g., 10,000 points per h) (29). The technique is based on time-domain thermoreflectance generated by a femtosecond pulsed laser. A thin Al film is first sputter-coated onto a sample to serve as a transducer to absorb the laser pulse and enable sensitive measurements of the temperature change through the relatively large thermoreflectance of Al at the wavelength of a Ti:sapphire laser, $\lambda = 770$ nm. The pulsed laser is split into two beams: the pump beam and the probe beam. The pump beam is modulated at ~10 MHz to limit lateral heat flow, and it heats the near surface region of the Al film.

As the heat diffuses through the sample, the surface temperature of the Al film decays and so does the Al reflectivity, which is monitored by the probe beam. The probe beam is modulated at audio frequencies to eliminate artifacts created by scattering of the pump beam by surface roughness. The thermal properties of the sample are evaluated by matching the temperature decay obtained from experiments with that calculated from heat flow models (D.G. Cahill, submitted). The thermal effusivity, the square root of the product of the thermal conductivity, Λ, and heat capacity per unit volume, *C*, are extracted from the ratio of in-phase to out-of-phase responses, $V_{in}(t)/V_{out}(t)$, of the thermoreflectance at the modulation frequency of the pump beam.

A Nb-Ti-Cr-Si diffusion multiple (Figure 7*a*) was coated with an Al film of 120 nm thickness. The thermal conductivity map taken of a Cr-Ti diffusion couple area of the diffusion multiple is shown in Figure 7*c* along with the corresponding EPMA Cr map shown in Figure 7*b*. The thermal conductivity data along the dashed line in Figure 7*c* are plotted in Figure 7*d*. It can be seen that Cr-rich side of the diffusion couple has a uniform conductivity. The intermetallic phase Cr_2Ti (Laves phase) has a smoothly varying conductivity (from 42 to 20 $Wm^{-1}K^{-1}$°$C =$ 2.9 $J\,cm^{-3}K^{-1}$ was estimated for Cr_2Ti). EPMA analysis showed that the Ti content increases from 33.5 at.% at the Cr-side to 36.1 at.% at the Ti-rich side of the Laves phase. This 2.6 at.% substitution of Ti for Cr reduces the thermal conductivity by a factor of about two. Whereas this is unlikely to be a simple solid-solution effect, the result might be caused by creation of point defects such as Ti interstitials/anti-site or Cr vacancies as Ti content increases. Note that even though the Cr_2Ti region has only 2.6 at.% variation in composition, it actually consists of two Laves phases. The Cr_2Ti adjacent to the Cr side is the C14-Laves with a hexagonal crystal structure (*h*P12), and the thickness of this phase is only about a micron. The Cr_2Ti at the Ti-rich side is the C15-Laves with a cubic crystal structure (*c*F24), and the change of thermal conductivity in the Cr_2Ti phase region of Figure 7*c*,*d* is essentially all for the C15-Laves phase. The two-phase microstructure on the Ti-rich side was the product of Laves phase precipitation from the bcc(Ti,Cr) matrix during cooling of the diffusion multiple from 1100°C to room temperature. The matrix phase on the Ti-rich side adjacent to Cr_2Ti is bcc(Ti,Cr) with a composition of 85.5 at.% Ti—14.5 at.% Cr. This bcc(Ti,Cr) alloy has the low thermal conductivity typical of Ti alloys, $\sim$7 $Wm^{-1}K^{-1}$.

Although the example here shows only the measurements of thermal conductivity of metals and intermetallic compounds, the technique can also be as easily applied to ceramics and polymers. It can be used to map any metallographic mount as long as a thin Al film can be sputtered onto the surface (similar to sputtering a carbon or gold film onto SEM samples). The accuracy of the technique is high, as shown in Figure 7*e* in which the thermal conductivity values measured for some common metals and ceramics are compared with the generally accepted values in the literature (X. Zheng, D.G. Cahill & J.-C. Zhao, submitted). Note that Figure 7*e* is not a calibration curve—the new technique makes the thermal conductivity measurements without the use of any standards.

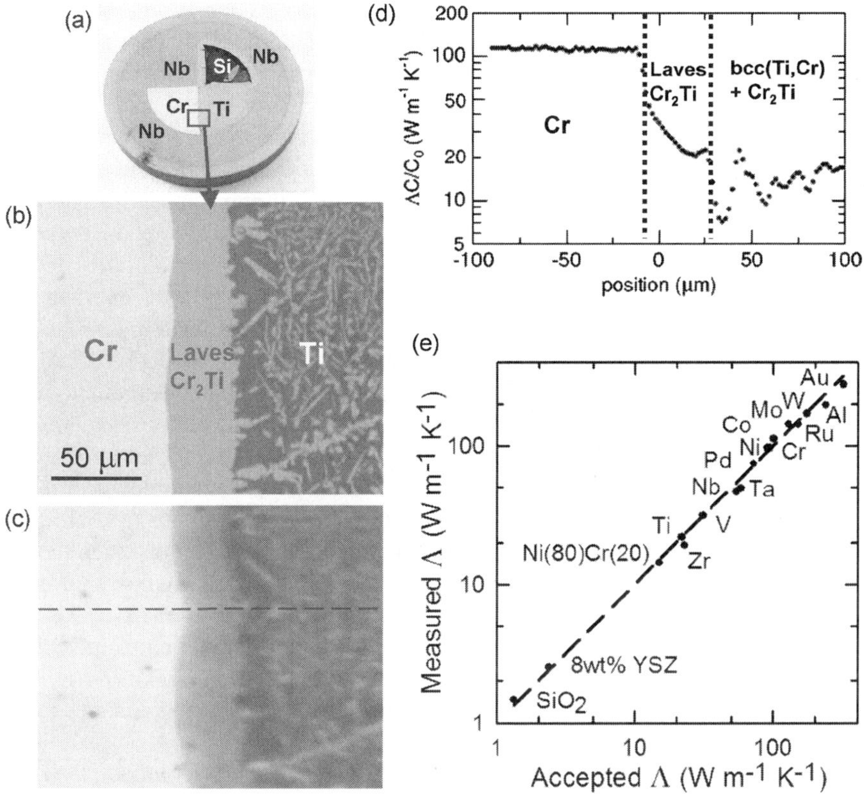

Figure 7 An accurate thermal conductivity mapping tool with micron-scale resolution (29): (*a*) optical image of a Nb-Ti-Si-Cr diffusion multiple; (*b*) EPMA Cr map of the Cr-Ti binary diffusion couple region showing the formation of Laves phases (Cr_2Ti) and precipitation of the C-15 Laves phase from bcc(Ti,Cr) during sample cooling from 1100°C to room temperature; (*c*) thermal conductivity map of the same region as that of (*b*); (*d*) thermal conductivity values along the line shown in (*c*); (*e*) comparison of thermal conductivity values of some common metals and ceramics measured using the new technique with well-accepted values (X. Zheng, D.G. Cahill & J.-C. Zhao, submitted).

The thermal conductivity of metals and intermetallic compounds is mainly carried by free electrons; thus it is highly sensitive to elemental substitution, point defect formation, and ordering. Both elemental substitution and point defect formation decrease the thermal conductivity because both increase the electron scattering. Chemical ordering in crystals decreases the electron scattering, thereby increasing the thermal conductivity. It is worth noting that thermal conductivity is far less sensitive to dislocation formation than the three effects discussed above. For instance, it has been shown that the thermal conductivity of Pt and several Pt alloys does not change much when they are cold rolled to 20% (even to 40% for the alloys) (30). This is simply because the dislocation density per unit volume is far less than that of point defects. Therefore, thermal conductivity measurements may be used to study elemental substitution, ordering, and perhaps even point defect formation in certain cases.

The thermal conductivity of NiAl varies sharply with composition (31) (as shown in Figure 8*a*), which can be correlated to point defect formation in non-stoichiometric compositions (which has been well characterized). There is a high concentration of Ni-vacancies in Al-rich compositions that varies from 0% at 50 at.% Al to 13% at 54 at.% Al (the results were based on samples water quenched from 1000°C). Similarly, the Ni anti-site point defects form in Al-lean compositions of NiAl with their concentration varying from 0% at 50 at.% Al to 4% at 48 at.% Al (32). Such high concentrations of point defects in non-stoichiometric NiAl are probably responsible for the thermal conductivity behavior shown in Figure 8*a*. Similarly, the thermal conductivity of CoAl varies appreciably with concentration, as shown in Figure 8*a*, which is also consistent with the varying amounts of point defects in non-stoichiometric compositions (33). Strangely, the thermal conductivity of FeAl changes little with composition, even though there are appreciable amounts of point defects in FeAl (34) depending on the compositions. It is speculated that even the stoichiometric composition of FeAl is not highly ordered—it has a significant number of vacancies; its thermal conductivity value is already so low that the point defect–induced thermal conductivity change no longer makes a difference. This interesting phenomenon is a subject of further study.

As a third element is added into a binary intermetallic compound, it often preferentially substitutes for one element or the other. Such elemental substitution (site preference) information is critical for designing the strength of intermetallics (35, 36), and it is also essential for selecting the right sublattice models for thermodynamic (CALPHAD) modeling of the intermetallics. For a stoichiometric compound or a compound with a limited single-phase region, the information can be directly read from the direction of the binary phase into the corresponding ternary phase diagram, i.e., the direction of the so-called solubility lobe. For a non-stoichiometric compound with wide single-phase region such as NiAl, the information cannot be deducted from the phase diagram. For instance, it is not possible to determine whether Ge substitutes for Ni or Al or both from the Ni-Al-Ge ternary phase diagram schematically shown in Figure 8*b*. For such cases, there have been many different ways to study the elemental substitution, including atomic probe field-ion microscopy (APFIM) (37, 38), channeling-enhanced microanalysis (ALCHEMI) in transmission electron microscopy (TEM) (39, 40), ion channeling (41, 42), and X-ray diffraction (XRD) (43, 44). These techniques often require stringent sample preparation or even single-crystal samples that are not easy to prepare for brittle intermetallics. An alternative way is to extract the site preference information from the thermal conductivity contour map, as shown in Figure 8*c*, for the Ge in the NiAl example (45–47). In the direction of elemental substitution, the thermal

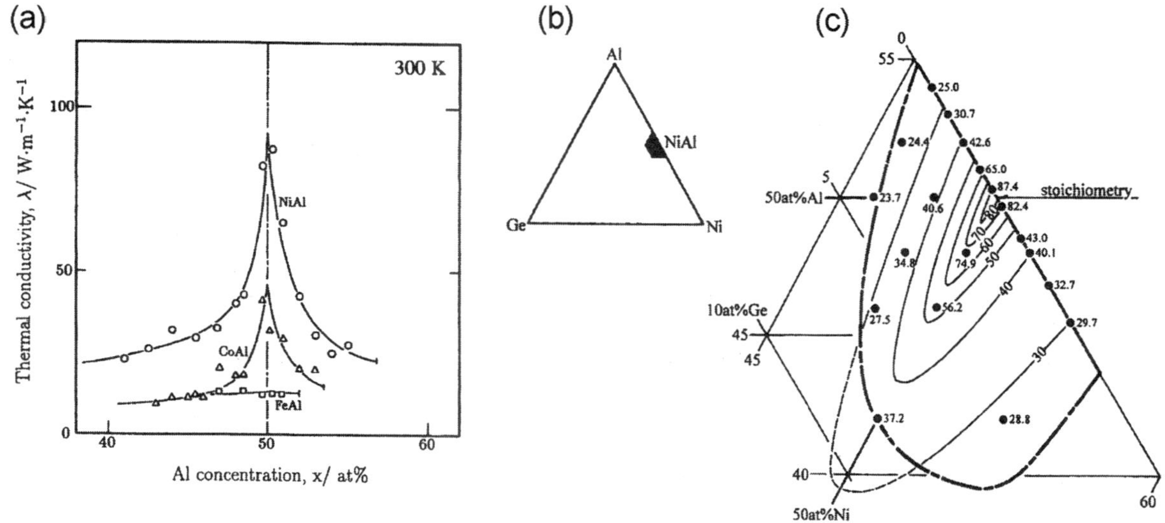

Figure 8 Variation of thermal conductivity with composition showing potential applications of thermal conductivity mapping: (*a*) thermal conductivity of NiAl and other aluminide phases; (*b*) a schematic Al-Ni-Ge phase diagram emphasizing the NiAl phase region; (*c*) a contour map of thermal conductivity for the NiAl phase doped with Ge, showing the ridge direction, thus indicating that Ge is substituting Al in the NiAl phase (31, 45–47).

conductivity displays a ridge in the contour map because the atom mixing in this direction is less random than other directions. Thus it can be clearly seen that Ge preferentially substitutes for Al in NiAl. Again, a combination of diffusion multiples to form intermetallic compounds and the new measurement technique to map thermal conductivity variation with composition can be used to effectively evaluate the site preference/elemental substitution in intermetallic compounds.

Thermal conductivity is generally correlated to electrical conductivity by the Wiedemann-Franz law (48, 49); thus thermal conductivity mapping can be used to get an indication of the electrical conductivity of the materials and also to map any ordered compositions that have increased thermal and electrical conductivity with respective to the disordered alloys.

Thermal conductivity measurement has rarely been used as a research tool to study alloy substitution and ordering simply because thermal conductivity is not a property easy to measure accurately and effectively. One exception is the work by Terada and co-workers (45–47), who effectively used thermal conductivity measurement to study elemental substitution/site preference. The development of this micro-scale technique may make thermal conductivity measurement a common tool for materials science research. As more micro-scale property tools are developed to enable fast measurements of certain properties as a function of composition and phases in diffusion multiples or other combinatorial libraries, new ways to do research will emerge to facilitate the development of more composition-structure-property relationships for faster materials design.

The new thermal conductivity probe can also be directly applied to the development of thermoelectrics, thermal barrier coatings (TBCs), and other functional materials for which thermal conductivity is a key design property.

SUMMARY

The key advantage of diffusion multiples is the formation of all equilibrium intermetallic compounds and solid-solution phases with a complete range of all single-phase compositions for binary and ternary systems. The local equilibrium at the phase interfaces allows accurate determination of phase diagrams. The compositions and phases created in the diffusion multiples allow mapping of phase properties when micro-scale measurement techniques are available. Several properties such as hardness, elastic modulus, and thermal conductivity can already be determined. With the development of micro-scale screening tools for determining optical, magnetic, mechanical, and other properties, numerous other properties can be directly determined for different phases and compositions. In this sense, diffusion multiples become a combinatorial synthesis tool for parallel generation of bulk (not thin film) intermetallic compounds and solid solutions.

Application of the diffusion-multiple approach in designing alloys and in performing effective materials research is summarized schematically in Figure 9, see color insert. The left-hand side shows the application of diffusion multiples

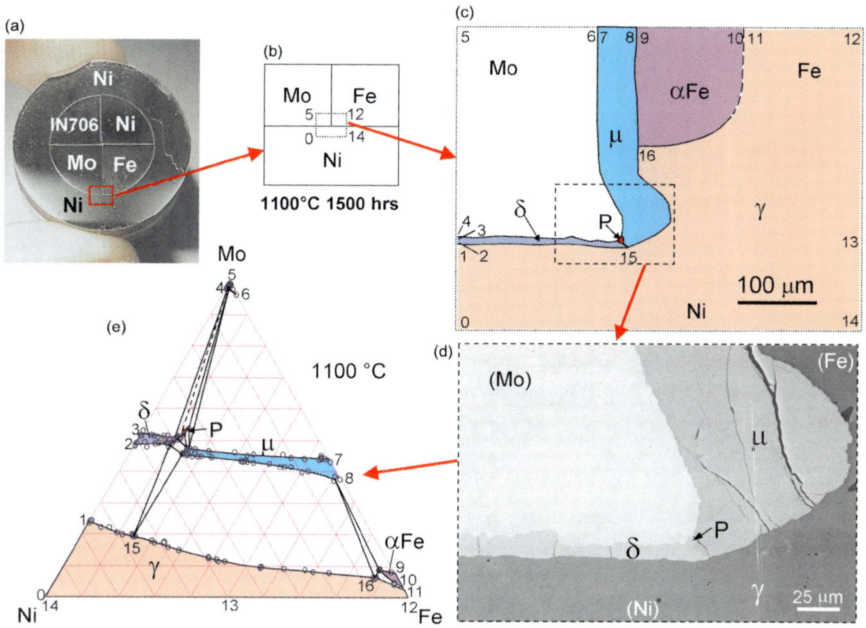

Figure 1 A diffusion multiple made up of Ni, Fe, Mo, and superalloy Inconel 706 (IN706) for rapid mapping of the Ni-Fe-Mo phase diagram and for studying the alloying effect of Mo addition to IN706. The phase diagram is plotted using atomic percent axes with numbering of the scales removed for simplicity (6, 7).

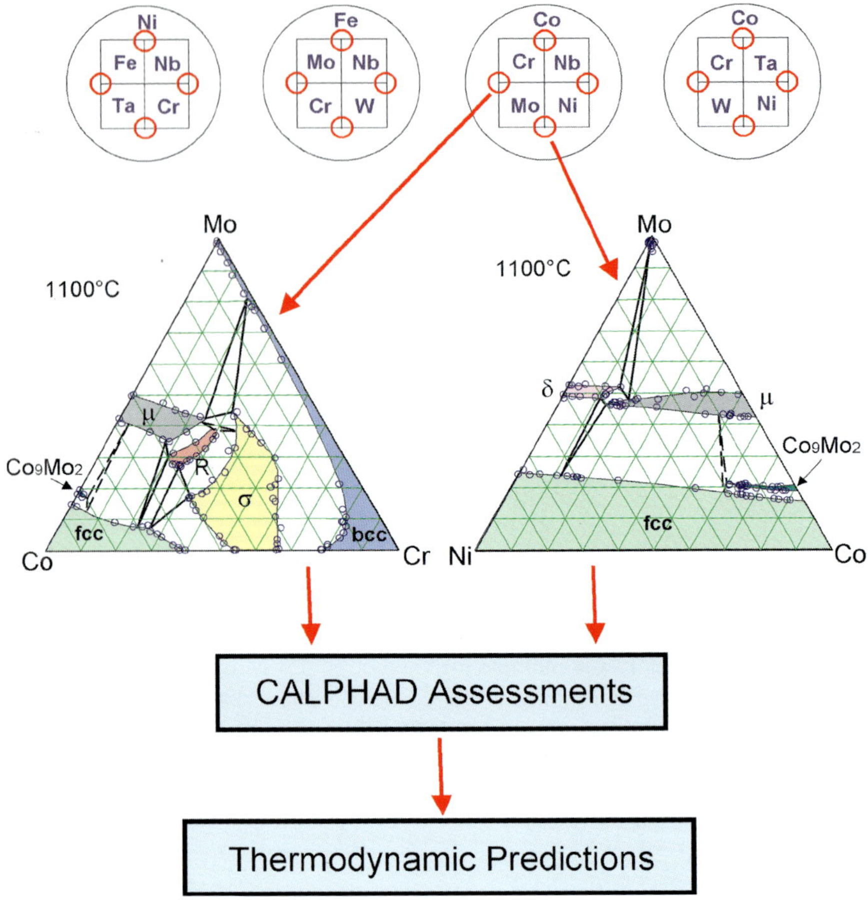

Figure 2 Application of diffusion multiples to obtain phase diagrams critical for accurate assessment of TCP phases in Ni-base superalloys. Four diffusion multiples were employed to obtain phase diagrams of 16 ternary systems (*circles*), two of which (Co-Cr-Mo and Co-Ni-Mo) are shown. These phase diagrams can then be used as input for thermodynamic modeling of the TCP phase stability (J.-C. Zhao & M.F. Henry, unpublished data).

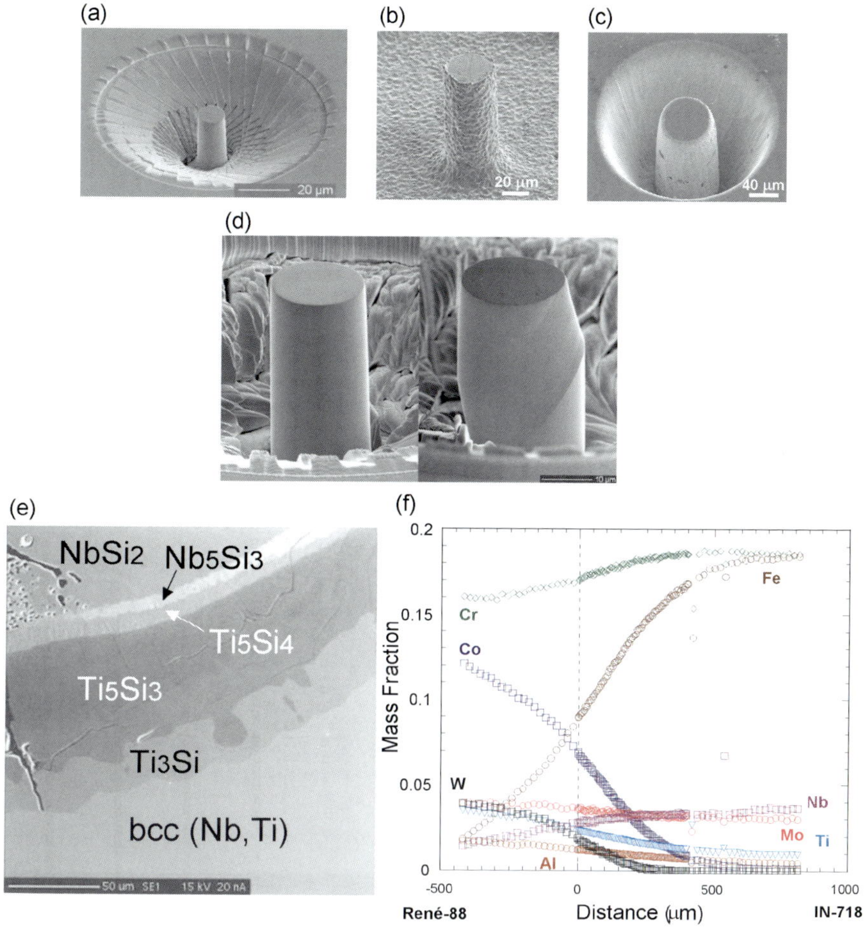

Figure 6 Potential screening for structural alloys and intermetallics using micro-compression tests on micro-pedestals: (*a*) a micro-pedestal machined from a sample using FIB (26, 27); (*b*) a micro-pedestal machined using micro-EDM (M. Uchic, unpublished data); (*c*) a micro-pedestal machined from a sample using a femtosecond laser (Q. Feng & T. Pollock, unpublished data); (*d*) SEM images showing a pedestal before and after a micro-compression test (26, 27); (*e*) intermetallic compound phases formed in a Nb-Ti-Si diffusion multiple showing their dimensions (9); (*f*) composition profiles from a diffusion multiple showing the wide interdiffusion region between Ni-base superalloys René 88 and IN718 (20). Micro-machining and micro-compression tests can be employed to screen new structural materials using diffusion multiples such as these shown in (*e*) and (*f*). Images (*a*, *b*, and *d*) are courtesy of M. Uchic. Image (*c*) is courtesy of Q. Feng & T. Pollock. Note that the pedestals shown in (*a*) to (*d*) are not machined from diffusion multiples shown in (*e*) and (*f*).

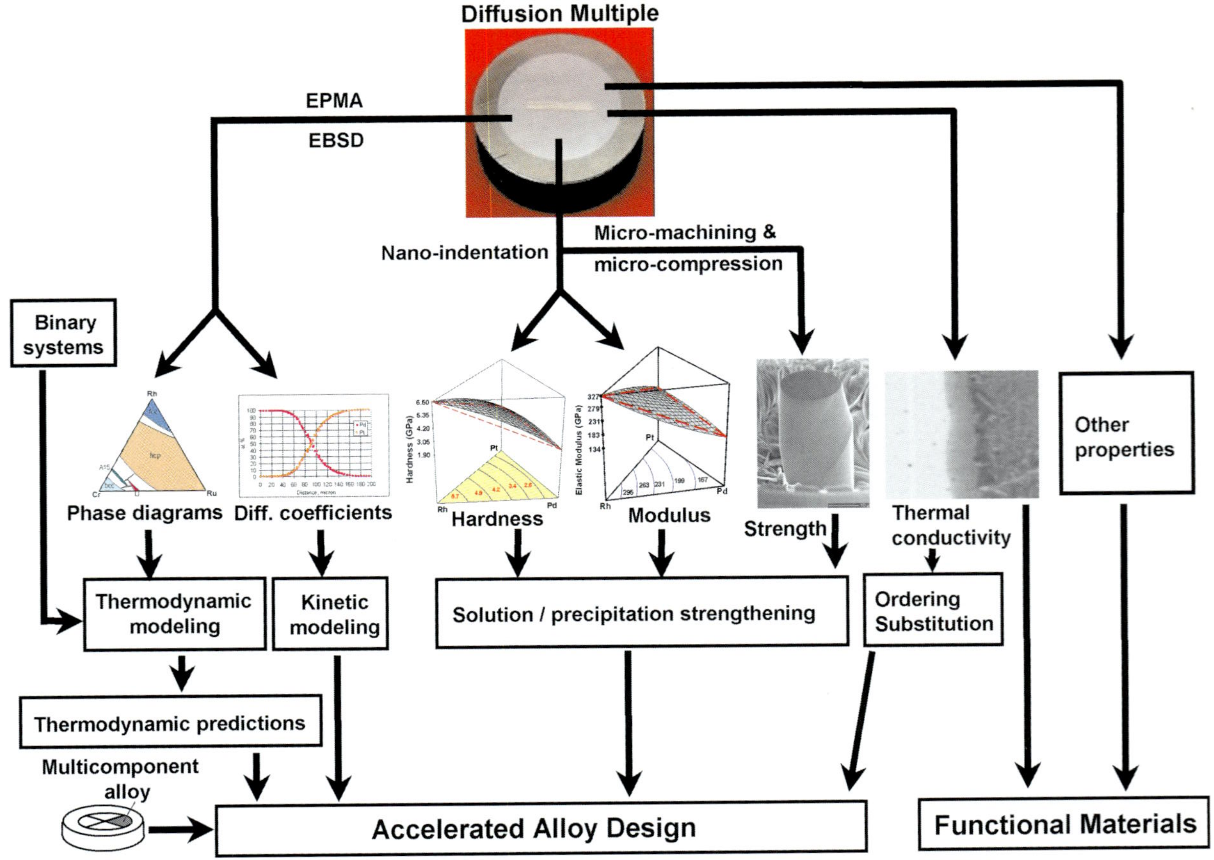

Figure 9 Summary of the diffusion-multiple approach to designing alloys and discovering functional materials.

to map phase diagrams as essential input to CALPHAD modeling to construct thermodynamic databases for prediction of phase equilibria in multicomponent alloys. This is like building a "global map (phase diagram)" for a particular multicomponent system. Because the thermodynamic data are extrapolated mostly from binary and ternary systems, the global map may not be very accurate (detailed). But, it will guide alloy designers to the proximity of a desired location (phases) in the multicomponent space. By using one or more multicomponent alloys as members of diffusion multiples, a detailed local map can be obtained to help to optimize the alloy and specify the composition windows. This is schematically shown at the bottom left of Figure 9. The diffusion-multiple approach can increase the efficiency of phase diagram mapping by orders of magnitude, thus making it possible to have phase diagrams available ahead of the alloy design process if the diffusion multiples are made early. This is particularly useful for development of completely new alloys. For development of existing alloy systems such as superalloys, the diffusion-multiple approach can be applied to obtain critical data (such as TCP phase equilibria) to improve the predictability of the thermodynamic database.

Many binary diffusion profiles can be obtained from diffusion multiples to allow extraction of chemical diffusion coefficients. Inert markers can be introduced when necessary to allow extraction of intrinsic diffusion coefficients. Because of the large number of profiles available, automated diffusivity extraction becomes imperative. It is important to develop robust algorithms to perform diffusivity data extraction from binary, multiphase diffusion profiles. Extraction of exact ternary diffusivity data from the tri-junction regions of diffusion multiples may not be an exercise worth doing because of the complications associated with the geometry and the difficulty in locating the initial interfaces. The quasi-binary diffusion profiles between two multicomponent alloys are useful both to check the reliability of diffusivity database predictions and to identify potential higher-order interactions. The diffusivity database is essential for modeling nucleation kinetics, precipitation kinetics, and grain growth.

In addition, diffusion multiples made up of multicomponent alloys can be used to generate new compositions by interdiffusion to allow direct screening of compressive strength using micro-compression tests on micro-pedestals machined with FIB, micro-EDM, or femtosecond laser techniques.

The middle of Figure 9 illustrates the importance of hardness and modulus mapping for structural materials design. The solution-hardening effect is directly correlated to solution-strengthening, which involves complex interactions. Mapping of hardness and modulus will provide critical data for alloy design and for testing first principles predictions. A solution-hardening database can be built to allow alloy developers to pick effective elements for strengthening the alloys. This is particularly useful for alloy design in new alloy systems when the elemental selection needs to be made. Precipitation kinetics and precipitation-hardening effects can also be found using diffusion multiples coupled with nano-indentation and microstructure analysis, as shown for the binary Ni-NiAl example.

An exciting approach for high-throughput screening of new alloys and new intermetallics consists of using (*a*) diffusion multiples to generate composition variations, (*b*) FIB, micro-EDM, or femtosecond laser to make micro-pedestals, and (*c*) nano-indentation to perform micro-compression tests. There are complications that will make the data analysis less straightforward, but the benefits far outweigh the complications arising from the dependence of strength on the pedestal diameter and the scale of the phases relative to the pedestal diameters. The compressive strength data obtained from such screening will certainly not be of high enough quality to replace bulk tests (also, tensile rather than compressive strength data are required for designing parts), but trends will have value in finding the right compositions for a new alloy. When the grain size is larger than the pedestal diameter, close to single-crystal-type compression data can be obtained. In this case, the EBSD orientation information will be very useful. These data can provide useful input for multi-scale constitutive modeling. It is worth noting that this method is especially important for the development of structural intermetallics. Most intermetallics are brittle and thus difficult to process to full density for macro-testing. A macro-scale compression test is a considerable amount of work for each composition. In contrast, many fully dense intermetallic compounds are formed in diffusion multiples to allow rapid screening of structural intermetallics. Some of the compounds may have cracks formed in certain regions owing to mismatch in thermal expansion coefficients of different compounds during cooling from the heat treatment temperature. Other un-cracked regions would allow micro-pedestals be made and tested.

Micro-scale thermal conductivity mapping on compositions and phases formed in diffusion multiples can provide information regarding elemental substitution (site preference) and ordering. Such information is also useful in the design of intermetallic compounds for various applications, especially for high-temperature strength.

The ability to focus X-ray beam to micron and sub-micron spots at the synchrotron beam lines (50–52) is an asset for the diffusion-multiple approach. With such a fine X-ray probe of sufficient intensity, it will be possible to measure lattice parameters as a function of composition and phases. Accurate lattice parameter measurements allow calculation of density for each composition with the local concentration being measured from EPMA. In addition, by measuring lattice parameters at different temperatures, the thermal expansion coefficients can be evaluated. All this information is going to accelerate alloy design and materials research.

The right-hand side of Figure 9 emphasizes the potential of employing diffusion multiples to discover functional materials. This will be made possible with the development of many micro-scale property probes to allow mapping of physical, thermal, optical, magnetic, and other properties. Robust micro-scale probes for electrical conductivity and thermodynamic quantities such as specific heat and heat of formation are among the most critical ones to be developed. Broader applications of the diffusion-multiple approach depend on the development of these probes.

The diffusion-multiple approach has been applied to the development of diffusion barriers for high-temperature coatings of jet engines (8) and hydrogen storage materials. In the first case, the diffusion multiples were made up of wedge-shaped diffusion barrier candidates sandwiched between superalloys and NiAl slabs. The diffusion multiples were used as a comprehensive design-of-experiments to examine many barrier concepts, i.e., compositions, thickness variations, and chemical (thermodynamic) stability/compatibility and kinetics at the same time. In the case of hydrogen storage materials, the diffusion multiples are used primarily to generate intermetallic compounds that are then charged with hydrogen to examine their suitability for its storage (J.-C. Zhao, J.P. Lemmon & V.S. Smentkowski, unpublished data). These two cases are examples of potential applications of the diffusion-multiple approach in development of various kinds of new materials.

ACKNOWLEDGMENTS

The author is grateful to S. Balsone, W.J. Boettinger, L.N. Brewer, D.G. Cahill, R.W. Cahn, C.E. Campbell, E.L. Hall, M.F. Henry, S. Huxtable, M.R. Jackson, J.P. Lemmon, I. Matthew, L.A. Peluso, A.M. Ritter, SD Sitzman, V.S. Smentkowski, J.A. Sutliff, J.H. Westbrook, and X. Zheng for help, support, and/or valuable discussions. Various parts of the work were supported by General Electric (GE) Company internal funding, the US Air Force Office of Scientific Research (AFOSR) under grant number F49620-99-C-0026 with C. Hartley as program manager, and Defense Advanced Research Projects Agency (DARPA) under the Accelerated Insertion of Materials (AIM) program (grant number F33615-00-C-5215) with Dr. L. Christodoulou as project manager.

LITERATURE CITED

1. Xiang X-D. 1999. Combinatorial materials synthesis and screening: an integrated materials chip approach to discovery and optimization of functional materials. *Annu. Rev. Mater. Sci.* 29:149–71
2. Koinuma H, Takeuchi I. 2004. Combinatorial solid-state chemistry of inorganic materials. *Nat. Mater.* 3:429–38
3. Amis EJ, Xiang X-D, Zhao J-C. 2002. Combinatorial materials science: what's new since Edison? *MRS Bull.* 27:295–300
4. Xiang X-D, Takeuchi I, eds. 2003. *Combinatorial Materials Synthesis*. New York: Dekker
5. Potyrailo RA, Amis EJ, eds. 2003. *High-Throughput Analysis: A Tool for Combinatorial Materials Science*. New York: Kluwer Academic/Plenum
6. Zhao J-C. 2001. A combinatorial approach for structural materials. *Adv. Eng. Mater.* 3:143–47
7. Zhao J-C. 2001. A combinatorial approach for efficient mapping of phase diagrams and properties. *J. Mater. Res.* 16:1565–78

8. Zhao J-C, Jackson MR, Peluso LA, Brewer LN. 2002. A diffusion multiple approach for accelerated design of structural materials. *MRS Bull.* 27:324–29
9. Zhao J-C. 2005. Diffusion multiple screening: phase diagram determination and related studies. In *Encyclopedia of Materials: Science and Technology—Updates*, ed. KHJ Buschow, RW Cahn, MC Flemings, EJ Kramer, S Mahajan, P Veyssiere, pp. 1–13. Oxford, UK: Elsevier
10. Zhao J-C. 2004. Reliability of the diffusion-multiple approach for phase diagram mapping. *J. Mater. Sci.* 39:3913–25
11. van Loo FJJ. 1990. Multiphase diffusion in binary and ternary solid-state systems. *Prog. Solid State Chem.* 20:47–99
12. Hasebe M, Nishizawa T. 1978. Analysis and synthesis of phase diagrams of the Fe-Cr-Ni, Fe-Cu-Mn and Fe-Cu-Ni systems. In *Application of Phase Diagrams in Metallurgy and Ceramics*, ed. GC Carter, 2:911–54. Gaithersburg, MD: NBS Special Publ. #496
13. Jin Z. 1981. A study of the range of stability of σ phase in some ternary systems. *Scand. J. Metall.* 10:279–87
14. van Loo FJJ, Bastin GF, Vrolijk JWQA, Hendriks JJM. 1980. Phase relations in the systems Fe-Ni-Mo, Fe-Co-Mo and Ni-Co-Mo. *J. Less-Common Met.* 72:225–30
15. Saunders N, Miodownik AP. 1998. *CALPHAD*. New York: Elsevier
16. Kaufman L, Bernstein H. 1970. *Computer Calculation of Phase Diagrams*. New York/London: Academic
17. Zhao J-C, Henry MF. 2002. CALPHAD—is it ready for superalloy design? *Adv. Eng. Mater.* 4:501–8
18. Zhao J-C, Jackson MR, Peluso LA, Brewer LN. 2002. A diffusion-multiple approach for mapping of phase diagrams, hardness, and elastic modulus. *J. Metals* 54(7):42–45
19. Campbell CE, Boettinger WJ, Kattner UR. 2002. Development of a diffusion mobility database for Ni-base superalloys. *Acta Mater.* 50:775–92
20. Campbell CE, Zhao J-C, Henry MF. 2004. Comparison of experimental and simulated multicomponent Ni-base superalloy diffusion couples. *J. Phase Equilib. Diff.* 25:6–15
21. Doerner MF, Nix WD. 1986. A method for interpreting the data from depth-sensing indentation instruments. *J. Mater. Res.* 1:601–9
22. Oliver WC, Pharr GM. 1992. An improved technique for determining hardness and elastic modulus using load and displacement sensing indentation experiments. *J. Mater. Res.* 7:1564–68
23. Schwartz AJ, Kumar M, Adams BL, eds. 2000. *Electron Backscatter Diffraction in Materials Science*. New York: Kluwer Academic/Plenum
24. Venkatesh TA, van Vliet KJ, Giannakopoulos AE, Suresh S. 2000. Determination of elasto-plastic properties by instrumented sharp indentation: guidelines for property extraction. *Scripta Mater.* 42:833–39
25. Giannakopoulos AE, Suresh S. 1999. Determination of elastoplastic properties by instrumented sharp indentation. *Scripta Mater.* 40:1191–98
26. Uchic MD, Dimiduk DM, Florando JN, Nix WD. 2003. Exploring specimen size effects in plastic deformation of Ni_3 (Al,Ta). Mater. Res. Soc. Symp. Proc. 753:BB1.4.1–6
27. Uchic MD, Dimiduk DM, Florando JN, Nix WD. 2004. Sample dimensions influence strength and crystal plasticity. *Science* 305:986–89
28. Feng Q, Picard YN, Liu H, Yalisove SM, Mourou G, Pollock T. 2004. Femtosecond laser micromachining of single-crystal superalloys. In *Superalloy 2004*, pp. 687–96. Warrendale, PA: TMS
29. Huxtable S, Cahill DG, Fauconnier V, White JO, Zhao J-C. 2004. Thermal conductivity imaging at micrometre-scale resolution for combinatorial studies of materials. *Nat. Mater.* 3:298–301
30. Terada Y, Ohkubo K, Mohri T, Suzuki T.

1999. Thermal conductivity in Pt-based alloys. *J. Alloys Comp.* 285:233–37
31. Terada Y, Ohkubo K, Nakagawa K, Mohri T, Suzuki T. 1995. Thermal conductivity of B2-type aluminides and titanides. *Intermetallics* 3:347–55
32. Kogachi M, Tanahashi T, Shirai Y, Yamaguchi M. 1996. Determination of vacancy concentration and defect structure in the B2 type NiAl β-phase alloys. *Scripta Mater.* 34:243–48
33. Kogachi M, Tanahashi T. 1996. Point defect behavior in the B2 type intermetallic compound CoAl. *Scripta Mater.* 35:849–54
34. Kogachi M, Tanahashi T. 1998. Random vacancy distribution in B2-type intermetallic compound FeAl. *Scripta Mater.* 39:159–65
35. Cotton JD, Noebe RD, Kaufman MJ. 1993. The effects of chromium on NiAl intermetallic alloys: part I. microstructures and mechanical properties. *Intermetallics* 1:3–20
36. Pike LM, Chang YA, Liu CT. 1997. Solid-solution hardening and softening by Fe additions to NiAl. *Intermetallics* 5:601–8
37. Miller MK, Horton JA. 1986. Site occupation determinations by APFIM for hafnium, iron, and cobalt in Ni_3Al. *Scripta Metall.* 20:1125–30
38. Almazouzi A, Numakura H, Koiwa M, Hono K, Sakurai T. 1997. Site occupation preference of Fe in Ni_3Al: an atom-probe study. *Intermetallics* 5:37–43
39. Chiba A, Shindo D, Hanada S. 1991. Site occupation determination of palladium in Ni_3Al by ALCHEMI. *Acta Metall. Mater.* 39:13–18
40. Munroe PR, Baker I. 1991. Ternary atom site location $L1_2$-structured intermetallic compounds. *J. Mater. Res.* 6:943–49
41. Stoloff NS, Koch CC, Liu CT, Izumi O, eds. 1987. *High-Temperature Ordered Intermetallic Alloys* II. Pittsburgh: Mater. Res. Soc. Symp. Proc. 81
41a. Bohn HG, Williams JM, Barrett JH, Liu CT. 1987. Lattice location of boron and hafnium dopants in an ordered nickel aluminide by use of ion channeling/nuclear reaction analyses. See Ref. 41, pp. 127–33
42. Lin H, Seiberling LE, Lyman PF, Pope DP. 1987. Lattice location of tantalum in Ni_3Al by ion channeling and nuclear reaction analysis. See Ref. 41, pp. 165–70
43. Masahashi N, Takasugi T, Izumi O. 1988. Atomistic defect structures of Ni_3Al containing carbon, boron and beryllium. *Acta Metall.* 36:1815–22
44. Lin H, Pope DP. 1990. The location of tantalum atoms in $Ni_3(Al,Ta)$. *J. Mater. Res.* 5:763–68
45. Terada Y, Ohkubo K, Mohri T. 2001. Site preference determination in intermetallic compounds by thermal conductivity measurement. *J. Mater. Res.* 16:2314–20
46. Terada Y, Ohkubo K, Mohri T, Suzuki T. 2002. Site preference in NiAl—determination by thermal conductivity measurement. *Mater. Sci. Eng.* A329–331:468–73
47. Terada Y, Ohkubo K, Mohri T, Suzuki T. 2002. Thermal conductivity of intermetallic compounds with metallic bonding. *Mater. Trans.* 43:3167–76
48. Klemens PG, Williams RK. 1986. Thermal conductivity of metals and alloys. *Inter. Met. Rev.* 31:197–215
49. Kumar GS, Prasad G, Pohl RO. 1993. Experimental determinations of the Lorenz number. *J. Mater. Sci.* 28:4261–72
50. Ice GE, Sparks CJ. 1999. Modern resonant X-ray studies of alloys: local order and displacements. *Annu. Rev. Mater. Sci.* 29:25–52
51. Larson BC, Lengeler B. 2004. High-resolution three-dimensional X-ray microscopy. *MRS Bull.* 29:152–56
52. Miao J, Ishikawa T, Johnson B, Anderson EH, Lai B, Hodgson KO. 2002. High resolution 3D X-ray diffraction microscopy. *Phys. Rev. Lett.* 89:088303–1–4

Annu. Rev. Mater. Res. 2005. 35:75–98
doi: 10.1146/annurev.matsci.35.082703.155604

First published online as a Review in Advance on March 2, 2005

SPATIAL ORDER AND DIFFRACTION IN QUASICRYSTALS AND BEYOND

Denis Gratias, Lionel Bresson, and Marianne Quiquandon
LEM-CNRS/ONERA, BP 72, F-92322 Châtillon cedex France;
email: denis.gratias@onera.fr, lionel.bresson@onera.fr, marianne.quiquandon@onera.fr

Key Words crystal, order, Bragg diffraction, diffuse scattering

■ **Abstract** This review is an introduction to the problem of determining the nature of the diffraction spectra of ordered solids. After explicit computation examples of Bragg diffraction spectra of quasicrystals (bulk, surface) in kinematical and dynamical (high-resolution electron microscopy) situations, we show on simple toy models that long-range order does not necessarily lead to Bragg peaks which, in turn, are not necessarily the signature of the unique long-range periodic order.

INTRODUCTION

The discovery of quasicrystals by Shechtman (1, 2) opened several basic questions in solid-state physics from the possible existence of non-periodic ground states for Hamiltonians with finite interaction distances to the simple mathematical question of which spatial distribution of point scatters has Fourier spectra containing Bragg reflections.

The existence of Bragg reflections in the diffraction spectra of solids has been long considered the signature of the sole crystalline state. The discovery in 1936 by Johansson and Linde (3) of an ordered structure with periodic antiphase boundaries in a Au-Cu alloy has been the starting point from the notion of a crystal to one of an incommensurate phase, where a parent crystal is perturbed by a modulation wave (in atomic position and/or chemical occupancy) of wavelength that is incommensurate with the crystal lattice. This generates a structure that is not periodic because of the irrationality between the crystal period and the modulation period (4). The fact that this type of structure still shows a set of Bragg peaks is easily understood by noting that the structure can be described by the product of two periodic functions (representing the parent crystal and the modulation), which leads to a convolution of the Fourier spectra of the two functions. This results in a dense, enumerable set of Bragg peaks: those corresponding to the Fourier spectrum of the parent crystal, which are called fundamental reflections, and those issued from the copy of the Fourier spectrum of the modulation wave at each fundamental peak, which are called satellite reflections. Although not periodic, these structures are still the result of a superimposition of two periodic signals. Both fundamental and associated satellite peaks distribute on the nodes of usual reciprocal lattices, and the basic relation between Bragg diffraction and periodicity remains.

The peculiar order found in quasicrystals, together with the progress in the 1970s and 1980 of our understanding of deterministic chaos (for example, see 6), made the paradigm of the Bragg diffraction as the unambiguous signature of long-range order periodicity critically questionable. New algorithms were studied that generated perfectly ordered objects with subtle Fourier spectra more complicated than simple periodic distributions of atomic masses (7, 8). The question of the relations between the kind of spatial deterministic order and the nature of the Fourier spectra became a non-trivial fundamental question as first expressed in early 1986 by Bombieri & Taylor in a seminal introductory paper: "Which distributions of matter diffract?" (9).

The goal of the present review is to address the question of spatial order and Bragg diffraction from the point of view of the crystallographer and the solid-state physicist, with no pretension of mathematical rigor. Our only ambition is to offer a bridge to material scientists for accessing some of the most efficient tools recently developed by mathematicians and theoretical physicists on the structural properties of solids. A substantial part of the examples presented here is extracted from several enlightening and didactic papers of Baake and coworkers (for example, see (10–14)) that the interested reader is strongly encouraged to consult for finding the details of the underlying mathematics.

We first discuss the basis of diffraction in quasicrystallography in the framework of the now-classical cut method. We use a convenient symbolic calculation of the diffraction spectra that has the advantage of simultaneously including the case of quasicrystals and approximants. We then discuss toy models—most of the information has been extracted from Baake's papers cited above—where the non-trivial spatial order shows different types of Fourier spectra.

DIFFRACTION AND IDEAL QUASICRYSTALS

The concept of an ideal quasicrystal (15) is similar to that of an ideal crystal: an idealized construct of an infinite solid ordered according to deterministic rules.[1] This model is a simplified image of what the local atomic structure of a quasicrystal actually is, irrespective of the (expected small) perturbations introduced by defects such as vacancies, interstitials, and phasons at thermal equilibrium.

Framework of Quasicrystallography

Ideal quasicrystals exhibit diffraction patterns with sharp diffraction peaks as do crystals but with specific unusual properties (for example, see (16, 17)): The pattern

[1]Triple periodicity for crystals.

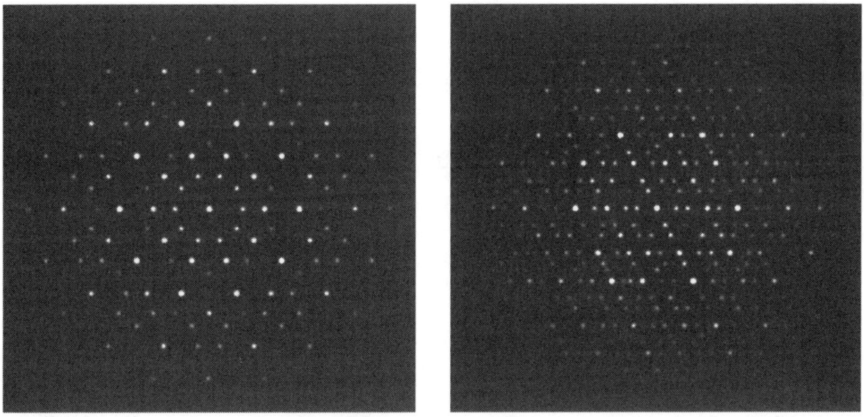

Figure 1 Example of transmission electron diffraction of *i*-AlPdMn along high-symmetry directions (on the left, fivefold; on the right, twofold). All diffraction spots can be unambiguously indexed with six integers, each associated to one of the fivefold directions of a regular icosahedron.

shows sharp, well-defined Bragg peaks that distribute in reciprocal space according to point symmetry generally not consistent with three-dimensional periodicity; the wave vectors of the diffraction peaks can be unambiguously written as a linear integer combination of $n > 3$ basic vectors, pointing on the vertices of a regular generally non-crystallographic polyhedron (see Figure 1):

$$\vec{q} = \sum_{i=1}^{n} n_i \vec{e}_i, \quad n_i \in \mathbb{Z} \qquad 1.$$

The set (Equation 1) forms a dense set of points called a $\mathbb{Z}$-module[2] introduced in crystallography by Janner & Janssen (4) in 1977 as the basic concept in the description of incommensurate structures.[3]

A $\mathbb{Z}$-module in a space of dimension d can be viewed as the irrational projection of a n-dimensional lattice $n > d$; n is the rank of the $\mathbb{Z}$-module and is larger than the dimension of the physical space. A simple example is given in Figure 2 where a (three-dimensional) cube is orthogonally projected on a plane along a generic direction. The three unit vectors of the cube, A, B, and C, project on the plane as a, b, and c, respectively. Because these vectors are coplanar, they are not linearly independent, but they are arithmetically independent: the integer linear

[2]A $\mathbb{Z}$-module is a vector space based on a ring. A base of a $\mathbb{Z}$-module of rank k is a set of vectors e_i, $(i = 1, \ldots, k)$ that are *arithmetically* linearly independent, i.e., such that there exist no non all zero n-uplets of *integers* $(n_1, n_2, \ldots, n_k)$ which make the sum $\sum n_i e_i$ the zero vector.

[3]Quasicrystals and incommensurate phases are both quasiperiodic structures.

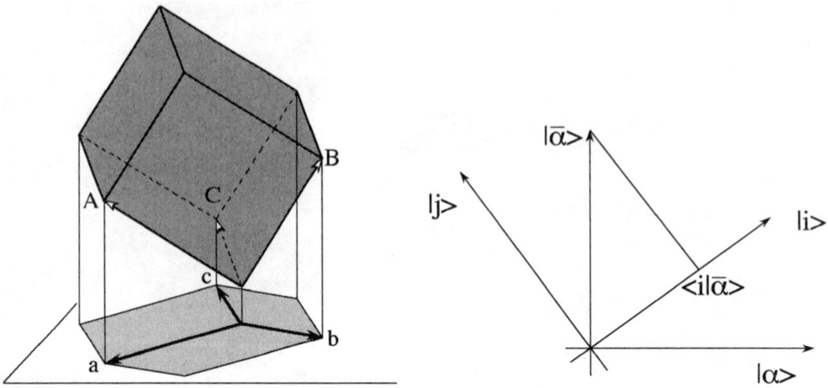

Figure 2 *Left*: an orthogonal projection of a cube along a generic direction defines a $\mathbb{Z}$-module of rank 3 on the projection plane, with generators a, b, and c projections of the three unit vectors A, B and C of the cube. The set of points $v(n, m, p) = na + mb + pc$ of the plane is an enumerable dense set. Each point is in one-to-one correspondence with a node of the three-dimensional cubic lattice. *Right*: the scalar products $\langle i|\alpha\rangle$ are the projection of $|i\rangle$ onto $|\alpha\rangle$ (or vice-versa); their values define the whole geometry of the n-dimensional aspect of quasicrystals.

combination leading to a null vector $na + mb + pc = 0$ is obtained for and only for $n = m = p = 0$. Here, the rank of the $\mathbb{Z}$-module defined by the set of points $na + mb + pc$ $(n, m, p \in \mathbb{Z})$ is 3, whereas the dimension of the representation space (the projection plane) is 2.

Basics of Quasicrystallography

The n-dimensional configurational space (18–21) $\mathbf{E}^n$ can be decomposed in $\mathbf{E}_\parallel$, the subspace of dimension d, which represents the physical space and $\mathbf{E}_\perp$ the subspace of dimension $n - d$ complementary to $\mathbf{E}_\parallel$ and orthogonal to it: $\mathbf{E}^n = \mathbf{E}_\parallel + \mathbf{E}_\perp$. As mentioned by Elser (21), the subspace $\mathbf{E}_\perp$ is best viewed as an order parameter space (or internal space in Janner & Janssen (4)). A $\mathbb{Z}$-module of rank n in a space of dimension $d < n$ is equivalent in the projection of a lattice of dimension n (irrationally) projected onto the d-dim space. This projection in reciprocal space corresponds to a cut in direct space. The geometry of the configurational space is determined by the relative orientations of $\mathbf{E}_\parallel$ and $\mathbf{E}_\perp$ in $\mathbf{E}^n$ (see Figure 2).

A convenient and compact way for manipulating the geometry of quasicrystals is to use Dirac-like notations in the Euclidean $\mathbf{E}^n$ space. We define these notations as

- the set of the n vectors $\{|i\rangle\}$ such that $\langle i|j\rangle = \delta_{i,j}$ forms a natural orthonormal basis of $\mathbf{E}^n$;

- the set of the d vectors $\{|\alpha\rangle\}$ is a d-dim orthonormal basis of $\mathbf{E}_\parallel$;
- the set of the $(n-d)$ vectors $\{|\bar{\alpha}\rangle\}$ is a $(n-d)$-dim orthonormal basis of $\mathbf{E}_\perp$;

from which the projection operators $\hat{\pi}^\parallel$ and $\hat{\pi}^\perp$ on $\mathbf{E}_\parallel$ and $\mathbf{E}_\perp$, respectively, can be written as

$$\hat{\pi}^\parallel = \sum_{\alpha=1}^{d} |\alpha\rangle\langle\alpha|, \quad \hat{\pi}^\perp = \sum_{\bar{\alpha}=1}^{n-d} |\bar{\alpha}\rangle\langle\bar{\alpha}|.$$

Let $\Lambda = \sum_{j=1}^{n} \lambda_j |j\rangle$ be a generic vector of $\mathbf{E}^n$. We can define the projections in $\mathbf{E}_\parallel$ and $\mathbf{E}_\perp$ of Λ in either internal or general bases as

$$\hat{\pi}^\parallel\Lambda = \sum_{\alpha=1}^{d} \lambda_\alpha^\parallel |\alpha\rangle = \sum_{j=1}^{n} \lambda_j^\parallel |j\rangle, \quad \hat{\pi}^\perp\Lambda = \sum_{\bar{\alpha}=1}^{n-d} \lambda_{\bar{\alpha}}^\perp |\bar{\alpha}\rangle = \sum_{j=1}^{n} \lambda_j^\perp |j\rangle,$$

which leads to the following explicit computations of the components:

$$\lambda_\alpha^\parallel = \sum_{j=1}^{n} \langle\alpha|j\rangle\lambda_j, \quad \lambda_i^\parallel = \sum_{j=1}^{n}\sum_{\alpha=1}^{d} \langle i|\alpha\rangle\langle\alpha|j\rangle\lambda_j,$$

$$\lambda_{\bar{\alpha}}^\perp = \sum_{j=1}^{n} \langle\bar{\alpha}|j\rangle\lambda_j, \quad \lambda_i^\perp = \sum_{j=1}^{n}\sum_{\bar{\alpha}=1}^{n-d} \langle i|\bar{\alpha}\rangle\langle\bar{\alpha}|j\rangle\lambda_j.$$

Thus the computation of the components of the projections of a generic vector of $\mathbf{E}^n$ on $\mathbf{E}_\parallel$ and $\mathbf{E}_\perp$ can be figured in two equivalent ways:

- expressed on the internal bases of $\mathbf{E}_\parallel(\mathbf{E}_\perp)$, in which case we use the $n \times d$ and $n \times (n-d)$ matrices $\hat{\mathrm{M}}_{\alpha,i}^\parallel = \langle\alpha|i\rangle$ and $\hat{\mathrm{M}}_{\bar{\alpha},i}^\perp = \langle\bar{\alpha}|i\rangle$;
- expressed of the basis of $\mathbf{E}^n$, using the $n \times n$ matrices:

$$\hat{\pi}_{i,j}^{\parallel(\perp)} = \sum_{\alpha(\bar{\alpha})=1}^{d} \langle i|\alpha(\bar{\alpha})\rangle\langle\alpha(\bar{\alpha})|j\rangle.$$

Let us illustrate this method with a simple example with $\mathbf{E}^n = \mathbb{R}^4$ and $\mathbf{E}_\parallel(\mathbf{E}_\perp) = \mathbb{R}^2$ where the four basic unit vectors in $\mathbb{R}^4$ project in $\mathbf{E}_\parallel(\mathbf{E}_\perp)$ according to the following $\hat{\mathrm{M}}^{\parallel(\perp)}$ matrices:

$$\hat{\mathrm{M}}^\parallel = \frac{1}{\sqrt{2}}\begin{pmatrix} 1 & 0 & \cos\alpha & -\sin\alpha \\ 0 & 1 & \sin\alpha & \cos\alpha \end{pmatrix}, \quad \hat{\mathrm{M}}^\perp = \frac{1}{\sqrt{2}}\begin{pmatrix} \cos\alpha & \sin\alpha & -1 & 0 \\ -\sin\alpha & \cos\alpha & 0 & -1 \end{pmatrix}.$$

The (internal) coordinates on $\mathbf{E}_\parallel$ and $\mathbf{E}_\perp$ of a four-dimensional-node $\lambda = (n_1, n_2, n_3, n_4)$ are

$$\lambda_\parallel = \frac{1}{\sqrt{2}}(n_1 + n_3\cos\alpha - n_4\sin\alpha; n_2 + n_3\sin\alpha + n_4\cos\alpha) \qquad 2.$$

$$\lambda_\perp = \frac{1}{\sqrt{2}}(n_1 \cos\alpha + n_2 \sin\alpha - n_3; -n_1 \sin\alpha + n_2 \cos\alpha - n_4). \qquad 3.$$

The angle α is a parameter between 0 and $\pi/4$. For $\alpha = 0$, using the cut method (see below) we obtain a simple square lattice, and for $\alpha = \pi/4$ we obtain the so-called octagonal tiling. In between, quasiperiodic and periodic tilings made of squares and rhombi are generated that represent a smooth transition from a square lattice to octagonal tiling as shown in Figure 3.

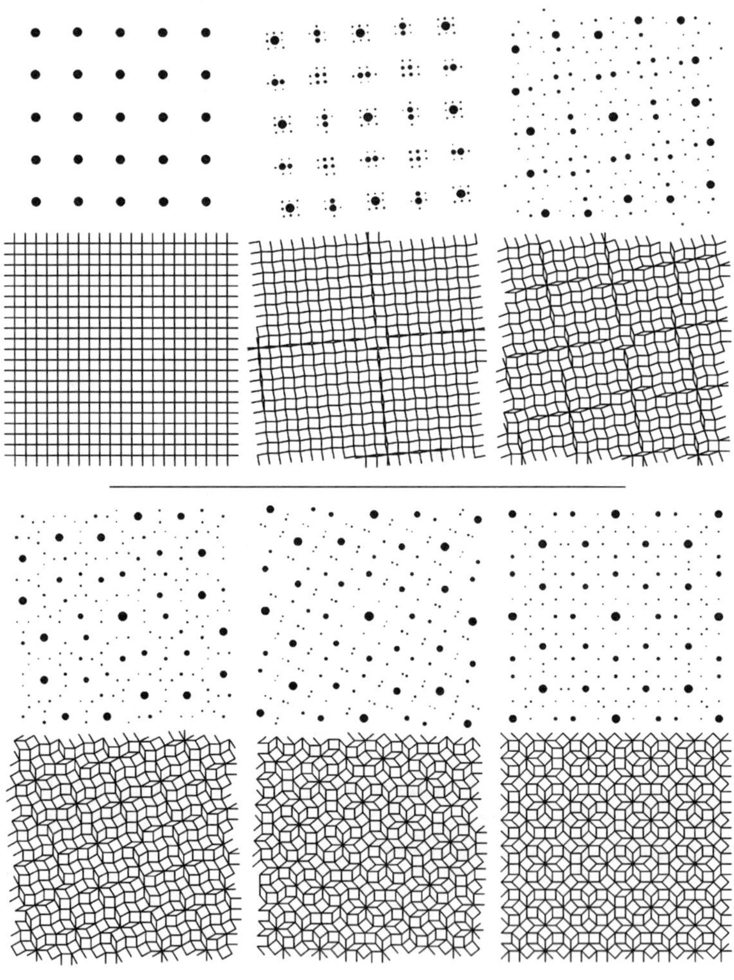

Figure 3 Sequence for $\alpha = 0, 10, 20, 30, 40, 45°$ (see Equation 3), respectively, of quasiperiodic tilings from the square lattice ($\alpha = 0$) to the octagonal tiling ($\alpha = 45°$) obtained from the cut and projection of a four-dimensional simple hypercubic lattice. The diffraction pattern is shown on top of each tiling.

The Cut Method

The natural and simplest method for generating quasiperiodic structures is the cut method introduced by Janner & Janssen (4) for describing incommensurate phases (for a complete discussion, see Bak (22–25)). It consists of cutting a $d_\perp$-manifold by the (Euclidean) physical space $\mathbf{E}_\parallel$. The intersection is an infinite series of points that represent the atomic positions of the structure. What differentiates quasicrystals from incommensurate structures is that in quasicrystals, this manifold is a collection of piece-wise linear, flat sections parallel to $\mathbf{E}_\perp$ instead of having a continuous sinuous surface (26). The atomic surface is noted by σ, and, for simplicity, we assume that the whole structure is generated by only one such atomic surface copied at each n-dimensional lattice node as shown in Figure 4. The actual structure is the collection of intersections of the σ atomic surfaces aligned along $\mathbf{E}_\perp$ with the physical cut space $\mathbf{E}_\parallel$.

The generalization to any finite dimension of the cut algorithm consists in copying in $\mathbb{R}^n$ a prototypic atomic surface σ (which is the bounded volume in a $(n - d_\parallel)$-dimensional vector subspace $\mathbf{E}_\perp$) at each node of a lattice Λ in $\mathbb{R}^n$ and selecting their intersection with the physical cut $\mathbf{E}_\parallel$ of dimension $d_\parallel$ perpendicular to the atomic surfaces and irrationally oriented with respect to Λ. To illustrate geometric properties of the various subspaces implied in the cut method, we use a slightly more general and flexible algorithm that is valid for both quasicrystals and approximant phases (27). With respect to the classical cut algorithm, we make the following generalization (see Figure 5):

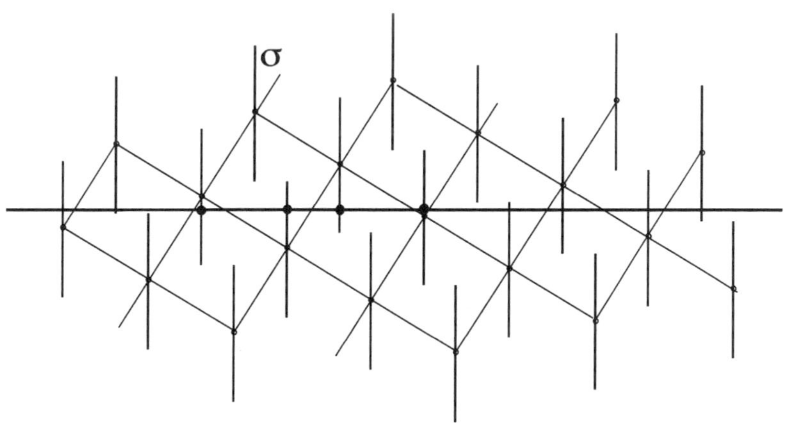

Figure 4 A one-dimensional quasiperiodic set of points can be generated by copying a given atomic surface, here a vertical segment σ, on the vertices of a two-dimensional lattice Λ and collecting the intersection points of these segments. The horizontal line represents the physical one-dimensional space $\mathbf{E}_\parallel$.

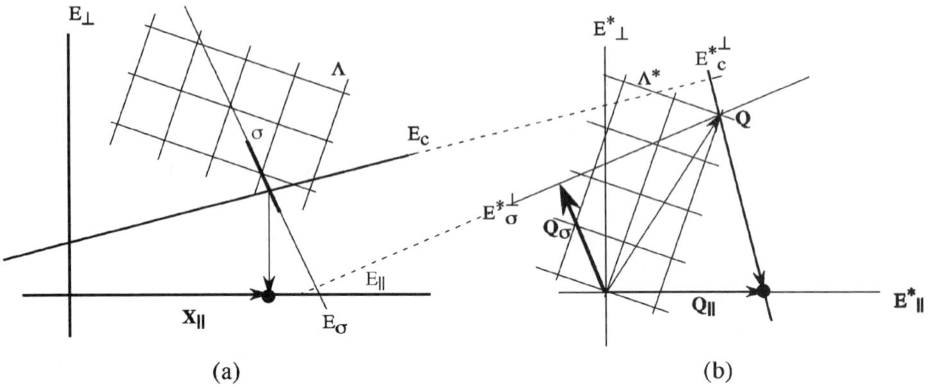

Figure 5 (*a*) A generalization of the cut method consists in differentiating $\mathbf{E}_\perp$ from the carrier $\mathbf{E}_\sigma$ of the atomic surfaces and $\mathbf{E}_\parallel$ from the cut space $\mathbf{E}_c$. (*b*) Reciprocal space geometry: the diffraction pattern is computed by calculating the Fourier transform of the characteristic function of the atomic surface σ, using as argument the projection of the Λ^* reciprocal node along $\mathbf{E}_\sigma^{\perp *}$ onto $\mathbf{E}_\sigma^*$. The location of the peak is determined by the oblique projection of Λ^* along $\mathbf{E}_c^{\perp *}$ onto $\mathbf{E}_\parallel^*$ (27).

1. The atomic surfaces, σ, which are still assumed to be flat, are no longer necessarily parallel to $\mathbf{E}_\perp$; we denote the space carrier of the atomic surfaces by $\mathbf{E}_\sigma$
2. The cut space, $\mathbf{E}_c$, used to collect the points that define the atomic positions in the final structure, is no longer necessarily parallel to the physical space $\mathbf{E}_\parallel$.

Here, the two additional subspaces $\mathbf{E}_\sigma$ and $\mathbf{E}_c$ are used to discuss the influence of slight deviations in the cut and project process from $\mathbf{E}_\perp$ and $\mathbf{E}_\parallel$, respectively. Hence $\mathbf{E}_\sigma$ and $\mathbf{E}_\perp$ have the same dimensions, and $\mathbf{E}_\sigma$ and $\mathbf{E}_c$ are properly oriented with respect to $\mathbf{E}_\parallel$ in the sense that $\mathbf{E}_\sigma$ is transverse to $\mathbf{E}_c$ which, in turn, projects one-to-one onto $\mathbf{E}_\parallel$ along $\mathbf{E}_\perp$. The standard technique is defined by $\mathbf{E}_\sigma \equiv \mathbf{E}_\perp$ and $\mathbf{E}_c \equiv \mathbf{E}_\parallel$.

We generate the structure using both a cut and a projection, which is arbitrary but has the advantage of clearly defining what kind of transformations can be made on quasicrystals and approximants and how they affect the Fourier transforms. The process is threefold: (*a*) We copy the prototypic atomic surface σ at each lattice node (the method generalizes to several atomic surfaces per unit cell); (*b*) we take the intersection of these surfaces with $\mathbf{E}_c$; and (*c*) we project the collected points onto $\mathbf{E}_\parallel$ along $\mathbf{E}_\perp$ (see Figure 5).

A rational orientation of $\mathbf{E}_c$ with respect to Λ generates a periodic structure (one, two, or three dimensions) made of a periodic collection of atomic clusters, where atoms are generally found at locations incommensurate with the lattice parameters. A rational orientation of $\mathbf{E}_\sigma$ leads to a quasiperiodic selection of

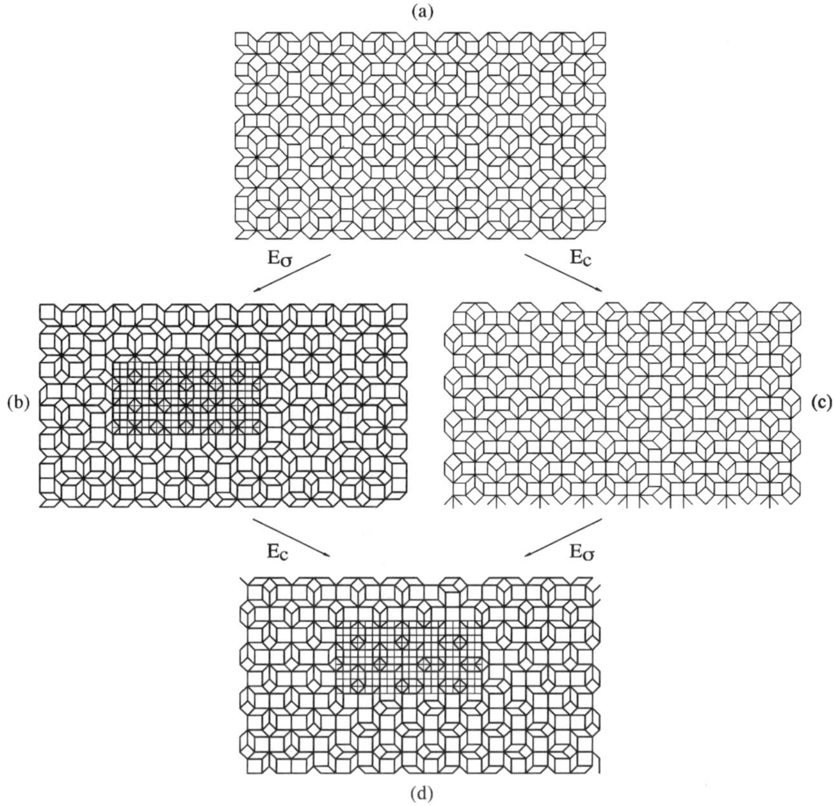

Figure 6 (*a*) Octagonal tiling after (*b*) a disorientation of $\mathbf{E}_\sigma$; (*c*) a disorientation of $\mathbf{E}_c$; (*d*) both.

atomic sites, all belonging to a host periodic lattice (see the octagonal example of Figure 6).

To compute the mass density ϱ associated with the corresponding structure in $\mathbf{E}_\parallel$ we introduce the following notations:

Λ is the n-D lattice, bearing one Dirac δ at each vertex;

$\eta = \eta_\sigma \otimes \delta_{\mathbf{E}_\sigma^\perp}$ is the measure carried by the atomic surface σ: It is the product of the characteristic function η_σ within $\mathbf{E}_\sigma$ (η_σ takes value 1 inside the atomic surface and 0 outside) by a Dirac δ along $\mathbf{E}_\sigma^\perp$;

$\mu_j = 1_j \otimes \delta_{j^\perp}$ is the Lebesgue measure carried by the subspace E_j.

Now, we follow the previous three step algorithm:

1. The prototypic atomic surface is copied at each lattice node of Λ. This is symbolically achieved by making the convolution product $\varrho = \Lambda * \eta$.

2. This set is cut by the subspace $\mathbf{E}_c$, which makes an additional simple product, μ_c: $\varrho = (\Lambda * \eta) \cdot \mu_c$. At this stage, we have defined a set of points in $\mathbf{E}_c$.
3. In order to perform the projection onto $\mathbf{E}_\parallel$, we copy $\mathbf{E}_\perp$ at each point of the previous set by making the convolution by $\mu_\perp$: $\varrho = ((\Lambda * \eta) \cdot \mu_c) * \mu_\perp$, and finally take the intersection with $\mathbf{E}_\parallel$, by performing a simple product by $\mu_\parallel$.

This leads to the following density function:

$$\varrho = (((\Lambda * \eta) \cdot \mu_c) * \mu_\perp) \cdot \mu_\parallel \qquad 4.$$

The standard cut method corresponds to $\mathbf{E}_\sigma \equiv \mathbf{E}_\perp$ and $\mathbf{E}_c \equiv \mathbf{E}_\parallel$ and to the density function reduced to the usual expression:

$$\varrho = (\Lambda * \eta) \cdot \mu_\parallel. \qquad 5.$$

The Fourier transform $\hat{\varrho}$ of Equation 4 is obtained by performing the standard formal transformations between convolution and simple products. Using these rules and noting that $\hat{\Lambda} = \Lambda^*$, $\hat{\eta} = \hat{\eta}_\sigma \otimes \mu_{\sigma_\perp^*}$, we obtain the restriction in $\mathbf{E}_\parallel^*$ of the Fourier transform $\hat{\varrho}$ of Equation 4:

$$\hat{\varrho}_\parallel = ((\Lambda^* \cdot (\hat{\eta}_\sigma \otimes \mu_{\sigma_\perp^*})) * \mu_{c_\perp^*}) \cdot \mu_{\parallel^*} \qquad 6.$$

The geometric interpretation of Equation 6, shown in Figure 5*b*, is as follows:

1. We first take the value of the Fourier transform $\hat{\eta}$ of the characteristic function η_σ using as argument the projections of the nodes of Λ^* along $\mathbf{E}_\sigma^{\perp *}$ onto $\mathbf{E}_\sigma^*$.
2. We make the oblique projection of Λ^* along $\mathbf{E}_c^{\perp *}$ onto $\mathbf{E}_\parallel^*$.
3. The final Fourier amplitude is obtained by multiplying the previous value of $\hat{\eta}$ by a correction term due to the fact that $\mu_{c_\perp^*}$ is not perpendicular to $\mu_{\parallel^*}$.

This analysis shows the main geometric features of the Fourier transform: (*a*) The locations of the peaks in the diffraction pattern depend only on the direction of the cut space $\mathbf{E}_c$. This is irrespective of the direction of $\mathbf{E}_\sigma$. Even if $\mathbf{E}_\sigma$ is rationally oriented, which corresponds to a quasiperiodic decoration of a usual (periodic) lattice, the carrier of the Fourier transform is a non-crystallographic[4] $\mathbb{Z}$-module; (*b*) the intensities of the peaks depend primarily on $\mathbf{E}_\sigma$ and on $\mathbf{E}_c$ as a secondary effect except if $\mathbf{E}_c$ is rationally oriented.

[4]The reason that the Fourier spectrum of a decorated square lattice can have a carrier that is not based on the reciprocal lattice of the host square lattice is because the quasiperiodic decoration does not result from the convolution of a finite pattern with the lattice.

Hence, the experimental determination of the direction of $\mathbf{E}_c$ requires only the determination of the locations of the diffraction peaks with respect to those of the parent quasicrystal irrespective of their intensities.

On the contrary, the determination of the direction of $\mathbf{E}_\sigma$ requires measuring both the locations and the integrated intensities of the diffraction peaks (or high-resolution imaging if the atomic displacements are large enough to be observed). This is similar to standard structure determination in usual crystals. The space group identification (irrespective of possible inversion symmetry) is achieved by purely geometric analysis of the diffraction patterns (point symmetry and possible systematic extinction of certain reflections); the relative locations of the atoms are determined by a further quantitative analysis of the integrated intensities of the diffraction peaks.

Surface Diffraction Amplitude Calculation

For a second example of explicit calculation of quasicrystalline Fourier spectrum, we consider the diffracted amplitude of a quasicrystalline surface (28). We assume the bulk quasicrystal to be well described by the standard density function (Equation 5). Here, the problem is to define $\mu_\parallel$ for the case of a semi-infinite quasicrystal. The direction perpendicular to the cut surface is z; the cut plane (x, y) is at level $z = c$. Let $H(z)$ be the Heaviside function $H(z)$ that takes value 1 for the half space $z \leq 0$ and 0 for $z > 0$. We introduce an amplitude damping function $\phi(z)$ that depends on the absorption coefficient κ, in the bulk defined by $\phi(z) = e^{2\pi\kappa z}$.

With these notations, the characteristic measure $\mu_\parallel$ associated with (half) a quasicrystal truncated by a plane perpendicular to z at level c is given by

$$\mu_\parallel = \delta_\perp \otimes 1_{(x,y)} \otimes \left((H(z)e^{2\pi\kappa z}) * \delta_c\right).$$

The Fourier transform $\hat{\rho}$ of the density function ρ is therefore given by

$$\hat{\rho} = (\hat{\sigma} \cdot \Lambda^*) * \hat{\mu}_\parallel,$$

with

$$\hat{\mu}_\parallel = 1_\perp \otimes \delta_{(x,y)} \otimes \frac{i}{2\pi} \frac{e^{2i\pi q_z c}}{(q_z + i\kappa)}$$

instead of $\hat{\mu}_\parallel = 1_\perp \otimes \delta_{(x,y,z)}$ for the perfect infinite quasicrystal.

We finally obtain the diffraction amplitude characteristic function:

$$\hat{\rho} = (\hat{\sigma} \cdot \Lambda^*) * \left(1_\perp \otimes \delta_{(x,y)} \otimes \frac{i}{2\pi} \frac{e^{2i\pi q_z c}}{(q_z + i\kappa)}\right),$$

which differs from the standard bulk spectrum only by the continuous component along z, which corresponds, in crystals, to the well-known rods of diffuse intensity typical of surface diffraction spectra.

Calculating the diffraction amplitude along the rods goes as follow. We first calculate the usual diffraction amplitude of the bulk for the wave vectors of Λ^*, which project on the $\mathbb{Z}$-module of Λ^* in parallel (reciprocal) space. Each peak is then convoluted by the rod function whose carrier is a line along the z-direction. If, as is generally the case, this direction is a quasicrystallographic direction, the convolution leads to a discrete sum of contributing terms issued from the wave vectors of the $\mathbb{Z}$-module that are aligned along z.

Let $\{q^j\}$ be the infinite set of reflections aligned along the z-direction passing through a given chosen reflection and let q be a generic wave vector along that direction. The diffracted amplitude is then given by

$$\hat{\rho}(q) = \frac{i e^{2i\pi q_z c}}{2\pi} \sum_j \frac{\hat{\sigma}(q^j) e^{-2i\pi q_z^j c}}{\left(q_z - q_z^j + i\kappa\right)} \qquad 7.$$

where $\hat{\sigma}(q^j)$ are the usual structure factors of the bulk quasicrystal. An example of rod profiles calculation using Equation 7 in i-AlPdMn is shown in Figure 7.

The cut level c introduces a quasiperiodically varying phase in the sum that cannot be factorized in the calculation of the intensity. Indeed, contrary to the case of crystals, there are no possible c values for which the phase factors $q_z^j c$ are simultaneously integers for all wave vectors of $\{q^j\}$. Because the termination planes are not periodically stacked, different intensities can be obtained depending on where the quasicrystal is cut.

DYNAMICAL ELECTRON DIFFRACTION AND HIGH-RESOLUTION IMAGES The formulation of dynamical electron diffraction of quasicrystals poses the problem

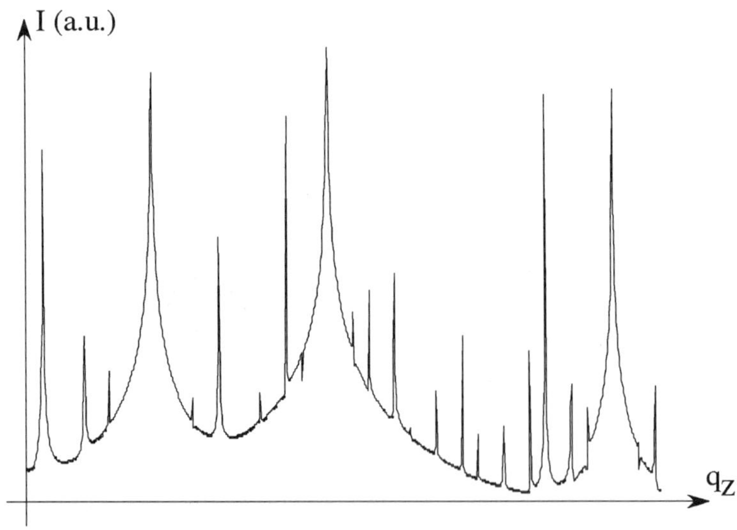

Figure 7 Calculated intensities of rods in i-AlPdMn (after Alvarez et al. (28)).

of diffraction by quasiperiodic potentials. The characterization of the spectra of quasiperiodic Hamiltonians is a non-trivial mathematical problem that goes far beyond our present subject. For the practical case of simulating high-resolution images (for example, see (29)) of quasicrystals, several approximate methods have been successfully used (30). The conceptually most simple one consists of embedding a finite part of the quasicrystal in a supercell by introducing an artificial (large) periodicity, as is often done in defects in crystals. The calculation reduces then to the standard techniques used for crystals.

A second method is based on a rewriting of the dynamical equation, à la Schrödinger, by assuming that the z parameter represents the propagation of the electrons in the solid as a time-like parameter (31–34). Under such an assumption (no backward scattering), the dynamical diffraction process satisfies an evolution equation that can be written as

$$i\frac{d}{dz}|\psi(z)\rangle = \Big(\sum_q |q\rangle(q^2-\chi^2)\langle q| + \sum_{q,q'} |q\rangle V(z)_{q-q'}\langle q'|\Big)|\psi(z)\rangle.$$

Here $|\psi\rangle$ is the wave function of the diffracting electron, χ and q are the projections on a plane perpendicular to the propagation direction z of the incident and diffracted beams (see Figure 8), and $V(z)_{q-q'}$ is the two-dimensional scattering potential for the reflection $G = q - q'$ at level z. In the case of the projected potential approximation, $V(z)_{q-q'}$ is replaced by its average value along z, $\bar{V}_{q-q'}$, which becomes independent of z. In that approximation, the wave function at level z is given by

$$|\psi(z)\rangle = \sum_j |j\rangle e^{i\gamma_j z}\langle j|\chi\rangle$$

where γ_j and $|j\rangle$ are the eigenvalues and eigenvectors of the scattering Hamiltonian $\hat{\mathrm{H}}$:

$$\hat{\mathrm{H}} = \sum_q |q\rangle(q^2-\chi^2)\langle q| + \sum_{q,q'} |q\rangle \bar{V}_{q-q'}\langle q'|. \qquad 8.$$

The main advantage of writing the usual scattering equations this way is that it explicitly shows the q-basis of diffracting wave vectors is not necessarily periodic but only enumerable. Thus it applies to quasicrystals[5] and is performed with a finite number of beams; each simulation is characterized by two lengths in reciprocal space limiting the size of the parallel and perpendicular components of the diffraction vectors, respectively. Typical test calculations (see Figure 9) show that the computed images do eventually stabilize only after a consequent number of beams have been included in the scattering matrix of Equation 8.

[5] Although in this case there is no proof that a finite size Hamiltonian will give converging eigenvalues and eigenvectors by increasing the number of beams, there are infinitely many beams in any finite region of the reciprocal space (the Fourier carrier is not a discrete set of points).

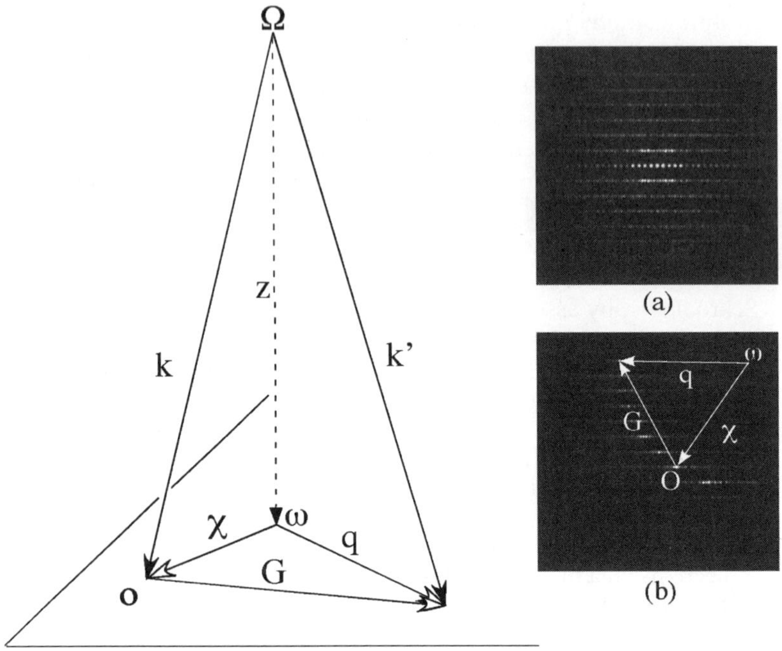

Figure 8 *Left*: In the approximation of forward diffraction of fast electrons, the scattering parameters are the χ and q vectors that represent the projection in the plane P perpendicular to the propagation direction z of the incident beam k and a diffracted beam k', respectively. The point O is the origin of the reciprocal lattice and ω is the orthogonal projection on P of the center of the Ewald sphere. *Right*: (*a*) typical (simulated) diffraction pattern of a large unit cell crystal oriented in exact symmetric Laue orientation ($\chi = 0$); (*b*) the same diffraction after a substantial disorientation showing the dynamical parameters defined on the left for an arbitrarily chosen vector q.

BEYOND QUASICRYSTALS: BRAGG REFLECTIONS AND ORDER

All examples examined so far are quasiperiodic, a long-range order that gives rise to Bragg diffraction-type spectra. To generalize the notion of long-range geometric order, we use the following definition of order: A set of point scatterers is long-range ordered if it can be generated by a deterministic algorithm of finite complexity. This definition is certainly approximate and perfectible with respect to the large notion of order, but it should be sufficient in our present context. It essentially states that an ordered object can be described by the (infinite) repetition of a deterministic construct that is made of a finite number of steps. This, of course,

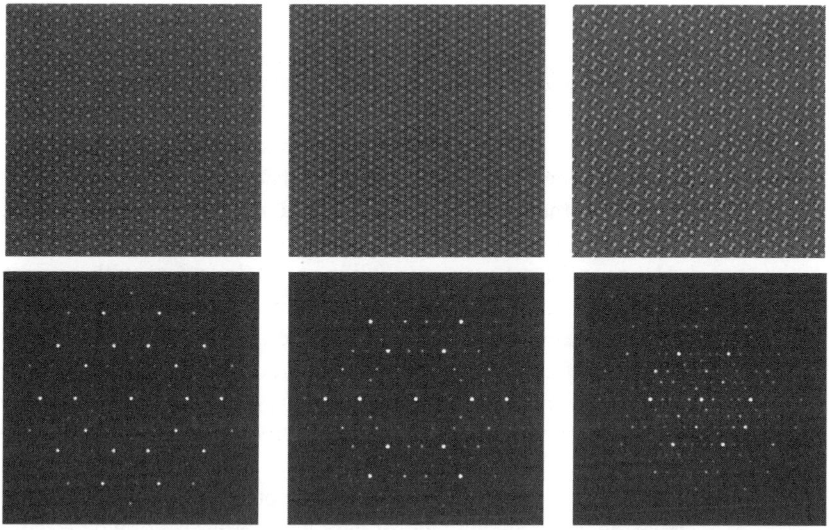

Figure 9 High-resolution image simulation in weak phase object condition of *i*-AlPdMn along the five-, three- and twofold directions with, respectively, 171, 145, and 417 beams (*top row*). The corresponding electron diffraction patterns are shown in the bottom row.

includes the case of periodic order that is generated by copying, side-by-side, the same prototypic (unit) cell parallel to itself.

Here we consider an infinite set of points Λ. Each point of Λ is a scatterer characterized by a scattering function, say f_λ at point λ, that is independent of the diffracting q direction.

There are relatively few but important concepts to consider when studying spatial order and Fourier transforms. The first one deals with the types of distribution in space of the point scatterers; the second considers the type of function to be defined in Fourier space that best characterizes the spectrum with respect to the long-range order.

Among the types of set Λ points in $\mathbb{R}^3$ (or any finite dimension) the following are relevant to solid-state physics:

- discrete;
- uniformly discrete: $\exists\, r > 0$ such that any ball of radius r contains at most one point of Λ;
- relatively dense: $\exists\, R > 0$ such that any ball of radius R contains at least one point of Λ;
- Delone set: Λ is uniformly discrete *and* relatively dense;
- Meyer set: Λ and $\Delta = \Lambda - \Lambda$ are Delone sets.

Here $\Delta = \Lambda - \Lambda$ is a short-hand notation for the set of all vectors between any two points of Λ. In crystals for example, this set is the collection of the points where the pair correlation (or Patterson) function has its maxima.

The fact that these definitions are relevant to solid-state physics is trivial. Indeed, atoms in a solid define a discrete set of points. This set is uniformly discrete in the sense that, because of short-range repulsion, the atoms are at a minimum distance from each other everywhere in the solid. Also, if the solid is homogeneous, the set is relatively dense, meaning that there are no diverging large holes in the solid. Hence, all homogeneous solids are Delone sets with respect to their atomic distribution in space. However, they are not necessarily Meyer sets. In general, Δ, which represents the set of the interatomic vectors, can well be non-uniformly discrete, for example, when the interatomic distances are distributed in a quasi-continuous fashion, as in glasses. On the other hand, atoms of a crystal do form a Meyer set: The pair correlation function shows peaks that distribute on a uniformly discrete set of points.

Hence, the set $\Delta = \Lambda - \Lambda$ is the basic concept to use for discussing long-range order because the physical concept for geometric order is not where the atoms are located but where they are located with respect to each other. In terms of Fourier analysis, this means that the key function to look at is the pair correlation function γ, if it exists, for one-dimensional Delone sets defined $\Lambda = \{f_1, f_2, \ldots, f_n, \ldots\}$, by

$$\gamma^{\Lambda}(h) = \lim_{N\to\infty} \frac{1}{V} \sum_{j=0}^{N-1} f_j f_{j+h} \qquad 9.$$

where f_j is the scattering function of the atom at site j, and the sum runs over all atoms of Λ (discrete set). By its very definition, the function $\gamma^{\Lambda}(h)$ gives the overlap degree of the sequence of itself. The nature of the Fourier spectrum is entirely determined by the way the $\gamma^{\Lambda}(h)$ behaves to infinity, i.e., the way the local order propagates to infinity. If the limits $\gamma^{\Lambda}(h)$ exist for all $h \in \mathbb{Z}$, then an important theorem of Bochner & Herglotz states the existence of a bounded increasing function σ^{Λ} such that

$$\gamma^{\Lambda}(h) = \int_0^1 e^{2i\pi hx} d\sigma^{\Lambda}(x).$$

The quantity $d\sigma^{\Lambda}(x)$ is known as the spectral measure of the sequence Λ. The theorem of Lebesgue's decomposition states that σ^{Λ} is the sum of three increasing functions:

$$\sigma^{\Lambda} = \sigma^{\Lambda}_{st} + \sigma^{\Lambda}_{ac} + \sigma^{\Lambda}_{sg}$$

where σ^{Λ}_{st} is a step function (that corresponds to the Bragg peaks in the spectrum), σ^{Λ}_{ac} is an absolutely continuous function (that corresponds to the diffuse scattering part of the spectrum), and σ^{Λ}_{sg} is a singular continuous spectrum (the derivative of which vanishes almost everywhere with respect to the Lebesgue measure). This last case corresponds to the so-called singular continuous part of the spectrum.

Deciphering which of the three parts is present in the spectrum of a given sequence Λ is not a trivial task. Some fundamental rules apply (see (6)):

1. If σ^{Λ} reduces to a step function, the spectral measure is discrete, a sum of Bragg peaks (pure point diffractive spectrum); this is the case for periodic and almost periodic sequences.
2. If $\lim_{N\to\infty} \frac{1}{N}\sum_{n<N} |\gamma^{\Lambda}(n)|^2 = 0$, then the spectrum is continuous (diffuse scattering).
3. If in addition to the former relation, $\limsup_{n\to\infty} |\gamma^{\Lambda}(n)| > 0$, then the spectrum is continuous singular.

Condition 2 shows a decrease to zero at infinity of the self-correlation function; the diffraction phase coherency vanishes at infinity. The singular continuous spectrum is a rather subtle intermediate between Bragg diffraction and diffuse scattering. The phase coherence with respect to distance is not strong enough to build Bragg reflections, but the correlation function itself still has a non-zero value limit to infinity. Order does propagate to infinity but such that it is not sufficient to generate Bragg peaks.

Some Examples

THE SIERPINSKY GASKET The Sierpinsky gasket is a well-known example of a fractal-type structure built by the following simple inflation iteration. Two copies of a triangle with unit length edges are placed at two of the vertices of the initial triangle thus forming a triangle twice as large as the original. This resulting triangle is used to iterate the inflation process, as exemplified in Figure 10. After n steps, a larger triangle is obtained of linear size 2^n.

We assume the initial triangle to be the placeholder of an elementary motif with scattering amplitude $f_0(q)$. If $F_n(q)$ is the diffraction amplitude per elementary motif at the n-th step, the diffraction amplitude at the $(n+1)$-th step is simply given by

$$F_{n+1}(q) = \frac{1}{3}F_n(q)(1 + e^{2i\pi q_x 2^n} + e^{2i\pi q_y 2^n}) = f_0(q)\prod_{k=0}^{n}\frac{1}{3}(1 + e^{2i\pi q_x 2^k} + e^{2i\pi q_y 2^k}).$$

At each iteration, the diffraction amplitude is multiplied by $\varrho_k(q) = \frac{1}{3}(1 + e^{2i\pi q_x 2^k} + e^{2i\pi q_y 2^k})$ that has a modulus smaller than one, except for the zero q-vector and for those of the form $q = (n/2^k, n'/2^{k'})$ with $n, k, n', k' \in \mathbb{Z}^+$. Rational fractions with powers of 2 as denominator are called diadic numbers. In this case, the multiplicative term ϱ at the iteration level $\sup(k, k')$ is equal to 1 and the product has reached a fixed value. Therefore, we conclude that the diffraction pattern of the Sierpinsky gasket is made of an enumerable set of Bragg peaks localized at the reciprocal points with diadic numbers as coordinates: $q = (n/2^k, n'/2^{k'})$.

Figure 10 *Top*: Inflation rule for generating the Sierpinsky gasket. *Bottom*: Diffraction pattern at increasing sensitivity (see (7)).

Irrespective of the variation of the elementary form factor $f_0(q)$, the Bragg intensities distribute in a hierarchical way according to the (k, k') values: the higher $\sup(k, k')$ the lower is the intensity as shown in Figure 10.

THE ONE- AND TWO-DIMENSIONAL THUE-MORSE SEQUENCES Our second example is an ordered sequence of two letters, say a and b, that is generated by inflation with the following rule: $S_n \to S_{n+1} = S_n\bar{S}_n$ where $\bar{S}_n$ denotes the sequence S_n and where as are transformed into bs and vice versa. For example, on a linear periodic chain, ± 1 spins ($a = 1$ and $b = \bar{1}$) are distributed according to the inflation rule:

$$1 \to 1(\bar{1}) \to 1\bar{1}(\bar{1}1) \to 1\bar{1}\bar{1}1(\bar{1}11\bar{1}) \to 1\bar{1}\bar{1}1\bar{1}11\bar{1}(\bar{1}11\bar{1}1\bar{1}\bar{1}1) \to \cdots.$$

The parentheses are here as a visual guide in the construction of the sequence. The Fourier transform of the linear periodic chain decorated by ± 1 spins is calculated by iteration, using the construction $S_{n+1} = S_n\bar{S}_n$, which leads to the diffracted amplitude per spin:

$$F_{n+1}(q) = \frac{1}{2}F_n(q)(1 - e^{2i\pi q 2^n}) = -iF_n e^{-i\pi q 2^n}\sin(\pi q 2^n),$$

i.e., up to a non-essential global phase factor

$$F_{n+1}(q) \sim \prod_{k=0}^{n} \sin(\pi q 2^k). \qquad 10.$$

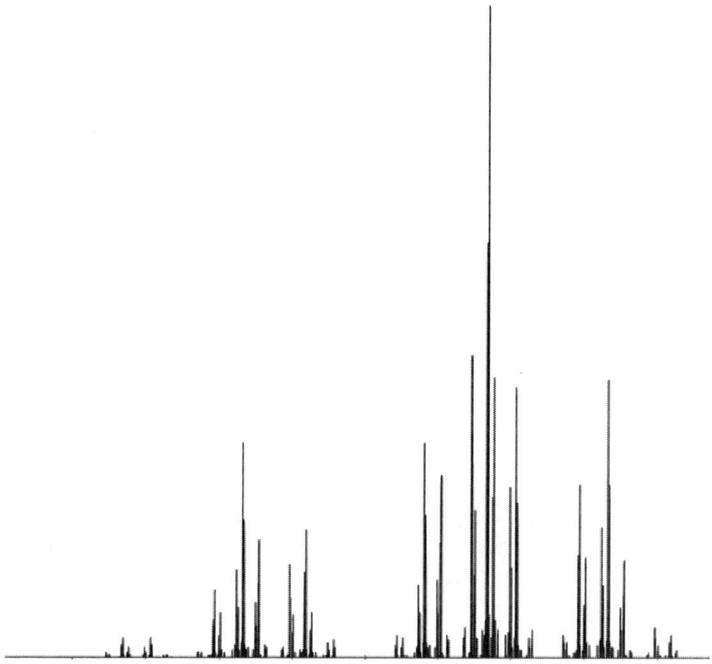

Figure 11 A schematic view of the diffraction spectrum of the ±1 Thue-Morse one-dimensional sequence.

This product, called a Riesz product, and its square converge to zero but its associated correlation function is satisfied by the relation $\limsup_{n\to\infty} |\gamma_n| > 1/3$, which leads to a singular continuous function (6, 8). In addition the ±1 Thue-Morse sequence can be proven to be pure singular continuous; there are neither Bragg nor continuous parts in the spectrum. We obtained a substantially different result from the previous case.[6] The diffraction intensity per spin at level n is a pure continuous singular function with zeros at all the diadic numbers and where one of the sine functions in the product vanishes at some level (8). Figure 11 gives an idea of what this type of spectrum looks like.

To have a better view of a singular continuous spectrum, we can generate a two-dimensional Thue-Morse tiling by using the substitution rule (see Figure 12):

$$a \to \begin{pmatrix} a & b \\ b & a \end{pmatrix}, \quad b \to \begin{pmatrix} b & a \\ a & b \end{pmatrix}.$$

[6]However, if we consider the same sequence but made of 0s and 1s, then a Bragg spectrum shows up in superposition with the singular continuous part on the diadic wave vectors.

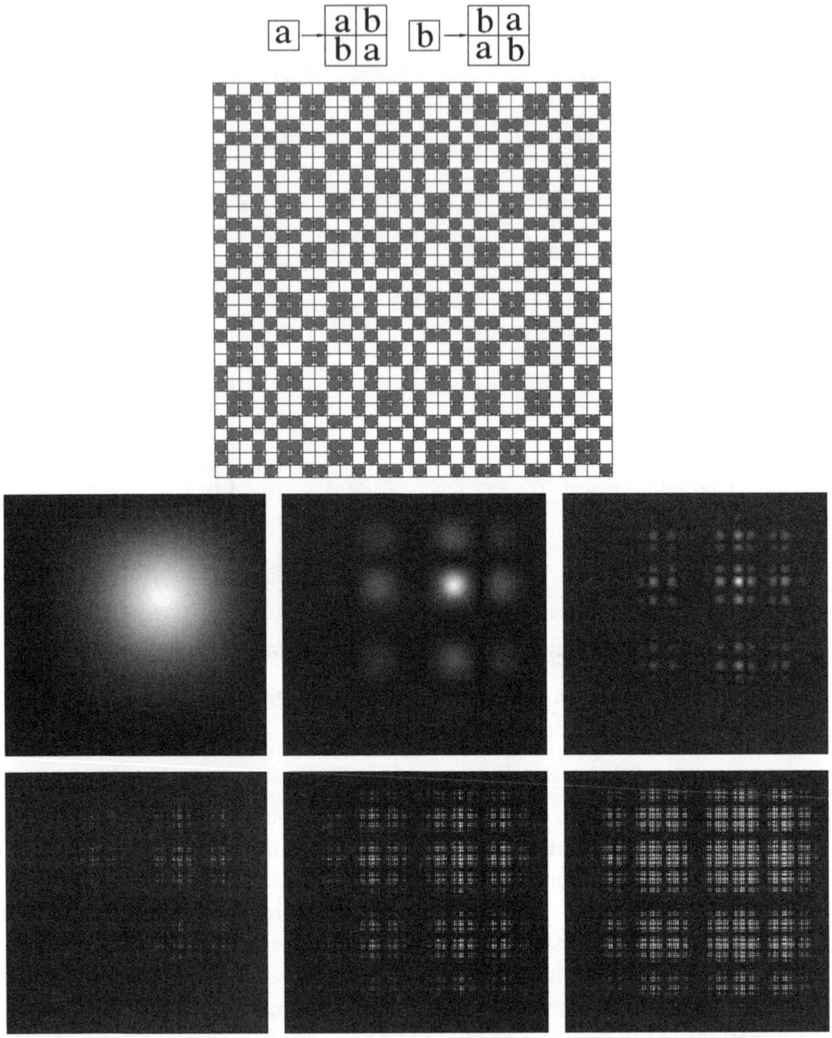

Figure 12 Generation of the two-dimensional Thue-Morse pattern as an ordered decoration of ±1 spins on a square lattice. Construction of the diffraction spectrum of the two-dimensional Thue-Morse pattern after the first six iterations. As in the one-dimensional case, the Fourier spectrum is a singular continuous spectrum with zero values at each q-vector with coordinates equal to diadic numbers.

By replacing a by 1 and b by $\bar{1}$, it is easy to see that the diffraction spectrum is a simple product of the one-dimensional ±1 Thue-Morse sequences along x and y. The construction of the diffraction is shown in Figure 12. Observe that here again the tiling is nothing but a decoration by ±1 spins on the nodes of a square lattice. Nevertheless, the Fourier carrier does not reduce to the simple reciprocal square lattice.

General Results

The previous examples have shown that a diffraction spectrum made of pure Bragg peaks is not the unambiguous signature of long-range order. Other types of long-range order exist beyond periodicity and almost periodicity that lead to the singular continuous spectrum without Bragg components. Some known extreme cases of ordered sequences are homometric to perfectly disordered sequences (for example, see (8, 12)). Such is the case between the regularly spaced linear chain with two types of atoms at stoichiometry, randomly distributed, and the Rudin-Shapiro sequence defined by the site occupancy variables $\sigma(k)$:

$$\sigma(0) = 1, \quad \sigma(-1) = 0, \quad \sigma(4m) = \sigma(4m+1) = \sigma(m)$$
$$\sigma(4m+\ell) = \frac{1}{2}(1 - (-1)^{m+\ell+\sigma(m)}) \quad \ell = 2, 3.$$

These two sequences realized on the same chain have the same diffraction intensity spectrum (*homometric*) $\omega = \frac{1}{4} + \frac{1}{4}\sum\delta$, although the first is totally (chemically) disordered, whereas the second is perfectly ordered!

Some general results have already been established (see (12)). In the case of a distribution Λ of points, the self-correlation function γ^{Λ} can be written as

$$\gamma^{\Lambda}(h) = \lim_{n\to\infty} \frac{1}{V_n} \sum_{\ell,m\in\Lambda\cap B_n; m-\ell=h} f(m)\bar{f}(\ell),$$

where B_n is a ball of volume V_n increasing with n. The sum runs over all pairs of points ℓ and $m = \ell + h$ of Λ that are inside the ball B_n. This expression represents the limit of the overlapping ratio-per-unit volume of Λ itself after a translation of h and is the natural extension of Equation 9. The manner in which the overlapping propagates with increasing translation is well described by the ensembles P_ε of the so-called ε-almost periods defined as

$$P_{\varepsilon>0} := \left\{ h \in \mathbb{R}^3; \left| 1 - \frac{\gamma^{\Lambda}(h)}{\gamma^{\Lambda}(0)} \right| < \varepsilon \right\}.$$

This definition can be used with $\varepsilon = 0$ to define the usual periods t of a crystal where $\gamma^{\Lambda}(t) = \gamma^{\Lambda}(0)$; in that case, P_0 contains the translations of the crystal lattice.

In the case where Δ is uniformly discrete, the following theorems have been established (see (12)):

- The diffraction pattern is a pure point diffractive if and only if the self-correlation measure γ_Λ is such that for all $0 < \varepsilon < 1$, P_ε is relatively dense.
- Under the additional assumption that Λ is relatively dense, the self-correlation of Λ exists and the diffraction of Λ contains a relatively dense set of Bragg peaks (Strungaru theorem).

The first result is a generalization to sets of points of the famous Bohr theorem of 1925 (see (35–38)) on almost periodic systems for continuous functions. It essentially states that a solid with Δ uniformly discrete and exhibiting pure Bragg diffraction (no diffuse scattering) is almost periodic (or trivially periodic).

The Strungaru theorem is of the greatest importance for solid-state physics. Bragg diffraction will be observable in any point scatter distribution Λ, the self-correlation function of which admits maxima on a uniformly discrete set Δ of points. This is trivially the case, for example, for a disordered binary alloy *AB*, where atoms *A* and *B* distribute randomly on the nodes of a lattice. The Δ function is the lattice itself and is therefore uniformly discrete. The theorem states that the diffraction spectrum will contain a set of Bragg peaks[7] with the addition, of course, of a background Laue diffuse intensity.

CONCLUSION

We have shown here that Bragg diffraction and long-range order are two distinct concepts that do not necessary imply each other. Several important properties have emerged over the past years concerning the characteristics of distributions of points that may or may not generate Bragg diffraction. One of the key properties of the diffraction spectrum is the nature of the ensemble Δ (a set of all difference vectors between any two points of the diffracting set): If it is a uniformly discrete set, then Bragg diffraction will show up in the pattern (not excluding possible absolute and/or singular continuous components). The set Δ being uniformly discrete is clearly not a necessary condition: A counterexample is given by the modulated structures where the pair distances involving the atoms displaced by the modulation do not distribute according to a uniformly discrete set. The general problem of Bragg diffraction connected to long-range order is far from being fully understood.

LITERATURE CITED

1. Shechtman D, Blech I, Gratias D, Cahn JW. 1984. Metallic phase with long-range orientational order and no translational symmetry. *Phys. Rev. Lett.* 53:1951
2. Shechtman D, Blech I. 1985. The microstructures of rapidly solidified Al_6Mn. *Metall. Trans.* 16A:1005–12
3. Johansson CH, Linde JO. 1936. Roentgenographic and electrical investigations of the Cu-Au system. *Ann. Phys.* 25:1–48
4. Janner A, Janssen T. 1977. Symmetry of periodically distorted crystals. *Phys. Rev. B* 15:643–58

[7]This result is often established in usual crystallography by replacing each atom by the average atom scatter $c_A f_A + c_B f_B$ and convoluting it with the host lattice.

5. Axel F, Gratias D, eds. 1995. *Beyond Quasicrystals*. Les Houches (France): Springer
6. Allouche JP, Mendès France M. 1995. Automata and automatic sequences. See Ref. 5, pp. 293–367
7. Janssen T. 1995. From quasiperiodic to more complex systems. See Ref. 5, pp. 75–140
8. Queffelec M. 1995. Spectral study of auromatic and substitutive sequences. See Ref. 5, pp. 369–414
9. Bombieri E, Taylor JE. 1986. Which distributions of matter diffract? An initial investigation. *J. Phys. Colloq.* 47(7 Suppl.):C3:19–28
10. Baake M, Höffe M. 2000. Diffraction of random tilings: some rigorous results. *J. Stat. Phys.* 99:219–61
11. Baake M, Moody RV, Schlottmann M. 1998. Limit-(quasi)periodic point sets as quasicrystals with p-adic internal spaces. *J. Phys. A* 31:5755–65
12. Baake M, Moody RV. 1998. Diffractive point sets with entropy. *J. Phys. A* 31:9023–9039
13. Baake M, Moody RV. 2004. Pure point diffraction. *J. Non-Cryst. Solids* 334-335: 77–82. In press
14. Baake M. 2002. A guide to mathematical quasicrystals. In *Quasicrystals*, ed. J-B Suck, M Schreiber, P Häussler, pp. 17–48. Berlin: Springer
15. Janssen T. 1986. Crystallography of quasicrystals. *Acta Cryst. A* 42:261–71
16. Cahn JW, Gratias D. 1987. Quasicrystals: a revolution in crystallography. In *Advancing Materials Research*, ed. PA Psaras, HD Langford. Washington, DC: National Academy Press
17. Gratias D. 1989. Quasicrystals: a new kind of ordered solids. *Physica Scripta* T29:38–45
18. Duneau M, Katz A. 1985. Quasiperiodic patterns. *Phys. Rev. Lett.* 54:2688–91
19. Katz A, Duneau M. 1986. Quasiperiodic patterns and icosahedral symmetry. *J. Physique* 47:181–196
20. Kalugin PA, Kitaev AY, Levitov LS. 1985. 6-dimensional properties of $Al_{0.86}Mn_{0.14}$ alloy. *J. Physique Lett.* 46:L601–L607
21. Elser V. 1986. The diffraction pattern of projected structures. *Acta Cryst. A* 42:36–43
22. Bak P. 1985. Phenomenological theory of icosahedral incommensurate ('quasiperiodic') order in Mn-Al alloys. *Phys. Rev. Lett.* 54:1517–19
23. Bak P. 1985. Symmetry, stability, and elastic properties of icosahedral incommensurate crystals. *Phys. Rev. B* 32:5764–72
24. Bak P. 1986. Icosahedral crystals: where are the atoms? *Phys. Rev. Lett.* 56:861–64
25. Bak P. 1986. Icosahedral crystals from cuts in six-dimensional space. *Scripta Met.* 20:1199–204
26. Katz A. 1990. On the distinction between quasicrystals and modulated crystals. *Proceedings of the Anniversary Adriatico Research Conference on Quasicrystals*, Trieste (Italy), World Scientific, pp. 200–217
27. Gratias D, Katz A, Quiquandon M. 1995. Geometry of approximant structures in quasicrystals. *J. Phys. D: Condens. Matter* 7:9101–25
28. Alvarez J, Calvayrac Y, Joulaud JL, Capitan M. 1999. Characterization of the fivefold plane surface of an $Al_{70.4}Pd_{21.4}Mn_{8.2}$ quasi-crystal by means of surface X-ray diffraction. *Surface Sci.* 423(2-3):L251–57
29. Hiraga K. 1991. High-resolution electron microscopy of quasicrystals. In *Quasicrystals: the State of the Art, Directions in Condensed Matter Physics*, ed. D Divincenzo, P Steinhardt, 11:95–110. Singapore: World Scientific
30. Feng J, Dai M, Wang R, Zou H. 1992. A theoretical and experimental analysis of higher-order Laue zone line splittings caused by a dislocation in an icosahedral quasi-crystal. *J. Phys. Condens. Matter* 4(47):9247–54
31. Gratias D, Portier R. 1983. Time-like perturbation method in high-energy electron diffraction. *Acta Cryst. A* 39:576–84

32. Cornier M, Portier R, Gratias D. 1985. Dynamical calculations of quasi-periodic structures. *Inst. Phys. Conf. Ser.* 78:Chapter 3:91–94
33. Cornier M, Zhang KYu, Portier R, Gratias D. 1986. High resolution electron microscopy of icosahedral phases. In *International Workshop on Aperiodic Crystals*, ed. D Gratias, L Michel, 47:C3, 447–56. Les Houches, FR: J. Phys.
34. Gratias D, Cornier M, Portier R. 1988. Symmetries and beam reduction in the matrix formulation of the dynamical theory of fast electrons. *Acta Cryst. A* 44:789–98
35. Bohr H. 1924. Zur theorie der fastperiodischen funktionen, I. *Acta Math.* 45:29–127
36. Bohr H. 1925. Zur theorie der fastperiodischen funktionen, II. *Acta Math.* 46:101–214
37. Bohr H. 1926. Zur theorie der fastperiodischen funktionen, III. *Acta Math.* 47:237–81
38. Besicovitch AS. 1932. *Almost Periodic Functions*, Cambridge University Press, Cambridge

Annu. Rev. Mater. Res. 2005. 35:99–126
doi: 10.1146/annurev.matsci.33.041002.094657

First published online as a Review in Advance on March 1, 2005

INFLUENCE OF INTERFACE ANISOTROPY ON GRAIN GROWTH AND COARSENING

Gregory S. Rohrer
Department of Materials Science and Engineering, Carnegie Mellon University, Pittsburgh, Pennsylvania 15213-3890; email: gr20@andrew.cmu.edu

Key Words grain boundaries, microstructure, polycrystal, crystal growth, interface energy

■ **Abstract** It has recently become possible to measure the anisotropic distribution of interfaces in polycrystals and composites. Because classical theories for grain growth and coarsening assume isotropic interface properties, they are incapable of explaining how these distributions arise. The purpose of this paper is to review the results of recent experiments, simulations, and theories that document the effects of anisotropy on the capillarity-driven evolution of granular systems. The results suggest that meaningful predictions of evolving microstructural characteristics can be made using models that incorporate the anisotropy of the interfacial energy and mobility.

INTRODUCTION

When a polycrystalline compact is heated at a temperature that is an appreciable fraction of its melting point, the sizes of the individual crystals change in such a way that the average grain size increases. This process is driven by a reduction in the total interfacial energy that occurs as the total interfacial area decreases. Because mass is conserved, some grains get smaller and disappear whereas others grow larger. In single-phase crystalline materials, where the grains are in intimate contact, this phenomenon is referred to as grain growth. When this process occurs in the midst of a distinct intervening phase that separates the crystals, it is usually referred to as coarsening. Because grain growth and coarsening are similar processes motivated by the same driving force, the two terms are frequently interchanged. In fact, without assigning a limiting value for the fractional volume of the intervening phase, a distinct classification may not be possible and, in many circumstances, the two processes may occur simultaneously. However, it should be recognized that distinct mechanisms are at work. In grain growth, the fundamental process is transferring an atom across a boundary from one grain to another. In coarsening, it is the dissolution of material from one crystal, its transport through the intervening phase, and precipitation onto another crystal that underpins the process.

There is an extensive literature on grain growth and coarsening that goes back more than 50 years and several reviews are available (1–3). One of the basic assumptions of the classical work is that the interface energies are isotropic. However, recent experiments and simulations have shown that anisotropy can have important consequences, particularly on the distribution of interface types. Furthermore, by classifying interfaces in a microstructure according to their crystallography, new metrics can be established to serve as a quantitative basis for understanding the relationship between the processing, structure, and properties of a material. In other words, accounting for anisotropy in growth processes not only has the potential to make more accurate predictions of polycrystalline structure, it also offers a way to characterize structural differences that have measurable influences on properties. Therefore, this review focuses on the influence of anisotropy on coarsening and grain growth and is constructed in the following way. The next section provides a summary of phenomenological observations of growth kinetics and size distributions. Classical theories and simulations of grain growth and coarsening are summarized in the third section. In the fourth section, the influence of anisotropic grain boundary and surface properties are discussed and in the fifth section, the impact of singular interfaces on growth is described. The final section identifies unanswered questions and likely trends for future research.

BRIEF ASSESSMENT OF EXPERIMENTALLY OBSERVED PHENOMENA

In a system of growing crystals, experimental observations consistently show that at a sufficiently high temperature, the average grain size increases with time until it approaches the sample size. For samples with anisotropic shapes, such as wires, films, or foils, the smallest dimension is the relevant size. The present discussion is limited to growth in three-dimensional systems where capillary forces dominate. Under these conditions, the time dependence of the average radius, $\langle r \rangle$, is

$$\langle r \rangle - \langle r_0 \rangle = (Kt)^n, \qquad 1.$$

where K is a positive constant, $\langle r_0 \rangle$ is the initial average radius, and n is a value less than or equal to $1/2$ (4–7). The upper limit of $1/2$ occurs for ideal conditions where the boundary velocity scales linearly with the capillary driving force. Smaller exponents are more frequently reported, and these results can be explained if there is a more complex driving force-velocity relationship determined by the presence of impurities (8, 9), pores, or second-phase precipitates (10). For the case of diffusion-limited coarsening, the exponent is $1/3$ (11). Although it has been convincingly demonstrated that in a sample of sufficient purity, grain growth at high temperature follows a kinetic law where $n = 1/2$, it should be recognized that this is an exceptional case and values of n less than $1/2$ are more typical (12).

Experimental observations also indicate that the rate of growth is inversely proportional to the size of the crystals (7):

$$\frac{d\langle r\rangle}{dt} = \frac{K}{\langle r\rangle}. \tag{2}$$

Note that integration of Equation 2 gives Equation 1 with an exponent of 1/2. Because boundaries have been observed to move toward their centers of curvature, and the boundary velocity has been observed to be linearly dependent on the curvature, this phenomenon can be explained in terms of the capillary driving force, $\gamma\kappa$, where γ is the energy per unit area of the interface (assumed to be isotropic) and κ is the interface curvature (7, 13–15). Assuming uniform curvature, $\kappa = 2r^{-1}$, the boundary velocity decreases as $\langle r\rangle$ increases and, therefore, the rate of growth decreases with increasing average grain size.

In the late stages of grain growth and coarsening, it is believed that there is a characteristic distribution of crystal sizes that remains constant even as the average radius increases. In other words, the distribution at two different times is identical except for a change in the magnification. This is referred to as statistical self-similarity or scaling. Although there is some experimental evidence for this assertion, it is difficult to demonstrate conclusively (16–18). As an example, grain size distributions (determined from true three-dimensional data) from four studies are compared in Figure 1*a* (19–22). The crystal size distributions are symmetric on a semilog plot and are most frequently reported to be lognormal:

$$f_L(\rho) = \frac{1}{(2\pi)^{1/2}}\frac{1}{\beta\rho}\exp[-(\ln\rho - \alpha)^2/2\beta^2], \tag{3}$$

where $\rho = r/\langle r\rangle$, $\langle r\rangle$ is the average grain size, and α and β are fitting parameters. The lognormal distribution is shown in Figure 1*c*. Three reported distributions resulting from coarsening are illustrated in Figure 1*b* (18, 23, 24). Note that the first two distributions are similar, do not appear to be lognormal, and both result from the coarsening of a solid within a solid. The third, which is much broader and symmetric, results from coarsening in a liquid. It should also be noted that bimodal distributions are frequently observed in anisotropic systems; this phenomenon is usually referred to as abnormal growth. For example, during the coarsening of $BaTiO_3$, two distinct distributions of grains with average sizes in the ranges of 1 μm and 100 μm have been observed to coexist (25). The exact forms of these distributions and the kinetics of their evolution have not yet been studied in detail.

Topological characteristics in systems of coarsening particles vary widely depending on the volume fraction and characteristics of the intervening phase. For example, the coarsening phase may be fully faceted or bounded by curved interfaces, depending on the anisotropy of the interface energy. For grain growth in single-phase systems, on the other hand, the topology is determined by the simultaneous requirements that the grains fill all space while maintaining local equilibrium at the interfacial junctions (26). Therefore, grains are approximately polygonal volumes whose faces have slight curvatures. When the number of faces per grain is large enough, the interfacial equilibrium constraint forces the faces to be concave so that the grain expands at the expense of its neighbors.

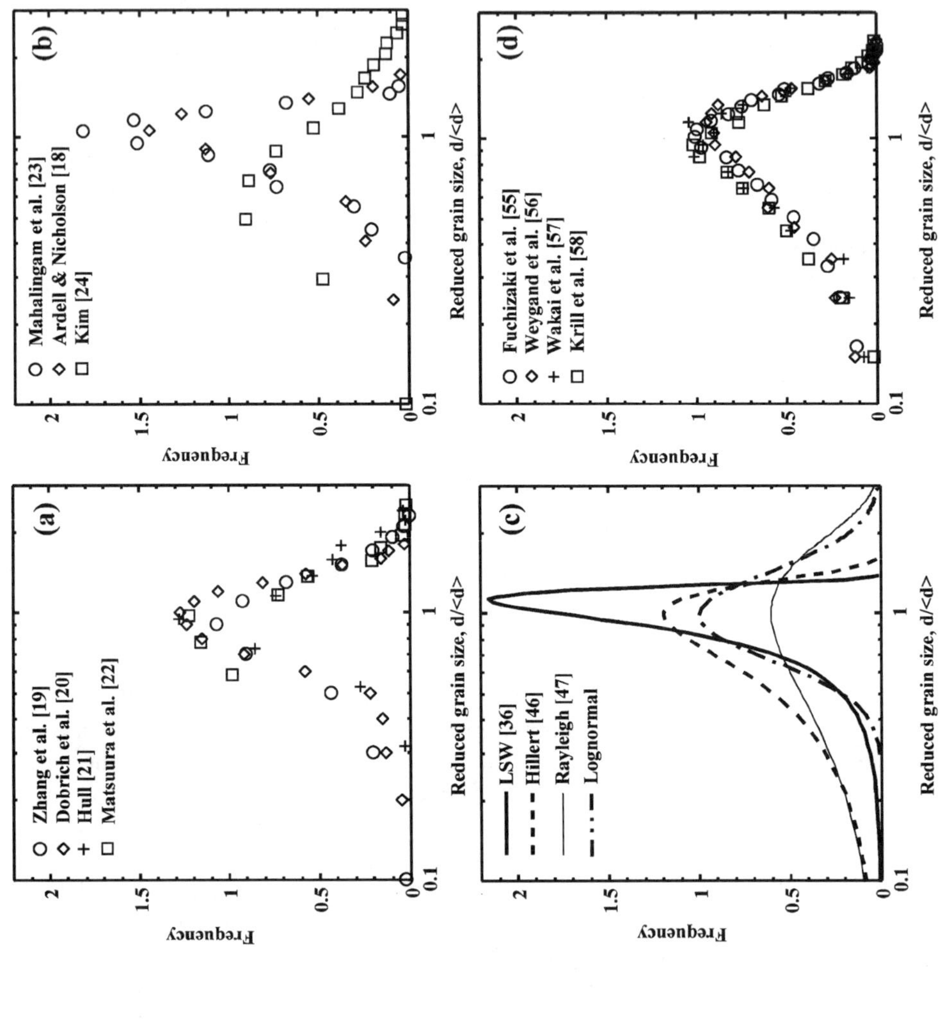
(a)
Zhang et al. [19]
Dobrich et al. [20]
Hull [21]
Matsuura et al. [22]
(b)
Mahalingam et al. [23]
Ardell & Nicholson [18]
Kim [24]
(c)
LSW [36]
Hillert [46]
Rayleigh [47]
Lognormal
(d)
Fuchizaki et al. [55]
Weygand et al. [56]
Wakai et al. [57]
Krill et al. [58]
Frequency
Reduced grain size, d/<d>
0.1
1
0
0.5
1.5
2

The implication is that larger, growing grains have more faces than smaller shrinking grains. This idea is supported by the observation that the average number of faces increases with the average crystal size, as illustrated in Figure 2. Measurements of the average number of faces per grain have yielded values between 12 and 14, whereas the average number of edges per face is very close to 5 (19–22, 27).

Grain growth in nanostructured materials has been an active area of research during the past decade (28–30). A review of these experimental results suggests that for the most part, grain growth phenomena in the nanometer size range is much the same as it is in the micrometer size range: growth follows the kinetics of Equation 1 (29) and the crystal size distribution is roughly lognormal (31). Furthermore, the same things that slow growth in microcrystalline specimens, such as porosity and impurities, also retard the rates of grain growth in nanostructured materials. One difference is that the onset of growth usually occurs at lower temperatures. Furthermore, in at least some systems, anomalous kinetic behavior has been reported. For grain sizes below 150 nm, a linear growth exponent ($n = 1$) was observed in nanocrystalline Fe (32). Also, when growth data for microcrystalline Fe is extrapolated to the nanometer size range, a grain growth rate is predicted that is several orders of magnitude greater than what is actually observed (30). This and similar observations have led to speculation that alloying elements can stabilize nanocrystalline structures. Not only do they provide a drag force on the boundary, but adsorption can reduce the grain boundary energy, which leads to a diminished driving force for growth (33, 34).

In summary, experimental observations lead to the following conclusions about grain growth and coarsening. As a sample is annealed, the grain size increases in

←

Figure 1 Equal area histograms for reduced grain sizes from experiment (*a, b*), theory (*c*), and simulation (*d*). In each case, the histograms are normalized to the same area and plotted on the same domain to simplify the comparison. (*a*) Three-dimensional grain size measurements of a relatively large number of grains ($793 \leq N \leq 4966$). In each case, the linear dimension d is proportional to the cube root of the volume and normalized by the average of this value, $\langle d \rangle$. Zhang et al. (19), 793 Fe grains measured by serial sections: Döbrich et al. (20), 4966 grains of Al in Al-Sn by X-ray microtomography; Hull (21), 941 grains of brass, separated by mercury infiltration; Matsuura et al. (22), 1000 grains of 304 stainless steel separated by boundary corrosion. (*b*) Grain size distributions in coarsening systems, from observations of planar sections: Mahalingam et al. (23), Al 2.4w/o Li (Al_3Li in Al-Li) after 168 h at 200°C; Ardell & Nicholson (18), Ni 6.71%Al (Ni_3Al in Ni-Al) after 6 h at 775°C; Kim (24), WC in liquid Co. (*c*) Theoretical distributions: LSW (coarsening) (36), Hillert (grain growth or coarsening by surface attachment limited kinetics) (46), Louat (grain growth) (47), lognormal (Equation 3). (*d*) Grain size distribution predicted from three-dimensional simulation: Fuchizaki et al. (55), vertex dynamics model; Wegand et al. (56), vertex dynamics model; Wakai et al. (57), surface evolver; Krill et al. (58), phase field model.

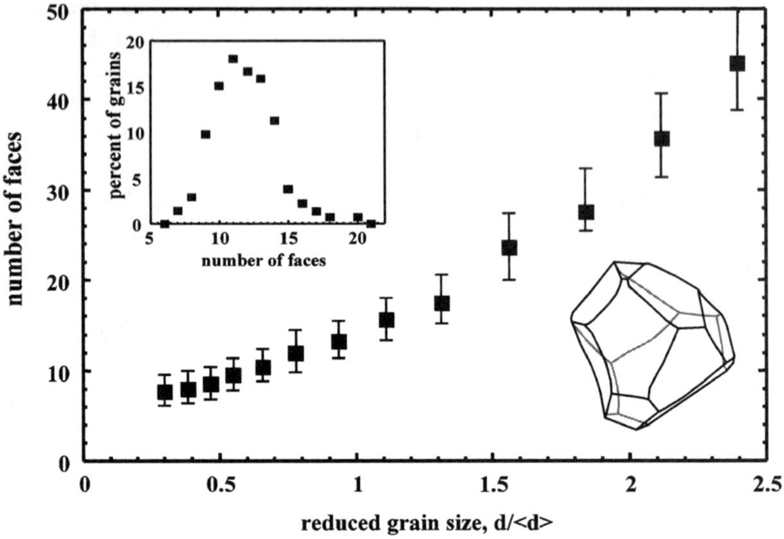

Figure 2 The number of grain faces as a function of reduced grain size (21). For each reduced grain size, the point represents the mean number of faces and the bars represent plus and minus one standard deviation. The upper inset shows the percentage of grains with different numbers of faces, and the lower inset shows a typical shape for a polygonal grain (27). Black lines show edges on the front of the grain and gray lines show edges on the back.

proportion to t^n, where n is less than or equal to 1/2. Boundaries migrate toward their centers of curvature at a rate that is linearly proportional to the curvature. There is a characteristic distribution of crystal sizes that is constant with time in the late stages of growth, which is frequently described as lognormal. However, in numerous situations bimodal distributions are also observed. In cases where the growth is unimodal, the maximum grain size is usually two to three times the average. In dense polycrystals, the grains have on average 12 to 14 faces and 5 edges per face.

ASSESSMENT OF CLASSICAL THEORIES FOR GRAIN GROWTH AND COARSENING

In coarsening systems, crystals grow (or shrink) by the addition (or removal) of atoms that are transported to (or away from) the crystal by capillary-driven diffusion. In the classical theory, it is assumed that if the coarsening crystals are sufficiently far apart, then the intervening phase will serve as a reservoir at a fixed chemical potential of $\mu_\infty = 2\gamma/r^*$, where r^* is the radius of a crystal that neither shrinks nor grows in contact with this reservoir (35, 36). The driving force for each crystal to grow or shrink is given by the free-energy difference between the atoms on the surface of a crystal with size r and the reservoir:

$$\Delta G = 2\gamma\Omega\left[\frac{1}{r^*} - \frac{1}{r}\right], \qquad 4.$$

where Ω is the molecular volume. The growth rate (dr/dt) of a crystal is proportional to this driving force and equal to either the rate of diffusion or the rate of the surface attachment/detachment process. When the evolution of the system is calculated, a kinetic law similar to Equation 1 is obtained where the exponent is $1/3$ for diffusion-limited kinetics and $1/2$ for surface attachment-limited kinetics. The time independent part of the distribution of crystal sizes resulting from diffusion-limited growth is (36)

$$f_{\text{diff}}(\rho) = \rho^2\left(\frac{3}{3+\rho}\right)^{7/3}\left(\frac{3/2}{3/2-\rho}\right)^{11/3}\exp\left[\frac{-\rho}{3/2-\rho}\right]. \qquad 5.$$

For the case of surface attachment-limited kinetics, the distribution is (36)

$$f_{\text{salk}}(\rho) = \rho\left(\frac{2}{2-\rho}\right)^5\exp[-3\rho/(2-\rho)]. \qquad 6.$$

Both distributions are plotted in Figure 1*c*. The classical coarsening theory was originally formulated under the assumption that the coarsening crystals were so far apart that their diffusion fields do not overlap. This occurs as the crystal volume fraction approaches zero. Real systems have finite volume fractions of the coarsening phase and several modifications have been made to the original theory to account for this (36–39). The corrections do not influence the growth exponent, but they do lead to broader and more symmetric particle size distributions that are a closer approximation to experimental observations.

The elementary process during grain growth is the change in the position of a grain boundary as volume is transferred from one grain to another. The theories to account for this process fall into two broad classes: those rooted in topology and those rooted in stochastic behavior. The fundamental premise of the topological models is the von Neumann-Mullins (40) rule for growth in two-dimensional systems. As long as the grain boundary energies are isotropic, then the dihedral angles where three grain boundaries meet must all be 120°. Therefore, grains with more than six sides have concave boundaries and will grow and grains with fewer than six sides will have convex boundaries and shrink. This leads to the so-called N-6 rule for grain growth in two dimensions:

$$\frac{dA}{dt} = M\gamma\frac{\pi}{3}(N-6), \qquad 7.$$

where M is the boundary mobility, A is the grain area, and N is the number of sides. Theorists have sought a similar rule for three-dimensional growth (41, 42, 45). Assuming that the number of faces in three dimensions is analogous to the number of sides in two dimensions, expressions have been produced that yield a zero growth rate when the number of faces is between 13 and 14. However, the complication is that there is no three-dimensional polyhedron that simultaneously fills space and satisfies the interfacial equilibrium constraints at the junctions. The assumed existence of such a critical polyhedron, and the correlation between size and number of faces (see Figure 2), led Hillert (46) to propose that there is a characteristic grain size above which all grains grow and below which all grains shrink. On this basis, he proposed a driving force in analogy to Equation 4 and developed a model that yields a parabolic growth law and the crystal size distribution, as illustrated in Figure 1*c*:

$$f_H(\rho) = (2e)^3 \frac{3\rho}{(2-\rho)^5} \exp[-6/(2-\rho)]. \qquad 8.$$

This distribution is identical (except for a scale factor) to that resulting from coarsening with surface attachment-limited kinetics (see Equation 6). It is less symmetric than the observed grain sized distributions in Figure 1*a* and predicts a much smaller maximum grain size.

Stochastic theories of grain growth assume grain sizes change randomly (47), i.e., crystals execute a random walk through the space of sizes. More elaborate theories combine a random term with a curvature-driven term that biases the random walk so that larger grains are more likely to grow and smaller grains are more likely to shrink (48–51). Depending on the exact formulation of the model, growth exponents varying from 1/4 to 1/2 arise, and the distribution of crystal sizes is usually given by the Rayleigh distribution, shown in Figure 1*c*:

$$f_R(\rho) = \frac{\alpha\rho}{\beta} \exp[-\rho^2/4\beta]. \qquad 9.$$

Although it is plausible that the motions of all boundaries, when averaged over a long time, behave randomly, these models are counterintuitive because experimental observations illustrate that individual boundaries move in a predictable way toward their centers of curvature. Furthermore, the purely stochastic model has been criticized because it does not conserve volume (49, 50, 51a), and it has been argued that, in combined models, random fluctuations in size are incompatible with the condition for continuity in self-similar growth (52).

It must be noted that observed crystal size distributions do not (in general) match well with the theoretical predictions (see Figure 1). However, it has been pointed out that the Hillert distribution is predicated on the basis of an assumed linear relationship between the grain size and the average number of faces. If this distribution has modest nonlinearities (as appears to be the case in Figure 2), the crystal size distribution has been shown to be quite different (53).

Three-dimensional grain growth simulations are expected to produce a more realistic account of the process, and numerous results have been published in recent years. Calculations based on the Monte Carlo Potts model (54), vertex tracking (55, 56), boundary tracking (57), phase field (58), and gradient-weighted moving finite elements (59) have been used to simulate the structure and evolution of grain boundary networks in three-dimensional systems. Until very recently, the simulations have used only isotropic grain boundary properties. The grain size distributions resulting from these models are illustrated in Figure 1*d*. Note that whereas the distributions are all similar, they are also distinct from the experimental distributions. All exhibit the shape characteristic of the theoretical distributions (Figure 1*c*).

For the theories described above, it is assumed that boundary motion is the limiting step in growth. More recently, however, several theories have emerged for growth limited by other factors. One is the vacancy drag model (60–62). Grain boundaries have an excess volume per area on the order of 0.02 nm. As grain growth occurs and interfacial area is eliminated, the excess volume must also be eliminated, presumably in the form of vacancies. If the diffusion of vacancies through the bulk is the rate-limiting step, then linear ($n = 1$) growth kinetics are expected. The possibility that vacancy elimination is rate limiting is expected only at nanometer-scale grain sizes and might explain the nonclassical kinetics observed in nanocrystalline Fe (32). A second possibility is that the motion of triple junctions becomes rate limiting (60). The usual assumption is that in equilibrium, the resolved force on the moving triple junction is zero. However, if the triple junction has a low intrinsic mobility, or if its motion is limited by impurities, then there is a net retarding force on the moving junction that will make the dihedral angles differ from the equilibrium value. Because of these changes in the dihedral angles, there is no unique topological class that divides growing grains from shrinking grains in two dimensions. This influence of the triple junctions will be greatest at the smallest grain sizes (2, 63). It has also been shown that at very small grain sizes, the random characteristics of atomic motion may influence grain growth (64). For the smallest grain sizes, stochastic atomic jumps determine the kinetics of grain growth and an exponent of $1/4$ is predicted, with a transition of $1/2$ by the time the grains reach an average size of 20 nm.

INTERFACE ANISOTROPY

One of the assumptions underpinning all of the theories described in the previous section is that the interface properties are isotropic. Whereas the surface energies of metals near their melting points might be reasonably approximated as being isotropic, it is well known that the surface energies of many ceramics (65, 66) and the grain boundary energies and mobilities of almost all materials vary with the character of the interface (67, 68). The grain boundary character distribution, $\lambda(\Delta g, \mathbf{n})$, can be defined as the relative areas of distinguishable grain boundaries characterized by their lattice misorientation (Δg) and boundary plane orientation ($\mathbf{n}$). Because the misorientation depends on three independent parameters, and the boundary inclination on two, the domain of grain boundary types is five-dimensional, and the number of distinguishable grain boundaries is large (69).

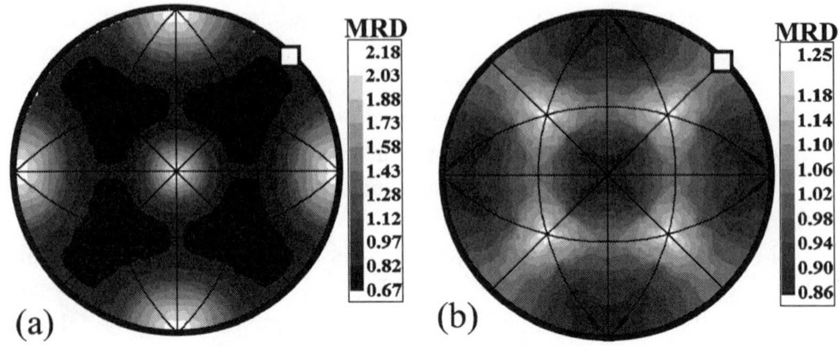

Figure 3 The distribution of grain boundary planes averaged over all misorientations for (*a*) MgO (70) and (*b*) Al (74). In the case of MgO, {100} planes are preferred; in the case of Al, the preference is for {111} planes. The data are shown in stereographic projection along [001], which is in the center of each projection. The [110] direction is marked with a square in each projection. The units are multiples of a random distribution (MRD).

The grain boundary character distribution can be thought of as a metric by which the structure of a polycrystal can be quantified.

Recent measurements of the grain boundary character distribution in a diverse range of ceramics and metals, including MgO (70), $SrTiO_3$ (71), TiO_2 (72), $MgAl_2O_4$ (73), Al (74), and Fe-1%Si (75), have led to three important conclusions. (*a*) The first is that the distribution of grain boundary planes is anisotropic (see Figure 3). (*b*) Second, the preferred habit planes for grains within polycrystals correspond to the same low-energy, low-index planes that dominate the external growth forms and equilibrium shapes of isolated crystals of the same phase (72). This finding amounts to a paradigm shift in our understanding of grain boundaries, which has been dominated for decades by consideration of crystal lattice orientation relationships instead of the grain surface relationships that have been found to be more influential. Furthermore, a correlation has been identified between the sum of the surface free energies that make up the boundaries and the grain boundary energy, $\gamma(\Delta g, \mathbf{n})$ (71, 76). This finding, derived from observations in a wide range of materials, provides an extremely useful simplifying principle: Knowledge of the surface energy anisotropy, which depends on only two parameters, is sufficient for predicting the grain boundary energy. (*c*) The third important finding is that the grain boundary character distribution is correlated to grain boundary energies, $\gamma(\Delta g, \mathbf{n})$ (76, 77). This is illustrated in Figure 4, which shows that low-energy boundaries occur more frequently in the distribution than do high-energy boundaries.

Whereas it is now clear that anisotropic boundary properties influence the grain boundary character distribution, it is not yet clear how the distribution forms during growth. One particularly striking example of the importance of anisotropy to growth was described in a recent attempt to deterministically predict the evolution

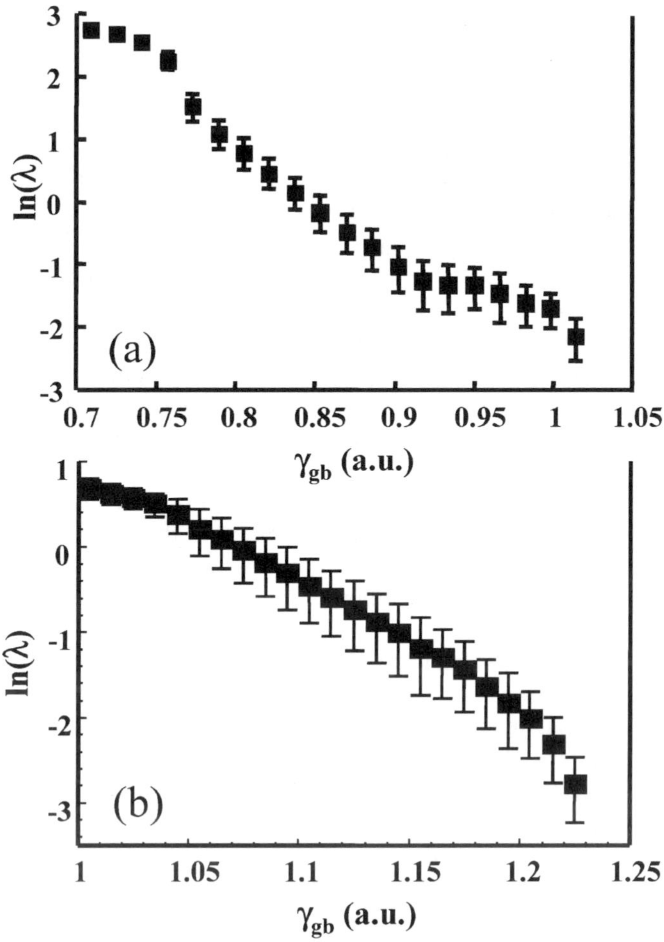

Figure 4 The logarithm of the grain boundary population (λ) as a function of the grain boundary energy (γ_{gb}). The mean is represented by the point; the bars indicate one standard deviation. Panel (*a*) is based on experimental observation (76) and panel (*b*) results from three-dimensional grain growth simulations with anisotropic boundary properties (77).

of a specific grain boundary network (78). The grain boundary network structure of a columnar Al sample was mapped using orientation imaging microscopy over an area consisting of more than 5000 grains. The sample was then annealed to allow the microstructure to evolve, and the orientation mapping was repeated. The initial grain structure was then used as the starting configuration for a curvature-driven grain growth simulation. Simulations were performed assuming isotropic grain boundary properties in one case and anisotropic properties in other cases. The simulations based on measured anisotropic properties reproduced the observed microstructural evolution much better than the isotropic properties. The implications of this experiment are clear: To realistically predict how a microstructure will evolve, it is necessary to include the correct properties of individual boundaries.

Although this example demonstrates the importance of anisotropy in influencing the movement of specific boundaries, it is perhaps more important to understand the effect of anisotropy on statistically averaged properties of the microstructure that are more accessible to experiment. Mullins (79) showed that as long as the crystal size distribution maintains statistical self-similarity during growth and the interface energy is differentiable at all orientations, then the anisotropy of the interface energy does not influence the normal kinetic laws for coarsening and grain growth. Furthermore, a one-dimensional capillary grain growth model demonstrated that for anisotropic boundary properties, a statistically similar particle size distribution is obtained and the distribution is different from that which results from uniform boundary properties (80). Two-dimensional coarsening models have also indicated that the crystal size distribution is influenced by interface energy anisotropy (81, 82). These results reproduced Mullins' predictions that anisotropy does not affect growth kinetics, but it does weakly influence the distribution of crystal sizes. Recent experimental studies of shape evolution during coarsening have shown a trend toward a greater population of low-index surfaces as growth proceeds. For $SrTiO_3$ coarsening in a TiO_2-rich eutectic liquid and PMN/PT coarsening in a PbO-rich liquid, it was observed that crystals with approximately isotropic initial shapes evolve to more polygonal forms bounded by low-energy surfaces (83, 84).

Experimentally, it is not usually possible to separate the comparative influence of the interface energy and mobility on the development of the interface character. However, simulations are capable of doing this. A number of two-dimensional simulations of capillary-driven grain growth indicate that the grain boundary energy plays a pivotal role in determining the grain boundary character distribution. Holm et al. (85) showed that the steady-state misorientation distributions in anisotropic systems vary with the assumed boundary energy anisotropy, which depended only on the lattice misorientation. The anisotropy of the mobility was not an important factor in this simulation. Grain boundaries with low energy were enhanced in the population when the data were weighted by length. However, the number densities of the boundaries were not changed by the energy anisotropy. In this case, it was concluded that the lower energy boundaries lengthened to achieve local equilibrium configurations at the grain boundary triple junctions. Simulations by Upmanyu et al. (86) used more complex forms of the energy and mobility anisotropy derived from molecular dynamics simulations, but with a simplified (one angular parameter) model for the grain misorientation. The results were similar to those of Holm et al. (85): A steady-state distribution of misorientations developed that was inversely related to the assumed grain boundary energy and relatively unaffected by the grain boundary mobility.

In both of the simulations described above, the energy of the grain boundary depended on only the lattice misorientation and not on the orientation of the interface. Kinderlehrer et al. (87) used a two-dimensional simulation to examine the effects of both lattice misorientation and boundary plane inclination on the distribution. When the grain boundary energy depends on only the lattice misorientation, the population and energy are related by the Boltzman distribution. This is reminiscent of the results in Figure 4, which shows that the logarithm of the population decreases linearly with grain boundary energy. When the grain boundary energy also depends on the orientation of the interface plane, the distribution appears to depend on both the energies and the gradients in the energies. Note that the orientation dependence of the energy and population that results from the model (Figure 5*a*) is similar to the trends in the observed values (Figure 5*b*). This

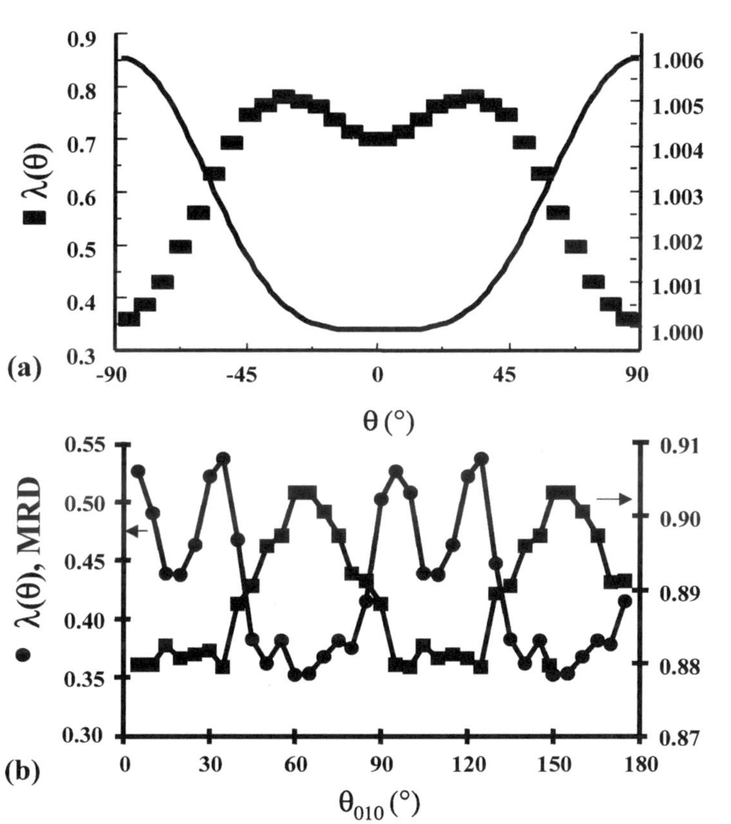

Figure 5 (*a*) Comparison of a simulated grain boundary character distribution and the assumed energy anisotropy as a function of the boundary normal (87). (*b*) Comparison of measured grain boundary energies (squares) and the observed population distribution (*circles*) for Σ5 tilt boundaries in MgO (76). The quantities are plotted in 5° intervals as a function of θ_{010}, the angle between [010] and the grain boundary normal. Note that in both cases, the population reaches a maximum at the positions where the gradient changes from being steeply negative to approximately zero.

sensitivity to the gradient of the energy is likely the source of the distribution in the populations at each energy value in Figure 4. As in the other simulations, it was found that the mobility has a subordinate effect on the distribution.

During microstructural evolution, the distribution changes most significantly when a boundary is annihilated in a neighbor switching or grain collapse event. It is known that as a grain loses sides, it is the shortest boundary that is lost first. If length is inversely correlated to energy, as suggested by Holm et al. (85), then we see that higher-energy boundaries are preferentially annihilated, and low-energy boundaries will form a greater portion of the population, as illustrated in Figures 4 and 5. Simultaneously, there are new boundaries being created, but as long as there is little texture, these should simply add a random component to the population. The abrupt creation and annihilation of boundaries means that triple junctions must adjust to accommodate these changes. If the energy depends on only misorientation, then the boundary plane can rotate without changing energy. However, when boundary plane anisotropy is included, these adjustments in position will occur in proportion to the torque (the differential of energy with respect to orientation) on the plane; this explains why the population is also affected by the gradients in the energy (see Figure 5).

The reorientation events that change the boundary character without changing the lattice misorientation are most likely to occur at the triple junctions, where boundaries are free to rotate to achieve a balance of the interfacial tensions and torques. When a simulation was run with anisotropic grain boundary energies, but also with the boundary condition that all triple-junction dihedral angles be $2\pi/3$ (the condition for isotropy), the grain boundary character distribution remained isotropic. From this result, it can be concluded that it is local adjustments at the triple junctions that lead to the anisotropy in the steady-state grain boundary character distribution.

The first simulations of anisotropic grain growth in three dimensions, using boundary properties that depend on all five crystallographic parameters and a statistically significant number of grain boundaries, have recently been completed. Using Grain3D, Gruber et al. (77) measured the evolution of the grain boundary character distribution as a function of time. In the initial state, there were more than 41,000 grains, and the grain boundary character distribution was completely isotropic. During the course of the simulated growth, approximately 30% of the total interfacial area and 75% of all the grains were eliminated. However, as illustrated in Figure 6, the distribution of interface planes reaches a steady state relatively quickly. In Figure 7, the distribution of grain boundary planes that develops from the initially random distribution, averaged over all misorientations, is compared with the assumed energy anisotropy. The approximate inverse relationship between energy and population evident in Figures 4 and 5 is also obvious here. Note that the distribution in Figure 7*a* is a statistically self-similar steady state that remains constant even as the network continues to evolve by normal growth. This is analogous to the self-similar grain size distributions predicted by the classical growth models.

Whereas mobility anisotropies do not appear to affect the grain boundary character distribution in materials with random orientation distributions, recent

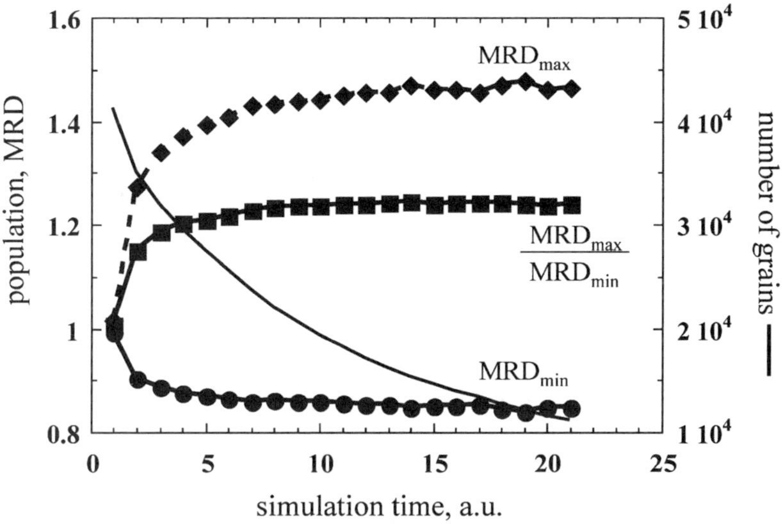

Figure 6 Simulated changes in the grain boundary character distribution with time (77). In the initial distribution, all boundaries occurred with the same frequency. As the number of grains is reduced by grain growth, the maxima and minima of the distribution (see Figure 7) reach steady-state values.

three-dimensional Monte-Carlo simulations of grain growth in a deformed and textured matrix suggest that mobility anisotropy can influence the development of grain orientation texture and, therefore, the grain boundary character distribution (88). In the initial condition, the model polycrystal was assigned a conventional rolling texture normally associated with an fcc metal deformed in plane strain; it

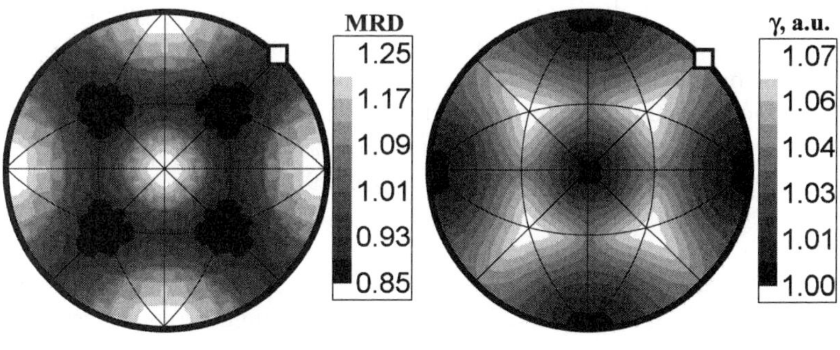

Figure 7 (*a*) The simulated distribution of grain boundary planes averaged over all misorientations, after the distribution has reached a steady state (time = 10 in Figure 6) (77). (*b*) The assumed variation of the grain boundary energy as a function of grain boundary plane orientation, provided as input for the simulation. The reference frames for the projections are the same as in Figure 3.

was also "seeded" with a small volume fraction of grains with the cube texture, $\{001\}\langle 100\rangle$. Mobility anisotropies with peaks for boundaries with a 40° misorientation about [111] were assigned to reflect experimentally measured anisotropies (89). The simulations showed that the greater the mobility of the $40°/\langle 111\rangle$ boundary, the greater the volume fraction of material with cube texture in the resulting microstructure. From this it was concluded that the presence of a peak in the mobility centered on the 40° about $\langle 111\rangle$ boundary type leads to the growth of the cube texture component.

To summarize, the steady-state kinetics of growth processes are not expected to be influenced by interface anisotropy; the grain size distribution is weakly influenced by anisotropy, and the interface character distribution is significantly influenced. Experiments and simulations suggest that steady-state interface character distributions are obtained during growth and, for the case of grain boundaries, the distribution is related to both the energy and the gradients of the energy with respect to the interface orientation. Boundary mobilities have little influence on the interface character distribution in randomly oriented systems but apparently do influence the development of the grain boundary character distribution in systems that have some texture in the initial state. In comparison with the distribution of grain sizes, the interface character distribution is a superior metric for differentiating microstructures and is expected, therefore, to provide a clearer link to materials properties.

INFLUENCE OF SINGULAR INTERFACES ON GROWTH PHENOMENA

Classical grain growth and coarsening theories assume isotropic interface energies and that all interfaces move by the same mechanism. However, it has been recognized for many decades that the mechanisms by which interfaces move depend to some extent on the interface structure. On atomically rough surfaces (assumed in classical theories), atoms can be added or removed from kink sites, and the rate of growth is generally limited by the rate at which reactants reach (or attach to) the surface. Singular or nonrough surfaces, which correspond to cusps in the plot of the surface energy as a function of orientation, advance or retreat via the lateral motion of steps that add or subtract atomic layers (90). If the singular surface has no steps or step sources (such as screw dislocations or twin plane reentrant edges), its motion is then limited by the rate of two-dimensional nucleation. One characteristic of materials exhibiting abnormal growth phenomena is that their microstructures contain a mixture of curved (rough) and flat (singular) interfaces. Therefore, it is reasonable to assume that the rates at which these interfaces move are controlled by different mechanisms. The important point here is that if these two mechanisms occur at very different rates, then it is possible to imagine how bimodal grain size distributions arise in some materials. This point has been the subject of numerous recent studies and is described further in the remainder of this section.

We begin with a consideration of how nucleation on a singular interface affects the coarsening of crystals bounded by singular interfaces. To simplify the situation, we begin with a collection of cube-shaped crystals with surface energy γ and a distribution of edge lengths, L. The crystals are dispersed in an intervening phase and the chemical potential of the material dissolved in that phase is $\mu_\infty = 2\gamma/r^*$. Therefore, there is a crystal size $L^* = 2r^*$ that is in equilibrium with the intervening phase. We can now consider the energy involved in transferring material from the intervening phase to the surface of the crystal. To do this, we place a square nucleus of area s^2 and height a on the flat facet, where the chemical potential is zero. The energy for this transfer is

$$\varepsilon(s) = 4as\gamma - as^2\frac{2\gamma}{r^*} \tag{10.}$$

for $0 \le s \le L$. In Equation 10, the first term on the right hand side is the energy of the perimeter of the nucleus and the second term is the change in the free energy on moving a volume of material (as^2) from the intervening phase to the flat surface. Equation 10 shows that the energy change on adding a complete layer ($s = L$) to a crystal with size $L = 2r^*$ is zero, as expected; however, before the layer is completed, the crystal must pass through a relatively higher energy state that contains a partial layer. By differentiation, we see that the maximum occurs at $s = r^*$ and that the barrier to addition is $\varepsilon_+ = 2a\gamma r^*$, where $0 < r^* < L$. It is interesting to note that the barrier for adding layers is independent of the size of the growing crystal and mainly governed by r^*, which is a measure of the average size of crystals in the system.

To remove material from the crystal and dissolve it in the intervening phase, the change in energy is

$$\varepsilon(s) = 4as\gamma - 4aL\gamma + (aL^2 - as^2)\frac{2\gamma}{r^*}. \tag{11.}$$

In the expression above, the first two terms on the right hand side represent the energy difference between a complete layer and a partial layer; this is ≤ 0 for all valid values of s. The final term on the right hand side is the energy increase associated with moving a volume of material from the flat surface to the reservoir. The maximum of Equation 11 is at $s = r^*$ and the barrier to removing a layer, ε_-, is

$$\varepsilon_- = 2a\gamma\left[r^* + L\left(\frac{L}{r^*} - 2\right)\right], \tag{12.}$$

where r^* is between $L/2$ and L. Note that for a crystal of size $L = 2r^*$, $\varepsilon_+ = \varepsilon_- = 2a\gamma r^*$, as expected for the equilibrium case. For smaller crystals, the barrier decreases (see Figure 8) and vanishes completely for crystals with $r = (^1\!/_2)r^*$. More generalized forms of these barriers have been derived previously, but the simplified forms given above are acceptable for the present purposes and do not lead to different conclusions (91).

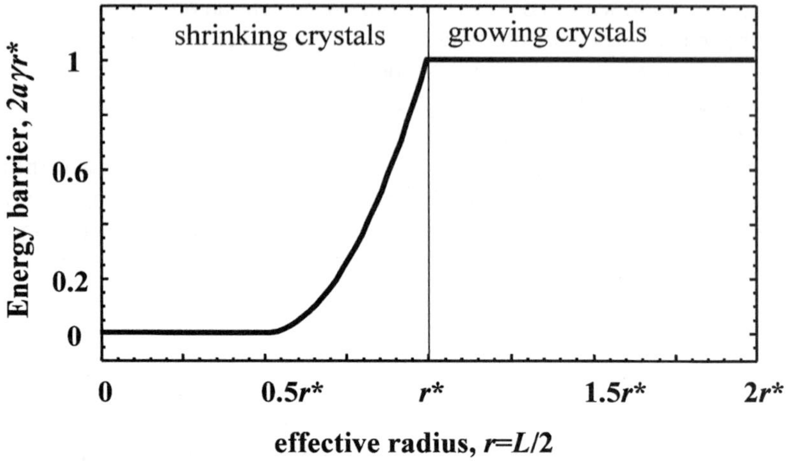

Figure 8 Plot of the nucleation energy barrier for crystals of different sizes in contact with a reservoir with a chemical potential $2\gamma/r^*$.

It is first important to compare the size of these barriers with the available thermal energy and the potential energy supplied by capillarity. The most important barrier is the one for the growth of crystals with sizes $L \geq 2r^*$ (this is the largest barrier). Even though small crystals can shrink without a barrier, the constraint of mass conservation will restrict their ability to do so if the barrier for growth is too large. In a typical situation that might be realized in the laboratory, we can take a to be 2×10^{-10} m, γ to be 1 J/m^2, and r^* to be 500 nm. To a first approximation, the choice of r^* is set to the average size of the crystals in the coarsening compact, and 500 nm is taken as a lower limit for most practical cases; this leads to a barrier to growth of 2×10^{-16} J. The relevant energies for comparison are the energy change resulting from capillarity-driven growth (taken to be $2\gamma a^3/r^* = 3 \times 10^{-23}$ J) and the available thermal energy ($kT = 2 \times 10^{-20}$ J at 1500°C). Although these calculations are admittedly approximate, several things are clear. First, because the barrier energy for growth on micron-sized crystals is more than 10^6 times the capillary driving force, we can be certain that capillarity cannot drive nucleation. In fact, for all cases where the grain size is larger than a few nm, if nucleation occurs, it must be driven by thermal fluctuations. The second point is that even thermal fluctuations are unlikely to drive nucleation on large crystals. According to the estimates above, the nucleation energy barrier is 10^4 kT. It has been estimated that barriers greater than 40 kT are not overcome in reasonable time periods (92). It should be noted that the step edge energy is usually thought to be a fraction of the surface energy, so the barrier height given above is probably overestimated. However, the nucleus perimeter energy would have to be reduced by a factor of 100, to 0.01 J/m^2 before the barrier would be reduced to 40 kT. Since this is considered unlikely, the size of the barrier energy and the extremely small driving force implies that singular interfaces without defects should never advance outward during coarsening (93).

For comparison, we should also consider the conditions under which two-dimensional nucleation can be expected to occur and has been observed experimentally. The theory for two-dimensional nucleation shows that the nucleation rate decreases exponentially as the ratio of the driving force to the available thermal energy decreases and as the ratio of the perimeter energy to the available thermal energy increases (94). In accordance with the theory, two-dimensional nucleation has been observed in cases where driving forces comparable to the thermal energy are provided by supersaturated liquids or vapors, or by cooling a melt below the crystallization temperature. For example, growth by two-dimensional nucleation has been reported during the solidification of liquid Ga, where the driving force is supplied by the latent heat of crystallization, which is greater than the available thermal energy (95, 96). In this case, it is also relevant to note that solid-liquid interface energy is very low (in the range of 0.04 to 0.07 J/m^2) and that the nucleus perimeter energy was determined to be in the range of 0.01 to 0.02 J/m^2. So, in addition to a high driving force, the nucleation energy barrier is significantly reduced by the low interfacial energy. The transition from dislocation controlled growth to two-dimensional nucleation controlled growth has been observed during the vapor deposition of NaCl (97). It was reported that the transition to two-dimensional nucleation occurs when the driving force exceeds 0.09 eV at 347°C. At this point, the driving force is approximately twice the thermal energy. In summary, two-dimensional nucleation has been observed in the past when the driving force is relatively large and comparable to the available thermal energy. In contrast, the capillary driving forces that arise during the coarsening of micron-sized crystals are only a small fraction of the thermal energy.

Several authors have argued that the growth mechanisms associated with singular interfaces can be used to explain abnormal growth phenomena. Through the analysis of microstructural data from many different systems, Yoon and coworkers (98) have noted that abnormal growth phenomena occur when singular interfaces are present in the system and that systems with fully rough interfaces grow normally. Their explanation is that abnormal growth is attributed to the step migration growth mechanisms of singular surfaces. Two different mechanisms have been invoked. The first, articulated in Reference 99 and citations therein, is that some grains can become large enough that the capillary driving force is higher than the critical value for the onset of nucleation, and this leads to abnormal growth. However, on the basis of the numerical estimates made in the paragraph above, this explanation is not viable. One reason is that under typical experimental conditions, the energies associated with thermal fluctuations are much larger than the capillarity driving force so that nucleation will be driven by thermal fluctuations or not at all. Another reason that the explanation is not viable is that all growing crystals are subject to the same barrier ($\varepsilon_+ = 2a\gamma r^*$) so that the size of the growing crystal is simply not a relevant factor; if two-dimensional nucleation by thermal fluctuations is possible on one growing crystal then it will occur on all growing crystals exposed to the same mean field chemical potential. The second possible mechanism is that step-producing structural defects, expected to occur in a subset of the crystals, allow the defective population to grow without a barrier and thus the average size of this subset becomes much larger than the more perfect crystals. It seems clear that having two distinct populations growing at very different rates is a plausible explanation for abnormal growth. However, a subset of large crystals growing by capillarity driven two-dimensional nucleation is not plausible.

The defect-controlled coarsening mechanism has been studied in detail and numerical simulations based on this model have demonstrated its feasibility (91). In the case where the growth of mixed populations is modeled (some crystals are limited by the barrier and others are assumed to have defects), bimodal populations are observed where the defect-containing subset of crystals grows much faster than the others, as illustrated in Figure 9. Note that because the smallest crystals in the slow population are able to shrink without a barrier (see Equation 12), they sustain the growth of the minority population of defective crystals.

The nucleation-limited coarsening model makes a number of predictions that can be compared with those of experiments. The first is that abnormal growth should be transient. In other words, at some point the defective (fast growing) grains have completely consumed the ideal (slow growing) population; because none of the remaining population has an advantage, growth should then proceed normally. The second is that during the period of bimodal growth, the number density of the large grains should remain constant, whereas the number of small grains should decrease. Although detailed measurements of the evolution of the grain size distribution are not yet available, some reports in the literature are consistent with this model (25).

The dislocation density in large $SrTiO_3$ seed crystals has been correlated with the rate at which they grow into a polycrystalline compact (100). Although there were only two observations, the seed with a higher dislocation density had a higher growth rate. Orientation imaging microscopy data suggest that all abnormal grains in PMN/PT have a twin (a boundary with a 60° rotation about $\langle 111 \rangle$). Twins are not found in the smaller matrix grains (101, 102), and these observations have been used to suggest that the reentrant twin plane edge provides a nucleation site with a greatly reduced barrier that allows the twinned crystals to grow much faster than the untwinned crystals. However, it should be noted that by observing only the final state, it is not possible to determine if the twin caused the grain to become large or if twins arise naturally in all grains when they become large. It can be reasoned that any grain that grows large enough will have encountered grains of many other orientations. In a random orientation distribution, approximately 1 in 50 misorientations will be close to the twin relationship (103). Thus, if a grain grows so that it has encountered many more than 50 differently oriented grains, it is likely that a twin will be formed, and this may be the reason that all large grains have twins. The independence of twinning and high growth rates was demonstrated in a recent paper where it was shown that there is no measurable difference in the growth rates of twinned and untwinned seeds (104). Since these

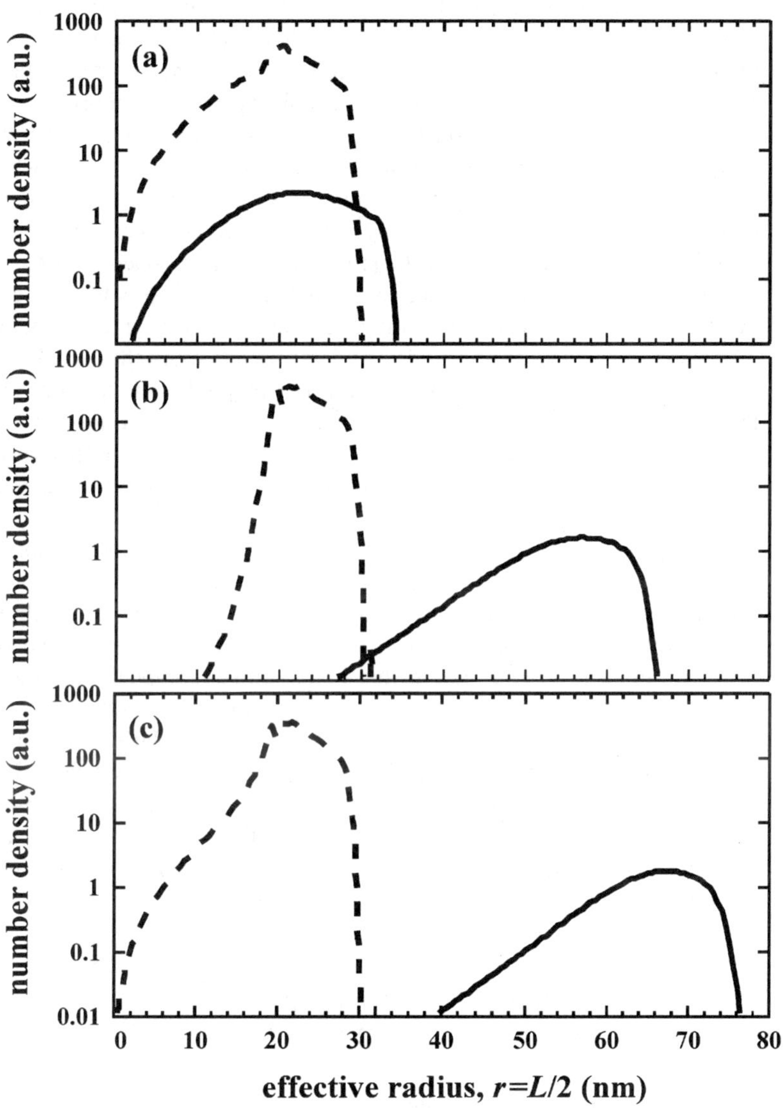

Figure 9 Simulated crystal size distribution for the defect-free crystals (*dashed line*) and the crystals with a persistent step source (*solid line*) coarsening in the same system: (*a*) after 1 arbitrary unit of time, (*b*) after 8, (*c*) after 11. The crystals with the step source grow without a barrier (91).

seeds were relatively large, it is likely that they all contained screw dislocations that could act as persistent step sources.

Whereas the habit plane of the twin in PMN/PT is incoherent, $BaTiO_3$ forms straight coherent twins on {111} planes. When $BaTiO_3$ compacts are seeded with twinned grains, the twinned seeds outgrew the untwinned seeds, suggesting that a reentrant twin plane edge nucleation mechanism promotes abnormal growth (105). This is also suggested by observations of $BaTiO_3$ coarsened at temperatures in the range of 1350°C (where an intergranular liquid is expected to form), which showed that twinned grains grow larger than others (106). Rehrig et al. (25) independently measured the grain sizes of the large and small populations and showed that within experimental error, the sizes of the small grains had stopped increasing, whereas the large grains continued to increase in size. It should be noted that although the ideal version of the nucleation limited coarsening theory suggests that the growth of ideal (smaller) crystals should completely cease when they reach a size where the nucleation energy barrier becomes insurmountable (and this size should be in the nanometer size range), it does not account for finite volume fraction effects and growth from coalescence that is expected even in the absence of nucleation. In other words, there may be some growth by grain boundary migration, even in the absence of coarsening.

Finally, it should also be recognized that mechanisms for abnormal growth phenomena have also been proposed that have nothing to do with whether the interface is singular or rough. MacLaren et al. (107) observe AGG in alumina where the rapidly moving boundaries are clearly rough and have accumulated a disordered layer of impurities. They argue that at particular compositions, the impurities induce a structural transformation to a highly mobile boundary structure. It has also been demonstrated by simulation that grain boundaries with high mobility misorientations can be sustained if they grow into a highly textured matrix and that this can lead to abnormal growth (108). On the basis of these observations, it seems likely that a variety of possible explanations exist for abnormal growth in different systems. However, in systems that are sufficiently anisotropic, evidence pointing to the importance of the interface structure is accumulating.

CURRENT DIRECTIONS IN RESEARCH ON THE EFFECTS OF ANISOTROPY ON GROWTH

Now that it is possible to measure the anisotropy of the grain boundary character distribution, there are a number of questions that can be asked. First, does the grain boundary character distribution actually reach a statistically self-similar state, as indicated by simulation (77)? If so, in what way is it connected to the statistically self-similar grain size distribution? To answer these questions experimentally, the evolution of the distribution will have to be measured at different points in time. Assuming that a steady-state distribution does exist, it should be predictable by simulations using the grain boundary properties as input. It may also be possible to develop an anisotropic theory for grain growth, in analogy to the classical grain growth theories, that allows the grain boundary character distribution to be predicted on the basis of anisotropic boundary properties. With such a theory, it might also be possible to work the inverse problem and determine the grain boundary properties from the experimentally observed grain boundary character distribution.

Assuming that the steady-state grain boundary character distribution is, in fact, a "fingerprint" of the microstructure, then it is sensible to ask how it is related to materials properties. If there is a direct correlation between certain materials properties and a predictable grain boundary character distribution, knowledge of this link should provide a path toward designing microstructures with predictable properties. In this case, it seems certain that research into the mechanisms of normal and abnormal growth phenomena will continue. The availability of three-dimensional anisotropic growth models will greatly aid this research because it will be possible to independently explore the influence of the grain boundary energy and mobility on the development of the grain boundary character distribution (77). It will also be possible to study the topology of the grain boundary network and the influence of initial orientation texture on the microstructures produced by growth. The recent emergence of new experimental tools will also advance this research by making it possible to collect the statistical data required to verify or refute hypothetical mechanisms for growth processes (69). For example, although abnormal growth remains a puzzle, with the proper data it should be possible to quantitatively assess the predictions of theories such as the one for nucleation limited coarsening described above.

ACKNOWLEDGMENT

This work was supported primarily by the MRSEC program of the National Science Foundation under award number DMR-00,79996 and partially by NASA under grant number 8-1674. I thank my colleagues at CMU who participated in this work, including W.W. Mullins, A.D. Rollett, D. Kinderlehrer, D. Saylor, T. Sano, C.-S. Kim, and J. Gruber. Helpful discussions with D.T. Wu and C.E. Krill, III, are also acknowledged.

LITERATURE CITED

1. Atkinson HV. 1988. Theories of normal grain growth in single phase systems. *Acta Metall.* 36:469–91
2. Fradkov VE, Udler D. 1994. Two-dimensional normal grain growth: topological aspects. *Adv. Phys.* 43:739–89
3. Voorhees PW. 1992. Ostwald ripening of two-phase mixtures. *Annu. Rev. Mater. Sci.* 22:197–215
4. Beck PA, Kremer JC, Demer L. 1947. Grain growth in high purity aluminum. *Phys. Rev.* 71:555

5. Beck PA, Holzworth ML, Hu H. 1948. Instantaneous rates of grain growth. *Phys. Rev.* 73:526–27
6. Burke JE. 1947. Grain growth in alpha-brass. *J. Appl. Phys.* 18:1028
7. Burke JE. 1949. Some factors affecting the rate of grain growth in metals. *Trans. Metall. Soc. AIME* 180:73–91
8. Lucke K, Detert K. 1957. A quantitative theory of grain boundary motion and recrystallization in metals in the presence of impurities. *Acta Metall.* 5:628–37
9. Cahn JW. 1962. The impurity-drag effect in grain boundary motion. *Acta Metall.* 10:789–98
10. Zener C. 1948. Grains, phases and interfaces: an interpretation of microstructure. *Trans. Metall. Soc. AIME* 175:15–51
11. Greenwood GW. 1956. The growth of dispersed precipitates in solutions. *Acta Metall.* 4:243–48
12. Grey EA, Higgins GT. 1973. Solute limited grain boundary migration: a rationalization of grain growth. *Acta Metall.* 21:309–21
13. Sun RC, Bauer CL. 1970. Measurements of grain boundary mobilities through magnification of capillary forces. *Acta Metall.* 18:635–48
14. Sun RC, Bauer CL. 1970. Tilt boundary migration in NaCl bicrystals. *Acta. Metall.* 18:639–47
15. Sun RC, Bauer CL. 1973. Kinetics of grain boundary migration in copper bicrystals with [001] rotation axes. *Acta Metall.* 21:1099–1109
16. Okazaki K, Conrad H. 1972. Grain size distribution in recrystallized alpha-titanium. *Trans. Jpn. Inst. Metals* 13:198–204
17. Hu H. 1974. Grain growth in zone refined iron. *Can. Metal. Q.* 13:275–86
18. Ardell AJ, Nicholson RB. 1966. The coarsening of gamma′ in Ni-Al Alloys. *J. Phys. Chem. Solids* 27:1793–804
19. Zhang C, Suzuki A, Ishimaru T, Enomoto M. 2004. Characterization of three-dimensional grain structure in polycrystalline iron by serial sectioning. *Metall. Mater. Trans.* 35A:1927–33
20. Döbrich KM, Rau C, Krill CE III. 2004. Quantitative characterization of the three-dimensional microstructure of polycrystalline Al-Sn using X-ray microtomography. *Metall. Mater. Trans* 35A:1953–61
21. Hull FC. 1988. Plane section and spatial characteristics of equiaxed β-brass grains. *Mater. Sci. Technol.* 4:778–85
22. Matsuura K, Itoh Y, Ohmi T, Ishii K. 1994. Evaluation of grain shape distribution in polycrystalline materials. *Mater. Trans. JIM* 35:247–53
23. Mahalingam K, Gu BP, Liedl GL, Sanders TH. 1987. Coarsening of delta (Al_3Li) precipitates in binary Al-Li alloys. *Acta Metall.* 35:483–98
24. Kim C-S. 2004. *Microstructural-mechanical property relationships in WC-Co composites*. PhD thesis. Carnegie Mellon Univ. 214 pp.
25. Rehrig PW, Messing GL, Trolier-McKinistry S. 2000. Templated grain growth of barium titanate single crystals. *J. Am. Ceram. Soc.* 83:2654–60
26. Smith CS. 1952. The shapes of metal grains, with some other metallurgical applications of topology. In *Metal Interfaces*, p. 65. Cleveland: ASM
27. Williams WM, Smith CS. 1952. A study of grain shape in an aluminum alloy and other applications of stereoscopic microradiaography. *J. Metals* 4:755–65
28. Marlow TR, Koch CC. 1996. Grain growth of nanocrystalline materials—a review. In *Synthesis and Processing of Nanocrystalline Materials*, ed. DL Bourell, pp. 33. Warrendale, PA: TMS
29. Gleiter H. 2000. Nanostructured materials: basic concepts and microstructure. *Acta Mater.* 48:1–29
30. Krill CE, Birringer R. 2001. Grain growth kinetics and thermal stability of nanocrystalline materials. In *Recrystallization and Grain Growth*, ed. G Gottstein, DA Molodov, pp. 205–18. New York: Springer Verlag

31. Krill CE, Birringer R. 1998. Estimating grain-size distributions in nanocrystalline materials from X-ray diffraction profile analysis. *Philos. Mag. A* 77:621–40
32. Krill CE, Helfen RL, Michels D, Natter H, Fitch A, et al. 2001. Size-dependent grain-growth kinetics observed in nanocrystalline Fe. *Phys. Rev. Lett.* 86:842–45
33. Weissmuller J. 1994. Alloy thermodynamics in nanostructures. *J. Mater. Res.* 9:4–7
34. Krill CE, Klein R, Janus S, Birringer R. 1995. Thermodynamic stabilization of grain boundaries in nanocrystalline alloys. *Mater. Sci. Forum* 179–181:443–48
35. Lifshitz IM, Slyozov VV. 1961. The kinetics of precipitation from saturated solids solutions. *J. Phys. Chem. Solids* 19:35–50
36. Wagner C. 1961. Theorie der Alterung Niederschlagen durch Umlosen. *Z. Elektrochem.* 65:581–91
37. Ardell AJ. 1972. The effect of volume fraction on particle coarsening: theoretical considerations. *Acta Metall.* 20:61–71
38. Davies CKL, Nash P, Stevens RN. 1980. The effect of volume fraction of precipitate on Ostwald ripening. *Acta Metall.* 28:179–89
39. Voorhees PW, Glicksman ME. 1984. Solution to the multi-particle diffusion problem with applications to Ostwald ripening—I. Theory. *Acta Metall.* 32:2001–11
40. Mullins WW. 1956. Two dimensional motion of idealized grain boundaries. *J. Appl. Phys.* 27:900–4
41. Mullins WW. 1989. Estimation of the geometrical rate constant in idealized three dimensional grain growth. *Acta Metall.* 37:2979–84
42. Hilgenfeldt S, Kraynik AM, Koehler SA, Stone HA. 2001. An accurate von Neumann's law for three-dimensional foams. *Phys. Rev. Lett.* 86:2685–99
43. Deleted in proof
44. Deleted in proof
45. Glicksman M. 2004. Capillarity-mediated grain growth in 3-D. *Mater. Sci. Forum* 467–470:1025–32
46. Hillert M. 1965. On the theory of normal and abnormal grain growth. *Acta Metall.* 13:227–38
47. Louat NP. 1974. On the theory of normal grain growth. *Acta Metall.* 22:721–24
48. Pande CS. 1987. On a statistical theory of grain growth. *Acta Metall.* 35:2671–78
49. Ryum N, Hunderi O. 1989. On the analytic description of normal grain growth. *Acta Metall.* 37:1375–79
50. Thorvaldson A. 1993. Grain growth as a stochastic process. *Acta Metall. Mater.* 41:1347–57
51. Mulheran PA. 1992. Mean field simulations of normal grain growth. *Acta Metall. Mater.* 40:1827–33
51a. Wu DT. 1994. Mass conservation in grain growth. *Mater. Res. Symp. Proc.* 343:61–64
52. Mullins WW. 1998. *Models of idealized grain growth: critique of stochastic theories and implications of scaling*. Presented at Proc. Int. Conf. Grain Growth, Pittsburgh, PA, p. 353
53. Mullins WW. 1998. Grain growth of uniform boundaries with scaling. *Acta Mater.* 46:6219–26
54. Anderson MP, Grest GS, Srolovitz DJ. 1989. Computer simulation of normal grain growth in three dimensions. *Philos. Mag. B* 59:293–329
55. Fuchizaki K, Kusaba T, Kawasaki K. 1995. Computer modelling of three-dimensional cellular pattern growth. *Philos. Mag. B* 71:333–57
56. Wegand D, Brechet Y, Lepinoux J, Gust W. 1999. Three-dimensional grain growth: a vertex dynamics simulation. *Philos. Mag. B* 79:703–16
57. Wakai F, Enomoto N, Ogawa H. 2000. Three-dimensional microstructural evolution in ideal grain growth-general statistics. *Acta Mater.* 48:1297–311
58. Krill CE, Chen L-Q. 2002. Computer simulation of 3-D grain growth using a phase field model. *Acta Mater.* 50:3057–73

59. Kuprat AP. 2000. Modeling microstructure evolution using gradient-weighted moving finite elements. *SIAM J. Sci. Comput.* 22:535–60
60. Gottstein G, King AH, Shvindlerman LS. 2000. The effect of triple-junction drag on grain growth. *Acta Mater.* 48:397–403
61. Estrin Y, Gottstein G, Shvindlerman LS. 1999. Intermittent 'self-locking' of grain growth in fine-grained materials. *Scripta Mater.* 41:385–90
62. Estrin Y, Gottstein G, Rabkin E, Shvindlerman LS. 2000. On the kinetics of grain growth inhibited by vacancy generation. *Scripta Mater.* 43:141–47
63. Adams BL, Kinderlehrer K, Livshits I, Mason DE, Mullins WW, et al. 1999. Extracting grain boundary and surface energy from measurement of triple junction geometry. *Interface Sci.* 7:321
64. Helfen L, Wu DT, Birringer R, Krill CE. 2003. The impact of stochastic atomic jumps on the kinetics of curvature-driven grain growth. *Acta Mater.* 51:2743–54
65. Saylor DM, Mason DE, Rohrer GS. 2000. Experimental method for determining surface energy anisotropy and its application to magnesia. *J. Am. Ceram. Soc.* 83:1226–32
66. Kitayama K, Glaeser AM. 2002. The Wulff shape of alumina: III, undoped alumina. *J. Am. Ceram. Soc.* 85:611–22
67. Hasson GC, Goux C. 1971. Interfacial energies of tilt boundaries in aluminum. Experimental and theoretical determination. *Scripta Metall.* 5:889–94
68. Gottstein G, Shvindlerman LS. 1999. *Grain Boundary Migration in Metals.* Boca Raton, FL: CRC Press
69. Rohrer GS, Saylor DM, El-Dasher BS, Adams BL, Rollett AD, Wynblatt P. 2004. The distribution of internal interfaces in polycrystals. *Z. Metall.* 95:197–214
70. Saylor DM, Morawiec A, Rohrer GS. 2003. Distribution of grain boundaries in magnesia as a function of five macroscopic parameters. *Acta Mater.* 51:3663–74
71. Saylor DM, El-Dasher BS, Sano T, Rohrer GS. 2004. Distribution of grain boundaries in $SrTiO_3$ as a function of five macroscopic parameters. *J. Am. Ceram. Soc.* 87:670–76
72. Saylor DM, El-Dasher BS, Pang Y, Miller HM, Wynblatt P, et al. 2004. Habits of grains in dense polycrystalline solids. *J. Am. Ceram. Soc.* 87:724–26
73. Miller HM, Saylor DM, El-Dasher BS, Rollett AD, Rohrer GS. 2004. Crystallographic distribution of internal interfaces in spinel polycrystals. *Mater. Sci. Forum* 467–470:783–88
74. Saylor DM, El-Dasher BS, Rollett AD, Rohrer GS. 2004. Distribution of grain boundaries in aluminum as a function of five macroscopic parameters. *Acta Mater.* 52:3649–55
75. Bennett TA, Kim C-S, Rohrer GS, Rollett AD. 2004. Five-parameter grain boundary character distribution in Fe-1%Si. *Mater. Sci. Forum* 467–470:1057–62
76. Saylor DM, Morawiec A, Rohrer GS. 2003. The relative free energies of grain boundaries in magnesia as a function of five macroscopic parameters. *Acta Mater.* 51:3675–86
77. Gruber J, George DC, Kuprat AP, Rohrer GS, Rollett AD. 2004. Effect of anisotropic interfacial energy on grain boundary distributions during grain growth. *Mater. Sci. Forum* 467–470:733–38
78. Demirel MC, Kuprat AP, George DC, Rollett AD. 2003. Bridging simulations and experiments in microstructure evolution. *Phys. Rev. Lett.* 90:016106
79. Mullins WW. 1986. The statistical self similarity hypothesis in grain growth and particle coarsening. *J. Appl. Phys.* 59: 1341–49
80. Mullins WW, Vinals J. 1993. Scaling in linear bubble models of grain growth. *Acta Mater.* 41:1359–67

81. Rutenberg AD, Vollmayr-Lee BP. 1999. Anisotropic coarsening: grain shapes and nonuniversal persistence. *Phys. Rev. Lett.* 83:3772–75
82. Brosh E, Shneck RZ. 2000. Anisotropic coarsening: the effect of interfacial properties on the shape of grains. *Model. Simul. Mater. Sci. Eng.* 8:815–23
83. Sano T, Kim C-S, Rohrer GS. 2005. Shape evolution of $SrTiO_3$ crystals during coarsening in a titania-rich liquid. *J. Am. Ceram. Soc.* In press
84. Gorzkowski EP, Chan HM, Harmer MP, Sano T, Kim C-S, Rohrer GS. 2005. Changes in the distribution of interfaces in PMN-35 mol% PT as a function of time. *Z. Metall.* 96:207–210
85. Holm EA, Hassold GN, Miodownik MA. 2001. On misorientation distribution evolution during anisotropic grain growth. *Acta Mater.* 49:2981–91
86. Upmanyu U, Hassold GN, Kazaryan A, Holm EA, Wang Y, et al. 2002. Boundary mobility and energy anisotropy effects on microstructural evolution during grain growth. *Interface Sci.* 10:201–16
87. Kinderlehrer D, Livshits I, Rohrer GS, Ta'asan S, Yu P. 2004. Mesoscale simulation of the evolution of the grain boundary character distribution. *Mater. Sci. Forum* 467–470:1063–68
88. Rollett AD. 2004. Crystallographic texture changes during grain growth. *J. Met.* 56:63–68
89. Yang C-C, Rollett AD, Mullins WW. 2001. Measuring relative grain boundary energies and mobilities in an aluminum foil from triple junction geometry. *Scripta Met.* 44:2735–40
90. Burton WK, Cabrera N, Frank FC. 1951. The growth of crystals and the equilibrium structure of their surfaces. *Philos. Trans. R. Soc. London Ser. A* 243:300–58
91. Rohrer GS, Rohrer CL, Mullins WW. 2002. Coarsening of faceted crystals. *J. Am. Ceram. Soc.* 85:675–82
92. Mullins WW, Rohrer GS. 2000. Nucleation barrier for volume conserving shape changes of faceted crystals. *J. Am. Ceram. Soc.* 83:214–16
93. Sheldon BW, Rankin J. 2002. Step-energy barriers and particle shape changes during coarsening. *J. Am. Ceram. Soc.* 85:683–90
94. Weeks JD, Gilmer GH. 1979. Dynamics of crystal growth. *Adv. Chem. Phys.* 40:157–228
95. Peteves SD, Abbaschian R. 1991. Growth kinetics of solid-liquid Ga interfaces: Part I. Experimental. *Metall. Mater. Trans.* 22A:1259–70
96. Peteves SD, Abbaschian R. 1991. Growth kinetics of solid-liquid Ga interfaces: Part II. Theoretical. *Metall. Mater. Trans.* 22A:1271–86
97. Keller KW. 1986. Modes of crystal growth revealed by electron microscopy on NaCl. *J. Cryst. Growth* 74:161–71
98. Yoon D-Y, Cho YK. 2005. Roughening transition of grain boundaries in metals and oxides. *Interface Sci.* In press
99. Oh K-S, Jun J-Y, Kim D-Y, Hwang N-M. 2000. Shape dependence of the coarsening behavior of niobium carbide grains dispersed in a liquid iron matrix. *J. Am. Ceram. Soc.* 83:3117–20
100. Chung S-Y Kang S-JL. 2000. Effect of dislocations on grain growth in $SrTiO_3$. *J. Am. Ceram. Soc.* 83:2828–32
101. Wallace JS, Huh J-M, Blendell JE, Handwerker CA. 2002. Grain growth and twin formation in 0.74PMN·0.26PT. *J. Am. Ceram. Soc.* 85:1581–84
102. Chung U-J, Park J-K, Hwang N-M, Lee H-Y, Kim D-Y. 2002. Effect of grain coalescence on the abnormal grain growth of $Pb(Mg_{1/3}Nb_{2/3})O_3$ 35 mol% $PbTiO_3$ ceramics. *J. Am. Ceram. Soc.* 85:965–68
103. Warrington DH, Boon M. Ordered structures in random grain boundaries; some geometrical probabilities. *Acta Metall.* 23:599–607
104. Rockosi DJ, Gorzkowski EP, King PT, Scotch AM, Chan HM, Harmer MP. 2004. Seeded growth from twinned

and untwinned abnormal grains of Pb $(Mg_{1/3}Nb_{2/3})O_3$ 35 mol% $PbTiO_3$ in a matrix containing PbO additions. *J. Am. Ceram. Soc.* 87:1339–42
105. Kang M-K, Yoo Y-S, Kim D-Y, Hwang N-M. 2000. Growth of $BaTiO_3$ seed grains by the twin-plane reentrant edge mechanism. *J. Am. Ceram. Soc.* 83:385–90
106. Lee H-Y, Kim J-S, Kim D-Y. 2002. Effect of twin plane reentrant edge on the coarsening behavior of barium titanate grains. *J. Am. Ceram. Soc.* 85:977–80
107. Maclaren I, Cannon RM, Gulgun MA, Voytovych R, Popescu-Pogrion N, et al. 2003. Abnormal grain growth in alumina: synergistic effects of yttria and silica. *J. Am. Ceram. Soc.* 86:650–59
108. Holm EA, Miodownik MS, Rollett AD. 2003. On abnormal subgrain growth and the origin of recrystallization nuclei. *Acta Mater.* 51:2701–16

Annu. Rev. Mater. Res. 2005. 35:127–42
doi: 10.1146/annurev.matsci.35.102303.140758

First published online as a Review in Advance on March 1, 2005

THE PREPARATION OF SKELETAL CATALYSTS*

A.J. Smith
School of Chemical Engineering and Industrial Chemistry, University of New South Wales, Sydney, 2052, Australia; email: andrew.smith@unsw.edu.au

D.L. Trimm
CSIRO Petroleum, Clayton, Victoria, 3168 Australia; email: david.trimm@csiro.au

Key Words Raney, alloy, leaching, structure, formation

■ **Abstract** The preparation of skeletal catalysts is considered, with particular reference to the preparation of skeletal nickel and skeletal copper. Skeletal catalysts have physical properties that are highly desired for some industrial processes, and the means of controlling these properties to optimize performance is reviewed. The preparation can be affected by the initial alloy composition, the quenching process, the leaching procedure, aging and by the addition of promoters. These processes may affect either or both the chemical and physical characteristics of the catalysts. Preparation procedures need to be adjusted for individual reactions if optimal performance is required. The effect of the preparation conditions on the structure and catalytic performance of these systems is reviewed.

INTRODUCTION

The properties of an industrial catalyst depend on the chemistry of the system and on the physical properties. In general, the activity, selectivity, and life of a catalyst are dictated by the chemical nature of the catalyst and the reaction being promoted. However, the physical nature of the solid is of equal importance, with mass transfer, heat transfer, porosity, and strength being critical parameters. Particularly in tubular reactors, the crush strength of the catalyst pellet must be such as to bear the weight of the overlying material (1). In slurry reactors, abrasion resistance is equally important.

Various methods have been used to improve the physical characteristics of catalysts. The development of ceramic-coated metallic supports for vehicle exhaust catalysis (2) is a good example of recent advances. However, some of the more interesting high-strength catalysts were first developed in the 1920s by Raney (3, 4). These consisted of a range of skeletal or sponge metal catalysts, prepared by

*The Australian Government has the right to retain a nonexclusive, royalty-free license in and to any copyright covering this paper.

alloying the desired metal with a reactive metal such as aluminum. The aluminum was subsequently leached out with strong alkaline solution under controlled conditions to leave a framework of active catalyst containing lesser or greater amounts of aluminum.

A number of different metal catalysts, including nickel, copper, cobalt, iron, and ruthenium, have been manufactured using these techniques (5–8). The catalysts are generally used for hydrogenation and dehydrogenation reactions (8) but are also active for hydrogenolysis (9, 10) and hydrolysis (11, 12). Skeletal nickel is by far the most popular skeletal catalyst and is used widely in slurry bed applications and in tubular beds, where pellet strength is an important factor (1). Skeletal copper has also received much attention due to the fact that selective hydrogenation reactions are favored over the catalyst (6). Other skeletal catalysts are available commercially (8).

The advantages of the skeletal catalysts revolve around their mechanical strength and the improved heat transfer offered by the metallic framework. Controlled leaching can produce reasonable surface areas and porosities that can be improved by the addition of promoters (see below). The main disadvantages lie with the disposal of the waste alkali lixiviant and with deactivation during use, although this may be reversible in some cases by further leaching. The problem of the waste alkali is handled by subsequent concentration and reuse or by catalyst preparation in association with an aluminum production facility, where disposal of waste alkali is inherent to the process.

It quickly became clear that the catalyst preparation conditions define the catalyst performance. From the precursor alloy composition and the preparation method through to leaching and catalyst storage, the effective properties of the solids and their catalytic performance can be controlled by careful manipulation. The present paper reviews the understanding of skeletal catalyst preparation with particular reference to skeletal nickel and skeletal copper—where selectivity issues can be more important.

PRECURSOR ALLOYS

The precursor alloy consists of the desired catalytic metal (or metals) alloyed with a reactive metal that can be leached out. Aluminum is the most commonly used reactive metal and gives the best results, although other metals can and have been used, including Raney's original silicon (13). The composition is usually around 50 wt% of reactive metal, but the optimum composition depends on which catalytic metal is being alloyed because different phases offer different advantages. Whatever the composition, it must still fall within dealloying limits. Insufficient reactive metal will cause leaching to come to a stop after a short time, with the surface completely covered in the catalytically active metal but with low porosity and low surface area.

Different intermetallic compounds (alloy phases) have various strengths and leaching resistance and lead to different catalytic activities (14–16). For example, the phase $NiAl_3$ leaches readily, but it is friable and disintegrates (5); the phase Ni_2Al_3 is also readily leached and forms a tight skeletal nickel structure (5), whereas the phase NiAl is only slightly soluble in 20% w/w NaOH (13, 17). Although the NiAl phase is not leachable, it is a tough compound (13) and thus can impart mechanical strength to the final catalyst. The ideal composition for the precursor alloy for skeletal nickel has been determined to be close to Raney's original alloy of ~50 wt% nickel (5). For skeletal copper, $CuAl_2$ (~50 wt% copper) produces an active and selective copper catalyst (18–20) with sufficient strength.

More than one catalytically active metal can be incorporated into an alloy to give a skeletal catalyst consisting of several active metals. The alloy composition again depends on the intermetallic phases that give the desired properties; however, the amount of aluminum is still around 50 wt% in these precursor alloys.

Other alloy components may be added either to improve the catalytic performance of the end-use material (21, 22) or to alter the leachability of the alloy (23). These need not always act in the same direction. Thus the ternary phase Cu_3Al_3Zn produces a catalyst with a very high specific activity (catalytic activity per unit surface area), but it is highly resistant to caustic leaching and therefore develops only low surface area material (23). Additive species that improve the overall activity of the final catalyst are known as dopants or promoters. These materials are discussed in some detail below.

It is impossible to generalize about the desired nature of the alloy because this varies widely and depends on the phase diagrams of the mixture. It is essential to plan the alloy composition with reference to the relevant phase diagrams. A typical example of this procedure, based on the preparation of skeletal nickel, is summarized in Reference 5.

Once the precursor alloy composition is identified, the alloy preparation involves standard foundry practice with the inclusion of quenching. A good mixture of the components is made and heated to ~1400°C, often in an induction furnace where small quantities are desired. Rapid quenching ensures homogeneity of the alloy and produces small crystals within the alloy (24). These small crystals are desirable because they provide a large number of grain boundaries through which the leach liquor can penetrate the alloy. Rapid quenching might also be used to produce a specific intermetallic phase, although this is less common. Yamauchi et al. (25–27) have obtained supersaturation of additive species by using a particularly fast quenching method.

Recent interest has focused on rapid quenching techniques designed to produce amorphous metal glasses. Hu et al. (28) have investigated the textural and structural properties of catalysts produced by leaching a rapidly quenched $Ni_{50}Al_{50}$ alloy, which produces lower surface areas, higher pore diameters, and larger crystallite sizes. Deng et al. (29) review progress in the design of amorphous alloys involving Ni-P, Ni-Co, Ni-B, Ni-Co-B, and Ni. Catalytic activity for the hydrogenation of various organic compounds is found to be higher, owing in part to an increase in surface area (Ni-P) and in part to chemical promotion (Ni-Co-B). At least some of the benefits of rapid quenching appear to result from aluminum species that remain on or near the surface and act as a matrix to stabilize nickel (30).

The quenched alloy is usually ground and screened to a required particle size. Catalytic activity can depend on particle size (31)—only one of many factors. It may be advantageous to cast an alloy into a form and leach it whole—e.g., catalytically active fuel cell electrodes, although modern methods use powdered catalysts bonded in PTFE (32–38).

LEACHING

Leaching is usually carried out using strong alkali, often at low temperatures. Much work has been published dealing with optimizing particle size, lixiviant (leach liquor) concentration, and temperature, in addition to the the effect of stirring and that of additives.

For the common case of leaching aluminum in concentrated sodium or potassium hydroxide, the general leaching equation can be written as

$$2\ MAl_{x(s)} + 2x\ OH^- + 6x\ H_2O \rightarrow 2\ M_{(s)} + 2x\ Al(OH)_4^- + 3x\ H_{2(g)}, \qquad 1.$$

where M is the catalytic metal such as nickel, copper, cobalt, etc.

Some workers have tried acid-leaching of alloys (39–41); however, this tends to give catalysts with smaller surface areas and hence lower activities.

The leaching procedure involves complex liquid-solid reactions. At first sight, it would appear that the process can be described by five steps:

1. diffusion of liquid hydroxide through the liquid film surrounding the particle to the surface of the solid;
2. penetration and diffusion of hydroxide through a rim of skeletal catalyst to the unreacted core;
3. reaction of hydroxide with aluminum to form sodium aluminate and hydrogen;
4. diffusion of products through the rim of skeletal catalyst to the exterior surface of the particle;
5. diffusion of the products into the bulk of the liquid.

As such, the models described by Levenspiel (42) for a fluid-solid reaction can be expected to describe the leaching kinetics. The kinetics of leaching are often found to agree with the predictions of the shrinking core model (21, 24).

Not surprisingly, however, the reactions involved in leaching are more complex than this simple model. The hydroxide dissolves aluminum, but, to a small extent, it can also dissolve the catalytically active metal, which subsequently reprecipitates (43). Because the dissolution process consumes hydroxide, there is a local fall of pH at the alloy-solution interface that may interfere with the dissolution and reprecipitation processes. If there is more than one catalytically active metal present, or if a promoter material is added to the leach liquor (21, 22), the dissolution and precipitation processes become even more complicated.

The relative importance of the various effects observed during leaching is hard to generalize and is best described in terms of the structure and reactivity of individual catalysts.

After leaching, the residual alloy must be kept moist. If the alloy is allowed to dry and become exposed to the atmosphere, rapid oxidation occurs to the point where the catalyst can be pyrophoric (13, 15, 16, 44).

The catalyst will slowly deactivate with time (16), even during the leaching process in some cases (45). Pore size and crystallite size result from a decreasing surface area (45), an effect that has been attributed to a rearrangement of the metal atoms (46). The rate of rearrangement increases at higher temperature and thus low temperatures are recommended during leaching and storage. The addition of stabilizers such as chromia can help (47–49) but may have a negative effect on catalyst selectivity.

The leaching of aluminum from a precursor alloy in the production of a skeletal catalyst can be likened to a dealloying reaction. Much research has focused on dealloying, investigating the kinetics, possible mechanisms, and development of the remaining structure. Most of this work has been in relation to corrosion studies. Skeletal catalyst preparation differs in that the environment of concentrated hydroxide is hostile compared with that of the relatively mild corrosion environments. Nevertheless, the theories of dealloying are relevant to the catalyst preparation reactions. Erlebacher et al. (50, 51) provide two recent reviews of this area.

CATALYST STRUCTURE

Catalyst activity is related to the structure that, in turn, is defined by the preparation conditions. No review of skeletal catalyst preparation would be complete without at least briefly visiting the structure of these catalysts.

A range of techniques have been used to gain an understanding of the catalyst structure including electron microscopy (52, 53), X-ray diffraction (52, 54–56), electron diffraction (57), Auger spectroscopy (17, 58) and X-ray photoelectron spectroscopy (XPS) (17, 58). The catalysts consist of a delicate structure and many are pyrophoric when dry (13, 15, 16, 44), which makes them difficult to image directly. Even with care, extremely high resolutions are required to see the very fine structure (52, 59). For many years structure was inferred by indirect means. For example, a structure of curved parallel rods was proposed for skeletal copper (45). This was inferred from electron microscope images of carbon replicas of the catalyst surface and from kinetics and dealloying theories in vogue at the time. The microstructure of the precursor alloy is retained upon leaching with a sharp interface between the leached and unleached regions (14, 45, 60–62). Extensive cracking is observed on the surface, probably resulting from shrinkage upon removal of the aluminum (52). The activity of the catalyst is proportional to the surface area and hence to the degree of crystallite dispersion (17, 56). Methods such as ball-milling of the precursor alloy to obtain a higher concentration of defects, which results in smaller crystallite sizes and hence higher activity, have been investigated (63).

Skeletal nickel has a crystallite size range of about 1 to 20 nm (15, 52, 54). The size decreases with decreasing temperature (54, 58, 64) or caustic concentration (54, 64) during leaching. The amount of residual aluminum can also affect the nickel crystallite size (65, 66). By comparison, skeletal copper crystallite size ranges from about 10 to 100 nm (25, 52, 67). Fasman and coworkers (41, 67, 68) have clarified this size range into copper crystal grains around 10–13 nm that formed into larger agglomerates of about 50 to 80 nm.

Skeletal copper has a narrow pore size distribution (45, 69), which has been shown to be bimodal (70) with maxima centered at about 20 nm and 1.3 μm. Increased leaching temperature produces a coarser structure; the pore spacing at 1°C is about 40 nm, whereas at 93°C it is significantly larger at about 110 nm (45). Similarly, skeletal nickel has a larger fraction of small pores when leached at 50°C compared with 110°C (61). Additives such as zinc in the skeletal copper system can decrease the pore size with a corresponding increase in surface area (69).

Nickel oxide has been detected among the ultrafine nickel particles of skeletal nickel, and it has been shown that the oxide arises from oxidation of nickel by water and not from dissolved oxygen in the leaching solution or from residual oxygen inside the electron microscope (71). The possibility of oxidizing nickel in the presence of hydrogen has been attributed to the ultradisperse nature of the skeletal nickel.

With the emergence of the focused ion beam (FIB) electron microscope, it became possible to slice the catalyst in situ inside the microscope under vacuum and to examine the fresh catalyst directly. A uniform, three-dimensional structure of fine metal ligaments was observed (Figure 1*a*,*b*) (22, 43). By milling thin (~100 nm) slices with the FIB, it was possible to image skeletal copper at very high resolutions in a transmission electron microscope (TEM) and directly visualize the clusters of copper forming the ~30-nm ligaments making up the structure (Figure 1*c*) (72).

Some FIB machines available today have two columns that combine the high-resolution power of an FESEM (field emission scanning electron microscope) with the FIB's ability to cut through the sample in situ, thus simplifying the technique above and providing an excellent opportunity to study the structure of these catalysts even further.

AGING

The activity of skeletal nickel and copper decrease with time (16, 18, 46, 73). Skeletal catalysts are known to age under operating conditions (18, 74–76). Causes include the oxidation of the highly reactive metal (52, 77) and potential fouling of the catalyst surface with by-products (10, 78, 79). Activity loss also occurs while the catalysts are still being leached, which causes the surface area to pass through a maximum with leach depth (18, 46, 80). This led to the proposal that metal rearrangement (or sintering) occurs under both leaching and operating conditions (20, 54, 75, 81, 82). The sintering process also occurs, although to a lesser extent,

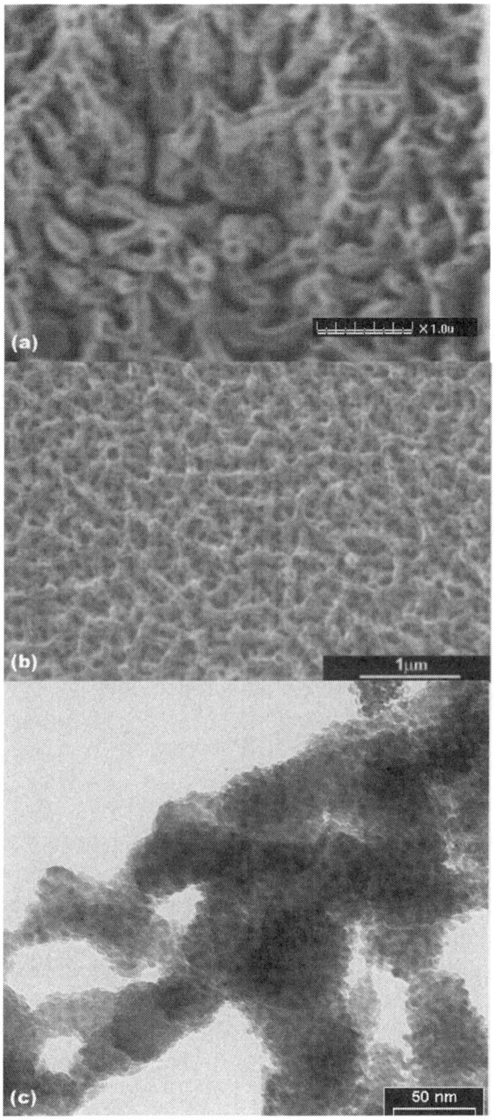

Figure 1 Structure of skeletal copper. (*a*) Undoped copper (FIB image), (*b*) zincate-doped copper (FIB image), and (*c*) zincate-doped copper (TEM image). Adapted from References 22, 43, 72.

with catalysts stored dry or under water (80, 83). Elevated leaching temperatures can accelerate aging (18, 83–85).

During the sintering-type aging, pore size and crystallite size both increase with the decreasing surface area; however, the pore volume remains constant (46). Pearce & Lewis (84) likened the aging to an annealing-type process. Defects (microstrains and stacking faults) are reduced, resulting in increased crystallite size and a corresponding decrease in surface area and hence activity. No such annealing process could be detected in cold-worked copper and nickel (which show some catalytic behavior owing to an increased number of structural defects) until the recrystallization temperature was reached. The low-temperature recrystallization in the skeletal systems may be the result of the extremely high concentration of defects in the leached residue. Sato & Ohta (80) found that low temperature and mild dissolution gave high-activity catalysts; however, the activity was lost rapidly. More severe leaching conditions gave lower-activity catalysts that were much more stable.

The rate of rearrangement decreases at lower temperatures and has a stronger dependence on temperature than on the rate of leaching (86). This means that lower temperatures will favor higher surface areas at given leach depths. The coarsening process is also accelerated by increased caustic concentrations (87). A mechanism consisting of dissolution and reprecipitation of copper has been proposed, and analysis of trace copper in the leach liquor is in agreeement with this idea (88). This mechanism has now been shown to operate in undoped skeletal copper as well as promoted copper systems (22, 43). Ma et al. (49) found the addition of small amounts of chromium prevents the coarsening of the structure during extended leaching, and thus maintains a high surface area and hence activity.

Controlled oxidation of the leached catalyst can passivate the material to allow transport of it in a dry state without any aging problems (52, 89). The activity can be restored in the reactor under a hydrogen flow.

ADDITIVES

The addition of a third component to skeletal catalysts may be associated with catalytic activity or with structural integrity. For example, the addition of zinc to copper produces a much more active catalyst for methanol synthesis (23, 90, 91), whereas alumina (18, 19) or chromia (73, 92, 93) can act as structural promoters, giving both greater surface areas and more long-term stability. The range of additives examined in skeletal catalyst systems has grown quite large and includes metallic as well as organic species. Recent advances have even seen the production of enantioselective skeletal catalysts through modification with tartaric acid (94–96).

The additive can be present in the original alloy or added to the leaching solutions. For example, leaching with a sodium hydroxide solution that contains zinc or chromium salts can lead to deposition of the metals as oxides:

$$Zn(OH)_4^{2-} \rightarrow Zn(OH)_{2(s)} + 2\ OH^-, \qquad 2.$$

$$2\ CrO_4^{2-} + 5H_2O + 6e^- \rightarrow Cr_2O_{3(s)} + 10\ OH^-. \qquad 3.$$

Both methods of addition affect the final catalyst, but the properties of the systems affect the choice. Thus, for example, zinc is quite soluble in the $CuAl_2$ phase (97); however, chromium is barely soluble (98) and must be added via the leaching solution. As a result, the two additives provide excellent examples of both methods of addition. Zinc may be added via either method, and Curry-Hyde et al. (70, 86, 99) found similar results with each method, i.e., whether the additive was part of the original alloy or present in the leaching solution. Recent work with mechanical alloying techniques has shown the feasibility of producing leachable alloys of metals that are alloy incompatible (100).

If zinc is added to the alloy, both dissolution and reprecipitation occur, the latter resulting from a localized fall in pH at the reaction front.

$$Cu(Zn)Al_{2(s)} + 12\ OH^- \rightarrow Cu_{(s)} + Zn(OH)_4^{2-} + 2\ Al(OH)_4^- + 8e^-, \qquad 4.$$

$$2\ H_2O + 2e^- \rightarrow H_{2(g)} + 2\ OH^-, \qquad 5.$$

$$Al(OH)_4^- \rightarrow Al(OH)_{3(s)} + OH^-, \qquad 6.$$

$$Zn(OH)_4^{2-} \rightarrow ZnO_{(s)} + OH^- + 2\ H_2O. \qquad 7.$$

Dissolved zincate may also react with aluminum still present in the alloy (101), although any metallic zinc produced will be readily attacked by hydroxide back to zincate.

$$CuAl_{2(s)} + 3\ Zn(OH)_4^{2-} \rightarrow Cu_{(s)} + 3\ Zn_{(s)} + 2\ Al(OH)_4^- + 4\ OH^-. \qquad 8.$$

Increasing the hydroxide concentration or the temperature was found to increase the leaching rate of $CuAl_2$ and $Cu(Zn)Al_2$ alloys, with leaching controlled by kinetics at the reaction front and by liquid phase diffusion in the leached residue region (20). Addition of either chromate or zincate to the leaching solution had the effect of reducing the rate of reaction (21, 86). Addition of chromate was found to reduce the leaching rate logarithmically, whereas hydroxide concentration had only a small effect.

Added chromate becomes reduced to chromic oxide during leaching (21) with deposition depending on reaction conditions (Equation 3). The total amount deposited increases with increasing chromate concentration and decreases with increasing hydroxide, probably as a result of the reaction between chromic oxide and strong alkali:

$$Cr_2O_{3(s)} + 2\ OH^- \rightarrow 2\ CrO_2^- + H_2O, \qquad 9.$$

$$Cr_2O_{3(s)} + 6\ OH^- \rightarrow 2\ CrO_3^{3-} + 3\ H_2O. \qquad 10.$$

The presence of sodium zincate was found to retard leaching and to enhance the surface area of the catalyst, but it did not inhibit copper rearrangement (102). Sodium chromate had less influence on leaching, but the structure and surface area of the catalyst were maintained for longer, and the surface area of the produced catalyst was much higher (Table 1).

The above observations on kinetics and final catalyst properties can be explained through the different modes of operation of these two promoters. Precipitation of either additive as an oxide species blocks the exposure of the $CuAl_2$ at the leach

TABLE 1 Specific catalytic activities

	Surface area (m^2 g^{-1})		Specific activity (× 10^3)					
			(mol/m^2 BET h)			(mol/m^2 Cu h)		
Catalyst	**Total**	**Cu**	**WGS**	**MS**	**RWGS**	**WGS**	**MS**	**RWGS**
RCu	19.49	8.16	2.44	1.56	2.63	5.83	3.73	6.28
RCu-ZnO	19.81	8.69	4.02	1.67	2.40	9.16	3.80	5.47
RCu-Cr_2O_3	40.65	14.11	3.31	1.11	1.47	9.52	2.92	4.23
RCu-ZnO-Cr_2O_3	38.61	10.45	5.72	1.19	1.48	21.13	2.97	5.45
ZnO	4.84	0.014	6.85	0	0	23.67	0	0
Cr_2O_3	27.29	0.22	1.45	0	0	18.00	0	0

WGS, water gas shift reaction: $CO + H_2O = CO_2 + H_2$; RWGS, reverse WGS: $CO_2 + H_2 = CO + H_2O$; MS, methanol synthesis: $CO + 2\ H_2 = CH_3OH$.

Activities reported under standard condition at 513 K in terms of mol m^{-2} surface area h^{-1}.

Adapted from Reference 105.

front, causing the decreased leaching rate, and also blocks copper dissolution, hindering the copper dissolution/reprecipitation mechanism. Chromate precipitates as chromium(III) oxide through electrochemical reduction coupled to the aluminum dissolution. Zincate precipitates as zinc oxide as a result of lowering of the pH at the reaction front through the consumption of hydroxide. Because the zincate precipitation is concentrated at the leach front, it shows a stronger effect on the leaching rate by blocking more of the exposed $CuAl_2$. However, the zinc oxide is also capable of redissolving after the leach front has passed, allowing the copper to rearrange once again through dissolution/reprecipitation.

FIB examination of the doped catalysts revealed the cause of the increased surface area for zinc- and chromium-doped skeletal copper (22, 104). The structure is similar to undoped skeletal copper except on a much finer scale (Figure 1*b*). Further examination of these catalysts by TEM of finely milled sections from the FIB gave an elemental map showing the location of dopant species within the structure (72). This information allowed the mechanism of operation to be deduced.

Whereas the copper dissolution/reprecipitation mechanism has been shown to be operating as the primary mode of structure rearrangement, copper diffusion appears to play a minor role in the rearrangement of the structure in the chromate system; however, it is mostly absent in the zincate system. The difference is because the zinc oxide is caught up and well dispersed inside the structural ligaments, thereby preventing the diffusion of the copper atoms as a rearrangement process. This location of the zinc oxide also confirms a nucleation-type role of the oxide, providing dispersed sites for the copper to precipitate on and resulting in a finer structure. Such a nucleation role was not operating in the chromate system.

Although Birkenstock et al. (52) claimed no morphological change with the addition of dopants, Fasman et al. (103) found that molybdenum addition facilitated creation and preservation of the porous structure and very high surface area. Klein & Hercules (58) also found that increasing the molybdenum concentration in the presursor alloy for skeletal nickel increased the amount of surface nickel, although only up to about 2 wt% molybdenum. After that, the surface nickel fell again; this profile matches the catalyst activity versus molybdenum content measurements exactly, proving that activity is related to the amount of surface nickel.

CATALYST PERFORMANCE

The performance of any catalyst is judged in terms of activity, selectivity, and life, and these parameters are affected by chemical and physical factors. The chemistry of catalysis essentially depends on what is on the surface and on the nanostructure of the surface. It is clear that additives can have a significant effect. For example, the addition of zinc or chromium to skeletal copper used to catalyze the water gas shift (WGS) reaction and the reverse WGS reaction (where the reactants are of similar size and if mass/heat transfer affects both reactions equally) have very different effects for the two systems (Table 1).

The catalyst activity is also dictated by the concentration of reactants at the active surface, which, in turn, is dictated by mass transfer effects. Skeletal catalysts, with their pronounced pore structure, usually offer excellent access but may be associated with smaller surface areas. However, this shortcoming is offset by excellent heat transfer properties of the sponge metal and by the mechanical strength that may be built into the catalyst.

The net result of these properties is the need to assess the use of skeletal catalysts for individual reactions. Certainly the historical use of skeletal nickel for hydrogenation of oils, for example, continues because of the advantages of the catalysts. On the other hand, skeletal copper is not used industrially for methanol synthesis despite the fact that the reactivity is about the same as current industrial systems (1).

It is also clear that skeletal catalyst preparation is complex and that several factors can affect the reactivity of the final catalyst. The control of preparation of skeletal nickel and copper catalysts is well understood and the ability to design catalyst specifications exists. Regrettably, this is not the case for other skeletal catalysts, and as such can be expected to limit the application of the systems.

The *Annual Review of Materials Research* is online at http://matsci.annualreviews.org

LITERATURE CITED

1. Goodman DR. 1989. Handling and using catalysts in the plant. In *Catalyst Handbook*, ed. MV Twigg, pp. 140–88. London: Wolfe
2. Wright CJ, Butler G. 1983. Preparation of catalysts III. In *Studies in Surface Science and Catalysis*, ed. G Poncelet, P Grange, PA Jacobs, 16:159. Amsterdam: Elsevier

3. Raney M. 1925. *U.S. Patent No. 1,563,587*
4. Raney M. 1927. *U.S. Patent No. 1,628,190*
5. Bakker ML, Young DJ, Wainwright MS. 1988. Selective leaching of nickel-aluminium ($NiAl_3$ and Ni_2Al_3) intermetallics to form Raney nickels. *J. Mater. Sci.* 23:3921–26
6. Ma L, Tran T, Wainwright MS. 2003. Hydrogenation of CO_2 over skeletal copper surfaces containing precipitated ZnO. *Topics Catal.* 22:287–93
7. Orchard JP, Tomsett AD, Wainwright MS, Young DJ. 1983. Preparation and properties of Raney nickel-cobalt catalysts. *J. Catal.* 84:189–99
8. Montgomery SR. 1981. Functional group activity of promoted Raney nickel catalysts. In *Catalysis of Organic Reactions*, ed. WR Moser, pp. 383–409. New York: Dekker
9. Kyowa Hakko Kogyo Co. Ltd. 1966. *British Patent No. 1,029,502*
10. Huang X, Cant NW, Wainwright MS, Ma L. 2004. Mitigation of fouling of skeletal copper catalysts for the gas-phase hydrogenolysis of methyl formate to methanol by addition of chromia. *Catal. Today* 93–95:107–12
11. Allain RJ, Seale VL. 1975. *U.S. Patent No. 3,928,440*
12. Svarz JJ, Goretta LA, Seale VL. 1976. *U.S. Patent No. 4,000,195*
13. Raney M. 1940. Catalysts from alloys. *Ind. Eng. Chem.* 32:1199–203
14. Young DJ, Wainwright MS, Anderson RB. 1980. Raney nickel-copper catalysts I: structure and leaching properties. *J. Catal.* 64:116–23
15. Anderson JR. 1975. *Structure of Metallic Catalysts*, pp. 228–33. London: Academic
16. Wainwright MS. 1998. Skeletal metal catalysts. In *Handbook of Heterogeneous Catalysis*, ed. G Ertl, H Knözinger, J Weitkamp, 1:64–72. Weinheim: Wiley-VCH
17. Klein JC, Hercules DM. 1981. Surface analysis of Raney nickel alloys. *Anal. Chem.* 53:754–58
18. Stanfield JA, Robbins PE. 1960. *Raney copper catalysts*. Presented at Proc. 2nd Int. Cong. Catal., Paris, pp. 2579–99
19. Nadirov NK, Ashirov AM, Savel'ev AF, Zhusupova A. 1977. Effect of activation conditions on the surface state of Raney copper catalysts. *Russ. J. Phys. Chem.* 51: 82–84
20. Curry-Hyde HE, Young DJ, Wainwright MS. 1987. Preparation and properties of Raney copper-zinc catalyst pellets for methanol dynthesis. *Appl. Catal.* 29:31–41
21. Ma L, Smith AJ, Tran T, Wainwright MS. 2001. Development of skeletal copper-chromia catalysts II. Kinetics of leaching Al and chromia deposition. *Chem. Eng. Proc.* 40:59–69
22. Smith AJ, Tran T, Wainwright MS. 2000. The kinetics and mechanism of the formation of doped skeletal copper catalysts: the effect of zincate compared to un-doped and chromate-doped systems. *J. Appl. Electrochem.* 30:1103–8
23. Bridgewater AJ, Wainwright MS, Young DJ, Orchard JP. 1983. Methanol synthesis over Raney copper-zinc catalysts. III. Optimisation of alloy composition and catalyst preparation. *Appl. Catal.* 7:369–82
24. Hu H, Qiao M, Pei Y, Fan K, Li H, et al. 2003. Kinetics of hydrogen evolution in alkali leaching of rapidly quenched Ni–Al alloy. *Appl. Catal. A* 252:173–83
25. Ohnaka I, Yamauchi I, Itaya M. 1992. Effect of solidification rate of mother alloys on structure of Raney copper catalysts and their catalytic activity in hydration reaction of acrylonitrile. *J. Jpn. Inst. Metals* 56:973–81
26. Yamauchi I, Ohnaka I, Ohashi Y. 1993. Effect of solidification conditions and Pd content on structure of Pd added Raney copper catalysts. *J. Jpn. Inst. Metals* 57:1064–71
27. Ohnaka I, Yamauchi I. 1994. Formation of new phases by chemical leaching of rapidly solidified alloys. *Mater. Sci. Eng.* A181–182:1190–94

28. Hu H, Qiao M, Wang S, Fan K, Li H, et al. 2004. Structural and catalytic properties of skeletal Ni catalyst prepared from the rapidly quenched $Ni_{50}Al_{50}$ alloy. *J. Catal.* 221:612–18
29. Deng J-F, Li H, Wang W. 1999. Progress in design of new amorphous alloy catalysts. *Catal. Today* 51:113–25
30. Lei H, Song Z, Bao X, Mu X, Zong B, Min E. 2001. XRD and XPS studies on the ultra-uniform Raney-Ni catalyst prepared from the melt-quenching alloy. *Surf. Interface Anal.* 32:210–13
31. Aller BV. 1957. Raney cobalt hydrogenation catalysts. I. The preparation of the catalyst. *J. Appl. Chem.* 7:130–34
32. Mund K, Richter G, Von Sturm F. 1977. Titanium-containing Raney nickel catalyst for hydrogen electrodes in alkaline fuel cell systems. *J. Electrochem. Soc.* 124:1–6
33. Kenjo T. 1985. Chromium-doped Raney nickel catalyst for hydrogen electrodes in alkaline fuel cells. *J. Electrochem. Soc.* 132:383–86
34. Tomida T, Nakabayashi I. 1989. Spongy Raney nickel hydrogen electrodes for alkaline fuel cells. *J. Electrochem. Soc.* 136:3296–98
35. Jenseit W, Khalil A, Wendt H. 1990. Material properties and processing in the production of fuel cell components. I. Hydrogen anodes from Raney nickel for lightweight alkaline fuel cells. *J. Appl. Electrochem.* 20:893–900
36. Fuehrer O, Rieke S, Schmitz C, Willer B, Wollny M. 1994. Alkaline eloflux fuel cells and electrolysis cells using a new kind of gas diffusion electrode. *Int. J. Hydrogen Energy* 19:343–48
37. Al-Saleh MA, Sleem-Ur-Rahman, Kareemuddin SMMJ, Al-Zakri AS. 1998. Novel methods of stabilization of Raney-nickel catalyst for fuel cell electrodes. *J. Power Sources* 72:159–64
38. Tanaka S, Hirose N, Tanaki T. 2000. Evaluation of Raney-nickel cathodes prepared with aluminium powder and tin powder. *Int. J. Hydrogen Energy* 25:481–85
39. Reynolds PW, Mackenzie JA. 1949. *British Patent No. 624,035*
40. Mellor JR, Coville NJ, Durbach SH, Copperthwaite RG. 1998. Acid leached Raney copper catalysts for the water-gas shift reaction. *Appl. Catal. A* 171:273–81
41. Kalina MM, Berdongarova SK, Sadykova GA, Fasman AB. 1976. *Physicochemical characteristics of Raney catalysts prepared from γ-metallides of some binary and ternary systems.* Presented at Mater. Resp. Nauchno-Tekh. Konf. Molodykh Uch. Pererab. Nefti Neftekhim., 3rd, Sredneaziat. Nauchno-Issled. Inst. Neftepererab. Prom-sti.: Tashkent, USSR, pp. 100–1 (Chem. Abs. 89:204764z)
42. Levenspiel O. 1972. *Chemical Reaction Engineering*, pp. 372. Singapore: Wiley & Sons
43. Smith AJ, Tran T, Wainwright MS. 1999. Kinetics and mechanism of the preparation of Raney copper. *J. Appl. Electrochem.* 29:1085–94
44. Kefeli LM. 1952. Pyrophoricity of the nickel skeleton catalyst. *Doklady Akad. Nauk S.S.S.R.* 83:697–99 (Chem. Abs. 46: 6916g)
45. Szot J, Young DJ, Bourdillon A, Easterling KE. 1987. Cellular morphologies in a de-alloying residue. *Philos. Mag. Lett.* 55:109–14
46. Tomsett AD, Curry-Hyde HE, Wainwright MS, Young DJ, Bridgewater AJ. 1987. Structural changes during the leaching of copper-based Raney catalysts. *Appl. Catal.* 33:119–27
47. Mellor JR, Coville NJ, Sofianos AC, Copperthwaite RG. 1997. Raney copper catalysts for the water-gas shift reaction. I: preparation, activity and stability. *Appl. Catal. A* 164:171–83
48. Mellor JR, Coville NJ, Sofianos AC, Copperthwaite RG. 1997. Raney copper catalysts for the water-gas shift reaction. II: initial catalyst optimisation. *Appl. Catal. A* 164:185–95

49. Ma L, Trimm DL, Wainwright MS. 1998. *Promoted skeletal copper catalysts for methanol synthesis*. Presented at Adv. of Alcohol Fuels in the World, Beijing, China, pp. 1–7
50. Erlebacher J, Aziz MJ, Karma A, Dimitrov N, Sieradzki K. 2001. Evolution of nanoporosity in dealloying. *Nature* 410: 450–53
51. Erlebacher J, Sieradzki K. 2003. Pattern formation during dealloying. *Scr. Mater.* 49:991–96
52. Birkenstock U, Holm R, Reinfandt B, Storp S. 1985. Surface analysis of Raney catalysts. *J. Catal.* 93:55–67
53. Ermolaev VN, Fasman AB, Semiletov SA, Khitrova VI. 1988. Study of the genesis of thin-layer Raney nickel catalysts by electron-optical Methods. IV. Transformation of aluminium + nickel-aluminium ($NiAl_3$) films during leaching (according to the data of transmission electron microscopy with microdiffraction). *Izv. Akad. Nauk Kaz. SSR, Ser. Khim.* pp. 34–40 (Chem. Abs. 108:138528h)
54. Urazov GG, Kefeli LM, Lel'chuk SL. 1947. The structure of the nickel skeleton catalyst. *Compt. Rend. Acad. Sci.* 55:735–38 (Chem. Abs. 41:6800h)
55. Kefeli LM, Lel'chuk SL. 1953. An X-ray study of the structure of skeleton catalysts. *Rentgen. Metody Issledovan. i Primenenie v Khim. Prom. Goskhimizdat* pp. 78–84 (Chem. Abs. 50:3862d)
56. Kagan AS, Kagan NM, Ul'yanova GD, Mironov LG. 1973. Relation between structure and activity of Raney catalysts. *Russ. J. Phys. Chem.* 47:978–80
57. Talipov GS, Nalibaev TN, Fasman AB, Sultanov AS. 1974. Electron-diffraction investigation of the structure and phase composition of skeletal nickel catalysts. *Kinet. Katal.* 15:744–50
58. Klein JC, Hercules DM. 1984. Surface analysis by X-ray photoelectron spectroscopy and Auger electron spectroscopy of molybdenum-doped Raney nickel catalysts. *Anal. Chem.* 56:685–89
59. Hamar-Thibault S, Koscielski T, Damon JP, Masson J. 1989. Structure of Raney nickel catalysts modified by chromium hydroxide deposition. *Appl. Catal.* 56:57–71
60. Colin P, Hamar-Thibault S, Joud JC. 1992. Metallurgical structure and alkali leaching of the Al/Al_3Ni eutectic. *J. Mater. Sci.* 27:2326–34
61. Freel J, Pieters WJM, Anderson RB. 1969. The structure of Raney nickel I. Pore structure. *J. Catal.* 14:247–56
62. Friedrich JB, Young DJ, Wainwright MS. 1981. Caustic leaching of Al-Cu-Zn alloys to produce Raney catalysts I. Morphological development. *J. Electrochem. Soc.* 128:1840–44
63. Sun L, Liu H, Dou S. 1997. The effect of ball milling and rapid quenching on the catalytic activity of the $Cu_{30}Al_{70}$ alloy. *Mater. Lett.* 33:79–83
64. Kefeli LM, Lel'chuk SL. 1952. Structure of skeleton catalysts. *Doklady Akad. Nauk S.S.S.R.* 84:285–88 (Chem. Abs. 46: 10822c)
65. Tanaka S, Hirose N, Tanaka T. 1997. Influence of the residual aluminium in Raney-nickel on the catalytic ability for hydrogen evolution reaction. *Denki Kagaku* 65:1044
66. Kubomatsu T. 1957. Role of remaining aluminium in Raney nickel catalyst in its activity. *Kagaku Kôgyô (Osaka)* 31:190–94 (Chem. Abs. 51:15837d)
67. Kalina MM, Fasman AB, Ermolaev VN. 1980. Mechanism of the formation of skeletal copper. *Kinet. Katal.* 21:813 (Chem. Abs. 93:81065v)
68. Kalina MM, Fasman AB, Ermolaev VN. 1980. Formation mechanism for Raney copper. *Deposited Doc., VINITI* Doc. No. 1022–80. 15 pp. (Chem. Abs. 95: 66291p)
69. Friedrich JB, Young DJ, Wainwright MS. 1981. Caustic leaching of Al-Cu-Zn alloys to produce Raney catalysts II. Leaching kinetics. *J. Electrochem. Soc.* 128:1845–50

70. Curry-Hyde HE, Sizgek GD, Wainwright MS, Young DJ. 1993. Improvements to Raney copper methanol synthesis catalysts through zinc impregnation. IV. Pore structure and the influence on activity. *Appl. Catal.* 95:65–74
71. Ermolaev VN, Pushkareva GA, Fasman AB. 1988. Study of the mechanism of formation of Raney Ni catalysts. *Kinet. Katal.* 29:431–36
72. Smith AJ, Munroe P, Tran T, Wainwright MS. 2001. FIB preparation of a sensitive porous catalyst for TEM elemental mapping at high magnifications. *J. Mater. Sci.* 30:3519–24
73. Ma L, Trimm DL, Wainwright MS. 1999. Structural and catalytic promotion of skeletal copper catalysts by zinc and chromium oxides. *Topics Catal.* 8:271–77
74. Smith HA, Bedoit WC Jr, Fuzek JF. 1949. The preparation and aging of Raney nickel catalysts. *J. Am. Chem. Soc.* 71:3769–71
75. Kelley RD, Candela GA, Madey TE, Newbury DE, Schehl RR. 1983. Surface and bulk analysis of a deactivated Raney nickel methanation catalyst. *J. Catal.* 80:235–48
76. Mikkola J-P, Vainio H, Salmi T, Sjöholm R, Ollonqvist T, Väyrynen J. 2000. Deactivation kinetics of Mo-supported Raney Ni catalyst in the hydrogenation of xylose to xylitol. *Appl. Catal. A* 196:143–55
77. Pattison JN, Degering EF. 1951. Some factors influencing the activity of Raney nickel catalyst. II. The role of oxygen in the aging of Raney nickel catalyst. *J. Am. Chem. Soc.* 73:486–87
78. Huang X, Cant NW, Wainwright MS, Ma L. 2005. The dehydrogenation of methanol to methyl formate: Part II. The effect of chromia on deactivation kinetics for copper-based catalysts. *Chem. Eng. Proc.* 44:403–11
79. Hoffer BW, Crezee E, Devred F, Mooijman PRM, Sloof WG, et al. 2003. The role of the active phase of Raney-type Ni catalysts in the selective hydrogenation of D-glucose to D-sorbitol. *Appl. Catal. A* 253:437–52
80. Sato M, Ohta N. 1955. Decrease in the surface area of Raney-type catalysts in the aluminium-dissolution process. *Bull. Chem. Soc. Jpn.* 28:182–85
81. Tomsett AD, Young DJ, Wainwright MS. 1987. Modelling the development of pore structure during the preparation of Raney copper catalysts for fixed bed operation. *Appl. Catal.* 35:321–28
82. Kefeli LM. 1949. Roentgenographic study of the structure of skeleton catalysts. *Prob. Kinetika Kataliza, Akad. Nauk S.S.S.R.* 6:210–14 (Chem. Abs. 49:14452d)
83. Schwab G-M, Marhenkel H. 1957. *BET Surfaces and their variability*. Presented at 2nd Int. Cong. Surface Activity. London: Butterworths, p. 64
84. Pearce CE, Lewis D. 1972. X-Ray diffraction line broadening studies of Raney copper and nickel. *J. Catal.* 26:318–25
85. Zhdanov GS, Kefeli LM. 1958. A study of the texture of copper obtained upon the leaching of $CuAl_2$ by the pole method. *Zhur. Fiz. Khim.* 32:666–69 (Chem. Abs. 52:17020e)
86. Curry-Hyde HE, Wainwright MS, Young DJ. 1991. Improvements to Raney copper methanol synthesis catalysts through zinc impregnation. II. Surface area development. *Appl. Catal.* 77:89–94
87. Tomsett AD. 1987. *Pore development in Raney copper catalysts*. PhD Thesis. Univ. New South Wales, Sydney, Australia. 197 pp.
88. Laine J, Ferrer Z, Labady M, Chang V, Frías P. 1988. Structure and activity of chromium-promoted Raney copper catalysts for carbon monoxide oxidation. *Appl. Catal.* 44:11–22
89. Tomsett AD, Wainwright MS, Young DJ. 1984. Pore structure of Raney copper according to adsorption and mercury porosimetry. *Appl. Catal.* 12:43–48

90. Marsden WL, Wainwright MS, Friedrich JB. 1980. Zinc-promoted Raney copper catalysts for methanol synthesis. *Ind. Eng. Chem. Prod. Res. Dev.* 19:551–56
91. Friedrich JB, Wainwright MS, Young DJ. 1982. Development of Raney-type low temperature methanol synthesis catalysts. *Chem. Eng. Commun.* 14:279–88
92. Ma L, Wainwright MS. 1999. Development of skeletal copper-chromia catalysts I. Structure and activity promotion of chromia on skeletal copper. *Appl. Catal. A* 187:89–98
93. Ma L, Tran T, Wainwright MS. 2003. Methanol synthesis from CO_2 using skeletal copper catalysts containing co-precipitated Cr_2O_3 and ZnO. *Topics Catal.* 22:295–304
94. Kukula P, Cervený L. 2001. Preparation of tartaric acid modified Raney nickel catalysts: study of modification procedure. *Appl. Catal. A* 210:237–46
95. Kukula P, Cervený L. 2002. Effects of reaction variables on enantioselectivity of modified Raney nickel catalyst. *J. Mol. Catal. A* 185:195–202
96. Haruna N, Acosta DE, Nakagawa S, Yamaguchi K, Tai A, et al. 2004. Asymmetric hydrogenation of furan-containing ketones over tartaric aicd-modified Raney nickel catalyst. *Heterocycles* 62:375–86
97. Ghosh G, van Humbeeck J. 1992. Aluminium-copper-zinc. In *Ternary Alloys: A Comprehensive Compendium of Evaluated Constitutional Data and Phase Diagrams*, ed. G Petzow, G Effenberg, 5:92–112. New York: VCH
98. Ghosh G. 1991. Aluminium-chromium-copper. In *Ternary Alloys: A Comprehensive Compendium of Evaluated Constitutional Data and Phase Diagrams*, ed. G Petzow, G Effenberg, 4:311–19. New York: VCH
99. Curry-Hyde HE, Wainwright MS, Young DJ. 1991. Improvements to Raney copper methanol synthesis catalysts through zinc impregnation. I. Electron microprobe analysis. *Appl. Catal.* 77:75–88
100. Zeifert BH, Salmones J, Hernández JA, Reynoso R, Nava N, et al. 2000. Preparation of iron-nickel catalysts by mechanical alloying. *Mater. Lett.* 43:244–48
101. Robertson SG, Ritchie IM, Druskovich DM. 1995. A kinetic and electrochemical study of the zincate immersion process for aluminium. *J. Appl. Electrochem.* 25:659–66
102. Curry-Hyde HE, Young DJ, Wainwright MS. 1988. The precipitation and dissolution of zinc during leaching of Cu-Al-Zn Alloys in alkali solutions. *J. Electrochem. Soc.* 135:1902–7
103. Fasman AB, Bedel'baev GE, Maksimova NA, Ermolaev VN, Kuanyshev AS. 1988. Mechanism of promoting a skeletal nickel catalyst by molybdenum. *Kinet. Katal.* 29:437–42
104. Smith AJ, Ma L, Tran T, Wainwright MS. 2000. Effect of chromate additive on the kinetics and mechanism of skeletal copper formation. *J. Appl. Electrochem.* 30:1097–102
105. Wainwright MS, Trimm DL. 1995. Methanol synthesis and water gas shift reactions on Raney copper catalysts. *Catal. Today* 23:29–42

Annu. Rev. Mater. Res. 2005. 35:143–66
doi: 10.1146/annurev.matsci.35.100303.122029

First published online as a Review in Advance on March 8, 2005

HETEROGENEOUS ASYMMETRIC CATALYSTS: Strategies for Achieving High Enantioselection

Graham J. Hutchings
School of Chemistry, Cardiff University, Cardiff CF10 3TB, United Kingdom; email: hutch@cf.ac.uk

Key Words asymmetric synthesis, supported metals, zeolites

■ **Abstract** Heterogeneous asymmetric catalysis remains a topic of intense research interest because control of enantioselectivity represents one of the major synthetic challenges today. Successful strategies to achieve high enantioselection have been developed using three key approaches: (*a*) adsorption of chiral modifiers onto an active metal surface, which is found to be particularly effective for enantioselective hydrogenation; (*b*) covalent tethering of homogeneous catalysts; and (*c*) electrostatic interaction between a negatively charged framework and a cation. Other approaches have been advocated, e.g., encapsulation, but in general these lead to lower enantioselectivity for these heterogeneous asymmetric systems when compared with the equivalent homogeneous catalyst. This review covers the strategies for achieving high enantioselection as well as recent interesting features concerning confinement effects, the inversion of enantioselectivity, and specific cases in which enantioselection increases with conversion, all being factors that can guide researchers to identifying improved systems.

INTRODUCTION

The subject of asymmetric catalysis has been extensively studied for several decades. It is a topic that exercises the interest of the many sub-sections of chemistry from synthetic chemists to catalytic chemists. The main drive has been to find new, exciting routes to chiral molecules. In many cases, this involves elegant three-dimensional solids. Chiral molecules lie at the heart of many pharmaceuticals and agrochemicals, either as intermediates or final products. The purity of these molecules for their use in medical or veterinary applications is paramount. Hence, there is always a need to find synthetic strategies that can give both high yields and enantioselectivity for these important molecules, so that the subsequent purification processes can be minimized.

To date, there have been significant advances in the field of homogeneous asymmetric catalysis, which culminated in the award in 2001 of the Nobel Prize for Chemistry to Noyori, Sharpless, and Knowles for their contributions to homogeneous asymmetric hydrogenation and oxidation (1). Homogeneous catalysts have been developed for numerous C–H, C–C, C–O and C–N bond formation reactions, which give high enantioselection. The design has been predicated on the synthesis of chiral ligands for active catalyst centers that ensure that the desired chiral transition state is readily accessed. However, these ligands are often difficult to recover and reuse and, for this reason, chiral homogeneous catalysts have not had the significant commercial input that researchers had initially hoped for (2). For this reason, attention has been focused on the design of immobilized homogeneous chiral catalysts or heterogeneous chiral catalysts because these can be readily recovered by filtration for a slurry reactor, or can be used in the numerous fixed bed reactor options available. This introduces a different aspect of catalyst usability, namely the stability of the heterogeneous catalyst, because leaching of active components can represent a real problem concerning commercialization (3). Consequently, emphasis has to be given to the design of stable heterogeneous catalysts that are capable of high enantioselectivity, which is the topic of this review.

In a previous review, approaches to the immobilization of asymmetric catalysts were fully explored (4), and in this article we intend to focus on three strategies that are known to lead to very stable catalysts capable of high enantioselection:

- modification of metal interfaces with chiral molecules;
- tethering of active homogeneous catalysts to a three-dimensional structure;
- electrostatic interaction as a means of immobilization.

These three approaches provide contrasting methodologies that are each suitable for particular applications and provide the basis for stable reuseable catalysts.

To date there have been been numerous approaches to the design of heterogeneous asymmetric catalysts, since Schwab and coworkers first demonstrated that Cu and Ni could be supported on chiral silica surfaces (5, 6) and that the resulting catalysts could give low enantioselection in the dehydration of butan-2-ol. The approaches have been reviewed by Davis (7) and several hundred publications and several extensive reviews are available (8–11). Blaser & Pugin (12) have also reviewed the scope and limitations of heterogeneous enantioselective catalysts using three strategies, namely: modification of metal and metal oxides by chiral ligands, chiral organic polymers and tethering chiral catalysts by covalent bonding. They concluded that only a few catalysts offer a way forward in the design of enantioselective heterogeneous catalysts. In addition to the three approaches featured in the present review, two other methodologies have been used: adsorption of homogeneous catalysts on the surface and encapsulation of homogeneous catalysts. Both approaches do not lead to stable catalysts capable of high enantioselectivity (4) and, consequently, these areas are excluded from this review.

The field of research is rapidly expanding, and it is contemplated that stable heterogeneous asymmetric catalysts will soon find commercial applications. This review concentrates on the three strategies that can be expected to provide the basis for such systems.

MODIFICATION OF METAL SURFACES WITH CHIRAL MOLECULES

Heterogeneous-enantioselective hydrogenation of prochiral ketones has been extensively studied using cinchona-modified-supported Pt catalysts (13–16) and tartrate-modified-supported Ni catalysts (17–19). However, the hydrogenation of pyruvate esters using cinchonidine-modified Pt/SiO_2 and $Pt/\gamma\text{-}Al_2O_3$ represents the most-well-investigated reactions, and these are often studied as a model system. For this reaction, the enantioselectivity can be high and enantioselective excess (e.e.) in excess of 90% has been reported in many studies (13–16, 20–22).

The reaction (Figure 1) has several fascinating features that have yet to be fully explained, although similarities with other heterogeneous asymmetric catalysts exist that warrant a detailed description in this review. In particular, the observation of high enantioselection was initially considered to be coupled with an enhancement in the rate of reaction (Table 1). This is a design approach that can be useful because the chiral-promoted sites inherently lead to high e.e. However, many studies have shown that the two effects can be decoupled. In addition, the system demonstrates the features of enantio-enhancement with conversion and enantio-inversion, and these observations may play a role in aiding the design of improved heterogeneous asymmetric catalysts.

The structure of the cinchona alkaloids has been extensively studied (23–27) and in general, cinchonidine produces a higher *R*-excess than the *S*-excess produced by cinchonine. The solvent has also been found to be important because it affects the proportions of the different conformers of cinchonidine present in solution (28), and the highest e.e.s are observed when the proportion of the open 3 conformer is maximized, typically using acetic acid as solvent. The reaction is also of interest because it does not have a homogeneous counterpart, which is in contrast to the other two methodologies discussed below, which are based on the immobilization strategies of active homogeneous catalysts.

Cinchona-modified Pt catalysts are highly specific for particular substrates (23–34), and only a few types of substrate can be hydrogenated with high enantioselectivity. Recent studies have extended the use of these catalysts to other activated ketones. Vermeer et al. (29) have described the enantioselective hydrogenation of buta-2,3-dione to 3-hydroxy-buta-2-one with e.e. in the range 20–40% (*R*) using cinchonidine-Pt/SiO_2. Baiker and coworkers (20, 28, 30) reported e.e. of 79% (*R*) and 92% (*R*) for the hydrogenation of ketopantolactone and

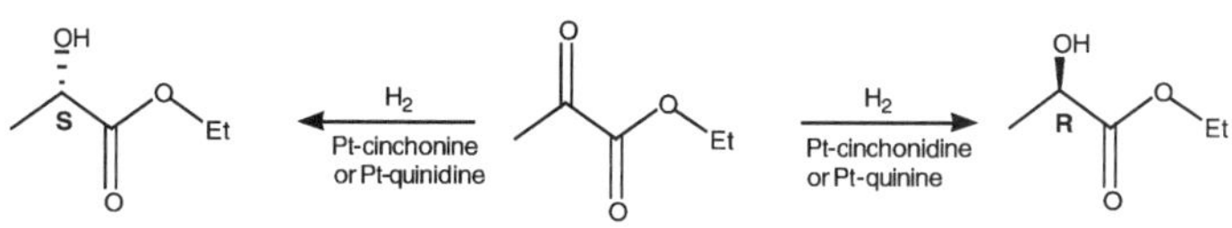

Figure 1 Reaction scheme for the hydrogenation of alkyl pyruvates.

TABLE 1 Effect of cinchona modifier on the rate of hydrogenation of ethyl pyruvate with Pt/Al_2O_3 catalysts[a]

Modifier[b]	Modification time/h	Solvent	Rate/mmol^{-1} g^{-1}
None	—	None	50
None	—	Ethanol	35
C	1	Ethanol	1100–1300
DHC	1	Ethanol	1100–1300
DHC	18	Ethanol	1490

[a]Data taken from Reference 21, H_2 pressure 10 bar, 20°C.
[b]C = cinchonidine, DHC = 10,11 dihydrocinchonidine.

methylketopantolactone, respectively, under optimized conditions using cinchonidine-modified Pt/γ-Al_2O_3. Blaser and coworkers (31) and Baiker and coworkers (30) have shown that α-keto acetals can be readily hydrogenated to the corresponding α-hydroxy acetals with e.e. up to 97%, indicating that these substrates are particularly well suited to the cinchona-modified Pt catalysts with respect to both enantioselectivity as well as activity.

To demonstrate the general effects of modifier concentration and solvent, a comparative study of the enantioselective hydrogenation of ethyl pyruvate and buta-2,3-dione has recently been carried out by Li et al. (35), and it is apparent that there are different optimal modifier concentrations for these two substrates (Figure 2), but that they show similar impact on the effect of the solvent medium on

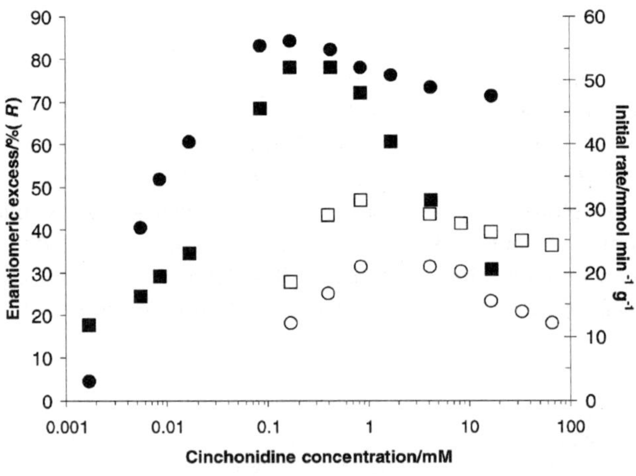

Figure 2 Effect of cinchonidine concentration on the e.e. and the initial rate of hydrogenation. Key: ■ e.e. ethyl pyruvate; • initial rate ethyl pyruvate; □ e.e. buta-2, 3-dione; ○ initial rate buta-2,3-dione. Reaction conditions: dichloromethane (12.5 ml), reactant (66 mmol), catalyst (0.25 g), 50 bar H_2, 20°C, 1200 rpm stirring.

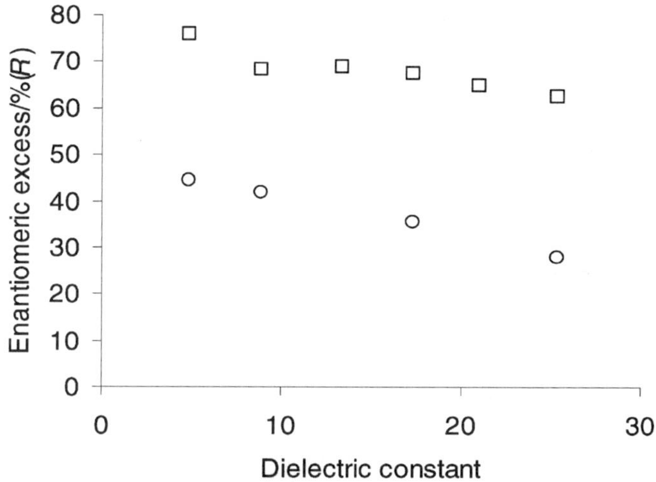

Figure 3 Effect of the dielectric constant of the solvent on e.e. Key: □ e.e. ethyl pyruvate, ○ e.e. buta-2,3-dione. Reaction conditions: solvent (12.5 ml), reactant (66 mmol), catalyst (0.25 g), 50 bar H_2, 20°C, 1200 rpm stirring.

the enantioselection, as shown by the relationship between e.e. and the dielectric constant of the solution (Figure 3). The difference in the optimal concentration of the modifier for the two reactants may be related to the competitive adsorption of the modifier and the substrate at the Pt surface.

One intriguing observation with some heterogeneous asymmetric catalysts, and in particular with the cinchona alkaloid-modified Pt/Al_2O_3 catalysts, is that the enantioselection is enhanced during the initial period of conversion, whereas, typically, the reverse is observed in many systems. Indeed the effect of this initial transient period has been most well studied for the cinchona-modified Pt catalysts (36–39). However, the origin of the enhanced enantioselection during the initial reaction remains a matter of considerable debate (40, 41). At present, two schools of thought exist concerning the origin of this effect. Baiker and coworkers (39, 41) consider the enhancement in e.e. to be caused by the removal of impurities present on the catalyst surface at the start of the reaction—for example, oxygen or the degradation products of dissociative chemisorption of the substrate. During the initial reaction in the presence of hydrogen, these impurities are removed. In contrast, Blackmond and coworkers (37, 40) consider the effect to be from a "reaction-driven equilibration" of the catalyst surface and have shown that the effect of rising rate and enantioselection in the initial reaction period is linked solely to substrate conversion and is independent of reaction conditions.

A typical example of the initial transient effect is shown in Figure 4. In an attempt to determine the origin of this intriguing effect, a sequential experiment was carried out in which the reaction of methyl pyruvate was carried out following the initial hydrogenation of ethyl pyruvate using chinona-modified Pt/Al_2O_3. In these experiments, the initial hydrogenation of ethyl pyruvate was carried out with

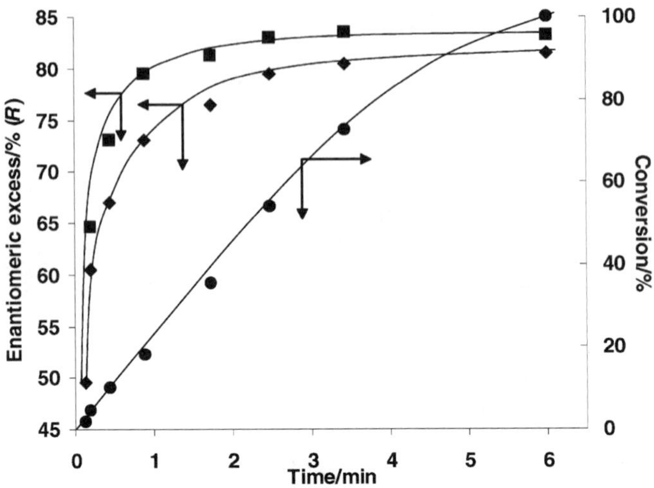

Figure 4 Dependence of enantiomeric excess and conversion of ethyl pyruvate hydrogenation. Key: • ethyl pyruvate conversion, ♦ e.e. (*R*)-ethyl lactate, ■ Δe.e. Conditions: 5% Pt/Al_2O_3 (0.25 g), CH_2Cl_2 (12.5 ml), cinchonidine (2.5 mg), ethyl pyruvate (66 mmol), P_{H_2} = 50 bar, 20°C, 1200 rpm stirring speed. Δe.e. is the mean value of the e.e. provided by the reaction between data points. As the e.e. is increasing with conversion and the initial turnovers have the lowest e.e., there is a cumulative effect for the e.e. at any time, which is the sum of all the earlier less enantioselective reactions. Δe.e. gives the e.e. for the conversion between the reaction time, at which it is recorded, and the preceding reaction time. This provides a better representation of the enantioselection achieved at the catalyst surface.

30 bar hydrogen pressure. Following this reaction, methyl pyruvate was rapidly added under hydrogen pressure, while the stirring was continued in such a way that neither oxygen nor other impurities could be admitted to the reaction mixture. The second hydrogenation was consequently conducted at a higher reaction pressure (50 bar) but, as the observation of the initial transient effect is independent of the reaction conditions (37, 40), this difference in pressure is not considered to be significant. The results of the sequential experiment (Figure 5) show that the initial

→

Figure 5 Sequential hydrogenation experiment using reaction procedure II: (*a*) ethyl pyruvate reacted to 100% conversion prior to addition of methyl pyruvate; (*b*) methyl pyruvate reacted to 100% conversion prior to addition of ethyl pyruvate; (*c*) ethyl pyruvate reacted to 50% conversion prior to addition of methyl pyruvate. Key: ■ ethyl pyruvate conversion, • e.e. in (*R*)-ethyl lactate, ▲ methyl pyruvate conversion, ♦ e.e. in (*R*)-methyl lactate. Conditions: 5% Pt/Al_2O_3 (0.25 g), CH_2Cl_2 (12.5 ml), cinchonidine (2.5 mg), ethyl pyruvate (33 mmol), methyl pyruvate (33 mmol), 20°C, 1200 rpm stirring speed, reaction pressure 30 bar H_2 for ethyl pyruvate hydrogenation, 50 bar H_2 for methyl pyruvate introduction.

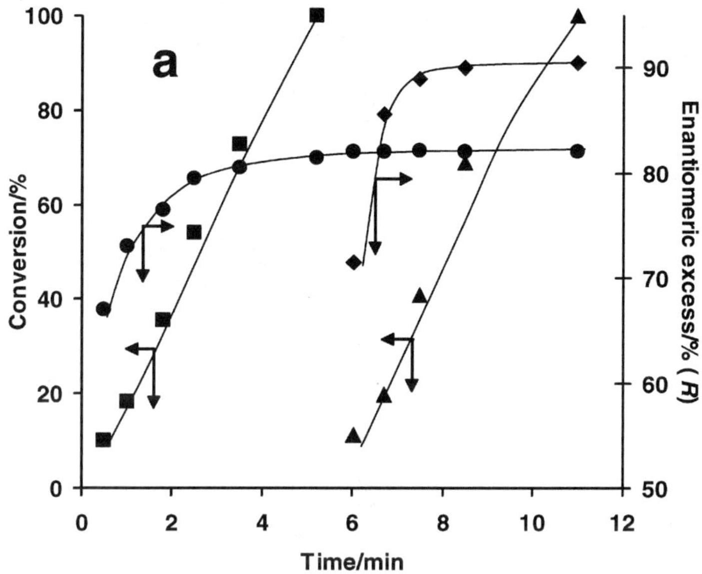
a
Conversion/%
Enantiomeric excess/% (R)
Time/min
0
20
40
60
80
100
50
60
70
80
90
0
2
4
6
8
10
12

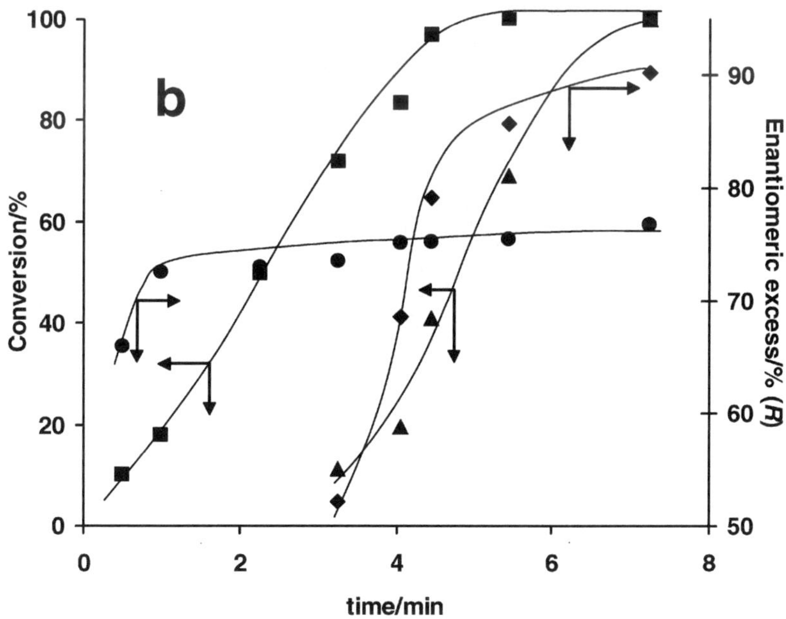
b
Conversion/%
Enantiomeric excess/% (R)
time/min
0
20
40
60
80
100
50
60
70
80
90
0
2
4
6
8

transient period is observed for the hydrogenation of methyl pyruvate and also for ethyl pyruvate. Experiments were also conducted with the order of reagent addition reversed, and the initial transient effects were observed as before. The observation that the initial transient effect is still observed with the sequential hydrogenation of ethyl and methyl pyruvates indicates that the increase in e.e. with conversion for this reaction is not attributable to impurity effects, because the optimal chiral active sites had been achieved from the reaction of ethyl pyruvate. The results were considered to support the ideas proposed by Blackmond and coworkers (37, 40) concerning the "reaction-driven equilibrium of the chiral surface environment." However, the experiments suggested that the cinchona modifier adsorbed on the metal surface may interact differently with the two reactants and the active chiral site is dependent upon the nature of the substrate. This may provide further scope to improve these catalyst systems.

In these earlier studies it was well established that with cinchonidine- and quinine-modified platinum catalysts, the reaction proceeds to form (*R*)-ethyl lactate predominantly, whereas with cinchonine- and quinidine-modified platinum catalysts, the reaction forms (*S*)-ethyl lactate preferentially (Figure 1). Considerable effort has been focused on the mechanism of these reactions, and three structural features of the cinchona and related modifiers that ensure they are effective in enantiodirection have been identified. They are (*a*) an aromatic moiety that enables adsorption on the platinum surface, (*b*) the absolute configuration at C_8, which controls the sense of the enantioselectivity, and (*c*) a basic nitrogen, which is considered to interact with the substrate, resulting in a 1:1 complex that is hydrogenated enantioselectively (16). The reaction is, however, complicated by effects observed in the early part of the reaction where enantioselection increases with conversion (36–41). Consequently, there has been limited progress in gaining an understanding of this fascinating reaction at the molecular level. Recently, Baiker and coworkers (42–44) and Bartók et al. (45) have shown a further intriguing aspect of this reaction; the sense of the enantioselection can be inverted for specific modifiers by changing the extent of reaction (42), the solvent (45), or the substituent at C_8 (43, 44). In all these cases the effect is induced by changes in the reaction conditions, but these experiments clearly demonstrate a further complexity of this heterogeneously catalyzed process. In a recent study, Balázsik & Bartók (46) have shown that reusing a cinchona-modified platinum catalyst can also affect the e.e., and thus it is clear that this catalyst and reaction are far from being fully understood.

The initial observation of inversion in enantioselectivity by Baiker and coworkers involved the hydrogenation of ethyl 4,4,4-trifluoroacetoacetate, which was conducted over Pt/alumina modified with altered cinchonidine containing a methoxy group at the C_9 position (42). It was reported that during the course of the reaction, the sense of enantio-differentiation changed with increasing yield of β-hydroxyester (Table 2). This phenomenon is accompanied by a rapid decrease in the rate of hydrogenation. The e.e. decreased from an initial constant value of 87% (*S*)- to 27% (*S*)-enantiomer at full conversion. However, the sense of

TABLE 2 Hydrogenation of ethyl 4,4,4-trifluoroaetoacetate over Pt/Al_2O_3 modified with cinchonidine (42)

Entry	Yield of β-hydroxyester (%)	Incremental enantiomeric excess Δee (%)	
		(*R*)-β-hydroxyester	(*S*)-β–hydroxyester
1	10		90
2	20		84
3	25		80
4	35		50
5	45	2	
6	55	12	
7	65	10	
8	75	8	
9	90	6	

enantio-differentiation was shown to invert at ~50% yield. Thus an incremental e.e. (Δee) value of 12% (*R*)-β-hydroxyester was reported (entry 9, Table 2). Inversion of the sense of enantioselectivity was suggested to arise from the hydrogenolysis of a hydrated form of the reactant at high conversion.

Recently Bartók et al. (45) have shown that the addition of an acid to the reaction mixture of 5% Pt/alumina, β-isocinchondine and ethyl pyruvate in toluene resulted in a change in the sense of enantiomeric excess of the lactate product (Table 3). The expected (*S*)-ethyl lactate enantiomer was generated in acetic acid; however, an excess of (*R*)-ethyl lacate was produced in the presence of toluene. This phenomenon was suggested to relate to a change of the intermediate responsible for chiral induction with different solvent composition.

Recently, von Arx & Hutchings (47) have shown that the supported platinum catalysts can be premodified with the cinchona alkaloid. These premodified

TABLE 3 Hydrogenation of ethyl pyruvate using Pt/Al_2O_3 modified with β-isocinchonidine

Entry	AcOH (%) in toluene	Enantiomeric excess (%)	
		(*R*)-ethyl lacate	(*S*)-ethyl lactate
1	0	50	
2	0.1	35	
3	1	5	
4	10		18
5	100		60

TABLE 4 Influence of reaction parameters on enantioselectivity for the gas phase enantioselective hydrogenation of ethyl pyruvate (48)[a]

Entry	H_2-concentration (%)	Amount of modifier[b]	Temperature/°C	AcOH additive[c]	(*R*)-methyl lactate e.e. (%)
1	2.5	0.5	40	—	25
2	5	0.5	40	—	31
3	10	0.5	40	—	37
4	25	0.5	40	—	42*[d]
5	50	0.5	40	—	43*
6	25	1	40	—	46 (54)[e]
7	25	2	40	—	47
8	25	1	25	—	29 (70)[e]
9	25	1	60	—	40* (35)[e]
10	25	2	40	5	48
11	25	2	40	20	51*

[a]Standard reaction conditions as specified below.

[b]Molar ratio of: modifier/Ms (metal surface atoms) at a metal dispersion of 0.25.

[c]Added during premodification (molar equiv. to CD).

[d]Values designated with * are initial values. ee decreases during time on line.

[e]Values in parentheses are the corresponding values from liquid phase batch reaction at 2 bar H_2-pressure using toluene as solvent.

catalysts can be dried and they still retain their activity. On the basis of this observation, the premodified catalysts were used for the gas phase heterogeneous asymmetric hydrogenation of ethylpyruvate (48). Enantioselective hydrogenation of methyl pyruvate was conducted at the gas-solid interface over premodified platinum catalysts (Table 4). The variation of experimental conditions was shown to be crucial to obtain successful reaction and high enantiomeric excess. An initial e.e. value of 51% (*R*)-methyl lactate (entry 11, Table 4) was obtained with the use of acetic acid as an additive during the premodification of the platinum catalyst with cinchonidine. Without the acid additive, the highest enantiomeric excess achieved was 47% (entry 7, Table 4).

TETHERING OF ACTIVE HOMOGENEOUS CATALYSTS TO A THREE-DIMENSIONAL STRUCTURE

At present this is one of the most popular approaches for the design of highly efficient heterogeneous asymmetric catalysts. The strategy employed tends to depend on the reaction being catalyzed, but overall this method has the potential for designing generic heterogeneous asymmetric catalysts applicable to many reaction types. The approaches are outlined in the following sections.

Carbon–Carbon Bond Formation

ALKYLATION OF ALDEHYDES Typically, the chiral auxiliary or complex is grafted onto the surface of a support material, most often the support of choice being a mesoporous silica. When the chiral auxiliary or the catalyst contains suitable functional groups, the chiral auxiliary is normally grafted directly onto the surface of the silica. If no such groups exist, then the ligand is modified to include $Si(OR)_3$ moieties, which can then be grafted by condensation onto the surface of the support.

One of the most widely studied systems using this approach is the immobilization of chiral auxiliaries for the addition of diethyl zinc to aldehydes to synthesize chiral secondary alcohols. For these reactions the chiral auxiliaries most often studied are ephedrine derivatives. This approach was first reported by Soai et al. (49) for the immobilization of a series of chiral *N*-alkylnorephedrines on silica gel, alumina, and silica gel coated with chloromethylated polystyrene. Reasonable yields and moderate e.e.s were obtained using these catalysts. A range of aldehydes were studied, although the study focused mainly on benzaldehyde, which gave the highest overall e.e. (59%). The alumina-supported catalysts tended to give higher e.e.s (40–59%) compared with the silica gel immobilized catalyst (25–39%), although little information concerning catalyst reuse was provided. A more in-depth study was carried out on the silica gel/chloromethylated polystyrene-immobilized catalyst. With this system, e.e.s up to 56% were reported with yields of 64%. Catalysts immobilized in this way could be recycled without the loss of enantioselectivity.

The surface grafting of ephedrine for the alkylation of benzaldehyde was studied in detail by Lasperas and coworkers (50), with particular emphasis on optimizing the support for the reaction. The mesoporous MCM-41 was used as the support rather than silica gel as this has well-defined pore dimensions, typically in the region of 40 Å, which help increase the rate of diffusion of the reactants to the active site. Initially, the same approach was used to graft the ephedrine chiral auxiliary to the support by modifying the support with (3-halopropyl)triethoxy silanes (Cl and I were studied) and then attaching the auxiliary to the support by displacement of the halide. Enhanced yields of the alkylation product were observed (>80%, calculated from conversion and selectivity) when compared with those in the studies of Soai et al. (49). The e.e.s were comparable for both studies (26–37%) but were inferior to the e.e.s obtained in the equivalent homogeneous reaction (67%). The unmodified support was also found to be active for the racemic reaction, although the rate was much lower than when the ephedrine was present. The residual activity of the support was concluded to be responsible for the lower e.e.s in the immobilized systems.

It was subsequently discovered that the ephedrine could directly interact with the surface either through the interaction of silanol groups with the N or O moiety or by the formation of Si-O-C moieties formed by condensation between silanols and the OH group present in ephedrine (50). This resulted in e.e.s similar to those obtained by the covalently bound ephedrine. Reuse experiments revealed the same decrease in enantioselectivity observed for the covalently bonded systems, suggesting that, even in the covalently bonded systems, the surface-bonded ephedrine constitutes an important part of the reaction system.

An alternative approach to the immobilization of ephedrine was reported by Chung & Rhee (51). Silica was treated with 3-aminopropyltriethoxysilane, which was used as an anchoring site for a dendrimer. After dendrimer formation, the ephedrine was immobilized onto the end of the dendrimer. This immobilized catalyst gave e.e.s comparable to those of other ephedrine-immobilized systems and could be reused three times without any loss of selectivity or e.e., which is a significant advantage over other methods mentioned above.

Proline (an alternative amino alcohol to ephedrine) has been immobilized on mesoporous silicas for the above reaction (52). A 3-chloropropyltriethoxysilane was first grafted onto the silica surface and the proline immobilized via an ether linkage. To study the effect of pore diameter on the reaction, amorphous silica (84 Å), MCM-41 (23–30 Å), and SBA-15 (85 Å) were used as supports. Two methods were used to carry out the reactions: method (*a*) where the catalyst was vacuum dried prior to reaction and method (*b*) where the catalyst was vacuum dried and pretreated with *n*-butyl lithium prior to reaction. The pretreatment with butyl lithium ensured all of the silanols on the catalyst surface were capped and all traces of water removed from the system. Generally, enhanced e.e.s were obtained using method (*b*). The e.e.s obtained for the MCM-41 and silica gel immobilized systems are broadly consistent with those obtained in the ephedrine system (20–50%). However, immobilization on SBA-15 resulted in slightly higher e.e.s (64 and 75%) but these are still lower than the e.e.s obtained for the homogeneous system (~90%).

FRIEDEL-CRAFTS HYDROXYALKYLATION An interesting example for this class of reaction is the alkylation of 1,3-dimethoxybenzene with 3,3,3-trifluoropyruvate using a silica- and MCM-41-immobilized copper-bis(oxazoline) (53). The results are significant because the e.e.s and conversions obtained with the immobilized catalyst were superior to those obtained in the homogeneous reaction. In this case, the bridge between the two oxazoline rings was modified to contain a terminal alkene moiety, which was used as the immobilization point. The homogeneous reaction gave an e.e. of 72% at 44% conversion, whereas the MCM-41-immobilized catalyst gave an e.e. of 82% at 77% conversion, and the silica-immobilized catalyst 92% e.e. at 72% conversion. Reuse of the MCM-41 immobilized material was possible, but a reduction in the rate of reaction was observed; however, the e.e. was maintained. Leaching studies were reported to reveal that the copper content of the solution was below detectable limits and, furthermore, that the reaction filtrate was inactive.

DIELS-ALDER REACTIONS Rechavi & Lemaire (54) reported the immobilization of bis(oxazoline) ligands on silica for Diels–Alder reactions. The Diels-Alder reaction between 3-acryloyl-2-oxazolidinone with cyclopentadiene was used as a test reaction. The silica-immobilized catalyst gave higher e.e.s (87% at −78°C and 65–70% e.e. at 0°C) than did the polyurethane-immobilized catalyst (51–56% e.e.

reaction resulted. These catalysts were also used for the hydrogenation of a variety of ketones, and excellent e.e.s (up to 98%) were reported.

ELECTROSTATIC INTERACTION AS A MEANS OF IMMOBILIZATION

Many porous solids, including zeolites, zeotypes, ordered mesoporous silicates, and layered materials, including clays, hydrotalcites, can act as ion exchangers. This presents a relatively facile mechanism for the immobilization of metal cations and complexes through electrostatic interaction. The approach has been used to produce stable heterogeneous enantioselective catalysts that can exhibit significantly improved catalytic performance compared with that of homogeneous counterparts. Mazzei et al. (63) were the first to use this approach for the immobilization of Rh complexes onto the interlayers of a range of clays. The interlayer distances of the clays were increased by treatment with water or alcohols to permit the ion exchange of the bulky Rh complexes. The immobilized catalyst gave $\leq$72% e.e. for the hydrogenation of (Z)-α-acetamidocinnamic acid. The catalyst could be reused, but the activity and selectivity decreased. Selke & Capka (64) subsequently immobilized cationic Rh chelates of 2,3-bis(O-diphenylphosphino)-β-D-glucopyranoside on silica using electrostatic interaction. The enantioselectivity for the hydrogenation of α-acetamidoacrylic acid ester with the immobilized catalyst was slightly higher (95% e.e.) than that for the homogeneous catalyst (91% e.e.). Catalysts were recycled up to 20 times, and while the enantioselectivity was retained, appreciable Rh leaching was noted. However, pretreatment of the silica support minimized leaching, and, in some cases leaching was eliminated.

We have used this immobilization strategy to design asymmetric heterogeneous catalysts for alcohol dehydration (65), aziridination (66), and epoxidation (67) of alkene, carbonyl- and imino-ene reactions (68). The design approach involves the ion exchange of suitable cations into the intracrystalline pores of zeolites (e.g., zeolite Y) or mesoporous solids (e.g., MCM-41) and the subsequent modification of the cation with a chiral ligand prior to reaction. Our initial studies used the acidic form of zeolite Y with chiral dithiane 1-oxides as ligands as a catalyst for the enantioselective dehydration of butan-2-ol. Although only one active center per zeolite supercage was modified, the modified sites were several orders of magnitude more active than the non-modified sites. As a result, the catalysts were effective for the enantioselective and catalyzed dehydration reaction. At low conversion, the chiral catalyst modified with the *R*-dithiane 1-oxide preferentially reacted with *S*-butan-2-ol in the presence of both enantiomers. Unfortunately, the enantioselective reaction was short-lived (69), but the enhancement in the rate of reaction with the modified catalyst was long-lived. Detailed spectroscopic studies of the active catalyst showed that the extra framework alumina, present in the pores of the zeolite, were an integral part of the active site. The rate enhancement was proposed to be from activation of the Brønsted OH group of the zeolite through interaction

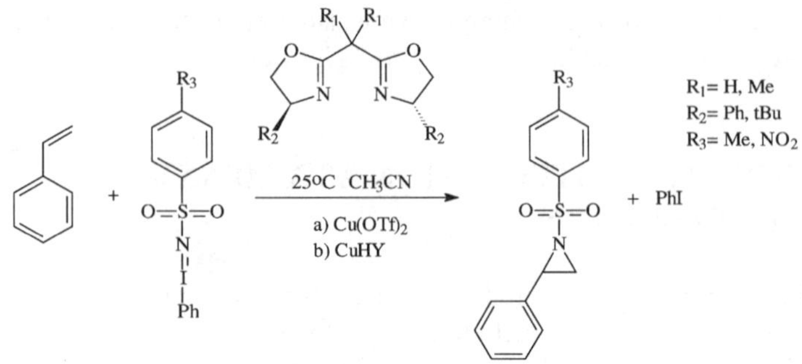

Figure 8 Reaction scheme for the enantioselective aziridination of styrene catalyzed by the copper/bis(oxazoline) catalysts (70).

with the oxygen of the dithiane 1-oxide and the extra framework alumina moiety (Figure 7, see color insert). These studies were considered as a proof-of-concept because it was shown that cations within zeolites could be modified to form chiral catalysts even though the enantioselectivity was short-lived.

Subsequently, we immobilized copper (II) bis(oxazoline) complexes with zeolite Y and Al-MCM-41 (66). Initially, the zeolite or mesoporous solid was partially ion exchanged to give a Cu^{2+}-ion-exchanged form with 3–5 wt% Cu. The aziridination of alkenes was investigated using [*N*-(*p*-tolylsulfonyl)imino]phenyliodinane (PhI = NTs) and [*N*-(*p*-nitrophenylsulfonyl)imino]phenyliodinane (PhI = NNs) as nitrene donors (Figure 8). Through careful control of the nitrene donor/substrate rates, e.e. in excess of 95% could be achieved with these catalysts with good yields. One of the key features with the immobilized catalysts is that they did not require excess substrate to drive the reaction to completion, a feature observed with the homogeneous catalyst. The catalysts were found to be very stable and although some Cu^{2+} was lost from the catalyst during use, the leached Cu^{2+} was inactived by the aziridination reaction (70), which is inhibited by the by-products, iodobenzene and sulfonamide. Variation in the bis(oxazoline) showed some interesting effects (Table 5). Typically, the heterogeneously catalyzed reaction gives much higher e.e.s than does the homogeneous catalyst reaction. However, less-hindered bis(oxazolines), which were particularly ineffective for the homogeneously catalyzed process, gave high e.e. with the heterogeneous catalyst. This, in effect, was considered to be from a containment effect within the pores of the zeolite. Epr spectroscopy (71) showed that the Cu^{2+} is chelated by the bis(oxazoline) in the zeolite pores as a square planar complex. Consequently, the zeolite cage structure places additional constraints on the approach of the substrate, an effect not possible in the homogeneously catalyzed process. As a result, the immobilized complex is capable of higher enantioselection than the non-immobilized complex. An interesting observation with the copper-bis(oxazoline) catalyzed aziridination reaction is that the enantioselection increases with the alkene conversion (Figure 9). The effect is likely from the interaction of the chiral aziridine product with the active copper-nitrene

TABLE 5 Effect of bis(oxazoline) on the aziridination of styrene (66)[a]

	PhI = NTs		PhI = NNs	
Bis(oxazoline)	**Yield (%)**	**e.e. (%)**	**Yield (%)**	**e.e. (%)**
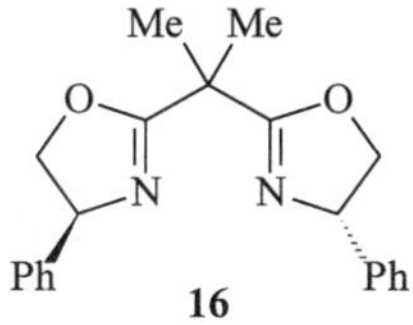 **16**	78 (91)	76 (73)	78 (96)	85 (81)
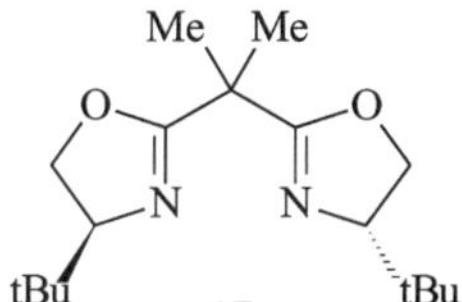 **17**	58 (78)	24 (35)	68 (79)	82 (43)
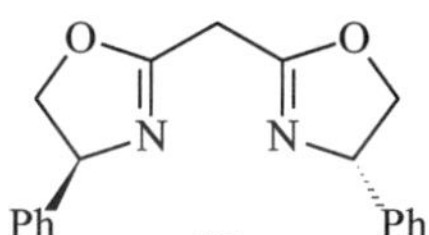 **18**	70 (74)	77 (28)	80 (85)	82 (54)
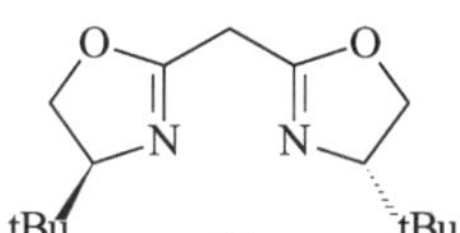 **19**	80 (85)	28 (8)	72 (42)	77 (31)
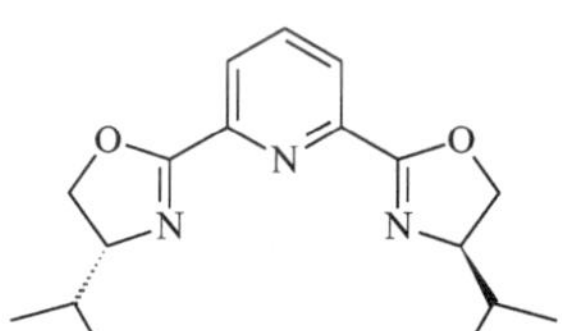 **20**	55 (70)	5 (11)	29 (42)	6 (15)

[a]Styrene:nitrene donor ratio = 1:1.5, 25°C, CuHY (0.3 g); figures in parenthesis are for the homogeneous catalyst $Cu(OTf)_2$ (0.015 mmol) under identical conditions. All reactions followed by dissolution of the nitrene donor, reaction times for all oxazolines similar for $Cu(OTf)_2$: 1 and 1.5 h for PhI = NTs and PhINNs, respectively; for CuHY: 3 and 5 h for PhI = NTs and PhINNs, respectively.

intermediate. It is, however, a feature that has been observed and discussed with the cinchona-modified platinum catalysts for the hydrogenation of alkyl privates (see above), which indicates that the active sites for heterogeneous asymmetric catalysts may be considerably more complex than originally envisaged.

Recently, we have extended the use of immobilized copper bis(oxazoline) catalysts to carbonyl-ene and imino-ene reactions (68). For these reactions, the catalyst

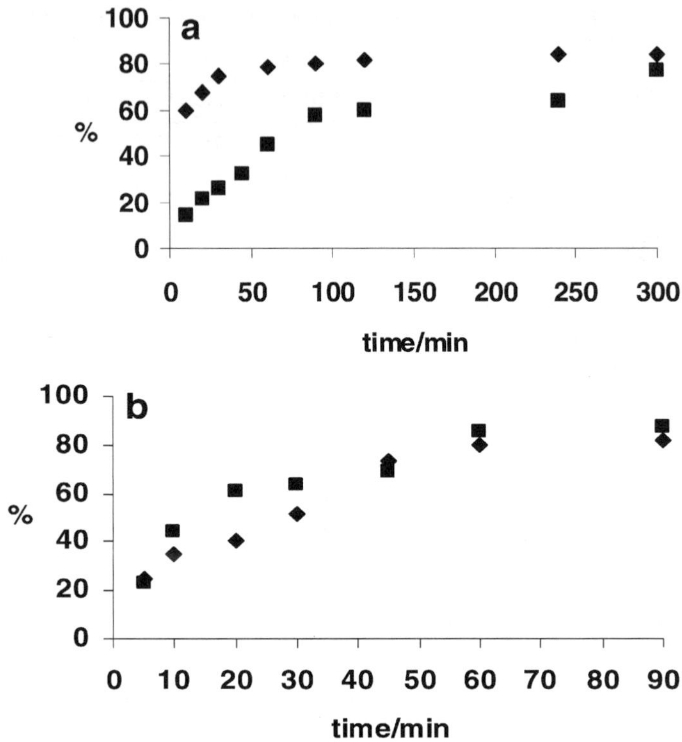

Figure 9 Effect of reaction time on yield and e.e. of (*R*)-aziridine [styrene (1 mmol) reacted with CuHY (0.3 g) or $Cu(OTf)_2$ (0.015 g), nitrene donor (1.5 mmol), with (*R*,*R*)-bis(oxazoline) 1 in CH_3CN at 25°C]. (*a*) CuHY with PhI = NNs; (*b*) $Cu(OTf)_2$ with PhI = NNs. Key: ■ aziridine yield, ♦ e.e.

reuse was demonstrated without loss of yield or enantioselection. In addition, the immobilized catalysts were found to be effective for substrates that were unreactive with the homogeneously catalyzed reaction. This approach was also used for the immobilization of manganese-salen complexes for the aziridination of alkenes (67) within the pores of Al-MCM-41. As observed with other attempts at immobilizing the salen ligand, this approach did not enhance the stability of the catalyst and, in general, lower e.e. and yields were observed with the immobilized catalysts.

A final feature of the well-studied Cu^{2+}-bis(oxazoline) immobilized within mesoporous materials concerns inversion in the enantioselectivity. This is a feature that has been initially considered in the cinchona-modified platinum catalyst system (see above). Recently, in the study of the Cu^{2+}-catalyzed Diels Alder reaction of *E*-ethyl-2-oxo-3-pentenoate with vinyl ether to form a dihydropyran, an interesting effect was observed with the heterogeneous catalyst that was absent with the homogeneously catalyzed reaction (72). The enantioselectivity was observed to

TABLE 6 Dihydropyran formation using Cu-AlSBA-15[a]

Time (h)	Conversion[b] %	Yield[c] %	e.e. %	Configuration
0.75	24.3	4.7	11.7	2*R*, 4*S*
3	28.5	8.1	9.4	2*R*, 4*S*
20.5	62.5	39.1	0.8	2*R*, 4*S*
23	63.2	37.8	1.7	2*S*, 4*R*
46	89.8	73.5	4.0	2*S*, 4*R*
120	90.4	75.7	4.1	2*S*, 4*R*

[a]2°C under standard conditions.

[b]Based on *E*-ethyl-2-oxo-3-pentenonate.

[c]Based on 2, isolated yield.

invert as the reaction proceeded (Table 6). Initially the 2*R*,4*S* product was seen, which subsequently switched to the 2*S*,4*R* product (the details are given in Table 5 for the Cu-AlSBA-15 catalyst), but the effect was also observed with the Cu-MCM-41 catalyzed reaction. It is likely that other catalyst systems will demonstrate similar inversions in enantioselectivity and will become an interesting research area.

COMMENTS ON ACHIEVING HIGH ENANTIOSELECTION

Of the many immobilization strategies developed to date for the preparation of asymmetric heterogeneous catalysts the three highlighted in this review have been shown to give high enantioselection, which is often achieved with high yields. The modification of metal surfaces using chiral modifiers does not provide systems that can be readily reused but ones that can be premodified and stored. This observation has led to the exciting observation that chiral reactions can be achieved at the gas-solid interface. The other two highlighted approaches, immobilization by covalent tethering and electrostatic interaction, have been shown to form reasonably stable catalysts that are capable of reuse. Covalent tethering will undoubtedly produce stable catalysts and can be considered the preferred methodology for catalysts that do not lend themselves to immobilization through electrostatic interaction. Immobilization via ionic interaction is a conceptually simple and facile method of immobilizing ionic catalysts. The approach can produce stable catalysts that are capable of reuse, but it was noted recently by Rechavi & Lemaire that methodology of catalyst synthesis is crucial if stable catalysts are to be obtained (73).

Many early studies concerned with immobilization of asymmetric homogeneous catalysts showed that, for the most part, lower yields and enantioselection were achieved with the immobilized catalysts. Hence, the immediate advantage of using a heterogeneous catalyst, namely, the facile recovery and reuse of the expensive catalyst/ligand system, could not be realized. However, recent studies have shown this need not be the case; immobilized catalysts can give higher e.e.s and yields than can non-immobilized catalysts. Indeed, some immobilized systems have been shown to be effective with immobilized systems for substrates that are non-selective with their homogeneous counterparts. Two reasons for this change of fortune for the immobilized catalysts are apparent. First, site isolation is achievable through the appropriate design of immobilized catalysts. This prevents the polymerization of some chiral ligands, which limits their effectiveness when used as homogeneous catalysts. Second, and more importantly, several groups have observed the containment effect, i.e., that the immobilized asymmetric catalyst is constrained through interaction with the supporting matrix (74). The additional constraints placed upon the chiral catalyst can induce increased enantioselection compared with that when using the homogeneous catalyst, and several examples have been given in this review. The additional constraint of the immobilized catalyst means that simpler chiral ligands may be more effective and, hence, the use of ligands specifically designed for use in homogeneously catalyzed processes may not necessarily be the best for use with the immobilized catalysts. Consequently, we can now embark on designing new ligand systems that can be effective asymmetric heterogeneous catalysts within the confined environment of the immobilized matrix, ones that can take full account of this highly effective containment effect. It is also likely that the containment effect observed with both the tethering and electrostatic interactions can be used effectively to design a new generation of immobilized asymmetric catalysts capable of both high yields and enantioselection. The effect provides a clear methodology by which high enantioselection can be obtained.

There are two additional features that require comment because both may suggest that there is a limit to the extent to which 100% enantioselection can be achieved. First, recent studies for both the chirally modified metal surfaces and the copper-bis(oxazolines) complexes immobilized within zeolite Y have shown that the sense of the enantiodirection can be changed by variation in reaction time or reaction conditions. This indicates that the heterogeneous catalysts contain a range of sites, some of which direct the chirality, for example, to the *R*-enantiomer, and others of which direct the reaction toward the alternative enantiomer. Secondly, a number of immobilized catalysts demonstrate the effect that the enantioselection is enhanced as the reaction proceeds. This indicates that the initial immobilized catalyst is probably modified during use. Although these catalysts give very high enantioselection, the initial reaction turnovers are not particularly enantioselective. This, again, will limit the overall e.e. that can be achieved. Clearly attention needs to be given to unraveling the nature of these effects using detailed spectroscopy. If these enantio-inversion and enanti-enhancement effects could be observed at the gas-solid interface, then the effect could be interrogated using a full range of surface-sensitive spectroscopies, which would aid in the design of improved catalysts.

The *Annual Review of Materials Research* is online at http://matsci.annualreviews.org

LITERATURE CITED

1. Nobel Lectures. 2002. *Angew. Chem. Int. Ed.* 41:1998
2. Blaser H-U. 2003. Enantioselective catalysis in fine chemicals production. *Chem. Commun.* 3:293–96
3. Sheldon RA, Wallau M, Arends IWCE, Schuchart U. 1998. Heterogeneous catalysts for liquid-phase oxidations: Philosophers' stones or Trojan horses? *Acc. Chem. Res.* 31:485–93
4. McMorn P, Hutchings GJ. 2003. Heterogeneous enantioselective catalysts strategies for the immobilization of homogeneous catalysts. *Chem. Soc. Rev.* 33:108–22
5. Schwab GM, Rudolph L. 1932. Catalytic cleavage of racemates by d- and l-quartz. *Naturwisserschaft* 20:363–64
6. Schwab GM, Rost F, Rudolph L. 1934. Optically asymmetric catalysis on quartz crystals. *Kolloid Z.* 68:157–59
7. Davis M. 1998. Zeolite-based catalysts for chemicals synthesis. *Microporous Mesoporous Mater.* 21:173–82
8. Blaser H-U, Pugin B. 1995. Scope and limitations of the application of heterogeneous enantioselective catalysts. In *Chiral Reactions in Heterogeneous Catalysis*, ed. G James, V Debuois, pp. 33–57. New York: Plenum
9. Sherrington DC. 2000. Polymer-supported metal complex alkene epoxidation catalysts. *Catal. Today* 57:87–104
10. Song CE, Lee S-G. 2002. Supported chiral catalysts on inorganic materials. *Chem. Rev.* 10:3495–524
11. Hutchings GJ, Bethel D, McGuire N, Page PCB, Robinson D, et al. 1999. Heterogeneous catalysts by design. *Curr. Top. Catal.* 2:39–58
12. Blaser H-U, Pugin B. 1998. Supported reagents and catalysts in chemistry, scope and limitations of heterogeneous enantioselective catalysts. Special Publication. *R. Chem. Soc.* 216:101–9
13. Baiker A. 2000. Transition state analogues—a guide for the rational design of enantioselective heterogeneous hydrogenation catalysts. *J. Mol. Catal. A* 163: 205–20
14. Blaser H-U, Jalett JP, Müller M, Studer M. 1997. Enantioselective hydrogenation of α-ketoesters using cinchona modified platinum catalysts and related systems: a review. *Catal. Today* 37:441–63
15. Wells PB, Wilkinson AG. 1998. Platinum group metals as heterogeneous enantioselective catalysts. *Top. Catal.* 5:39–50
16. von Arx M, Mallat T, Baiker A. 2002. Asymmetric hydrogenation of activated ketones on platinum: relevant and spectator species. *Top. Catal.* 19:75–87
17. Kukula P, Cerveny L. 2002. Effects of reaction variables on enantioselectivity of modified Raney nickel catalyst. *J. Mol. Catal. A* 185:195–202
18. Sugimura T, Nakayama S, Tai A. 2002. Over 98% optical yield achieved by a heterogeneous catalysis. Substrate design and analysis of enantio-differentiating factors of tartaric acid-modified Raney nickel hydrogenation. *Bull. Chem. Soc. Jpn.* 75:355–63
19. Osawa T, Sakai S, Harada T, Takayashu O. 2001. Highly durable enantio-differentiating nickel catalyst for repeated use for the hydrogenation of methyl acetoacetate. *Chem. Lett.* 392–99
20. Schürch M, Künzle N, Mallat T, Baiker A. 1998. Design of new modifiers for the enantioselective hydrogenation of ethyl pyruvate. *J. Catal.* 176:569–71
21. Hutchings GJ. 1999. New approaches to rate enhancement in heterogeneous catalysis. *Chem. Commun.* 301–6

22. Zuo X, Liu H, Liu M. 1998. Asymmetric hydrogenation of α-ketoesters over finely dispersed polymer-stabilized platinum clusters. *Tetrahedron Lett.* 39:1941–44
23. Baiker A. 1997. Progress in asymmetric heterogeneous catalysis: design of novel chirally modified platinum metal catalysts. *J. Mol. Catal.* A 115:473–93
24. Blaser H-U, Jallet HP, Monti DM, Baiker A, Wehrli JT. 1991. Enantioselective catalysis by chiral solids: approaches and results. *Stud. Surf. Sci. Catal.* 67:147–55
25. Webb G, Wells PB. 1992. Aysmmetric hydrogenation. *Catal. Today* 12:319–37
26. Baiker A. 1996. Towards molecular design. *Stud. Surf. Sci. Catal.* 101:51–61
27. Blaser H-U, Jallet HP, Muller M, Studer M. 1997. Enantioselective hydrogenation of α-ketoesters using cinchona modified platinum catalysts and related systems: a review. *Catal. Today* 37:441–63
28. Burgi T, Baiker A. 1998. Conformational behavior of cinchonidine in different solvents: a combined NMR and ab initio investigation. *J. Am. Chem. Soc.* 120:12920–26
29. Vermeer WAH, Fulford A, Johnston P, Wells PB. 1993. New enantioselective reactions catalyzed by cinchonidine-modified platinum. *J. Chem. Soc. Chem. Commun.* 1053–54
30. Stürch M, Schwalm O, Mallat T, Weber J, Baiker A. 1997. Enantioselective hydrogenation of ketopantolactone. *J. Catal.* 169:275–86
31. Studer M, Bukhardt S, Blaser H-U. 1999. Enantioselective hydrogenation of α-keto acetals with cinchona modified Pt catalyst. *Chem. Commun.* 1727–28
32. Török B, Felföldi K, Balàzsik K, Bartók M. 1999. New synthesis of a useful C3 chiral building block by a heterogeneous method: enantioselective hydrogenation of pyruvaldehyde dimethyl acetal over cinchona modified Pt/Al_2O_3 catalysts. *Chem. Commun.* 1725–26
33. Sutherland IM, Ibbotson A, Moyes RB, Wells PB. 1990. Enantioselective hydrogenation. I. Surface conditions during methyl pyruvate hydrogenation catalyzed by cinchonidine-modified platinum/silica (EUROPT-1). *J. Catal.* 125:77–88
34. Meheux PA, Ibbotson A, Wells PB. 1991. Enantioselective hydrogenation. II. Variation of activity and optical yield with experimental variables in methyl pyruvate hydrogenation catalyzed by cinchona-modified platinum/silica (EUROPT-1). *J. Catal.* 128:387–96
35. Li X, Dummer N, Jenkins R, Wells RPK, Wells PB, et al. 2004. Enantioselective hydrogenation using cinchona-modified Pt/α-Al_2O_3 catalysts: comparison of the reaction of ethyl pyruvate and butane-2,3-dione. *Catal. Lett.* 96:147–51
36. Singh UK, Landau RN, Sun Y, LeBlond C, Blackmond DG, et al. 1995. Enantioselective catalysis: influence of conversion and bulk diffusion limitations on selectivity in the hydrogenation of ethyl pyruvate. *J. Catal.* 154:91–97
37. Wang J, Sun Y, LeBlond C, Landau RN, Blackmond DG. 1996. Nature of initial transient period during enantioselective hydrogenation on Pt and Pd. *J. Catal.* 161:752–58
38. Sun Y, Wang G, LeBlond C, Landau RN, Blackmond DG. 1996. Asymmetric hydrogenation of ethyl pyruvate: diffusion effects on enantioselectivity. *J. Catal.* 161:759–65
39. Mallat T, Bodnar Z, Minder B, Borszeky K, Baiker A. 1997. Nature of initial transient period during enantioselective hydrogenation on Pt and Pd. *J. Catal.* 168:183–93
40. Blackmond DG. 1998. Comments on "Nature of initial transient period during enantioselective hydrogenation on Pt and Pd." *J. Catal.* 176:267
41. Mallat T, Baiker A. 1998. Reply to comment on "Nature of initial transient period during enantioselective hydrogenation on Pt and Pd." *J. Catal.* 176:271–74
42. von Arx M, Mallat T, Baiker A. 2001. Heterogeneous enantioselective hydrogenation of trifluoromethyl ketones. *Angew. Chem. Int. Ed.* 40:247–54

43. Hess R, Vargas A, Mallat T, Bürgi T, Baiker A. 2004. Inversion of enantioselectivity in the platinum-catalyzed hydrogenation of substituted acetophenones. *J. Catal.* 222:117–28
44. Diezi S, Szabo A, Mallat T, Baiker A. 2003. Inversion of enantioselectivity in the hydrogenation of ketopantolactone on platinum modified by ether derivatives of cinchonidine. *Tetrahedron Asymm.* 14:2573–77
45. Bartók M, Sutyinski M, Felföldi K, Szöllősi Gy. 2002. Unexpected change of the sense of the enantioselective hydrogenation of ethyl pyruvate catalyzed by a Pt-alumina-cinchona alkaloid system. *Chem. Commun.* 1130–31
46. Balázsik K, Bartók M. 2004. Heterogeneous asymmetric reactions. Part 40. New data on repeated use of Pt-alumina catalyst during enantioselective hydrogenation of ethyl pyruvate. *J. Mol. Catal. A* 219:383–89
47. von Arx M, Hutchings GJ. 2003. Molecularly modified noble metal catalysts for gas phase hydrogenation reactions. *New J. Chem.* 27:1367–70
48. von Arx M, Dummer N, Willock DJ, Taylor SH, Wells RPK, et al. 2003. Observation of high enantioselectivity for the gas phase hydrogenation of methyl pyruvate using supported Pt catalysts pre-modified with cinchonidine. *Chem. Commun.* 1926–27
49. Soai K, Watanabe M, Yamamoto A. 1990. Enantioselective addition of dialkylzincs to aldehydes using heterogeneous chiral catalysts immobilized on alumina and silica gel. *J. Org. Chem.* 55:4832–35
50. Abramson S, Bellocq N, Lasperas M. 2000. Enantioselective C–C bond formation with heterogenized catalysts. *Top. Catal.* 13:261–82
51. Chung Y-M, Rhee H-K. 2002. Dendritic chiral auxiliaries on silica: a new heterogeneous catalyst for enantioselective addition of diethylzinc to benzaldehyde. *Chem. Commun.* 238–39
52. Bae JS, Kim S-W, Hyeon T, Kim BM. 2000. New chiral heterogeneous catalysts based on mesoporous silica: asymmetric diethylzinc addition to benzaldehyde. *Chem. Commun.* 31–32
53. Corma A, Garcia H, Moussaif A, Sabatier MJ, Zniber R, Redouane A. 2002. Chiral copper(II) bisoxazoline covalently anchored to silica and mesoporous MCM-41 as a heterogeneous catalyst for the enantioselective Friedel-Crafts hydroxyalkylation. *Chem. Commun.* 1058–59
54. Rechavi D, Lemaire M. 2002. Enantioselective catalysis of Diels-Alder reactions by heterogeneous chiral bis(oxazoline) catalysts. *J. Mol. Catal. A* 182–83:239–47
55. Xiang S, Zhang Y, Xin Q, Li C. 2002. Asymmetric epoxidation of allyl alcohol on organic-inorganic hybrid chiral catalysts grafted onto the surface of silica and in the mesopores of MCM-41. *Chem. Commun.* 2696–97
56. Pini D, Mandoli A, Orlandi S, Salvadori P. 1999. First example of a silica gel-supported optically active Mn(III)-salen complex as a heterogeneous asymmetric catalyst in the epoxidation of olefins. *Tetrahedron: Asymmetry* 10:3883–86
57. Choudary BM, Chowdari NS, Kantam ML, Santhi PL. 2001. Silica gel anchored chiral Mn(III)salen complexes for enantioselective epoxidation of unfunctionalised olefins. *Catal. Lett.* 76:213–18
58. Baleizao C, Gigante B, Sabatier MJ, Garcia H, Corma A. 2002. On the activity of chiral chromium salen complexes covalently bound to solid silicates for the enantioselective epoxide ring opening. *Appl. Catal. A* 288:279–88
59. Burguete MI, Fraile JM, Garcia JI, Garcia-Verdugo E, Herrerias CI, et al. 2001. Bis(oxazoline)copper complexes covalently bonded to insoluble support as catalysts in cyclopropanation reactions. *J. Org. Chem.* 66:8893–901
60. Clarke RJ, Shannon IJ. 2001. Mesopore immobilized copper bis(oxazoline) complexes for enantioselective catalysis. *Chem. Commun.* 1936–37

61. Raynor SA, Thomas JM, Raja R, Johnson BFG, Bell RG, Mantle MD. 2000. A one-step, enantioselective reduction of ethyl nicotinate to ethyl nipecotinate using a constrained, chiral, heterogeneous catalyst. *Chem. Commun.* 1925–26
62. Adima A, Moreau JJE, Man MWC. 2000. Immobilization of rhodium complexes in chiral organic-inorganic hybrid materials. *Chirality* 12:411–20
63. Mazzei M, Marconi W, Riocci M. 1980. Asymmetric hydrogenation of substituted acrylic acids by Rh′-aminophosphine chiral complex supported on mineral clays. *J. Mol. Catal.* 9:381–87
64. Selke RM, Capka M. 1990. Methods for preparing immobilized chiral rhodium(I) complexes. *J. Mol. Catal.* 63:319–34
65. Feast S, Bethell D, Page PCB, King F, Rochester CH, et al. 1995. Gas-phase catalytic asymmetric reaction using chirally modified microporous catalysts. *J. Chem. Soc. Chem. Commun.* 2409–11
66. Taylor S, Gullick J, McMorn P, Bethell D, Bullman Page PC, et al. 2001. Catalytic heterogeneous aziridination of styrene using CuHY catalyst: an assessment of catalyst stability. *J. Chem. Soc. Perkin 2* 1714–23
67. Piaggio P, McMorn P, Murphy D, Bethell D, Bullman Page PC, et al. 2000. Enantioselective epoxidation of (Z)-stilbene using a chiral Mn(III)-salen complex: effect of immobilization on MCM-41 on product selectivity. *J. Chem. Soc. Perkin 2* 2008–15
68. Hancock FE, Hutchings GJ, Caplan NA. 2003. *Patent WO 03/018191*
69. Feast S, Siddiqui MRH, Wells RPK, Willock DJ, King F, et al. 1997. Enantioselective dehydration of butan-2-ol using zeolite Y modified with dithiane oxides. *J. Catal.* 167:533–42
70. Taylor S, Gullick J, McMorn P, Bethell D, Bullman Page PC, et al. 2001. Catalytic asymmetric heterogeneous aziridination of styrene using CuHY: effect of nitrene donor on enantioselectivity. *J. Chem. Soc. Perkin 2* 1724–28
71. Traa Y, Murphy DM, Farley RD, Hutchings GJ. 2001. An EPR study on the enantioselective aziridination properties of a CuNaY zeolite. *Phys. Chem. Chem. Phys.* 3:1073–80
72. Wan Y, McMorn P, Hancock FE, Hutchings GJ. 2003. Heterogeneous enantioselective synthesis of a dihydropyran using Cu-exchanged microporous and mesoporous aterials modified by bis(oxazoline). *Catal. Lett.* 91:145–48
73. Rechavi D, Lemaire M. 2002. Enantioselective catalysis using heterogeneous bis(oxazoline) ligands: which factors influence the enantioselectivity? *Chem. Rev.* 102:3467–93
74. Thomas JM, Maschmeyer T, Johnson BFG, Shephard DS. 1999. Constrained chiral catalysts. *J. Mol. Catal. A* 141:139–44

Annu. Rev. Mater. Res. 2005. 35:167–207
doi: 10.1146/annurev.matsci.35.101503.091652

THE OXIDATION OF NiAl: What Can We Learn from Ab Initio Calculations?

M.W. Finnis,[1] A.Y. Lozovoi,[1] and A. Alavi[2]
[1]*School of Mathematics and Physics, Queen's University Belfast, Belfast BT7 1NN, Northern Ireland, United Kingdom; email: m.finnis@qub.ac.uk, s.lozovoi@qub.ac.uk*
[2]*Chemistry Department, University of Cambridge, Lensfield Road, Cambridge CB2 1EW, United Kingdom; email: asa10@cam.ac.uk*

Key Words ordered compounds, selective oxidation, point defects, environmental parameters

■ **Abstract** We review here the theory of the early stages of oxidation of the (110) surface of $Ni_{1-x}Al_x$, based on ab initio calculations using a plane-wave pseudopotential method. The clean surface and several oxidized surfaces have been investigated, with oxygen coverages up to 2 ML of oxygen (1 ML = 3 O atoms per 2 surface Al atoms). The theory to date is a description in terms of equilibrium thermodynamics, with a comparison of the free energies of several surfaces of different composition, implemented at the atomic scale. Three environmental parameters are singled out as control variables in this treatment, namely the alloy composition x (assumed to be near 0.5), the temperature T and the partial pressure of oxygen p_{O_2}. With certain reasonable approximations an analytic formula for the surface energy σ is derived in terms of these variables and some constants that are calculated ab initio together with others that are derived from experimental thermodynamic tables. At oxygen pressures just above the threshold for bulk oxidation of NiAl, the calculations explain the observed formation of a thin film of alumina in place of NiAl surface layers, with the consequent dissolution of Ni into the bulk. Ab initio calculations illustrate how the energetics of supplying Al to the surface depends on bulk stoichiometry, which alters the relative stability of different surface oxidation states so as to favour oxidation more if the alloy is Al-rich than if it is Ni-rich.

INTRODUCTION

We now recognize that the oxidation of materials, once considered a problem similar to other kinds of corrosion, can sometimes be a process of enormous value. Silicon dioxide films for example are still at the heart of the electronics industry and will be for many years to come because they are the simplest useful gate oxides. One might well ask where that industry would be if silicon oxidized in the uncontrollable manner of iron. At the another extreme, aluminum oxide films occur spontaneously, not only on aluminum itself, but on many useful aluminum-containing alloys, including NiAl. These films are protective because they are relatively stable and continuous, and hence they are desirable, for example on turbine blades in power generation, or other components subject to high temperature corrosive environments (1). Their morphology and rate of growth are still rather poorly understood from a fundamental point of view. Such films can also play a quite different role, by acting as substrates for deposited films of other, catalytic materials (2). In this review we focus on NiAl as a well studied example.

Our purpose here is to illustrate how ab initio calculations can be used effectively to predict the relative stability of various oxidized surface structures as a function of externally imposed thermodynamic parameters (referred to here as environmental parameters), such as temperature, oxygen partial pressure, and chemical composition of the substrate. This latter aspect, in particular, adds considerable richness to the problem. Unlike elemental systems, our chosen substrate, being a two-component compound, has the freedom to vary the chemical composition of the surface layer (and more generally the surface region) as the oxidation proceeds: The compound will do so if it can lower its free energy (i.e., its surface excess free energy). This can lead to enrichment of the surface region with one of the components, thus exhibiting selective oxidation. This enrichment, in turn, requires the mass transport of material to and from the bulk. Therefore, the oxidation process at the surface becomes intimately related to bulk thermodynamics and kinetics in a manner whose description goes well beyond the standard treatment of oxidation processes of elemental systems. In particular, we show that the bulk point-defect behavior of NiAl is an essential physical ingredient. Furthermore, we show that the extension of the ab initio thermodynamic formalism, combined with the theory of point defects in NiAl, is a framework in which this selective oxidation process can be studied.

Considering oxidation at an atomistic level, one would hope to address at least three questions: (*a*) How would the oxidation proceed? (*b*) What is the rate of oxidation? (*c*) What is the final structure at which the system arrives? It is generally assumed that there is an initial, transient phase during which the first layer or layers of oxide are created, merging into a steady-state phase during which bulk diffusion events and interface reactions can be distinguished, one of which may be rate-determining. In steady state, the presence of grain boundaries in the oxide film and interface roughness, as well as the development and relief of stresses in the oxide film are complications that would make a fully atomistic simulation out of reach of current techniques, and thus continuum mechanics and diffusion theory would be the tools of choice. By studying the early stages of oxidation, however, or investigating particular events at any stage, ab initio calculations can now contribute to our understanding.

To answer the first question, an extreme and naive approach would be to try to monitor the evolution of the System, for example, by running a molecular dynamics (MD) simulation. Some accelerated way to do this (3) would be necessary, otherwise it would take unacceptably long for a single vacancy simply to jump into a neighboring site. Nevertheless, such an approach would be a formidable challenge to present computational resources, unless one used such a drastically simplified model of interatomic interactions that the computed outcomes would be of questionable value. This is not to say that the MD technique is of no value for any oxidation studies at present; on the contrary, in specific cases it might lead to results that would be hard to achieve in any other way. An impressive example is the study of the oxidation of Al(111) (4), which demonstrated that the energy released during the adsorption and dissociation of an O_2 molecule is sufficient to initiate the local precursor of an oxide phase, at an oxygen coverage as low as 0.5 ML.

An alternative approach, which is the method of choice in most oxidation studies, is to identify a number of consecutive atomic structures through which the system proceeds during the oxidation process. This approach relies on our intuition and even luck in selecting structures worth considering, so it is very much modeling rather than simulation, but it offers a much better ratio of insight to computation than does brute force MD. Given that the number of possible structures grows exponentially with the number of atoms in the model, the final result can be no more than a best-educated guess, no matter how accurate the calculations. The studies of oxidation based on first-principles calculations, which have started to appear in recent years, have therefore concentrated on low-index surfaces of single-component crystals. The systems considered are either late transition and noble metals (5–15), where the formation enthalpy of the oxides is relatively low, or metals that form highly stable oxides such as Mg (16, 17) and Al (18, 19). In the former, the oxidation typically proceeds through a number of ordered structures of the adsorbed oxygen (14, 20) and eventually begins introducing oxygen subsurface (15). Soon after that the surface oxide starts to resemble a stack of ordered oxygen-metal layers, which continues to thicken as more oxygen atoms arrive, subject to kinetic restrictions. On Mg(0001), on the other hand, oxygen atoms go straight under the surface layer (16). Oxygens tend to group together forming oxide islands (17), which grow laterally until the whole surface is covered. Oxidation of Al(111) is a little more complicated. The introduction of subsurface oxygens becomes energetically favorable only after 1 ML of oxygen atoms has settled on the surface. However, the same tendency of oxygen atoms to cluster together promotes oxide nucleation long before an average oxygen coverage of 1 ML is reached (18, 19). The oxide islands then continue to grow laterally, as on Mg(0001). In order to identify the system's oxidation path and the rate of oxidation (the second question), one needs not only to find out the sequence of surface structures through which the oxidizing system might pass, but also to calculate the energetic barriers separating these structures from each other. The search for saddle points on the energy surface is a time-consuming task, and to our knowledge has not been accomplished for the oxidation process for any metals. Fortunately, the elucidation of the exact sequence of atomic jumps occurring at and near the particular oxidizing surface is not really the only answer in which one is interested. It is also useful to understand the energy differences that drive the system through a particular path.

Recent groundbreaking work on calculating energy barriers for the oxidation of Si (21) illustrates the current state of the art in ab initio simulation. This study also shows the difficulties that still must be overcome even for single-component substrates. The authors created an ensemble of amorphous oxide structures on Si using classical interatomic potentials, and they relaxed these structures with an ab initio (density functional) code. They then studied in detail how a single oxygen molecule arriving at the interface would dissociate, breaking open the Si lattice and bonding to the Si. This was achieved by dragging the molecule to the interface, relaxing all the atomic positions to their minimum energy configuration at each stage on the journey. Note that a fully dynamic simulation of the inward diffusion of the molecule would have been too time consuming. Despite its size limitations, this model was clearly able to provide new insights, notably that oxygen dissociation and incorporation presents no significant energy barrier and is spontaneous when the molecule has arrived at the interface. Such a procedure for a multicomponent substrate, even for NiAl, would be considerably more difficult, because oxygen not only must diffuse to the interface, aluminum must be there to meet it, so a sequence of processes, or cooperative processes, must be identified. However, that is not to say it would be impossible, and this process represents another general strategy for future consideration.

The answer to the third question is again a matter of kinetics. Unless the temperature is very high or the partial pressure of oxygen is very low, thermodynamic equilibrium is not attained until the entire metal has become an oxide or a combination of oxides. This can be easily appreciated by comparing the heat of formation of NiAl, $H_f(\mathrm{NiAl}) = -61.8$ kJ/mol (22) with those of the respective oxides, $H_f(\mathrm{NiO}) = -489.5$ kJ/mol, and $H_f(\mathrm{Al_2O_3}) = -1675.7$ kJ/mol (23). For some noble and late transition metals, a thin oxide film may appear to be of lower energy than the equivalent amount of bulk oxide (6, 10, 13); however, the metal is still unstable against further oxidation. In practical terms this means that for any model of an oxidized surface constructed in the laboratory, another structure with lower free energy always exists, which could be created by adding more oxygen atoms. However, the timescale for further oxidation may be much longer than the timescale for attaining and equilibrating the oxide layer at its current thickness. Hence, any particular structure is at best metastable, but this is the normal situation in thermodynamics; one can make satisfactory predictions by excluding any states that are inaccessible on the timescales of interest. With this in mind, the general strategy we and others (10–14) have followed for predicting equilibrium configurations and free energies can be stated as follows: One calculates the interfacial excess Gibbs free energy for each structure of a finite set, which must be locally stable; that is, all the atomic coordinates are relaxed to equilibrium, although globally the system is only metastable. This set contains a manageable number of trial stoichiometries and atomic configurations, of order perhaps 10–100. It cannot be exhaustive but it should be representative. The best prediction of the structure will be the one in this set having the lowest excess Gibbs free energy. A key role in this analysis is played by the chemical potentials of the components, upon which the excess Gibbs free energies depend. Chemical potential is inherently a nonlocal property; that is, the chemical potential of a species of atom does not depend on any atom's immediate environment but on the system as a whole. The chemical potentials are computed separately from bulk calculations, and therefore they enter the expression for the excess Gibbs energy of a surface film a posteriori, after the detailed and time-consuming atomistic simulations of the surface have been made. This greatly extends the analysis because approximate calculations of excess Gibbs free energy versus temperature and versus chemical potential of the components can be made easily to reveal the limits of stability of a particular structure.

In the case of NiAl(110), the initial oxidation results in a continuous thin aluminum oxide film. According to the suggested structural model of the film (24, 25), it consists of two O–Al bilayers with every 3 oxygen atoms in each bilayer roughly corresponding to 2 aluminum atoms in a layer of the substrate. Hence, to model the initial oxidation of a NiAl(110) surface it suffices to consider the oxygen coverage up to 2 ML (1 ML = 3 O atoms per every 2 surface Al atoms). The following sections describe how the excess Gibbs free energy of the surface in its early stages of oxidation has been calculated as a function of the temperature, the stoichiometry of the metal, and the p_{O_2}. We hope to make clear how the same approach can be applied to other systems and how it can be improved and extended to make use of faster computers.

THEORY

In the following sections we outline the theory of the excess Gibbs free energy (for short: the surface energy) σ of a planar surface of arbitrary thickness at the atomistic level and explain how σ can be calculated by making use of ab initio total energy results for internal energies at 0 K. Surface energies of oxidized surfaces are very sensitive to temperature, and to the chemical potentials of the components including oxygen itself, which is why the ab initio internal energies are just one part of the story. Vibrational contributions are usually neglected in surface energy calculations, as they have been in studies of oxidation to date. On the other hand more important variations with temperature occur owing to the temperature dependence of the chemical potentials of the components, and these can be included analytically. The chemical potential of each component can be expressed in terms of its partial pressure (most appropriate for oxygen, or other naturally gaseous species) or solid concentration (most appropriate for alloy components or impurities). The latter step requires proper acknowledgment of the fact that the zero of chemical potential in atomistic models of the energy is different to the zero employed in chemistry.

The advantages of the approach just summarized have been widely appreciated. It has been referred to as ab initio thermodynamics (26–28) or first-principles atomistic thermodynamics (29), and it is currently used by a number of independent research groups. For example, the oxidation of the Ag, Rh, Pt, and Pd surfaces has recently been investigated using the above technique (10–14). It has now been applied for studying the equilibrium structure of several oxide surfaces: $Cr_2O_3(0001)$

(30), $Fe_2O_3(0001)$ (28), $V_2O_5(0001)$ (31), ZnO(0001) (32), $RuO_2(110)$ (33), and various surfaces of PdO (29) and α–Al_2O_3 (34–38).

Below we outline the approach in terms of surface excesses of the components (39), which makes the effect of non-stoichiometry at the surface explicit. Throughout we refer to NiAl as an illustration. For more detailed descriptions in other contexts the reader is referred to these references (5–19, 26–38, 40, 41). The formulae are readily applied to other systems.

Thermodynamics at the Atomic Scale

Our formulation of the surface energy is based on that of Cahn's (39) for homogeneous substances, which was generalized previously to multi-component crystals (42). The system we consider is always an infinite slab some tens of atomic planes in thickness, which is the kind of atomistic system that is simulated in most calculations. Three-dimensional periodic boundary conditions are applied to a piece of the slab, with a vacuum separating it typically by one nanometer from its periodically repeated parallel images. The basic repeat unit is called a supercell. How can we justify modeling the vapor phase as a vacuum? The point here is that the vacuum within the supercell does actually represent a vapor phase; however, the density of atoms in it is low. By regarding the supercell as a member of a grand canonical ensemble of systems at constant temperature and pressure, the chance of finding a gas molecule in such a small volume is indeed low, and does not significantly effect the internal thermodynamic variables E_s or G_s of the supercell. We mention that interfaces between crystals are treated in the same way as surfaces, except that a second crystal slab takes the place of the vacuum. We begin with the general thermodynamic formula for σ:

$$\sigma = \frac{1}{A_s}\left(G_s - \sum_i \mu_i N_i\right). \qquad 1.$$

Here G_s denotes the total Gibbs free energy of the contents of the supercell, A_s is the surface area of the slab within the supercell, index i runs over all the component species: Ni, Al, and O; μ_i is the chemical potential of these species, and N_i is the number of atoms of each species contained within the supercell. We are assuming here for simplicity that the surfaces of the slab are identical. Recasting Equation (1) in terms of excesses Γ_i,

$$\sigma = \frac{1}{A_s}\left(G_s - \frac{N_A}{N_A^b}\, G^b\right) - \sum_{i \neq A} \mu_i\, \Gamma_i. \qquad 2.$$

Here A represents the component with respect to which excesses are measured. This is possible at an interface with only one condensed phase; if there were two different condensed phases, two reference components would be required (42). N_A^b and G^b are the number of atoms of species A within an arbitrary representative quantity of the bulk phase and the Gibbs free energy of that quantity, respectively.

The surface excess Γ_i is given by

$$\Gamma_i = \frac{1}{A_s}\left(N_i - N_A \frac{\nu_i^b}{\nu_A^b}\right), \qquad 3.$$

where ν_i^b is the atomic fraction of component i in the bulk phase.

Now let us be more specific about the degrees of freedom under our control. We assume that the oxidized surface of the NiAl crystal (not necessarily stoichiometric) is in equilibrium with $Ni_{1-x}Al_x$ bulk and with oxygen gas at a certain partial pressure p_{O_2}. The equilibrium is

$$Ni_{1-x}Al_x \text{ (bulk)} \rightleftharpoons \{NiAl + O\} \text{ (surface)} \rightleftharpoons O_2 \text{ (gas)}.$$

How many phases are there in the equilibrium? It would be incorrect to think about even several layers of the surface oxide as a separate phase, as the system does not have a well-defined bulk oxide region. Fortunately, it does not matter how thick the surface region of the oxidized metal is as long as there is some well-defined bulk metal. As there are $r = 2$ phases with $c = 3$ components (Ni, Al, and O) in equilibrium, one has $f = 3$ degrees of freedom according to the Gibbs phase rule:

$$f = c - r + 2 = 3 - 2 + 2 = 3.$$

Hence, p_{O_2}, T, and x can be chosen as three variables fully characterizing the system. We call them the environmental parameters because in a sense they provide an environment for the surface.

If, for example, Ni is chosen as the reference component, the above equations become

$$\sigma(x, T, p_{O_2}) = \frac{1}{A_s}\left(G_s - N_{Ni}\frac{g_{NiAl}(x, T)}{2(1-x)}\right) - \mu_{Al}(x, T)\Gamma_{Al} - \mu_O(p_{O_2}, T)\Gamma_O, \qquad 4.$$

with the surface excesses of Al and O given by

$$\Gamma_{Al} = \frac{1}{A_s}\left(N_{Al} - N_{Ni}\frac{x}{1-x}\right), \quad \Gamma_O = \frac{1}{A_s}N_O, \qquad 5.$$

and where g_{NiAl} is the Gibbs free energy of the bulk $Ni_{1-x}Al_x$ per formula unit:

$$g_{NiAl}(x, T) = 2\left\{(1-x)\mu_{Ni}(x, T) + x\mu_{Al}(x, T)\right\}. \qquad 6.$$

Equations (4)–(6) are helpful for discussing which approximations need to be employed to make this a practical scheme. Recall that we intend to define a set of atomic configurations and stoichiometries of the supercell, calculate σ for each, and declare the one with the lowest σ to be our predicted equilibrium structure and corresponding surface energy. The most expensive part of the calculations is to obtain G_s because this involves the search for the atomic positions that minimize the free energy of a slab containing from several tens to hundreds of atoms, and we want to use an ab initio method in order to properly account for the strength of chemical bonds. If G_s can be assumed to be independent of the three environmental parameters, it needs to be calculated only once for each member of the pre-selected set of atomic configurations. First of all, we assume that the effect of the pressure on G_s is negligible. In practice we set the external pressure to zero. Second, we consider the oxygen pressure to be low enough such that no oxygen atoms need to appear in the vacuum part of the supercell; and third, we assume that the concentration of oxygen atoms dissolved in the bulk is sufficiently small so that no oxygens need to be introduced explicitly into the interior of the slab. All these approximations are reasonable given that p_{O_2} will never exceed 1 atm. Similarly, one can neglect the dependence of G_s on x if the deviation of the alloy composition from the exact stoichiometry is of the order or less than $1/N$, where N is the number of metal atoms in the supercell. This condition places an upper limit on the range of compositions, which is important to keep in mind in dealing with compounds such as $Ni_{1-x}Al_x$, which exists as a homogenous phase in the range $0.40 < x < 0.55$. The most serious approximation is to neglect the temperature dependence of G_s. This dependence comes through both the vibrations of the atoms and the expansion of the lattice constant of the underlying bulk crystal. As such, it can be evaluated by means of explicit ab initio calculations; a number of studies of this type already exist (43–45). On the other hand, the variation of the chemical potential of oxygen with temperature is much stronger and thus dominates in the $\sigma(T)$ dependence. In other words, the assumption that G_s is just the zero-temperature total energy still gives reasonably accurate results. There is considerable cancellation of the bulk contributions to G_s with the second purely bulk term in the parenthesis of Equation (2).

In practical calculations then, the dependence of surface energy on x, p_{O_2}, and T in Equation (4) enters through the chemical potentials of the species, which, as the equilibrium condition is assumed, are equal to the respective chemical potentials in the bulk phases. As such, they can be calculated separately. With a few additional assumptions, their dependence on the environmental parameters can thus be estimated. In particular, for μ_O we can make use of accurate experimental data, so the main uncertainties are found in the chemical potentials of the species in the solid. This whole scheme is illustrated in Figure 1.

Chemical Potentials

SOLID PHASE Let us assume that the oxidized surface of NiAl is in chemical equilibrium with the underlying crystal so that the chemical potentials of Ni and Al, μ_{Ni} and μ_{Al}, are equal to those of the bulk phase. These chemical potentials depend on the composition x of the bulk phase $Ni_{1-x}Al_x$, as well as on temperature T and external pressure p. The pressure dependence of chemical potentials in solids can, of course, be neglected under normal conditions. Before proceeding with the discussion of how chemical potentials can be obtained in actual calculations, some

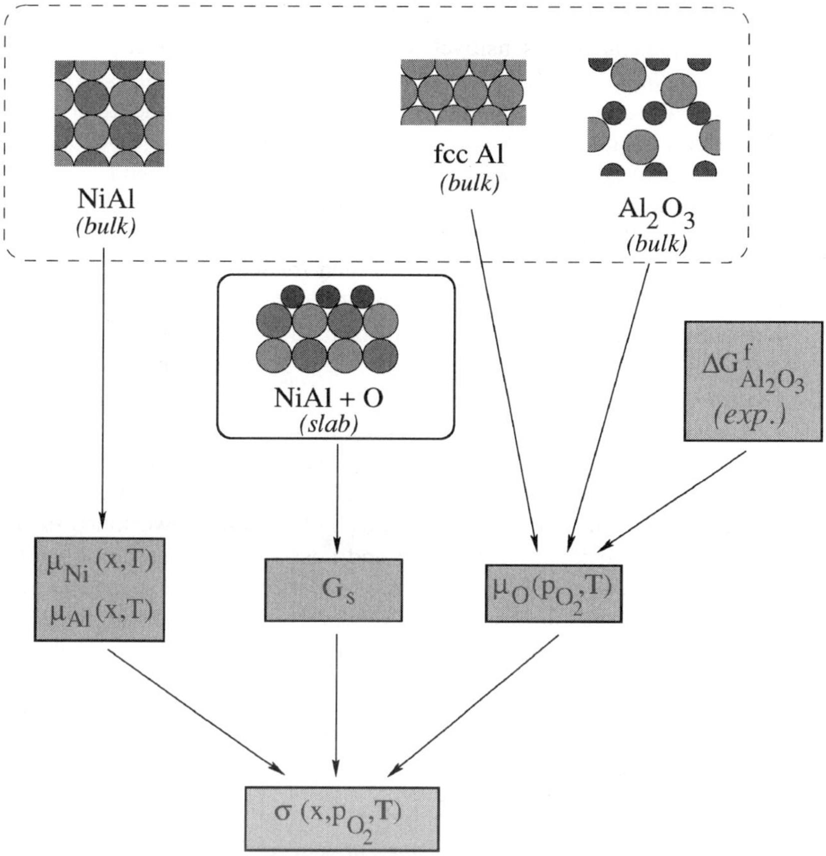

Figure 1 Schematic diagram describing the flow of data to obtain the surface energy σ as a function of the temperature T, partial pressure of oxygen gas p_{O_2}, and the composition x of the substrate $Ni_{1-x}Al_x$.

comment about their reference states is appropriate. It may be the case that chemical potentials used in semi-empirical calculations are not comparable with those used in ab initio calculations, and the latter in turn differ by orders of magnitude between all-electron and pseudopotential calculations. This is simply because in a semi-empirical calculation, isolated atoms (at rest) are taken to have a chemical potential of zero. On the other hand, in all-electron ab initio calculations, isolated nuclei and electrons may be assigned a zero value of chemical potential. In ab initio pseudopotential schemes the reference state would be isolated ions and valence electrons, and none of these corresponds to the chemical convention, which takes the enthalpy of an element at standard temperature and pressure as zero. Such dependence on the reference state should cancel out in the calculation of any physical observable such as σ, but it matters if data are transferred to one model from another or from experiment. The chemical potentials of components in a solid may depend very sensitively on small concentrations of point defects. NiAl is a good example of this because it is a strongly ordered alloy, which if slightly Ni-rich, will support a population of Ni antisite defects (rather than Al vacancies). On the other hand if it is slightly Al-rich, this compound supports a population of Ni vacancies (rather than Al antisites). If we add one Al atom to an Al-rich alloy, it will create on average one more Ni vacancy, whereas if we add one Al atom to a Ni-rich alloy it will fill on average one Al vacancy. The energy required for the second process will be much less than for the first (at constant *T*). But this energy is precisely the chemical potential of Al in the alloy, which is therefore a sensitive function of stoichiometry near the stoichiometric composition. The chemical potentials in the case of NiAl have been estimated theoretically by the following procedure.

One assumes a dilute or ideal solution model, in which interactions between point defects can be neglected and they are randomly distributed. This is often referred to as the Wagner-Schottky model (46). Within this model, the Gibbs free energy $g(x, T)$ of a compound $A_{1-x}B_x$ is the sum of two kinds of terms. The first terms are linear in each defect and host atom concentration, and represent the point defect energies and vibrational entropies. The second terms describe the configurational entropy of the point defects. If only substitutional defects and vacancies are considered (which is a very reasonable approximation for NiAl), then the Gibbs free energy per formula unit is given by (47)

$$g(x, T) = \sum_{i,j} c_{ij}\, g_{ij}(x, T) + k_B T \sum_{i,j} c_{ij}\, \log c_{ij}, \tag{7.}$$

where c_{ij} denotes the concentration (per sublattice site) of species I (A, B, or vacancy V) on sublattice J (A or B), g_{ij} are partial defect energies to be found from ab initio calculations, and k_B is Boltzmann's constant. Equilibrium concentrations c_{ij}^{eq} are those that minimize $g(x, T)$ given by (7), subject to some natural constraints: The sum of all the concentrations on each sublattice is unity, and the ratio of total concentrations of the A and B species is prescribed by the fixed composition x of the compound. If the defect parameters g_{ij} are known, the c_{ij}^{eq} can be obtained either by finding roots of a fourth-order polynomial (47) or by solving numerically the set of mass-action equations (48–50). Then the chemical potentials in question can be obtained directly. They are given by (47)

$$\mu_a = g_{aa} - g_{va} + k_B T \log\left(\frac{c_{aa}^{eq}}{c_{va}^{eq}}\right) \tag{8.}$$

$$\mu_b = g_{bb} - g_{vb} + k_B T \log\left(\frac{c_{bb}^{eq}}{c_{vb}^{eq}}\right). \tag{9.}$$

As $T \to 0\,\mathrm{K}$ these chemical potentials as a function of x approach constants with a step at $x = 0.5$. The defect parameters g_{ij} can be assembled from total energy calculations for perfect supercells and for supercells containing a single defect, in the approximation that the supercells are large enough for interactions between defects to be neglected. For example, if G_{va}^N is the Gibbs free energy of the supercell containing N unit cells with one vacancy on sublattice A, and g_0 is the free energy per unit cell of the ideal stoichiometric compound AB, then

$$g_{va}(T) - g_{aa}(T) = G_{va}^N(T) - N g_0(T). \qquad 10.$$

Note that the above Gibbs free energies might include the vibrations, but not the configurational entropy; the latter is already accounted for in Equation (7). Equations similar to Equation (10) can be obtained for all four types of point defects. Finally, we arbitrarily assign some value to g_{aa} and g_{bb} such that

$$g_{aa}(T) + g_{bb}(T) = g_0(T) \qquad 11.$$

to complete the system of linear equations, which is sufficient to calculate all g_{ij}. It can be shown that arbitrary division of g_0 between g_{aa} and g_{bb} does not affect any physical properties of the system (47). In particular, μ_a and μ_b are well defined. If the supercell Gibbs free energies are approximated by zero-temperature total energies, the partial defect energies are also temperature independent.

So far, it has been demonstrated that the chemical potentials of a compound are unambiguously defined by the reference state chosen and by the energies of point defects. The alternative approach is to put natural bounds on the chemical potentials (51, 52) and to study a particular problem either over the whole range of chemical potentials specified by these bounds or simply for the maximum and minimum values of the allowed interval. In our case of the AB compound, this method works as follows. If μ_a, the chemical potential of component A, were to exceed that of the pure A crystal, μ_a^0, component A would start to precipitate out of the compound. Hence

$$\mu_a \le \mu_a^0; \quad \mu_b \le \mu_b^0. \qquad 12.$$

We also need the expression for the Gibbs energy g_0 of bulk alloy, which is given by Equation (6) for $Ni_{1-x}Al_x$, but which for the present purpose is sufficiently accurate if we take $x = 0.5$.

$$g_0 = \mu_a + \mu_b. \qquad 13.$$

From Equations (12) and (13), one gets

$$\begin{cases} g_0 - \mu_b^0 \le \mu_a \le \mu_a^0 \\ \mu_b = g_0 - \mu_a \end{cases}, \qquad 14.$$

in which only one of two chemical potentials is an independent variable. The above inequality is often recast in the equivalent form:

$$\Delta G_f \le \Delta\mu_a \le 0, \qquad 15.$$

where $\Delta\mu_a = \mu_a - \mu_a^0$ and ΔG_f is the Gibbs free energy of formation of the compound AB. The method of bounds is often used in dealing with insulators or semiconductors where either the ab initio calculation of point defect statistics or the ideal solution model is more problematic than in metals.

GAS PHASE The last remaining ingredient needed in the expression for surface energy (4) is the chemical potential of oxygen $\mu_O(p_{O_2}, T)$. In fact, we need to know μ_O only for the standard pressure $p^0 = 1$ atm; one can then evaluate it accurately at any other pressure using the ideal gas expression:

$$\mu_O(p_{O_2}, T) = \mu_O^0(T) + \frac{1}{2}k_B T \log\left(p_{O_2}/p^0\right). \qquad 16.$$

The superscript 0 on any quantity denotes its value at the standard pressure p^0. Different authors have taken somewhat different approaches for calculating $\mu_O^0(T)$, but usually the information in research papers is insufficient for an uninitiated reader to follow the procedure with confidence. We therefore present our recommended approach in some detail.

The trick (34, 53) is to invoke a thermodynamic cycle corresponding to formation of some oxide, for which the standard Gibbs free energy of the oxide formation, $\Delta G^0_{\mathrm{oxide}}(T)$, is available from tabulated experimental data. Here, we employ the reaction for $\alpha-Al_2O_3$ formation at standard partial oxygen pressure $p_{O_2} = p^0$:

$$2\mathrm{Al(s)} + \frac{3}{2}\mathrm{O_2(g)} \rightleftharpoons \mathrm{Al_2O_3(s)},$$

in which Al_2O_3(s) is the bulk crystal of $\alpha-Al_2O_3$, Al(s) is the Al crystal, and O_2(g) is oxygen gas. The Gibbs free energy balance for this reaction is

$$g^0_{Al_2O_3}(T) = 2\mu^0_{Al}(T) + 3\mu^0_O(T) + \Delta G^0_{Al_2O_3}(T). \qquad 17.$$

Now we write our desired equation for $\mu^0_O(T)$ by turning this equation around, and we introduce an arbitrary reference temperature T_1 for reasons that will become clear:

$$\mu^0_O(T) = \frac{1}{3}\left\{g^0_{Al_2O_3}(T_1) - 2\mu^0_{Al}(T_1) - \Delta G^0_{Al_2O_3}(T_1)\right\} + \Delta\mu^0_O(T) \qquad 18.$$

where

$$\Delta\mu^0_O(T) = \mu^0_O(T) - \mu^0_O(T_1). \qquad 19.$$

The virtue of Equation (18) is that one does not have to calculate any absolute properties of an oxygen molecule directly; particularly, its internal energy at 0 K is subject to greater uncertainty than the energy of simple metals and their oxides, and one can now sidestep the problem of calculating it. The relative quantities $\Delta G^0_{Al_2O_3}(T_1)$ and $\Delta\mu^0_O(T)$ are readily obtained from tables of experimental data. Only the Gibbs free energy of bulk alumina per formula unit $g^0_{Al_2O_3}(T_1)$ and the Gibbs free energy per atom of a bulk fcc Al crystal $\mu^0_{Al}(T_1)$ must be calculated ab initio. These have been approximated by total energies at zero Kelvin. The lower

T_1, the more accurate will be the result, but standard room temperature T^0 is adequate, which is fortunate because experimental values of $\Delta G^0_{\text{oxide}}$ are generally available only at and above T^0. Notice that the value of $\mu^0_O(T)$ depends very much on the implicit reference states of the components in the ab initio calculation, which may be separated electrons and nuclei at rest (if all electrons are treated explicitly), or separated ions and nuclei at rest (in a pseudopotential calculation). However, this introduces no error because we are using the same convention for all the terms. Substitution of μ^0_O back into Equation (16) completes the scheme:

$$\mu_O(p_{O_2}, T) = \frac{1}{3}\left\{g^0_{Al_2O_3} - 2\mu^0_{Al} - \Delta G^0_{Al_2O_3}(T^0)\right\} + \Delta\mu^0_O(T) + \frac{1}{2}k_B T \log\left(p_{O_2}/p^0\right). \qquad 20.$$

One could choose any tabulated oxide instead of Al_2O_3 to close the thermodynamic cycle. The only requirement is that the chosen oxide and the oxide-forming element should be amenable to accurate first principles calculations. In theory, different oxides should lead to the same $\mu_O(T)$, but in practice there are small differences (41). Apart from experimental errors, the obvious reason for μ_O to differ from one oxide to another is the neglect of thermal contributions to the Gibbs energies of the solids, which are present in the experimental $\Delta G^0_{\text{oxide}}$. In principle this could be corrected by using the experimental data on the entropy and/or specific heat C_p of the bulk phases, if such information is at hand. Another source of error is intrinsic to density functional theory in which the errors in the treatment of exchange and correlation may differ from one system to another. Also, if pseudopotentials are used, they are not perfectly transferable.

Some observations on $\Delta\mu^0_O(T)$ are worth making. The best values of $\Delta\mu^0_O(T)$ are probably those based on combining experimental data with statistical mechanics for ideal gases (54). However, one could disregard experimental data and integrate from $\mu^0_O(T^0)$ to temperature T, using the formula for an ideal gas of rigid dumbells (41):

$$\Delta\mu^0_O(T) = -\frac{1}{2}\left\{\left(S^0_{O_2} - C^0_P\right)(T - T^0) + C^0_P T \ln(T/T^0)\right\}, \qquad 21.$$

where $C^0_P(T) \equiv 7k_B/2$ and $S^0_{O_2}$ (the entropy of oxygen gas at $T = T^0$ and $p_{O_2} = p^0$) is the only parameter taken from thermodynamic tables. As seen in Figure 2, this simple formula, Equation (21), predicts the temperature dependence of $\Delta\mu^0_O$ surprisingly well; the error at 1000 K is only 0.005 eV rising to 0.036 eV at 2000 K.

Finally let us comment on the way ab initio data is used in Equation (20). So far we have assumed the 0 K total energies are estimates of the Gibbs free energies of solid phases (38). What if, instead, we take them to be estimates of the enthalpies (10). It has been suggested (27) that in condensed phases the enthalpy change Δh^0 between 0 K and T^0 is much smaller than the contribution to the Gibbs energy $T^0 s^0$, hence the second option is preferable. To clarify this point, we have compared the

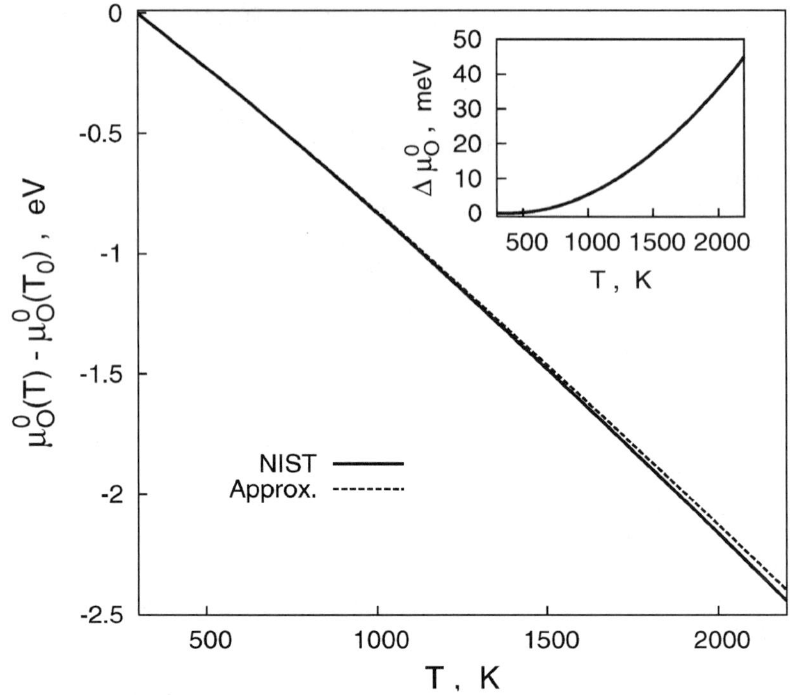

Figure 2 Chemical potential of oxygen μ_O^0 in the gas phase (per oxygen atom) relative to its value at $T_0 = 298.15$ K according to the NIST data (54) (*solid line*) and given by the approximate expression Equation (21) (*dashed line*). The difference between them is shown in the inset.

errors implied by either approach (see Table 1), using available experimental data. Although T^0s^0 indeed exceeds Δh^0 in most cases, these are of opposite sign. As a result, the differences $\Delta h^0 - T^0s^0$ are in many cases smaller in magnitude than Δh^0. Although the errors listed in Table 1 are rather small, using the 0 K ab initio total energies as estimates of the Gibbs free energies is generally better than using them as estimates of the enthalpies.

Analytical Approximation

It is possible (although not strictly necessary) to make a further simplification of the calculations in order to explicitly reveal the main effects of the environmental parameters. One can approximate the composition dependence of chemical potentials (Equations 8–9) with simple algebraic expressions (47, 48) applicable in the range of concentrations near $x = 0.5$ over which the B2 phase is stable but still excluding a small region around the stoichiometric composition $x = 0.5$. The basic idea behind this approximation is to divide point defects present in an

TABLE 1 Comparison of the errors (in meV per oxygen atom) generated if one assumes that total energies at 0 K approximate either enthalpies (Method *H*) or Gibbs free energies (Method *G*) of crystalline oxide A_nO_m and oxide-forming element *A* in Equation (17). The errors are calculated as $[\Delta h^0_{A_nO_m} - n\Delta h^0_A]/m$ (Method *H*) and $[\Delta h^0_{A_nO_m} - n\Delta h^0_A - T_0(s^0_{A_nO_m} - ns^0_A)]/m$ (Method *G*), where $\Delta h^0 = h^0(T_0) - h^0(0\,\mathrm{K})$ is the enthalpy change and $s^0 = s^0(T_0)$ is the entropy change between 0 K and standard temperature $T_0 = 298.15$ K. The data are taken from References (54–57)

Oxide A_mO_n	$\Delta h^0_{A_nO_m}$, kJ/mol	Δh^0_A, kJ/mol	$s^0_{A_nO_m}$, J/(mol K)	s^0_A, J/(mol K)	Error, meV/at	
					Method *H*	Method *G*
Al_2O_3	10.02	4.54	50.92	28.30	3.23	9.08
B_2O_3	9.30	1.22	53.97	5.90	23.69	−19.73
BaO	9.98	6.91	72.05	62.48	31.83	2.27
BeO	2.84	1.95	13.77	9.50	9.19	−4.00
CaO	6.75	5.74	38.1	41.59	10.51	21.29
CdO	8.41	6.25	54.85	51.80	22.42	13.00
Cr_2O_3	15.28	4.06	80.65	23.62	24.76	−9.64
GeO_2	7.23	4.64	39.71	31.09	13.44	0.13
MgO	5.16	5.00	26.95	32.67	1.68	19.35
NbO_2	9.27	5.26	54.50	36.46	20.81	−7.05
Nb_2O_5	22.29	5.26	137.30	36.46	24.41	−15.35
SiO_2	6.92	3.22	41.46	18.81	19.17	−15.81
SnO	8.74	6.32	57.17	51.18	25.01	6.51
SnO_2	8.38	6.32	49.04	51.18	10.68	13.99
SrO	8.67	6.36	55.42	55.68	23.98	24.78
Ta_2O_5	23.04	5.68	142.86	41.48	24.21	−12.80
TiO_2	8.68	4.82	50.62	30.72	19.98	−10.75
ThO_2	10.56	6.35	65.23	51.8	21.82	1.08
UO_2	11.28	6.36	77.03	50.20	25.48	−15.96
UO_3	14.59	6.36	96.11	50.20	28.40	−18.86
U_3O_8	42.74	6.36	282.55	50.20	30.64	−20.31
VO	6.91	4.64	39.10	28.93	23.50	−7.91
V_2O_3	17.48	4.64	98.16	28.93	28.32	−13.17
V_2O_4	18.10	4.64	103.52	28.93	22.85	−12.41
V_2O_5	21.51	4.64	130.40	28.93	25.34	−19.47
ZnO	6.93	5.66	43.65	41.63	13.22	6.99

ordered $A_{1-x}B_x$ compound into thermal and structural defects. The former are the result of thermal disorder; they disappear as $T \to 0$ K. The latter defects necessarily exist in the compound even at $T = 0$ K to maintain the given composition. The concentration of structural defects thus depends on the composition x but not on the temperature, and in a stoichiometric compound they are entirely absent. The ratio of thermal defects to structural defects is a small parameter, unless the composition is very close to stoichiometry; in real alloys the structural defects usually outnumber the thermal defects by orders of magnitude. The first term of a Taylor expansion of the chemical potentials with respect to this small ratio already turns out to be a good approximation. For NiAl in particular, there is good agreement with the exact formulae up to around 1400 K, as demonstrated in Reference (47) to which we refer the reader for more details. These expressions explicitly take into account that the constitutional defects in NiAl are Ni antisite atoms on the Ni-rich side and Ni vacancies on the Al-rich side, as found experimentally (58) and confirmed in a number of independent ab initio calculations (48, 49, 59, 60). Explicitly (47), on the Ni-rich side ($x < 0.5$):

$$\mu_a = \frac{1}{2}\{e_{aa} + e_{ab} + k_B T \log(1 - 2x)\}, \qquad 22.$$

$$\mu_b = e_{bb} + \frac{1}{2}\{e_{aa} - e_{ab} - k_B T \log(1 - 2x)\}, \qquad 23.$$

and on the Al-rich side ($x > 0.5$):

$$\mu_a = e_{aa} - e_{va} - k_B T \log \frac{2x - 1}{x}, \qquad 24.$$

$$\mu_b = e_{bb} + e_{va} + k_B T \log \frac{2x - 1}{x}, \qquad 25.$$

where a stands for Ni, b stands for Al, and x is the composition of the $Ni_{1-x}Al_x$ compound. In effect, Equations (22)–(25) approximate the dependence of chemical potentials on composition as being exclusively the result of the ideal configurational disorder of the structural defects on their own sublattices.

Another approximation used in Equations (22)–(25) is the neglect of terms of order $k_B T\delta$, compared with the $k_B T \log \delta$ terms, where $\delta = |x - 0.5|$ is the deviation of the composition of NiAl from the exact stoichiometry. We do not lose any essential physics in making this approximation, as the temperature dependence of surface energy σ is mainly from the oxygen chemical potential, even when compared with $k_B T \log \delta$ terms which we keep. To this order the formula for the surface energy Equation (4) simplifies to

$$\sigma(x, T, p_{O_2}) = \frac{1}{A_s}(G_s - N_{Ni} g_0) - \mu_{Al}(x, T)\Gamma'_{Al} - \mu_O(p_{O_2}, T)\Gamma_O, \qquad 26.$$

where g_0 is the (free) energy of the unit cell of bulk stoichiometric NiAl, Equation (11), the aluminum surface excess is now the stoichiometry of the supercell itself:

$$\Gamma'_{Al} = \frac{1}{A_s}(N_{Al} - N_{Ni}), \tag{27}$$

and $\Gamma_O = N_O/A_s$, as before. To make the surface energy dependence on all three environmental parameters explicit, we rewrite Equation (26) in the form

$$\sigma(x, T, p_{O_2}) = \sigma_0 + \Delta\sigma_0\theta(x - 0.5) - \frac{1}{2}k_BT\left\{\Gamma_O \log(p_{O_2}/p^0) + \Gamma'_{Al}\eta(x)\right\} - \Delta\mu_O^0(T)\Gamma_O, \tag{28}$$

where σ_0 and $\Delta\sigma_0$ absorb all ab initio numbers, $\theta(x)$ is the step function and $\eta(x)$ is defined by

$$\eta(x) = \begin{cases} -\log(1 - 2x), & x < 0.5 \text{ (Ni-rich NiAl)} \\ 2\log[(2x - 1)/x], & x > 0.5 \text{ (Al-rich NiAl)}, \end{cases} \tag{29}$$

and $\Delta\mu_O^0(T)$ can be parameterized as discussed above.

Stability Limits

Before we proceed with looking at the relative stability of various structures of a thin oxide film formed on the surface of NiAl, it is useful to estimate the range of stability of the respective bulk oxide phases. For example, if in the ab initio calculations we find a particular oxide layer to be stable at some p_{O_2} at which the NiAl substrate turns out to be unstable with respect to the formation of bulk Al_2O_3, then the predicted surface structure is at best metastable with respect to further oxidation. It might still take much longer than the experimental timescale for the material to convert itself into a bulk oxide, e.g., because the oxide film itself is a barrier to further oxidation.

To find the p_{O_2} above which NiAl starts decomposing to form some bulk oxide phase, or equivalently, the pressure below which any bulk oxide would be unstable in contact with NiAl, we take the following approach. Imagine a system comprising a large sample of NiAl in contact with some bulk oxide surrounded by its vapor. The amount of the oxide phase is assumed to be large enough to consider it as a bulk phase and to neglect any surface or interface effects, but small in comparison to that of NiAl such that one could think of NiAl as a reservoir of Ni and Al atoms. We next check, in turn, three possible oxides to derive the p_{O_2} at which the equilibrium in the system is established at given temperature T and NiAl composition x and then choose the smallest of these pressures, p_{min}, as the p_{O_2} in question. This leaves us with only three representative situations to examine. The first one corresponds to NiAl releasing Al atoms to form Al_2O_3:

$$2\text{Al}(\text{from } \text{Ni}_{1-x}\text{Al}_x) + \frac{3}{2}\text{O}_2(\text{g}) \rightleftharpoons \text{Al}_2\text{O}_3(\text{s}). \tag{30}$$

The pressure p_a corresponding to the above equilibrium can be found as in (53):

$$\log\left[p_a(x,T)/p^0\right] = \frac{2}{3k_BT}\left[\Delta G^0_{Al_2O_3}(T) - 2\Delta G_{Al}(x,T)\right], \qquad 31.$$

where ΔG_{Al} is the partial Gibbs free energy of Al atoms in NiAl. It can be taken from experiment (61), or calculated ab initio as $\Delta G_{Al}(x,T) = \mu_{Al}(x,T) - \mu^0_{Al}$. The values of p_a at 1273 K obtained in both ways are compared in Figure 3. The rather good agreement between them mirrors the similar level of consistency

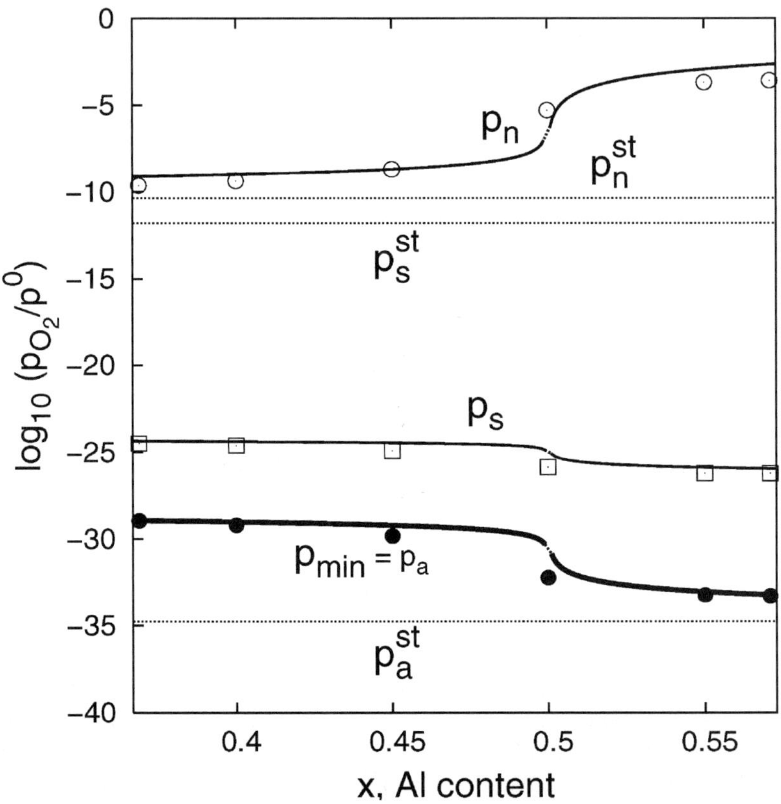

Figure 3 Composition dependence of p_a, Equation (31), p_n, Equation (34), and p_s, Equation (35) at $T = 1273$ K according to experiment (*symbols*) and ab initio calculations (*solid lines*) (62). The gaps on the curves in the vicinity of the stoichiometric composition appear because the analytical approximation for Ni and Al chemical potentials (Equations (22)–(25)), is not applicable there. Horizontal dotted lines indicate the boundary of the absolute stability of $\alpha-Al_2O_3$, p_a^{st}, Equation (37); NiO, p_n^{st}, Equation (38); and $NiAl_2O_4$, p_s^{st}, Equation (40). Experimental data are from Reference (23) ($\Delta G^0_{Al_2O_3}$ and ΔG^0_{NiO}), Reference (63) ($\Delta G^0_{NiAl_2O_4}$), and Reference (61) (ΔG_{Al} and ΔG_{Ni}).

between theoretical and experimental ΔG_{Al} explicitly shown in Reference (53).

The second and third reactions are similar to Equation (30) and describe the formation of nickel oxide NiO, and the spinel $NiAl_2O_3$, respectively:

$$\mathrm{Ni(from\ Ni_{1-x}Al_x)} + \frac{1}{2}\mathrm{O_2(g)} \rightleftharpoons \mathrm{NiO(s)}, \qquad 32.$$

$$2\mathrm{Al(from\,Ni_{1-x}Al_x)} + \mathrm{Ni(from\,Ni_{1-x}Al_x)} + 2\mathrm{O_2(g)} \rightleftharpoons \mathrm{NiAl_2O_4(s)}. \qquad 33.$$

The pressures p_n and p_s corresponding to the equilibrium in Equations (32) and (33) are given by

$$\log\left[p_n(x,T)/p^0\right] = \frac{2}{k_BT}\left[\Delta G^0_{NiO}(T) - \Delta G_{Ni}(x,T)\right], \qquad 34.$$

$$\log\left[p_s(x,T)/p^0\right] = \frac{1}{2k_BT}\left[\Delta G^0_{NiAl_2O_4}(T) - 2\Delta G_{Al}(x,T) - \Delta G_{Ni}(x,T)\right]. \qquad 35.$$

These pressures are also plotted in Figure 3, which shows that p_a is lowest over the whole range of B2 compositions: $p_{min} = p_a$.

To complete our analysis, let us check that at the pressures found the respective oxides would not decompose into their constituents. The pressure p_a^{st} below which Al_2O_3 decomposes into Al(s) and oxygen gas

$$\mathrm{Al_2O_3(s)} \rightarrow 2\mathrm{Al(s)} + \frac{3}{2}\,\mathrm{O_2(g)}, \qquad 36.$$

is defined by Equation (34):

$$\log\left[p_a^{st}(T)/p^0\right] = \frac{2}{3k_BT}\Delta G^0_{Al_2O_3}(T). \qquad 37.$$

Similarly, NiO decomposes into Ni(fcc) and oxygen below pressure p_n^{st} given by

$$\log\left(p_n^{st}/p^0\right) = \frac{2}{k_BT}\Delta G^0_{NiO}(T). \qquad 38.$$

Finally, at an oxygen partial pressure below p_s^{st}, $NiAl_2O_4$ becomes unstable with respect to decomposition into Al_2O_3 and Ni (64):

$$\mathrm{NiAl_2O_4(s)} \rightarrow \mathrm{Al_2O_3(s)} + \mathrm{Ni(s)} + \frac{1}{2}\mathrm{O_2(g)}. \qquad 39.$$

where

$$\log\left[p_s^{st}(T)/p^0\right] = \frac{2}{k_BT}\left[\Delta G^0_{NiAl_2O_4} - \Delta G^0_{Al_2O_3}\right]. \qquad 40.$$

The pressures p_a^{st}, p_s^{st} and p_n^{st} are shown in Figure 3 as horizontal lines. Importantly, p_a^{st} is lower than p_a over the whole composition range (which simply means that it is harder to extract an Al atom from NiAl than from pure Al). Hence, the $p_a(x)$ curve represents the equilibrium between NiAl and alumina; above $p_a(x)$, NiAl starts shedding Al atoms to form Al_2O_3. At $p_{O_2} = p_s$, NiAl might conceivably be in equilibrium with $NiAl_2O_4$ spinel, but because p_s is smaller than p_s^{st}, any $NiAl_2O_4$ formed should decompose into Ni and Al_2O_3. We observe that increasing the p_{O_2} above p_a appears to activate more channels for NiAl oxidation. However, a mixture of products (including perhaps more metastable phases not considered here) is conceivable only if the local concentration of Al is depleted down to the Ni_3Al phase boundary and beyond by the slow bulk diffusion in the NiAl. Another way of looking at this is to consider the sequence of solid phases that would be in equilibrium under increasing p_{O_2} if the NiAl were completely exhausted, assuming it remains homogeneous. The sequence, both according to Figure 3 and the experimental phase diagram (65), would be

$$\mathrm{NiAl} \rightarrow \mathrm{NiAl} + \mathrm{Al_2O_3} \rightarrow \mathrm{Ni_3Al} + \mathrm{Al_2O_3} \rightarrow \mathrm{Ni} + \mathrm{Al_2O_3}$$

$$\rightarrow \mathrm{Ni} + \mathrm{NiAl_2O_4} \rightarrow \mathrm{NiO} + \mathrm{NiAl_2O_4}.$$

Note that $NiAl_2O_4$ *does* appear as a transient phase in some experimental studies of NiAl oxidation (66).

The pressure p_{min} is estimated here by assuming that bulk $\alpha-Al_2O_3$ is the oxide phase formed reversibly by NiAl. $\alpha-Al_2O_3$ is the most stable form of alumina, hence it is plausible to suggest that the pressure at which a thin oxide layer on top of NiAl starts to form could not be less than p_{min}. Clearly, a thin Al_2O_3 layer alone cannot be of lower energy than the bulk phase, otherwise the bulk $\alpha-$alumina would decompose into such layers. However, one cannot rule out the theoretical possibility that the interaction of the layer with the substrate strongly stabilizes the system. This can be tested by ab initio calculations; so far the situation in which a thin oxide layer is more stable than bulk oxide has been found only for some noble and late transition metals: Ag (10), Ru (6), and Pd (13).

THE PATH OF NiAl OXIDATION

NiAl is a popular substrate for growing alumina films. It suits this purpose better than pure Al because the temperature required to convert the initially deposited oxygen atoms into an ordered film exceeds the melting temperature of Al. The experimental procedure to produce a 5 Å alumina film on NiAl(110) is well established (24). One must expose clean NiAl(110) to a dose 1000–1200 L ($1\,L = 10^{-6}$ Torr s) of oxygen at a temperature of about 600 K, after which the surface is annealed in ultra-high vacuum (UHV) at a temperature in excess of 1000 K. After the first stage, a layer of amorphous alumina is formed, which is believed to be locally ordered but which does not produce any diffraction spots (hence the name).

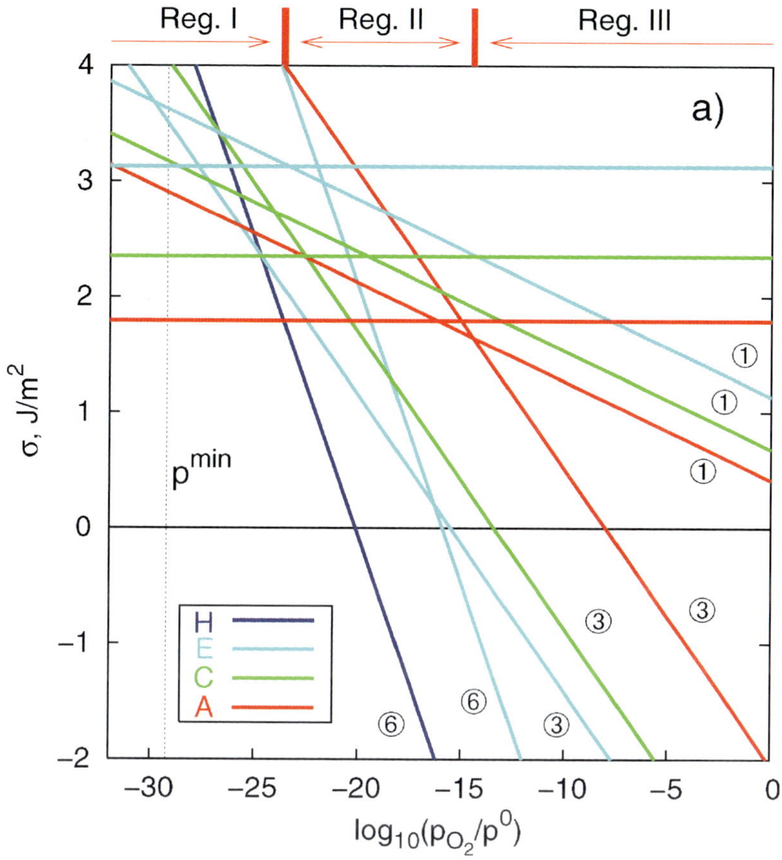

Figure 6 Surface energy σ, Equation (26), of configurations *A* (*red lines*), *C* (*green lines*), E (*light blue lines*), and *H* (*dark blue lines*) as a function of the oxygen partial pressure p_{O_2} at $T = 1273$ K: (*a*) assuming the bulk composition of the substrate $Ni_{0.55}Al_{0.45}$; (*b*, overleaf) assuming the bulk composition of the substrate $Ni_{0.45}$ $Al_{0.55}$. The thin vertical dotted line indicates the stability limit of bulk α–Al_2O_3 in contact with NiAl, Equation (31). Horizontal lines correspond to the $\Gamma_O A_s = 0$ family; lines corresponding to $\Gamma_O A_s = 1$, 3, and 6 are marked with the respective numbers; to convert to oxygen coverage in monolayers, these should be divided by 3. Oxidation regions: I (clean surface), II (oxidation with additional barriers), and III (down-hill oxidation) are indicated above the plot.

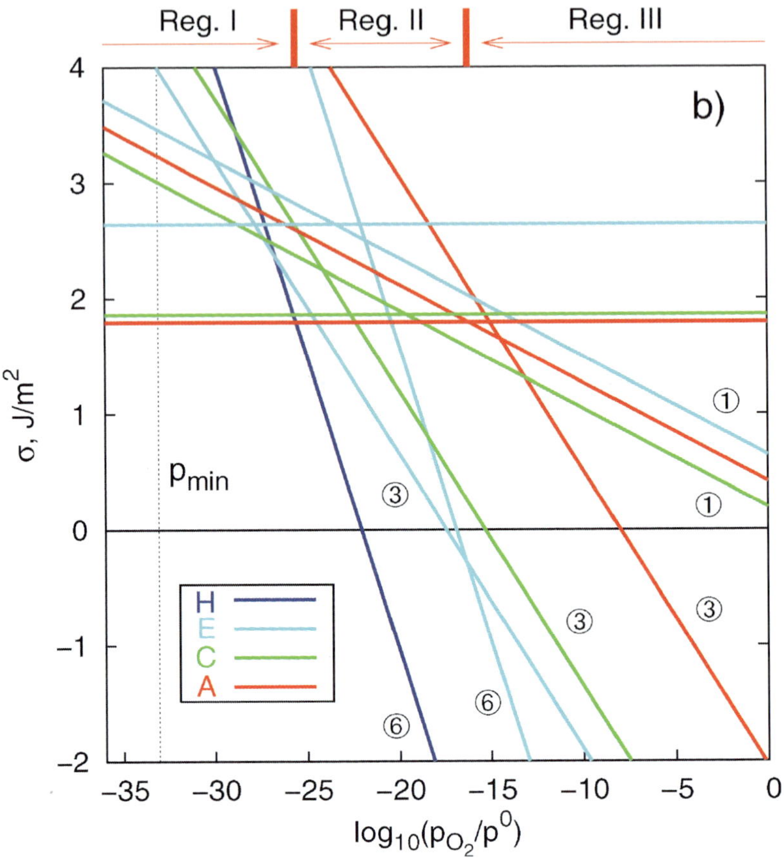

Figure 6 (*Continued*)

TABLE 2 Notation for different configurations considered in the present study. Point defects are denoted as $X_Y^{(i)}$,which means species X (Al, Ni, or vacancy V) on sublattice Y (Al or Ni) in layer i (layer 1 is the surface). Configuration H was designed to simulate experimentally observed stoichiometry of the oxide (two Al_2O_3 layers); it was used in combination with 6 oxygen atoms (H_6 in our notation). The last column represents the surface excess of Al atoms Γ_{Al} (assuming a stoichiometric substrate) multiplied by the cross-sectional area of the supercell A_s

Notation	Point defects present	$\Gamma_{Al} \times A_s$
A	none	0
B	$V_{Ni}^{(2)}$	1
C	$V_{Ni}^{(1)}$	1
D	$V_{Ni}^{(2)} + Ni_{Al}^{(2)} + Al_{Ni}^{(1)}$	1
E	$V_{Ni}^{(1)} + Ni_{Al}^{(2)} + Al_{Ni}^{(1)}$	1
F	$V_{Al}^{(2)} + Al_{Ni}^{(1)}$	1
G	$Ni_{Al}^{(2)} + Al_{Ni}^{(1)}$	0
H	$V_{Ni}^{(1)} + Ni_{Al}^{(2)} + Al_{Ni}^{(1)}$ + Al extra atom	2
I	$V_{Al}^{(1)}$	−1
J	$V_{Ni}^{(1)} + Ni_{Al}^{(1)}$	−1

motions of the atoms and can be good at finding natural pathways linking different basins of the potential energy surface. As a result,we managed to locate new configurations that would be impossible to guess otherwise. In some cases, these configurations were 1–2 eV/cell lower in energy than those obtained from initial plausible atomic arrangements. For more details and discussion of the MD simulations and all ab initio calculations, which were made with the plane-wave pseudopotential method using the CPMD code, the reader is referred to References (40, 53). Below we present only the results obtained within the local density approximation (LDA). The effect of the generalized gradient approximation (GGA) is discussed in (40).

The study of the adsorption of one O atom per supercell, $\Gamma_O A_s = 1$, reveals that oxygen atoms prefer to adsorb at the threefold hollow site formed by two Al and one Ni atoms. The second oxygen's preference is the bridge site between Al atoms, which is just 0.2 eV higher in energy. Strong distortion of the substrate was observed in both cases. According to structural models of the NiAl(110)/Al_2O_3 interface (24, 25), configurations with $\Gamma_O A_s = 3$ should correspond to one of the two Al + O bilayers of the Al_2O_3. However, the theoretical structure of the oxygen layer on the *ideal* NiAl substrate does not at all resemble any atomic arrangement typical of Al_2O_3. O atoms have arranged themselves there in a kind of butterfly structure, occupying an Al–Al bridge site and two (2Al + Ni) hollow sites per cell (configuration A_3 in Figure 4 (see color insert)), providing each O atom with two Al neighbors. However, it is problematic to call such a structure an oxide layer because the parallel O–Al chains do not appear to be linked to each other. Configuration G_3, in which we exchanged a surface Ni with subsurface Al atoms, would not seem particularly stable either (see Figure 4). Owing to its larger radius, the extra Al atom has been pushed out of the surface and forms a kind of pyramid with its four nearest-neighbor Al atoms. This pyramid is decorated with oxygens on three of its four sides in such a way that each O atom now has three Al neighbors. Nevertheless, the total energy of this structure is almost 2 eV lower, probably because of the increased number of O–Al bonds.

We checked next the effect of introducing a Ni vacancy in the top layer (configuration C_3). The energy again significantly decreased (by 1.92 eV for Ni-rich, and by 2.96 eV for Al-rich NiAl). During the MD run, atoms first vibrated about their initial positions, but after about 300 steps (0.36 ps) the Al atoms changed their striped structure by forming distorted hexagonal rings around Ni atoms, in which each pair of adjacent Al atoms became connected by an oxygen atom. Although each oxygen atom in this structure still had only two Al neighbors, the extra space provided by the vacancy allowed the system to form a number of short O–Al bonds (1.79 Å). We thus conclude that the effect of an antisite defect is to expose more Al atoms to the surface oxygen and thus to increase the number of Al–O bonds, whereas the vacancy makes space for the system to arrange atoms in a more optimal way.

Finally, trying both the exchange defect and the Ni vacancy in the top layer simultaneously, we arrived at our lowest energy atomic arrangement, which demonstrates two clearly separated oxygen and aluminum layers (configuration E_3 in Figure 4). Al atoms in the top layer arrange themselves in a distorted hexagonal packing and have almost lost their sublattice individuality, therefore the resulting O–Al bilayer can be considered as the first distinct precursor of the oxide phase.

How does the system arrive at the configuration E_3 starting from configuration A_3? Would it go through G_3 or C_3 (Figure 4)? To answer this, we have considered a number of intermediate configurations listed in Table 2, such that the system's trip from A_3 to E_3 can be decomposed into elementary steps. The resulting pathways are shown in Figure 5 (see color insert) together with the energy change at each step. For example, if a Ni bulk vacancy is initially absorbed by the second layer ($A \rightarrow B$), it may then jump to the surface layer ($B \rightarrow C$), and the creation of exchange defect ($C \rightarrow E$) completes the path *ABCE*. Alternatively, having a Ni vacancy in the second layer, the system can first create an exchange defect ($B \rightarrow D$) and then either allow the Ni vacancy to jump to the first layer immediately ($D \rightarrow E$), or first convert $V_{Ni}^{(2)}$ and $Ni_{Al}^{(2)}$ defects into $V_{Al}^{(2)}$ (i.e., the Ni antisite atom returns to its own sublattice, filling a Ni vacancy there). A nearest-neighbor Ni atom from the first layer can then jump into this vacancy ($D \rightarrow F \rightarrow E$). Another family of pathways arises if we assume that the system starts by creating an exchange defect ($A \rightarrow G$). Then a Ni vacancy arrives at the second layer ($G \rightarrow D$), and the subsequent history is as above.

We emphasize here the dependence of the relative energies on the substrate composition. In the zero-temperature limit, the surface energy dependence on composition is described by the step function, Equation (28). Hence the precise composition of the bulk substrate is not important, only whether it is Al- or Ni-rich. Looking at the pathways shown in Figure 5 we see that an additional free-energy barrier of at least 0.5 eV appears on the Ni-rich side compared with that on the Al-rich side (40). This is because configurations containing a Ni vacancy are higher in energy in Ni-rich NiAl simply because on the Al-rich side, these vacancies are the structural defects (see above).

Our zero-temperature analysis thus shows that the oxidized NiAl substrate tries to get rid of Ni atoms in the top layer(s) by either exchanging them with Al atoms beneath or absorbing Ni vacancies from the bulk. Both processes seem essential: The first allows the surface to expose more Al atoms to oxygen, whereas the second provides more space in the surface layer to accommodate larger Al atoms and subsequently to form a relatively open oxide structure (77). On the Ni-rich side, we expect larger barriers for the oxidation. For any substrate composition, the initial growth of the oxide film invokes the segregation and further consumption of Ni vacancies, therefore the substrate vacancies are destroyed, not injected.

This picture remains essentially valid in the finite temperature analysis (see below). Before proceeding, however, let us discuss how realistic the suggested mechanism is. In particular, according to the presented scheme one would expect that the substrate is first locally enriched with Ni near the metal-oxide interface, and this has not been observed experimentally (25). Of course, such enrichment is a transient effect; the excess Ni atoms should dissolve in the bulk to restore the overall equilibrium of the system by absorbing Ni vacancies. But is the diffusion in bulk NiAl fast enough to supply all the required vacancies?

To confirm that this is the case, one could make a routine estimate of the diffusion rates (24, 69), but we think more direct evidence exists. It comes from the elegant work by McCarty et al. (78), who observed the growth and shrinkage of surface steps at the NiAl(110) surface while cycling the temperature around 1000 K. The characteristic time of the cycle was of the order of several seconds. This turned out to be sufficient for the surface steps to respond to the temperature oscillation by changing their size. The essential point is that the size of the steps changed because they were emitting or absorbing vacancies to maintain the temperature-dependent equilibrium concentration of vacancies in the NiAl bulk.

Even more direct confirmation of the model comes from the X-ray study of the NiAl(110)/Al_2O_3 interface by Stierle et al. (71). The exposure of the NiAl specimen to oxygen gas at two different temperatures, 370 K and 570 K, followed by a 4 min annealing at 1020 K, resulted in drastically different concentrations of point defects in the first few layers of the substrate. At the lower exposure temperature, almost half of the aluminum sublattice was occupied by vacancies in the first two layers. In addition, the Ni sublattice of the top layer contained Ni and Al atoms in almost equal proportion. We understand that the temperature of 370 K was insufficient to activate the diffusion, and the oxidation removed a significant fraction of the Al atoms from the substrate leaving Al vacancies; the very open structure that resulted induced major chemical disorder on the sublattices. Some structural disorder is to be expected in such open structures, especially in the vicinity of the incommensurate oxide phase. At the higher temperature, $T = 570$ K, on the contrary, no vacancies were found in the top two layers, and only 18% of Al antisite atoms was observed. These atoms were significantly displaced outwards (by half of the NiAl(110) interplanar spacing). We consider the above result as evidence of some diffusion exchange already in operation at 570 K. The antisite Al atoms are probably the consequence of the strong chemical interaction between the oxide layer and the NiAl substrate. Surprisingly, introduction of 11% of Ni atoms in a Ni-Ni bridge site in the surface layer led to significant improvement of the fit to X-ray data (71). This might be a consequence of the above mentioned lateral distortion of the substrate atoms induced by the oxide film. Somewhat different results have been reported by Torrelles (79). In this study (1200 L at 600 K, then annealing to 1200 K for few minutes), neither structural nor additional chemical disorder in the top substrate layer was detected after the formation of the Al_2O_3 overlayer. The intermixing of the species between sublattices in the top layer of about 4% was the same as for the pure NiAl(110) surface. Such results are interpreted in (79) as the result of the diffusion of Al to the surface and Ni to the bulk.

Overall, we conclude that the point defect–mediated oxidation mechanism (40, 53), controlled by defect mobility as a function of temperature, is fairly well documented in NiAl. The rate of diffusion required for this mechanism is already noticeable at temperatures of about 600 K, and is surely sufficient at 1000–1200 K.

Surface Energy Dependence on the Environmental Parameters

In this section, we expand the analysis made above to finite temperatures. In doing this, we neglect any vibrational contributions to the Gibbs free energy, thus simply converting each of the previously calculated zero-temperature surface energies into a function of p_{O_2}, x, and T by using Equation (28). We concentrate on the following two issues: What is the lowest energy configuration at a given set of environmental parameters? Do any oxidation barriers appear in addition to those previously discussed? Notice in the following that a classical treatment of surface energies would distinguish the surface of the film from its interface with the substrate, and attribute a separate interface energy to the latter. We recognize that this distinction has no meaning when the films are too thin for their surface to be decoupled from the interface with the substrate, and we note that our surface energies include this coupling implicitly, together with the surface and interface contributions.

We plot in Figure 6 (see color insert) the surface energies σ of our configurations as a function of p_{O_2} at $T = 1273$ K, separately for a Ni-rich ($x = 0.45$) and an Al-rich ($x = 0.55$) composition. In this way we can determine the behavior of

σ at all intermediate compositions, as it should change smoothly from Figure 6*a* to Figure 6*b* as x goes from 0.45 to 0.55. The observation widely used in similar studies [(see, for example, References (26, 29)] is that changing the temperature roughly corresponds to rescaling the pressure axis if the latter is plotted in units of $\log(p_{O_2}/p^0)$ (34). This is because the composition-dependent term is small compared with the oxygen pressure term in curly brackets in Equation (28). The last term in Equation (28) defines the pressure about which the rescaling should be performed. For NiAl, this is a large pressure of about 10^{11} atm (53). Therefore, all essential information is already contained in Figure 6.

In Figure 6 we show only configurations A, C, E, and H as this is sufficient to capture the overall picture. Different configurations are shown by different colors, whereas the oxygen excess $\Gamma_O A_s = 1, 3$, or 6, is indicated by numbers on the plot. Horizontal lines correspond to zero oxygen excess, meaning an oxygen-free surface. Let us start with the Ni-rich system (Figure 6*a*). The structure of this busy plot is in fact quite simple. One can distinguish three regions indicated above the plot. Region I corresponds to the pressures at which the clean surface (A_0) is most stable. These are the pressures below $\sim 10^{-24}$ atm. Region III is the region in which the oxidation can proceed without any additional barriers. It corresponds to the pressures above $\sim 10^{-14}$ atm. Indeed, it is easy to appreciate that the following sequence decreases the surface energy at each step: $A_0 \rightarrow A_1$ (an O atom adsorbs at the clean surface), $A_1 \rightarrow A_3$ (two more O atoms adsorb), $A_3 \rightarrow C_3$ (a Ni vacancy arrives at the top layer), $C_3 \rightarrow E_3$ (an exchange defect is created), $E_3 \rightarrow E_6$ (three more O atoms adsorb), and $E_6 \rightarrow H_6$ (an Al atom is extracted from the substrate).

Correspondingly, Region II is the region in which an additional barrier for the oxidation appears owing to the non-monotonic dependence of surface energies on oxygen coverage. This situation arises if, for example, configurations with three oxygen atoms per cell are lower in energy than configurations with one oxygen atom, but clearly the surface cannot adsorb three atoms before it adsorbs one! This behavior is discussed below. The paths that the system can choose in such circumstances are $A_0 A_1 C_1 C_3 E_3 E_6 H_6$ or $A_0 C_0 C_1 C_3 E_3 E_6 H_6$ (Figure 6*a*).

If we look only at the lines representing the lowest σ, these are A_0 below the pressure 10^{-24} atm, and H_6 above it. Therefore, there is a range of pressures between p_{min} and 10^{-24} atm. in which the substrate should already be oxidizing (see above), but this does not happen according to the calculations of surface energy. There might be two reasons for this. First, the estimate refers to the formation of bulk α–Al_2O_3 by oxidation of NiAl, and the thin oxide film that forms on the surface may be less stable than bulk alumina. Second, in calculations, we can consider only a limited number of configurations; there might be a configuration of lower energy than H_6, which would intersect A_0 at a lower pressure. The low-energy configurations for each coverage for which calculations are available (A_0, A_1, E_3, and H_6) form a sequence in which configurations with progressively higher oxygen content intersect A_0 at lower and lower pressure. They should tend to a vertical line in the limit of infinite oxygen excess (a thick Al_2O_3 layer). Without extrapolating from the present configurations, we note that at a pressure of $\sim 10^{-21}$ atm, the H_6

line intersects the zero axis, which signals the instability of the system to catastrophic oxidation. We interpret $\sigma < 0$ to mean that the whole system is unstable to fragmentation into oxidized surfaces.

For the Al-rich substrate (Figure 6*b*), the situation is similar, except the vacancy-containing configurations (C, E, and H) are of lower energies because they have Al in excess. As a result, the boundaries for Regions I, II, and III shift slightly to the left: 10^{-26} and 10^{-16} atm instead of 10^{-24} and 10^{-14} atm, respectively. Also, the nominal barriers for paths $A_0A_1C_1C_3E_3E_6H_6$ and $A_0C_0C_1C_3E_3E_6H_6$ in Region II are lowered on the Al-rich side. This is the same trend we observed for transitions at fixed oxygen coverage (see above).

According to Figure 6, the experimental conditions under which the oxide film on NiAl is grown (for the initial oxygen deposition and even for the high-temperature annealing) fall into Region III for both Ni-rich and Al-rich substrate. Although the metal is already unstable with respect to oxidation (configuration H_6 gives a negative surface energy), surprisingly, a thin oxide film can still be produced. This is likely due to kinetic limitations: The diffusion of oxygen and/or aluminum through the compact oxide film is sufficiently slow to prevent its further growing. However, it remains unclear how to explain the following experimental observation: If the annealing temperature exceeds 1300 K, the oxide film starts to disintegrate (80). According to Figure 6, in order for the oxide film to dissolve at 1300 K, the p_{O_2} must be of the order of 10^{-24} atm or even lower. We can only speculate that lower values than this might prevail at the oxide–metal interface if the diffusion gradient is sufficiently steep.

'Inverted' Surface Energy Behavior and Relation to the Oxide Growth Mode

Previously, we pointed out a surprising feature of $\sigma(p_{O_2})$ lines as the p_{O_2} is raised: At the point where a configuration with oxygen excess $\Gamma_O A_s = 1$ becomes stable, i.e., crosses the clean surface line A_0, a configuration with $\Gamma_O A_s = 3$ is already more stable. We call this behavior inverted. Similarly, when $\Gamma_O A_s = 3$ becomes more stable than A_0, a $\Gamma_O A_s = 6$ configuration is already more stable. This is opposite to the picture that one would naively expect, thinking of adsorbed oxygen atoms as negatively charged ions that repel each other. Indeed, a monotonic decrease of binding energy per adatom as the coverage increases is often observed for other systems such as O/Ru(0001) and O/Rh(111) (6, 7), up to a monolayer coverage. A normal pattern would correspond to $\sigma(p_{O_2})$ plots shown on the left graph in Figure 7, in which the lines with higher oxygen content intersect the pure surface line at higher pressures. For NiAl(110), the situation is reversed, and this inverted behavior is shown on the right graph in Figure 7. Why do surface energies of oxidized NiAl show this counterintuitive trend? The answer is rather simple, but before we explain it, let us make one more observation. The lowest energy configurations for each oxygen coverage, A_0, A_1, E_3, and H_6 on the Ni-rich side, and A_0, C_1, E_3, and H_6 on the Al-rich side, both form the inverted sequence (Figure 6).

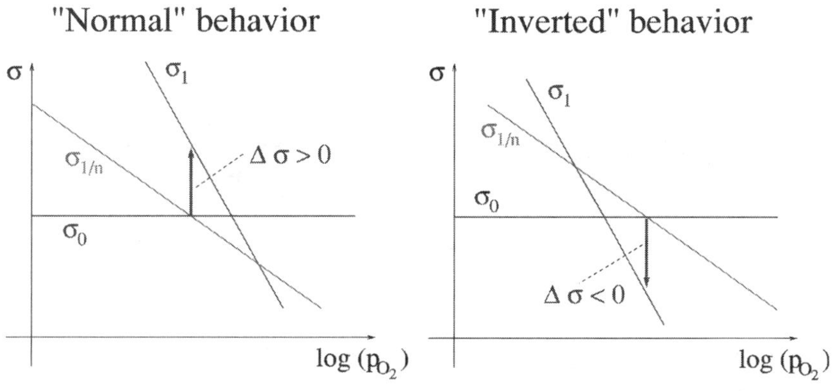

Figure 7 Schematic illustration of the normal and inverted dependence of surface energy on coverage. In this example, we show surface energies corresponding to coverages 0 ML, 1 ML, and some fractional coverage of $1/n$ ML, where $n > 1$. The normal type corresponds to σ_1 crossing σ_0 at higher oxygen partial pressure than $\sigma_{1/n}$; the type is inverted if this is the other way round. The normal (inverted) type is indicated by the positive (negative) sign of $\Delta\sigma$ defined by Equation (42) and has a simple geometric interpretation, as shown in the Figure.

However, by comparing only A_i configurations in which no point defects are allowed (Figure 6), we see that they form a normal sequence. Hence, it is the point defects created in the substrate that invert the surface energies. Now the picture we have is thus. Because of the strong attraction between adsorbed oxygen atoms and substrate Al atoms, the adsorbate tries to re-arrange the substrate to create the local configuration that would resemble stable alumina as closely as possible. Apart from simple atomic relaxation (which is also present in configurations A_i), such a rearrangement requires absorption of vacancies and interchange of Al and Ni atoms. If these movements can be achieved, the more oxygen atoms that combine forces, the lower the energy of the structure they create. The gain in energy overcomes the above mentioned mutual repulsion and the disadvantage of sharing substrate atoms. In other words, if point defects are available, the oxygen atoms can work together to design the substrate that suits them best. If point defects are not available, then the ability of the adsorbed oxygens to cooperate is reduced.

There is a direct relation between inverted versus normal behavior of the surface energies and the the stability of islands versus a continuous film, illustrated in Figure 8. Consider two configurations with surface energies σ_m and σ_n, where the subscript is an integer that measures the oxygen excess Γ_O, and $m < n$. The monolayer corresponds to $n = 1$, the double layer to $n = 2$ and so on. σ_0 denotes the energy of the pure surface. The islands in Figure 8 contain the same material as the continuous layer, but have a relative thickness n/m. They therefore occupy a fraction m/n of the surface and the combination of exposed surface and islands

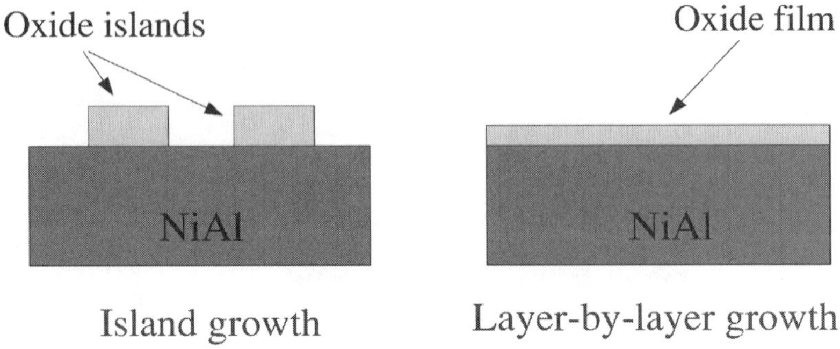

Figure 8 Illustration of the island growth and layer-by-layer growth modes of the surface oxide.

has an effective surface energy of

$$\sigma_{islands} = \frac{m}{n}\sigma_n + \left(1 - \frac{m}{n}\right)\sigma_0. \qquad 41.$$

Compare this with the surface energy σ_{layer} of the continuous layer on the right of Figure 8 and define

$$\Delta\sigma(x, T) = \sigma_{islands} - \sigma_{layer} = m \times \left\{\frac{1}{n}\sigma_n(x, T, p_{O_2}) + \left(\frac{1}{m} - \frac{1}{n}\right)\sigma_0(x, T) - \frac{1}{m}\sigma_m(x, T, p_{O_2})\right\}. \qquad 42.$$

All surface energies depend on three environmental parameters, except for σ_0, which does not depend on p_{O_2}. However, the dependence of σ_n/n and σ_m/m on p_{O_2} cancels out in Equation (42). Therefore, $\Delta\sigma$ does not depend on p_{O_2}, which we explicitly indicate in Equation (42). Now it is easy to see that the normal (inverted) regime corresponds to positive (negative) $\Delta\sigma$: one just needs to substitute into Equation (42) the pressure at which $\sigma_m = \sigma_0$. The theory we have just presented is a discrete version of a phenomenon known in the theory of adsorption, namely that if the surface energy as a function of coverage is convex, it is favorable for the adsorbate to partition into areas or phases of unequal coverage, and conversely. Our set of inclined lines of surface energy versus p_{O_2} are sampled from a much larger set of such lines, from which, in principle, a continuous curve of surface energy versus coverage could be constructed. That this curve would be convex can be seen by noting just the three lines drawn in the right hand graph of Figure 7.

Of course, there should be some additional penalty owing to the boundaries of the islands. This affects the minimum size of the islands beyond which such a phase separation lowers the surface energy. We conclude that the inverted behavior, which is demonstrated by the calculated surface energies of oxidized NiAl, should favor the oxide initially appearing as isolated islands, and not the layer-by-layer growth mode (Figure 8). The layer-by-layer mode would, however, be energetically favored if the supply of point defects from the substrate were suppressed, such as by lowering the temperature.

The above conjectures have some experimental support, although it is probably not conclusive. Recall here that the fabrication of the oxide layer on NiAl(110) is a two-step process (see above): 1200 L of oxygen is initially deposited at about 600 K, which results in a layer of amorphous alumina (α-Al_2O_3) covering the whole surface. In a second step, high-temperature annealing leads to the formation of well-ordered crystalline Al_2O_3 on top of NiAl. During the first step, the bright diffraction spots caused by the substrate gradually disappear, and no diffraction pattern is observed by the end of the process. No oxide islands are observed at this stage (80). In the second step, holes revealing clean surface re-appear (80), as the ordered crystalline alumina is more compact than α–Al_2O_3. Hence, another cycle of oxidation is required to fill these holes (81). The formation of the patches of clean surface is more likely related to the crystallization of the α–Al_2O_3 than to the growth mode, but the driving force would be the same. Direct observation of the growth of an ordered alumina film on an otherwise clean NiAl(110) would resolve this question. Perhaps, the most relevant observation is that of McCarty (80), who initially exposed a $Ni_{57}Al_{43}$ sample to a low dose of oxygen (360 L) at 325°C, after which low-energy electron microscopy (LEEM) micrographs showed that the oxygen adsorbed uniformly. After annealing at 925°C, most of the substrate became oxide free, with relatively few oxide islands being formed at terraces and step edges. (Interestingly, some of these oxide patches were of a type different from those produced by the standard procedure, an observation similar to that in Reference (71)). Thus McCarty's results suggest that the high-temperature, ordered Al_2O_3 does prefer to form islands rather than to spread itself over the surface, and that a transition between these two types of behavior should occur somewhere between 600 K and 1200 K. According to our interpretation, the transition temperature is the temperature at which diffusion mass-transport should be sufficient to supply the substrate surface with point defects, which are responsible for the inverted behavior. This complements our earlier observation drawn from Stierle's results (71) according to which at 570 K some diffusion in stoichiometric NiAl is already in operation. Note in passing that oxide islands were also observed at the NiAl(100) surface after incomplete oxidation and subsequent annealing at 1200 K (82).

FUTURE STUDIES

Structural Model of the Al_2O_3/NiAl(110) Oxide Film

The atomic structure of the interface between the Al_2O_3 film and the NiAl(110) surface is of major interest. However, even the question as to which of the Al_2O_3 polymorphs corresponds to the structure of the film is still under debate (see above). Another issue is the atomic structure of the NiAl substrate, including its point defects (71). A number of point defects (oxygen vacancies, for instance) are also suspected in the oxide layer (2). This is exactly the type of problem for which ab initio calculations have been helpful. To go beyond the early oxidation calculations described here, the obstacles are largely technical. The oxide film is incommensurate with the substrate, hence a large simulation cell is needed. Otherwise, additional constraints imposed by a small periodic supercell can easily change the structure of alumina and its orientation. Nevertheless, our feeling is that the power of modern computers has reached a level sufficient to tackle the problem. Taking the structural model of Jaeger et al. (24) as a guide, about 200–250 atoms of the substrate and 100 atoms of the oxide film would be a reasonable starting size for the system. We envisage a calculation in which the oxide film is of either γ- or κ-alumina type, and point defects are introduced into the two or three surface layers of the NiAl substrate. The bulk composition of $Ni_{1-x}Al_x$ should make its impact indirectly through the Ni and Al chemical potentials, but otherwise it need not be modeled explicitly. Given that even slight deviation from the stoichiometry in ordered alloys affects the composition of their surfaces (83, 84), one might expect the effect to show up in our case as well. In addition, there have recently appeared calculations of free surfaces of γ–Al_2O_3 (85) and κ–Al_2O_3 (86), thus a more comprehensive investigation of NiAl oxidation, or a similar study for other alloys, seems feasible if computationally demanding.

If the atomic structures of the growing oxide film, the substrate, and the interface between them have been established, an even more ambitious project would be a similar study of the boundaries formed between neighboring domains of the oxide film, in which there is much interest at present (67, 87, 88). The observed reflection and antiphase domain boundaries represent the simplest cases of crystal defects, namely special twin grain boundaries and stacking faults, which are straightforward to study in isolation but pose much more difficult problems for simulation where they meet the metal.

Film Thickness

The reason for the original oxide film on NiAl(110) to be exactly 5 Å thick is still a matter of speculation. It cannot be solely because the film, once formed, covers the whole surface and effectively stops the oxidation, for which two Al–O bilayers are sufficient. Indeed, on NiAl(100), the oxide film is twice as thick. It could be reasoned that there is some competition between the chemical energy gain owing to the film formation and the penalty from the elastic energy, which a strained film maintains to remain coherent with the substrate. Indeed, the lattice constants observed for the oxide film are noticeably larger than those in the bulk oxide (by 9.5% and 6.7% compared with bulk γ-alumina (24)). This might explain the dependence of the film thickness on the substrate orientation (the lattice mismatch for the NiAl(100)/γ–Al_2O_3 interface is only 3% (25)). A refinement of the elastic strain argument is the suggestion that the Madelung energy in thin two-dimensional films favors larger lattice constants. Thus the thickening of the film leads to an increase in the effective strain. The verification of such apparently reasonable arguments requires ab initio calculations, as the charges on the atoms at surfaces and interfaces do not correspond to their nominal ionic charges (89).

Stability of the Oxide Film

Another intriguing observation is the disappearance of the oxide film at about 1300 K. Is it because of enhanced diffusion, with the film growing so thick that it becomes too expensive to maintain the registry with the substrate, and thus pieces of the film spall? Such an effect is indeed observed during cyclic oxidation experiments. It is usually exploited to prepare a clean NiAl surface, free from impurities (a NiAl sample is heated up to 1500 K and than cooled). However, the alumina film was actually observed to gradually disappear, starting from the step edges, which hints at some sort of thermodynamic instability of the film rather than at any mechanical event (80). This is indeed puzzling because the calculations do not show any signs of the film instability at reported experimental temperature and oxygen partial pressure if static equilibrium is assumed. Resolving this paradox remains a challenge for both theoreticians and experimentalists working on the thermodynamics and kinetics of oxidation. Note in passing that the decomposition of the oxide film at sufficiently high temperature seems to be a rather general effect: It occurs at 1500 K for the Al_2O_3 film on NiAl(100) (90), whereas the Ga_2O_3 film formed on CoGa decomposes at 770 K on the (110) surface (91) and at 850 K on the (100) surface (73).

Free Energy of the Oxidized Surfaces

A significant step forward for simulations of the oxidation process would be a more explicit treatment of the finite temperature effects. Nevertheless, as mentioned above, both the effect of the vibration of substrate atoms and the configurational disorder of bulk point defects would likely be cancelled (to a large extent) if the equivalent amount of bulk material in the expression for the surface energy were subtracted (Equation 4). That is, the remaining effects should be small compared with the gas phase contribution to the free energy. Indeed, according to simple estimations Reference (33) for the RuO_2 (110) surface, the vibrational contribution to the surface energy should not exceed 10 meV/Å^2 at temperatures up to 1000 K. Our own surface energy calculations using the quasi-harmonic approximation for Al_2O_3(0001) indicated a contribution rising to about 10% at 2000 K. Any more detailed study of this important issue is still lacking to our knowledge. We do expect a somewhat larger contribution in less ordered surface structures due to configurational entropy of adsorbed species at relatively low coverages, and owing to the vibration of loosely arranged surface atoms at higher coverages. For the latter, standard harmonic approaches may not be adequate, as the potential energy surface of such configurations is rich in local minima separated by low barriers. For instance, even during the relatively short MD runs reported here, the A_3 configuration kept jumping between two closely spaced minima (53). For situations like this, the harmonic approximation fails and ab initio MD simulations (92) in combination with thermodynamic integration could be considered instead. Even if simply the entropy S^0 at some reference temperature T^0 could be reasonably estimated in such calculations, a substantial improvement of the estimated Gibbs free energy is expected by adding the linear $(T - T^0)S^0$ term. More generally, systematic studies of the magnitude and relative importance of various contributions to the free energy of an oxidizing substrate up to the coverage of a few monolayers would not only improve the accuracy of estimates of the relative stability of competing surface structures, but, more importantly, would provide general guidance for either neglecting or including particular contributions to the Gibbs free energy of oxidized surfaces.

Other Systems to Study

A large number of first-principles investigations of the oxidation of metallic surfaces have appeared in the past several years, confirming that this important topic is now within the reach of modern ab initio techniques. However, apart from the cited studies of the NiAl(110) surfaces (40, 53), these studies have been restricted to the oxidation of elemental crystals. As we have seen for NiAl, oxidation of multicomponent systems is a much more complex issue, the main principles of which we are just beginning to understand. It is also a rather common situation in practice. Any attempt to look at other systems would therefore be of interest, to see which of the discussed features are common, and which are peculiar to the oxidation of NiAl(110).

An immediate example would be the oxidation of the other low-index surfaces of NiAl, (100) and (111). Both surfaces can be either Ni- or Al-terminated, unlike the (110) surface, which contains half Al and half Ni sites. Hence, the composition of the bulk alloy (either Ni-rich, or Al-rich) may be even more important for such non-stoichiometric faces. A peculiar property of the oxidized NiAl(100) surface is that the structure of the thin Al_2O_3 film formed is highly sensitive to temperature. It is twice as thick as the (110) surface and proceeds through several allotropic modifications with temperature before disappearing at 1500 K (90). NiAl(111) upon oxidation at 1100–1200 K forms (110) facets. The oxide film is found to be located on the (110) surfaces of these facets, whereas (111) terraces remain oxide free. Upon further annealing at higher temperatures (1500 K), the (111) surface becomes flat again (67). The ab initio approach may throw light on the mechanism behind this delicate interplay of surface energies. CoAl, FeAl, and CoGa are among other high-temperature-ordered intermetallic compounds that exhibit selective oxidation and have been extensively studied experimentally (see 67, 72 and references therein). It would be interesting to see whether the surface energies in these compounds depend on oxygen coverage in the same inverted way as they do for NiAl(110), and whether the transition from layer-by-layer to island growth occurs if point defects are allowed to arrive from the substrate. Characteristic temperatures used for initial oxygen deposition and subsequent annealing steps vary quite significantly from one system to another, which might be helpful in decoupling the bulk diffusion and film-ordering events.

A valuable next step toward the theory of the oxidation of ordered multicomponent crystals would be a study of binary systems in which both components form oxides at comparable rates. TiAl might be an interesting and rather rich system for such an investigation. Upon oxidation, Al_2O_3 and various Ti oxides are formed depending on the oxidation conditions and the composition of the substrate (see, e.g., 93).

CONCLUSIONS

Oxidation of NiAl is a complex process, characterized by the strong interaction of oxygen with Al and the highly ordered nature of the alloy with its characteristic point defect structure. We have described a theoretical framework within which the surface energy of an ordered, but not necessarily stoichiometric, compound can be calculated as a function of external variables defining the state of the system (environmental parameters). The parameters are the temperature, composition of the substrate, and partial pressure of oxygen, with an emphasis in particular on how the surface energy dependence on the environmental parameters can be well approximated with simple analytical expressions.

Oxidation of NiAl proceeds through the arrival and re-arrangement of point defects in the substrate, the ultimate goal of which is to expose Al atoms to oxygen at the oxide-substrate interface and to transport Ni atoms away from it. This is done by exchanging Ni and Al atoms between the surface and subsurface layers and by absorbing Ni vacancies from the bulk. Thus vacancy injection into the substrate is unlikely to occur during the initial stages of the oxidation. On the contrary, the vacancies must be consumed at the oxide-substrate interface. Vacancy injection, however, might be possible at a later stage, after a mature oxide film has been formed through which Al atoms diffuse out to form more oxide. Our oxidation scenario is thus an extended version of the one suggested by Jaeger et al. (24), namely, the initial 5 Å-thick oxide film replaces the two outermost layers of the NiAl(110), whereas Ni atoms dissolve in the bulk and are partially replaced by Al; the latter together with the Al atoms from the two surface layers combine with oxygen to form the Al_2O_3 film.

The 5 Å-thick oxide film is not in thermodynamic equilibrium under most experimental conditions but rather is a transient state in which the system is held by energy barriers. It seems that thermodynamic barriers must be overcome during film formation, in addition to the energy barriers, in order for mass transport to proceed. These thermodynamic barriers are expected to be larger for Ni-rich than for Al-rich NiAl.

Some such thermodynamic barriers are a consequence of the inverted dependence of the surface energy σ on coverage: Under conditions where the surface with a particular coverage becomes stable, a surface with a higher coverage has an even lower energy. The opposite, normal type of behavior, is characterized by a monotonic increase in the adsorbed oxygen with p_{O_2}. We showed that because of the linear dependence of σ on log p_{O_2} (the ideal gas assumption), the inverted behavior of surface energies means the convexity of the dependence of surface energy on oxygen coverage. This implies that the differential binding energy of oxygen increases with coverage. The reason lies in the strong interaction between oxygen and aluminum atoms: Increasing, locally, the concentration of oxygen atoms enables them to create a more favorable local environment of aluminum. This, in turn, favors the oxide film growing as separate islands rather than as a uniform layer. Our calculations reveal an interesting link between the dependence of σ on coverage and the point defect mechanism of oxidation. An inverted behavior is observed as long as point defects are allowed in the immediate substrate. If, on the contrary, the immediate substrate is forced to stay perfect, then the surface energies depend on the coverage in a normal way. In other words, constraining the substrate to remain defect-free is sufficient to reduce dramatically the cooperative effect between the oxygen atoms. Whether this is a typical for selective oxidation in ordered compounds or peculiar to NiAl is not yet clear. We look forward to calculations on other systems that could clarify this point.

This picture of NiAl oxidation rationalizes a large body of experimental data. For instance, we now understand how the study of lattice point defects in the top substrate layers, on the one hand, and monitoring the oxide growth on the other hand, are looking at two sides of the same coin, namely, whether particular experimental conditions are sufficient to activate mass transport, mediated by point defects, in the bulk NiAl. This idea can be used to clarify why the established procedure to grow a thin alumina film with a well-controlled thickness is such as it is. Because no oxide islands were observed during the initial oxygen deposition at 600 K (80), we conclude that at this temperature the point defect mobility is too low to invert the dependence of surface energy on coverage, which would induce the island growth mode. This means that the type and number of point defects created at and near the substrate surface are not thermodynamically optimal. Subsequent high-temperature annealing, provided that it is long enough and/or the temperature is sufficiently high, removes this constraint. The difference in the annealing temperature might thus explain the apparent discrepancy between the similar studies (71) and (79) regarding the presence of lattice defects in the top substrate layer. The normal behavior of the surface energies is what is needed at the initial deposition stage to produce a uniform layer of adsorbed oxygen atoms that will later be converted into an ordered film. Therefore, raising the temperature at this stage is undesirable because further enhancement of the point defect mobility may induce the island growth mode. Decreasing the deposition temperature, on the other hand, would be unfavorable for a different reason. We recall here the experiment of

Isern & Castro (94), showing that during room temperature oxidation, the oxygen uptake saturates after 90 L exposure. If this is insufficient to build a 5 Å alumina film during the annealing stage, we can understand why low-temperature oxygen deposition results in a different oxide structure (71) and why the reduced dose of oxygen at 600 K leads to similar consequences (80). The reason for this early saturation of oxygen coverage may be that even the near-surface exchanges of Ni and Al necessary for further uptake of oxygen are too slow at room temperature. The existence of an optimal temperature with respect to the quality of the oxide film has been confirmed in experimental work by Lay et al. (70).

We are far from pretending to having reached an exhaustive description of the mechanism of the initial oxidation of NiAl. Surely, many points can also be elucidated by future experimental studies. However, we believe that the NiAl case study is a good illustration of how ab initio calculations, despite their limitations, can provide useful insight even into such a complex issue as the selective oxidation of ordered compounds.

LITERATURE CITED

1. Miracle DB. 1993. The physical and mechanical properties of NiAl. *Acta Metall. Mater.* 41:649–84
2. Bäumer M, Freund H-J. 1999. Metal deposits on well-ordered oxide films. *Prog. Surf. Sci.* 61:127–98
3. Voter AF, Montalenti F, Germann TC. 2002. Extending the time scale in atomistic simulation of materials. *Annu. Rev. Mater. Res.* 32:321–46
4. Ciacchi LC, Payne MC. 2004. "Hot-Atom" O_2 dissociation and oxide nucleation on Al(111). *Phys. Rev. Lett.* 92:176104
5. Stampfl C, Scheffler M. 1996. Theoretical study of O adlayers on Ru(0001). *Phys. Rev. B* 54:2868–72
6. Reuter K, Ganduglia-Pirovano MV, Stampfl C, Scheffler M. 2002. Metastable precursors during the oxidation of the Ru(0001) surface. *Phys. Rev. B* 65:165403
7. Ganduglia-Pirovano MV, Scheffler M. 1999. Structural and electronic properties of chemisorbed oxygen on Rh(111). *Phys. Rev. B* 59:15533–43
8. Bocquet M-L, Michaelides A, Sautet P, King DA. 2003. Initial stages in the oxidation and reduction of the 4×4 surface oxide phase on Ag{111}: A combined density-functional theory and STM simulation study. *Phys. Rev. B* 68:075413
9. Li W-X, Stampf C, Scheffler M. 2003. Subsurface oxygen and surface oxide formation at Ag(111): A density-functional theory investigation. *Phys. Rev. B* 67:045408
10. Li W-X, Stampf C, Scheffler M. 2003. Insights into the function of silver as an oxidation catalyst by ab initio atomistic thermodynamics. *Phys. Rev. B* 68:165412
11. Gustafson J, Mikkelsen A, Borg M, Lundgren E, Köhler L, et al. 2004. Self-Limited Growth of a Thin Oxide Layer on Rh(111). *Phys. Rev. Lett.* 92:126102
12. Li WX, Österlund L, Vestergaard EK, Vang RT, Matthiesen J, et al. 2004. Oxidation of Pt(110). *Phys. Rev. Lett.* 93:146104
13. Lundgren E, Kresse G, Klein C, Borg M, Andersen JN, et al. 2002. Two-dimensional oxide on Pd(111). *Phys. Rev. Lett.* 88:246103
14. Lundgren E, Gustafson J, Mikkelsen A,

Andersen JN, Stierle A, et al. 2004. Kinetic hindrance during the initial oxidation of Pd(100) at ambient pressures. *Phys. Rev. Lett.* 92:046101
15. Todorova M, Li WX, Ganduglia-Pirovano MV, Stampfl C, Reuter K, Scheffler M. 2002. Role of subsurface oxygen in oxide formation at transition metal surfaces. *Phys. Rev. Lett.* 89:096103
16. Schröder E, Fasel R, Kiejna A. 2004. O adsorption and incipient oxidation of the Mg(0001) surface. *Phys. Rev. B* 69:115431
17. Schröder E, Fasel R, Kiejna A. 2004. Mg(0001) surface oxidation: a two-dimensional oxide phase. *Phys. Rev. B* 69: 193405
18. Kiejna A, Lundqvist BI. 2001. First-principles study of surface and subsurface O structures at Al(111). *Phys. Rev. B* 63:085405
19. Kiejna A. 2003. Vacancy formation and O adsorption at the Al(111) surface. *Phys. Rev. B* 68:235405
20. Tang H, van der Ven A, Trout BL. 2004. Phase diagram of oxygen adsorbed on platinum(111) by first-principles investigation. *Phys. Rev. B* 70:045420
21. Bongiorno A, Pasquarello A. 2004. Reaction of the oxygen molecule at the Si(100)-SiO_2 interface during silicon oxidation. *Phys. Rev. Lett.* 93:086102
22. Nash P, Kleppa O. 2001. Composition dependence of the enthalpies of formation of NiAl. *J. Alloys Comp.* 321:228–31
23. Lide DR, ed. 1995. *CRC Handbook of Chemistry and Physics*. Boca Raton, FL: CRC Press. 75th ed.
24. Jaeger RM, Kuhlenbeck H, Freund H-J, Wuttig M, Hoffmann W, et al. 1991. Formation of a well-ordered aluminium oxide overlayer by oxidation of NiAl(110). *Surf. Sci.* 259:235–52
25. Yang JC, Schumann E, Müllejans H, Rühle M. 1996. Chemistry and bonding investigations of NiAl/γ-Al_2O_3 interfaces. *J. Phys. D* 29:1716–24
26. Reuter K, Scheffler M. 2003. First-principles atomistic thermodynamics for oxidation catalysis: surface phase diagrams and catalytically interesting regions. *Phys. Rev. Lett.* 90:046103
27. Zhang W, Smith JR, Wang X-G. 2004. Thermodynamics from ab initio computations. *Phys. Rev. B* 70:024103
28. Bergermayer W, Schweiger H, Wimmer E. 2004. Ab initio thermodynamics of oxide surfaces: O_2 on $Fe_2O_3(0001)$. *Phys. Rev. B* 69(19):195409
29. Rogal J, Reuter K, Scheffler M. 2004. Thermodynamic stability of PdO surfaces. *Phys. Rev. B* 69:075421
30. Wang X-G, Smith JR. 2003. Surface phase diagram for $Cr_2O_3(0001)$: ab initio density functional study. *Phys. Rev. B* 68:201402
31. Ganduglia-Pirovano MV, Sauer J. 2004. Stability of reduced $V_2O_5(001)$ surfaces. *Phys. Rev. B* 70:045422
32. Meyer B. 2004. First-principles study of the polar O-terminated ZnO surface in thermodynamic equilibrium with oxygen and hydrogen. *Phys. Rev. B* 69:045416
33. Reuter K, Scheffler M. 2002. Composition, structure, and stability of $RuO_2(110)$ as a function of oxygen pressure. *Phys. Rev. B* 65:035406
34. Batyrev IG, Alavi A, Finnis MW. 2000. Equilibrium and adhesion of Nb/sapphire: the effect of oxygen partial pressure. *Phys. Rev. B* 62:4698–706
35. Wang X-G, Chaka A, Scheffler M. 2000. Effect of the environment on α-$Al_2O_3(0001)$ surface structures. *Phys. Rev. Lett.* 84:3650–53
36. Lodziana Z, Nørskov JK, Stoltze P. 2003. The stability of the hydroxylated (0001) surface of α-Al_2O_3. *J. Chem. Phys.* 118: 11179–88
37. Wang X-G, Smith JR. 2004. Hydrogen and carbon effects on Al_2O_3 surface phases and metal deposition. *Phys. Rev. B* 70:081401
38. Marmier A, Parker SC. 2004. Ab initio morphology and surface thermodynamics of α-Al_2O_3. *Phys. Rev. B* 69:115409
39. Cahn JW. 1978. Thermodynamics of solid and fluid surfaces. In *Interfacial*

Segregation, ed. WC Johnson, JM Blakely, pp. 3–23. Ohio: ASM
40. Lozovoi AY, Alavi A, Finnis MW. 2000. Surface stoichiometry and the initial oxidation of NiAl(110). *Phys. Rev. Lett.* 85:610–13
41. Johnston K, Castell MR, Paxton AT, Finnis MW. 2004. $SrTiO_3(001)$ (2×1) reconstructions: first-principles calculations of surface energy and atomic structure compared with scanning tunneling microscopy images. *Phys. Rev. B* 70:085415
42. Finnis MW. 1998. Accessing the excess: an atomistic approach to excesses at planar defects and dislocations in ordered compounds. *Phys. Stat. Solidi* 166:397–416
43. Xie J, de Gironcoli S, Baroni S, Scheffler M. 1999. Temperature-dependent surface relaxations of Ag(111). *Phys. Rev. B* 59:970–74
44. Marzari N, Vanderbilt D, de Vita A, Payne MC. 1999. Thermal contraction and disordering of the Al(110) Surface. *Phys. Rev. Lett.* 82:3296–99
45. Lazzeri M, de Gironcoli S. 2002. First-principles study of the thermal expansion of Be(1010). *Phys. Rev. B* 65:245402
46. Wagner C, Schottky W. 1930. Theorie der geordneten Mischphasen. *Z. Phys. Chem. B* 11:163–210
47. Hagen M, Finnis MW. 1998. Point defects and chemical potentials in ordered alloys. *Philos. Mag. A* 77:447–64
48. Meyer B, Fähnle M. 1999. Atomic defects in the ordered compound B2-NiAl: a combination of ab initio electron theory and statistical mechanics. *Phys. Rev. B* 59:6072–82
49. Korzhavyi PA, Ruban AV, Lozovoi AY, Vekilov YuKh, Abrikosov IA, Johansson B. 2000. Constitutional and thermal point defects in B2 NiAl. *Phys. Rev. B* 61:6003–18
50. Mishin Y, Herzig C. 2000. Diffusion in the Ti-Al system. *Acta Mater.* 48:589–623
51. Qian G, Martin RM, Chadi DJ. 1988. Stoichiometry and surface reconstruction: an ab initio study of GaAs(100) surfaces. *Phys. Rev. Lett.* 60:1962–65
52. Zhang SB, Northrup JE. 1991. Chemical potential dependence of defect formation energies in GaAs: application to Ga self-diffusion. *Phys. Rev. Lett.* 67:2339–42
53. Lozovoi AY, Alavi A, Finnis MW. 2001. Surface energy and the early stages of oxidation of NiAl(110). *Comput. Phys. Commun.* 137:174–94
54. Linstrom PJ, Mallard WG, eds. 2003. *NIST Chemistry Web Book, NIST Standard Reference Database Number 69*. Gaithersburg, MD: Natl. Inst. Standard. Technol., http://webbook.nist.gov
55. Cox JD, Wagman DD, Medvedev VA. 1989. *CODATA Key Values for Thermodynamics*. New York: Hemisphere
56. JANAF Thermochemical Tables. 1974. Suppl. 1974. *J. Phys. Chem. Ref. Data* 3
57. JANAF Thermochemical Tables. 1975. Suppl. 1975. *J. Phys. Chem. Ref. Data* 4
58. Bradley AJ, Taylor A. 1937. An X-ray analysis of the nickel-aluminium system. *Proc. R. Soc. London Ser. A* 159:56–72
59. Fu CL, Ye Y-Y, Yoo MH, Ho KM. 1993. Equilibrium point defects in intermetallics with the B2 structure: NiAl and FeAl. *Phys. Rev. B* 48:6712–15
60. Alavi A, Lozovoi AY, Finnis MW. 1999. Pressure-induced isostructural phase transition in Al-rich NiAl alloys. *Phys. Rev. Lett.* 83:979–82
61. Desai PD. 1987. Thermodynamic properties of selected binary aluminum alloy systems. *J. Phys. Chem. Ref. Data* 16:109–24
62. In calculating the ab initio curves for p_n and p_s shown in Figure 3, we have employed an additional thermodynamic cycle that allowed us to circumvent explicit calculations of Ni(fcc) and NiO crystals (see Reference 53 for details).
63. Brandes EA, Brook GB, eds. 1992. *Smithells Metals Reference Book*. Oxford, UK: Butterworth-Heinemann. 7th ed.
64. Ustundag E, Clausen B, Bourke MAM. 2000. Neutron diffraction study of the reduction of $NiAl_2O_4$. *Appl. Phys. Lett.* 76:694–96

65. Elrefaie FA, Smeltzer WW. 1981. Thermodynamics of nickel-aluminum-oxygen system between 900 and 1400 K. *J. Electrochem. Soc.* 128:2237–42
66. Doychak J, Smialek JL, Mitchell TE. 1989. Transient oxidation of single-crystal β–NiAl. *Metall. Trans. A* 253:499–518
67. Franchy R. 2000. Growth of thin, crystalline oxide, nitride and oxynitride films on metal and metal alloy surfaces. *Surf. Sci. Rep.* 38:195–294
68. Stierle A, Renner F, Streitel R, Dosch H, Drube W, Cowie BC. 2004. X-ray diffraction study of the ultrathin Al_2O_3 layer on NiAl(110). *Science* 303:1652–56
69. Yoshitake M, Mebarki B, Lay TT. 2002. Crystallinity and thickness control of well-ordered ultra-thin Al_2O_3 film on NiAl(110). *Surf. Sci.* 511:L313–18
70. Lay TT, Yoshitake M, Mebarki B. 2002. Temperature effect on growth of well-ordered thin Al_2O_3 film on NiAl(110). *J. Vac. Sci. Technol. A* 20:2027–31
71. Stierle A, Renner F, Streitel R, Dosch H. 2001. Observation of bulk forbidden defects during the oxidation of NiAl(110). *Phys. Rev. B* 64:165413
72. Rose V, Podgursky V, Costina I, Franchy R. 2003. Growth of ultra-thin amorphous Al_2O_3 films on CoAl(100). *Surf. Sci.* 541: 128–36
73. Pan FM, Pflitsch C, David R, Verheij LK, Franchy R. 2001. Formation of an ordered oxide on the CoGa(100) surface by room temperature oxidation and annealing. *Surf. Sci.* 479:191–200
74. Frazer HL, Loretto MH, Smallman RE, Wasilewski RJ. 1973. Oxidation-induced defects in NiAl. *Philos. Mag.* 28:639–50
75. Eibner JE, Engell H-J, Schultz H, Jakobi H, Schlatte G. 1975. Annealing of point defects in quenched NiAl. *Philos. Mag.* 31:739–42
76. Parthasarathi A, Frazer HL. 1984. The annealing of vacancy defects in β–NiAl. II. The role of surface oxidation in vacancy loop growth in slowly cooled crystals. *Philos. Mag. A* 50:101–15
77. Some results presented in Reference 40 were misinterpreted in Reference 71, suggesting that there is a discrepancy between the results of these studies. The authors of the latter paper took the composition of the top layer in configuration E_3 as a prediction for the defect structure of the first substrate layer. However in Reference 40, the layer in question was treated as part of the oxide film, not the substrate. Also, in Reference 40, the defect concentration was not optimized in any layer, as the supercell is far too small for that. Thus there is actually no discrepancy between the results of the two studies.
78. McCarty KF, Nobel JA, Bartelt NC. 2001. Vacancies in solids and the stability of surface morphology. *Nature* 412:622–25
79. Torrelles X, Wendler F, Bikondoa O, Isern H, Moritz W, Castro GR. 2001. Structure of the clean NiAl(110) surface and the Al_2O_3/NiAl(110) interface by measurements of crystal truncation rods. *Surf. Sci.* 487:97–106
80. McCarty KF. 2001. Imaging the crystallization and growth of oxide domains on the NiAl(110) surface. *Surf. Sci.* 474:L165–72
81. Ceballos G, Song Z, Pascual JI, Rust H-P, Conrad H. 2002. Structure investigation of the topmost layer of a thin ordered alumina film grown on NiAl(110) by low temperature scanning tunneling microscopy. *Chem. Phys. Lett.* 359:41–47
82. Blum R-P, Ahlbehrendt D, Niehus H. 1998. Growth of Al_2O_3 stripes in NiAl(001). *Surf. Sci.* 396:176–88
83. Ruban AV. 2002. Surface composition of ordered alloys: An off-stoichiometric effect. *Phys. Rev. B* 65:174201
84. Blum V, Hammer L, Schmidt C, Meier W, Wieckhorst O, et al. 2002. Segregation in strongly ordering compounds: a key role of constitutional defects. *Phys. Rev. Lett.* 89:266102
85. Pinto HP, Nieminen RM, Elliott SD. 2004. Ab initio study of γ-Al_2O_3 surfaces. *Phys. Rev. B* 70:125402
86. Ruberto C, Yourdshahyan Y, Lundqvist BI.

2003. Surface properties of metastable alumina: a comparative study of κ- and α-Al_2O_3. *Phys. Rev. B* 67:195412
87. Kulawik M, Nilius N, Rust H-P, Freund H-J. 2003. Atomic structure of antiphase domain boundaries of a thin Al_2O_3 film on NiAl(110). *Phys. Rev. Lett.* 91:256101
88. Nilius N, Kulawik M, Rust H-P, Freund H-J. 2004. Defect-induced gap states in Al_2O_3 thin films on NiAl(110). *Phys. Rev. B* 69:121401
89. Finnis MW. 1999. Why is InSb(111) reconstructed? *CCP3 Newslett.* 24:2–11. http://www.Ccp3.Ac.uk/newsletters.Shtml
90. Gassmann P, Franchy R, Ibach H. 1994. Investigation on phase transitions within thin Al_2O_3 layers on NiAl(001)—HREELS on aluminium oxide films. *Surf. Sci.* 319:95–109
91. März A, Franchy R. 2000. Oxidation of CoGa(110). *Surf. Sci.* 466:54–65
92. Walker BG, Molteni C, Marzari N. 2004. Ab initio molecular dynamics of metal surfaces. *J. Phys. Condens. Matter* 16:S2575–96
93. Schmiedgen M, Graat PCJ, Baretzky B, Mittemeijer EJ. 2002. The initial stages of oxidation of γ-TiAl: an X-ray photoelectron study. *Thin Solid Films* 415:114–22
94. Isern H, Castro GR. 1989. The initial interaction of oxygen with a NiAl(110) single crystal: a LEED and AES study. *Surf. Sci.* 211/212:865–71

Annu. Rev. Mater. Res. 2005. 35:209–38
doi: 10.1146/annurev.matsci.35.012704.142050

First published online as a Review in Advance on March 17, 2005

ENGINEERED POROUS CATALYTIC MATERIALS

Ferdi Schüth
Max-Planck-Institut für Kohlenforschung, 45470 Mülheim, Germany;
email: schueth@kofo.mpg.de

Key Words catalysis, single site, solid, oxide

■ **Abstract** In recent years substantial progress has been made in the control of the properties of porous materials on the nano-, meso-, and macroscale. The feature of solids can, in favorable cases, be tailored to such an extent that one is justified in using the term engineering of porous catalytic materials. This review highlights recent developments concerning (*a*) the control of the pore structure of zeolites, (*b*) the creation of single site type catalysts based on ordered mesoporous oxides, (*c*) the catalytic potential of porous coordination compounds, and (*d*) the advances with respect to direct production of desired shapes of porous catalytic materials.

INTRODUCTION

Solid catalysts, in contrast to molecular catalysts, which are used in homogeneous catalysis, are often not very well-defined systems. They are mostly complex solids consisting of various components and phases and have inhomogeneous surface properties. The fact that one often describes the production of solid catalysts as preparation and not as synthesis already suggests a comparably low degree of rational control over the properties of the resulting materials. Although solid catalysts can typically be produced in a highly reproducible manner, as impressively demonstrated by the various commercial heterogeneously catalyzed processes with capacities of several hundred thousand tons per year per plant, we are in most cases far away from a true design of solid catalytic materials. This is different for molecular catalysts. Here the structures of the catalyst are typically well defined, even if there is disagreement in some cases with respect to the state of the catalyst during reaction, such as in the case of the Heck-reaction (1). By tailoring the ligands around a metal center, one can fine-tune the performance of molecular catalysts often in a quasi-rational manner. In addition, computational tools are nowadays developed to such a level that it is possible to describe molecular catalysts by quantum chemical methods, which also helps to understand the catalyst action and to support the design of improved catalysts.

However, some developments of the past years have also brought a more rational tailoring of solid catalysts within reach in special cases. Acid catalysis over zeolites is probably as close as one can get toward single sites in heterogeneous catalysis, at least in those cases where it is clear that the reaction proceeds on the strong Brønsted acidic sites of the zeolite. In addition, a more rational control of specific features of zeolitic materials has become possible recently. These developments are addressed in the first section of this review.

Somewhat less defined, but still highly regular, are ordered mesoporous materials (2). They also provide the opportunity to control properties in a rational manner and also allow the detailed analysis of the resulting materials, which is as important as the control of the properties. Modified ordered mesoporous materials are discussed in the following section of the review.

The third development of the past several years, which is relevant here, is that of porous coordination compounds, which have thoroughly been reviewed recently (3–6). Molecular catalysts are often isolated coordination compounds in solution, and porous solid coordination compounds can be considered, with some justification, as simply the extended solid-state version of such complexes. If change of coordination of the metal centers is possible, such materials could be considered as polymeric, single-site catalysts. Although at present not many catalytic applications have been reported, a subsequent section four alerts readers interested in novel catalytic materials to this emerging class.

Finally, as important as it is to tailor the catalytic sites, the macroscopic form of the catalysts can also help to realize novel process options. Methods to control this property are covered in the final section.

ZEOLITES BY DESIGN

Zeolite Structure Design by Tailoring Templates

Templates are often used in the synthesis of zeolites in order to direct the formation process to a specific structure. The word template suggests a one-to-one correspondence between the shape of the molecules and the cavity formed, which has been proven with a high level of confidence only in few cases, such as for instance in the case of ZSM-18, where the tri-quat molecule with a D_{3h} point symmetry is located in a cage with C_{3h} symmetry. The interactions with the framework are so strong that rotation of the guest molecule is not possible, which suggests not only a mere space filling but a specific interaction defining the shape of the cavity (7). Because convincing evidence for true templating has not been given in most cases, it is preferable to use the term structure-directing agent (SDA) or organic additive. However, even if a true templating is not in operation in most zeolite synthesis processes, one can still use the correspondence between size and shape of molecules used as additives, and size and shape of the cavities formed by them to rationally design crystalline porous materials. The basis for this type of work was laid by Gies & Marler, who had demonstrated that in cases where Coulomb interactions were absent, the van der Waals interactions between organic molecule and the walls of the silica determine the pore system of the resulting porous silica polymorphs (8).

The earliest approach for a computational design of template molecules was published by Lewis et al. (9). In short, first the cavity to be templated is defined, and then an organic molecule is computationally grown in this cavity from predefined fragments until van der Waals overlap with the cavity wall occurs. The program then tries to remove this conflict by a set of operations such as moving the organic molecule, rotating it, etc. If the conflict cannot be removed, the fragment is rejected and growth stops. The best templates are selected on the basis of maximum van der Waals contact between molecule and cavity walls. This technique had subsequently been used to design novel templates that were indeed successful in the synthesis of microporous crystalline materials (10, 11). In more recent years, however, the computational de novo design of template molecules seems to be used to a lesser extent, and more emphasis is placed on rationalizing the action of SDAs during the formation of specific zeolitic materials. This may be due to the fact that the interactions often do seem to be more complex than simply the optimum space filling. A recent study on the efficiency of different fluorine-substituted benzylpyrrolidine derivates as SDAs in the formation of the aluminophosphate $AlPO_4$-5 has shown that subtle differences in packing can lead to strongly differing templating abilities (12). Without accounting for packing effects, the computationally determined templating ability did not correspond to the experimentally observed trends. The van der Waals interaction, which is still the dominating contribution, is rather similar for all molecules studied. However, if SDA-SDA interactions are taken into account, strong differences for the different SDAs investigated become obvious. An especially important observation was the fact that the SDA-SDA interaction changes the density of the templates in the structure. So even if the interaction energy per SDA molecule does not substantially change for different derivates, the total system energy is lowered for more densely packed SDAs. This is normally not taken into account if the stabilization of microporous structures by molecular templates is investigated.

Another complication arises from the fact that in many cases it is not the most stable structure that is formed in the presence of a particular template, but rather the one that nucleates most easily. This has been investigated in a study by Sastre et al. (13). In a system where several different zeolites can form, i.e., beta, EU-1, ZSM-11, or ZSM-12, both the most stable structures and most stable nuceli were calculated (approximated by considering only the energy contribution from close-range interactions, which are considered to best model the gel situation during formation of the material). The authors showed that depending on conditions, the system could be either thermodynamically or kinetically controlled and the influence of the different SDAs could be rationalized. For the different structures they also included the effect of different packing density per tetrahedral unit. Water was still excluded in these calculations, but the computational results of the study coincided well with the experimental trends. A similar rationalization of the selectivity of the synthesis for one of several possible structures (based on energetic considerations) has also been possible for systems containing germanium (see below for the influence of germanium). In situations where the template does not provide a good fit with the zeolite, several other factors contribute to directing the synthesis toward a certain structure, such as the fraction of trivalent atoms incorporated, the water/silicon ratio, or the presence of germanium or fluoride (14).

Considering the level of understanding that has been reached with respect to the influence of organic additives in a zeolite synthesis, it can be stated that an ex-post rationalization of the SDA action seems to be possible in many cases. The ex-ante planning of a zeolite synthesis by design of the template through computational methods is still in the early stages, but several successful attempts have been reported (see above). One should also keep in mind that a more qualitative design of template molecules to produce certain zeolite structures has been successful in the past and has resulted in the synthesis of highly interesting zeolitic materials, which could be considered as designed materials.

High silica zeolite A, which was recently described (15) for the first time, is a prominent example of this strategy. The requirements of a successful template are these: A successful template should not form strong complexes with the solvent. It should fit the cage to be templated with as many van der Waals contacts as possible and have minimum deformation, and it should be sufficiently hydrophobic to allow the formation of a high silica zeolite. For the sodalite cage, which has to be templated in the synthesis of zeolite A, it proved very difficult to find a suitable organic molecule. Corma et al. therefore decided to search for compounds that would form aggregates by π-π-interactions. Eventually, 4-methyl-2,3,6,7-tetrahydro-1H,5H-pyrido[3.2.1-ij]quinolinium iodide was found to be successful in templating high silica zeolite A. The molecule was not incorporated as such, but in the form of a dimer that fulfilled the requirements formulated for the template (Figure 1, see color insert). The material can be produced with high Si/Al ratios up to a fully aluminum-free sample. The first sorptive and catalytic data reported for this material make it interesting for the processing of small hydrocarbon molecules.

Zeolite Structure Design by Controlling Ring Sizes

The last example in the previous section, the synthesis of high silica zeolite A, relied not only on a tailored template, but also on another method to rationally control the formation of certain structures: the introduction of germanium into the framework to prefer certain ring sizes, a method introduced by the Corma group (16). The importance of heteroatoms in directing the synthesis of microporous materials to certain structures had previously been discussed by Davis and coworkers (17, 18), especially in connection with the synthesis of VPI-8 (19), which is made possible by the addition of zinc compounds to the synthesis gel, where one may speculate (20) that zinc favors the formation of three rings. These are present in VPI-8 and other structures with related structural elements that are formed by the incorporation of Zn, such as RUB-17 (21). However, whereas in previous work the incorporation of heteroelements was used on a more empirical basis to discover novel zeolites, Corma et al. deliberately used the concept to stabilize certain ring sizes by heteroelement incorporation to favor the formation of desired structures.

In quantum chemical studies it was discovered that germanium is able to stabilize double four-membered rings unlike those in the all-silicon variety (22). The calculations were carried out at the Hartree-Fock level. For both hypothetical situations (formation of double four rings in aqueous media, neglecting solvation) or from a hypothetical solid-state reaction), an increase in the number of germanium atoms in the rings made the rings more stable, irrespective of the relative position of the germanium atoms, as long as no Ge-O-Ge bonds were formed. This observation was the starting point for a series of studies on the synthesis of germanium-substituted zeolite materials, in which germanium incorporation was found to facilitate the crystallization of materials with a high fraction of small rings, such as double four rings. It was thus possible to prepare the pure polymorph C of zeolite beta with a variety of different templates, which gave three different zeolite structures (beta, ITQ-4, and ZSM-12) in the absence of germanium (16). For ITQ-7 a strong increase in the formation rate has been observed upon addition of germanium to the synthesis gel (22). In addition, several novel zeolitic materials with interesting pore systems and high fraction of double four ring units could be produced. ITQ-22 is a material with three different pore sizes (8-, 10-, and 12-membered rings forming a three-dimensional pore system) that is prepared by the combination of a special SDA and germanium as one component of the synthesis gel (23). A similar approach led to the formation of ITQ-21, a large pore material with 12-membered ring pores and a large cavity (24), and the high silica zeolite A mentioned above (15). Interesting from an applications point of view is ITQ-15, which also was synthesized from germanium-containing gels (25); its structure has recently been determined (26). The material has a unique pore system with 14-membered ring pores perpendicular to 12-membered rings. At the same time, the structure of a material (IM-12) that is probably identical to ITQ-15 has been described by the Mulhouse group (27). The Rietveld analysis of both teams supports the notion that the germanium atoms are preferentially, possibly even exclusively, located in the double four rings in the structure.

Replacement of silicon with germanium does not lead to changes in the framework charges and, therefore, in order to create Brønsted acidic sites, incorporation of trivalent elements is necessary if the materials are to be used in acid catalysis. An important additional effect of the presence of germanium is the facilitated incorporation of aluminum in the structure.

Creating Transport Pore Systems in Zeolites

As much progress as has been made with respect to increasing the pore size of zeolites (see above), diffusion is still a highly restricted process in zeolites. Therefore, it is essential for many processes either to decrease zeolite particle size in order to maximize catalyst efficiency or to provide an additional pore system with larger pore sizes to facilitate mass transfer. Just minimizing particle sizes is in most cases not sufficient, because the colloidal zeolites tend to grow to larger

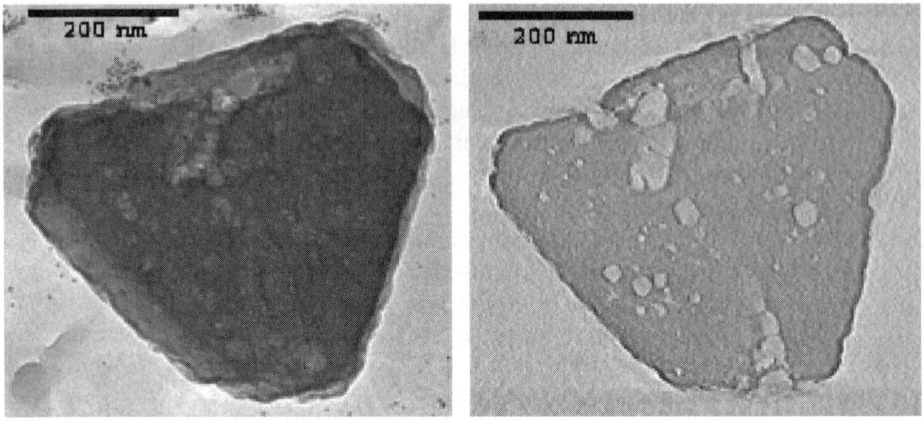

Figure 2 Conventional transmission electron microscopy (TEM) (*left*) and a two-dimensional section from a tomographic reconstruction (*right*) of an ultrastabilized zeolite Y crystal. Reproduced with permission from Reference 28.

particles if one tries to isolate them, and additional precautions have to be taken to maintain the high dispersion achieved in the synthesis process.

The creation of a mesopore system in zeolites has long been known and is one beneficial effect of the ultrastabilization process in the production of ultrastabilized zeolite Y (USY) zeolites for catalytic cracking. Using electron tomography, it is now possible to directly visualize the mesopore system generated in a faujasite during high-temperature steam treatment (28). Figure 2 shows one cut generated from the reconstruction of the electron micrographs, and the transport pore system can clearly be seen.

Beyond the generation of pore systems within zeolite crystals by the ultrastabilization process, several methods have been reported in recent years that are suitable to produce zeolitic materials with a designed mesopore system. One such technique is the use of different carbon materials as endotemplate (29) in zeolite synthesis (30, 31). The general procedure consists of the impregnation of a carbon source, such as carbon black or carbon nanotubes, with a zeolite precursor solution or gel, hydrothermal synthesis and subsequent calcination, which removes both the molecular template used for structure direction of the zeolite and the carbon template for the generation of the mesopore system. Interestingly, the carbon is incorporated into zeolite crystals, which appear as single crystals, but are penetrated by holes that are either spherical if carbon blacks are used as additives (30) or channel-like if carbon nanotubes are used (31). Visualization of the pore system

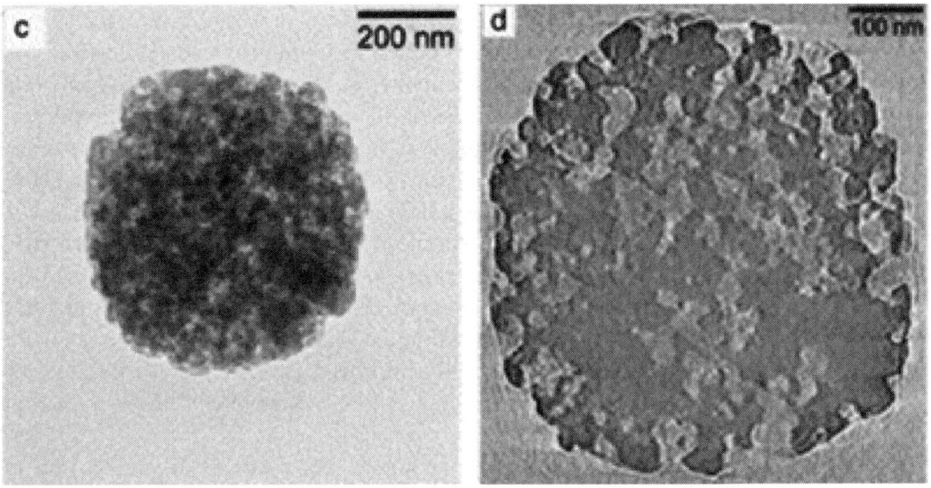

Figure 3 Conventional TEM (*c*) and two-dimensional section from a tomographic reconstruction (*d*) of a ZSM-5 type zeolite synthesized with carbon black as *porogen*. Reproduced with permission from Reference 32.

by electron microscopy is given in Figure 3 (32), and comparison with Figure 2 reveals that the pore system is more regular than the pores created by steaming. The reverse situation, where carbon is not incorporated in a growing zeolite crystal, but zeolite is grown in the pore system of porous carbon aerogel or an ordered mesoporous carbon to create zeolites with a mesopore system, can also be realized (33, 34). Due to the possibilities for tuning the texture of the carbon aerogels, the textural properties of the zeolite prepared in the carbon aerogels can be tuned to some extent as well. However, due to the current high costs of carbon aerogel, the synthesis involving cheap carbon blacks at present appears to be more attractive.

A rather surprising method—which is probably not relevant—to create mesopores in zeolite crystals was discovered accidentally. Kato et al. (35) discovered that platinum particles present on the external surface of ZSM-5 crystals would under DeNOx reaction conditions tunnel into the crystal and thus produce a pronounced mesopore system with the size of the tunnels determined by the size of the platinum particles. The mesopores seem to be "cut out" of the structure and have remarkably precisely defined edges. As a means for practically controlled generation of a mesopore system, this pathway is probably not useful, but more precise analysis of the data may induce ideas for easier methods to achieve similar results.

Another possibility to create a mesopore system in a zeolitic material is the organization of colloidal zeolites. Such colloidal zeolites can be synthesized following procedures developed by Schoeman et al. (36), but if isolated from solution, they are often difficult to obtain as solids consisting of separated colloidal particles. One way to organize colloidal zeolites in order to create a mesopore system is cross linking with a silicone and subsequent calcination. The silicone precursor is converted to silica by the calcination process that separates the colloidal zeolite particles and thus forms a mesopore system. This makes the MFI-type zeolite easily accessible for reagents (37). A test of such materials in the Beckmann rearrangement of cyclohexanone oxime to ε-caprolactam, a reaction that is severely transport-limited in MFI materials or may even occur on the external surface, revealed a substantially improved performance of these cross-linked colloidal zeolites compared with the performance of conventional zeolitic catalysts.

Alternatively, colloidal zeolites or zeolite "seeds" can be organized into an ordered mesoporous material or coated onto the surface of ordered mesoporous materials of the MCM-41 or SBA-15 type. Several attempts have been made to create ordered mesoporous materials with zeolitic structural elements in the walls. After first reports, where probably only composites were formed (38, 39), the first successful attempts likely came from the groups of Pinnavaia (40) and Zhang et al. (41). Both groups used zeolite synthesis gels at stages before crystalline material was detected and added surfactant to create the mesostructure. Because wall crystallinity cannot be detected in such materials owing to the thin walls, evidence for zeolitic structural elements is circumstantial. However, the catalytic activity in acid-catalyzed reactions was superior to regular MCM-41, which provides evidence for an altered wall structure. Subsequently, Kaliaguine et al. developed methods by which zeolite particles are coated onto the walls of ordered mesoporous materials, mostly onto SBA-15-type materials due to the larger pore sizes (42, 43). Here improved stability and altered catalytic behavior also give evidence for success of this strategy.

ORDERED MESOPOROUS SOLIDS

Ordered mesoporous solids are not only useful as supports for zeolites. These materials, the most investigated members being MCM-41 (44) and SBA-15 (45) silicas, extend the size range accessible with systems with a sharp pore size distribution into the mesopore range. A pore system with equal pores and a homogeneous surface is one requirement for the creation of solid single-site catalysts. Even if molecular catalysts are anchored to a solid surface, site homogeneity not only depends on the molecular catalyst but also on its environment in a pore system, and only a pore system in which all pores are rather similar provides the opportunity for solid single-site systems.

Engineering Surface Properties

One requirement for a solid single-site catalyst is a high degree of surface uniformity, because surface heterogeneity would immediately imply heterogeneity of the properties of a molecular species owing to different interactions with the surface. A truly homogeneous surface probably does not exist, with the possible exception of single-crystal surfaces with very low defect concentration. Such materials are not useful in technical catalysis because single-crystal surfaces are not stable under technical conditions and have surface areas that are too low. In addition, even on such perfect surfaces, multiple different sites exist for the adsorption of molecules, such as the on-top or bridge-bonded adsorption mode of CO on noble metal surfaces.

However, ordered mesoporous silica is a material that could come close to providing a homogeneous surface. Unmodified ordered mesoporous silicas have some distribution of surface sites, because the silanol density of most materials is around 2 nm^{-2}, i.e., about 50% of the silanol density of regular hydroxylated silica (46, 47). For SBA-15, a silanol density of about 1 nm^{-1} has been reported, which is about the expected value for a silica pretreated at 550°C (48). It is not clear how the regular hydroxylated silicas differ from the surfaces of MCM-41-type or SBA-15-type silicas, because in many cases the pretreatment conditions are not fully specified and it is not clear to which extent the surface is indeed rehydroxylated after the calcination step. There is evidence for surface heterogeneity of such materials with hydrophobic and hydrophilic patches, at least for the MCM-41-type materials (49). However, irrespective of whether the silanol groups are homogeneously distributed over the surface or clustered, the surface is invariably composed of silicon atoms that are connected to all neighbors via siloxane bonds and silicon atoms bearing a silanol group. Thus there is an intrinsic heterogeneity of the surface. Compared with conventional silica, at least the pore sizes and thus the spatial restrictions are identical for all sites.

Surface modification allows the passivation of the silanol groups by silylation to obtain a more homogeneous surface functionality, and additional surface functionality can also be induced. A whole wealth of methods are available for surface modification of ordered mesoporous materials that are often based on the techniques developed for the surface functionalization of conventional silica. These techniques had been developed for use in chromatography, for example, and they have been reviewed in detail in several relatively recent papers (50–52).

Surface silylation by trialkylchloro- or trialkylalkoxysilanes is relatively straightforward and was early on described by the Mobil group (53). Since then, numerous papers have appeared describing the formation of organically modified ordered mesoporous materials. A novel development was the direct formation of such solids by co-condensation of tetraalkoxysilanes with organically functionalized silanes, which was introduced in 1996 almost simultaneously by Burkett et al. (54) and Macquarrie (55). The co-condensation routes have several disadvantages: (*a*) the need for surfactant extraction instead of removal by calcination and the altered synthesis parameters for formation of the mesostructure [which can, in turn, be used to direct the formation of specific structures (56)]; and (*b*) the need for compatibility of the functional group of the organosilane with the mesophase synthesis and certain problems created by different hydrolysis rates of the tetraalkoxysilane and the functional silane. There are also some advantages. First, the synthesis proceeds in a single step, possibly even accompanied with the introduction of the desired functional group (57). In addition, especially if low loadings are desired, the two-step procedure does not seem to be optimal because it has been reported to lead to an inhomogeneous modification of the surface, with high concentration of the organic groups on both the external surface and near the pore mouths as well as depletion in the center of the particles (58). Such inhomogeneous surface modifications are not desirable in the creation of immobilized single-site catalysts.

On the contrary, capping of most of the silanol groups is necessary in order to leave only a few free for modification with functional silanes to which the molecular catalysts can be anchored, thereby creating truly isolated sites. Otherwise the ligand spheres and/or the metal centers of neighboring catalyst species may influence each other, and substantial inhomogeneity of the site may result.

Creating Mesoporous Solids with Single Sites

If molecular catalysts are anchored to relatively homogeneous surfaces, solid single-site catalysts can be created. Here, a single-site catalyst is defined as a catalyst having exclusively catalytic centers that are identical with respect to their environment and thus identical with respect to their catalytic performance. It is, of course, difficult to prove the existence of a single-site system, but normally high selectivity in a catalytic transformation would be a feature expected from a single-site system. Molecular catalysts heterogenized on ordered mesoporous silica have been reviewed recently (2, 59–61), and therefore this section focuses on materials in which the mesostructure has a special effect indicating the formation of a homogeneous environment of the catalytically active species. These aspects were also recently highlighted by Thomas & Raja (62). With respect to the creation of solid single-site catalysts, ordered mesoporous silicas offer unique possibilities compared with other supports. In these materials, the surface can be made chemically somewhat homogeneous by the methods discussed above. Moreover, beyond chemical homogeneity spatial homogeneity can also be an issue in catalysis. Ordered mesoporous silicas provide pore systems with constant curvature of the surfaces throughout, something that is never realized in silica gel. Moreover, the homogeneous pore size distribution also guarantees that the distance of the anchored molecular catalyst to the next pore wall is equal for all species immobilized if the chosen loading is sufficiently low. However, one should keep in mind that it is not always clear whether the immobilized molecular catalysts really protrude freely into the pore space, as suggested by many of the published drawings of such

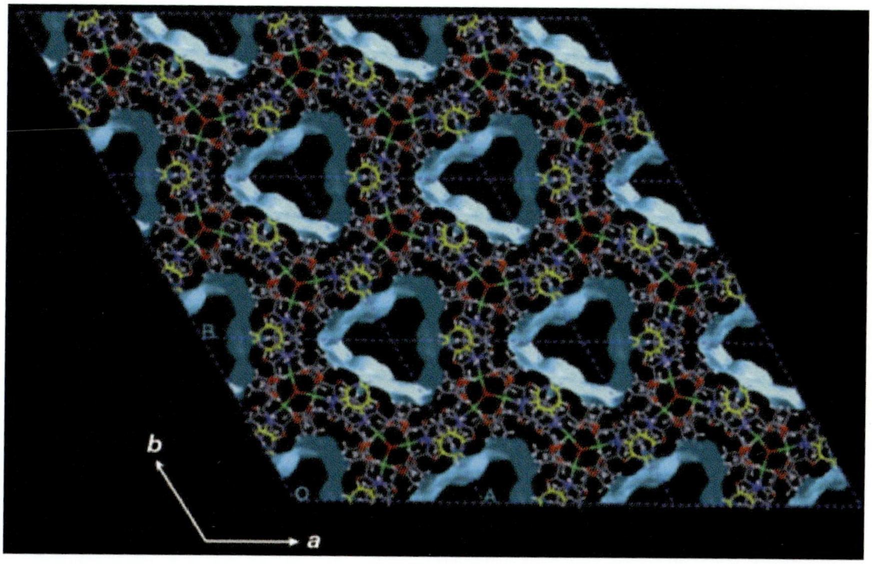

Figure 8 Pore system of the chiral coordination compound, which is active for a transesterification reaction. Racemic alcohols are converted with a slight preference for one enantiomer. Reproduced with permission from Reference 96.

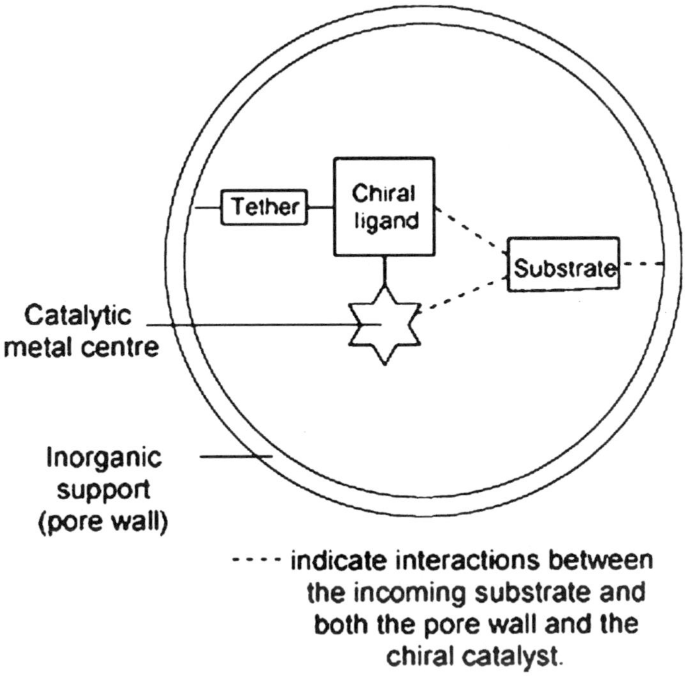

Figure 5 Schematic drawing of the additional spatial restriction imposed by the pore system on the substrate in the vicinity of the catalytic center. Reproduced with permission from Reference 64.

ligands, a tripodally bonded, isolated titanium center formed. Other methods of titanium introduction tend to lead to formation of polynuclear titanium species on the surface. In a similar way, isolated molybdenum sites can be generated from molybdenocene dichloride at low loading around 1%, whereas at higher loading, polymerization of the molybdenum species is observed (68).

Isolated metal centers have also been produced by Tilley et al. for copper (69), iron (48), and titanium (70). This group uses (*tert*-butoxy)siloxy metal complexes as precursors to graft metal centers onto the surface of ordered mesoporous silica. After removal of the ligands, stable, isolated metal centers are obtained, and the materials are active and selective catalysts in a variety of different reactions.

Nanoengineering Mesoporous Catalysts

With the creation of ordered mesoporous materials, it was possible to modify them spatially and analyze the modifications with techniques such as diffraction and electron microscopy, which revealed many more details in these materials than is possible for disordered, amorphous mesoporous systems. The X-ray diffraction (XRD) pattern, for instance, is strongly dependent on the location of guest species in the pore system (71), and nowadays sophisticated models are available to fit the experimental pattern with structural models so that XRD can provide detailed insight into the structure resulting after modification (72). TEM also allows precise localization of guest species introduced into the materials and therefore helps to control the success of a desired modification step (73–76).

Several recently developed methods permit spatial control of the modification. Selective passivation of the silanol groups on the external surface of ordered mesoporous silica by reaction with small amounts of diphenyldichlorosilane was the first method of spatially controlled modification (77). After such a passivation step, the interior of the pore system can be selectively functionalized. This method has been applied especially for the creation of enantioselective catalysts based on ordered mesoporous silica (63, 78).

Alternatively, instead of passivating the external surface of ordered mesoporous solids, the pore system can be blocked reversibly, the external surface can then be treated, and subsequently, after removal of the blocking agent from the pore system, the inner surface of the pores is accessible for further steps. Such techniques have been used by Lu et al. for the creation of both magnetic ordered mesoporous carbons (79) and ordered magnetic mesoporous silica (80). Figure 6, see color insert, describes the general approach to achieve such a selective functionalization. From an ordered mesoporous material a composite is created in which the pores are blocked. The external surface is then modified, in this case by deposition of magnetic cobalt nanoparticles. After removal of one of the components from the composite, a material selectively functionalized on the external surface is formed. The blocking agent can be the silica, for example, which is already used as a hard template in the formation of an ordered mesoporous carbon, and thus no additional step is needed. A blocking agent, which can be removed by simple thermal treatment, was used in the case of the magnetic silica. Here, the pores were reversibly blocked by poly(methylmetacrylate). This polymer is easily and almost tracelessly decomposed upon heating at 500°C and therefore provides a convenient reversible blocking agent for a pore system.

Two other recent developments provide the potential for spatially selective functionalization on the smaller scale of the pores. Yang et al. have discovered that the template from triblockcopolymer-templated silicas can be removed in a sequential manner, thus liberating first the mesopores and then the micropores (81, 82). These materials are templated with polyethylenoxide-polypropyleneoxide-polyethyleneoxide triblockcopolymers. In the as-made materials, the polypropylene block is located predominantly in the mesopores, whereas the polyethylene blocks are embedded in the silica walls. Upon calcination, the removal of the polyethyleneoxide moieties from the walls creates micropores that connect the mesopores (83–85). The method for sequential removal of the template relies on an ether cleavage, using around 50% sulfuric acid. This cleaves the accessible polymer in the mesopore system (predominantly polypropyleneoxide), whereas the polyethyleneoxide part is essentially unaffected as shown by ^{13}C NMR spectroscopy. Mild calcination at 200°C then also removes the template from the silica walls and makes micropores accessible. Thus after the ether cleavage, the walls of the mesopores can selectively be modified, whereas after removal of the polyethyleneoxide block, the micropores can be modified.

Another discovery that provided the opportunity for spatially controlled modification was the first synthesis of CMK-5 (86). In this material two different pore systems are also created in separate steps. CMK-5 is formed by nanocasting of carbon from SBA-15. However, the pores of the SBA-15 are not completely filled with a carbon precursor, they are only coated. Thus a pore system inside the carbon tubes is created first. After removal of the silica, the second independent pore system is generated. By control of the synthesis process, the pore systems in such materials can be tailored to some extent (87). Because the two pore systems are created independently of each other, it should also be possible to modify them independently, for instance, by creating a hydrophilic pore system by oxidation of the inner surface of the carbon tubes and leave the other pore system hydrophobic after removal of the silica.

The developments that seem possible by using spatially selective functionalization on the nanometer scale (nanoengineering) will probably not lead immediately to the development of novel industrial catalysts due to the complexity of the processes. However, it may be possible to design catalysts in a highly controlled manner and then study the influence of the different components forming the catalytic material.

POROUS COORDINATION COMPOUNDS

Porous coordination compounds are a class of materials that have metal centers linked to each other by organic linkers, such as bicarboxylates, to create an open framework material. The pore system is crystallographically defined as in zeolites, but there is a much wider range of framework constituents. Catalytic applications of these materials have not been well investigated so far, but considering the properties of these materials, there could be many options for their use in catalysis. In the following, the properties most relevant in catalytic applications are briefly addressed, and then the available data on catalysis over this class of materials is reviewed.

General Properties of Porous Coordination Compounds

Several recent reviews cover the synthesis, properties, and applications of porous coordination compounds (3–6, 88). These materials are composed of metal centers, often linked to clusters, that are linked to each other by organic molecules bearing at least two binding groups, such as bicarboxylates or bipyridines. Copper, zinc, and nickel are probably the most abundant metal centers for which porous coordination compounds are known so far. Linking of the metal centers in various geometries leads to the formation of network structures that are porous in certain cases. Intuitively, one would expect the porosity to increase with increasing length of the linkers. However, interpenetration or interweaving of the structures counteracts this trend. Nevertheless, even for strongly interpenetrating networks, high porosity is possible as demonstrated for various materials (89).

Porosity is one of the most remarkable features for this class of materials and possibly the most interesting feature with respect to catalysis. Aperture sizes range up to 1.5 nm, while the diameter of cavities accessible through these apertures can reach close to 3 nm. Pore volumes can appreciably exceed 1 ml/g. Surface areas are only ill defined for such materials, but values of several thousand square meters per gram are quoted. Figure 7, see color insert, gives examples for such high porosity structures.

A special property with respect to the porosity of these materials is the flexibility/adaptability of the framework upon removal or introduction of guest species. These properties are discussed in detail by Kitagawa et al. (4) and Rosseinsky (5). Most remarkable is the existence of a "gate pressure," which means that the gas offered is not adsorbed up to a certain pressure threshold and then suddenly an adsorption step occurs, giving evidence for expansion of the structure (90). A gate pressure is typically accompanied by a hysteresis, which may be exploited to adsorb hydrogen at high pressure that can then be stored at much lower pressure, as recently reported (91).

For materials with organic framework constituents, thermal stability is always a possible problem. However, many porous coordination compounds are thermally remarkably stable, because many of the organic linker molecules also have high intrinsic stability, which is further increased by linking. Thermal stabilities of the materials exceeding 200°C are therefore quite common; in exceptional cases, stability up to 400°C has been reported (92, 93).

However, even if the thermal stabilities are rather high, catalytic applications will typically be envisaged for solution-based processes. Therefore, solubility of the compounds is an important issue. In general, the frameworks are composed of charged metal centers and charged ligands. Therefore, they are typically synthesized from polar solvents, and the solubility in unpolar solvents is usually very low. Even for polar solvents, many of the porous coordination compounds are not very soluble. With respect to catalysis, though, even slight solubility makes it difficult to distinguish between true heterogeneous catalysis and homogeneous catalysis by low concentration of dissolved species or fragments from the crystals. Thus great care has to be taken in the interpretation of catalytic results.

A final property that is relevant in the context of catalysis is the degree of coordinative saturation of the metal centers. In order for a metal to function as a catalyst, it is generally assumed that it needs to have a free coordination site or at least sufficiently labile ligands so that coordination of a substrate is possible. This, however, is problematic for porous coordination compounds. Normally, all coordination sites are used in such materials because this provides the high structural stability. Removal of a coordinating ligand or of many coordinating ligands for the extended structure may strongly destabilize such materials. Fortunately, however, many of the transition metal ions have a rather flexible coordination sphere. For example, some compounds have free coordination sites that do not compromise structural integrity. In $Cu_2ATC \cdot 2\,H_2O \cdot \{4H_2O\}$ (ATC = adamantyl tetracarboxylate), the copper coordination is square pyramidal by strong bonds, which is complemented by an axial water ligand. This water ligand can be removed without strong distortion of the structure so that a free copper coordination site exists in the dehydrated material (94). However, examples like this are rare, and frequently the framework collapses if ligands are removed from the metal centers. Another option to create coordinatively unsaturated sites is similar to zeolite chemistry, where aluminum centers substituting silicon can create a catalytically active OH-group. If one substitutes a metal center in a porous coordination compound with another one with different requirements with respect to coordination environment, accessible coordination sites may result, as has been demonstrated by Kitagawa et al. (95). Finally, one should keep in mind that catalytic activity does not necessarily need to be located at the metal center; basic or acidic sites of the linker molecules can also be used to catalyze chemical transformations such as transesterifications (see below) (96).

Perspectives in Catalysis

Reports on catalytic applications of porous coordination compounds are still scarce, which may possibly be attributed to the fact that it took some time before the catalysis community widely realized the existence and properties of such materials. In addition, the problem of instability upon change of coordination environment of the metal centers may limit their use in catalysis, at least for many of the materials. In principle, however, the materials should allow the design of catalytic sites when the interest moves more toward the possible applications of these materials and away from the synthesis and structures.

Most interesting in this respect, at least from a fundamental point of view, is the observation of Seo et al. (96) that a transesterification with a racemic alcohol proceeds preferably with one enantiomer over a chiral porous coordination compound (Figure 8, see color insert), even if the selectivity is low with an ee of 8%. Catalysis in this case does not occur on the metal centers but on pyridyl groups of the linker, which protrude into the channels of the material.

The first catalytic application of a porous coordination compound, however, dates more than ten years back, and was described by Fujita et al. These authors used $Cd(4,4'\text{-}BPY)_2(NO_3)_2$ as a catalyst for the cyanosilylation of aldehydes (97). As in zeolites, shape selectivity was observed for this reaction: 2-tolualdehyde was converted more rapidly than the bulkier 3-tolualdehyde, which was tentatively attributed to the effect of the cages in the material. Cyanosilylation activity was also reported for a lamellar lanthanide phosphate (98). Aldehydes of different sizes could be processed with similar efficiency, which led to the conclusion that the catalysts swell under reaction conditions to allow access of the substrates. The material was also active in ring opening of meso-carboxylic anhydrides. Although the framework of the material is chiral, no enantioselectivity was observed in the catalytic transformations.

In the previous case, incorporation of the substrates led to swelling of already existing pores. Some reports on catalytic activity of coordination compounds describe reactions that do not proceed in a pore system as such. However, the reagents seem to lead to expansion of the layers in such compounds with corresponding incorporation of the substrate molecules in the layers before the catalytic transformation can take place (99, 100).

Hydrogenation reactions and hydrogen exchange were observed to occur in the micropores of a rhodium dicarboxylate (101, 102). Interestingly, the reaction proceeds at very low temperature (around 200 K), and the authors present evidence that a bimolecular hydrogen exchange between two propene molecules takes place exclusively in the micropores of the compound. Over a rhodium porphyrin coordination polymer, high activities for the hydrogenation of ethane and propene were reported; the materials were active at 194 K (103). For these materials, prophyrin tetracarboxylate was used as linker, which provides a four-connector to the rhodium centers. In addition to these reports, there are scattered patents resulting from the collaboration between BASF and Yaghi's group (104, 105) in which organometallic framework materials are claimed as essential components of the catalysts.

In addition, a number of transformations appear to be catalyzed by porous coordination compounds that do not seem to occur in pores of the materials and are thus not as relevant in the context of this discussion (106, 107). There was a recent report of a catalytic oxidation reaction occurring in the pores of an amorphous coordination polymer gel containing palladium sites (108).

Intensive research on catalytic properties of porous coordination compounds has started only recently, and it is therefore expected that many more applications in catalysis will be reported within the next years. Use of functional groups of the linkers for catalysis instead of the metal centers seems an attractive approach, especially in the light of the advent of organocatalysis (109). Another attractive option seems to be the guest-responsive pore system, which may be used to trigger certain catalytic functions by changing concentrations of gas phase species in a highly nonlinear manner. Because of the almost unlimited structural and compositional options that may be realized with these materials, and the emerging possibility to computationally predict the structures of such solids (110), there may be many more discoveries than can presently be imagined.

ENGINEERING SHAPES

Shaped bodies for catalytic applications are conventionally prepared by methods such as tableting, extrusion, or pelletizing. However, it can be advantageous to produce directly shaped bodies, such as has been demonstrated in chromatography by the introduction of the Nakanishi columns (111). In the field of porous catalytic materials, several recent interesting developments have made the direct synthesis of shaped materials possible.

Creating Membranes

Membranes are useful for catalysis, if catalytic function and separation can be incorporated in the membrane simultaneously (112, 113). Zeolitic membranes are especially attractive for two reasons: (*a*) Other than sol-gel membranes, zeolite membrane pore size distribution is sharp if the membranes are perfect, and thus they have molecular sieving properties; and (*b*) the catalytic function is normally directly present in the zeolite. Substantial progress has been made in the synthesis of zeolitic membranes. Reviews covering earlier work are available (114, 115), and here only some highlights concerning special types of membranes are mentioned.

One crucial point in the synthesis of a zeolite membrane is the production of membranes with oriented crystals. Lai and coworkers have successfully synthesized a very thin, perfect membrane based on MFI-type zeolite, which provided high flux and good separation properties (116). They used a seed layer composed of colloidal zeolite crystals and a trimeric form of the tetrapropylammonium template normally employed for the formation of MFI-type zeolites. This combination led to the growth of a b-oriented membrane (Figure 9), which provided both high flux and high separation factors of several 100, depending on temperature, for the o- and p-isomers of xylene. In subsequent papers, a more detailed account on the synthesis

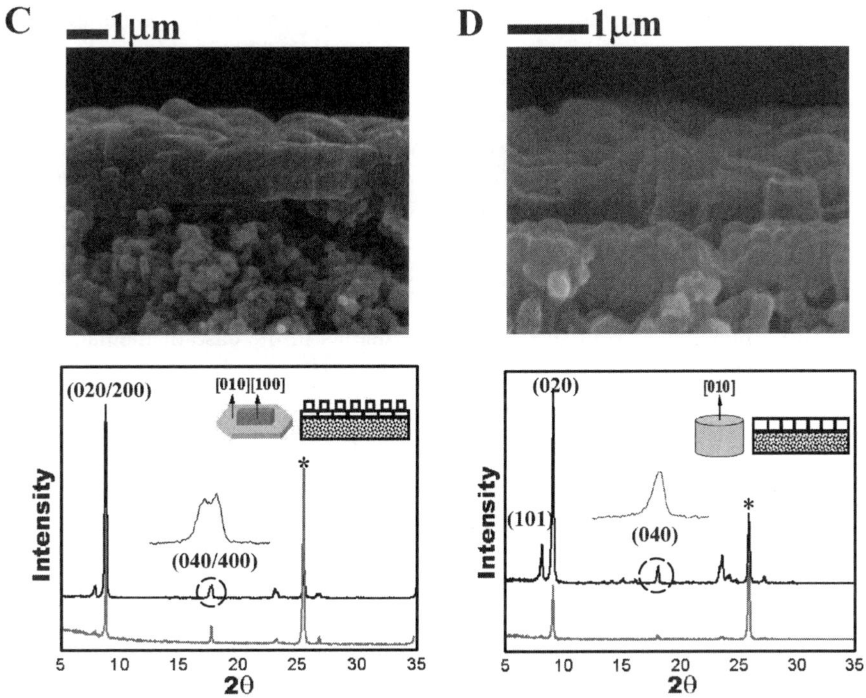

Figure 9 TPA$^+$-templated, a/b-oriented MFI-membrane starting from a seed layer (*c*) and membrane-templated with a trimeric form of TPA$^+$ (*d*). Under the scanning electron microscopy (SEM), the XRD traces are shown. The absence of the (040) reflection in the right pattern demonstrates the almost perfect orientation. Reproduced with permission from Reference 116.

and characterization of the membranes (117) and further results on good separation of other molecules, such as benzene/cyclohexane and benzene/2,2-dimethylbutane (118), were published.

The perfect pore system present in zeolites provides an additional possibility for controlling the membrane properties. Because the pore sizes of the zeolites are of the same order of magnitude as molecular sizes, small changes in the pore space can control the flow through the membranes. Such changes can be brought about by temperature or by switchable molecules with differently bulky states located in the pore system.

The fluxes and selectivities through membranes are to some degree always temperature dependent. However, temperature-induced changes of the pore system could provide an additional means to influence the performance of the membranes. The adsorptive properties of ETS-4, a titanosilicate, can be altered strongly by temperature ("molecular gating") (119). It has been reported that ETS-4 membranes can be prepared (120–122). So far, the molecular gating effect does not seem to have been exploited in connection with a membrane system, but it could provide interesting options for creating a switchable membrane.

Proof-of-concept, however, has been achieved for the adsorption of switchable molecules in the zeolite crystals that form the membrane. Such an approach at this point does not seem to be immediately applicable to solve practical separation problems, but the concept is highly interesting. This approach has been used by Weh et al. (123). These authors have adsorbed *trans*-azobenzene into a zeolite film composed of MFI- or FAU-type crystals from the gas phase. Upon irradiation, the azobenzene isomerizes to the *cis*-state, which has different spatial requirements in the zeolite crystals and thus changes the flux of gases through the membrane. Changes in permeance of selected gases between the *cis*- and *trans*-states of the azobenzene were up to a factor of four. Measured changes in separation factors for two different molecules were modest for most gases but came close to a factor of two in selected cases, such as in the case of methane and carbon dioxide.

Producing Monoliths by Novel Pathways

Monoliths are used in many processes and there are well established techniques of applying catalytically active compounds on preformed ceramic monoliths (124). The direct synthesis of monolithic particles, which can be used without further shaping, saves one synthesis step and can provide particles with improved mechanical properties. Also, recent progress in this field makes it possible to engineer ready-to-use shaped bodies in a single step. Scalability of such approaches to technical dimensions is in many cases still in question, but further progress in the field may also bring that within reach.

A recently developed method for the direct formation of monolithic materials is the hard templating from suitable sacrificial exotemplates (125), such as carbons or silicas, which can easily be obtained in monolithic form by conventional methods,

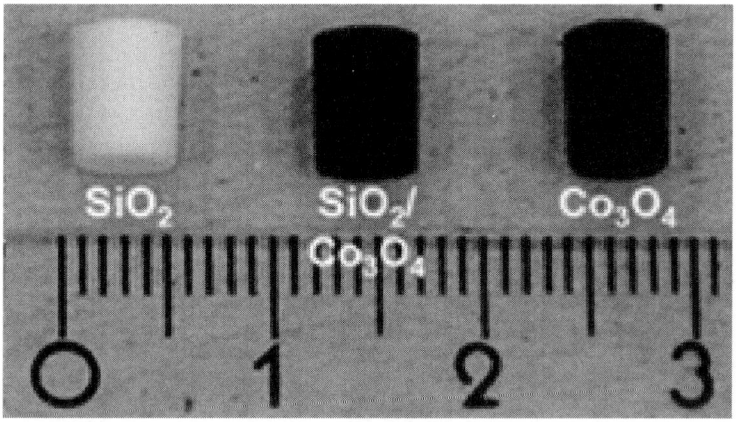

Figure 10 Photographs of monolithic macro/mesoporous samples. Right: ordered mesoporous SiO_2; middle: composite of SiO_2 and Co_3O_4; left: Co_3O_4 after leaching of silica. Reproduced with permission from Reference 126.

for instance, by the sol-gel process. The advantage of this method is the direct replication of the parent material, which enables the pore system of the resulting cast material to be tuned. Thus Co_3O_4 was produced in the form of a monolithic material by casting from a mesoporous silica monolith (126). The resulting cobalt oxide monoliths had surface areas of 30–70 m^2/g and inherited the macropore system of the parent silica (Figure 10). In the same manner, carbons can also be casted from silica (127–130).

Carbon monolithic materials have been used as hard templates to a lesser extent, and there are so far few reports on the synthesis of larger monolithic materials based on a carbon templating. Kaneko et al. succeeded in the synthesis of 3–5-mm ZSM-5 granules from carbon aerogel monoliths by a hard templating pathway (131). High surface area amorphous alumina granules can also be prepared by nanocasting from carbon aerogels and subsequent combustion of the aerogel template (132). Whether larger shaped bodies are accessible from carbon monolithic templates remains to be seen.

An alternative for the direct preparation of the monolith from zeolitic material or the coating of a ceramic monolith from a zeolite slurry (133) is the synthesis of a zeolite directly onto a ceramic foam. This approach has been reported by Schwieger et al. (134). Cellulose membranes have been coated with zeolites by synthesizing the zeolite onto the membrane (135). Possibly the most interesting development in this respect is the synthesis of pure, large zeolite monoliths by first coating polyurethane (PU) foams with the zeolite by direct synthesis onto the foam and subsequent removal of the PU by calcination (136). The presence of the PU surface facilitates nucleation of the zeolite so that the PU is coated, and mechanically stable monoliths are obtained in various shapes and sizes (Figure 11).

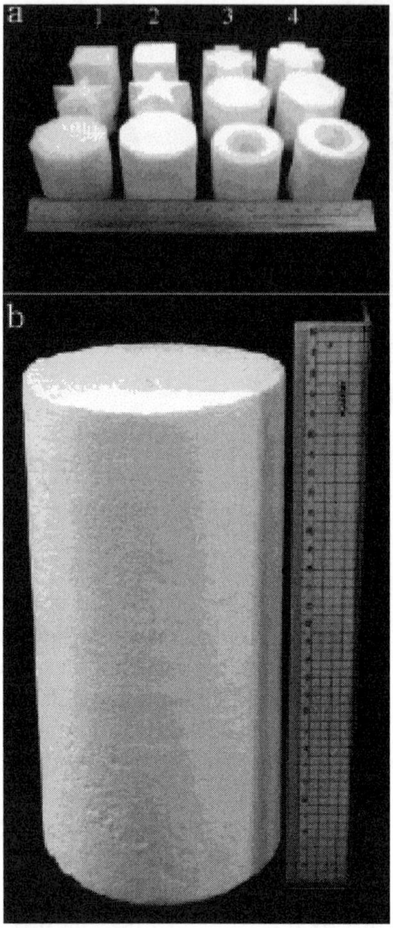

Figure 11 Photographs of PU foams (*top*, rows 1 and 3); zeolite monoliths synthesized with these PU foams as templates (*top*, rows 3 and 4). Bottom: large sized-zeolite monolith synthesized from PU foam. Reproduced with permission from Reference 136.

PERSPECTIVES

There has been immense progress in the design of solids and control of their properties on the atomic and mesoscales. However, there is often a lack of communication between the synthesis-oriented groups and the more catalysis-oriented groups unless the research teams are working on the border line between synthesis and catalysis. Nonetheless, as the achievements in the controlled synthesis of solids become better known, more and more groups will use such materials and employ them in various catalytic reactions. The results of these studies will be fed back to the synthesis groups, and it can be expected that for selected cases a true engineering of catalytic materials on various length scales will become possible over the next few years. One should bear in mind, however, that the methods used to produce such engineered catalytic materials are very demanding and the associated costs are high. Industrial applications therefore do not seem to be within easy reach.

The *Annual Review of Materials Research* is online at http://matsci.annualreviews.org

LITERATURE CITED

1. Reetz MT, de Vries JS. 2004. Ligand free Heck reactions using low Pd-loading. *Chem. Commun.* 2004:1559–63 and references therein
2. Taguchi A, Schüth F. 2005. Ordered mesoporous materials in catalysis. *Micropor. Mesopor. Mater.* 77:1–45
3. Yaghi OM, O'Keeffe M, Ockwig NW, Chae HK, Eddaoudi M, Kim J. 2003. Reticular synthesis and the design of new materials. *Nature* 423:705–14
4. Kitagawa S, Kitaura R, Noro S. 2004. Functional porous coordination polymers. *Angew. Chem. Int. Ed.* 43:2334–75
5. Rosseinsky MJ. 2004. Recent developments in metal-organic framework chemistry: design, discovery, permanent porosity and flexibility. *Micropor. Mesopor. Mater.* 73:15–30
6. Kesanli B, Lin W. 2003. Chiral porous coordination networks: rational design and applications in enantioselective processes. *Coord. Chem. Rev.* 246:305–26
7. Davis ME, Lobo RF. 1992. Zeolite and molecular sieve synthesis. *Chem. Mater.* 4:756–68
8. Gies H, Marler B. 1992. The structure-controlling role of organic templates for the synthesis of porosils in the system SiO_2/template/H_2O. *Zeolites* 12:42–49
9. Lewis DW, Willock DJ, Catlow CRA, Thomas JM, Hutchings GJ. 1996. De novo design of structure-directing agents for the synthesis of microporous solids. *Nature* 382:604–6
10. Lewis DW, Sankar G, Wyles JK, Thomas JM, Catlow CRA, Willock DJ. 1997. Synthesis of a small-pore microporous material using a computationally designed template. *Angew. Chem. Int. Ed.* 36:2675–77
11. Willock DJ, Lewis DW, Catlow CRA, Hutchings GJ, Thomas JM. 1997. Designing templates for the synthesis of microporous solids using de novo molecular design methods. *J. Mol. Catal. A* 119:415–24
12. Gomez-Hortiguela L, Cora F, Catlow CRA, Perez-Pariente J. 2004. Computational study of the structure-directing effect of benzylpyrrolidine and its fluorinated derivatives in the synthesis of the aluminophosphate AlPO-5. *J. Am. Chem. Soc.* 126:12097–102
13. Sastre G, Leiva S, Sabater MJ, Gimenez I, Rey F, Valencia S, Corma A. 2004. Computational and experimental approach to the role of structure-directing agents in the synthesis of zeolites: the case of cyclohexyl alkyl pyrrolidinium salts in the synthesis of beta, EU-1, ZSM-11, and ZSM-12 zeolites. *J. Phys. Chem. B* 107:5432–40
14. Sastre G, Pulido A, Castaneda R, Corma A. 2004. Effect of the germanium incorporation in the synthesis of EU-1, ITQ-13, ITQ-22, and ITQ-24 zeolites. *J. Phys. Chem. B* 108:8830–35
15. Corma A, Rey F, Rius J, Sabater MJ, Valencia S. 2004. Supramolecular self-assembled molecules as organic directing agent for synthesis of zeolites. *Nature* 431:287–90
16. Corma A, Navarro MT, Rey F, Rius J, Valencia S. 2001. Pure polymorph C of zeolite beta synthesized by using framework

isomorphous substitution as a structure-directing mechanism. *Angew. Chem. Int. Ed.* 40:2277–80
17. Lobo RF, Zones SI, Davis ME. 1995. Structure direction in zeolite synthesis. In *Inclusion Chemistry with Zeolites: Nanoscale Materials by Design*, ed. N Herron, DR Corbin, 1:47–78. Dordrecht/Boston/London: Kluwer. 340 pp.
18. Helmkamp MM, Davis ME. 1995. Synthesis of porous silicates. *Annu. Rev. Mater. Sci.* 25:161–92
19. Freyhardt CC, Lobo RF, Khodabandeh S, Lewis JE, Tsapatsis M, et al. 1996. VPI-8: a high-silica molecular sieve with a novel "pinwheel" building unit and its implications for the synthesis of extra-large pore molecular sieves. *J. Am. Chem. Soc.* 118:7299–310
20. Corma A, Davis ME. 2004. Issues in the synthesis of crystalline molecular sieves: towards the crystallization of low framework-density structures. *Chem. Phys. Chem.* 5:304–13
21. Röhrig C, Gies H. 1995. A new zincosilicate zeolite with 9-ring channels. *Angew. Chem. Int. Ed.* 34:63–65
22. Blasco T, Corma A, Diaz-Cabanas MJ, Rey F, Vidal-Moya JA, Zicovich-Wilson CM. 2002. Preferential location of Ge in the double four-membered ring units of ITQ-7 zeolite. *J. Phys. Chem. B* 106: 2634–42
23. Corma A, Rey F, Valencia S, Jorda JL, Rius J. 2003. A zeolite with interconnected 8-, 10- and 12-ring pores and its unique catalytic selectivity. *Nat. Mater.* 2:493–97
24. Corma A, Diaz-Cabanas M, Martinez-Triguero J, Rey F, Rius J. 2002. A large-cavity zeolite with wide pore windows and potential as an oil refining catalyst. *Nature* 418:514–17
25. Corma A, Diaz-Cabanas MF, Rey F. 2000. *Spanish Patent No. ES2186487*
26. Corma A, Diaz-Cabanas MF, Rey F, Nicolopoulos S, Boulahya K. 2004. ITQ-15: the first ultra large pore zeolite with a bi-directional pore system formed by intersecting 14- and 12-ring channels, and its catalytic implications. *Chem. Commun.* 2004:1356–57
27. Paillaud JL, Harbuzaru B, Patarin J, Bats N. 2004. Extra-large-pore zeolites with two-dimensional channels formed by 14 and 12 rings. *Science* 304:990–92
28. Janssen AH, Koster AJ, de Jong KP. 2001. Three-dimensional transmission electron microscopic observations of mesopores in dealuminated zeolite Y. *Angew. Chem. Int. Ed.* 40:1102–4
29. Schüth F. 2003. Endo- and exotemplating to create high surface area materials. *Angew. Chem. Int. Ed.* 42:3604–22
30. Jacobsen CJH, Madsen C, Houzvicka J, Schmidt I, Carlsson A. 2000. Mesoporous zeolite single crystals. *J. Am. Chem. Soc.* 122:7116–17
31. Schmidt I, Boisen A, Gustavsson E, Stahl K, Pehrson S, et al. 2001. Carbon nanotube templated growth of mesoporous zeolite single crystals. *Chem. Mater.* 13: 4416–18
32. Janssen AH, Schmidt I, Jacobsen CJH, Koster AJ, de Jong KP. 2003. Exploratory study of mesopore templating with carbon during zeolite synthesis. *Micropor. Mesopor. Mater.* 65:59–75
33. Tao Y, Kanoh H, Kaneko K. 2003. ZSM-5 Monolith of uniform mesoporous channels. *J. Am. Chem. Soc.* 125:6044–45
34. Yang ZX, Xia YD, Mokaya R. 2004. Zeolite ZSM-5 with unique supermicropores synthesized using mesoporous carbon as a template. *Adv. Mater.* 16:727–32
35. Kato H, Minami T, Kanazawa T, Sasaki Y. 2004. Mesopores created by platinum nanoparticles in zeolite crystals. *Angew. Chem. Int. Ed.* 43:1251–54
36. Schoeman BE, Sterte J, Otterstedt JE. 1993. Synthesis and size tailoring of colloidal zeolite particles. *Chem. Commun.* 1993:994–95
37. Agren P, Thomson S, Ilhan Y, Zibrowius B, Schmidt W, Schüth F. 2002. Chemical

linking of MFI-type colloidal zeolite crystals. *Stud. Surf. Sci. Catal.* 142:159–66
38. Karlsson A, Stöcker M, Schmidt R. 1999. Composites of micro- and mesoporous materials: simultaneous syntheses of MFI/MCM-41 like phases by a mixed template approach. *Micropor. Mesopor. Mater.* 27:181–92
39. Verhoef MJ, Kooyman PJ, van der Waal JC, Rigutto MS, Peters JA, van Bekkum H. 2001. Partial transformation of MCM-41 material into zeolites: Formation of nanosized MFI type crystallites. *Chem. Mater.* 13:683–87
40. Liu Y, Zhang WZ, Pinnavaia TJ. 2001. Steam-stable MSU-S aluminosilicate mesostructures assembled from zeolite ZSM-5 and zeolite beta seeds. *Angew. Chem. Int. Ed.* 40:1255–58
41. Zhang ZT, Han Y, Zhu L, Wang RW, Yu Y, et al. 2001. Strongly acidic and high-temperature hydrothermally stable mesoporous aluminosilicates with ordered hexagonal structure. *Angew. Chem. Int. Ed.* 40:1258–62
42. On DT, Kaliaguine S. 2002. Ultrastable and highly acidic, zeolite-coated mesoporous aluminosilicates. *Angew. Chem. Int. Ed.* 41:1036–40
43. On DT, Nossov A, Springuel-Huet MA, Schneider C, Bretherton JL, et al. 2004. Zeolite nanoclusters coated onto the mesopore walls of SBA-15. *J. Am. Chem. Soc.* 126:14324–25
44. Kresge CT, Leonowicz ME, Roth WJ, Vartuli JC, Beck JS. 1992. Ordered mesoporous molecular-sieves synthesized by a liquid-crystal template mechanism. *Nature* 359:710–12
45. Zhao DY, Feng JL, Huo QS, Melosh N, Fredrickson GH, et al. 1998. Triblock copolymer syntheses of mesoporous silica with periodic 50 to 300 angstrom pores. *Science* 279:548–52
46. Widenmayer M, Anwander R. 2002. Pore size control of highly ordered mesoporous silica MCM-48. *Chem. Mater.* 14:1827–31
47. Landmesser H, Kosslik H, Storek W, Fricke R. 1997. Interior surface hydroxyl groups in ordered mesoporous silicates. *Solid State Ionics* 101:271–77
48. Nozaki C, Lugmaier CG, Bell AT, Tilley TD. 2002. Synthesis, characterization, and catalytic performance of single-site iron(III) centers on the surface of SBA-15 silica. *J. Am. Chem. Soc.* 124:13194–203
49. Ottaviani MF, Galarneau A, Desplantiers-Giscard F, Di Renzo F, Fajula F. 2001. EPR investigations on the formation of micelle-templated silica. *Micropor. Mesopor. Mater.* 44:1–8
50. Anwander R. 2001. SOMC@PMS. Surface organometallic chemistry at periodic mesoporous silica. *Chem. Mater.* 13: 4419–38
51. Sayari A, Hamoudi S. 2001. Periodic mesoporous silica-based organic-inorganic nanocomposite materials. *Chem. Mater.* 13:3151–68
52. Schüth F, Wang Y, Yang CM, Zibrowius B. 2005. Mesostructured silica and organically functionalized silica—status and perspectives. In *Organosilicon Chemistry VI*, ed. N Auner. Weinheim: Wiley-VCH. In press
53. Beck JS, Vartuli JC, Roth WJ, Leonowicz ME, Kresge CT, et al. 1992. A new family of mesoporous molecular-sieves prepared with liquid-crystal templates. *J. Am. Chem. Soc.* 114:10834–43
54. Burkett SL, Sims SD, Mann S. 1996. Synthesis of hybrid inorganic-organic mesoporous silica by co-condensation of siloxane and organosiloxane precursors. *Chem. Commun.* 1996:1367–68
55. Macquarrie DJ. 1996. Direct preparation of organically modified MCM-type materials. Preparation and characterisation of aminopropyl-MCM and 2-cyanoethyl-MCM. *Chem. Commun.* 1996:1961–62
56. Wang Y, Yang CM, Zibrowius B, Spliethoff B, Linden M, Schüth F. 2003. Directing the formation of SBA-15 or

large pore Ia3d structure of vinyl-functionalized mesoporous silica. *Chem. Mater.* 15:5029–36
57. Yang CM, Zibrowius B, Schüth F. 2003. A novel synthetic route for negatively charged ordered mesoporous silica SBA-15. *Chem. Commun.* 2003:1772–73
58. Lim MH, Stein A. 1999. Comparative studies of grafting and direct syntheses of inorganic-organic hybrid mesoporous materials. *Chem. Mater.* 11:3285–95
59. de Vos DE, Dams M, Sels BF, Jacobs PA. 2002. Ordered mesoporous and microporous molecular sieves functionalized with transition metal complexes as catalysts for selective organic transformations. *Chem. Rev.* 102:3615–40
60. Trong On D, Desplantier-Giscard D, Danumah C, Kaliaguine S. 2003. Perspectives in catalytic applications of ordered mesoporous materials. *Micropor. Mesopor. Mater.* 253:545–602
61. Witht AP, Davis ME. 2002. Design and preparation of organic-inorganic hybrid catalysts. *Chem. Rev.* 102:3589–614
62. Thomas JM, Raja R. 2004. Significance of mesoporous crystals for catalytic application. *Stud. Surf. Sci. Catal.* 148:163–211
63. Johnson BFG, Raynor SA, Shephard DS, Maschmeyer T, Thomas JM, et al. 1999. Superior performance of a chiral catalyst confined within mesoporous silica. *Chem. Commun.* 1999:1167–68
64. Raynor SA, Thomas JM, Raja R, Johnson BFG, Bell RG, Mantle MD. 2000. A one-step, enantioselective reduction of ethyl nicotinate to ethyl nipecotinate using a constrained, chiral, heterogeneous catalyst. *Chem. Commun.* 2000:1925–26
65. Jones MD, Raja R, Thomas JM, Johnson BFG, Lewis DW, et al. 2003. Enhancing the enantioselectivity of novel homogeneous organometallic hydrogenation catalysts. *Angew. Chem Int. Ed.* 42:4326–31
66. Raja R, Thomas JM, Jones MD, Johnson BFG, Vaughan DEW. 2003. Constraining asymmetric organometallic catalysts within mesoporous supports boosts their enantioselectivity. *J. Am.Chem. Soc.* 125:14982–83
67. Maschmeyer T, Rey F, Sankar G, Thomas GM. 1997. Heterogeneous catalysts obtained by grafting metallocene complexes onto mesoporous silica. *Nature* 378:159–62
68. Shannon IJ, Maschmeyer T, Oldroyd RD, Sankar G, Thomas JM, et al. 1998. Metallocene-derived, isolated Mo-VI active centres on mesoporous silica for the catalytic dehydrogenation of methanol. *J. Chem. Soc. Faraday Trans.* 94:1495–99
69. Fujdala KL, Drake IJ, Bell AT, Tilley TD. 2004. Atomic level control over surface species via a molecular precursor approach: Isolated Cu(I) sites and Cu nanoparticles supported on mesoporous silica. *J. Am. Chem. Soc.* 126:10864–66
70. Jarupatrakorn J, Tilley TD. 2002. Silica-supported, single site titanium catalysts for olefin epoxdiation: a molecular precursor strategy for control of catalyst structure. *J. Am. Chem. Soc.* 124:8380–88
71. Sauer J, Marlow F, Schüth F. 2001. Simulation of powder diffraction patterns of modified ordered mesoporous materials. *Phys. Chem. Chem. Phys.* 3:5579–84
72. Solovyov LA, Kirik SD, Shmakov AN, Romannikov VN. 2001. X-ray structural modeling of silicate mesoporous mesophase material. *Micropor. Mesopor. Mater.* 44:17–23
73. Sauer J, Marlow F, Spliethoff B, Schüth F. 2002. Rare earth oxide coating of the walls of SBA-15. *Chem. Mater.* 14:217–24
74. Vradman L, Landau MV, Herskowitz M, Ezersky V, Talianker M, et al. 2003. High loading of short WS2 slabs inside SBA-15: promotion with nickel and performance in hydrodesulfurization and hydrogenation. *J. Catal.* 213:163–75
75. Anderson MW, Ohsuna T, Sakamoto Y, Liu Z, Carlsson A, Terasaki O. 2004. Modern microscopy methods for the structural study of porous materials. *Chem. Commun.* 2004:907–16

76. Thomas JM, Midgley PA. 2004. High-resolution transmission electron microscopy: the ultimate nanoanalytical technique. *Chem. Commun.* 2004:1253–67
77. Shephard DS, Zhou WZ, Maschmeyer T, Matters JM, Roper CL, et al. 1998. Site-directed surface derivatization of MCM-41: use of high-resolution transmission electron microscopy and molecular recognition for determining the position of functionality within mesoporous materials. *Angew. Chem. Int. Ed.* 37:2719–23
78. Thomas JM, Johnson BFG, Raja R, Sankar G, Midgley PA. 2003. High-performance nanocatalysts for single-step hydrogenations. *Acc. Chem. Res.* 36:20–30
79. Lu AH, Schmidt W, Matoussevitch N, Bönnemann H, Spliethoff B, et al. 2004. Nanoengineering of a magnetically separable hydrogenation catalyst. *Angew. Chem. Int. Ed.* 43:4303–6
80. Lu AH, Li WC, Kiefer A, Schmidt W, Bill E, et al. 2004. Fabrication of magnetically separable mesostructured silica with an open pore system. *J. Am. Chem. Soc.* 126:8616–17
81. Yang CM, Schmidt W, Schüth F. 2003. Consecutive generation of mesopores and micropores in SBA-15. *Chem. Mater.* 15: 3739–41
82. Yang CM, Zibrowius B, Schmidt W, Schüth F. 2004. Consecutive removal of the copolymer template from mesopores and micropores in SBA-15. *Chem. Mater.* 16:2918–25
83. Ryoo R, Ko CH, Kruk M, Antochshuk V, Jaroniec M. 2000. Block-copolymer-templated ordered mesoporous silica: Array of uniform mesopores or mesopore-micropore network? *J. Phys. Chem.* B 104:11465–71
84. Lukens WW, Schmidt-Winkel P, Zhao DY, Feng JL, Stucky GD. 1999. Evaluating pore sizes in mesoporous materials: a simplified standard adsorption method and a simplified Broekhoff-de Boer method. *Langmuir* 15:5403–9
85. Miyazawa K, Inagaki S. 2000. Control of the microporosity within the pore walls of ordered mesoporous silica SBA-15. *Chem. Commun.* 2000:2121–22
86. Joo SH, Choi SJ, Oh I, Kwak J, Liu Z, et al. 2001. Ordered nanoporous arrays of carbon supporting high dispersions of platinum nanoparticles. *Nature* 412:169–72
87. Lu A, Schmidt W, Spliethoff B, Schüth F. 2003. Synthesis of ordered mesoporous carbon with bimodal pore system and high pore volume. *Adv. Mater.* 15:1602–6
88. Forster PM, Cheetham AK. 2003. Hybrid inorganic-organic solids: an emerging class of nanoporous catalysts. *Top. Catal.* 24:79–86
89. Eddaoudi M, Kim J, Rosi N, Vodak D, Wachter, et al. 2002. Systematic design of pore size and functionality in isoreticular MOFs and their application in methane storage. *Science* 295:469–72
90. Kitaura R, Seki K Akiyama G, Kitagawa S. 2003. Porous coordination-polymer crystals with gated channels specific for supercritical gases. *Angew. Chem. Int. Ed.* 42:428–31
91. Zhao XB, Xiao B, Fletcher AJ, Thomas KM, Bradshaw D, Rosseinsky MJ. 2004. Hysteretic adsorption and desorption of hydrogen by nanoporous metal-organic frameworks. *Science* 306:1012–15
92. Forster PM, Cheetham AK. 2002. Open-framework nickel succinate, $[Ni_7(C_4H_4O_4)_6(OH)_2(H_2O)_2]2H_2O$: a new hybrid material with three-dimensional Ni-O-Ni connectivity. *Angew. Chem. Int. Ed.* 41:457–59
93. Pan L, Liu H, Lei X, Huang X, Olson DH, et al. 2003. RPM-1: a recyclable nanoporous material suitable for ship-in-bottle synthesis and large hydrocarbon sorption. *Angew. Chem. Int. Ed.* 42:542–46
94. Chen B, Eddaudi M, Reineke TM, Kampf JW, O'Keefe M, Yaghi OM. 2000. $Cu_2(ATC)6H_2O$: design of open

metal sites in porous metal-organic crystals (ATC:1,3,5,7-adamantane tetracarboxylate). *J. Am. Chem. Soc.* 122:11559–60

95. Noro S, Kitagawa S, Yamashita M, Wada T. 2002. New microporous coordination polymer affording guest-coordination sites at channel walls. *Chem. Commun.* 2002:222–23
96. Seo JS, Whang D, Lee H, Jun SI, Oh J, et al. 2000. A homochiral metal-organic porous material for enantioselective separation and catalysis. *Nature* 404:982–86
97. Fujita M, Kwon YJ, Washizu S, Ogura K. 1994. Preparation, clathration ability and catalysis of a two-dimensional square network material composed of cadmium(II) and 4,4′-bipyridine. *J. Am. Chem. Soc.* 116:1151–52
98. Evans OR, Ngo EL, Lin W. 2001. Chiral porous solids based on lamellar lanthanide phosphonates. *J. Am. Chem. Soc.* 123:10395–96
99. Tannenbaum R. 1994. Three-dimensional coordination polymers of ruthenium^{2+} with 1,4-diisocyanobenzene ligands and their catalytic activity. *Chem. Mater.* 6: 550–55
100. Tannenbaum R. 1996. Radiation enhancement of the catalytic properties of three-dimensional coordination polymers of Ru(II) with diisocyanide ligands. *J. Mol. Catal. A* 107:207–15
101. Mori W, Takamizawa S, Kato CN, Ohmura T, Sato T. 2004. Molecular-level design of efficient microporous materials containing metal carboxylates: inclusion complex formation with organic polymer, gas-occlusion properties, and catalytic activities for hydrogenation of olefins. *Micropor. Mesopor. Mater.* 73:31–46
102. Naito S, Tanibe T, Saito E, Miyao T, Mori W. 2001. A novel reaction pathway in olefin-deuterium exchange reaction inside the micropores of Rh(II) dicarboxylate polymer complexes. *Chem. Lett.* 2001:1178–79
103. Sato T, Mori W, Kato CN, Ohmura T, Sato T, et al. 2003. Microporous rhodium(II) 4,4′,4″,4‴-(21H,23H-porphine-5,10,15,20-tetrayl)tetrakisbenzoate. Synthesis, nitrogen adsorption properties, and catalytic performance for hydrogenation of olefin. *Chem. Lett.* 2003:854–55
104. Müller U, Luinstra G, Yaghi OM. 2003. *U.S. Patent No. 6.617.467 2003*
105. Müller U, Lobree L, Hesse M, Yaghi OM, Eddaoudi M. 2003. *U. S. Patent No. 6.624.318 2003*
106. Gomez-Lor B, Gutierrez-Puebla E, Iglesias M, Monge MA, Ruiz-Valero C, Snejko M. 2002. $In_2(OH)_3(BDC)_{1.5}$ (BDC = 1,4-benzenedicarboxylate: An In(III) supramolecular 3D framework with catalytic activity. *Inorg. Chem.* 41:2429–32
107. Feinstein-Jaffe I, Efraty A. 1986. New palladium$_0$ and platinum$_0$ 4,4′-diisocyanobiphenyl matrices for heterogeneous catalytic-hydrogenation of alkenes and alkynes. *J. Mol. Catal.* 35:285–302
108. Xing B, Choi MF, Xu B. 2002. Design of coordination polymer gels as stable catalytic systems. *Chem. Eur. J.* 8:5028–32
109. List B. 2004. Enamine catalysis is a powerful strategy for the catalytic generation and use of carbanion equivalents. *Acc. Chem. Res.* 37:548–57
110. Ferey G, Serre C, Mellot-Draznieks C, Millange F, Surble S, et al. 2004. A hybrid solid with giant pores prepared by a combination of targeted chemistry, simulation and powder diffraction. *Angew. Chem. Int. Ed.* 43:6296–301
111. Nakanishi K, Soga N. 1991. Phase-separation in gelling silica organic polymer-solution-systems containing poly-(sodium stryrenesulfonate). *J. Am. Ceram. Soc.* 74:2518–30
112. Dittmeyer R, Svajda K, Reif M. 2004. A review of catalytic membrane layers for gas/liquid reactions. *Top. Catal.* 29:3–27

113. Liu SM, Tan XY, Li K, Hughes R. 2001. Methane coupling using catalytic membrane reactors. *Catal. Rev. Sci. Eng.* 43:147–98
114. Noack M, Kolsch P, Schafer R, Toussaint P, Caro J. 2002. Molecular sieve membranes for industrial application: problems, progress, solutions. *Chem. Eng. Tech.* 25:221–30
115. Coronas J, Santamaria J. 2004. State-of-the-art in zeolite membrane reactors. *Top. Catal.* 29:29–44
116. Lai ZP, Bonilla G, Diaz I, Nery JG, Sujaoti K, et al. 2003. Microstructural optimization of a zeolite membrane for organic vapor separation. *Science* 300:456–60
117. Lai ZP, Tsapatsis M, Nicolich JR. 2004. Siliceous ZSM-5 membranes by secondary growth of b-oriented seed layers. *Adv. Funct. Mater.* 14:716–29
118. Lai ZP, Tsapatsis M. 2004. Gas and organic vapor permeation through b-oriented MFI membranes. *Ind. Eng. Chem. Res.* 43:3000–7
119. Kuznicki SM, Bell VA, Nair S, Hillhouse HW, Jacubinas RM, et al. 2001. A titanosilicate molecular sieve with adjustable pores for size-selective adsorption of molecules. *Nature* 412:720–24
120. Braunbarth CM, Boudreau LC, Tsapatsis M. 2000. Synthesis of ETS-4/TiO_2 composite membranes and their pervaporation performance. *J. Membr. Sci.* 174:31–42
121. Jeong HK, Krohn J, Sujaoti K, Tsapatsis M. 2002. Oriented molecular sieve membranes by heteroepitaxial growth. *J. Am. Chem. Soc.* 124:12966–68
122. Guan GQ, Kusakabe K, Morooka S. 2001. Synthesis and permeation properties of ion-exchanged ETS-4 tubular membranes. *Micropor. Mesopor. Mater.* 50:109–20
123. Weh K, Noack M, Hoffmann K, Schroder KP, Caro J. 2002. Change of gas permeation by photoinduced switching of zeolite-azobenzene membranes of type MFI and FAU. *Mircopor. Mesopor. Mater.* 54:15–26
124. Nijhuis TA, Beers AEW, Vergunst T, Hoek I, Kapteijn F, Moulijn JA. 2001. Preparation of monolithic catalysts. *Catal. Rev. Sci. Eng.* 43:345–80
125. Lu AH, Schüth F. 2005. Nanocasting pathways to create ordered mesoporous solids. *C.R. Chimie* 8:609–20
126. Smatt JH, Spliethoff B, Rosenholm JB, Linden M. 2004. Hierachically porous nanocrystalline cobalt oxide monoliths through nanocasting. *Chem. Commun.* 2004:2188–89
127. Lu AH, Smatt JH, Backlund S, Linden M. 2004. Easy and flexible preparation of nanocasted carbon monoliths exhibiting a multimodal hierarchical porosity. *Micropor. Mesopor. Mater.* 72:59–65
128. Taguchi A, Smatt JH, Linden M. 2003. Carbon monoliths possessing a hierarchical, fully interconnected porosity. *Adv. Mater.* 15:1209–13
129. Shi ZG, Feng YQ, Xu L, Da SL, Zhang M. 2004. A template method to control the shape and porosity of carbon materials. *Carbon* 42:1677–82
130. Shi ZG, Feng YQ, Xu L, Da SL, Zhang M. 2003. Synthesis of a carbon monolith with trimodal pores. *Carbon* 41:2677–79
131. Tao Y, Kanoh H, Kaneko K. 2003. ZSM-5 monolith of uniform mesoporous channels. *J. Am. Chem. Soc.* 125:6044–45
132. Li WC, Lu AH, Schmidt W, Schüth F. 2005. High surface area mesoporous glassy alumina with controllable pore size by nanocasting from carbon areogels. *Chem. Eur. J.* 11:1658–64
133. Buciuman FC, Kraushaar-Czarnetzki B. 2001. Preparation and characterization of ceramic foam supported nanocrystalline zeolite catalysts. *Catal. Today* 69:337–42
134. Zampieri A, Colombo P, Mabande GTP, Selvam T, Schwieger W, Scheffler F. 2004. Zeolite coatings on microcellular

ceramic foams: a novel route to microreactor and microseparato. *Adv. Mater.* 16:819–23
135. Wang YJ, Tang Y, Dong AG, Wang XD, Ren N, et al. 2002. Self-supporting porous zeolite membranes with sponge-like architecture and zeolitic microtubes. *Adv. Mater.* 14:994–97
136. Lee YJ, Lee JS, Park YS, Yoon KB. 2001. Synthesis of large monolithic zeolite foams with variable macropore architectures. *Adv. Mater.* 13:1259–63

Annu. Rev. Mater. Res. 2005. 35:239–314
doi: 10.1146/annurev.matsci.35.102303.091623

First published online as a Review in Advance on March 17, 2005

ANALYTICAL TRANSMISSION ELECTRON MICROSCOPY

Wilfried Sigle
Max-Planck-Institut für Metallforschung, D-70569 Stuttgart; email: sigle@mf.mpg.de

Key Words electron spectroscopy, EELS, EDXS, energy-filtered TEM, electron scattering

■ **Abstract** Chemical analysis at high spatial resolution is the domain of analytical transmission electron microscopy. Owing to rapid instrumental developments during the past decade, electron energy-loss spectroscopy offers now a spatial resolution close to 0.1 nm and an energy resolution close to 0.1 eV. This development has been accompanied by the introduction of numerous new techniques and methods for data acquisition and analysis, which are outlined in the present article. Recent results for a wide range of material systems are addressed. These comprise first-principles calculations, which have contributed to enormous progress in the calculation of near-edge fine structures, and fingerprinting methods, which are still important for the interpretation of experimental data.

INTRODUCTION

In ancient times the development of new materials took place on a time-scale of thousands of years. This long time-scale was not only related to the lack of appropriate processing techniques (e.g., high temperatures) but also to the ignorance of the nature of materials. The discovery of the light microscope in the late sixteenth century by the Dutch optician Zacharias Janssen offered the possibility to study the outer shape of small pieces of materials. However, it was not before the late nineteenth century that the microscope was used to study the polycrystalline microstructure of metals (1). Although the majority of the elements of the periodic table were already known at this time, the internal structure of materials was unknown. Suitable experimental techniques were invented some decades later [X-ray diffraction (2) and electron diffraction (3)] that made structural analysis of the interior of materials possible. Shortly after, the electron microscope was developed (4), which soon offered a spatial resolution superior to optical microscopes. Another 15 years passed until it was possible to prepare sections from materials thin enough for the transmission of fast electrons (5, 6). This was the advent of spatially resolved characterization of materials. The ongoing technical improvement of transmission electron microscopes (TEMs) has nowadays reached the level of atomic resolution.

Electrons can not be used only for imaging. Along their path through the specimen, many electrons undergo inelastic scattering processes which result in an energy loss. Relaxation processes can lead to the emission of characteristic X-ray quanta, Auger electrons, or light. It is these inelastic processes that make chemical analysis of materials possible in the TEM. These electron spectroscopic methods are the main scope of this article. However, numerous other techniques make use of the Z-dependence of elastic scattering (Z contrast), of phase shifts of electrons passing the specimen (electron holography), of chemically sensitive Bragg reflections (high-resolution TEM), or of the intensity of atom columns in high-resolution TEM micrographs. All these methods therefore contain chemical information, although less direct than found in spectroscopic measurements, but in some cases with a spatial resolution that is difficult to achieve by spectroscopy. Therefore, the potentials of these methods are also shortly addressed.

THE ELECTRON MICROSCOPE

The Electron Source

Apart from a long lifetime, the main requirements of the electron source are high brightness (electron current density per solid angle) and small spread of the kinetic electron energy. This requires materials with a small work function. Modern TEMs are equipped with either LaB_6 or field emission guns (FEGs) (Table 1). LaB_6 filaments offer a high current, but the brightness is by three orders of magnitude lower than that of FEGs. Whereas the high current is well suited for energy-filtering TEM (EFTEM), the low brightness leads to impractically small currents if a small electron probe of nanometer size is formed. This is the domain of field-emission sources, which consist of pointed tungsten wires exposed to a high electric field. In the case of heated FEGs (Schottky emitters), the tip of the wire is coated with ZrO_2. Cold FEGs have the main advantage of a small energy spread and a very long lifetime, but the total current is small and the long-term current stability is poor.

TABLE 1 Parameters of different electron emitters

		Field emission source (W)	
	LaB_6	**Schottky type**	**Cold**
Brightness (Acm^{-2} sr^{-1}) (at 100 kV)	10^6	10^8–10^9	10^7–10^9
Energy spread (eV)	0.7–0.9	0.7–0.9	0.3–0.5
Total current (nA)	500	300	3
Current density (Acm^{-2})	10	10^5	10^4–10^5

In order to improve the energy spread of the source, monochromators have recently been introduced. There are two basic approaches: the Wien filter with crossed electric and magnetic fields and the electrostatic Omega filter. Both types improve the energy width to about 0.2 eV at a yet acceptable beam current of several 100 pA. The single Wien filter limits the probe size to about 2 nm, whereas there is no such limitation in the case of the Omega filter. This spatial aberration of the Wien filter can be improved by using double Wien filters (7).

Electron Lenses

Owing to their electric charge, electrons can be deflected by electric or magnetic fields. Modern electron microscopes are equipped with round magnetic lenses. The magnetic flux (typically of the order of 1 T) is concentrated in a narrow circular gap between two pole pieces. Because the focal length of such lenses is of the order of millimeters, the specimen is immersed in the magnetic field. This is done by using a long specimen rod that is inserted either into the pole piece gap (side-entry lens) or from the top of the lens (top-entry lens). The former offers more flexibility for in situ experiments (heating, cooling, straining, etc.), the latter is less influenced by external influences such as variation of room temperature and air pressure, or noise.

Round magnetic lenses suffer from severe aberrations, which are the resolution-limiting factor in TEMs as long as other instabilities are small (e.g., those of lens currents, high voltage, external vibrations, or AC electric and magnetic fields). Because the strongest aberration is the spherical aberration (aberration coefficient C_S), a longstanding dream of TEM was the implementation of C_S correctors. Owing to the need for high mechanical precision and lens current stability, as well as for computer control, this was achieved in only the past decade (8, 9), mostly on the basis of suggestions by Rose (10, 11). In scanning TEM, the resolution-limiting factor is the size of the focused electron probe. With use of C_S correctors, probes below 0.1 nm can be achieved (12–14), and values below 0.1 nm were recently demonstrated (15). An important advantage of C_S correctors is that larger acceptance angles can be used that increase the probe current considerably. In the TEM, C_S correctors allow for a much more straightforward interpretation of high-resolution images. Recent examples (16) demonstrate how the application of a negative C_S value simplifies the detection of light atoms.

Illumination Modes

In conventional TEMs the specimen is illuminated by an almost parallel electron beam. The objective lens is used to form a magnified (intermediate) image of the illuminated area, which is further magnified by postspecimen lenses onto the image detector. Modern FEG-TEMs are often equipped with scanning coils that allow the microscopes to be used in a second imaging mode, the scanning mode. In this case the electron beam is focused on the specimen by the condenser lenses and the objective-lens prefield. The small probe is scanned across the area of interest in the specimen plane. The image is obtained either (*a*) from electrons that have not been scattered or are only under small angles or (*b*) from electrons that have undergone large-angle scattering events. The first mode gives a bright-field image, the second one an annular-dark-field image. If electrons are detected that were scattered under angles above about 70–100 mrad, the second mode is called high-angle annular-dark-field (HAADF) mode. Microscopes operating only in the scanning mode are called scanning TEMs (STEMs). Some STEMs (e.g., the VG HB501) have no lenses behind the specimen, which makes their design rather intuitive (Figure 1).

Important quantities for spectroscopy are the convergence angle, β, and the collection angle, α, because they enter into the formulae for the partial inelastic scattering cross-sections (see below). The convergence angle is defined as the half angle of the cone illuminating the specimen. The collection angle is the maximal scattering angle that is accepted by the spectrometer. In a dedicated STEM, β is determined by the size of the objective aperture, whereas α is determined by the size of the collection aperture. In TEMs and in STEMs with postspecimen lenses the situation is more complicated because the postspecimen lenses have the

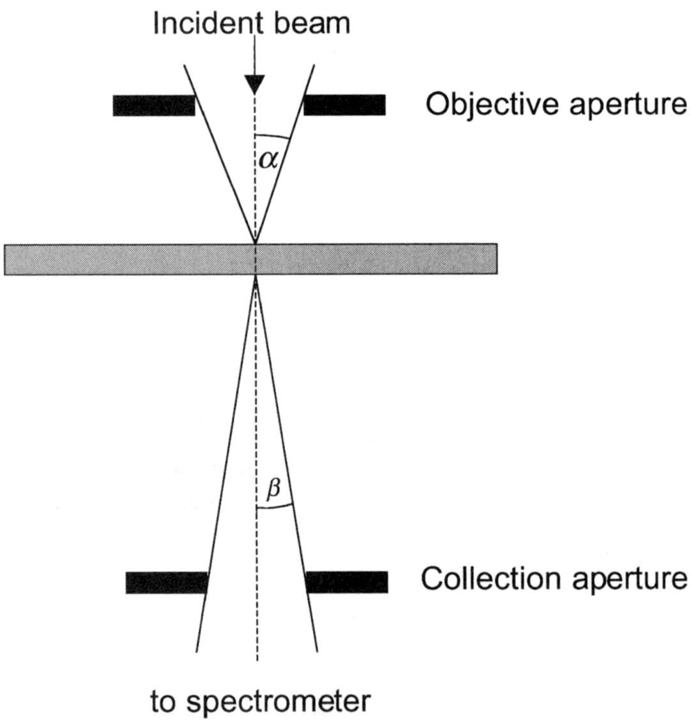

Figure 1 Electron beam path in a dedicated STEM. The objective aperture determines the angular range of illumination of the specimen, whereas the collection aperture is used to limit the angle of the scattered electrons entering the spectrometer.

TABLE 2 How apertures determine the collection and convergence angles in EELS in the case of a dedicated STEM and a TEM

	STEM (with no lenses after the specimen)	TEM	
		Image mode	Diffraction mode
Convergence angle α	OA	Condenser aperture	Condenser aperture
Collection angle β	CA	OA	The smaller of SEA and OA (for SEA: β depends on camera length)
Selected area	Probe size	Size of image on SEA, but affected by chromatic aberration	SAD aperture or probe size if beam is focused

OA, objective aperture; CA, collection aperture; SEA, spectrometer entrance aperture; SAD, selected area aperture.

property of changing beam angles. In TEMs, one has further to distinguish between image mode (image on fluorescent screen) and diffraction mode (diffraction pattern on fluorescent screen). Table 2 shows how the apertures in the STEM and TEM determine α and β. Note that for $\alpha > \beta$, the geometrically given collection angle should be replaced by an effective one (17) (see below).

Electron Spectrometers

Electron spectroscopy in the STEM was pioneered by Crewe et al. (18, 19). The basic electron spectrometer type is the prism spectrometer (Figure 2). Electrons are deflected by about 90° in a perpendicular static magnetic field. According to the Lorentz force, the deflection angle depends on the electron velocity. Using quadrupole lenses the electrons are focused in the energy-dispersive plane of the spectrometer where a position-sensitive detector is located. This detector can be an array of photodiodes or a charge-couple device (CCD), in both cases combined with a suitable combination of a scintillator and a light-guide. The CCD has the advantage of a better (electron-to-counts) conversion efficiency, better spatial resolution (narrower point-spread function), and often a higher dynamic range (16 bit). High conversion efficiency is important if the signal of interest is low. The spatial resolution is directly linked with the energy resolution. A high dynamic range is essential for the detection of the low-energy part of the energy-loss spectrum.

Energy Filters

The purpose of the energy filter is to form images from electrons that have suffered a specific energy loss [energy-filtered TEM (EFTEM) or electron spectroscopic imaging (ESI)] (20). An energy filter consists of an electron spectrometer with a mechanical slit in the energy-dispersive plane that is used for the selection of

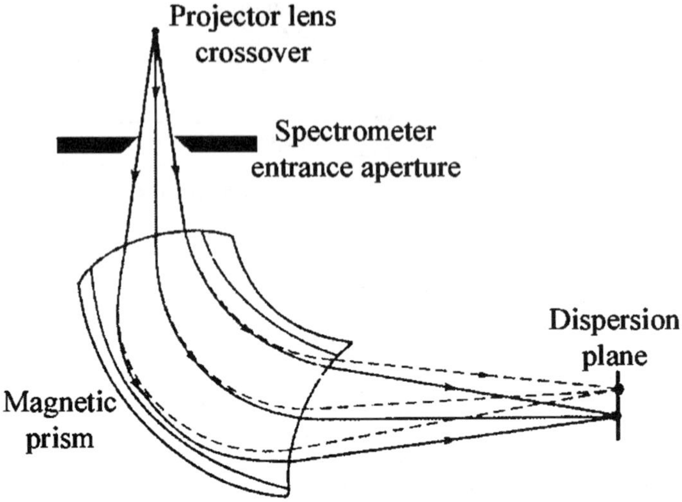

Figure 2 Design of a magnetic prism electron spectrometer. Electrons of various kinetic energies (depending on the energy losses they suffered within the specimen) enter the spectrometer. Because the magnetic force acting on the electrons depends on their velocity, an energy-loss spectrum is formed in the energy-dispersive plane. The spectrum is recorded by a CCD array located in this plane.

the energy loss of interest (Figure 3, see color insert). Behind the slit either an energy-filtered image (or diffraction pattern) or an image of the energy-dispersive plane [the electron energy-loss (EEL) spectrum] is formed by additional electron lenses. These lenses can be considered as a second projector lens system of the microscope.

There are basically two types of energy filters: the in-column filter and the post-column filter. The in-column filter is a symmetric filter incorporated in the microscope column below the usual projector lenses (Figure 3). Owing to the symmetry, a number of aberrations vanish for this type of filter. The pre- and post-filter electron trajectories are parallel, which makes the in-column filter design an obvious choice. However, a drawback is that an in-column filter cannot be attached to an existing TEM column. This is possible with the postcolumn filter (21) which has a 90° design and thus lacks the symmetry of the in-column filter. The corresponding higher number of aberrations is corrected by additional multipole lenses behind the filter.

An energy filter can be characterized by its transmissivity, which is a measure of the volume in phase-space that can be transferred (22, 23). A filter with a high transmissivity accepts electrons with large angles with respect to the optical axis, and it gives a large field of view even if narrow energy-selecting slits are used. Unfortunately no round-robin tests have been performed yet in order to compare the different filter designs.

Energy-Dispersive X-Ray Detector

For the detection of characteristic X-rays, energy-dispersive semiconductor detectors are used. Although they have a rather poor energy resolution, their detection efficiency is far superior to wavelength-dispersive detectors, which is an important requirement in the TEM owing to the low count rate. At present, mainly Li-drifted silicon detectors are used. They have an energy resolution of about 140 eV at 5.9 keV. Using digital pulse-processors, count rates of 10,000 counts $\cdot$ s^{-1} can be handled. Despite the poor energy resolution, energy-dispersive X-ray spectroscopy (EDXS) is often used because quantification is more straightforward than in electron energy-loss spectroscopy (EELS) and the detection of high-Z elements is often easier. However, the detection of low-energy quanta is considerably affected by absorption effects in the specimen and in the detector window. Therefore, elements of low atomic number should be more easily detected by EELS. A further disadvantage of EDXS spectrometers is the small collection solid angle (at most a few tenths of a steradian). This is limited by the size of the detector crystal, as well as by the specimen–detector distance.

Recently, microcalorimeters have been employed for the energy-dispersive detection of X-rays. They measure the heat generated by the X-rays in a superconducting material (24, 25). The energy resolution is below 10 eV and thus comparable to wavelength-dispersive spectrometers. However, the maximal count rate is lower by a factor of 10 compared with that found with conventional EDXS detectors. Furthermore, such detectors are at present extremely expensive. Figure 4 shows

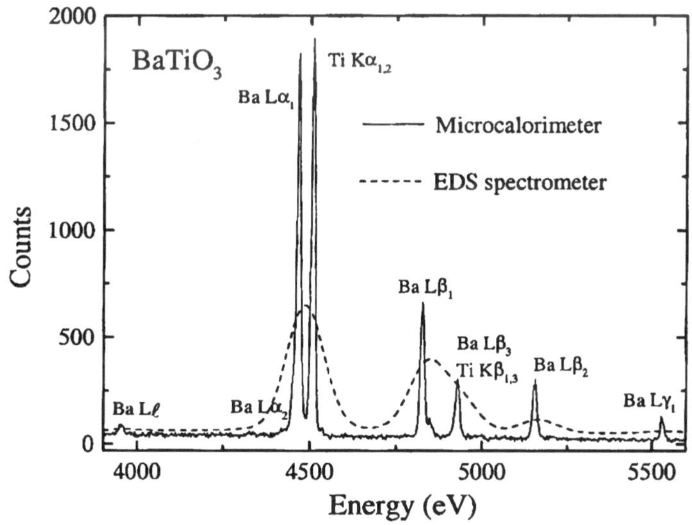

Figure 4 Comparison of EDXS spectra of $BaTiO_3$ obtained by a conventional semiconductor detector (*broken line*) and a microcalorimeter detector (*solid line*). Reprinted from Wollmann et al. (24).

the richness of information obtained with such detectors in comparison to that obtained with semiconductor detectors (dashed line) (24).

Whereas calorimeters have the advantage of high-energy resolution, silicon drift detectors offer a factor of 10 increase of count rate compared with that of Si(Li) detectors (26). This makes very fast EDXS mapping possible if the count rate is high enough. Furthermore, such detectors can be made with larger areas, which increases the collection solid angles, probably by a factor of 2 or 3. The space required for these large detectors becomes available in C_s corrected microscopes with large pole-piece gaps. For reviews on EDXS analysis the reader is referred to (27, 28).

ELECTRON-SPECIMEN INTERACTION

The interaction of electrons with matter is much stronger than that of X-rays with matter. This is a natural consequence of the electrical charge, and it is this strong interaction that makes the study of small volumes of material possible and thus is the basis for high spatial resolution. On the other hand, multiple scattering effects often complicate quantitative analysis.

Electron Scattering

Although the total potential of an atom (generated by the charged nucleus and the electron cloud) is responsible for the scattering process, one can distinguish limiting cases (see Table 3) where the scattering center can be ascribed to a single nucleus or an electron, or to a multiple of both. Because of the same mass, electron–electron scattering leads to significant energy losses that makes this type of scattering the most significant for analytical TEM. Because the electron–nucleus interaction depends on the atomic number Z, it can also be used for chemical analysis. However, quantification is complicated in this case (see below).

TABLE 3 Interactions of electrons in a solid

	Interaction of electron with		**Interaction of electron with**		
	1 electron	**Many electrons**	**1 nucleus**	**Many nuclei**	
Type of scattering	Inelastic	Inelastic	(Quasi) elastic	Elastic	Inelastic
Scattering effect	Electron Compton effect, electron excitation	Plasmon excitation, Cerenkov effect	Rutherford scattering, phonon scattering	Bragg scattering	Bremsstrahlung

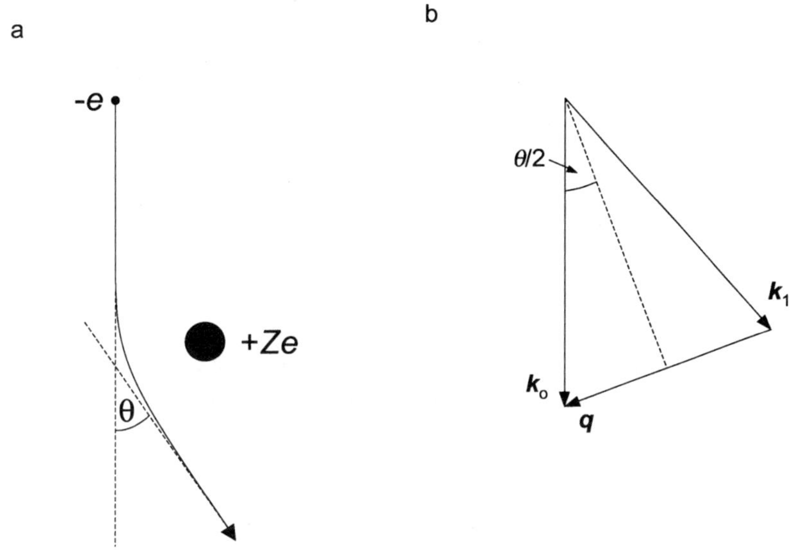

Figure 5 (*a*) Scattering of an electron at a nucleus. (*b*) Initial ($\boldsymbol{k}_0$) and final ($\boldsymbol{k}_1$) wave vector of an elastically scattered electron and momentum transfer $\boldsymbol{q}$.

ELASTIC SCATTERING Let us assume an electron with kinetic energy E and velocity v undergoing a scattering event with an atom of atomic number Z and atom mass A (see Figure 5). The potential of the atom nucleus is screened by the surrounding electrons. Therefore only approximate analytical expressions of the potential are available. In the simple case of a Wentzel potential, which uses an exponential decay of the potential with a decay length r_0 ($\approx a_0 Z^{-1/3}$), the differential scattering cross-section is given by

$$\frac{d\sigma}{d\Omega} \approx \frac{4\gamma^2 Z^2}{a_0^2 k_0^4} \frac{1}{\left(\theta^2 + \theta_0^2\right)^2}. \qquad 1.$$

Here, a_0 is the Bohr radius (0.0529 nm), $\gamma = (1 - v^2/c^2)^{-1/2}$, with c the speed of light, and θ the electron scattering angle. k_0 is the modulus of the wave vector of the incident electron. $\theta_0 = (k_0 r_0)^{-1}$ is the characteristic angle for elastic scattering.

In the limit of small impact parameter we obtain the expression for Rutherford scattering. In this case, the scattering angle is large and in the extreme case the electron is back scattered. The energy transfer to the atom can be of the order of 10s or 100s of eV, which may be sufficient for a displacement of the atom and thus to radiation damage (see below). This type of a scattering event can no more be considered elastic. It is often neglected in TEM because of the small scattering cross-section compared with that of other inelastic scattering events.

In crystalline materials elastic scattering can lead to the formation of strong Bragg reflections, described as originating from the constructive interference of spherical secondary electron waves emanating from the crystal atoms. An alternative approach is to solve the Schrödinger equation. The solution of the equation in a periodic potential of a crystal is characterized by Bloch functions, which are usually called Bloch waves in the case of fast incident electrons. Following the solution for an atom potential, the Bloch waves are denoted as s-, p-type etc., depending on their symmetry. Whenever lattice planes are close to the Bragg orientation certain Bloch waves are strongly excited, e.g., if $\boldsymbol{k}_0$ of the incident electron wave is parallel to a row of atoms (zone axis orientation), electrons are concentrated ("channelled") in the vicinity of this atom row of positive ions. This corresponds to an s-type Bloch wave. The strong excitation of certain Bloch waves is correlated with an inhomogeneous electron distribution within the unit cell of the crystal. Such conditions should be avoided for quantitative chemical analysis unless it is done intentionally as in the ALCHEMI method (29, 30). In the case of simultaneous *Z*-contrast imaging (see below) and EELS analysis along a zone axis, one should be aware of artifacts that may arise owing to the channelling effect.

INELASTIC SCATTERING A typical spectrum of energy losses suffered by an electron penetrating through a thin slice of a material is shown in Figure 6. The zero-loss peak includes all elastically scattered electrons, but also electrons with energy losses less than the energy resolution of the TEM (mostly phonon scattering). The energy-loss range up to about 100 eV is dominated by collective electron excitations (plasmons), but it also contains interband transitions from the valence to the conduction band (Figure 7). This regime includes the optical transitions and can therefore be used for the determination of optical properties (17). At higher energy

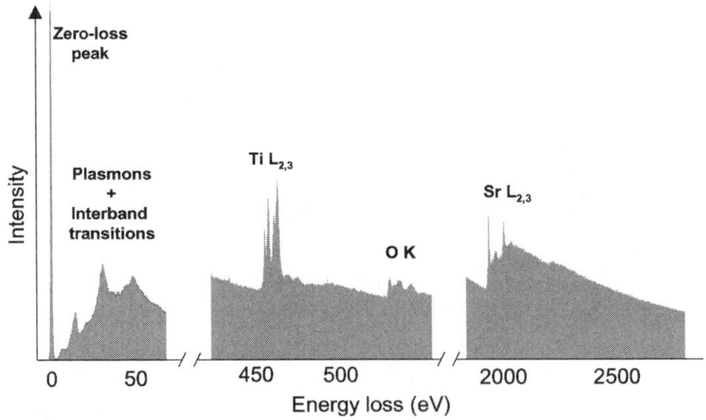

Figure 6 Electron energy-loss spectrum of $SrTiO_3$. The middle and right parts of the spectrum are strongly magnified compared with those in the low-loss spectrum.

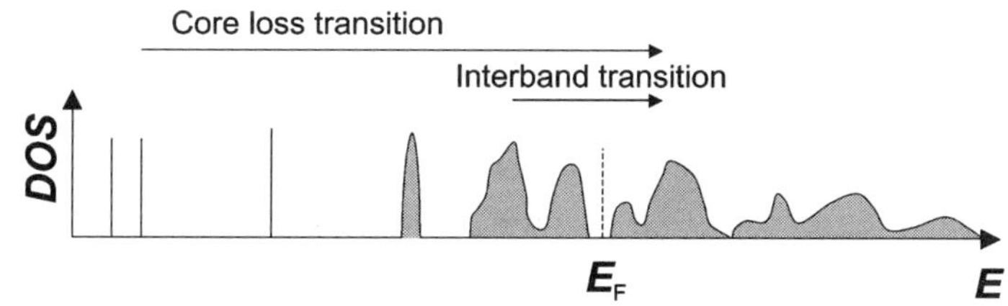

Figure 7 Electron transitions in a solid material. Core levels can be considered as atomic-like levels, whereas the levels are band-like near and above the Fermi level E_F.

losses excitations from core levels to unoccupied states above the Fermi level are found (Figure 7), which are superimposed on a background signal stemming from the low-loss events. These core-level excitations lead to edge-like features with an onset energy corresponding to the lowest unoccupied state. They are classified according to the initial state of the excited electron (Figure 8). Because the initial

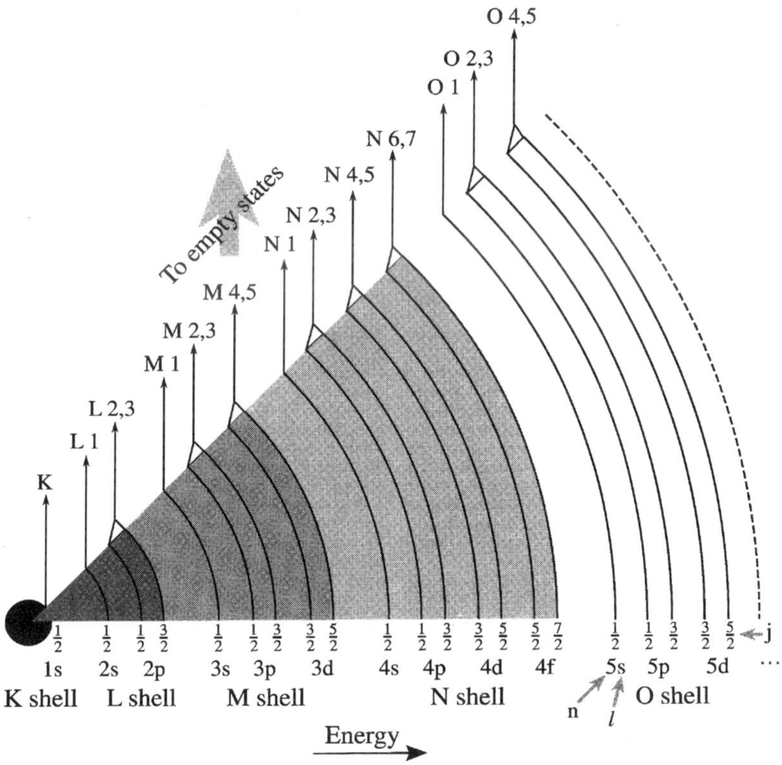

Figure 8 Classification of core transitions in EELS. Reprinted with permission from Williams & Carter (97).

states are hardly modified from atomic levels even in a solid, the shape of the absorption edge is mainly determined by the density of unoccupied states (see below) and thus includes information about the bonding state. Elemental concentrations can be extracted from the area underneath the absorption edge.

According to Lenz (31) the differential inelastic cross-section for electron–electron scattering can be written as (17)

$$\frac{d\sigma_i}{d\Omega} \approx \frac{4\gamma^2 Z}{a_0^2 q^4}\left\{1 - \frac{1}{\left[1+(qr_0)^2\right]^2}\right\}, \qquad 2.$$

with the scattering vector $q^2 = k_0^2(\theta^2 + \bar{\theta}_E^2)$ (Figure 5). $\bar{\theta}_E = \bar{E}/(\gamma m_0 v^2)$ is a scattering angle corresponding to a mean energy loss $\bar{E}$ and is usually denoted as the characteristic angle for inelastic scattering. Because the majority of energy losses is well below 100 eV, the characteristic angle is of the order of a few tenths of a milliradian, which is much smaller than typical Bragg angles. This means that the Lorentzian distribution of inelastic scattering is much more forward-directed than elastic scattering. Note, however, that for high-energy losses the characteristic scattering angle can be quite appreciable (e.g., 2.5 mrad at $\Delta E = 1000$ eV for 200 keV electrons). In the limit of "ballistic electron–electron collisions" (or electron Compton scattering (32, 33), a broad energy-loss peak occurs at high-energy losses (Figure 9). Because such electrons are deflected under relatively large

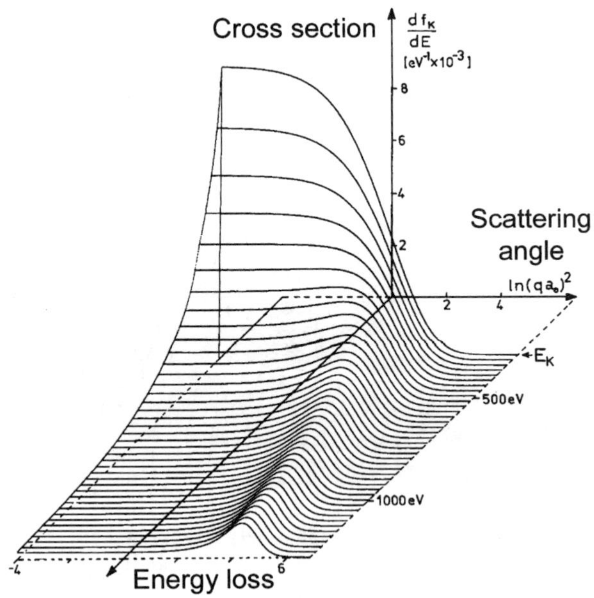

Figure 9 Electron scattering intensity as a function of energy loss and scattering angle. At large scattering angles, the Bethe ridge occurs. Reprinted with permission from Egerton (17).

angles (known as the Bethe ridge) (17), they are often not observed when low-angle scattering is studied.

As can be seen from Equations 1 and 2, inelastic electron–electron scattering shows a weaker dependence on the atomic number than does elastic scattering. Therefore, the ratio of inelastic to elastic scattering is particularly strong for light atoms and vice versa. In addition to electron–electron scattering, electron–phonon interaction is an important contribution to inelastic scattering, especially at high scattering angles. This contribution is particularly strong for heavy elements, whereas for light elements electron–electron scattering dominates at high angles (34).

Equation 2 describes the inelastic electron–electron scattering process as a Rutherford-type scattering, which is modified for small scattering angles by the second term in the bracket. In Equation 2 this factor is approximated by an empirical expression. In a concise quantum-mechanical treatment (35, 36) this term contains the probability that a crystal electron can be excited from an initial eigenstate ψ_i to a final eigenstate ψ_f, i.e., the transition-matrix element. Taking this into account, Equation 2 can be rewritten as

$$\frac{\mathrm{d}\sigma_i}{\mathrm{d}\Omega} \approx \frac{4\gamma^2}{a_0^2 q^4} \left| \left\langle \psi_f \left| \sum_j \exp(i\boldsymbol{q} \cdot \boldsymbol{r}_j) \right| \psi_i \right\rangle \right|^2, \qquad 3.$$

where $\boldsymbol{r}_j$ are the electron coordinates. The atomic number Z is now implicitly included in the sum. In a solid, the unoccupied final states are not discrete energy levels but can be described as energy bands with energy-dependent densities of states (DOS). Hence, the DOS enters into Equation 3 as an energy-dependent weighting factor. Because the initial states have discrete energies, the shape of absorption edges in EELS is determined by the product of the transition-matrix element and the density of unoccupied states. Schattschneider et al. (37) pointed out that Equation 3 is valid only under kinematic scattering conditions. In the case of strong excitations of Bloch waves, the expression has to be modified to the so-called mixed dynamic form factor (38).

The sharpness of the features observed in EEL spectra is limited by three effects: (*a*) Although the core level energy is discrete, the energy-loss spectrum will be broadened by the limited lifetime τ_i of the initial state. In other words, the core hole left behind by the excited electron is filled by other electrons from higher energy levels, either directly or via emission of Auger electrons. According to the uncertainty principle, the limited lifetime corresponds to a finite width of the measured energy level. The core hole lifetime limitation has long been known and was reported by Krause & Oliver and others (39–41) (Figure 10). The lifetime broadening owing to the core hole becomes stronger as Z increases and the angular momentum decreases. (*b*) Second, further broadening occurs by the limited mean free path (typically a few nanometers) (see Figure 11) of the excited electron, mainly by inelastic scattering events. This leads to a limited lifetime of the excited electron and thus again to an energy broadening. This effect is zero

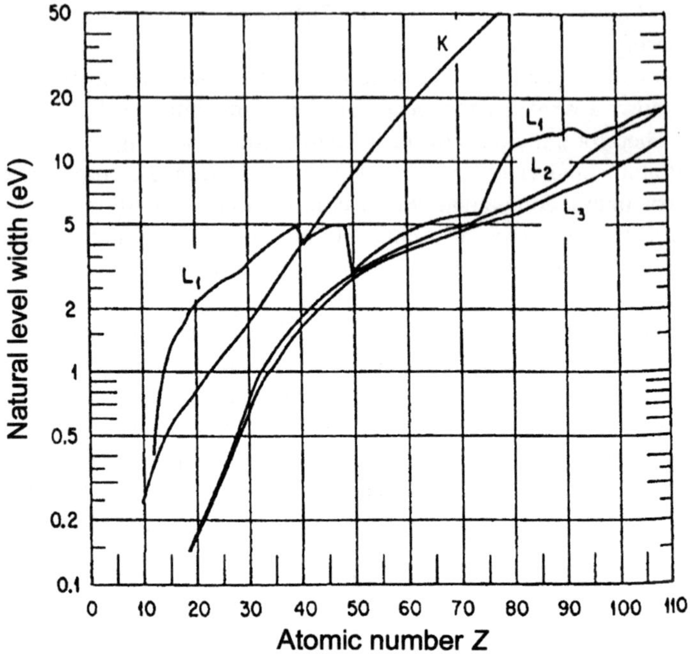

Figure 10 Line widths of electron transitions due to the limited lifetime of the core hole. Reprinted with permission from Krause & Oliver (39).

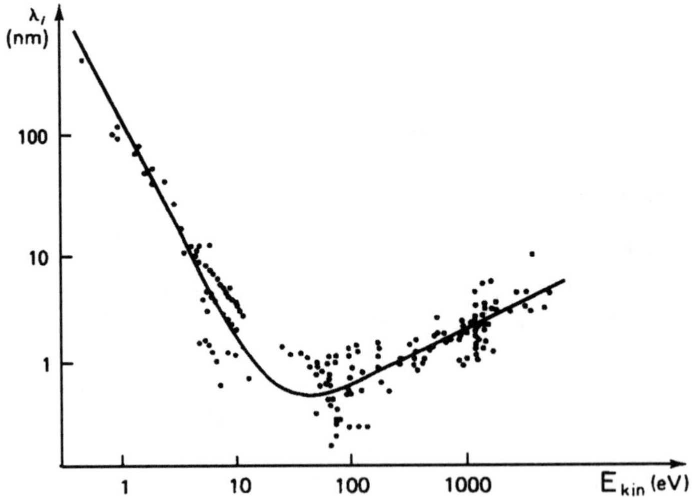

Figure 11 Inelastic mean free path of electrons as a function of their kinetic energy. Reprinted with permission from Egerton (17).

for electrons excited just to the edge onset energy and increases with energy loss (17). Well above the edge onset this effect dominates over the core-hole effect and leads to the broad wiggles in the extended region of the absorption edge (extended energy-loss fine structure, EXELFS) (42). (*c*) Finally, the instrumentation for the detection of the absorption edge can introduce a limitation to the energy resolution.

For $q \cdot r_j \ll 1$, i.e., if the length of the scattering vector is small compared with the average radius of the initial orbital, the exponential function in Equation 3 can be approximated by $(1 + i\boldsymbol{q} \cdot \boldsymbol{r}_j)$, and higher-order terms can be neglected (Dipole regime). The unity term cancels because the initial and final state wave functions are orthogonal. The vector product is non-zero only if the symmetry (angular momentum ℓ) of initial and final states is different.[1] In the case of dipole transitions ($\Delta\ell = \pm 1$), transitions from s- to p-states (e.g., K edges) or from p- to d-states (e.g., $L_{2,3}$-edges) are allowed. Therefore an absorption edge in EELS can be described by a site- and angular-momentum-projected density of states. The dipole selection rule may be violated if the DOS of dipole-forbidden transitions is much higher than the dipole-allowed states (44–46). Essex et al. (46) recommend not using the dipole approximation for energy losses below 50 eV. For low-energy losses the use of a small collection aperture (objective aperture in the TEM) ensures that non-dipole transitions are excluded from the EEL spectrum.

An alternative interpretation of the energy-loss spectrum is to consider it as the dielectric response of the material to the impact of the incident electron. The incident electron polarizes the material thus creating an electric field that acts back on the electron. In this formulation, the differential inelastic cross-section is given as (17)

$$\frac{\mathrm{d}\sigma_i}{\mathrm{d}E} = \frac{2\,\mathrm{Im}(-1/\varepsilon(E))}{\pi a_0 m_0 v^2 n_a} \ln\left[1 + \left(\frac{\beta}{\theta_E}\right)^2\right], \qquad 4.$$

where $\varepsilon(E)$ is the complex energy-dependent dielectric function, m_0 the rest mass of the electron, and n_a the atom density of the material. For a comparison with the experiment, the measured EEL spectrum is deconvoluted to obtain the single-scattering energy-loss function, *ELF*. This function compares with Equation 4 by $ELF = I_0 n_a t(\mathrm{d}\sigma_i/\mathrm{d}E)$, where I_0 is the intensity of the zero-loss peak and t is the specimen thickness. Because t can be calculated from the ratio of the total intensity of the spectrum and the zero-loss intensity, the dielectric function $\varepsilon(E)$ can be obtained from the EEL spectrum. The real and imaginary parts of the dielectric function $\varepsilon(E) = \varepsilon_1(E) + i\varepsilon_2(E)$ can be obtained by a Kramers–Kronig analysis. The real part is a measure of the polarizability of the material. The imaginary part

[1]A concise treatment of allowed and forbidden transitions, including the local symmetry around the excited atom, is given by Nufer et al. (43).

of $\varepsilon(E)$ describes absorption effects and is directly related with the joint density of states (JDOS) by

$$\varepsilon_2 E \propto JDOS \cdot \sum_i |\langle \psi_f | v_i | \psi_i \rangle|, \qquad 5.$$

where the sum describes the matrix element. This means that for low-energy losses, electron transitions from the valence to the conduction band are involved, i.e., the DOS of both bands is important.

Each inelastic scattering process is not only related to an energy loss but also to a change of the momentum of the scattered electron (Figure 12). This can be an additional source of information about the material. However, in most EELS experiments this dependence is not examined, and only electrons with scattering angles ranging from zero to β are selected by the collection aperture. Experimentally, larger scattering angles can be inspected by tilting the incident beam. From Figure 12 it is easy to show that for small scattering angles θ

$$q_{II} = k_0 \frac{\Delta E}{2E_0} \quad \text{and} \quad q_\perp \cong k_0 \theta. \qquad 6.$$

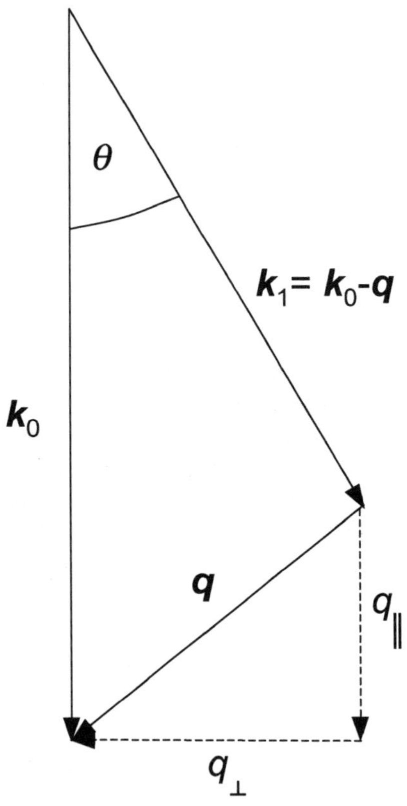

Figure 12 Electron wave vectors $\boldsymbol{k}_0$ and $\boldsymbol{k}_1$ and transferred momentum $\boldsymbol{q}$ in an inelastic electron scattering event. The momentum transfer can be decomposed into components parallel and perpendicular to the incident electron.

This means that for low-energy losses the momentum transfer is mainly perpendicular to the incident electron momentum, whereas for high-energy losses the parallel contribution can be quite appreciable.

Due to the scattering geometry and the limited size of the collection aperture (STEM) or objective aperture (TEM), only a certain angular range of scattering vectors enters the spectrometer, which can be used to obtain angle-resolved EEL spectra. Leapman et al. (47) proposed a method that was subsequently called the 45° method (48). The collection aperture is placed remote from the incident beam direction such that the magnitude of the perpendicular ($q_\perp$) and parallel ($q_\parallel$) components of the scattering vector $\boldsymbol{q}$ are equal (Figure 13). According to the above equations this position depends on the studied energy loss. In addition, the specimen is tilted so that a certain crystallographic direction $\boldsymbol{c}$ makes an angle of 45° with the incident electron beam. This means that the direction of the scattering vector $\boldsymbol{q}$ is then parallel to this crystallographic direction. By displacing the aperture to the opposite side of the incident beam, a spectrum with $\boldsymbol{q}$ perpendicular to this direction is obtained. In an alternative technique proposed by Browning et al. (49), EEL spectra are obtained with different collection angles. From these the parallel and perpendicular components are extracted.

A method to increase the angular resolution was presented by Midgley (50). After focusing the electron beam onto the specimen, the specimen is raised above the original position thereby resulting in a diffraction pattern with very high magnification (or angular resolution). Using a slit, a certain fraction of this pattern

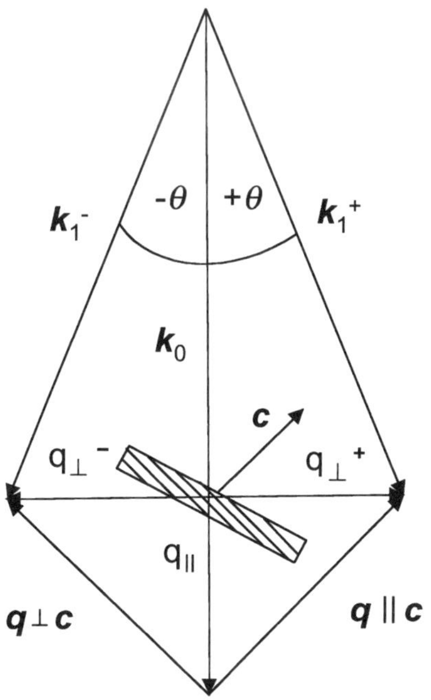

Figure 13 Scattering geometry used in the 45° method in order to obtain momentum-resolved EEL spectra (for details see text).

(corresponding to a certain range of scattering vectors) is allowed to enter the electron spectrometer. This permits direct imaging of the dispersion of features in the low-loss regime of the EEL spectrum, which helps to disentangle the physical origin of these features.

DATA ACQUISITION AND ANALYSIS IN EDXS AND EELS

The Signals Obtained by EDXS

The main advantage of EDXS over EELS is based on the fact that EDXS spectra are composed of peaks on a relatively low background originating mainly from bremsstrahlung. The removal of this rather smooth background is straightforward either by applying the second derivative of the spectrum or by fitting an analytical function to the background. Subsequently, peak intensities are determined by a fit of either a Gaussian function or peaks obtained from library standards, which are stored in a reference library in the computer. Although this library is often supplied by the manufacturer, it may be advisable to use one's own standards in order to optimize the quantification results.

Other minor contributions to the background are more difficult to separate from the real spectrum, such as

1. X-rays from areas not illuminated by the incident electron beam (e.g., from the pole piece of the microscope lens or, if the specimen is placed on a grid, from the grid material). These spurious X-rays can be excited by (back-) scattered electrons or by X-ray fluorescence. This signal is minimized but not eliminated by manufacturing the specimen stage from low-*Z* materials such as Be. It is partly possible to extract the remaining background signal by acquiring a hole spectrum where the incident electrons pass through the TEM specimen hole.
2. Similar to the above, secondary X-rays, which can be excited within the specimen itself.
3. Coherent bremsstrahlung, which can produce peaks in the bremsstrahlung background with shapes similar to X-ray peaks but with energy positions dependent on the energy of the incident electrons and the periodicity of the crystal lattice along the path of the incident electrons (51). Coherent bremsstrahlung is particularly strong if the crystalline material is oriented close to zone axes. These peaks cannot occur in amorphous materials, but they do occur in quasi-crystals (52).

Among the three contributions, X-ray fluorescence is probably the strongest effect. However, the prerequisite of strong X-ray fluorescence is that the energies of the original and fluorescing X-rays are close and that the original energy is the higher one. An example would be a steel specimen containing Cr and Fe, where Fe-K radiation can strongly excite Cr-K electrons. Clearly this effect changes the relative intensities of X-ray peaks in the spectrum and thus leads to wrong quantitative results, especially when quantifying Cr composition in dilute Fe-Cr steels. It is not trivial to consider this effect in quantification procedures because it depends on the specimen geometry (also called geometric factor) (53). As mentioned above, X-ray absorption within the specimen can be a major problem if low-energy X-rays (<3 keV in typical TEM specimens) are used for quantification. Because this effect depends on the specimen geometry, it is very difficult to correct for it by means of spectrum-processing software. Despite this problem, methods have been developed to consider the absorption effects in the analysis (e.g., 54, 55). Apart from the low X-ray yield in low-*Z* materials, it is mainly this absorption effect that makes EELS the preferred method in light-element analysis. Watanabe (56) proposed a new quantification procedure (the ζ-factor method) in which absorption and fluorescence effects are incorporated. It might replace the traditional Cliff–Lorimer method (57).

Signal Processing in EFTEM

Three images are sufficient to obtain a distribution map of a certain element: two from the pre-edge energy-loss range and one image from the edge area (58, 59). The pre-edge images are used for background subtraction (Figure 14). When crystalline materials are studied, a frequent problem is the preservation of diffraction contrast in inelastic images (60–65, 67). If only the amplitudes of diffracted beams are considered, this effect occurs when an area of the specimen is under strong

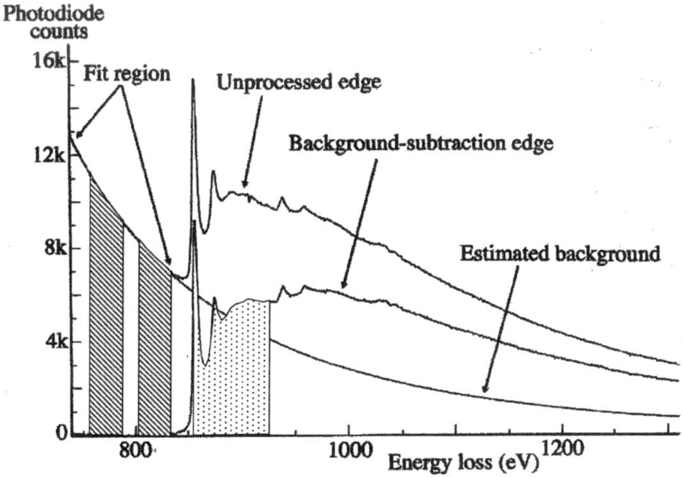

Figure 14 Three-window method used in EFTEM in order to obtain elemental distribution maps. Images using energy losses from the hatched areas are used for background subtraction. The signal is obtained from the third window above the edge onset.

diffraction conditions and most of the electrons are scattered under angles that are beyond the allowed range determined by the objective aperture. There are methods such as beam rocking (68) that reduce the effect; however, this is at the expense of lower intensity and thus a lower signal-to-noise ratio (SNR). The preservation of diffraction contrast is particularly pronounced under conditions used for atom-resolved imaging, i.e., zone-axis orientations of the crystal. Under such conditions, Bragg diffraction is strongly pronounced and can be described by strongly excited Bloch waves (see above). The spatial information contained in these Bloch states is partly preserved even after an inelastic scattering process (69). Navidi-Kasmai & Kohl (70) showed that the thicker the specimen the stronger the effect and that even for thin specimens this effect cannot be prevented, which means that interpretation of atomically resolved EFTEM images (71) is always difficult (see also below).

In order to reduce diffraction contrast in energy-filtered images the jump-ratio method (58, 72, 73) can be used instead of the three-window method. In the jump-ratio method only two images are acquired: those from energy losses below and above the edge onset (Figure 15). In the ratio image, diffraction effects are eliminated because the elastic diffraction effect included in the two energy-loss windows is almost the same due to their close proximity. Moore et al. (74) showed that the elimination of diffraction contrast is not complete, particularly at low-energy losses. In order to improve this, Crozier (59) suggested dividing ratio images by an image taken at very low-energy loss. Unfortunately, ratio images give only qualitative information about the chemical composition of the specimen, i.e., they cannot be used for quantification.

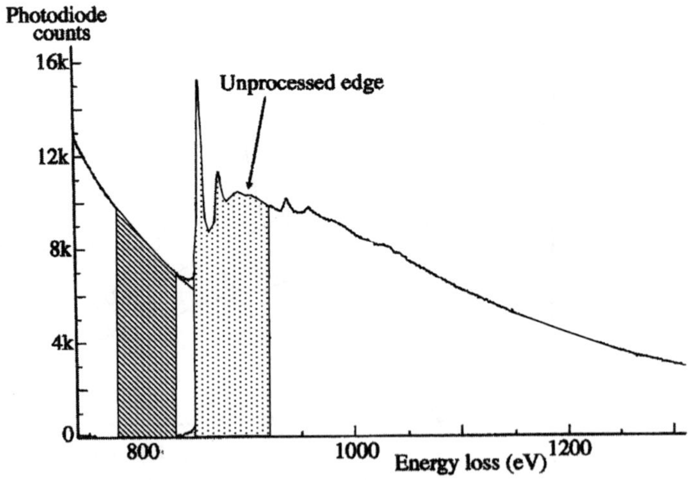

Figure 15 Energy-loss windows used in the jump-ratio method. The final image is obtained by a division of the dotted and hatched windows. Background subtraction is not done in this case.

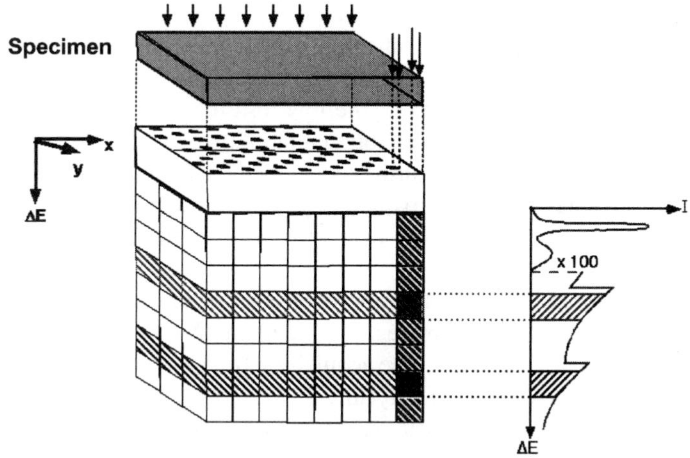

Figure 16 Data cube used to illustrate the collected data in a three-dimensional way. In a STEM, a full EEL spectrum is obtained from the illuminated area. In EFTEM, a full image is obtained from electrons that have suffered certain energy losses. This requires an energy filter as shown in Figure 2.

Acquisition of Data Series by EELS and EFTEM

The information content available from EELS or EFTEM can be represented through use of the data cube (75) (Figure 16). The x- and y-axes determine the lateral position of the investigated area and the z-axis is the energy-loss axis. In a normal elemental map obtained by EFTEM, slices parallel to the x/y-plane are obtained with thicknesses given by the width of the mechanical slit behind the energy filter. An EEL spectrum obtained from a selected position on the specimen corresponds to a line parallel to the z-axis. The entire data cube can be filled with data in two ways (Figure 16).

STEM SPECTRUM IMAGING Subsequent EEL spectra are recorded in the STEM from individual positions by scanning the focused electron beam (76–79). Each EEL spectrum has the full energy resolution determined by the spectrometer and the electron source. The acquisition time can be long. Owing to specimen drift, image distortions can be introduced that are almost impossible to correct after the data acquisition. The individual spectra can be analyzed after the acquisition using standard EELS analysis software.

EFTEM IMAGE SPECTROSCOPY Images are recorded from electrons of subsequent energy loss ranges. In this mode, each image contains the full image information; however, this occurs at the expense of lower energy resolution in contrast to that of the STEM approach (80–89). Post-acquisition specimen drift compensation is possible in this case, although it is complicated if strong contrast variations are present in the images. Then image processing such as the application of Laplacian filters may simplify the drift-compensation process (86). Using a combination of energy-filtered series of the low-loss region and of the core-loss region, Mayer et al. (85) were able to obtain quantitative results by EFTEM image spectroscopy. This method was further developed by Thomas & Midgley (89) who acquired an EFTEM series across a continuous energy range including the zero-loss peak. This method makes full data analysis possible, including plural-scattering deconvolution.

In order to compare the two approaches let us assume a data cube of dimension $N^2 \times M$, i.e., the area of interest is divided into $N \times N$ pixels and the energy loss range contains M elements. The acquisition time for each pixel in STEM is τ_S, the acquisition time of each energy-filtered image in EFTEM image spectroscopy is τ_E. The total beam currents are I_S and I_E, respectively. If we want to have the same statistics within each pixel (i.e., $\tau_S I_S = \tau_E I_E/N^2$), we need an M times higher total current in EFTEM image spectroscopy than in EELS spectrum imaging. If we assume a typical total current of 200 pA in EELS spectrum imaging and of 1 μA in EFTEM image spectroscopy, an area of 100×100 pixels would therefore give a data cube with comparable statistics. The total acquisition time in EELS spectrum imaging would be $T_S = N^2 \cdot \tau_S = (I_E/I_S) \cdot \tau_E \approx 10{,}000\ \tau_E$ and in EFTEM image spectroscopy $T_E = M \cdot \tau_E$. Because typical values of M are well below 10,000, EFTEM image spectroscopy is the faster approach. This example shows that EFTEM image spectroscopy is superior if large areas with many pixels are recorded, whereas EELS spectrum imaging has advantages for small pixel numbers because the energy resolution does not depend on N. The total current per pixel is $\tau_S \cdot I_S$ ($= \tau_E \cdot I_E/N^2$ in the case of the same statistics as in EFTEM image spectroscopy) in EELS spectrum imaging and $M \cdot \tau_E \cdot I_E/N^2$ in EFTEM image spectroscopy. Therefore, the total electron dose is larger by a factor of M in EFTEM image spectroscopy. In the case of radiation-sensitive materials, M should be small in EFTEM image spectroscopy.

A special technique to obtain a continuous series of EEL spectra along a line was demonstrated by Reimer et al. (90) and was applied later on by several groups (91–96). This so-called spectral profile technique makes use of a slit above the energy filter oriented perpendicular to the energy-dispersive direction of the filter. In the energy-dispersive plane of the filter an EEL spectrum is visible for each position along the slit. The advantage of this technique is the simultaneous acquisition of all data. Thus specimen drift or energy instabilities are almost negligible. Walther (96) reported that for a TEM with high-energy stability and a high stability of the spectrometer currents, chemical shifts of 0.01 eV should be detectable. An interesting application of the method is the study of interfaces and grain boundaries by placing the slit perpendicular to the interface. A rotation specimen holder is advantageous in order to orient the interface exactly with respect to the fixed slit orientation.

Quantification

The intensity of a peak in an energy-dispersive X-ray (EDX) spectrum or under an absorption edge in EELS can be written as

$$I_X \propto I_0 \cdot \sigma_X \cdot t \cdot C_X \cdot n, \qquad 7.$$

where I_0 is the intensity of the incident electron beam, t is the specimen thickness, σ_X the cross-section for inelastic electron scattering of the element X under study, C_X is the concentration of element X in the illuminated volume, and n is the atom density. I_0, t, and n can be eliminated by using ratios of peak or edge intensities (e.g., of elements A and B):

$$\frac{I_A}{I_B} \propto \frac{\sigma_A \cdot C_A}{\sigma_B \cdot C_B}. \qquad 8.$$

This is the standard technique in EDXS. Together with the knowledge of $C_A + C_B = 1$ (in a binary system) the concentrations of elements A and B can be calculated. The proportionality factor between intensities and concentrations is called Cliff–Lorimer factor (57). For details the reader is refered to Reference 97.

In EELS, I_0 can be directly measured from the low-loss part of the EEL spectrum. This also allows quantification of single edges (in units of atoms per unit area) by dividing the core-loss signal by $(\sigma_X \cdot I_0)$. The inelastic cross-sections for K edges can be calculated using the SigmaK program of Egerton (17, 98), which gives a typical accuracy of about 10% [see also Hofer (99)]. The accuracy for L edges (and even more so for M edges, etc.) is lower, typically 10–20%. A source of error in EELS quantification of elemental concentrations is multiple inelastic scattering, which becomes important at specimen thicknesses above about 0.3 inelastic mean free paths. Multiple scattering basically leads to a convolution of the absorption edge with the low-loss part of the EEL spectrum. Deconvolution algorithms can be used to obtain the single-scattering intensity (17). Quantification errors introduced by the background subtraction are discussed below.

It has been shown by Knippelmeyer et al. (100) that relativistic effects become important at kinetic energies of the incident electrons, E_0, above 200 kV and that these should be considered in the quantification process. In the case of K edges, they found deviations between 5 and 20% at $E_0 = 400$ keV and between 10 and 70% at $E_0 = 1200$ keV.

Spatial Resolution

If a focused electron probe is used (e.g., in the STEM), spatial resolution is primarily determined by the size of the electron probe. As shown above, the probe size is typically a few Ångstrøm for field-emission TEMs and close or below 1 Å if a C_s corrector for the illumination system is used. Nowadays it is possible to focus the electron probe onto a single atom column in a zone-axis orientation of a crystalline material. As mentioned previously, under such conditions a strong 1s-type Bloch wave is excited that is concentrated close to the nuclei of the atom column (axial channeling) (101–103). This strong localization of the electron intensity allows one to obtain an EELS signal largely originating from a single atom column. Lupini & Pennycook (104) showed that large collection apertures are necessary to achieve atomically resolved EELS. This will increase the fraction of nondipole transitions.

It has been argued that even a small shift of the electron probe away from the very column position leads to considerable loss of intensity to the neighboring atom columns (105). Furthermore, the electron intensity spreads with increasing specimen thickness away from the central atom column (dechanneling) (101, 106). In C_s-corrected TEMs, one must further consider that the convergence angle is much higher than that in uncorrected instruments. This large convergence leads to an increased dechanneling of the electrons away from the central atom column. Dwyer & Etheridge (107) calculated the EELS signal for different probe sizes and crystal orientations and found that atomically resolved EELS is possible for small specimen thicknesses and small probes. However, the strong variation of the electron intensity through the crystal by dynamic scattering can lead to misinterpretations in cases where atoms are not distributed homogeneously along the electron trajectory. On the other hand, this effect might be used to detect the z-location of a dopant atom within an atom column (106).

Under arbitrary specimen orientations and in thicker specimen regions (typically >30–70 nm depending on Z), the beam broadening owing to elastic scattering within the specimen (54, 108–112) becomes the resolution-limiting factor. Alber et al. (112) have shown that high-resolution chemical information can still be obtained if this broadening is quantitatively taken into account. Unlike in EDXS, in EELS the whole broadening is not relevant because the limited collection aperture β cuts off the tails of the broadened electron probe. On the basis of purely geometric arguments Egerton (17) showed that the broadening in EELS can be approximated by $2r \approx \beta t$.

Several authors have described spatial resolution in inelastic scattering by the convolution of the electron probe function with the so-called inelastic object function (113, 114). The latter is equal to the inelastic response of an atom for a perfectly localized electron beam.

At low-energy losses, the spatial resolution, d, is limited by the delocalization of inelastic scattering (115–118). This effect has its origin in the small characteristic inelastic scattering angle θ_E at low-energy loss (see above), which, according to the Rayleigh criterion $d \approx 0.6\,{}^{\lambda}\!/_{\theta_E}$, limits the spatial resolution. However, as Muller & Rose (119) point out, about 50% of the electrons is scattered at angles larger than θ_E, which diminishes the delocalization effect. This is supported by results of van Benthem et al. (120) who found that the effect of delocalization at energy losses below 20 eV is less severe than those predicted by the models of Pennycook & Narayan (116) and Muller & Silcox (118).

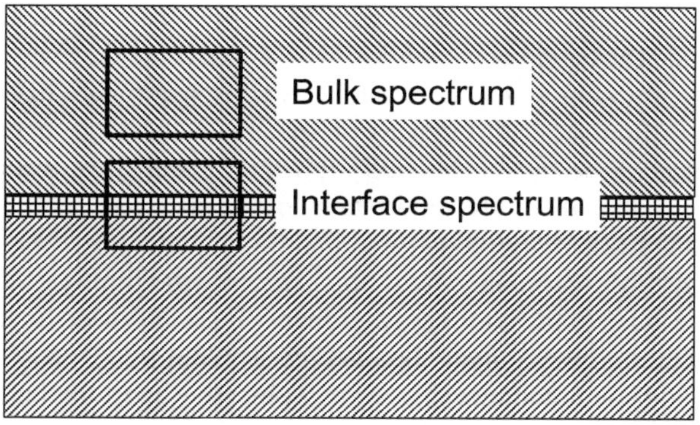

Figure 17 Illustration of the spatial difference technique. By comparing two spectra obtained in the bulk or on the interface, an interface-specific spectrum can be obtained.

In order to detect the interface-specific energy-loss near-edge structure (ELNES) or segregation at interfaces, the so-called spatial difference (SD) method has been applied in many studies (121, 122). With this technique spatial resolution is obtained by subtracting a spectrum from an area (usually rectangular) containing the interface, I_{IF}, from a bulk spectrum, I_{bulk}: $I_{SD} = I_{IF} - \xi I_{bulk}$ (Figure 17). The method is based on the assumption that the signals stemming from the bulk and the interface region overlap incoherently. Because no focused probe is used, the spatial resolution of this technique is not limited by the probe size. A main advantage of this technique compared with that of spot measurements is that the position of the beam can be monitored during the EELS or EDXS data acquisition, i.e., specimen drift can actively be compensated for. Furthermore, beam damage can be reduced if dose rate is a problem for the material investigated. The weighting factor ξ corresponds to the fraction of bulk-like information contained in the interface spectrum. This factor is often determined empirically during the subtraction procedure. Nufer (123) determined ξ by calculating the ELNES for atoms at different distances from the interface. Using this "interfacial width" he found excellent agreement between experiment and theory demonstrating the validity of the spatial difference approach. This validity was also demonstrated by Scheu et al. (124, 125) who measured the difference signal for different positions of the interface within the rectangular area. The same results were obtained in all cases. The SD method has been criticized by Muller (126), who pointed out that subtraction of spectra with small energy shifts with respect to one another (e.g., due to chemical shifts or energy instabilities) can give completely artificial results in the difference spectrum. Although this poses a problem for the technique, careful data acquisition and analysis can minimize such artifacts.

In another approach, the interface-specific EEL or EDX spectra are obtained by multivariate statistics (127, 128, 128a) by sorting spectra from a line scan or a two-dimensional map of the interface region according to their similarity

(129–131). The N spectra, each with M energy channels, are treated as an $N \times M$ matrix and the eigenvectors of this matrix constitute the axes of an N-dimensional space. Within this space the position of all N spectra are displayed. This method is able to find weak differences of spectra in the large data sets. Furthermore, it is objective because it basically relies on a statistical evaluation. It should be mentioned that there are other statistical approaches such as the use of neural networks (132).

In EFTEM, chromatic aberration is usually the limiting factor for spatial resolution because of the wide range of energy losses allowed by the selecting slit. For a slit width δ the chromatic broadening is given by $C_c \cdot \beta \cdot \delta / E$. In principle this effect can be reduced by using a small objective aperture, but this can introduce a blurring by diffraction ($= 0.61\ \lambda/\beta$ with the electron wavelength λ). The final resolution in EFTEM is thus determined by these contributions added in quadrature (133). Figure 18 shows the individual contributions and the total EFTEM resolution as a function of the objective aperture size for a 200 keV TEM with C_c = 2 mm and C_s = 2 mm. This example shows that there is an optimal size of the diffraction aperture. However, in practice it may happen that larger apertures have to be used in order to increase the SNR. In Figure 19 the energy-loss dependence of the resolution-limiting factors is shown. In fact, only the delocalization of inelastic scattering is energy-loss dependent. In the case of best resolution at high-energy

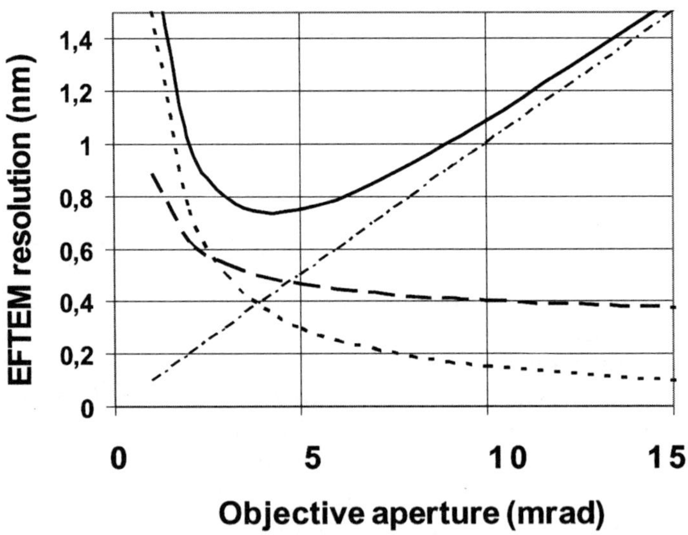

Figure 18 Spatial resolution in EFTEM as a function of the objective aperture size (which limits the collection angle) in a TEM. Limitations result from diffraction (· · ·), delocalization (– – –), and chromatic aberration (– · – ·). These can be summed in quadrature to obtain the total resolution limit (—). Parameters: electron energy 200 keV, $C_s = C_c = 2$ mm, 10 eV width of the energy-selecting slit, energy loss 285 eV.

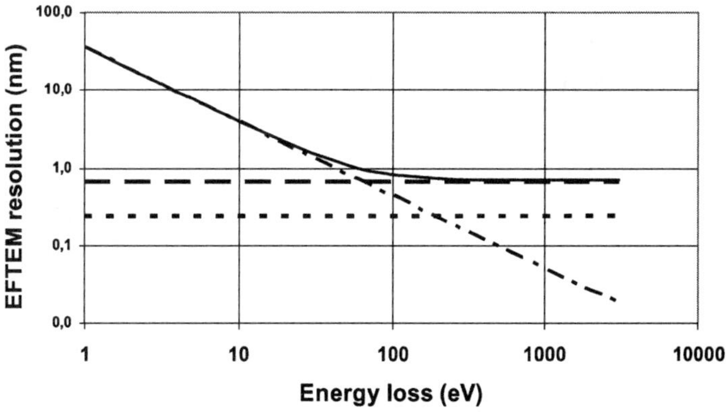

Figure 19 Spatial resolution in EFTEM as a function of the energy loss. Limitations are due to chromatic aberration (– – –), delocalization (– · – ·), and diffraction (· · ·). These are summed in quadrature to obtain the total resolution limit (—). Parameters: objective aperture 6.5 mrad, $C_c = 2$ mm, electron kinetic energy 200 keV, width of the energy-selecting slit 10 eV.

losses, the horizontal lines for chromatic aberration and diffraction limit overlap. Delocalization is important only for energy losses below 100 eV; in the case of large slit widths ($\geq$20 eV) only below 20 eV. A way to reduce the effect of chromatic aberration in EFTEM is to use a series of energy-filtered images across the energy range of interest using a narrow energy-selecting slit.

In a number of articles the possibility of atom-column resolution in EFTEM has been addressed. Lattice images can be formed from inelastically scattered electrons, e.g., by using electrons which have undergone a plasmon loss (69, 134–140). The resolution is affected by the width of the energy-selecting slit due to chromatic aberration. Unfortunately it is usually not possible to directly interpret such images as atomically resolved elemental maps. This is due to the preservation of contrast from elastic scattering (141), i.e., the elastic part of the image contrast cannot simply be subtracted from the image. Notwithstanding this problem, there are examples which are likely to show atomic resolution in EFTEM images (71, 96). Kimoto et al. (142, 143) showed how the contrast of lattice fringes changes with energy loss because the energy loss leads to an effective defocus change.

Egerton (34) discussed spatial resolution in EELS and EFTEM taking into account both elastic and inelastic scattering.

Signal-to-Noise Ratio and Detection Limits

Characteristic X-ray peaks as well as absorption edges in EELS are always superimposed on a background signal. Furthermore, experimental data always suffer from statistical noise. Both effects lead to a lower limit of detectability. Generally, the presence of a peak or an edge can be regarded with 98% certainty if the peak (edge) intensity exceeds three times the standard deviation of the background signal. Often the Rose criterion (144) is used, which states that a factor of five instead of three should be used.

In EDXS, the typical detection limit in terms of a concentration (minimum mass fraction, MMF) is in the range 0.1 to 1 at%. However, because small volumes are probed, this can correspond to the detection of a small number of atoms (minimum detectable number of atoms, MDN). Watanabe & Williams (145) demonstrated a MDN of two atoms for the detection of Mn in a Cu matrix using EDXS. They state that this can be achieved only in thin specimens in order to minimize beam spreading and thus to keep the probed volume small. One should keep in mind that the best experimental conditions for small MMF and small MDN are not the same. For small MMF, high beam current is mandatory, which may be advisable to illuminate across a large area to prevent specimen damage. For small MDN, a small probe is important, which poses a limit to the total current.

The problem of SNR in EELS was discussed (among others) by Egerton (17), Berger & Kohl (146), and Overwijk & Reefman (147). The main difference between EELS and EDXS in SNR is that it is much more difficult to remove the background signal in EELS than in EDXS. Due to the better signal collection in EELS, MMF can be in the range of some 10 ppm in favorable cases. Near-single atom detection was reported by Mory & Colliex (148) and Krivanek et al. (149). Suenaga et al. (150) demonstrated single-atom detection by EELS spectrum imaging. Leapman (151) studied biological specimens and found that the four Fe atoms in a hemoglobin molecule are detectable, and in the case of calcium, even single-atom detection was found to be possible, both cases having a SNR of 5. These results were made possible by the use of a modern CCD exhibiting high detection quantum efficiency. Feng et al. (152) proposed using ratios of differences of three EEL spectra, which were taken in a series. These ratios are independent of gain variations of the detector and thus improve detection of trace elements.

Recently Menon & Krivanek (153) presented a software package to model entire EEL spectra. This can be used to find detection limits and to optimize the experimental conditions.

Noise is a frequent problem in EFTEM images (58) because the number of electrons within the selected energy-loss range is small (mainly at high energy losses) and because the total intensity is distributed over N by N image pixels. In principle, the SNR can be improved by increasing the exposure time, but this can introduce artifacts from specimen drift or radiation damage. This problem was treated by Berger et al. (154) and Kohl & Berger (155). Berger et al. state that the SNR can be improved by placing one pre-edge window close to the edge onset, with the second window shifted as much as possible to lower energies, taking into account that the parameters A and r are only constant in a limited energy-loss range. This was stated more precisely by Hofer et al. (58) and Kothleitner & Hofer (156): For saw-tooth-like edges (like many K edges) and L edges with white lines, the window close to the edge should be 20–30 eV wide. For delayed edges, the post-edge window should be positioned at the maximum of the edge and should

More Sophisticated Analysis of EFTEM Series and STEM Spectrum Images

Unfortunately, an automized analysis of EEL spectra is at present not available. This would be particularly useful for the analysis of a large number of spectra as in energy-filtered series or in STEM spectrum images. However, Kundmann & Krivanek (175) have described a way for automatic detection of edges in an EEL spectrum. It is based on the derivative of spectra that is sensitive to the sharp onset of absorption edges. This is followed by a selection of those peaks having intensities that are statistically significant above the noise level. Using this approach, elemental occurrence maps can be extracted from the original data cube (176). For quantification, correction for multiple scattering is necessary in thicker specimen areas. Clearly this also requires data from the low-loss region including the zero-loss peak. Once the elements are detected, their concentration can be determined by the usual approach mentioned above. Because the edge positions of the elements are known, the window widths for background subtraction and signal integration can be determined automatically. However, this fails if edges are too close. In this case the spectra can be synthesized using reference spectra as shown above.

By comparing different elemental maps, phases can be identified in a material. A simple method is to use colors for the different elemental maps and to overlap them, which results in a certain color value for each phase. In another approach a scatter diagram is used in which the axes correspond to the intensities of the maps. Suppose there are two elemental maps (e.g., from C and N) with $N \times N$ pixels. The intensity pairs of each N^2 pixels are transferred into the scatter diagram. All pixels of a certain phase will have a similar combination of intensity values in the two images and will therefore be located in the same area of the scatter diagram. By selecting areas with high pixel density in the scatter diagram and back tracing them into the elemental maps, the location of the phases can be determined (58). To circumvent diffraction effects, such phase maps are better performed from jump-ratio maps (see above).

Radiation Damage

During electron exposure the studied material may undergo a change in its atomic structure. This can be from (*a*) the direct knock-on process between electron and nucleus or (*b*) inelastic excitation (electron–electron scattering). In the first case the hit atom directly receives momentum from the electron (which is back scattered). It is the dominant damage process in metals and probably in semiconductors and ceramics. The knock-on damage has a lower threshold, i.e., it can be avoided by using sufficiently low electron energies. In the case of inelastic electron–electron scattering, atom displacement occurs during the relaxation process of the excited atom. The emission of secondary and Auger electrons leads to a positive charge. Because of the small illuminated region, a high-electric field strength develops that can expel positive ions (177). It is most pronounced in biological specimens and polymers but also in inorganic materials such as ceramics. This type of damage cannot be prevented by lowering the electron energy because the inelastic cross-section would also increase. An improvement could be obtained by cooling the specimen or by coating it with a conducting film.

Depending on the material, the amount of damage may depend not only on the dose but also on the dose rate. In this case the size of the illuminated area should be increased in order to reduce the damage. In the study of interfaces, a strongly astigmatic beam aligned parallel to the interface can be used to probe a larger volume; in the STEM, a narrow rectangular area can be illuminated. In any case, it is advisable to acquire spectra at different doses and dose rates in order to determine whether damage actually occurs before starting with the real experiments.

An example of radiation damage in α-Al_2O_3 is shown in Figure 20. Electron irradiation of this material can lead to the formation of metallic Al and molecular oxygen (178). In Figure 20 sharp peaks at 531 eV and 15.3 eV appear after irradiation, corresponding to molecular oxygen and metallic aluminum (123). This damage occurs only with specimens that have been Ar ion–thinned at ion energies above about 3 keV. The ion bombardment introduces point defects that act as nucleation centers for the formation of oxygen bubbles during the electron irradiation in the TEM. The damage can be minimized by using lower ion energies or by heating the specimen after ion milling to anneal the point defects.

Radiation damage also occurs during the specimen preparation by ion sputtering. Apart from the atoms that are removed from the surface, atoms in the near-surface regions can be displaced from their regular lattice sites. The structural disordering can lead to an amorphization. Even worse for quantification in analytical TEM, different atoms in multi-element specimens are displaced with different cross-sections leading to a nonstoichiometry of the near-surface regions. This can be improved (but not avoided) by using low-energy ions and small inclination angles of the ion beam (179). Such surface nonstoichiometry can be checked by comparing the specimen composition at different specimen thicknesses.

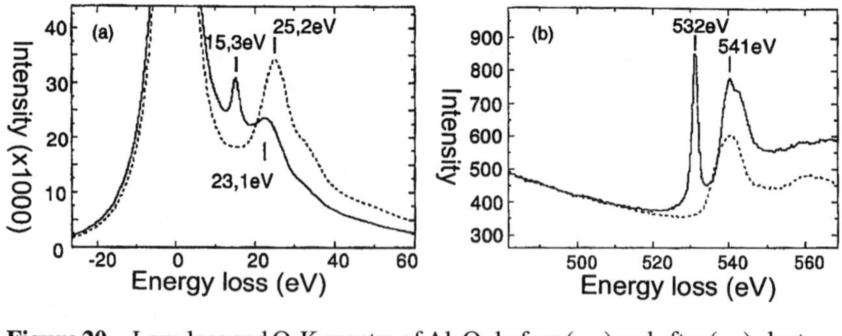

Figure 20 Low-loss and O-K spectra of Al_2O_3 before (· · ·) and after (—) electron irradiation. Reprinted with permission from Nufer (123).

For a recent review about irradiation-induced effects, including heating, charging, and contamination, the reader is referred to Reference 180.

CALCULATION OF THE ELNES

There are a number of approaches to calculate the fine structure of absorption edges based on first principles (181–187). The main purpose of such calculations is to obtain a relationship between the ELNES and structural properties in materials. This is not easily derived directly from EEL spectra. Due to the complexity of the problem, fingerprinting methods are frequently applied for the interpretation of EELS data. These methods rely on the comparison of the data with spectra obtained from known materials. Despite the strength of the fingerprinting approach there is always room for misinterpretations. Therefore reliable ab initio calculations will be of utmost importance in the future.

The theoretical methods can roughly be divided into real-space and reciprocal-space methods. Both should give the same results if properly performed. In real space the excited electron is treated as a spherical wave emanating from the atom. For excitation energies up to ~10 eV above the Fermi level, the inelastic mean free path of the excited electron is larger than the next-nearest neighbor distance (188) (Figure 11). Therefore, the electron wave is scattered at the potentials of several surrounding atoms. The interference of all waves leads to the fine structure of the absorption edge, which thus contains information about the type and position of surrounding atoms. The advantage of this multiple scattering approach is that the ELNES can be calculated for small clusters of atoms, i.e., no long-range periodicity is required as in the ***k***-space methods. The origin of ELNES features can be ascribed readily to different atom shells around the excited atom. In reciprocal space the eigenvalues of the system are calculated at several positions in the Brillouin zone. From the calculated band structure the density of occupied and unoccupied states is calculated, which directly reflects the ELNES. Apart from the information about the DOS, in this approach a (most probable) relaxed structure of minimum total energy can be determined. In the multiple-scattering approach, this is not possible. Note that, according to Equation 3, the transition matrix elements must also be considered for a quantitative comparison with EEL spectra; however, their energy dependence is usually rather smooth compared with the variations of the DOS.

The ELNES calculations are significantly complicated by two effects: (*a*) Calculations within the density-functional approximation give the ground-state electronic structure. In reality the excited electron and the core hole interact with each other and with all other electrons leading to changes in the band structure (final state effects) (189). In fact, the partial unscreening of the nucleus by the missing core electron leads to a contraction of the upper bands. This can be particularly strong in ionic solids, often leading to sharp features at the edge onset. In extreme cases excitonic states can appear near the edge onset, i.e., the excited electron and the core hole can be treated as an electron–hole pair. There are methods to take final-state effects into account; for example, the Z + 1 approach (190–193), which treats the excited atom as an atom with increased nuclear charge (e.g., replacing an O atom by a F atom). Alternatively suitable pseudopotentials can be employed (frozen core hole) (181, 183, 193, 194). The core hole often leads to a localization of the wave functions in real space. The treatment of the core hole makes $\boldsymbol{k}$-space first-principles calculations time-consuming because large supercells have to be employed in order to prevent interaction of the core hole atoms (periodic boundary conditions!). Paxton et al. describe the use of Slaters' transition state theory to include final-state effects (195). (*b*) Most first-principles methods assume that the exchange energy density is locally the same as in a homogeneous gas of the same density (local density approximation). However, this can fail in systems with strong exchange and correlation effects, such as in magnetic materials or generally in materials with highly localized electron states (e.g., d- and f-orbitals in transition metals and rare-earth elements).

Köstlmeier (196) showed that the onset energy of K (ground state 1s) and L_1 edges (ground state 2s) can be calculated reliably only up to about $Z = 15$ (phosphorous). Elsässer & Köstlmeier (197) showed that the $Z + 1$ approximation accounts well for the final-state effects if the total-energy difference between two neighboring elements in the periodic table is similar to the excitation energy of the process under study.

ADDITIONAL METHODS

In this section, several TEM techniques are addressed that do not directly make use of spectral information such as EELS or EDXS, but still allow the retrieval of chemical information.

Z Contrast

Equation 1 shows that the differential cross section for elastic scattering is proportional to the square of the atomic number. This is true only for large scattering angles, typically well above 100 mrad for electron energies normally used in TEM. Therefore, by scanning the electron beam across the region of interest and using a HAADF detector, the resulting image contains chemical information. If the illuminating beam is small and oriented parallel to a zone axis of a crystalline material, the electron intensity can be confined almost to a single-atom column, which gives atomically resolved Z-contrast images (102, 103). For quantitative chemical analysis, comparison with image simulations is indispensable. The reader is referred to References 198–201 for some examples of this technique.

Z-contrast images can also be obtained by hollow-cone illumination and detection of electrons scattered to the proximity of the optical axis (202). However, the technique suffers from very low intensity because only a small fraction of the scattered electrons is detected, whereas HAADF detectors detect all electrons scattered within a certain angular range.

ALCHEMI

It was shown above that the electron intensity varies within the unit cell of a crystal if Bragg reflections are excited. By intentionally varying the Bragg conditions by carefully tilting the specimen and using a simultaneous detection of EELS or EDXS signals, the occupancies of particular sites within the unit cell by specific elements can be determined. Spence & Taftø (29) introduced the acronym ALCHEMI (atom location by channeling-induced enhanced microanalysis) for this technique. The importance of delocalization of the inelastic scattering process for ALCHEMI was shown by Pennycook (117), Nüchter & Sigle (203), and Horita et al. (204). Oxley & Allen (205) show that the approximate expression given by Pennycook has to be modified. Important developments were made by Rossouw's group on the basis of their initial work (206) using a multivariate statistical approach.

Chemical Information from Elastic HRTEM Images

The contrast in high-resolution transmission electron microscopy (HRTEM) images is usually not directly interpretable in terms of chemical contrast, although there are exceptions. Ourmazd et al. (207) showed that by extracting the contrast originating from chemically sensitive reflections, one can determine the composition of each single unit cell. The method relies on pattern recognition and has been further developed (208–212).

Fresnel Contrast

Fresnel contrast occurs at edges where the average inner potential of the specimen changes. This can be applied to the study of thin films at grain boundaries that exhibit compositions different from the surrounding bulk material. By varying the defocus of the objective lens, one observes bright or dark contrast along the boundary. A comparison with calculation allows determination of the shape and magnitude of the potential across the boundary (213).

Tomography

In order to obtain three-dimensional information of an object shape, taking a series of images from various directions is a prerequisite. Several papers have demonstrated that this can also be performed by the use of EFTEM or STEM images (214, 215). Using this approach it is possible to get an element-specific three-dimensional image with a spatial resolution of a few nanometers. The available resolution and field of view is compared with other tomography techniques in Figure 21.

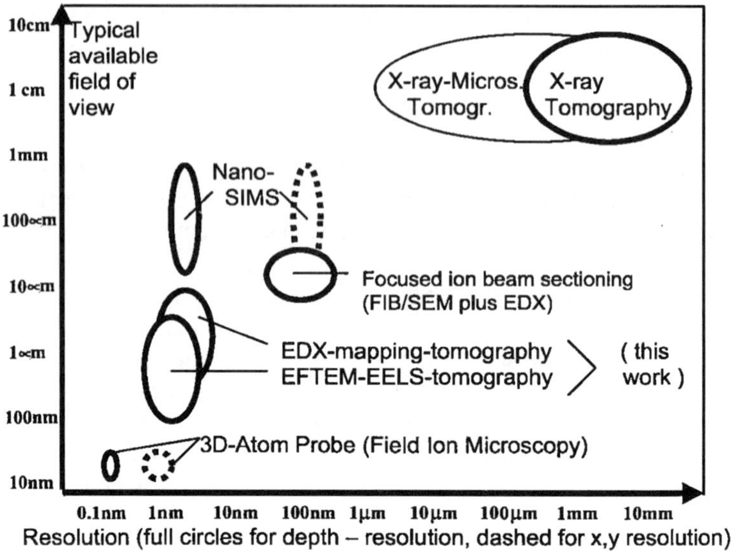

Figure 21 Comparison of different tomography approaches in terms of field of view and depth or lateral resolution. Reprinted with permission from G. Möbus et al. (214).

APPLICATIONS OF EELS AND EFTEM

The second part of this review concentrates on applications of analytical TEM, with the intention of giving the reader an idea of the type of information that can be extracted for particular material systems. The given references usually contain citations for further reading, and may thus be helpful as a starting point. An attempt to review the whole literature on this subject is simply not possible. We begin with some general remarks on low-loss EELS and anisotropic materials. This is followed by a discussion of core-loss studies of particular material systems, generally in the order of atomic number. Finally, special studies on interfaces, grain boundaries, and dislocations are addressed.

Low-Loss EELS

It was shown in Equation 4 that the EEL spectra contain the entire dielectric information $\varepsilon(E)$ necessary to calculate optical properties. The data analysis starts with a deconvolution algorithm [typically a Fourier-log deconvolution (17)] in order to obtain the single-scattering intensity. This is followed by several simple mathematical operations. Finally a Kramers–Kronig analysis separates real and imaginary parts of the dielectric function. Some examples have been given by Wang (216), Dorneich et al. (217), Müllejans & French (218), French et al. (219, 220), Ryen et al. (221), Rafferty & Brown (222), and van Benthem et al. (223).

In the formalism of dielectric theory the energy loss of the impinging electron polarizes the dielectric medium. The electric field connected with this polarization exerts a force on the electron that, integrated over the path of the electron, results in a certain energy loss. A general introduction to plasmons in solids can be found in Reference 224. Theoretical and experimental studies of the low-energy-loss region for special specimen geometries and crystal symmetries have been done by several groups. von Festenberg (225) calculated plasmon losses in systems containing thin films. Ferrell & Echenique (226) used classical dielectric theory to describe the dielectric loss for specimens of arbitrary geometry.

Ritchie (227) and Otto (228) pointed out that the free surfaces of a thin foil may lead to a reduced polarization and thus to a lower energy loss than in the bulk. Using the Maxwell equations, Economou (229) took account of the retardation effects that have a significant influence on the plasmon modes. A classical treatment of surface excitations was performed by Howie (230). This was extended to relativistic electrons moving parallel to the interface of two media with different dielectric functions by Garcia-Molina et al. (231). Howie (232, 233) points out that the sensitivity to the interface-specific dielectric function can be increased by using electrons scattered to higher angles. The energy loss function of an electron passing a multilayer material (both perpendicular and parallel to the layers) was studied by Bolton & Chen (234), who found interface plasmons in addition to bulk plasmons. The interface plasmons are weak for boundaries between two similar materials, for example, between two semiconductors. These interface features are more pronounced in the case of parallel incidence. Another study of dielectric losses in multilayers can be found in Reference 235.

Batson (236) studied the energy loss of an electron passing an Al sphere covered with alumina. He found a coupling between two close spheres that resulted in a loss peak at 3.2 eV, which is well below the surface plasmon peak at 6.7 eV. The effects of the dielectric coupling of small spheres to a substrate were theoretically treated by Ouyang & Isaacson (237). They found a reduced surface plasmon intensity. The same authors (238) extended the theory to arbitrary particle shapes. Echenique et al. (239) applied the theory of Ferrell & Echenique (226) to oxide-coated and uncoated spheres. Ugarte et al. (240) used EELS and EFTEM to study plasmon excitation in small Si spheres coated with SiO_2. A surface mode was found at an energy loss of 3–4 eV and this was imaged by energy-filtered imaging in the STEM (241, 242).

Chu et al. (243) performed a non-relativistic study of the energy loss of electrons passing a hollow cylinder parallel to the cylinder axis. For narrow cylinders, mainly bulk plasmons are excited, whereas the plasmon peak shifts to lower energies toward the surface plasmon peak with increasing cylinder radius. This was extended to the relativistic case by De Zutter & De Vleeschauwer (244) (see also 245, 246).

The energy loss close to the edge of a wedge (which is a typical geometry found in TEM specimens) is discussed in References 247, 248.

The work mentioned above assumed bulk-like dielectric functions. This assumption may not be applicable in structures of atomic dimensions. Ekardt (249) calculated the dielectric response of small metal particles in the time-dependent local density approximation and found that bulk plasmon excitation disappears below a critical cluster size. The bulk plasmon energy increases with decreasing cluster size owing to the quantization of electronic states, whereas the surface plasmon energy decreases because of an increasingly diffuse surface. Mitome et al. (250) found a d^{-2} dependence of the volume plasmon energy in small Si clusters (d, cluster diameter), which is opposite to the case of metal clusters. They argue that the effect results from the d dependence of the energy gap.

The change of the optical properties across an Al–SiO_2–Si field emitter structure was performed by Turowski & Kelly (251, 252). Recently, a similar approach was used for the study of a grain boundary in $SrTiO_3$ (120, 253). From local changes of the electronic structure dispersion forces can be quantified from the low-loss EEL spectrum. van Benthem et al. found a reduction in the number of transitions in the grain boundary region. From this finding a reduced atomic density and thus a reduced valence electron density in this region was deduced. With the assignment of the spectral features to specific transitions, it was concluded that the lower density was mainly from a loss of oxygen atoms. This is consistent with results by Browning and coworkers (254). It is argued that the London dispersion makes an important contribution to the grain boundary energy. The role of the Landau dispersion for the stability of intergranular films in silicon nitride is discussed by French et al. (220). It was found that the film thickness scales inversely with the Hamaker constant. Lo et al. (255) used EFTEM image spectroscopy in the low-loss region to obtain two-dimensional information about optical properties in a SiO_2-based low-k material. Because the sampling of the energy scale was too rough in EFTEM, they applied a FFT interpolation algorithm. This was followed by the normal Fourier-log deconvolution and Kramers–Kronig analysis.

A critical step in the low-loss data analysis is the subtraction of the zero-loss peak because there is no simple mathematical function that would fit to the whole tail of the peak. Therefore, an analytical function is fitted to only a certain range of the tail. Alternatively, Fourier-ratio deconvolution can be applied (256). Also non-Fourier space methods were reported (257, 258) that use the zero-loss peak recorded without the specimen as a reference function. Bangert et al. (259) found that the noise introduced by deconvolution procedures does not allow unraveling the information on band gaps in semiconductors. They recommended the use of an analytical function derived from the natural energy spread of the field-emission tip (260) convoluted by a Gaussian function, and the point-spread function of the detector.

In materials with a band gap above the valence band, the zero-loss peak is followed by a narrow range with zero intensity (assuming that no states are present in the gap region). Unfortunately this range is often not visible due to the width of the zero-loss peak or the poor point spread function of the EELS detector. Despite this difficulty, several groups have attempted to analyze the band-gap region (222, 256, 259). Recent technological developments, including electron monochromators, will facilitate the study of the band-gap region considerably, even with narrow-gap semiconductors. In Equation 5 the JDOS is proportional to $(E\text{-}E_g)^{0.5}$, where E_g is the band gap. The matrix element is constant for direct band gaps and proportional to $(E\text{-}E_g)$ for indirect band gaps. Therefore, the imaginary part of the energy-loss function is proportional to $(E\text{-}E_g)^n$ with $n = 0.5$ in the case of direct band gaps and $n = 1.5$ for indirect band gaps (222). Thus, using EELS, the nature of the band gap can be determined. This was done in (222) for MgO, diamond, h-BN, c-BN, GaAs, and ZnO, and by Alexandrou et al. (261) for carbon (amorphous, tetrahedral amorphous, and C_{60}).

In Equation 4 it was shown that the energy-loss spectrum can be described by $\text{Im}\{-1/\varepsilon E\}$. This expression, which can also be written as $\varepsilon_2/(\varepsilon_1{}^2 + \varepsilon_2{}^2)$, has maxima if $\varepsilon_1 = 0$ and ε_2 is small. These maxima correspond to plasmons, which are collective oscillations of free electrons in a metal or bound valence electrons, e.g., in semiconductors. The plasmon energy E_p is a material property (for a compilation, see 17). In the case of free electron oscillations, $E_p{}^2 = (\hbar^2 ne^2/4\pi^2 m)$, where n is the free electron density. A shift of the plasmon energy appears if bound electron oscillations are relevant (262) or if there are absorption edges by interband transitions close to the plasmon peak (158, 263). Recently it was demonstrated that by acquiring energy-filtered series across the plasmon peak, the peak position can be determined for each pixel image with an accuracy of 0.05 eV (264) at a spatial resolution of about 2 nm. This is shown for a Si-B-C-N ceramic material in Figure 22 (see color insert) where Si_3N_4 grains, SiC grains, and the surrounding matrix can be clearly distinguished by their color-coded plasmon peak energies.

Edge Loss

ANISOTROPIC MATERIALS As mentioned previously, the inelastic electron scattering process does not probe the whole solid angle but is concentrated to a certain angular range of the scattering vector q, which depends on energy loss. Hence, it is possible to obtain EEL spectra with different directions of momentum transfer. This is of particular interest for the study of materials with anisotropic electronic structure. According to Equation 3 the scattering vector enters into the ELNES as a scalar product with the electron coordinate r_j. Thus the ELNES is the projection of the matrix element onto the scattering vector and therefore the material anisotropy is also reflected in the ELNES. This effect is particularly pronounced in graphite (47, 48, 265) or BN (45, 266), which leads to a dependence of the π^* and σ^* peak intensities, e.g., of the C-K edge, on the size of the collection and convergence apertures (Figure 23). The relative intensities of these two peaks is of interest because it allows the determination of the amount of sp^3 bonding in a material. In a number of articles conditions were discussed under which the mentioned dependencies vanish. Berger et al. (271) obtained an sp^2 reference spectrum from polycrystalline graphite by illuminating a large area in order to average out the orientation effect. Browning et al. (267) calculated this averaging effect and suggested using these values because this would not require the illumination of large areas,

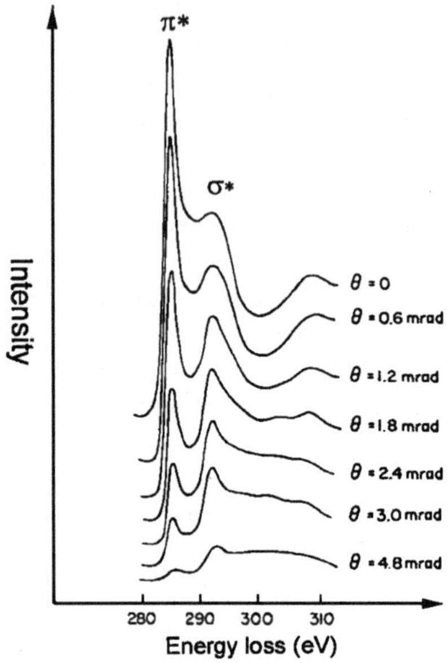

Figure 23 Carbon K edge for different scattering angles of the incident electrons. The relative intensities of the π^* and σ^* peaks change owing to the anisotropic electronic structure of graphite. Reprinted with permission from Leapman et al. (47).

thus making high spatial resolution studies possible. Some authors (268, 269) reported the so-called "magic angle" for the convergence and collector angles, for which the dependence of relative peak heights with specimen orientation vanishes. Values of two and four times the characteristic angle for inelastic scattering, $\bar{\theta}_E$, were proposed. Recently Hébert et al. (270) pointed out that $2\bar{\theta}_E$ is correct and that the apparent discrepancy was due to the fact that inelastic cross sections were not calculated relativistically. Further studies on the determination of the π^*/σ^* ratio can be found in References 271–278. Mapping of the π^*/σ^* ratio was done by both STEM techniques (279) and EFTEM (165, 280).

BORON The detection limit of B in a carbon matrix was shown to be 0.2 at% (281). Experimental studies of BN were performed by Schmid using EELS (282) and Moscovici et al. using X-ray absorption near edge structure (XANES) (283). The B-K edge of hexagonal BN (h-BN) is dominated by a strong π^* peak at 190 eV energy loss. This peak disappears if the momentum transfer $\boldsymbol{q}$ to the excited electron is perpendicular to the *c*-axis, showing that it is entirely due to $2p_z$-type orbitals. The σ^* peak at 198 eV is mostly due to 2s and $2p_{x,y}$ orbitals. Although a dipole transition into s states is forbidden for the B-K edge, such transitions contribute to the edge intensity due to strong 2s–2p hybridization. Further smaller peaks at higher energy losses can be correlated with transitions into 3s and 3p orbitals (284). The core hole effect strongly influences peak intensities. The N-K edge also shows two main peaks at 400 and 408 eV, which are similarly related to N $2p_z$ and $2p_{x,y}$ orbitals, as in the case of the B-K edge. The peaks are broader, which has not been explained yet. The B-K ELNES of cubic BN (c-BN) and wurtzite

BN (w-BN) shows a broad peak centered around 196 eV, which results from σ^* interactions. This peak is followed by small features at about 204 and 213 eV, the latter originating from B-3s and -3p contributions (284). Schmid (282) pointed out that the only difference between c-BN and w-BN is the presence of a low-energy shoulder at the 213 eV peak in the case of w-BN. Terauchi et al. (285) compared low-loss spectra of h-BN with those of BN nanotubes. They found distinct differences in the energy of interband transitions and plasmons.

CARBON The C-K edge in materials formed from elemental C shows several features (Figure 24). The π^* peak at 284.5 eV is due to transitions from the 1s to the antibonding 2pπ^* state. This peak occurs only if unsaturated π bonds are present in the material, as e.g., in graphite and amorphous carbon. At 290 eV, transitions to the antibonding σ^* state form a small peak. A peak appearing between π^* and σ^* was found by Fischer et al. (286) and assigned to interlayer electronic states.

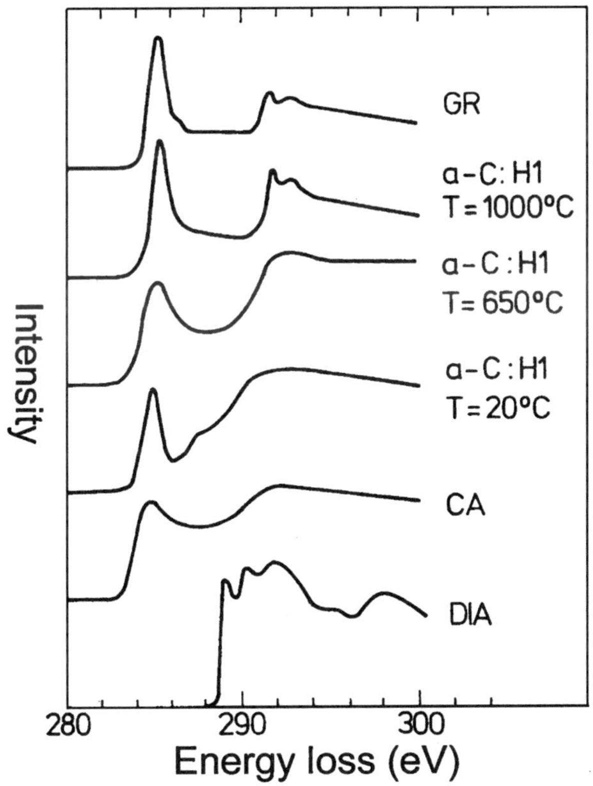

Figure 24 Carbon K edge of graphite (GR), evaporated amorphous carbon before (CA) and after (a-C) heat treatments, and diamond (DIA). Reprinted with permission from Fink et al. (287).

Fink et al. (287) assign a similar peak in this region to the presence of hydrogen, thus forming C:H bonds. In diamond, no π^* peak is present because all C atoms are tetrahedrally coordinated. The absorption starts at 289.1 eV with a sharp 1s core excitonic peak.

In the low-loss regime, the oscillations of π electrons occur at a plasmon energy of 5–6 eV, whereas a combined oscillation of π and σ electrons occurs at about 23 eV. From the $(\pi + \sigma)$ plasmon energy, the valence electron density and thus the atomic density can be determined, but the anisotropy of the material, which can also shift plasmon peaks, should be considered (288).

There are several features in EEL spectra of carbon nanotubes that differ from those of bulk carbon materials. Menon & Yuan (289) found that the π^* peak is more pronounced if the electron beam passes the tube near the surface instead of near the center, which can be understood from the anisotropy of the material (see above). The σ^* peak disappears with increasing curvature of graphene planes (290). In multiwalled nanotubes the $(\pi + \sigma)$ plasmon energy is similar to that in graphite, whereas it is considerably lower in single-walled nanotubes (291–293). In contrast to graphite, C nanotubes exhibit π-to-π^* interband transitions between 1 and 3 eV (294–297). Kociak et al. (298) reported on surface plasmon modes in carbon nanotubes, excited with the electron beam remote from the tube surface. They found two modes at 12–13 eV and 17–18 eV energy loss, and assigned these losses to in-plane and out-of-plane components of the dielectric tensor. Suenaga et al. (299) found that in nanotubes composed of B, C, and N, the individual layers are composed of either C or BN, with C forming the outermost and innermost layers.

Elementary excitations in C_{60} molecules studied by high-resolution EELS are discussed by Lucas et al. (300). Both in the low-loss regime and between the π^* and the σ^* peaks in the C-K edge, C_{60} molecules show distinct peaks not observed in graphite. For larger sphere diameters, as in "carbon onions," the electronic structure resembles that of graphite (301).

In thin films of diamond-like carbon, Yan et al. (165) showed that the near-surface regions exhibited a higher sp^2 bonding than the film interior.

POLYMERS AND BIOLOGICAL MATERIALS Apart from C, these materials also contain other elements, e.g., H and O. The bonding of C with these elements leads to additional features in the ELNES. C–H bonds and C=O bonds (299 eV) lead to peaks close to the π^* peak (287). A mapping of C–C- and C–O-bonds by energy-filtered series was shown by Martin et al. (302).

A main issue in the study of polymers is radiation damage induced by the incident electrons. In the case of PET (polyethylene terephthalate), a critical dose of 10^3 cm^{-2} was found (303, 304), which is at the lower limit for the study of the ELNES of core-loss edges. A factor of about 10 can be gained by cooling the specimen to liquid nitrogen temperature (305).

In polymers the O-K edge displays peaks at 531.5 eV (π^* transition), 534.6 eV, and 541.6 eV (σ^* transitions), all originating from O=C bonds (304).

MAGNESIUM OXIDE Lindner et al. (306) compared experimental Mg-L and O-K spectra with multiple-scattering calculations and concluded that the inclusion of the core hole effect is particularly important for the K edge. Using first-principles molecular orbital calculations for Mg-K, Mg-L, and O-K edges, Mizoguchi et al. (183) found a good correspondence with experimental ELNES studies only after taking into account the effect of the core hole. They assigned spectral features in the Mg spectra to Mg*-Mg interactions, those in the O-K edge to Mg-Mg and Mg-O* interactions (Mg* and O* denote the excited atoms with core holes). Using the mixed-basis pseudopotential approach, Elsässer & Köstlmeier (197) found a good validity of the $Z+1$ approximation for the O-K and Mg-L_1 edges, whereas for the Mg-$L_{2,3}$ edge the correlation with experimental data is less satisfactory. According to Köstlmeier & Elsässer (193) the peaks in the O-K ELNES can be understood simply by the presence of the strongly scattering oxygen ions. Therefore, the O-K ELNES can generally be used as a fingerprint for the structure of the anion sublattice in metal oxides. The influence of the cation sublattice is comparatively weak.

ALUMINUM AND ITS COMPOUNDS Figure 25 shows the Al-$L_{2,3}$ edge for different Al compounds. Bouchet & Colliex (307) identified 10 peaks a–j between 77 and 86 eV. Peaks at higher energy losses are from the transitions to 3d states, between 85 and 95 eV to 3p states (dipole-forbidden) (308), and the peaks near threshold are transitions to 3s states (193), which are related to the formation of core excitons induced by the core hole (194). Owing to the screening of a core hole, no such sharp features are observed in metallic Al near threshold. According to Bouchet & Colliex (307) peaks g–j are related with the medium-range order around the Al atoms. The positions of peaks a–d can be used for an identification of Al compounds. Nufer et al. (43) pointed out that under channeling conditions (i.e., in a zone axis orientation) dipole-forbidden, but symmetry-allowed, transitions can occur. Nufer et al. (309) also reported on calculations of core-loss edges in α-Al_2O_3 for the Al-K and -L_1 edges. For both edges, transitions to the 2p states occur, implying that they have a similar shape. However, this is not the case because of different core holes (1s for the K edge, 2s for the L edge). The authors (309) find a good match with experimental data in the $Z+1$ approximation for the Al-K edge, whereas for the Al-L_1 edge better results are obtained without this approximation. It is argued that the core hole effect is stronger the more localized is the core hole (i.e., for the 1s core hole) because in such cases, the perturbation of the system is particularly strong.

A simple way to distinguish metallic Al from Al compounds is the sharp plasmon peak of metallic Al at 15 eV. Figure 26 (see color insert) shows an energy-filtered plasmon image of an α-Al_2O_3 specimen after electron irradiation at high temperatures. Bright circular areas clearly display the metallic regions.

SILICON In a number of articles Batson (e.g., 310, 311) used spatially resolved EELS to measure the electronic structure of the band gap in Si-based semiconductor

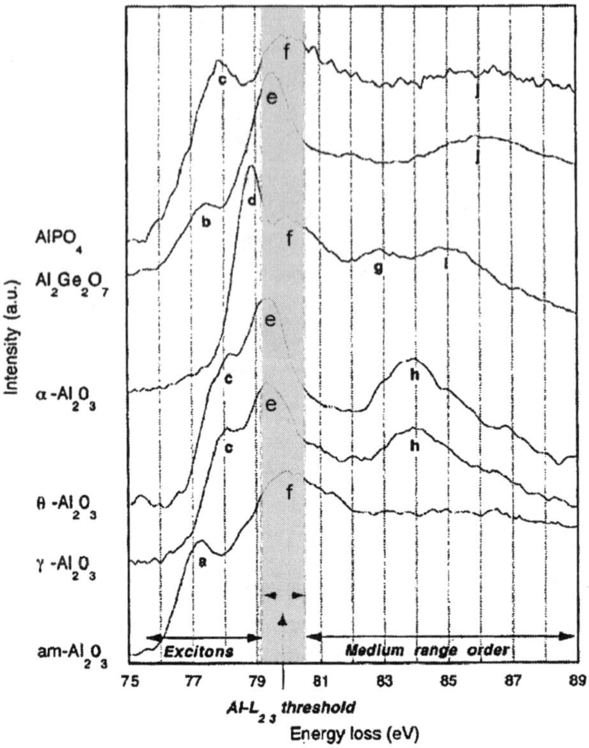

Figure 25 Al-L edge in different Al compounds. The peaks a–j are discussed in the text. Reprinted with permission from Bouchet et al. (307).

devices. After deconvolution with the low-loss signal and subtraction of transitions from the $2p_{1/2}$ ground state (the L_2 edge), the remaining Si-L_3 edge is fitted to the band structure. In SiGe alloys Batson was able to correlate strain with a change of the shape of the Si-L edge. In a dissociated 60° dislocation in SiGe he found distinct differences between the Si-L ELNES in the partial dislocation cores and the intrinsic stacking fault in between the cores (312). He ascribed these differences to optically active electronic states in the band gap in the partial dislocation cores.

Duscher et al. (313) show that the Si-L edge is influenced by the core hole effect not only in oxides but also in pure Si.

TRANSITION METALS Transition metals with unoccupied d-states show two sharp peaks (white lines) in the $L_{2,3}$ absorption edge originating from transitions from $2p_{1/2}$ and $2p_{3/2}$ initial states. These two states have different energies because of the coupling between the spin and angular magnetic moments of the 2p core hole. With increasing *Z*, this energy difference becomes larger. The width of the white

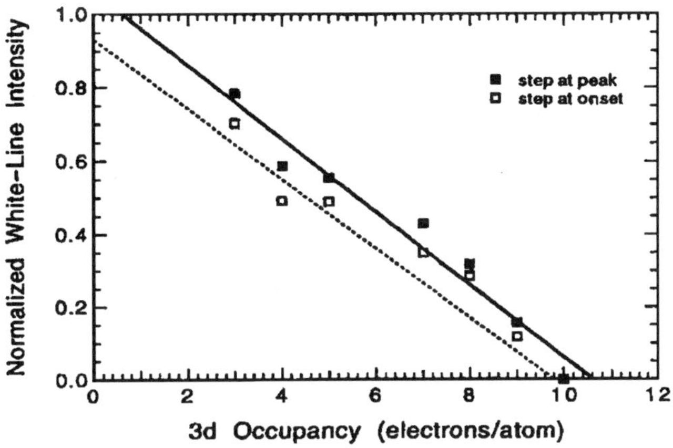

Figure 27 White line intensity for transition metals with different occupancy of 3d states. Reprinted with permission from Pearson et al. (315).

lines decreases with Z owing to the steady filling of the 3d band, and it is narrower for oxides than it is for the metal. For most of the transition metals (except Cu) the edge onset is shifted toward higher energy by oxidation. The total intensity of the white lines compared with the continuum following the edge is directly related to the d-band occupancy (314, 316) (Figure 27). The intensity ratio of the white lines is not equal to the simple statistical ratio of 2:1 expected from the $2j + 1$ degeneracy ($j = \ell + s$ with spin s and orbital momentum ℓ) of the two initial states (317, 318). Rather, it is a function of the d-band occupancy and thus of the local magnetic moment and the oxidation state of the atom (318–320). An improved analysis of L_2/L_3 ratios was shown by Pease et al. (321). Murakami et al. (322) detected a change of the d-band occupancy in a TiNiFe alloy after the martensitic transformation.

In transition metal (TM) oxides the oxygen 2p states hybridize with the TM s-, p-, and, in particular, d-states. For this reason the ELNES of the O-K edges show features that resemble the $L_{2,3}$ edges of the TMs, as e.g., peaks from the t_{2g}- and e_g-orbitals (323–326).

The sharp white lines make the detection of transition metals relatively easy. It was shown that 100 ppm can be detected (327).

The magnetic-dichroic effect can be used to characterize magnetic anisotropy and is normally performed using X-ray absorption spectroscopy (328, 329). The circular magnetic dichroism (difference in absorption of right- and left-circularly polarized photons) has been demonstrated until now by using X-rays, which allows determination of the magnitude and direction of spin and orbital magnetic moments. See Reference 330 for a discussion of possible EELS experiments. Schattschneider et al. (331) point out that it is the chirality of the $\Delta m = \pm 1$ transition that is measured in EELS because circular polarization is not possible for electrons. Linear magnetic dichroism (difference in absorption for momentum transfer parallel and perpendicular to the magnetization) also can be detected by EELS (49, 289, 332, 333). It basically leads to a variation of the $L_{2,3}$ intensities by changing the orientation of the specimen with respect to the electron beam. Owing to the hybridization of O and TM orbitals, the magnetic-dichroic effect is even visible in the O-K edge (333). The linear dichroism measures the charge anisotropy of the valence states.

Murakami et al. (334) have reported a broad peak about 30–40 eV above the O-K edge of transition metal oxides and perovskites and ascribe it to a resonance effect of the ejected core electrons. They found a linear dependence of the energy position above the edge onset and $1/R^2$, where R denotes the distance to the next O atom. Thus the energy position of this peak allows a simple measure of the O–O separation.

Titanium oxides The shape of the Ti-$L_{2,3}$ edge is dominated by four peaks a–d (Figure 28). These arise from the dipole-allowed excitation of electrons from the inner $2p_{3/2}$ (L_3 edge, peaks a and b) and $2p_{1/2}$ levels (L_2 edge, peaks c and d) to the narrow unoccupied 3d band, resulting in two white lines. Both white lines are doubled by the action of the Coulomb field imposed by the surrounding O atoms (ligand-field splitting). A perfect octahedral ligand field causes the five degenerate

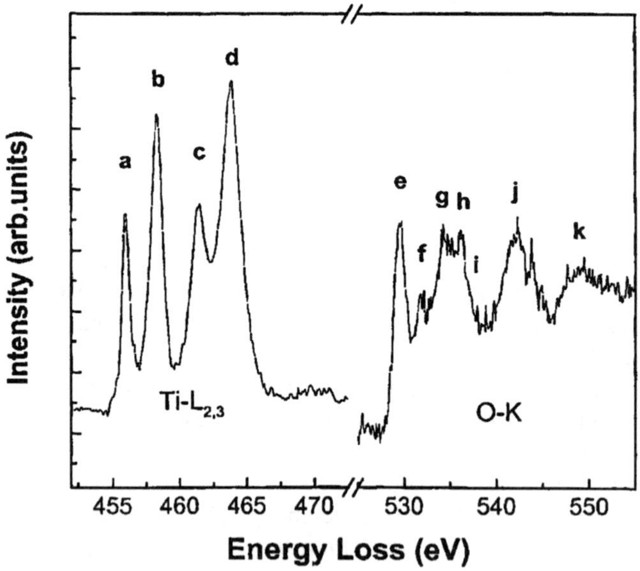

Figure 28 Ti-L and O-K edges in $SrTiO_3$. The peaks a–k are discussed in the text. Reprinted figure with permission from Zhang et al. (379).

d-states in the conduction band to split into twofold e_g orbitals that are directed at the ligands (corresponding to peaks b and d), and the threefold t_{2g} orbitals that are directed between the ligands (corresponding to peaks a and c) (335). Whereas a deviation from the octahedral symmetry will create new states in the conduction band, a change of the magnitude of the crystal field affects the energetic position and relative intensities of the transitions (336). However, these changes are generally weak and are often not observable as additional peaks. Rather, a broadening of the white lines or a change of the peak separations is detected. The e_g:t_{2g} intensity ratio increases and their separation decreases by a reduction of the crystal field (337). A chemical shift of 1.4 eV toward lower energies is observed in the Ti metal compared with that in TiO_2 (317). The same authors report a L_3/L_2 ratio of 0.8 for both Ti and TiO_2. The deviation from the statistical ratio of 2:1 (obtained from the $2j + 1$ degeneracy of the initial states) is mainly due to the excitation process, which leaves a core hole in one of the 2p states and an excited electron in the 3d band (319). Taking this into account, Ogasawara et al. (338) recently succeeded in approximately reproducing the experimental ratio by nonempirical ab initio cluster calculations. By ab initio calculations, van Benthem (339) has shown that due to the influence of the core hole, the measured white line separations deviate from the values in the electronic ground state.

The O-K edge originates from transitions from the O 1s-shell to unoccupied 2p states and would thus not be visible in a purely ionic compound where all 2p states are filled. In transition metal (TM) oxide compounds, the density of unoccupied oxygen 2p states is significantly modified by the strong hybridization with metal 3d states. The O-K edge in bulk $SrTiO_3$ is shown in Figure 28. Seven peaks labeled e–k are visible. It is generally accepted that peaks e and f are correlated with the interaction of the oxygen atom with the two nearest-neighbor Ti atoms; e can be correlated with the Ti t_{2g} orbital and f with the Ti e_g orbital. Brydson et al. (340) made the point that a large e/f intensity ratio is typical for a linear chain of Ti-O–Ti tight bonds, as is true for $SrTiO_3$. Any distortion of this linearity increases the intensity of f. For tetrahedral coordination, both peaks have similar intensities (340, 341). By comparing different TM oxides, de Groot et al. (325) concluded that the total intensity of e and f is inversely proportional to the d-band filling and thus to the covalency of the Ti–O bonding. The weakness of peak f has also been attributed to its broadening by the strong e_g-e_g interaction (341, 342). It is argued that this interaction even leads to a splitting of the e_g band leading to the separate peak g. From this, Tanaka et al. (342) conclude that the e_g intensity and the separation of f and g should sensitively depend on the Ti–O bonding angle. Peaks h and i most probably are related to the Sr–O bonding (340–342). There is no unambiguous assignment of peak j to structural features. Whereas de Groot et al. (341) assumes this peak to originate from the Sr 4p band, Tanaka assigns it to the Ti 4s band. Finally, peak k is generally assumed to be related to the Ti 4p band. Wallis & Browning (343) and Browning et al. (344) performed multiple scattering calculations using clusters extracted from the perovskite structure. By removing atoms from the cluster they assigned spectral features to structural features. Apart

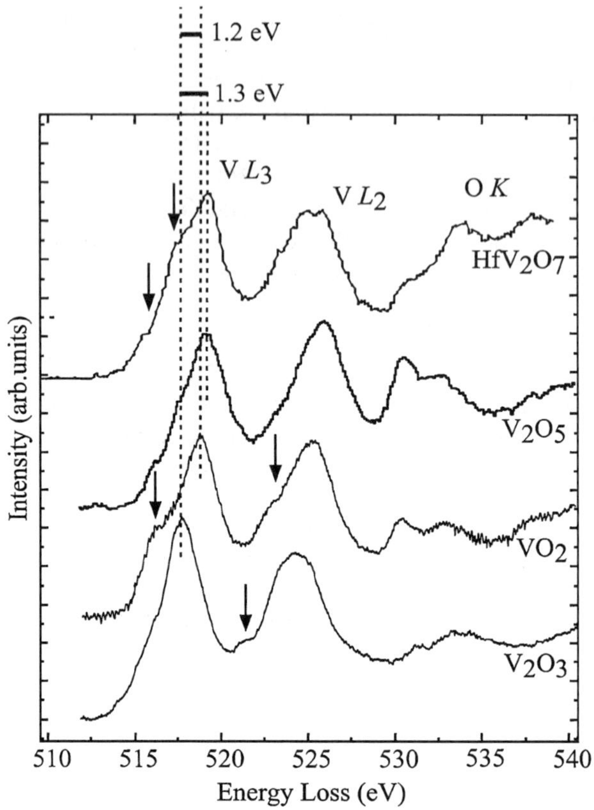

Figure 29 V-L edge in various vanadium oxides. Reprinted with permission from Gloter et al. (346).

from peaks e and f, they assigned the other peaks mainly to the scattering from the surrounding oxygen atoms.

Vanadium In Figure 29 the V-$L_{2,3}$ edge is shown for different vanadium oxides (345, 346). There is a gradual decrease of the L_3/L_2 white line ratio with increasing oxidation state. This is accompanied by an increase of the oxygen-K edge intensity at the onset of the edge (530 eV).

Manganese The Mn-$L_{2,3}$ and O-K ELNES were measured by Kurata & Colliex (347) for oxides with five different nominal oxidation states of the Mn ions. The results are shown in Figure 30. The O-K edge is composed of features a–d. Peak a is attributed to transitions to O 2p states hybridized with Mn 3p states. The structure of this peak varies significantly in the different oxides. The Mn-L edge shows two prominent white lines. With increasing oxidation state the L_3 peak shifts toward higher energy and the L_2/L_3 intensity ratio increases.

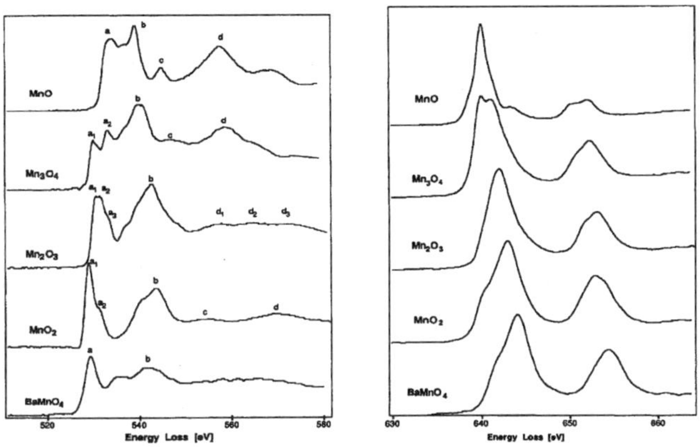

Figure 30 O-K and Mn-L edges in different manganese oxides. Reprinted with permission from Colliex (347).

Iron The structure of the O-K and Fe-$L_{2,3}$ edges of various Fe oxides was discussed by Colliex et al. (348). Similar to other transition metal oxides, it was found that the oxygen-K edge is sensitive to the local bonding and coordination, and the Fe-L edge is strongly influenced by electron Coulomb and exchange interactions between the 2p core hole and 3d electrons. The latter was confirmed by Kurata et al. (349) using ab initio cluster calculations. In addition, a high-spin state of α-Fe_2O_3 (parallel spins of the 5d electrons, $S = 5/2$) had to be assumed. The crystal field from the oxygen ligands causes a shoulder at the onset of the L_3 white line. The L_3/L_2 ratio increases with decreasing crystal field owing to a transition from a high-spin to a low-spin state (Figure 31) (319, 320, 347, 350). An overview of Fe $L_{2,3}$ edge shapes in many Fe-containing minerals is shown by van Aken & Liebscher (351). From a detailed analysis they determined the relative content of Fe^{3+} ions with an absolute error of $\pm\,0.02$. This is possible because the L_3 white line of Fe^{2+} is located at 707.8 eV, whereas it is shifted to 709.5 eV in the case of Fe^{3+}.

Copper Owing to the filled d-states, metallic copper does not show white lines in the $L_{2,3}$ edge. However, in the presence of oxidizing elements, charge can be removed from the Cu atoms, which makes the white lines appear. White line intensity can be used to determine the oxidation state (314, 315, 317). Cu oxide spectra of different compounds (different oxidation states) are shown in Figure 32.

Zinc From the shape of the Zn-L edge, Mizoguchi et al. (354) were able to distinguish different oxygen coordination numbers of Zn atoms.

Intermetallic alloys Muller & Silcox (352) studied the Ni-L and the Al-L edges in $Ni_{1-x}Al_x$, both experimentally and by theory. For the Ni-L edge they found a

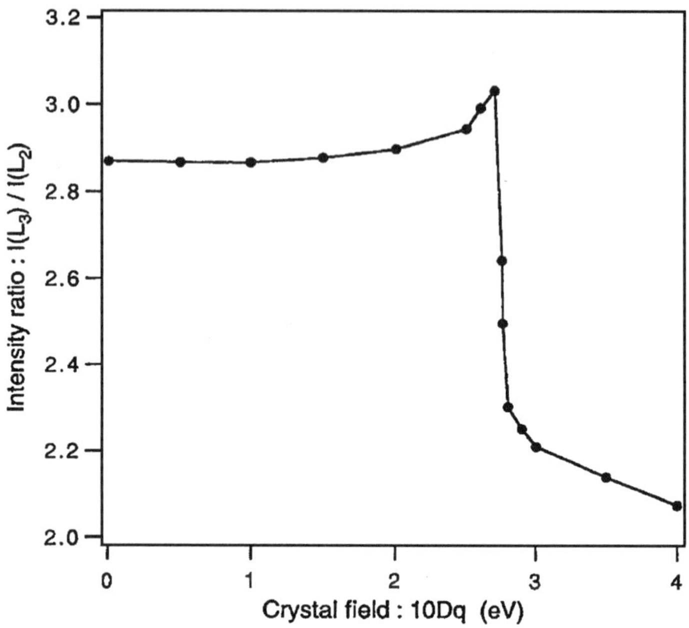

Figure 31 Intensity ratio of Fe L_3 and L_2 white lines as a function of the magnitude of the crystal field acting on the Fe atom by surrounding oxygen ions. Reprinted with permission from Kurata et al. (349).

steady decrease of the L_3 white line with increasing Al content (Figure 33) because of hybridization of Al 2s with Ni 3d orbitals. The intensity close to the onset of the Al-L edge was found to decrease with increasing Ni content. This was attributed to a pseudogap, also originating from s–d hybridization. Changes of the Ni-$L_{2,3}$ edge by the formation of Ni-based intermetallic compounds were measured by Potapov et al. (353).

GALLIUM NITRIDE The N-K edge shows several well pronounced peaks (Figure 34) (161). Their positions are almost the same in c-GaN and h-GaN because in both systems the N coordination is tetrahedral. Lazar et al. (161) also studied the low-loss regime of GaN and detected the band gap, as well as several interband transitions. They were able to fit the intensity above the band gap to that expected from a direct semiconductor.

The variation of the N-K and Al-K edges with N content in $Al_xGa_{1-x}N$ is shown in Figure 35 (355). The strong variation of the N-K edge is attributed to the fact that, through a variation of x, the composition of nearest-neighbor atoms of N changes. For Al atoms only the second-nearest neighbors change with x. The slight peak shifts of the Al-K edge are attributed to an increase in interatomic distances.

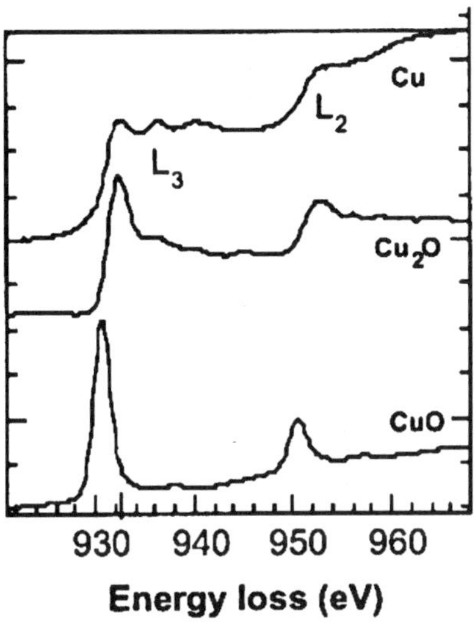

Figure 32 Cu-L edge in Cu and two Cu oxides. Reprinted with permission from Imhoff et al. (47).

RARE-EARTH ELEMENTS In lanthanides the $M_{4,5}$ edge is dominated by white lines originating from transitions $3d_{3/2} \rightarrow 4f_{5/2}$ (M_4) and $3d_{5/2} \rightarrow 4f_{7/2}$ (M_5). The M_4:M_5 intensity ratio was found to decrease with increasing 4f orbital occupancy (Figure 36) (358) and could therefore be used to determine the oxidation state (356).

Buck & Fortner (357) reported the detection of less than 200 ppm of U and Pt in nuclear waste glass by analyzing the $M_{4,5}$-edges at about 3.7 keV. A similar

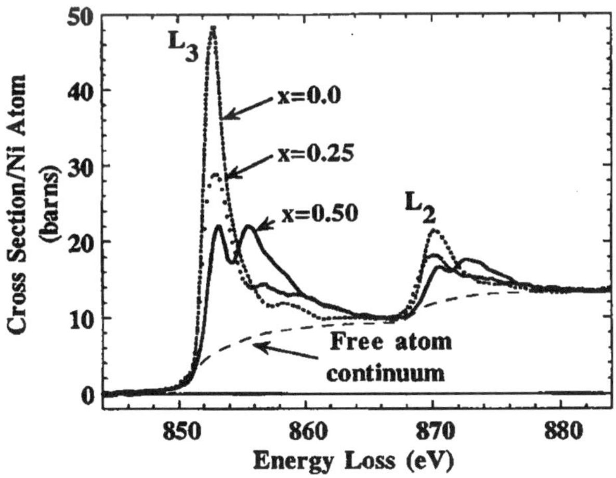

Figure 33 Ni-L edge in $Ni_{1-x}Al_x$ for different values of x. Reprinted with permission from Muller & Silcox (352).

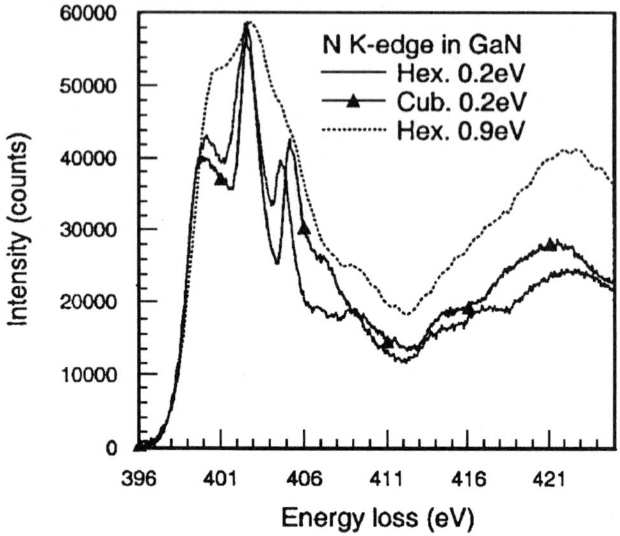

Figure 34 N-K edge in hexagonal and cubic GaN. The spectra with 0.2 eV energy resolution were obtained using an electron monochromator. Reprinted with permission from Lazar et al. (161).

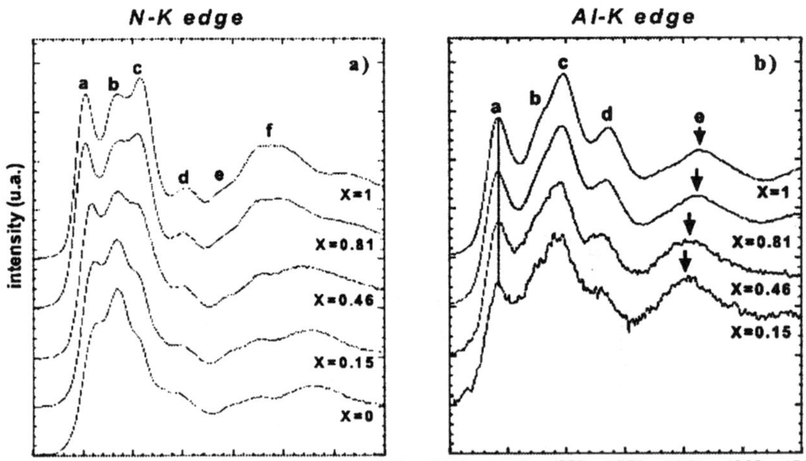

Figure 35 N-K and Al-K edges in $Al_xGa_{1-x}N$ for different values of x. Reprinted with permission from Radtke et al. (355).

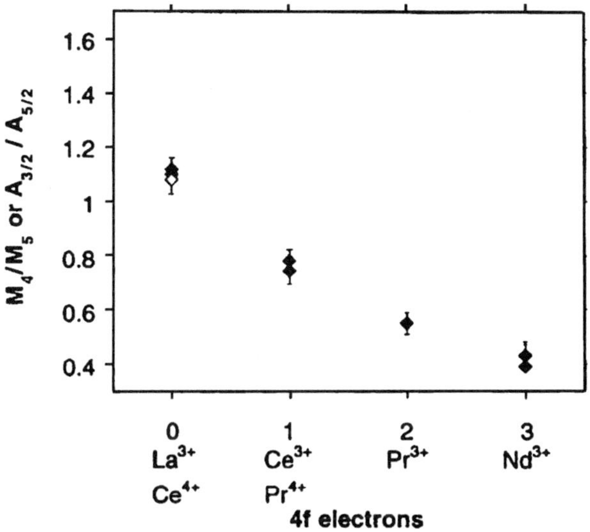

Figure 36 Intensity ratio of M_4 and M_5 white lines for different lanthanides. Reprinted with permission from Fortner et al. (358).

trend of the M_4:M_5 intensity ratio with 5f orbital occupancy was found as in the lanthanides (Figure 37) (358).

SPINELS Spinels are compounds with the composition AB_2O_4, where A and B are transition metals and O is oxygen. Depending on the site occupancy of the metal atoms, normal and inverse spinels are distinguished. In normal spinels all octahedral sites are occupied by B^{3+} ions, whereas in the inverse spinel these sites are occupied by A^{2+} and B^{3+} ions in equal proportions.

The O-K edge was studied for a series of normal spinels with $B{=}Cr^{3+}$ (chromites) by Docherty et al. (359) both experimentally (Figure 38) and by multiple scattering simulation using the FEFF8 code. They identified six peaks denoted by A, B, C, D (with a shoulder, denoted as D*), and E. From the fact that peaks A–C are not present in aluminate spinels it was concluded that these peaks are related to the vacant d-states of the Cr ions because these d-states hybridize with O 2p-states. The intensity of peak C changes with the nature of the A ion. No marked changes were observed in a variation of the A ions, both in aluminates and chromites. The FEFF calculations showed good correspondence with experimental spectra for the Mg aluminate but poor agreement for all chromates. It was suggested that this is due to the influence of magnetic interactions.

The intensity of the low-energy shoulder in the aluminates (left side of Figure 38) was found to be related to the degree of inversion, being particularly

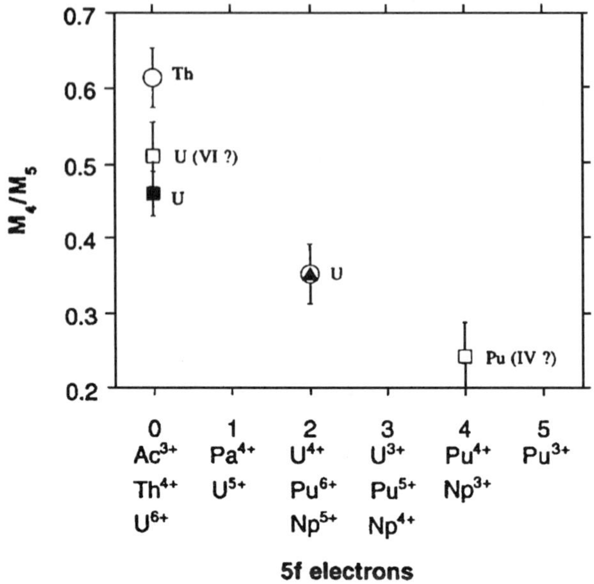

Figure 37 Intensity ratio of M_4 and M_5 white lines for different actinides. Reprinted with permission from Fortner et al. (358).

strong at full inversion (193, 360). Thus this shoulder is related to the tetrahedral nearest-neighbor coordination (361, 362).

PEROVSKITES (SEE ALSO Ti-OXIDES) Perovskites are oxides with composition ABO_3, where A and B are metal ions. Using the combination of atom-column-resolved Z-contrast and EELS, Klie et al. (363) studied different perovskites with [A = (La,Sr)/B = (Fe,Cr); A = Sr/B = Co; A = Sr/B = Ti]. Using EEL spectra acquired from different atom columns, they determined the valency of the transition metals and from this deduced the number of oxygen vacancies. In the case of $SrTiO_3$, they showed evidence for the accumulation of oxygen vacancies near a grain boundary.

HIGH-TEMPERATURE SUPERCONDUCTORS A prepeak of the O-K edge at about 529 eV in $La_{2-x}Sr_xCuO_{4-\delta}$ (364), $YBa_2Cu_3O_{7-\delta}$ (365), and $Bi_2Sr_2Ca_{1-x}Y_xCu_2O_8$ (366) was reported for p-doped material. This peak indicates the presence of holes on O sites. From the orientation dependence of the shape of the O-K edge, it was possible to determine the symmetry of these hole states (367–369).

METAL–CERAMIC INTERFACES Owing to the different interatomic bonds in metals and ceramics (covalent and/or ionic), atoms close to interfaces between these

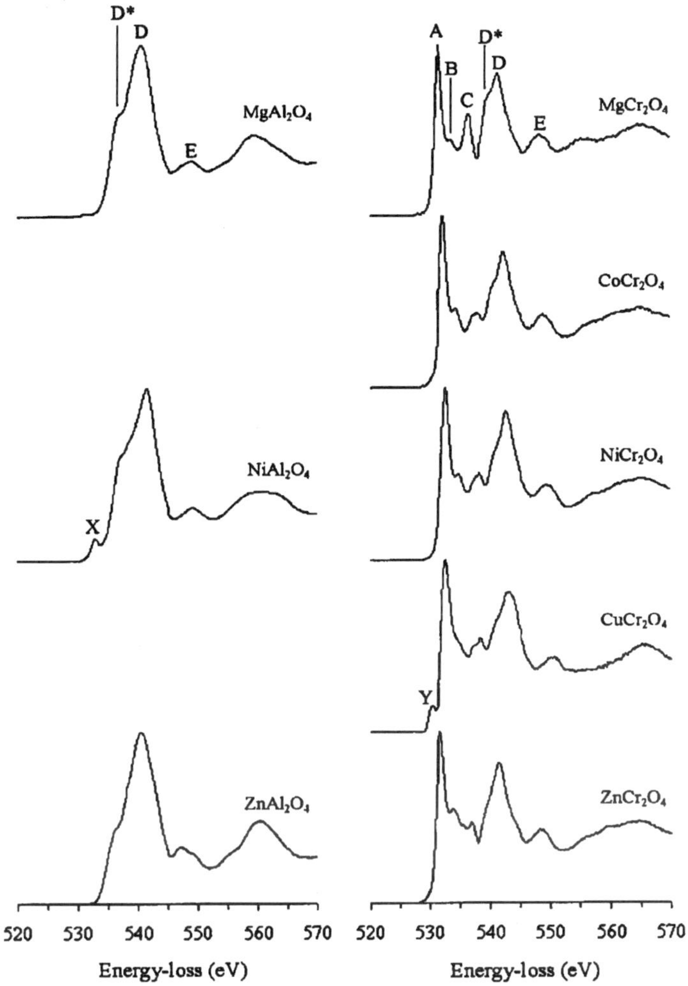

Figure 38 O-K edges of various Al- and Cr-spinels. Reprinted with permission from Docherty et al. (359).

material systems may differ in oxidation state and in coordination from the parent crystals or may even form new phases along the interfaces. Unfortunately, at present it is not possible to make predictions about the interfacial electronic and atomic structure from the basic properties of the materials. Experimentally, the most frequently used analytical technique in this field is the spatial difference technique as well as EELS line scans across the interface.

MgO–Cu Using the spatial difference method Imhoff et al. (370) found a CuO-like Cu-$L_{2,3}$ ELNES indicating transfer of charge from Cu to MgO. From EELS line scan data, Muller et al. (371) concluded that metal-induced gap states exist in the MgO. These originate from the Cu wave function extending into the band gap of MgO.

Cu–α-Al₂O₃ Dehm et al. (372) and Scheu et al. (373) studied Cu films deposited on the basal plane of sapphire by molecular beam epitaxy (MBE) at 473 K. The spatial difference spectrum showed Cu_2O-type bonding, i.e., the Cu atoms are reduced to the Cu^{1+} state by charge transfer to the neighboring O atoms. MBE films grown at 800°C on the $(11\bar{2}0)$ plane show a mixed Al-Cu interface plane facing an oxygen layer in the sapphire substrate (374).

Nb–α-Al_2O_3 From a change of the interface-specific ELNES of the Al-L edge, Bruley et al. (375) concluded that diffusion-bonded material showed Al–Nb bonding. From the absence of this signal in MBE-grown interfaces, Nb–O bonding was deduced.

Al–α-Al_2O_3 Scheu et al. (376) found changes of both Al-L and O-K edges at the interface between Al and sapphire. Therefore it is concluded that the sapphire is O-terminated.

Cr,Ni/$SrTiO_3$ At interfaces between Ni or Cr and $SrTiO_3$ bonding between metal and oxygen atoms was detected by van Benthem et al. (377).

DISLOCATION CORES The experimental study of the ELNES in dislocation cores is particularly challenging because the dislocation has to be viewed along the dislocation line and thus appears as a one-dimensional defect. Batson (312) studied the electronic structure of dislocations in SiGe (see above). Zhang et al. (378, 379) detected a change of the Ti-L and O-K ELNES and attributed this to a distortion of the TiO_6 octahedra in the dislocation core. This distortion was accompanied by a loss of oxygen in the core region.

Arslan & Browning (380) and Xin et al. (381) obtained N-K ELNES from dislocation cores in GaN. By comparing multiple scattering calculations they concluded that in pure GaN dislocations do not induce states in the band gap.

GRAIN BOUNDARIES As in the study of interfaces, the spatial difference method is particularly powerful for the study of the electronic structure or segregation at grain boundaries (GBs) (see above). Nufer et al. (382) studied the electronic structure at the rhombohedral twin grain boundary in α-Al_2O_3. They found good correspondence between band structure calculations, EELS, and HRTEM results.

Until the 1970s, grain boundary segregation studies were the domain of Auger electron spectroscopy. The first to measure segregation by TEM were Doig & Flewitt (108). The quantitative measure of the GB excess is possible by illuminating a narrow area containing the GB and comparing the EELS or EDXS signal with that of an area remote from the GB. This is basically a spatial difference method.

It can be done most effectively in a STEM. A field-emission electron source is almost mandatory because the relative signal from the GB is higher the smaller the illuminated area. Usually a width of a few nanometers perpendicular to the boundary is used. In thick specimen areas, beam broadening can introduce errors into the analysis. This was taken into consideration by Alber et al. (112). Another approach also applicable in a TEM was described by Walther et al. (383). They measured the signal from the segregated element using different sizes of the focused electron beam. From the interpolation to zero beam size they obtain the GB excess. Using the ζ-factor method (56), Watanabe et al. obtained compositional maps of the Cr-, Ni-, and Mo-segregation at GBs in Fe (383a). A method to measure GB segregation at heterointerfaces was shown by Kooi et al. (384).

With the availability of highly stable microscopes it has become possible to probe the ELNES of almost single-atom columns on the GB or interface. Several examples are shown in (385) and references therein.

CONCLUSIONS

With the advent of new instrumentation (C_s correctors, electron monochromaters, energy filters, CCD detectors) it has now become possible to obtain chemical information with a spatial resolution well below 1 nm and an energy resolution better than 0.2 eV in analytical TEM. This unique combination makes analytical TEM an indispensable technique in materials science. The instrumental achievements have been accompanied by a steady development of data evaluation and interpretation, the aim of which is to extract the utmost information from the available experimental data. Owing to this rapid progress, the present article can at best give a snap-shot of the present state in this exciting field of research.

ACKNOWLEDGMENT

I thank Professor M. Rühle for inspiring me to write this article and for his careful comments. Critical reading of a part of the manuscript by Professor C. Elsässer and Dr. Masashi Watanabe is gratefully acknowledged.

LITERATURE CITED

1. Sorby H. 1887. On the microscopical structure of iron and steel. *J. Iron Steel Inst.* 1:255–88
2. Friedrich W, Knipping P, Laue M. 1912. *Interferenz-Erscheinungen bei Röntgenstrahlen.* Presented at Mathematisch-Physikalische Klasse der königlich bayerischen Akademie der Wissenschaften

3. Davisson C, Germer L. 1927. Diffraction of electrons by a crystal of nickel. *Phys. Rev.* 30:705–40
4. Knoll M, Ruska E. 1932. Das Elektronenmikroskop. *Zeitschrift für Physik* 78:318–39
5. Bollmann W. 1956. Interference effects in the electron microscopy of thin crystal foils. *Phys. Rev.* 103:1588–9
6. Hirsch P, Horne R, Whelan M. 1956. Direct observations of the arrangement and motion of dislocations in aluminium. *Philos. Mag.* 1:677–84
7. Martínez G, Tsuno K. 2004. Design of Wien filters with high resolution. *Ultramicroscopy* 100:105–14
8. Haider M, Rose H, Uhlemann S, Schwan E, Kabius B, Urban K. 1998. A spherical-aberration-corrected 200 kV transmission electron microscope. *Ultramicroscopy* 75:53–60
9. Krivanek OL, Dellby N, Lupini AR. 1999. Towards sub-Å electron beams. *Ultramicroscopy* 78:1–11
10. Rose H. 1990. Outline of a spherically corrected semiaplanatic medium-voltage transmission electron-microscope. *Optik* 85:19–24
11. Rose H. 1999. Prospects for realizing a sub-Å sub-eV resolution EFTEM. *Ultramicroscopy* 78:13–25
12. Haider M, Uhlemann S, Zach J. 2000. Upper limits for the residual aberrations of a high-resolution aberration-corrected STEM. *Ultramicroscopy* 81:163–75
13. Krivanek O, Nellist P, Dellby N, Murfitt M, Szilagyi Z. 2003. Towards sub-0.5 Å electron beams. *Ultramicroscopy* 96:229–37
14. Batson P. 2003. Aberration correction results in the IBM STEM instrument. *Ultramicroscopy* 96:239–49
15. Nellist P, Chisholm M, Dellby N, Krivanek O, Murfitt M, et al. 2004. Direct sub-Angstrom imaging of a crystal lattice. *Science* 305:1741
16. Urban K, Kabius B, Haider N, Rose H. 1999. A way to higher resolution: spherical-aberration correction in a 200 kV transmission electron microscope. *J. Electron Microsc.* 48:821–26
17. Egerton RF. 1996. *Electron Energy-Loss Spectroscopy in the Electron Microscope*. New York/London: Plenum. 485 pp.
18. Crewe A, Isaacson M, Johnson D. 1969. A simple scanning electron microscope. *Rev. Sci. Instrum.* 40:241–46
19. Crewe A, Isaacson M, Johnson D. 1971. High resolution electron spectrometer for use in transmission scanning electron microscopy. *Rev. Sci. Instrum.* 42:411–20
20. Reimer L. 1994. *Energy-Filtering Transmission Electron Microscopy*. Berlin/Heidelberg: Springer
21. Gubbens AJ, Krivanek OL. 1993. Applications of a post-column imaging filter in biology and materials science. *Ultramicroscopy* 51:146–59
22. Krivanek OL, Friedman SL, Gubbens AJ, Kraus B. 1995. An imaging filter for biological applications. *Ultramicroscopy* 59:267–82
23. Uhlemann S, Rose H. 1996. Acceptance of imaging energy filters. *Ultramicroscopy* 63:161–67
24. Wollmann D, Irwin K, Hilton G, Dulcie L, Newbury D, Martinis J. 1997. High-resolution, energy-dispersive microcalorimeter spectrometer for X-ray microanalysis. *J. Microsc.* 188:196–223
25. Newbury D, Wollman D, Irwin K, Hilton G, Martinis J. 1999. Lowering the limit of detection in high spatial resolution electron beam microanalysis with the microcalorimeter energy dispersive X-ray spectrometer. *Ultramicroscopy* 78:73–88
26. Zaluzec N. 2004. *XEDS systems for the next generation analytical electron microscope*. Presented at Microscopy and Microanalysis 2004, Savannah, GA
27. Goldstein J. 2002. *Scanning Electron Microscopy and X-Ray Microanalysis*. The Netherlands: Kluwer

28. Williams DB. 1995. *X-Ray Spectrometry in Electron Beam Instruments*: New York: Plenum
29. Spence J, Tafto J. 1983. ALCHEMI—a new technique for locating atoms in small crystals. *J. Microsc.* 130:147–54
30. Taftø J, Spence JCH. 1982. Atomic site determination using the channeling effect in electron-induced X-ray emission. *Ultramicroscopy* 9:243–47
31. Lenz F. 1954. Zur Streuung mittelschneller Elektronen in kleinste Winkel. *Z. Naturforsch. A* 9:185–204
32. Williams BG, Parkinson GM, Eckhardt CJ, Sparrow JMT. 1981. A new approach to the measurement of the momentum densities in solids using an electron microscope. *Chem. Phys. Lett.* 78:434–38
33. Exner A, Schattschneider P. 1996. Asymmetries in electron Compton profiles of silicon—a coherence effect. *Ultramicroscopy* 65:131–45
34. Egerton R, Wong K. 1995. Some practical consequences of the Lorentzian angular-distribution of inelastic-scattering. *Ultramicroscopy* 59:169–80
35. Bethe H. 1928. Theorie der Beugung von Elektronen an Kristallen. *Ann. Phys.* 4:55–129
36. Inokuti M. 1971. Inelastic collisions of fast charged particles with atoms and molecules—Bethe theory revisited. *Rev. Mod. Phys.* 43:297–347
37. Schattschneider P, Hébert C, Jouffrey B. 2001. Orientation dependence of ionization edges in EELS. *Ultramicroscopy* 86:343–53
38. Nelhiebel M, Luchier N, Schorsch P, Schattschneider P, Jouffrey B. 1999. The mixed dynamic form factor for atomic core-level excitations in interferometric electron-energy-loss experiments. *Philos. Mag. B* 79:941–53
39. Krause M, Oliver J. 1979. Natural width of atomic K-levels and L-levels, K-alpha X-ray-lines and several KLL Auger lines. *J. Phys. Chem. Ref. Data* 8:329–38
40. Fuggle JC, Inglesfield JE. 1992. Introduction. In *Unoccupied Electronic States*, ed. JC Fuggle, JE Inglesfield, pp. 1–23. Berlin: Springer
41. Fuggle J, Alvarado S. 1980. Core-level lifetimes as determined by X-ray photoelectron-spectroscopy measurements. *Phys. Rev. A* 22:1615–24
42. Rehr J, Albers R. 2000. Theoretical approaches to X-ray absorption fine structure. *Rev. Mod. Phys.* 72:621–54
43. Nufer S, Gemming T, Köstlmeier S, Rühle M. 2001. Validity of the dipole-selection rule for the Al-$L_{2,3}$ edge of -Al_2O_3 under channeling conditions. *Ultramicroscopy* 88:253–63
44. Grunes L, Leapman R. 1980. Optically forbidden excitations of the 3s subshell in the 3d transition-metals by inelastic-scattering of fast electrons. *Phys. Rev. B* 22:3778–83
45. Hébert-Souche C, Louf P-H, Blaha P, Nelhiebel M, Luitz J, et al. 2000. The orientation-dependent simulation of ELNES. *Ultramicroscopy* 83:9–16
46. Essex D, Nellist P, Whelan C. 1999. Limitations of the dipole approximation in calculations for the scanning transmission electron microscope. *Ultramicroscopy* 80:183–92
47. Leapman R, Fejes P, Silcox J. 1983. Orientation dependence of core edges from anisotropic materials determined by inelastic-scattering of fast electrons. *Phys. Rev. B* 28:2361–73
48. Botton G, Boothroyd CB, Stobbs WM. 1995. Momentum dependent energy loss near edge structures using a CTEM: the reliability of the methods available. *Ultramicroscopy* 59:93–107
49. Browning N, Yuan J, Brown L. 1991. Real-space determination of anisotropic electronic-structure by electron-energy loss spectroscopy. *Ultramicroscopy* 38: 291–98
50. Midgley PA. 1999. A simple new method to obtain high angular resolution ω–q patterns. *Ultramicroscopy* 76:91–96

51. Spence J, Reese G. 1986. Pendellösung radiation and coherent bremsstrahlung. *Acta Crystallogr. A* 42:577–85
52. Sigle W, Carstanjen H. 1992. Observation of coherent bremsstrahlung in quasicrystalline Al-Cu-Co-Si. *Philos. Mag. B* 66:533–40
53. Anderson I, Bentley J, Carter C. 1997. Secondary fluorescence correction formulae for X-ray microanalysis. 1. Parallel-sided thin foil, wedge, and bulk specimens. *Ultramicroscopy* 68:77–94
54. Goldstein J, Costley J, Lorimer G, Reed S. 1977. *Quantitative X-ray analysis in the electron microscope*. Presented at 10th Annu. Scanning Electron Microscopy Symp. Chicago
55. Watanabe M, Horita Z, Nemoto M. 1996. Absorption correction and thickness determination using the ζ factor in quantitative X-ray microanalysis. *Ultramicroscopy* 65:187–98
56. Watanabe M, Williams DB. 1999. *Microscopy and Microanalysis* 5:88 (Suppl. 2)
57. Cliff G, Lorimer G. 1975. Quantitative-analysis of thin specimens. *J. Microsc.* 103:203–7
58. Hofer F, Grogger W, Kothleitner G, Warbichler P. 1997. Quantitative analysis of EFTEM elemental distribution images. *Ultramicroscopy* 67:83–103
59. Crozier P. 1995. Quantitative elemental mapping of materials by energy-filtered imaging. *Ultramicroscopy* 58:157–74
60. Craven A, Gibson J, Howie A, Spalding D. 1978. Study of single-electron excitations by electron-microscopy. 1. Image-contrast from delocalized excitations. *Philos. Mag. A* 38:519–27
61. Cundy S, Metherel A, Whelan M. 1967. Contrast preserved by elastic and quasi-elastic scattering of fast electrons near Bragg beams. *Philos. Mag.* 15:623–30
62. Cundy S, Howie A, Valdrè U. 1969. Preservation of electron microscope image contrast after inelastic scattering. *Philos. Mag.* 20:147–63
63. Watanabe H. 1964. Energy selecting microscope. *Jpn. J. Appl. Phys.* 3:480–85
64. Kuwabara S, Uefuji T. 1975. Variation of electron-microscopic thickness fringes of aluminium single-crystals with energy-loss. *J. Phys. Soc. Jpn.* 38: 1090–97
65. Castaing R, Henoc P, Henry L, Natta M. 1967. Degree of coherence of electron scattering by electron-phonon interaction. *C. R. Acad. Sci. B* 265:1293–96
66. Deleted in proof
67. Reimer L, Fromm I, Rennekamp R. 1988. Operation modes of electron spectroscopic imaging and electron energy-loss spectroscopy in a transmission electron microscope. *Ultramicroscopy* 24:339–54
68. Hofer F, Warbichler P. 1996. Improved imaging of secondary phases in solids by energy-filtering TEM. *Ultramicroscopy* 63:21–25
69. Howie A. 1963. Inelastic scattering of electrons by crystals. 1. Theory of small-angle inelastic scattering. *Proc. R. Soc. London Ser. A* 271:268–87
70. Navidi-Kasmai T, Kohl H. 2000. Computation of contrasts in atomic resolution electron spectroscopic images of planar defects in crystalline specimens. *Ultramicroscopy* 81:223–33
71. Freitag B, Mader W. 1999. Element specific imaging with high lateral resolution: an experimental study on layer structures. *J. Microsc.* 194:42–57
72. Krivanek OL, Gubbens AJ, Kundmann MK, Carpenter GC. 1993. *Elemental mapping with an energy-selecting imaging filter*. Presented at 51st EMSA Meet., San Francisco
73. Johnson D. 1979. Energy loss spectrometry for biological research. In *Introduction to Analytical Electron Microscopy*, ed. J Hren, J Goldstein, D Joy, pp. 245–58. New York: Plenum
74. Moore KT, Howe JM, Elbert DC. 1999.

Analysis of diffraction contrast as a function of energy loss in energy-filtered transmission electron microscope imaging. *Ultramicroscopy* 80:203–19
75. Krivanek O, Gubbens A, Dellby N. 1991. Developments in EELS instrumentation for spectroscopy and imaging. *Microsc. Microanal. Microstruct.* 2:315–32
76. Jeanguillaume C, Colliex C. 1989. Spectrum-image: The next step in EELS digital acquisition and processing. *Ultramicroscopy* 28:252–57
77. Hunt J, Williams D. 1991. Electron energy-loss spectrum-imaging. *Ultramicroscopy* 38:47–73
78. Botton G, L'Esperance G. 1994. Development, quantitative performance and applications of a parallel electron energy-loss spectrum imaging system. *J. Microsc.* 173:9–25
79. Bouchet D, Colliex C, Flora P, Krivanek O, Mory C, Tencé M. 1990. Analytical electron-microscopy at the atomic level with parallel electron-energy loss spectroscopy. *Microsc. Microanal. Microstruct.* 1:443–54
80. Jeanguillaume C, Trebbia P, Colliex C. 1978. About use of electron energy-loss spectroscopy for chemical mapping of thin foils with spatial-resolution. *Ultramicroscopy* 3:237–42
81. Lavergne J, Martin J, Belin M. 1992. Interactive electron-energy-loss elemental mapping by the imaging-spectrum method. *Microsc. Microanal. Microstruct.* 3:517–28
82. Lavergne J, Foa C, Bongrand P, Seux D, Martin J. 1994. Application of recording and processing of energy-filtered image sequences for the elemental mapping of biological specimens—Imaging-spectrum. *J. Microsc.* 174:195–206
83. Martin J, Lavergne J, Vacher B, Inoue K. 1995. Interactive image-spectrum EELS—Application to elemental mapping of lubricant colloids. *Microsc. Microanal. Microstruct.* 6:53–63
84. Mayer J, Plitzko J. 1996. Mapping of ELNES on a nanometre scale by electron spectroscopic imaging. *J. Microsc.* 183:2–8
85. Mayer J, Eigenthaler U, Plitzko J, Dettenwanger F. 1997. Quantitative analysis of electron spectroscopic imaging series. *Micron* 28:361–70
86. Plitzko JM, Mayer J. 1999. Quantitative thin film analysis by energy filtering transmission electron microscopy. *Ultramicroscopy* 78:207–19
87. Hofer F, Kothleitner G. 1993. Quantitative microanalysis using electron-energy-loss spectrometry. 1. Li and Be in oxides. *Microsc. Microanal. Microstruct.* 4:539–60
88. Egerton R. 1974. The electron energy loss spectrum and band structure of diamond. *Philos. Mag.* 30:739–49
89. Thomas PJ, Midgley PA. 2001. Image-spectroscopy—I. The advantages of increased spectral information for compositional EFTEM analysis. *Ultramicroscopy* 88:179–94
90. Reimer L, Fromm I, Hirsch P, Plate U, Rennekamp R. 1992. Combination of EELS modes and electron spectroscopic imaging and diffraction in an energy-filtering electron-microscope. *Ultramicroscopy* 46:335–47
91. Kimoto K, Sekiguchi T, Aoyama T. 1997. Chemical shift mapping of Si L and K edges using spatially resolved EELS and energy filtering TEM. *J. Electron Microsc.* 46:369–74
92. Bentley J. 1997. *Interfacial segregation studied with modern AEM techniques.* Presented at Microscopy and Microanalysis, Cleveland, OH
93. Flaitz PL. 1999. *Characterization of a TiN/Ti/CoSi$_2$ interfacial reaction layer by spectrum line imaging.* Presented at Electron Microscopy and Analysis, Sheffield. pp. 605–8. Bristol, UK: Inst. Phys. Conf. Ser. 161
94. Walther T, Mader W. 1999. *Application of spatially resolved electron energy-loss spectroscopy to the quantitative*

analysis of semiconducting layer structures. Presented at MSM-11, Oxford. pp. 121–18. Bristol, UK: Inst. Phys. Conf. Ser. 164

95. Sekiguchi T, Kimoto K, Aoyama T, Mitsui Y. 1998. Nitrogen distribution and chemical bonding state analyses in oxynitride film by spatially resolved electron energy loss spectroscopy (EELS). *Jpn. J. Appl. Phys. Lett.* 37: L694–96
96. Walther T. 2003. Electron energy-loss spectroscopic profiling of thin film structures: 0.39 nm line resolution and 0.04 eV precision measurement of near-edge structure shifts at interfaces. *Ultramicroscopy* 96:401–11
97. Williams DB, Carter B. 1996. *Transmission Electron Microscopy*. New York: Plenum
98. Egerton R. 1993. Oscillator-strength parameterization of inner-shell cross sections. *Ultramicroscopy* 50:13–28
99. Hofer F. 1994. In *Energy-Filtering Transmission Electron Microscopy*, ed. L Reimer, pp. 225–68. Berlin: Springer
100. Knippelmeyer R, Wahlbring P, Kohl H. 1997. Relativistic ionisation cross sections for use in microanalysis. *Ultramicroscopy* 68:25–41
101. Kambe K, Lehmpfuhl G, Fujimoto F. 1974. Interpretation of electron channeling by the dynamical theory of electron diffraction. *Z. Naturforsch.* 29a:1034–44
102. Pennycook S, Jesson D. 1990. High-resolution incoherent imaging of crystals. *Phys. Rev. Lett.* 64:938–41
103. Pennycook SJ, Jesson DE. 1991. High-resolution Z-contrast imaging of crystals. *Ultramicroscopy* 37:14–38
104. Lupini AR, Pennycook S. 2003. Localization in elastic and inelastic scattering. *Ultramicroscopy* 96:313–22
105. Plamann T, Hÿtch MJ. 1999. Tests on the validity of the atomic column approximation for STEM probe propagation. *Ultramicroscopy* 78:153–61
106. Möbus G, Nufer S. 2003. Nanobeam propagation and imaging in a FEG-TEM/STEM. *Ultramicroscopy* 96:285–98
107. Dwyer C, Etheridge J. 2003. Scattering of Å-scale electron probes in silicon. *Ultramicroscopy* 96:343–60
108. Doig P, Flewitt PEJ. 1978. The influence of temper embrittlement on the stress corrosion susceptibility of Fe-3 wt.% Ni alloys. *Acta Metall.* 26:1283–91
109. Michael J, Williams D, Klein C, Ayer R. 1990. The measurement and calculation of the X-ray spatial resolution obtained in the analytical electron microscope. *J. Microsc.* 160:41–53
110. Doig P, Londsdale D, Flewitt P. 1980. The spatial resolution of X-ray microanalysis in the scanning transmission electron microscope. *Philos. Mag. A* 41: 761–75
111. Reed S. 1982. The single-scattering model and spatial-resolution in X-ray-analysis of thin foils. *Ultramicroscopy* 7:405–9
112. Alber U, Müllejans H, Rühle M. 1997. Improved quantification of grain boundary segregation by EDS in a dedicated STEM. *Ultramicroscopy* 69:105–16
113. Maslen V, Rossouw C. 1983. The inelastic scattering matrix element and its application to electron energy loss spectroscopy. *Philos. Mag.* 47:119–30
114. Rafferty B, Pennycook SJ. 1999. Towards atomic column-by-column spectroscopy. *Ultramicroscopy* 78:141–51
115. Bourdillon A, Self P, Stobbs W. 1981. Crystallographic orientation effects in energy dispersive X-ray analysis. *Philos. Mag. A* 44:1335–50
116. Pennycook S, Narayan J. 1985. Atomic location by axial-electron-channeling analysis. *Phys. Rev. Lett.* 54:1543–46
117. Pennycook SJ. 1988. Delocalization corrections for electron channeling analysis. *Ultramicroscopy* 26:239–48

118. Muller DA, Silcox J. 1995. Delocalization in inelastic scattering. *Ultramicroscopy* 59:195–213
119. Müller H, Rose H. 2003. Electron scattering. In *High-Resolution Imaging and Spectrometry of Materials*, ed. F Ernst, M Rühle, pp. 9–68. Berlin: Springer
120. van Benthem K, French R, Sigle W, Elsässer C, Rühle M. 2001. Valence electron energy loss study of Fe-doped $SrTiO_3$ and a $\Sigma 13$ boundary: electronic structure and dispersion forces. *Ultramicroscopy* 86:303–18
121. Bruley J. 1992. Detection of nitrogen at (100) platelets in a type IAA/B diamond. *Philos. Mag. Lett.* 66:47–56
122. Müllejans H, Bruley J. 1994. Improvements in detection sensitivity by spatial difference electron energy-loss spectroscopy at interfaces in ceramics. *Ultramicroscopy* 53:351–60
123. Nufer S. 2001. *Ortsaufgelöste analytische Elektronenmikroskopie an Korngrenzen in alpha-Al_2O_3*. PhD thesis. Univ. Stuttgart. 161 pp.
124. Scheu C. 2002. Electron energy-loss near-edge structure studies at the atomic level: reliability of the spatial difference technique. *J. Microsc.* 207:52–57
125. Gao M, Scheu C, Tchernychova E, Rühle M. 2003. Successful application of the spatial difference technique to electron energy-loss spectroscopy studies of Mo/$SrTiO_3$ interfaces. *J. Microsc.* 210:94–101
126. Muller DA. 1999. Why changes in bond lengths and cohesion lead to core-level shifts in metals, and consequences for the spatial difference method. *Ultramicroscopy* 78:163–74
127. Trebbia P, Bonnet N. 1990. EELS elemental mapping with unconventional methods I. Theoretical basis: image analysis with multivariate statistics and entropy concepts. *Ultramicroscopy* 34:165–78
128. Bonnet N, Simova E, Thomas X. 1991. Application of multivariate statistical-analysis to time-dependent spectroscopy. *Microsc. Microanal. Microstruct.* 2:129–42
128a. Kotula PG, Keenan MR, Michael JR. 2003. Automated analysis of SEM X-ray spectral images: a powerful new microanalysis tool. *Microsc. Microanal.* 9:1–17
129. Titchmarsh J, Dumbill S. 1996. Multivariate statistical analysis of FEG-STEM EDX spectra. *J. Microsc.* 184:195–207
130. Titchmarsh J, Dumbill S. 1997. Multivariate statistical analysis of STEM-EDX data from radiation-induced sensitization in stainless steel. *J. Microsc.* 188:224–36
131. Titchmarsh JM. 1999. Detection of electron energy-loss edge shifts and fine structure variations at grain boundaries and interfaces. *Ultramicroscopy* 78:241–50
132. Gatts C, Duscher G, Müllejans H, Rühle M. 1995. Analyzing line scan EELS data with neural pattern recognition. *Ultramicroscopy* 59:229–39
133. Krivanek O, Kundmann M, Kimoto K. 1995. Spatial resolution in EFTEM elemental maps. *J. Microsc.* 180:277–87
134. Howie A. 1979. Image-contrast and localized signal selection techniques. *J. Microsc.* 117:11–23
135. Humphreys C, Whelan M. 1969. Inelastic scattering of fast electrons by crystals I. Single electron excitations. *Philos. Mag.* 20:165–72
136. Craven A, Colliex C. 1977. High-resolution energy filtered images in STEM. *J. Microsc. Spectrosc. Electron.* 2:511–22
137. Endoh H, Hashimoto H, Makita Y. 1994. Theoretical and observed electron microscope images of impurity atoms in thin crystals formed by L-shell ionization electrons. *Ultramicroscopy* 56:108–20
138. Wang ZL, Shapiro AJ. 1995. Energy-filtering and composition-sensitive

imaging in surface and interface studies using HREM. *Ultramicroscopy* 60: 115–35
139. Wang Z. 1997. Lattice imaging using plasmon energy-loss electrons in an energy-filtered transmission electron microscope. *Ultramicroscopy* 67:105–11
140. Krivanek O, Gubbens A, Dellby N, Meyer C. 1992. Design and 1st applications of a post-column imaging filter. *Microsc. Microanal.* 3:187–99
141. Stallknecht P, Kohl H. 1996. Computation and interpretation of contrast in crystal lattice images formed by inelastically scattered electrons in a transmission electron microscope. *Ultramicroscopy* 66:261–75
142. Kimoto K, Matsui Y. 2001. Fourier images feature of lattice fringes formed by low-loss electrons as observed using spatially-resolved EELS technique. *J. Electron Microsc.* 50:377–82
143. Kimoto K, Matsui Y. 2003. Experimental investigation of phase contrast formed by inelastically scattered electrons. *Ultramicroscopy* 96:335–42
144. Rose A. 1948. The sensitivity performance of the human eye on an absolute scale. *J. Opt. Soc. Am.* 38:196–208
145. Watanabe M, Williams DB. 1999. Atomic-level detection by X-ray microanalysis in the analytical electron microscope. *Ultramicroscopy* 78:89–101
146. Berger A, Kohl H. 1993. Optimum imaging parameters for elemental mapping in an energy filtering transmission electron-microscope. *Optik* 92:175–93
147. Overwijk M, Reefman D. 2000. Maximum-entropy deconvolution applied to electron energy-loss spectroscopy. *Micron* 31:325–31
148. Mory C, Colliex C. 1989. Elemental analysis near the single-atom detection level by processing sequences of energy-filtered images. *Ultramicroscopy* 28:339–46
149. Krivanek O, Mory C, Tencé M, Colliex C. 1991. EELS quantification near the single-atom detection level. *Microsc. Microanal. Microstruct.* 2:257–67
150. Suenaga K, Tencé M, Mory C, Colliex C, Kato H, et al. 2000. Element-selective single atom imaging. *Science* 290:2280–82
151. Leapman R. 2003. Detecting single atoms of calcium and iron in biological structures by electron energy-loss spectrum-imaging. *J. Microsc.* 210:5–15
152. Feng J, Ho R, Shao Z, Somlyo A. 1999. Parallel electron energy-loss spectroscopy free from gain variation. *Ultramicroscopy* 76:221–31
153. Menon N, Krivanek O. 2002. Synthesis of electron energy loss spectra for the quantification of detection limits. *Microsc. Microanal.* 8:203–15
154. Berger A, Mayer J, Kohl H. 1994. Detection limits in elemental distribution images produced by energy-filtering TEM—case-study of grain-boundaries in Si_3N_4. *Ultramicroscopy* 55:101–12
155. Kohl H, Berger A. 1995. The resolution limit for elemental mapping in energy-filtering transmission electron microscopy. *Ultramicroscopy* 59:191–94
156. Kothleitner G, Hofer F. 1998. Optimization of the signal to noise ratio in EFTEM elemental maps with regard to different ionization edge types. *Micron* 29:349–57
157. Terauchi M, Kuzuo R, Satoh F, Tanaka M, Tsuno K, Ohyama J. 1991. Performance of a new high-resolution electron-energy-loss spectroscopy microscope. *Microsc. Microanal. Microstruct.* 2:351–58
158. Fink J. 1989. Recent developments in energy-loss spectroscopy. *Adv. Electron. Electron Phys.* 75:121–232
159. Batson P. 1999. Advanced spatially resolved EELS in the STEM. *Ultramicroscopy* 78:33–42
160. Mitterbauer C, Kothleitner G, Grogger

W, Zandbergen H, Freitag B, et al. 2003. Electron energy-loss near-edge structures of 3d transition metal oxides recorded at high-energy resolution. *Ultramicroscopy* 96:469–80

161. Lazar S, Botton G, Wu M-Y, Tichelaar F, Zandbergen H. 2003. Materials science applications of HREELS in near edge structure analysis and low-energy loss spectroscopy. *Ultramicroscopy* 96:535–46
162. Brydson R, Scott A, Brown A. 2003. Some thoughts on source monochromation and the implications for electron energy loss spectroscopy. *Z. Metall.* 94:277–81
163. Gloter A, Douiri A, Tencé M, Colliex C. 2002. Improving energy resolution of EELS spectra: an alternative to the monochromator solution. *Ultramicroscopy* 96:385–400
164. Overwijk M, Reefman D, Batson P. 1998. *Maximum entropy deconvolution applied to EELS*. Presented 14th Int. Congress on Electron Microscopy, Cancun, Mexico
165. Yan J, Chen F, Kai J. 2002. Mapping of sp^2/sp^3 in DLC thin film by signal processed ESI series energy-loss image. *J. Electron Microsc.* 51:391–400
166. Haking A, Troester H, Richter K, Crucifix C, Spring H, Trendelenburg MF. 1999. An approach to an objective background subtraction for elemental mapping with core-edges down to 50 eV: description, evaluation and application. *Ultramicroscopy* 80:163–82
167. Egerton R, Malac M. 2002. Improved background-fitting algorithms for ionization edges in electron energy-loss spectra. *Ultramicroscopy* 92:47–56
168. Chan H, Williams D. 1985. Quantitative-analysis of lithium in Al-Li alloys by ionization-energy loss spectroscopy. *Philos. Mag. B* 52:1019–32
169. Leapman RD, Hunt JA, Buchanan RA, Andrews SB. 1993. Measurement of low calcium concentrations in cryosectioned cells by parallel-EELS mapping. *Ultramicroscopy* 49:225–34
170. Tencé M, Quartuccio M, Colliex C. 1995. PEELS compositional profiling and mapping at nanometer spatial resolution. *Ultramicroscopy* 58:42–54
171. Shuman H, Somlyo A. 1987. Electron energy loss analysis of near-trace-element concentrations of calcium. *Ultramicroscopy* 21:23–32
172. Leapman R, Hunt J. 1991. Comparison of detection limits for EELS and EDX. *Microsc. Microanal. Microstruct.* 2:231–44
173. Verbeeck J, Van Aert S. 2004. Model based quantification of EELS spectra. *Ultramicroscopy* 101:207–24
174. Chapman J, Steele J, Paterson J, Titchmarsh J. 1985. *The separation of characteristic signals from complex EELS spectra*. Presented Electron Microscopy and Analysis, Newcastle-upon-Tyne
175. Kundmann M, Krivanek O. 1991. Automated processing of parallel-detection EELS data. *Microsc. Microanal. Microstruct.* 2:245–56
176. Kothleitner G, Hofer F. 2003. Elemental occurrence maps: a starting point for quantitative EELS spectrum image processing. *Ultramicroscopy* 96:491–508
177. Cazaux J. 1995. Correlations between ionization radiation damage and charging effects in transmission electron microscopy. *Ultramicroscopy* 60:411–25
178. Berger S, Salisbury I, Milne R, Imeson D, Humphreys C. 1987. Electron energy-loss spectroscopy studies of nanometre-scale structures in alumina produced by intense electron-beam irradiation. *Philos. Mag. B* 55:341–58
179. Scheu C, Gao M, van Benthem K, Tsukimoto S, Schmidt S, et al. 2003. Advances in EELS spectroscopy by using new detector and new specimen preparation technologies. *J. Microsc.* 210:16–24
180. Egerton R, Li P, Malac M. 2004.

Radiation damage in the TEM and SEM. *Micron* 35:399–409
181. Paxton A, M vS, McKenzie M, Craven A. 2000. The near-edge structure in energy-loss spectroscopy: many-electron and magnetic effects in transition metal nitrides and carbides. *J. Phys. Condens. Matter* 12:729–50
182. Lie K, Hoier R, Brydson R. 2000. Theoretical site- and symmetry-resolved density of states and experimental EELS near-edge spectra of AlB_2 and TiB_2. *Phys. Rev. B* 61:1786–94
183. Mizoguchi T, Tanaka I, Yoshiya M, Oba F, Ogasawara K, Adachi H. 2000. Core hole effects on theoretical electron-energy-loss near-edge structure and near edge X-ray absorption fine structure of MgO. *Phys. Rev. B* 61:2180–87
184. Rez P, MacLaren J, Saldin D. 1998. Application of the layer Korringa-Kohn-Rostoker method to the calculation of near-edge structure in X-ray-absorption and electron-energy-loss spectroscopy. *Phys. Rev. B* 57:2621–27
185. Muller D, Singh D, Silcox J. 1998. Connections between the electron-energy-loss spectra, the local electronic structure, and the physical properties of a material: a study of nickel aluminum alloys. *Phys. Rev. B* 57:8181–202
186. Brydson R. 1991. Interpretation of near-edge structures in the electron energy-loss spectrum. *EMSA Bull.* 21:2:57–67
187. Rehr JJ, Albers RC. 2000. Theoretical approaches to X-ray absorption fine structure. *Rev. Mod. Phys.* 72:621–54
188. Seah M, Dench W. 1979. Quantitative electron spectroscopy of surfaces: a standard data base for electron inelastic mean free paths in solids. *Surf. Interface Anal.* 1:2–11
189. Pantelides S. 1975. Electronic excitation-energies and soft-X-ray absorption-spectra of alkali-halides. *Phys. Rev. B* 11:2391–411
190. Fujikawa T. 1983. Theory of the X-ray absorption near edge structure (XANES) at a deep $L_{2,3}$ edge studied by the short-range order multiple scattering theory. *J. Phys. Soc. Jpn.* 52:4001–7
191. Durham P, Pendry J, Hodges C. 1982. Calculation of X-ray absorption near-edge structure, XANES. *Computer Phys. Commun.* 25:193–205
192. Vvedensky D, Saldin D, Pendry J. 1986. An update of DLXANES, the calculation of X-ray absorption near-edge structure. *Computer Phys. Commun.* 40:421–40
193. Köstlmeier S, Elsässer C. 1999. Ab initio calculation of near-edge structures in electron-energy-loss spectra for metal-oxide crystals. *Phys. Rev. B* 60:14025–34
194. Tanaka I, Adachi H. 1996. Calculation of core hole excitonic features on Al L_{23}-edge X-ray-absorption spectra of alpha-Al_2O_3. *Phys. Rev. B* 54:4604–8
195. Paxton A, Craven A, Gregg J, McComb D. 2003. Bandstructure approach to near edge structure. *J. Microsc.* 210:35–44
196. Köstlmeier S. 2001. Success and limits of common final-state approximations. *Ultramicroscopy* 86:319–24
197. Elsässer C, Köstlmeier S. 2001. Density-functional modelling of core hole effects in electron energy-loss near-edge spectra. *Ultramicroscopy* 86:325–37
198. Kaiser U, Muller D, Grazul J, Chuvilin A, Kawasaki M. 2002. Direct observation of defect-mediated cluster nucleation. *Nat. Mater.* 1:102–5
199. Muller D, Schorsch T, Moccio S, Baumann F, Evans-Lutterodt K, Timp G. 1999. The electronic structure at the atomic scale of ultrathin gate oxides. *Nature* 399:758–61
200. Erni R, Heinrich H, Kostorz G. 2003. Quantitative characterisation of chemical inhomogeneities in Al-Ag using high-resolution Z-contrast STEM. *Ultramicroscopy* 94:125–33
201. Yamazaki T, Watanabe K, Kikuchi Y, Kawasaki M, Hashimoto I, Shiojiri M. 2000. Two-dimensional distribution of

As atoms doped in a Si crystal by atomic-resolution high-angle annular dark field STEM. *Phys. Rev. B* 61:13833–39

202. Schmitz G, Ewert JC, Hartung F. 1999. Chemical analysis by high-angle hollow cone illumination. *Ultramicroscopy* 77:49–63
203. Nüchter W, Sigle W. 1995. Electron channelling: a method in real-space crystallography and a comparison with the atomic location by channelling-enhanced microanalysis. *Philos. Mag. A* 71:165–86
204. Horita Z, Matsumura S, Baba T. 1995. General formulation for ALCHEMI. *Ultramicroscopy* 58:327–35
205. Oxley MP, Allen LJ. 1999. Impact parameters for ionization by high-energy electrons. *Ultramicroscopy* 80:125–31
206. Rossouw C, Turner P, White T, O'Connor A. 1989. Statistical analysis of electron channelling microanalytical data for the determination of site occupancies of impurities. *Philos. Mag. Lett.* 60:225–32
207. Ourmazd A, Baumann FH, Bode M, Kim Y. 1990. Quantitative chemical lattice imaging: theory and practice. *Ultramicroscopy* 34:237–55
208. Schwander P, Kisielowski C, Seibt M, Baumann F, Kim Y, Ourmazd A. 1993. Mapping projected potential, interfacial roughness, and composition in general crystalline solids by quantitative transmission electron-microscopy. *Phys. Rev. Lett.* 71:4150–53
209. De Jong A, Van Dyck D. 1990. Composition determination from HREM images of substitutional alloys. *Ultramicroscopy* 33:269–79
210. Thoma S, Cerva H. 1991. The influence of non-linear interference processes on the HREM contrast of AlGaAs in ⟨100⟩ projection. *Ultramicroscopy* 35:77–97
211. Walther T, Gerthsen D, Carius R, Forster A, Urban K. 1993. Correlation between the structural and optical-properties of AlGaAs/GaAs quantum-well structures. *Inst. Phys. Conf. Ser.* 134:449–54
212. Hillebrand R, Werner P, Hofmeister H, Gösele U. 1996. A fuzzy logic approach to image analysis of HREM micrographs of III–V compounds. *Ultramicroscopy* 66:73–88
213. Kara F, Dunin-Borkowski RE, Boothroyd CB, Little JA, Stobbs WM. 1996. Fresnel contrast analysis of composition changes and space charge at grain boundaries in mullite. *Ultramicroscopy* 66:59–71
214. Möbus G, Doole R, Inkson B. 2003. Spectroscopic electron tomography. *Ultramicroscopy* 96:433–51
215. Midgley PA, Weyland M. 2003. 3D electron microscopy in the physical sciences: the development of Z-contrast and EFTEM tomography. *Ultramicroscopy* 96:413–31
216. Wang Z. 1996. Valence electron excitations and plasmon oscillations in thin films, surfaces, interfaces and small particles. *Micron* 27:265–99
217. Dorneich A, French R, Müllejans H, Loughin S, Rühle M. 1998. Quantitative analysis of valence electron energy-loss spectra of aluminium nitride. *J. Microsc.* 191:286–96
218. Müllejans H, French R. 1996. Interband electronic structure of a near-Sigma 11 grain boundary in alpha-alumina determined by spatially resolved valence electron energy-loss spectroscopy. *J. Phys. D* 29:1751–60
219. French R, Müllejans H, Jones D. 1998. Optical properties of aluminum oxide: Determined from vacuum ultraviolet and electron energy-loss spectroscopies. *J. Am. Ceram. Soc.* 81:2549–57
220. French R, Müllejans H, Jones D, Duscher G, Cannon R, Rühle M. 1998. Dispersion forces and Hamaker constants for intergranular films in silicon nitride from spatially resolved-valence electron energy loss spectrum imaging. *Acta Mater.* 46:2271–87

221. Ryen L, Wang X, Helmersson U, Olsson E. 1999. Determination of the complex dielectric function of epitaxial $SrTiO_3$ films using transmission electron energy-loss spectroscopy. *J. Appl. Phys.* 85:2828–34
222. Rafferty B, Brown L. 1998. Direct and indirect transitions in the region of the band gap using electron-energy-loss spectroscopy. *Phys. Rev. B* 58:10326–37
223. van Benthem K, Elsässer C, French R. 2001. Bulk electronic structure of $SrTiO_3$: experiment and theory. *J. Appl. Phys.* 90:6156–64
224. Raether H. 1988. *Excitation of Plasmons and Interband Transitions by Electrons.* Berlin/Heidelberg: Springer
225. von Festenberg C. 1969. Energieverlustmessungen an III/IV-Verbindungen. *Z. Phys.* 227:453–81
226. Ferrell T, Echenique P. 1985. Generation of surface excitations on dielectric spheres by an external electron-beam. *Phys. Rev. Lett.* 55:1526–29
227. Ritchie R. 1957. Plasma losses by fast electrons in thin films. *Phys. Rev.* 106:874–81
228. Otto A. 1965. Plasmaschwingungen gebundener Elektronen: Der charakteristische Energieverlust von 3,6 eV in dünnen Silberfolien. *Z. Phys.* 185:232–62
229. Economou E. 1969. Surface plasmons in thin films. *Phys. Rev.* 182:539–54
230. Howie A. 1983. Surface reactions and excitations. *Ultramicroscopy* 11:141–48
231. Garcia-Molina R, Gras-Marti A, Howie A, Ritchie R. 1985. Retardation effects in the interaction of charged particle beams with bounded condensed media. *J. Phys. C* 18:5335–45
232. Howie A, Milne R. 1985. Excitations at interfaces and small particles. *Ultramicroscopy* 18:427–34
233. Howie A, Milne R. 1984. Electron energy loss spectra and reflection images from surfaces. *J. Microsc.* 136:279–85
234. Bolton J, Chen M. 1995. Electron energy loss in multilayered slabs. *Ultramicroscopy* 60:247–63
235. Chase JB, Kliewer KL. 1970. Electron energy loss due to surface modes in a thin ionic crystal film. *Phys. Rev. B* 2:4389–400
236. Batson P. 1982. Surface-plasmon coupling in clusters of small spheres. *Phys. Rev. Lett.* 49:936–40
237. Ouyang F, Isaacson M. 1989. Accurate modeling of particle substrate coupling surface-plasmon exctation in EELS. *Ultramicroscopy* 31:345–50
238. Ouyang F, Isaacson M. 1989. Surface plasmon excitation of objects with arbitrary shape and dielectric constant. *Philos. Mag. B* 60:481–92
239. Echenique P, Howie A, Wheatley D. 1987. Excitation of dielectric spheres by external electron beams. *Philos. Mag.* 56:335–49
240. Ugarte D, Colliex C, Trebbia P. 1992. Surface-plasmon and interface-plasmon modes on small semiconducting spheres. *Phys. Rev. B* 45:4332–43
241. Echenique P, Bausells J, Rivacoba A. 1987. Energy-loss probability in electron-microscopy. *Phys. Rev. B* 35: 1521–24
242. Batson P. 1980. Damping of bulk plasmons in small aluminum spheres. *Solid State Commun.* 34:477–80
243. Chu Y, Warmack R, Ritchie R, Little J, Becker R, Ferrell T. 1984. Contribution of the surface plasmon to energy losses by electrons in a cylindrical channel. *Particle Accelerators* 16:13–17
244. De Zutter D, De Vleeschauwer D. 1986. Radiation from and force acting on a point charge moving through a cylindrical hole in a conducting medium. *J. Appl. Phys.* 59:4146–50
245. Zabala N, Rivacoba A, Echenique P. 1989. Energy-loss of electrons traveling through cylindrical holes. *Surf. Sci.* 209:465–80
246. Walsh C. 1989. Analysis of electron

energy-loss spectra from electron-beam-damaged amorphous AlF_3. *Philos. Mag. A* 59:227–46
247. Dobrzynski L, Maradudin A. 1972. Electrostatic edge modes in a dielectric wedge. *Phys. Rev. B* 6:3810–15
248. Davis LC. 1976. Electrostatic edge modes of a dielectric wedge. *Phys. Rev. B* 14:5523–25
249. Ekardt W. 1985. Size-dependent photoabsorption and photoemission of small metal particles. *Phys. Rev. B* 31:6360–70
250. Mitome M, Yamazaki Y, Takagi H, Nakagiri T. 1992. Size dependence of plasmon energy in Si clusters. *J. Appl. Phys.* 72:812–14
251. Turowski M, Kelly T. 1992. Profiling of the dielectric function across Al/SiO_2/Si heterostructures with electron-energy loss spectroscopy. *Ultramicroscopy* 41: 41–54
252. Moreau P, Brun N, Walsh C, Colliex C, Howie A. 1997. Relativistic effects in electron-energy-loss-spectroscopy observations of the Si/SiO_2 interface plasmon peak. *Phys. Rev. B* 56:6774–81
253. van Benthem K, Tan G, Denoyer L, French R, Rühle M. 2004. Local optical properties, electron densities and London dispersion energies of atomically structured grain boundaries. *Phys. Rev. Lett.* 93:227201/1–4
254. Browning N, Pennycook S. 1996. Direct experimental determination of the atomic structure at internal interfaces. *J. Phys. D* 29:1779–98
255. Lo S, Kai J, Chen F, Chang L, Chen L, et al. 2002. Four-dimensional dielectric property image obtained from electron spectroscopic imaging series. *J. Electron Microsc.* 50:497–507
256. Rafferty B, Pennycook S, Brown L. 2000. Zero loss peak deconvolution for bandgap EEL spectra. *J. Electron Microsc.* 49:517–24
257. Reed BW, Sarikaya M. 2002. Background subtraction for low-loss transmission electron energy-loss spectroscopy. *Ultramicroscopy* 93:25–37
258. Reed B, Chen J, MacDonald N, Silcox J, Bertsch G. 1999. Fabrication and STEM/EELS measurements of nanometer-scale silicon tips and filaments. *Phys. Rev. B* 60:5641–52
259. Bangert U, Harvey A, Keyse R. 1997. Assessment of electron energy-loss spectroscopy below 5 eV in semiconductor materials in a VG STEM. *Ultramicroscopy* 68:173–80
260. Fowler R, Nordheim L. 1928. Electronic emission in intense electric fields. *Proc. R. Soc. A* 119:173–81
261. Alexandrou I, Papworth A, Rafferty B, Amaratunga G, Kiely C, Brown L. 2001. Calculation of the electronic structure of carbon films using electron energy loss spectroscopy. *Ultramicroscopy* 90:39–45
262. Fink J. 1992. Transmission electron energy-loss spectroscopy. In *Unoccupied Electronic States*, ed. J Fuggle, JE Inglesfield. Berlin: Springer
263. Knupfer M, Widder K, Sing M, Knauff O, Fink J. 1998. Influence of the core polarization on the dielectric properties of polyvalent metals. *Eur. Phys. J. B* 6:323–28
264. Sigle W, Krämer S, Varshney V, Zern A, Eigenthaler U, Rühle M. 2003. Plasmon energy mapping in energy-filtering transmission electron microscopy. *Ultramicroscopy* 96:565–71
265. Rosenberg R, Love P, Rehn V. 1986. Polarization-dependent C(K) near-edge X-ray-absorption fine structure of graphite. *Phys. Rev. B* 33:4034–37
266. Souche C, Jouffrey B, Nelhiebel M. 1998. Orientation sensitive EELS-analysis of boron nitride nanometric hollow spheres. *Micron* 29:419–24
267. Browning N, Yuan J, Brown L. 1993. Theoretical determination of angularly-integrated energy-loss functions for anisotropic materials. *Philos. Mag. A* 67: 261–71

268. Menon NK, Yuan J. 1998. Quantitative analysis of the effect of probe convergence on electron energy loss spectra of anisotropic materials. *Ultramicroscopy* 74:83–94
269. Daniels H, Brown A, Scott A, Nichells T, Rand B, Brydson R. 2003. Experimental and theoretical evidence for the magic angle in transmission electron energy loss spectroscopy. *Ultramicroscopy* 96:523–34
270. Hébert C, Schattschneider P, Jouffrey B. 2004. *Magic angle: an unsolved mystery*. Presented at 13th European Microscopy Congr., Antwerp
271. Berger S, McKenzie D, Martin P. 1988. EELS analysis of vacuum arc-deposited diamond-like films. *Philos. Mag. Lett.* 57:285–90
272. Pappas D, Saenger K, Bruley J, Krakow W, Cuomo J. 1992. Pulsed laser deposition of diamond-like carbon films. *J. Appl. Phys.* 71:5675–84
273. Davis C, Amaratunga G, Knowles K. 1998. Growth mechanism and cross-sectional structure of tetrahedral amorphous carbon thin films. *Phys. Rev. Lett.* 80:3280–83
274. Alexandrou I, Scheibe H, Kiely C, Papworth A, Amaratunga G, Schultrich B. 1999. Carbon films with an sp^2 network structure. *Phys. Rev. B* 60:10903–7
275. Papworth A, Kiely C, Burden A, Silva S, Amaratunga G. 2000. Electron-energy-loss spectroscopy characterization of the sp^2 bonding fraction within carbon thin films. *Phys. Rev. B* 62:12628–31
276. Ferrari A, Libassi A, Tanner B, Stolojan V, Yuan J, et al. 2000. Density, sp^3 fraction, and cross-sectional structure of amorphous carbon films determined by X-ray reflectivity and electron energy-loss spectroscopy. *Phys. Rev. B* 62:11089–103
277. Serin V, Beche E, Berjoan R, Abidate O, Dorignac D, et al. 1997. *XAES, XPS, EELS and Raman spectroscopy of polycrystalline to amorphous carbon films with various sp^3 to sp^2 bondings*. Electrochem. Soc. Proc. 5th Int. Symp. Diamond Materials, Paris, pp. 126–42
278. Zhang S, Zeng X, Xie H, Hing P. 2000. A phenomenological approach for the I_d/I_g ratio and sp^3 fraction of magnetron sputtered a-C films. *Surf. Coat. Technol.* 123:256–60
279. Muller D, Tzou Y, Raj R, Silcox J. 1993. Mapping sp^2 and sp^3 states of carbon at subnanometer spatial-resolution. *Nature* 366:725–27
280. Sigle W, Zern A, Hahn K, Eigenthaler U, Rühle M. 2001. Advances in energy-filtering transmission electron microscopy. *J. Electron Microsc.* 50: 509–15
281. Zhu Y, Egerton RF, Malac M. 2001. Concentration limits for the measurement of boron by electron energy-loss spectroscopy and electron-spectroscopic imaging. *Ultramicroscopy* 87:135–45
282. Schmid H. 1995. Phase identification in carbon and BN systems by EELS. *Microsc. Microanal. Microstruct.* 6:99–111
283. Moscovici J, Loupias G, Parent P, Tourillon G. 1996. Polarization-dependent boron and nitrogen K NEXAFS of hexagonal BN. *J. Phys. Chem. Solids* 57:1159–61
284. Tanaka I, Araki H, Yoshiya M, Mizoguchi T, Ogasawara K, Adachi H. 1999. First-principles calculations of electron-energy-loss near-edge structure and near-edge X-ray-absorption fine structure of BN polytypes using model clusters. *Phys. Rev. B* 60:4944–51
285. Terauchi M, Tanaka M, Matsumoto T, Saito Y. 1998. Electron energy-loss spectroscopy study of the electronic structure of boron nitride nanotubes. *J. Electron Microsc.* 47:319–24
286. Fischer D, Wentzcovitch R, Carr R, Continenza A, Freeman A. 1991. Graphitic interlayer states: A carbon K near-edge X-ray-absorption fine-structure study. *Phys. Rev. B* 44:1427–29

287. Fink J, Mullerheinzerling T, Pfluger J, Bubenzer A, Koidl P, Crecelius G. 1983. Structure and bonding of hydrocarbon plasma generated carbon-films—an electron-energy loss study. *Solid State Commun.* 47:687–91
288. Reed B, Sarikaya M. 2002. TEM/EELS analysis of heat-treated carbon nanotubes: experimental techniques. *J. Electron Microsc.* 51:S97–105
289. Menon NK, Yuan J. 1999. Towards atomic resolution EELS of anisotropic materials. *Ultramicroscopy* 78:185–205
290. Suenaga K, Sandré E, Colliex C, Pickard C, Kataura H, Iijima S. 2001. Electron energy-loss spectroscopy of electron states in isolated carbon nanostructures. *Phys. Rev. B* 63:165408–14
291. Kuzuo R, Terauchi M, Tanaka M. 1992. Electron energy-loss spectra of carbon nanotubes. *Jpn. J. Appl. Phys. Lett.* 31: L1484–87
292. Ajayan P, Iijima S, Ichihashi T. 1993. Electron-energy-loss spectroscopy of carbon nanometer-size tubes. *Phys. Rev. B* 47:6859–62
293. Bursill L, Stadelmann P, Peng J, Prawer S. 1994. Surface-plasmon observed for carbon nanotubes. *Phys. Rev. B* 49:2882–87
294. Saito. 1998. *Physical Properties of Carbon Nanotubes*. London: Imperial College
295. Mintmire J, White C. 1998. Universal density of states for carbon nanotubes. *Phys. Rev. Lett.* 81:2506–9
296. Pichler T, Knupfer M, Golden M, Fink J, Rinzler A, Smalley R. 1998. Localized and delocalized electronic states in single-wall carbon nanotubes. *Phys. Rev. Lett.* 80:4729–32
297. Reed B, Sarikaya M. 2001. Electronic properties of carbon nanotubes by transmission electron energy-loss spectroscopy. *Phys. Rev. B* 64:195404–13
298. Kociak M, Henrard L, Stephan O, Suenaga K, Colliex C. 2000. Plasmons in layered nanospheres and nanotubes investigated by spatially resolved electron energy-loss spectroscopy. *Phys. Rev. B* 61:13936–44
299. Suenaga K, Willaime F, Loiseau A, Colliex C. 1999. Organisation of carbon and boron nitride layers in mixed nanoparticles and nanotubes synthesised by arc discharge. *Appl. Phys. A* 68:301–8
300. Lucas A, Gensterblum G, Pireaux J, Thiry P, Caudano R, et al. 1992. Elementary excitations of C_{60} from the far infrared to the far vacuum ultraviolet studied by high-resolution electron-energy-loss spectroscopy. *Phys. Rev. B* 45:13694–702
301. Pichler T, Knupfer M, Golden M, Fink J, Cabioc'h T. 2001. Electronic structure and optical properties of concentric-shell fullerenes from electron-energy-loss spectroscopy in transmission. *Phys. Rev. B* 63:155415–19
302. Martin JM, Vacher B, Ponsonnet L, Dupuis V. 1996. Chemical bond mapping of carbon by image-spectrum EELS in the second derivative mode. *Ultramicroscopy* 65:229–38
303. Hitchcock A, Urquhart S, Rightor E. 1992. Inner-shell spectroscopy of benzaldehyde, terephthalaldehyde, ethyl benzoate, terephthaloyl chloride, and phosgene: models for core excitation of poly(ethylene terephthalate). *J. Phys. Chem.* 96:8736–50
304. Varlot K, Martin JM, Quet C, Kihn Y. 1997. Towards sub-nanometer scale EELS analysis of polymers in the TEM. *Ultramicroscopy* 68:123–33
305. Grubb D, Groves G. 1971. Rate of damage of polymer crystals in the electron microscope: dependence on temperature and beam voltage. *Philos. Mag.* 24:815–28
306. Lindner T, Sauer H, Engel W, Kambe K. 1986. Near-edge structure in electron-energy-loss spectra of MgO. *Phys. Rev. B* 33:22–24
307. Bouchet D, Colliex C. 2003. Experimental study of ELNES at grain boundaries

in alumina: intergranular radiation damage effects on Al-L_{23} and O-K edges. *Ultramicroscopy* 96:139–52
308. Hansen P, Brydson R, McComb D. 1992. p-p-like transitions at the silicon $L_{2,3}$-edges of silicates. *Microsc. Microanal. Microstruct.* 3:213–19
309. Nufer S, Gemming T, Elsässer C, Köstlmeier S, Rühle M. 2001. Core hole effect in the ELNES of alpha-Al_2O_3: experiment and theory. *Ultramicroscopy* 86:339–42
310. Batson P. 1995. Conduction band-structure in strained silicon by spatially-resolved electron-energy-loss spectroscopy. *Ultramicroscopy* 59:63–70
311. Batson P. 2000. Near-edge conduction band electronic states in SiGe alloys. *J. Electron Microsc.* 49:267–73
312. Batson P. 2000. Structural and electronic characterization of a dissociated 60° dislocation in GeSi. *Phys. Rev. B* 61:16633–41
313. Duscher G, Buczko R, Pennycook S, Pantelides S. 2001. Core hole effects on energy-loss near-edge structure. *Ultramicroscopy* 86:355–62
314. Pearson D, Fultz B, Ahn C. 1988. Measurements of 3d state occupancy in transition-metals using electron-energy loss spectrometry. *Appl. Phys. Lett.* 53: 1405–7
315. Pearson D, Ahn C, Fultz B. 1993. White lines and d-electron occupancies for the 3d and 4d transition-metals. *Phys. Rev. B* 47:8471–78
316. Pearson D, Ahn C, Fultz B. 1994. Measurements of the 3d occupancy from Cu L(2,3) electron-energy-loss spectra of rapidly quenched CuZr, CuTi, CuPd, CuPt, and CuAu. *Phys. Rev. B* 50:12969–72
317. Leapman RD, Grunes LA, Fejes PL. 1982. Study of the $L_{2,3}$ edges in the 3d transition-metals and their oxides by electron-energy-loss spectroscopy with comparisons to theory. *Phys. Rev. B* 26:614–35
318. Sparrow T, Williams B, Rao C, Thomas J. 1984. L_3/L_2 white-line intensity ratios in the electron energy-loss spectra of 3d transition-metal oxides. *Chem. Phys. Lett.* 108:547–50
319. Waddington WG, Rez P, Grant IP, Humphreys CJ. 1986. White lines in the $L_{2,3}$ electron-energy-loss and X-ray absorption spectra of 3d transition metals. *Phys. Rev. B* 34:1467–73
320. Thole B, van der Laan G. 1988. Branching ratio in X-ray absorption spectroscopy. *Phys. Rev. B* 38:3158–71
321. Pease DM, Fasihuddin A, Daniel M, Budnick JI. 2001. Method of linearizing the 3d L_3/L_2 white line ratio as a function of magnetic moment. *Ultramicroscopy* 88:1–16
322. Murakami Y, Shindo D, Otsuka K, Oikawa T. 1998. Electronic structure changes associated with a martensitic transformation in a $Ti_50Ni_{48}Fe_2$ alloy studied by electron energy-loss spectroscopy. *J. Electron Microsc.* 47:301–9
323. Krivanek OL, Paterson JH. 1990. ELNES of 3d transition-metal oxides: I. Variations across the periodic table. *Ultramicroscopy* 32:313–18
324. Paterson JH, Krivanek OL. 1990. ELNES of 3d transition-metal oxides: II. Variations with oxidation state and crystal structure. *Ultramicroscopy* 32:319–25
325. de Groot F, Grioni M, Fuggle J, Ghijsen J, Sawatzky G, Petersen H. 1989. Oxygen 1s X-ray-absorption edges of transition-metal oxides. *Phys. Rev. B* 40:5715–23
326. Craven AJ. 1995. The electron energy-loss near-edge structure (ELNES) on the N K-edges from the transition metal mononitrides with the rock-salt structure and its comparison with that on the C K-edges from the corresponding transition metal monocarbides. *J. Microsc.* 180:250–62
327. Leapman R, Newbury D. 1993. Trace

elemental analysis at nanometer spatial-resolution by parallel-detection electron-energy-loss spectroscopy. *Anal. Chem.* 65:2409–14
328. Thole B, van der Laan G, Fuggle J, Sawatzky G, Karnatak R, Esteva J. 1985. 3d X-ray-absorption lines and the $3d^9 4f^{n+1}$ multiplets of the lanthanides. *Phys. Rev. B* 32:5107–18
329. Laan GVD, Thole BT, Sawatzky GA, Goedkoop JB, Fuggle JC, et al. 1986. Experimental proof of magnetic X-ray dichroism. *Phys. Rev. B* 34:6529–31
330. Hébert C, Schattschneider P. 2003. A proposal for dichroic experiments in the electron microscope. *Ultramicroscopy* 96:463–68
331. Schattschneider P, Rubino S, Hébert C. 2004. *Observation of circular dichroism in the TEM*. Presented at Microscopy and Microanalysis, Savannah, GA, p. 836–37 (on CD)
332. Yuan J, Menon N. 1997. Magnetic linear dichroism in electron energy loss spectroscopy. *J. Appl. Phys.* 81:5087–89
333. van Aken P, Lauterbach S. 2003. Strong magnetic linear dichroism in Fe $L_{2,3}$ and O K electron energy-loss near-edge spectra of antiferromagnetic hematite alpha-Fe_2O_3. *Phys. Chem. Miner.* 30 (8):469–77
334. Murakami Y, Shindo D, Kikuchi M, Zuo J, Spence J. 2002. Resonance effect in ELNES from perovskite-type manganites $BiMnO_3$ and $LaMnO_3$. *J. Electron Microsc.* 51:99–103
335. Borg RJ, Dienes GJ. 1992. Quantum mechanical principles and the covalent bond. In *The Physical Chemistry of Solids*, pp. 179–227. New York: Academic
336. Brydson R, Garvie LAJ, Craven AJ, Sauer H, Hofer F, Cressey G. 1993. $L_{2,3}$ edges of tetrahedrally coordinated d_0 transition-metal oxyanions XO_4N^-. *J. Phys. Condens. Matter* 5:9379–92
337. Laan Gvd, Kirkman IW. 1992. The 2p absorption spectra of 3d transition metal compounds in tetrahedral and octahedral symmetry. *J. Phys. Condens. Matter* 4:4189–204
338. Ogasawara K, Iwata T, Koyama Y, Ishii T, Tanaka I, Adachi H. 2001. Relativistic cluster calculation of ligand-field multiplet effects on cation L-2,L-3 x-ray-absorption edges of $SrTiO_3$, NiO, and CaF_2. *Phys. Rev. B* 64:115413–15
339. van Benthem K. 2002. *Electron Microscopic Investigations of the bonding behaviour of metals on $SrTiO_3$ substrates*. PhD thesis. Univ. Stuttgart. 206 pp.
340. Brydson R, Sauer H, Engel W, Hofer F. 1992. Electron energy-loss near-edge structures at the oxygen K-edges of titanium(IV) oxygen compounds. *J. Phys. Condens. Matter* 4:3429–37
341. de Groot FMF, Faber J, Michiels JJM, Czyzyk MT, Abbate M, Fuggle JC. 1993. Oxygen 1s X-ray absorption of tetravalent titanium oxides: a comparison with single-particle calculations. *Phys. Rev. B* 48:2074–80
342. Tanaka I, Nakajima T, Kawai J, Adachi H, Gu H, Rühle M. 1997. Dopant-modified local chemical bonding at a grain boundary in $SrTiO_3$. *Philos. Mag. Letters* 75:21–27
343. Wallis DJ, Browning ND. 1997. Probing 3-d structural distortions and coordination changes at $SrTiO_3$ grain boundaries using electron energy loss spectroscopy. *J. Am. Ceram. Soc.* 80:781–85
344. Browning ND, Moltaji HO, Buban JP. 1998. Investigation of three-dimensional grain-boundary structures in oxides through multiple-scattering analysis of spatially resolved electron-energy-loss spectra. *Phys. Rev. B* 58:8289–300
345. Lin X, Wang Y, Dravid V, Michalakos P, Kung M. 1993. Valence states and hybridization in vanadium oxide systems investigated by transmission electron-energy-loss spectroscopy. *Phys. Rev. B* 47:3477–81

346. Gloter A, Serin V, Turquat C, Cesari C, Leroux C, Nihoul G. 2001. Vanadium valency and hybridization in V-doped hafnia investigated by electron energy loss spectroscopy. *Eur. Phys. J. B* 22: 179–86
347. Kurata H, Colliex C. 1993. Electron-energy-loss core-edge structures in manganese oxides. *Phys. Rev. B* 48:2102–8
348. Colliex C, Manoubi T, Ortiz C. 1991. Electron-energy-loss-spectroscopy near-edge fine structures in the iron-oxygen system. *Phys. Rev. B* 44:11402–11
349. Kurata H, Hojou K, Uozumi T. 1998. Cluster model calculations for the Fe L_2,L_3-edge fine structure of alpha-Fe_2O_3. *J. Electron Microsc.* 47:293–99
350. Kurata H, Tanaka N. 1991. Iron $L_{2,3}$ white line ratio in nm-sized gamma-iron crystallites embedded in MgO. *Microsc. Microanal. Microstruct.* 2:183–90
351. van Aken P, Liebscher B. 2002. Quantification of ferrous/ferric ratios in minerals: new evaluation schemes of Fe $L_{2,3}$ electron energy-loss near-edge spectra. *Phys. Chem. Miner.* 29:188–200
352. Muller DA SD, Silcox J. 1998. Connections between the electron-energy-loss spectra, the local electronic structure, and the physical properties of a material: a study of nickel aluminum alloys. *Phys. Rev. B* 57:8181–202
353. Potapov P, Kulkova S, Schryvers D. 2003. Study of changes in $L_{3,2}$ EELS ionization edges upon formation of Ni-based intermetallic compounds. *J. Microsc.* 210:102–9
354. Mizoguchi T, Yoshiya M, Li J, Oba F, Tanaka I, Adachi H. 2001. Electron-energy-loss near edge structures of six-fold-coordinated Zn in MgO. *Ultramicroscopy* 86:363–70
355. Radtke G, Bayle-Guillemaud P. 2004. *EELS study of near edge fine structure in Al(x)Ga(1−x)N alloys*. Presented at 13th Eur. Microscopy Congr., Antwerp, Belgium, pp. 239–40
356. Fortner J, Buck E. 1996. The chemistry of the light rare-earth elements as determined by electron energy loss spectroscopy. *Appl. Phys. Lett.* 68:3817–19
357. Buck E, Fortner J. 1997. Detecting low levels of transuranics with electron energy loss spectroscopy. *Ultramicroscopy* 67:69–75
358. Fortner JA, Buck EC, Ellison AJG, Bates JK. 1997. EELS analysis of redox in glasses for plutonium immobilization. *Ultramicroscopy* 67:77–81
359. Docherty F, Craven A, McComb D, Skakle J. 2001. ELNES investigations of the oxygen K-edge in spinels. *Ultramicroscopy* 86:273–88
360. van Benthem K, Kohl H. 2000. Methods for ELNES-quantification: characterization of the degree of inversion of Mg-Al-spinels. *Micron* 31:347–54
361. Köstlmeier S, Elsässer C, Meyer B. 1999. Ab initio analysis of electron energy loss spectra for complex oxides. *Ultramicroscopy* 80:145–51
362. Brydson R. 1996. Multiple scattering theory applied to ELNES of interfaces. *J. Phys. D* 29:1699–708
363. Klie RF, Ito Y, Stemmer S, Browning ND. 2001. Observation of oxygen vacancy ordering and segregation in Perovskite oxides. *Ultramicroscopy* 86: 289–302
364. Nücker N, Fink J, Renker B, Ewert D, Politis C, et al. 1987. Experimental electronic-structure studies of $La_{2-x}Sr_xCuO_4$. *Z. Phys. B* 67:9–14
365. Yarmoff J, Clarke D, Drube W, Karlsson U, Talebibrahimi A, Himpsel F. 1987. Valence electronic-structure of $Y_1Ba_2Cu_3O_7$. *Phys. Rev. B* 36:3967–70
366. Matsuyama H, Takahashi T, Katayamayoshida H, Kashiwakura T, Okabe Y, et al. 1989. Impurity-state-like nature

of fermi-liquid states in $Bi_2Sr_2CaCu_2O_8$ observed by photoemission and X-ray absorption. *Physica C* 160:567–70
367. Himpsel F, Chandrashekhar G, McLean A, Shafer M. 1988. Orientation of the O 2p holes in $Bi_2Sr_2Ca_1Cu_2O_8$. *Phys. Rev. B* 38:11946–48
368. Nücker N, Romberg H, Xi X, Fink J, Gegenheimer B, Zhao Z. 1989. Symmetry of holes in high-Tc superconductors. *Phys. Rev. B* 39:6619–29
369. Kuiper P, Grioni M, Sawatzky GA, Mitzi D, Kapitulnik A, et al. 1989. Symmetry of the oxygen hole states in $Bi_2Sr_2CaCu_2O_8$ investigated by XAS. *Physica C* 157:260–62
370. Imhoff D, Lauren S, Colliex C, Backhaus-Ricoult M. 1999. Determination of the characteristic interfacial electronic states of {111} Cu-MgO interfaces by ELNES. *Eur. Phys. J. AP* 5: 9–18
371. Muller D, Subramanian S, Batson P, Sass S, Silcox J. 1998. Near atomic scale studies of electronic structure at grain boundaries in Ni_3Al. *Phys. Rev. Lett.* 80:4741–44
372. Dehm G, Scheu C, Möbus G, Brydson R, Rühle M. 1997. Synthesis of analytical and high-resolution transmission electron microscopy to determine the interface structure of Cu/Al_2O_3. *Ultramicroscopy* 67:207–17
373. Scheu C, Dehm G, Rühle M, Brydson R. 1998. Electron-energy-loss spectroscopy studies of Cu-alpha-Al_2O_3 interfaces grown by molecular beam epitaxy. *Philos. Mag. A* 78:439–65
374. Scheu C, Stein W, Rühle M. 2000. Electron energy-loss near-edge structure studies of a Cu/(11–20)-Al_2O_3 interface. *Phys. Status Solidi B* 222:199–211
375. Bruley J, Brydson R, Müllejans H, Mayer J, Gutekunst G, et al. 1994. Investigations of the chemistry and bonding at niobium-sapphire interfaces. *J. Mater. Res.* 9:2574–83
376. Scheu C, Dehm G, Müllejans H, Brydson R, Rühle M. 1995. Electron-energy-loss near-edge structure of metal-alumina interfaces. *Microsc. Microanal. Microstruct.* 6:19–31
377. van Benthem K, Scheu C, Sigle W, Rühle M. 2002. Electronic structure investigations of Ni and Cr films on (100)$SrTiO_3$ substrates using electron energy-loss spectroscopy. *Z. Metall.* 93: 362–71
378. Zhang Z, Sigle W, Kurtz W, Rühle M. 2002. Electronic and atomic structure of a dissociated dislocation in $SrTiO_3$. *Phys. Rev. B* 66:214112–17
379. Zhang Z, Sigle W, Rühle M. 2002. Atomic and electronic characterization of the a[100] dislocation core in $SrTiO_3$. *Phys. Rev. B* 66:094108–18
380. Arslan I, Browning N. 2002. Intrinsic electronic structure of threading dislocations in GaN. *Phys. Rev. B* 65:075310–11
381. Xin Y, James E, Arslan I, Sivananthan S, Browning N, et al. 2000. Direct experimental observation of the local electronic structure at threading dislocations in metalorganic vapor phase epitaxy grown wurtzite GaN thin films. *Appl. Phys. Lett.* 76:466–8
382. Nufer S, Marinopoulos A, Gemming T, Elsässer C, Kurtz W, et al. 2001. Quantitative atomic-scale analysis of interface structures: transmission electron microscopy and local density functional theory. *Phys. Rev. Lett.* 86:5066–69
383. Walther T, Daneu N, Recnik A. 2004. A new method to measure small amounts of solute atoms on planar defects and application to inversion domain boundaries in doped zinc oxide. *Interface Sc.* 12:267–75
383a. Watanabe M, Williams DB. 2003. Quantification of elemental segregation to lath and grain boundaries in low-alloy steel by STEM X-ray mapping combined with the ζ-factor method. *Z. Metall.* 94:307–16

384. Kooi B, Wouters O, de Hosson J. 2001. Analysis of Gibbsian segregation at heterophase interfaces using analytical transmission electron microscopy: a novel approach. *Acta Mater.* 50:223–35

385. Browning N, Arslan I, Ito Y, James E, Klie R, et al. 2001. Application of atomic scale STEM techniques to the study of interfaces and defects in materials. *J. Electron Microsc.* 50:205–18

BIOCATALYSTS AND ENZYME MIMICS

It is widely acknowledged that biocatalysts and enzymes have many advantages but also some major limitations. In a recent, authoritative text, Bommarius & Riebel (32) tabulated these as follows (Table 1).

The key point to remember about enzymes is that their supremely good selectivity has to be weighed against their lack of stability. There are other general considerations. First, notwithstanding Klibanov's work (33), where it was shown that many enzymes can function perfectly satisfactorily in nonaqueous solvents, they are usually far better in their general performance in aqueous media. Second, some reactions cannot be catalyzed by enzymes: This is particularly true of those that play a central role in industrial manufacture. Third, enzymes generally require the involvement of expensive cofactors (such as NAD(P)H). Ideally, it would be better to have just one essential catalytic agent [and this can certainly be achieved in some of the enzymes bred by Arnold et al. (3–6), as well as by our own (10–12, 30) and others' inorganic enzyme mimics (13)]. The requirement for a cofactor is quite limiting because it precludes the use of enzymes in vitro without addition of a coupled, cofactor-regenerating system (34).

Enzymes by Darwinian Evolution

Quite apart from the subtle (and comprehensive) changes that may be made to the amino acid sequence in more or less any protein by the process of site-selected mutagenesis (35), there are other ways of developing powerful enzyme catalysts, and perhaps the most important of these is that of Darwinian evolution (3–9). In effect, the ethos behind this approach is based on the fact that, since one does not really know what sequence of amino acids will confer the required catalytic activity and selectivity, it is best, instead, to breed the proteins (by forced evolution) so as to achieve novel and desirable functions. In the words of Arnold (36)...

> This approach circumvents our profound ignorance of how the amino acid sequence encodes protein function and exploits the ability of biological systems to evolve and innovate.... Structure-guided recombination of homologous proteins generates libraries of diverse sequences, a large fraction of which retain the parent fold.

TABLE 1 Advantages and disadvantages of biocatalysts and enzymes

Advantages	Disadvantages
Very high enantioselectivity	Often low specific activity
Very high regioselectivity	Instability at extremes of temperature and pH
Transformation under mild conditions	Availability for selected reactions only
Solvent often water	Long development times for new enzymes

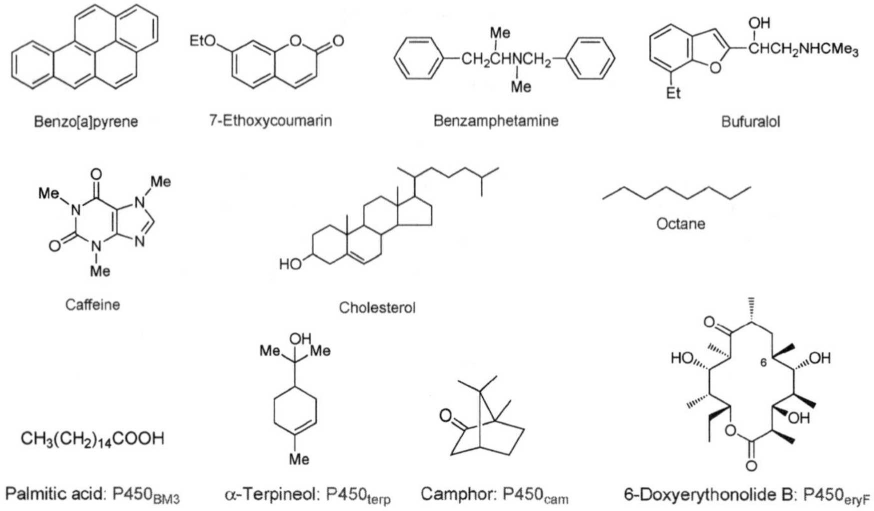

Figure 2 A selection of hydrocarbons that may be oxidized by cytochrome P450 enzymes (9).

Arnold and her coworkers have used this approach to generate thousands of properly folded cytochrome P450s, which differ from the parent sequence by up to 100 amino acid substitutions. These laboratory-grown enzymes possess many useful characteristics in the pursuit of clean technology, as we outline below. A measure of the vast scope that exists for breeding or otherwise fine-tuning the structures (and hence the catalytic performances) of cytochrome P450s is reflected by the wide range of compounds that are known to be processed by such enzymes. Figure 2, taken from the work of Wong (9), illustrates the point, and Figure 3 is the summary

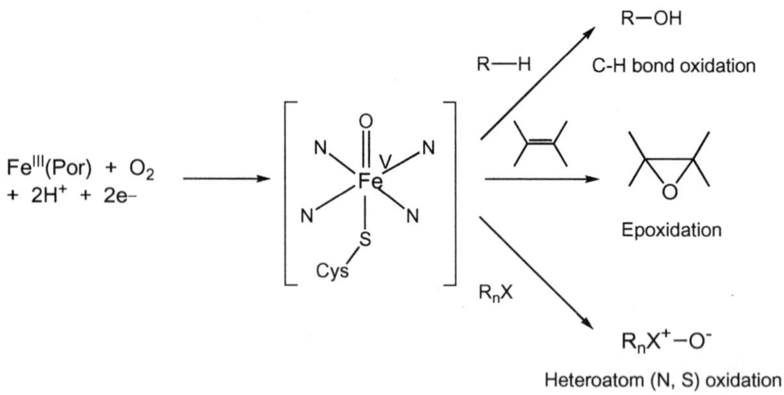

Figure 3 Summary of the overall action of the O_2-utilizing P450 enzyme (9).

TABLE 2 Biocatalysts by directed Darwinian evolution (3–8)

Wild-type enzyme (e.g., lipase from *Pseudomonas aeruginosa* or P450 from *Pseudomonas putida*)
Gene that encodes the enzyme is first subjected to random mutagenesis: a 'library' of mutant genes is created in a test tube.
The library is then inserted into a suitable microorganism and expressed as a library of mutant enzymes (screened for their activity, even enantioselectivity).
Inferior mutants are discarded. Mutant genes of the enzyme exhibiting optimal activity (enantioselectivity) are then subjected to further cycles of mutagenesis, expression and screening.
In each round, ~2000–5000 mutant enzymes are created. Therefore high-throughput screening (e.g., some 1000 to 10,000 *ee* values to be determined per day) is required.

of the overall action of the enzyme in utilizing O_2, two protons and two electrons, to create the highly reactive catalytically active center.

To explain the essence of Darwinian breeding of enzymes, we cite the work of Reetz et al. (7, 8) on lipases, which they evolved so as to effect enantioselective conversions. The essence of the Darwinian method is outlined in Table 2.

The starting point of Reetz's biocatalysis is a wild-type enzyme that catalyzes a reaction of interest, but not enantioselectively (37). An example is the lipase from the bacterium *Pseudomonas aeruginosa* that catalyzes the hydrolysis of esters. The gene that encodes the enzyme is first subjected to random mutagenesis to create a library of mutant genes in a test tube. The library is then inserted into a suitable microorganism and expressed as a library of mutant enzymes, which are screened for enantioselectivity. Inferior mutants are discarded, and the mutant gene of the enzyme exhibiting optimal enantioselectivity is then subjected to further cycles of mutagenesis, expression, and screening.

As about 2000 to 5000 mutant enzymes are created, it is necessary to employ high-throughput screening for the developed enantioselectivity (Reetz has developed NMR and mass-spectroscopic techniques to enable them to determine up to 10,000 enantiomeric excess (ee) values per day). As a particular example, Reetz et al. (37) have dramatically enhanced the enantioselectivity of a lipase in the hydrolytic kinetic resolution of a chiral ester; they improved the selectivity factor from 1.1 to more than 51. In addition to catalyzing the hydrolysis mentioned above, Reetz et al. (37) produced, by directed Darwinian evolution, an enantioselective cyclohexanone monooxygenase for the Baeyer-Villiger (see below) oxidation of 4-hydroxycyclohexanone to form two lactone enantiomers.

Some Selected Examples of the Value of Darwinian Evolution in the Benign Oxidation of Hydrocarbons

All the examples cited in this section are taken from the work of Arnold et al. (3–6, 36).

1. First, we recall the essence of their laboratory evolution of peroxide-mediated cytochrome P450 hydroxylation (4). By directed evolution of the P450 from *Pseudomonas putida* to create mutants that hydroxylate naphthalene in the absence of cofactors, they achieved more than a 20-fold higher activity than the native enzyme. (It is believed that this system should allow the development of mono- and di-oxygenases into practically useful biocatalysts for the hydroxylation of a wide range of aromatic compounds.)
2. The Darwinian evolution of a soluble, self-sufficient alkane hydroxylase was effected by Glieder et al. (38), who converted cytochrome P450 BM-3 from *Bacillus megaterium*, a medium-chain (C_{12} to C_{18}) fatty acid monooxygenase, into a highly efficient catalyst for the conversion of alkanes to alcohols. The turnover rates (see Figure 4) are higher than any reported biocatalyst for the selective oxidation of hydrocarbons of small-to medium-chain length (C_3 to C_8). Unlike naturally occurring alkane hydroxylases, the best known of which are large complexes of methane monooxygenase and membrane-associated nonheme iron alkane monooxygenases, the Darwinian-evolved enzyme is monomeric, soluble, and requires no additional proteins (cofactors) for catalysis. The evolved alkane hydroxylase was found to be even

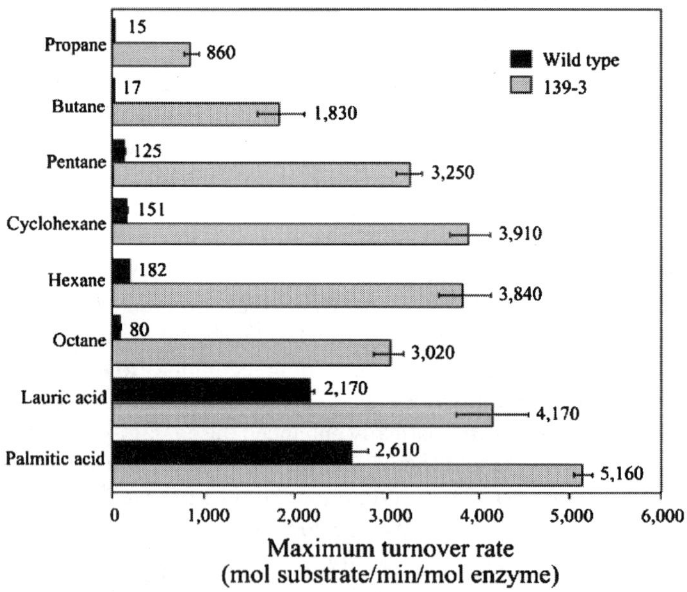

Figure 4 Maximum turnover rates (mol substrate/min/mol enzyme) for P450 BM-3 wild-type and 139-3. Substrate concentrations for maximum initial rates of 139-3 (*shaded bars*) and wild-type (*black bars*) are as follows: propane (saturated solution), butane (saturated solution), pentane (2.5 and 2.5 mM, respectively), hexane (2.5 and 2.5 mM), cyclohexane (2.5 and 2.5 mM), octane (2.5 and 5.0 mM), lauric acid (0.5 and 1.0 mM), palmitic acid (0.5 and 0.5 mM).

more active on fatty acids than wild-type BM-3, which was already one of the most efficient fatty acid monooxygenases known. A broad range of reactants (substrates), including the gaseous alkane propane induces the low- to high-spin shift that activates the enzyme.

This bio-catalyst for alkane hydroxylation at room temperature opens up new opportunities for clean, selective hydrocarbon activation for chemical synthesis (and manufacture) as well as bioremediation. The soluble P450 BM-3 enzyme not only catalyzes the hydroxylation of subterminal C–H bonds of medium-length chains (C_{12} to C_{18}) of fatty acids and amides (with initial turnover frequencies of 3000 m^{-1}), it also can epoxidize unsaturated hydrocarbons at high rates (5).

Alkene Epoxidation by Cytochrome P450 BM-3 139-3

Arnold et al. (6) have shown that the 139-3 mutant of the cytochrome P450 BM-3, developed by Darwinian evolution, is active in the selective oxidation of styrene, cyclohexene, 1-hexene and propylene (Figure 5). It also converts benzene to

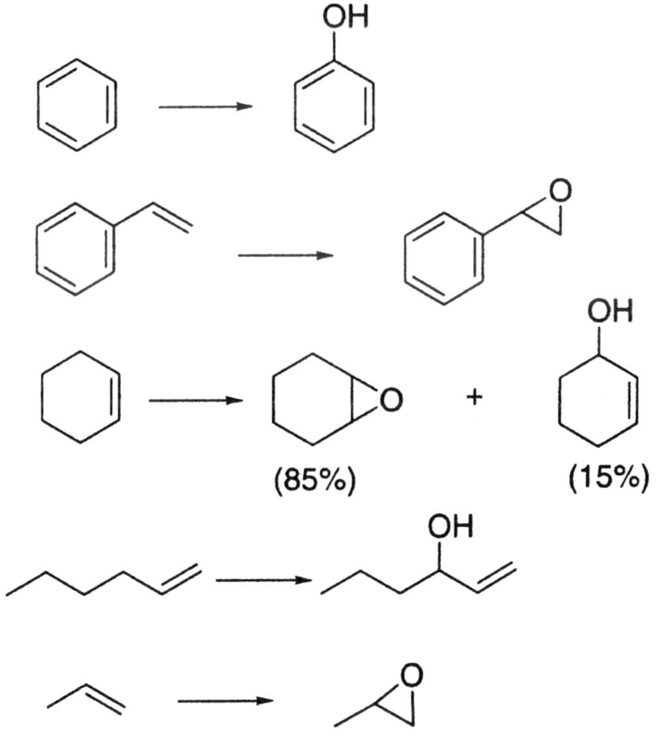

Figure 5 Substrates oxidized by BM-3 variant 139-3 and the corresponding products. For benzene, styrene, cyclohexene, or 1-hexene oxidations, the reaction mixture contained enzyme, substrate, and methanol in potassium phosphate buffer. For propylene oxidation, the buffer was first saturated with propylene.

phenol. Its far greater activity than that of the wild-type P450 BM-3 makes it a convenient platform for further directed evolution to optimize epoxidation of specific organic reactants.

Wong and his group in Oxford (9) are also engineering P450 enzymes so as to selectively oxidize a wide variety of non-natural reactants (substrates). Already this group has arrived at a new P450 that oxidizes butane and propane at rates comparable to that of camphor, and they, like Arnold et al. (3–6), have succeeded in converting ethane to ethanol. One of the great goals of this work is to produce methanol from methane.

Solid Inorganic Enzyme Mimics (Zeozymes)

Metalloenzymes having a metal active site owe their high selectivities in chemical reactions primarily to the protein tertiary structure. This structure wraps around the active site and performs several crucial functions such as protecting the active site from deleterious reactions via bimolecular pathways, sieving the substrate molecules so that only those capable of passing through the protein channels can gain access to the embedded prosthetic group, and providing a stereochemically demanding void space, in the vicinity of the active site, where the substrate resides during the catalytic reaction.

The open framework of microporous and other zeolitic solids has a number of structural similarities strikingly similar to the protein mantle of natural enzymes. They have pores, cavities, and channels of molecular dimensions that select only those molecules with the proper size and shape to pass through the pore. They also orient those molecules (based on their physical dimensions) that gain access to the internal voids of the crystallite, where the active site is located. By taking advantage of these similarities, new catalysts that combine the essential features of the robust, chemically inert molecular sieve with the high activity and selectivity of natural enzymes may be developed. Moreover, as much is now known about ways of designing the molecular dimensions (39), degree of hydrophobicity (40), and other features of open-structure inorganic catalysts, genuine inorganic mimics of certain enzymes may be prepared. In addition, a range of highly refined techniques has been evolved to probe, under operating conditions (41), the behavior of reactant species and the active sites inside the cavities of open-structure inorganic solids.

Herron (42), Raja & Ratnasamy (10–12), Jacobs et al. (13), Bein et al. (43) and Bäckvall et al. (44) have designed zeolite-encapsulated metal complexes of porphyrins, phthalocyanines, salens, salophens, bipyridines, and trimethyl-triazocyclononane (tmtacn) for a number of oxidation reactions. Such single-site-encapsulated catalysts allow access of only those reactant molecules small enough to diffuse through the windows of the zeolitic cage and prevent any of those products, which, if formed, are too large, from leaving the sphere of reaction. More importantly, the encapsulation protects the active site entity from losing its intrinsic catalytic performance because of dimerization or other complicating side-reactions.

These catalysts, frequently termed zeozymes or enzyme mimics for obvious reasons, have proved effective in oxyfunctionalizing the terminal methyl groups of linear alkanes (2), for converting methane to methanol (11) and propane to isopropanol (45), for epoxidizing alkenes with H_2O_2 (46) with high yields and selectivities, and for the aerobic oxidation of alcohols. Apart from highlighting the usual advantages of being able to remove and reuse the ship-in-bottle catalyst by simple filtration, these workers showed that these single-site heterogeneous catalysts had a higher specific rate than their homogeneous analogues.

Specific Examples of Enzyme Mimics

TYROSINASE MIMIC Copper-containing monooxygenase enzymes, such as tyrosinase (EC 1.14.18.1), reversibly bind O_2 and catalyze two different reactions—the hydroxylation of monophenols to ortho-diphenols (monophenolase activity) and the oxidation of the ortho-diphenols to ortho-quinones (diphenolase activity)—using molecular oxygen as the oxidant. The active site contains a pair of antiferromagnetically coupled copper ions. Raja & Ratnasamy (47) have shown that dimeric copper-acetate complexes, encapsulated within the microcavities of porous solids (zeolites X, Y and MCM-22), were effective in converting L-tyrosine to L-DOPA (see Scheme 1). Their catalysts (enzyme mimics) also displayed high

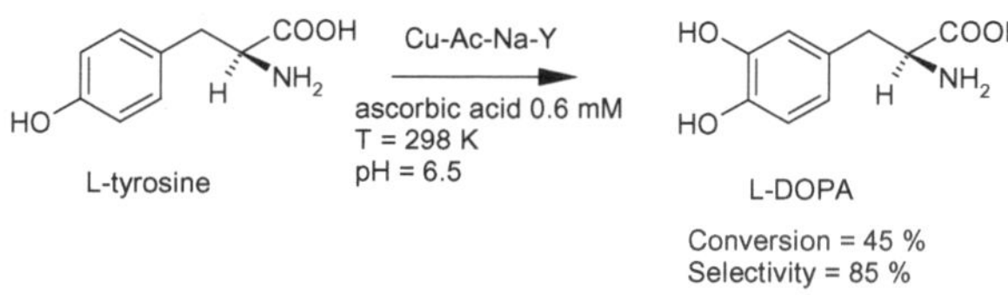

Scheme 1

substrate specificity (only monophenols were oxidized) and regioselectivity (hydroxylation always occurred at the position ortho to the –OH group). The turnover frequencies for the encapsulated catalysts are higher than that of their neat analogues, suggesting that these copper dimers are well-isolated in the molecular sieve matrix.

METHANE MONOOXYGENASE MIMIC Methane monooxygenases are a group of enzymes that hydroxylate methane to yield methanol, an extremely valuable commodity chemical (produced mainly on an industrial scale by steam-reforming of natural gas). Raja & Ratnasamy (11) have shown that ship-in-bottle, single-site heterogeneous catalysts, consisting of encapsulated complexes of Fe and Cu phthalocyanine (see Figure 6), in which all or most of the hydrogen atoms have been replaced by electron-withdrawing substituents (halogens or nitro groups), may convert methane in air at 273 K to methanol and formaldehyde as principal products (Scheme 2).

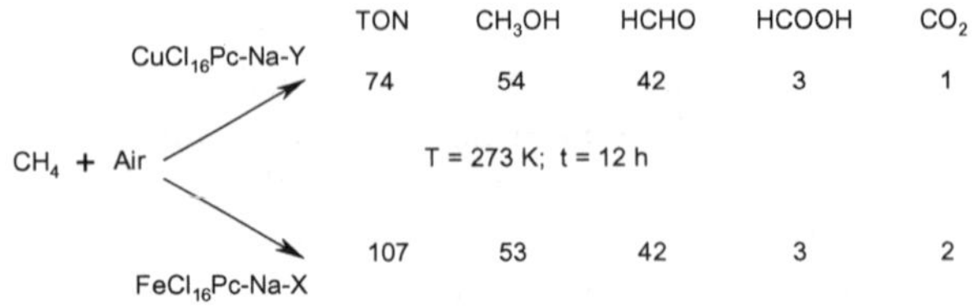

Scheme 2

HALOPEROXIDASE MIMIC Haloperoxidases, which occur widely in marine organisms, are enzymes that catalyze the oxidation of a halide (Cl^-, Br^- or I^-), by hydrogen peroxide, in such as manner as to achieve concomitant halogenation of certain organic substrates. $CuCl_{16}Pc$ complexes, encapsulated within the supercages of zeolites X and Y, are effective for the oxychlorination and oxybromination of a wide range of aromatic compounds, at ambient conditions, using molecular oxygen as the oxidant, in the presence of a suitable alkali halide (such as KBr) (see Table 3). The in situ formation of Br_2 from KBr (established by spectroscopic methods) and the pH dependence of the catalytic activity suggest that the halogenating agent is an electrophilic species (48).

HYDROXYLASES Herron & Tolman (42) made significant progress in designing a completely inorganic mimic of the alkane ω-hydroxylases for the regioselective hydroxylation of terminal methyl groups of linear alkanes by using encapsulated Pd(0) and Fe(II) entities inside zeolite A. This ship-in-bottle design succeeds in producing hydrogen peroxide from a O_2/H_2 mixture, but they were unable to selectively oxyfunctionalize only the terminal methyl group of *n*-octane with this

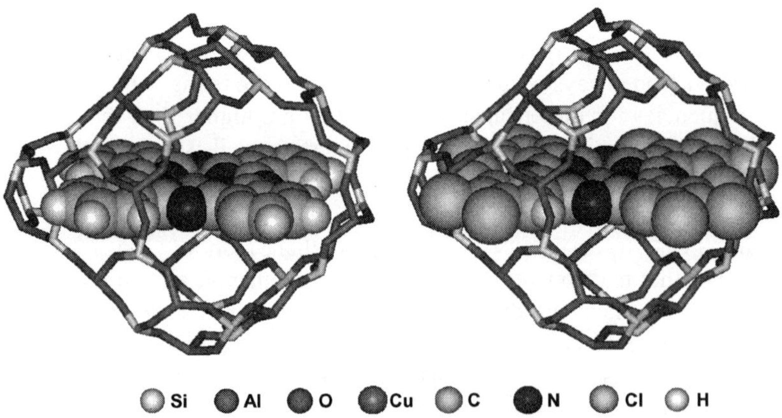

Figure 6 Computer graphic images of copper phthalocyanine (*left*) and copper chlorophthalocyanine (*right*) encapsulated within the supercages of zeolite X.

TABLE 3 Oxyhalogenation of aromatics in air using $CuCl_{16}Pc$-Na-X and KBr

Substrate	Halide	Temp. (K)	Conv. (%)	Halogenated products (wt.%) Mono-	Di-	Tri-	Side-chain oxidation
Benzene	KBr	338	6.2	100	—	—	—
Toluene	KBr	338	16.8	69.5	5.5	—	25.0
Phenol	KBr	323	18.1	62.0	25.0	13.0	—
Aniline	KBr	338	10.3	63.0	15.5	21.5	—
Anisole	KBr	338	6.7	25.5	55.0	19.5	—
Resorcinol	KBr	323	18.2	100	—	—	—

Reaction conditions: Catalyst ≈ 0.5 g; T ≈ 338 K; t ≈ 10 h; halide = KBr; Air ≈ 400 psig; solvent (CH_3CN: H_2O) ≈ 2:1 mol%; initiator ≈ 2 mol% TBHP (tert butyl hydroperoxide).

catalyst. Open-structure single-site heterogeneous catalysts (such as M^{III}AlPO-18 where M ≡ Co, or Mn) are, however, effective (49) in preferentially hydroxylating the terminal methyl groups of linear alkanes (C_nH_{2n+2}, n = 6–12) in air or oxygen (see Figure 7, see color insert).

EXAMPLES OF SHIP-IN-BOTTLE CATALYSTS FOR THE EPOXIDATION OF OLEFINS AND FOR THE OXIDATION OF ALCOHOLS Jacobs et al. (46) encapsulated *cis*-[Mn $(bpy)_2]^{2+}$ complexes (bpy, bipyridine) within the micropores of zeolite Y and used the resulting catalyst for the epoxidation of alkenes with H_2O_2. This proceeded without complications of competing processes, such as the decomposition of the peroxide. High yields of the epoxides of hex-1-ene, cyclohexene, dodec-1-ene and cyclododecene were obtained. It is being argued that the H_2O_2 activates the Mn ions so as to produce a $Mn^{(IV)}=O$ function, which releases its oxygen in the critical act of addition across the double bond to form the epoxide. Bein et al. (43) also used a ship-in-bottle alkene epoxidation catalyst formed inside the cages of zeolite Y when extraframework Mn^{II} cations were complexed with a trimethyl-triazocyclononane ligand (tmtacn), the role of the methyl groups being chiefly to protect the amine against oxidation by H_2O_2. In situ electron spin resonance (ESR) measurements show that the catalytically active site is a Mn^{III}-Mn^{IV} dinuclear complex.

The great practical advantages of ship-in-bottle catalysts are well illustrated in the recent work of Bäckvall et al. (44) on ruthenium-catalyzed aerobic oxidation of alcohols {where encapsulated cobalt Salophen was the active, single-site catalyst: H_2 Salophen is 2.2′-[benzene-1,2-diil-bis(nitrilomethylidyne)]diphenol}. This catalyst, which exhibits shape-selectivity in the oxidation of secondary alcohols, had a higher specific rate than that of the homogeneous analogue, and the zeolite (bottle) serves as a water acceptor, thereby removing the necessity to introduce a separate one.

The Adipic Acid Story

In 1999, the worldwide annual production of adipic acid (AA) was 2.5 million metric tons, and since that time it has increased further. The main use of AA is in the preparation of Nylon 6,6. Indeed some 80% of AA consumption is for the manufacture of Nylon 6,6 fibers and resins. Large amounts are also converted to esters for use in plasticizers, lubricants, and in a variety of polyurethane resins. AA is also added to gelatins and jams as a acidulant and to other foods as a buffering or neutralizing agent. Other uses include adhesives, insecticides, tanning and drying agents, paper manufacture, as well as uses in the textile industry.

Currently, most commercial processes for the production of AA involve multiple steps starting from either cyclohexane (or cyclohexanol)—or from benzene or phenol that are first hydrogenated—and entail the nitric acid oxidation of cyclohexanol, cyclohexanone, or a mixture of the two known as KA oil (see Figure 8) (14).

The environmentally undesirable features of this multi-step mode of manufacture are

- the generation of voluminous quantities of N_2O (a greenhouse gas that also depletes the ozone layer),
- the massive consumption of concentrated nitric acid (which is used as a stoichiometric oxidant), and
- the several disadvantages associated with many of the multiple steps (use of Cr^{III} salts and their problematic disposal).

Other industrial methods of producing AA, used by various multinational companies, entail the use of butadiene (see Figure 9). For example, BASF used a carboxymethoxylation of the latter with CO and methanol.

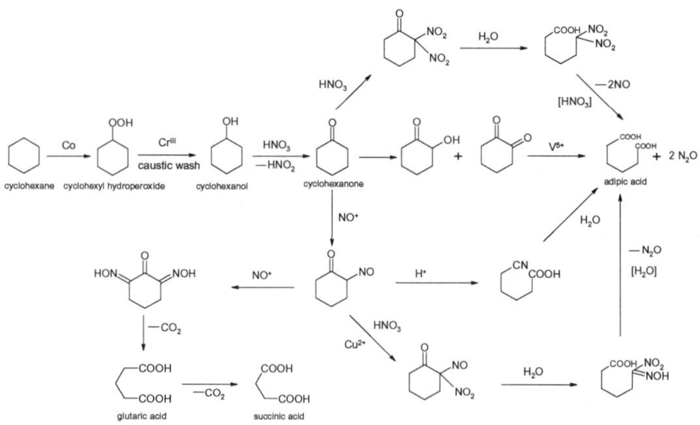

Figure 8 The production of KA oil (cyclohexanol + cyclohexanone) and adipic acid from cyclohexane.

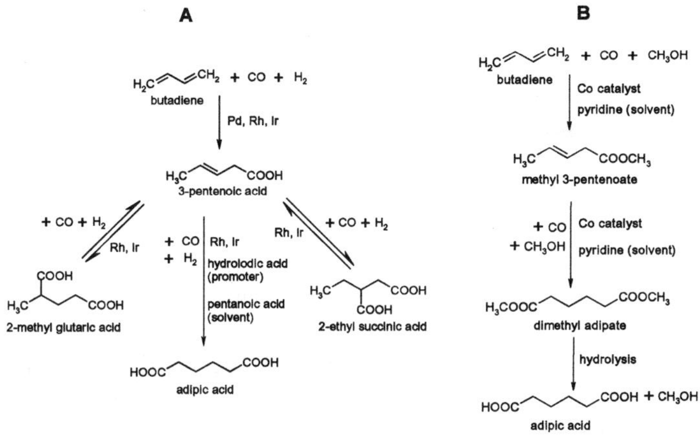

Figure 9 Production of adipic acid from butadiene by (*A*) dihydrocarboxylation and (*B*) carbomethoxylation of butadiene.

New, Clean, and Green Methods of Producing Adipic Acid

In 1998, Sato et al. (50) published a green route to AA that involved the direct oxidation of cyclohexene with 30% H_2O_2. The reaction pathway for this conversion is shown in Figure 10. We and our colleagues (51) used a similar procedure by replacing the phase-transfer catalyst employed by the Japanese workers with a Ti^{IV}-substituted microporous aluminophosphate (so-called TAPO-5). The detailed

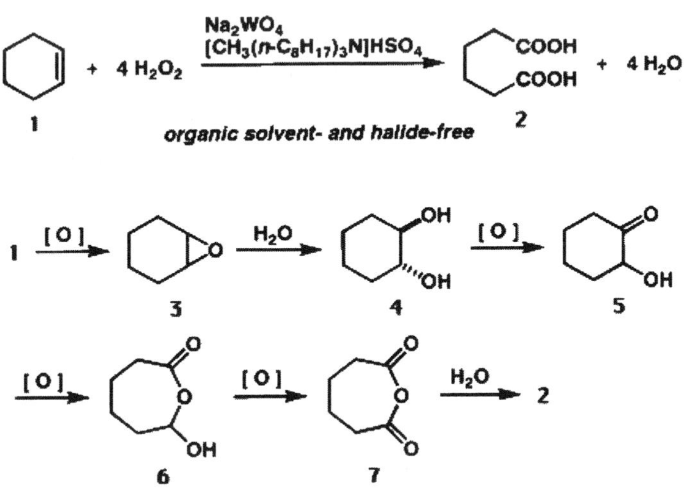

Figure 10 The reaction pathway for the direct oxidation of cyclohexene to adipic acid with aqueous H_2O_2 (50).

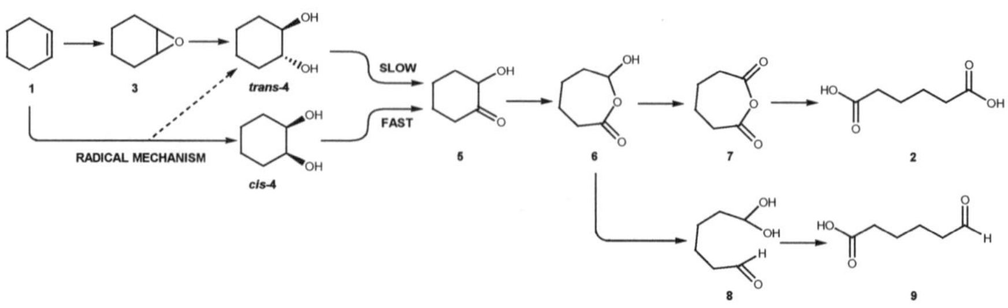

Figure 11 Mechanistic pathway for the oxidation of cyclohexene to adipic acid (51).

course of the overall reaction, determined by a combination of multinuclear NMR, chromatographic, and mass spectrometric measurements, is shown in Figure 11.

Earlier, we had devised a benign method of producing AA from cyclohexane using an Fe^{III}AlPO-31 catalyst, as shown in Figure 12 (52). Here, using dry air only as the oxidant, the FeAlPO-31 shape-selective catalyst converts the reactant with 65% selectivity to AA at 100 to 130°C.

We have also devised appropriate catalysts to produce AA in other benign ways. First, using *n*-hexane as reactant (53), AA is generated again with air as an oxidant in the constricted space separating two catalytically active (redox) centers inside Co^{III}AlPO-18 and Co^{III}AlPO-34 (see Figure 13). Second, we produced (Figure 14) AA by selective catalytic hydrogenation of muconic acid (54), an intermediate that may be generated biocatalytically from D-glucose [as demonstrated by Frost (55)].

The key significance of the second of these methods is that it obviates the need for a petroleum (or fossil) source for the reactant. [For these and possibly other reasons the editor of *Science* entitled his piece on our work "Renewable Nylons"] (56).

Solvent-Free Production of ε-Caprolactam by Generating Hydroxylamine In Situ using a Bifunctional Catalyst

The conversion of cyclohexanone (1) to the oxime (2) and its subsequent Beckmann rearrangement to ε-caprolactam (3) are vital stepping stones in the manufacture of Nylon 6 (Scheme 3). On an industrial scale, one popular procedure in converting

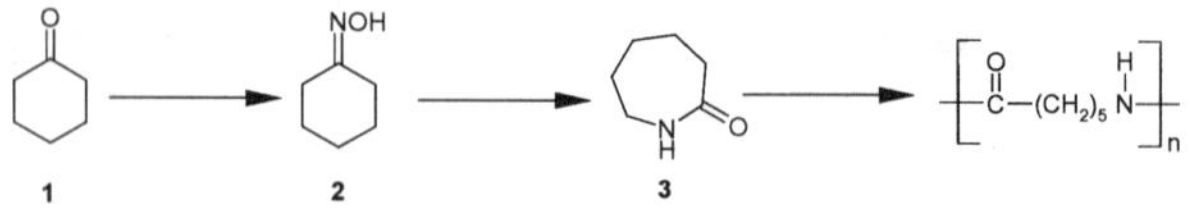

Scheme 3

(1) to (2) is to employ hydroxylamine sulfate; the sulfuric acid is thus liberated being neutralized by ammonia, with the consequential production of large quantities of (low value) ammonium sulfate. The traditional industrial route for effecting the Beckmann rearrangement of (2) to (3) is by use of a strong mineral acid such as oleum (see Scheme 4).

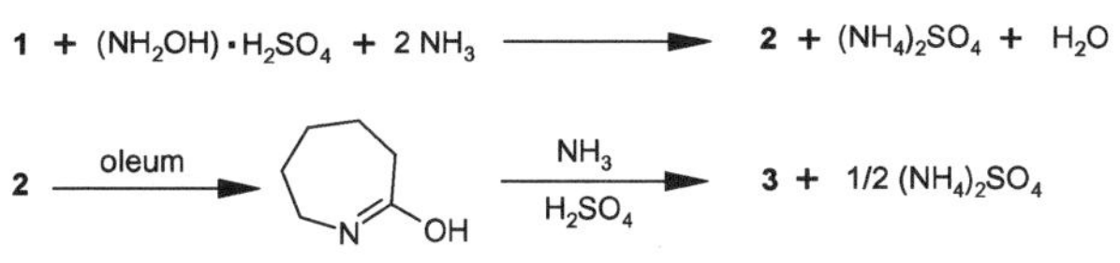

Scheme 4

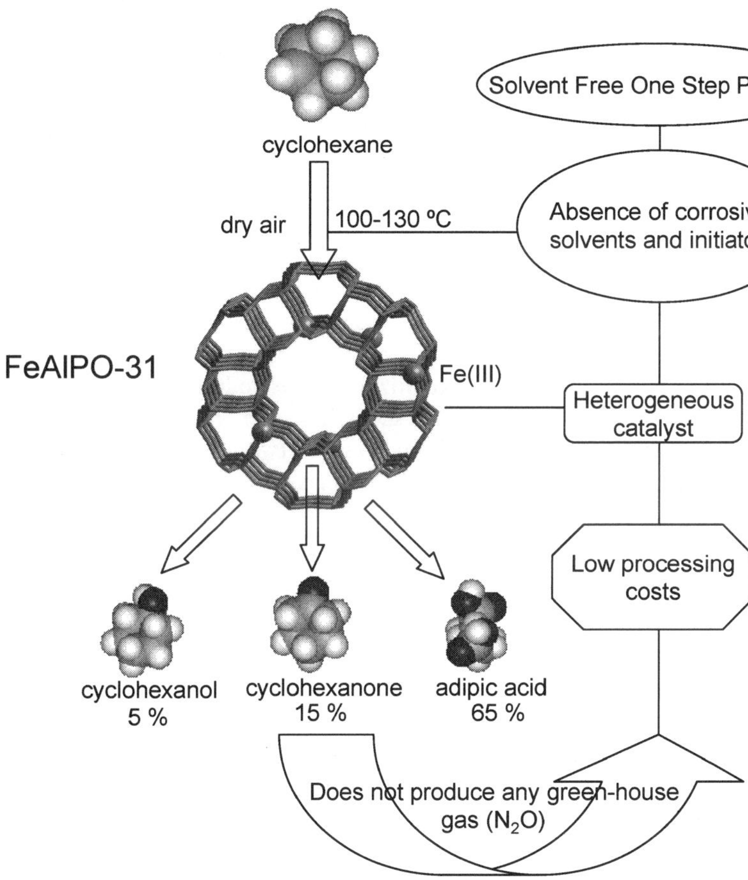

Figure 12 Adipic acid is at present produced in a costly multi-step process involving homogeneous catalysts, aggressive oxidants such as concentrated HNO_3, and the liberation of large quantities of N_2O, a greenhouse gas (see Figure 8). An alternative, one-step, solvent-free process using air and an inexpensive microporous catalyst is shown.

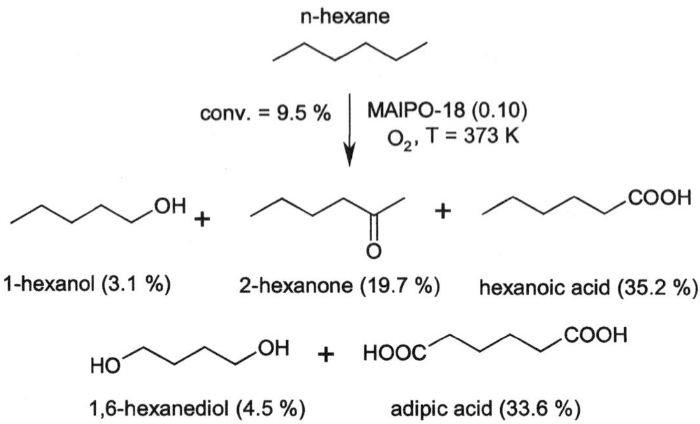

Figure 13 The direct conversion of *n*-hexane to adipic acid using CoAlPO-18 (0.10).

Workers at Enichem Co. showed that the titanosilicate molecular sieve TS-1 offered an attractive, more environmentally benign route of production, a particular advantage being the use of H_2O_2 as the oxidant (57) in aqueous solution. Other hydrogen peroxide-based processes employ organic solvents such as methanol, ethanol, or isopropanol, and a stabilizer such as sodium EDTA so as to generate 1,1′-dihydroxydicyclohexyl peroxide. Earlier, Armor (58) demonstrated a direct (gas phase) route to (2) and (3) from (1) using two distinct (double-bed) catalysts consisting of a variety of silicas and aluminas in the range 120–250°C (58).

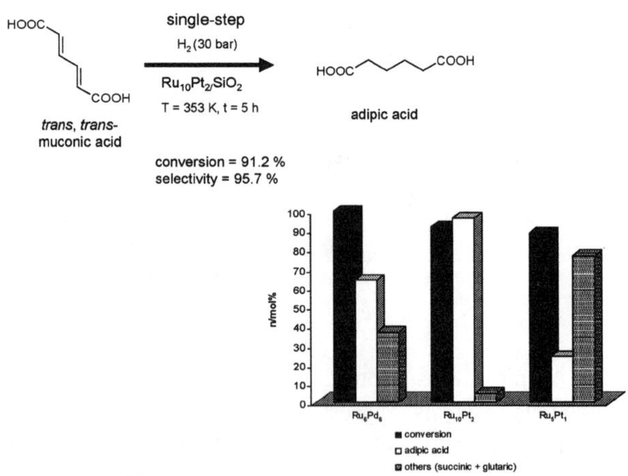

Figure 14 The hydrogenation of *trans,trans*-muconic acid to adipic acid using a $Ru_{10}Pt_2$ nanoparticle catalyst.

Our own program, which aims at designing new catalysts for clean technology, green chemistry, and sustainable development, has led us to assemble bifunctional transition-metal-ion (framework)-substituted aluminophosphate molecular sieve catalysts for converting (1) to (2) and (3) in a one-step, solvent-free manner, in the liquid phase, using oxygen (air) and ammonia. Full details have been given in (1) and a patent (59). In essence, the catalysts are all based on the formula [($M^{II}M^{III}$)AlPO-36, M ≡ Co, Mn]. They are structurally well-defined, possessing pore apertures of 6.5 × 7.5 Å, and a surface area (overwhelmingly internal) of ~700 $m^2\ g^{-1}$. X-ray absorption spectroscopy has established that of the 4 at% of the framework Al^{III} isomorphously replaced by M ions, approximately 50% are in the M^{II} and 50% in the M^{III} state (Figure 15, see color insert). M^{II} ions, because they have protons loosely bound to an adjacent framework oxygen atom, are the loci of Brönsted acid active sites. The M^{III} framework ions, on the other hand, are redox active sites, capable of activating hydrocarbons and oxygen. The pore dimensions of MAlPO-36 are just large enough to permit ingress of any of the molecules (1), (2), or (3). In ($M^{II}M^{III}$)AlPO-18, M ≡ Co, which we have also studied (see below) for the purpose of elucidating the nature of the catalysis, all the Co ions are in the Co^{III} state, and the pore diameter is so small that only air, H_2O_2, and ammonia (or hydroxylamine when formed) may gain access to the interior surface of the sieve.

Our designed bifunctional catalysts, $M^{II}M^{III}$AlPO-36, perform well in consecutively converting (1) to (2) to (3) because (see below and Table 4):

TABLE 4 Solvent-free ammoximation of cyclohexanone: performance of various catalysts[a]

I.D.	Solid molecular sieve	Oxidant	t (h)	Conv. mol%	TON[c]	Product selectivity[b] (mol%) Oxime	ε-caprolactam	Others
A	$Co^{II}Co^{III}$AlPO-36	O_2	6	13.1	217	73.9	17.7	10.2
			20	20.2	322	54.2	21.1	24.5
B	$Mn^{II}Mn^{III}$AlPO-36	O_2	6	15.0	245	75.8	16.9	7.3
			20	23.6	381	57.0	20.2	22.8
C	$Mg^{II}Co^{III}$AlPO-36	O_2	6	12.4	108	44.0	36.2	19.5
			20	21.5	212	33.9	42.9	22.7
D	$Mg^{II}Mn^{III}$AlPO-36	O_2	6	13.7	136	48.4	31.3	19.5
			20	23.0	262	36.2	45.2	18.0

[a]Catalyst = 0.5 g; air = 3.5 MPa; cyclohexanone: NH_3 1:3 (mol); T = 328 K; cyclohexanone ≅ 50 g; mesitylene (internal standard) ≡ 2.5 g.

[b]Oxime ≡ cyclohexanone oxime (2); others = mainly high molecular weight conjugated products formed through aldol condensation of cyclohexanone plus some polycaprolactam and water.

[c]TON = mol of substrate converted per mol of metal (Co, Mn and/or Mg) in the catalyst.

1. hydroxylamine (NH_2OH) is readily formed in situ inside the pores from NH_3 and O_2 at the M^{III} active (redox) sites;
2. the NH_2OH converts (1) to (2) both inside and outside the pores, and, likewise, (2) is isomerized at the Brönsted active sites to (3) inside the pores of the molecular sieve catalyst; and
3. O_2 yields NH_2OH more efficiently than H_2O_2 or TBHP (tert butyl hydroperoxide) at the redox active sites.

This eminently clean method of making first the oxime of cyclohexanone and then its important isomerized product (and precursor for the manufacture of nylon) is deemed a highly attractive method of a potential manufacturing practice (59).

CATALYSTS FOR EPOXIDATION AND THE SCOPE FOR SUSTAINABLE PRODUCTION

Epoxidation entails addition of an oxygen atom across a carbon-carbon double bond in an unsaturated hydrocarbon (such as propylene). Epoxides are important building blocks for the fine chemical and pharmaceutical industries, and also for the production of glycols, which are widely used as antifreeze and other agents such as surfactants (60). For many of the industrially produced epoxides, the so-called chlorohydrin stoichiometric process (see Scheme 5) is used—a distinctly bad procedure in this environmentally conscious age.

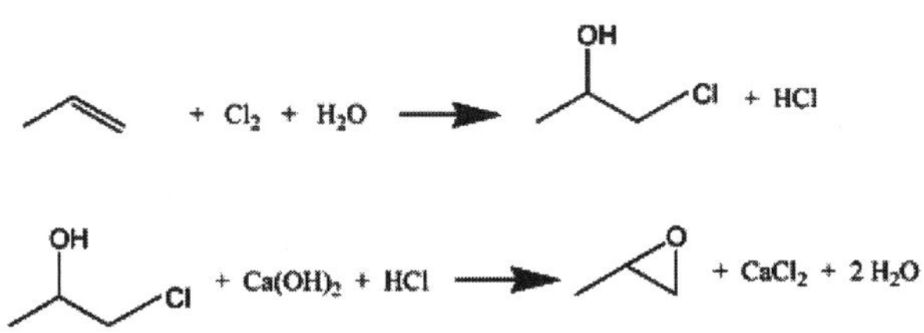

Scheme 5

On the laboratory scale, organic chemists tend to use iodosyl benzene (PhIO) as the selective oxidant, not an ideal choice. The commercial production of propylene oxide (61), carried out on a massive industrial scale by the Shell Co. and others, uses a Ti^{IV}-based heterogeneous catalyst involving silica supports and alkyl hydroperoxides (typically cumene or tert butyl hydroperoxide, CHBP and TBHP, respectively) as oxidants.

With H_2O_2-based catalytic epoxidation, the catalysts have been the open-structure titanosilicate TS-1 (63, 64), methyltrioxorhenium (62), tungsten-containing compounds, and the silicotungstate compound devised by Mizuno et al. (62): $[\gamma\text{-}SiW_{10}O_{34}(H_2O)_2]^{4-}$. This material is synthesized by protonation of a divacant lacunary, Keggin-type polyoxometalate of $[\gamma\text{-SiW}_{10}O_{36}]^{8-}$ and exhibits high catalytic performance for the epoxidation of various olefins, including propylene, with H_2O_2 as oxidant at 305 K (see Table 5). $[\gamma\text{-SiW}_{10}O_{34}(H_2O)_2]^4$ is a remarkably good catalyst in that it reaches 99% selectivity to the epoxide, an efficiency of greater than 99% in the utilization of H_2O_2, high stereospecificity, and easy recovery of the catalyst from the homogeneous reaction mixture.

TABLE 5 Epoxidation of various olefins catalyzed by $(Bu_4N)_4\bullet 1$* with H_2O_2

Entry	Substrate	Time (h)	Yield %	Product [selectivity (%)]	H_2O_2 efficiency (%)
1*		8	90	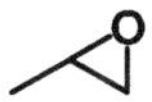 (>99)	>99
2*		8	88	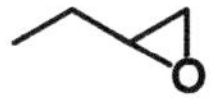 (99)	>99
3*		9	91	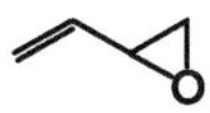 (99)	99
4		10	90	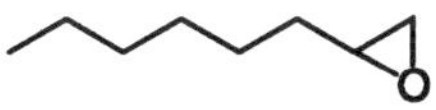 (99)	>99
5	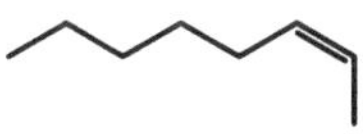	3	>99	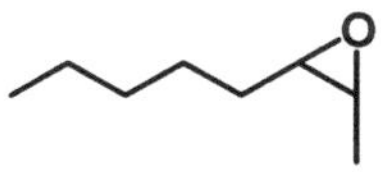 (>99)	>99
6		14	91	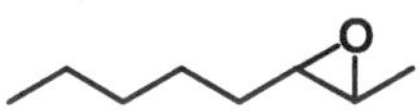 (99)	99
7		6	84	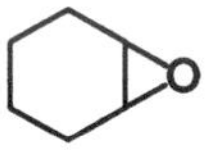 (>99)	>99
8	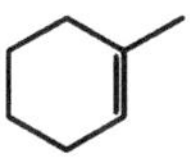	4	95	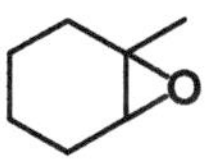 (>99)	99
9		4	>90	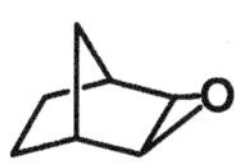 (99)	>99
10	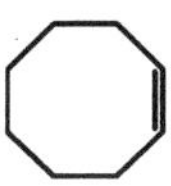	2	99	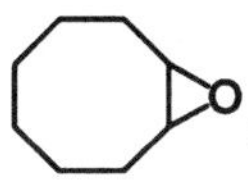 (>99)	>99
11†	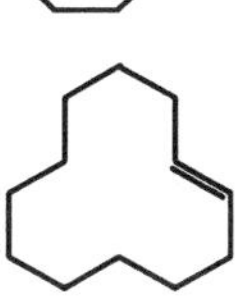	4	97	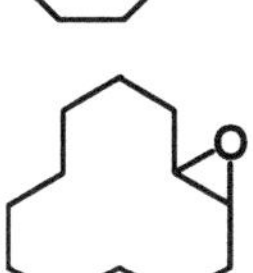 (>99)	>99

*Propylene (6 atm), 1-butane (3 atm), 1,3-butadiene (2.5 atm). †MeCN (9 mL).

TABLE 6 Liquid-phase epoxidation of olefins and terpenes using APB and a redox catalyst

					Product selectivity			
Substrate[a]	**Catalyst**	**Time (h)**	**Conv.**[b] **Mol%**	**Peroxide Eff. (%)**[c]	**Epoxide**	**Diol**	**C_6H_5CHO**	**Others**
Styrene	MnAlPO-5	1	31.7	95.2	100	—	—	—
		4	32.9	98.8	86.5	13.3	—	—
Styrene	FeAlPO-5	1	28.0	84.1	100	—	—	—
		4	33.0	99.1	87.3	12.5	—	—
Styrene	$CuCl_{16}$Pc-NaX	1	24.3	73.0	97.3	2.5	—	—
		4	29.8	89.5	95.5	4.4	—	—
α-pinene	MnAlPO-5	1	25.9	77.8	100	—	—	—
		4	29.5	88.6	68.7	—	—	31.2
	$CuCl_{16}$Pc-NaX	1	11.7	35.1	96.2	—	—	3.5
		4	21.5	64.6	91.3	—	—	8.5
(+)-limonene	MnAlPO-5	1	26.0	78.0	100	—	—	—
		4	31.2	93.7	61.3	—	—	38.5
	$CuCl_{16}$Pc-NaX	1	13.0	39.0	100	—	—	—
		4	22.6	67.9	78.4	—	—	21.5

[a]Styrene ≡ 2.8 g; α-pinene, (+)-limonene ≡ 3.7 g; MeAlPO-5 ≡ 0.25 g; $CuCl_{16}$Pc-NaX ≡ 0.5 g; T = 65° C; solid APB ≡ 3.49 g (shown by titration studies to liberate 0.701 g of PAA and 0.045 g of H_2O_2 immediately on dissolution in 20.5 g of double-distilled (D-d) water;

[b]Conv. % = [(mol of initial substrate—mol of residual substrate)/(mol of initial substrate)] × 100;

[c]Peroxide efficiencies calculated on the basis of a substrate:oxidant mol ratio of 3:1.

Recently, we and our colleagues have shown that a new solid oxidant, acetyl peroxy borate (R. Raja, J.M. Thomas, M. Xu, K.D.M. Harris, M. Greenhill-Hooper & K. Quill, manuscript submitted), in the dissolved state (and in the presence of MAlPO-5 catalysts, M ≡ Mn^{III}, Co^{III}) is extremely effective in epoxidation. Styrene, for example, is converted to styrene oxide with high efficiency under environmentally mild conditions (see Table 6).

We and others took early advantage of large-pore (mesoporous) silicas (such as those illustrated in Figure 16) that would enable us to prepare single-site, ion-centered heterogeneous catalysts. A significant advance was made when Ti^{IV} active centers were grafted on to the inner walls of mesoporous silica using an organometallic precursor, in particular titanocene dichloride [$Ti(Cp)_2Cl_2$], as was done by Maschmeyer et al. (65). The key steps in the introduction of the isolated, single-site active centers on the inner walls of the so-called MCM-41 silica are shown in Figure 17, see color insert. The detailed course of this heterogenization of a Ti^{IV} active center was followed by in situ X-ray absorption spectroscopy

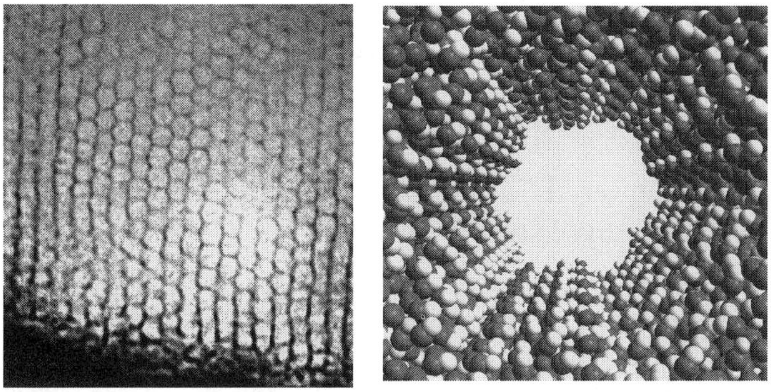

(a) (b)

Figure 16 *A high-resolution electron micrograph of an ordered mesoporous silica* where the diameter of the pores is $\simeq$100 Å and the walls are only a few atoms thick. Each pore is lined with pendant silanol ($\equiv$Si$-$OH) groups.

combined with in situ X-ray diffractometry (DFT) (66). DFT calculations yield important supplementary information. In particular they show that the titanyl group ($>$Ti $=$ O), once postulated (67) as the active site for epoxidation of alkenes, is energetically unfavorable.

A tripodally grafted Ti(OH) group constitutes the isolated active site, very resistant to leaching, in this instance. Note that the van der Waals envelopes (Figure 18, see color insert) of the cyclopentadienyl groups attached to the surface-bound ("half-sandwich") intermediate, shown in Figure 17, ensure that no two Ti^{IV} active centers are closer than $\simeq$7 Å, thereby allowing the generation of a genuine, isolated, single-site catalyst. (Other methods of introducing Ti^{IV} active centers involving salts of titanium or in-grown titanium alkoxide additives to the siliceous precursor do not guarantee production of isolated, tetrahedral Ti^{IV} centers and a tendency to form Ti$-$O$-$Ti linkages exists with these methods.) This single-site catalyst is an exceptionally good epoxidation catalyst that uses alkyl hydroperoxides (described below).

An altogether different, but equally effective approach has been pioneered by Tilley et al. (68, 69) in which a molecular precursor is taken to yield a series of active catalysts on mesoporous (and certain other) supports. The metal ions in question are those of Ti, Cr, Fe, and vanadyl; the essence of their preparation is that the desired atomic environment aimed at in the final catalyst (e.g., Ti$-$$(OSi)_4$ or Ti$-$$(OSi)_3$ is already present in the thermolytic precursor. Thus, by taking $(^iPrO)Ti[OSi(O^tBu)_3]_3$, the tris(*tert*-butoxy)siloxy titanium complex, the environment ultimately achieved in the single-site catalyst is Ti$-$$(OSi)_3$, and from $Ti[OSi(O^tBu)_3]_4$ it is Ti$-$$(OSi)_4$. Tilley et al. have shown that both isolated Ti- and Fe-centered active sites at mesoporous silica surfaces are good epoxidation catalysts.

Because the Ti-SiO_2 catalysts that Thomas et al. (70) have prepared are spatially isolated and structurally well defined (as determined by in situ XAFS measurements), they are ideally suited for computational (DFT) studies, which in turn shed light on the mechanism of epoxidation.

Grafted Ti^{IV}-Centered Catalysts for the Epoxidation of Fatty Acid Methyl Esters (71)

Epoxidized fatty acids and their derivatives have been used for many commercial applications such as plasticizers and stabilizers in chlorine-containing resins, as additives in lubricants, as components in thermosetting plastics, in urethane foams, and as wood impregnants. Vegetable oils and fats are renewable sources of two popular unsaturated fatty methyl esters: methyl-(Z)-9-octadecanoate (methyl oleate, structure 1 in Figure 19 and methyl-(E)-9-octadecanoate, methyl elaidate, structure 2). In the past, an environmentally unfriendly "peracid" method was used to epoxidize the naturally occurring unsaturated compounds.

Now, however, as Ravasio and her coworkers have shown (71), the Ti^{IV}-grafted active site on mesoporous silica via $Ti(Cp)_2(Cl)_2$ (65) is an excellent and environmentally friendly method for converting the fatty acid methyl esters (FAME) into their epoxides. These workers have recently shown that the Ti^{IV}-grafted catalyst also effectively converts the (doubly) unsaturated components of soya

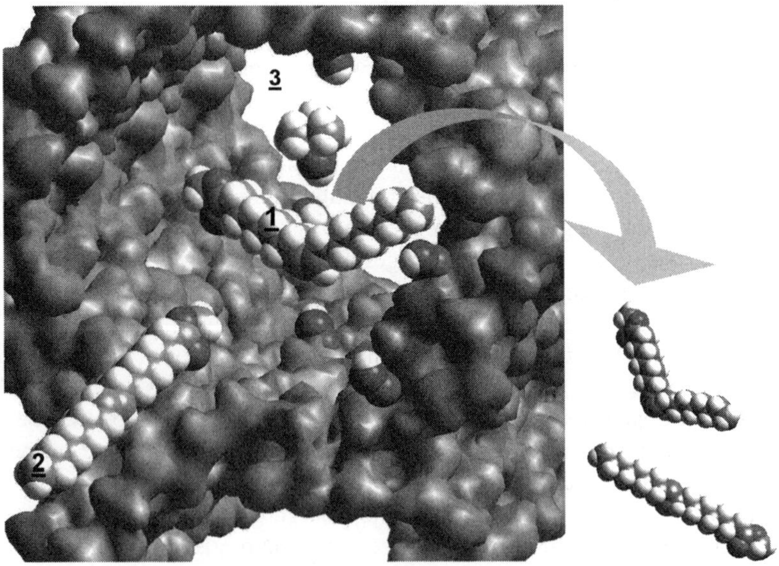

Figure 19 Both methyl oleate (1) and methyl elaidate (2) are completely epoxidized using single-site Ti^{IV}-mesoporous silica catalysts and tert butyl hydroperoxide (3) as oxidant.

TABLE 7 Epoxidation of terpenes using Ti^{IV} catalysts on silica under mild conditions

	TOF (h^{-1})	
Substrate	**Ti→**	**Ti↑**
α-terpineol	2	20
Carveol	15	33
Limonene	4	20

T = 85°C; CH_3CN (solvent); 30% wt catalyst; TBHP:terpene mol ratio = 1.

bean oil into useful epoxides—another important step toward sustainable development.

Grafted Ti^{IV}-Centered Catalysts for the Epoxidation of Terpenes

The major sources of terpenes (which are natural products, the structure of which is built up from isoprene units) are balsams, natural resins, and essential oils, but they are also by-products of lemon and orange juice production, as well as of the pulp and paper industries. Some terpenes, notably (−)-α-pinene and (+)-limonene are among the more readily (naturally) available optically active substances and are therefore used for the syntheses of other optically active products. Here again it is obvious that catalysts capable of efficiently functionalizing terpenes are of value in the context of sustainable development.

The work of Ravasio and her collaborators (72) has shown that the Ti↑MCM-41 (grafted) catalyst [derived from $Ti(Cp)_2(Cl)_2$] (65) is particularly good in epoxidizing such important terpenes as α-terpinol, carveol and limonene (see Table 7) under mild conditions [i.e., at ~85°C using tert butyl hydroperoxide (TBHP) in CH_3CN]. Indeed, in harmony with earlier work on the epoxidation of cyclohexene also using TBHP, the Ti↑MCM-41 surpasses the activity of the sol-gel grown Ti → MCM-41 by a factor of 10 in the case of the α-terpinol, the main constituent of pine oil.

Judging by the results of other workers who have compared the catalytic performance of Ti↑MCM-41 (i.e., the grafted variety) with that of the Ti → MCM-41 (sol-gel preparation), there is no doubt of the superiority of the former. Thus in their study of the hydroxylation of benzene in the liquid phase (using aqueous H_2O_2) He et al. (73) found both higher activity and enhanced selectivity to phenol (as well as greater chemical stability) with the grafted catalyst.

ACID-CATALYZED REACTIONS

Acid catalysts have been used on an industrial scale for several decades, but the need to develop new ones has recently intensified in order to minimize the disadvantages associated with the present extensive use of corrosive and environmentally dangerous liquids such as concentrated phosphoric, sulfuric, and hydrofluoric acids. Very strong acids—capable of catalyzing "demanding" reactions such as alkylations (e.g., of isobutene by butane) to yield octane fuels or the isomerization of C_5 to C_9 n-alkanes—such as the liquids mentioned above, are often supported on siliceous or clay minerals. But there is an urgent need to develop new solid catalysts that do not suffer from the environmental disadvantages and adverse corrosive properties of such liquids. There is also a need for new, intrinsically weaker, acidic catalysts that would be useful for the production of ethers and small hydrocarbons by the dehydration of alkanols under rather mild conditions. A brief discussion of approaches to these problems has been given elsewhere (74).

In all so-called pentasil zeolites (e.g., ZSM-5 and ZSM-11), where the Si/Al ratios are seldom less than 10, and in numerous other aluminosilicate and aluminophosphate open-structure solids (such as faujasite, ferrierite, erionite, and SAPO-34), a large variety of (Brönsted) acid-catalyzed reactions may be carried out (75). Table 8 enumerates some of these important reactions.

One extremely important acid-catalyzed reaction merits further comment. It is the dehydration of methanol to (predominantly) ethylene and propylene (with traces of the butanes). These are feedstocks for the polyolefin market, which is enormous. Polyethylene production is projected (76) to increase from 55×10^6 metric tons in 2002 to 87×10^6 metric tons in 2010, and polypropylene production is projected to increase from 35×10^6 to 60×10^6 metric tons in the same period.

One of us has been involved in the design of shape-selective solid acid catalysts that are extremely efficient in producing light olefins (ethylene and propylene mainly) in the dehydration of methanol. Two important examples are MAlPO-18 ($M \equiv Zn^{II}$, Mg^{II}, Mn^{II}, Co^{II}, ...) (77) and DAF-4 (Davy Faraday microporous solid No. 4) (78).

TABLE 8 Economically significant catalytic conversions effected by zeolitic solids

A	Hydrocarbons produced (often shape-selectively) from acid-catalyzed reactions
	Xylenes from isomerizations and disproportionation of toluene
	Methanol to light olefins (ethene and propene) and methanol to gasoline
	Ethyl benzene from benzene and ethene
	Oligomerization of olefins
	Cumene from benzene plus propene
	Detergent alkylate from benzene plus C_6 to C_{14} olefins
	Isomerization of methyl naphthalenes
	Vinylcyclohexene from butadiene by dimerization
B	Reactions involving functionalized hydrocarbons
	Ester hydrolysis
	Alkylation of phenol
	Glycol ethers by ethoxylation of lower aliphatic alcohols
	Primary amines from attack of ammonia on olefin, and secondary amines by elimination of ammonia from primary amine
	Alkyl carboxylates from carboxylic acids and olefins
	Nitration of chlorobenzene

If, in place of substituting Al^{III} ions by M^{II} ions, one replaces P^{V} with Si^{IV}, the resulting SAPO microporous solids are also good solid acid catalysts and are now in commercial use as the preferred means of producing the light olefins required for the polymer industry.

GENERATING POLYMERS AND SMALL MOLECULES FROM SUSTAINABLE SOURCES

Above and in Figure 14 we drew attention to the fact that adipic acid (a key component in many commercial polymers) may be produced by hydrogenating muconic acid, which, in turn, may be extracted from plant (botanical) sources using an adroit choice of biocatalyst. We also referred to the epoxidation of fatty acid methyl esters and terpenes, each of which may be extracted from renewable plants (sunflower and soya bean oil). An increasing number of such possibilities exist that pose new challenges for the catalyst merchant. In this regard, polyoxygenates such as polyesters and polycarbonates, derived from readily renewable resources such as lactide from corn and carbon dioxide, are attractive as environmentally friendly alternatives to polymers derived from fossil fuels. Very recently (79), Chisholm & Zhou produced an important summary of the role likely to be played in the near future by metal-oxygen bonds to alkoxide ligands as catalysts for ring-opening polymerizations (ROP), ring-opening copolymerizations, and transesterifications (see Figure 20).

These workers, among others, draw attention to the joint venture of Cargill-Dow in the production of the lactide monomer as a bulk commodity material (16). Ring-opening polymerizations of lactide yield polylactide, a polyester that can be used for packaging and control release as a biodegradable and biocompatible material. The use of CO_2 to replace phosgene or its equivalent in the production of polycarbonates is clearly desirable. We now proceed to highlight a few issues

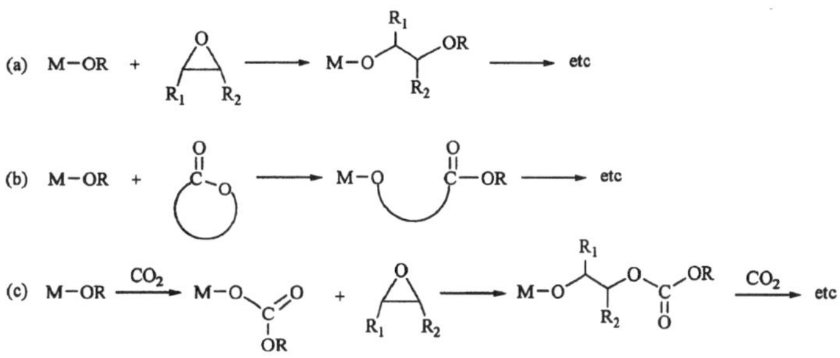

Figure 20 Fundamental reactions involving metal–oxygen (alkoxide) bonds in (*a*) the ring opening of epoxides and (*b*) the ring opening of cyclic esters, and (*c*) for the alternating copolymerization of an epoxide and CO_2.

(79) discussed in the article by Chisholm & Zhou (which should be consulted for fuller information).

Polylactides and Related Polyesters

When lactic acid is heated, water is given off and a gummy polymer/oligomer, polylactide, is formed. If this polymer/oligomer is heated in the presence of zinc dust or zinc oxide at ~150°C, lactide (LA), the cyclic ester, can be distilled or sublimed from the molten polymer. Lactide comes in the three forms as shown in Scheme 6: *L* (or *S,S*)-lactide, *D* (or *R,R*)-lactide, and *meso* (or *R,S*)-lactide,

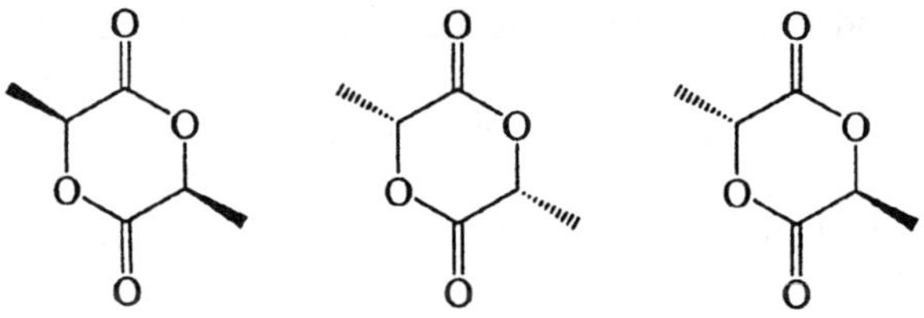

Scheme 6

and the 50:50 mixture of *L* and *D* is known as *rac*-lactide. Closely related to lactide is a family of cyclic esters that include glycolide (GL), methylglycolide, and capro-, valero-, and butyro-lactones. With the exception of γ-butyrolactone, all these cyclic esters and β-butyrolactone are subject to ring-opening polymerization under various conditions.

The presence of the two stereocenters in lactide affords many interesting possibilities for the microstructure of polylactides (PLAs). The polymer microstructure of PLA has been investigated by NMR studies following the pioneering work of Kricheldorf (80).

A reaction pathway for the ring opening of LA is shown in Scheme 7. Coordination of the LA substrate to the metal center is clearly an important step.

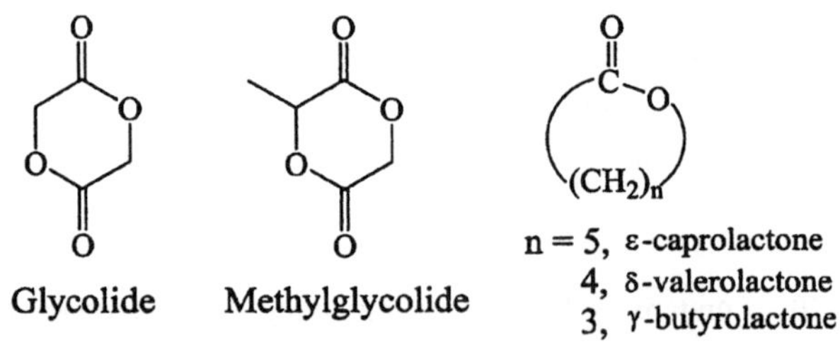

Scheme 7

This brings the ketonic carbon and the nucleophilic alkoxide group into close proximity, and furthermore activates the ketonic carbon toward nucleophilic attack. With this model in mind it is clearly evident that a polar M—OR bond, $M^{\delta+}$ and $RO^{\delta-}$, will be more active than a more covalent and less polar one. Furthermore, steric factors associated with the OR group, the metal center, and the ring itself will be important. A similar mechanism is operating for lactones which for $n = 5$ more readily undergo ROP than lactide. At least one theoretical study modeling the ring opening of lactide by Me_3SnOMe has already appeared in the literature (82).

Comparison of ROP of Lactides and Epoxides

Well-defined metal complexes that are highly active in polymerizing lactides and related cyclic esters by a single-site mechanism have thus far not proved to be similarly effective in the polymerization of epoxides. There is an abundance of literature to support the view that the ring opening of epoxides by nucleophiles occurs by backside attack and thus lends itself to a bimetallic mechanism (see Scheme 8). There is, however, now unequivocal evidence from Chisholm's work

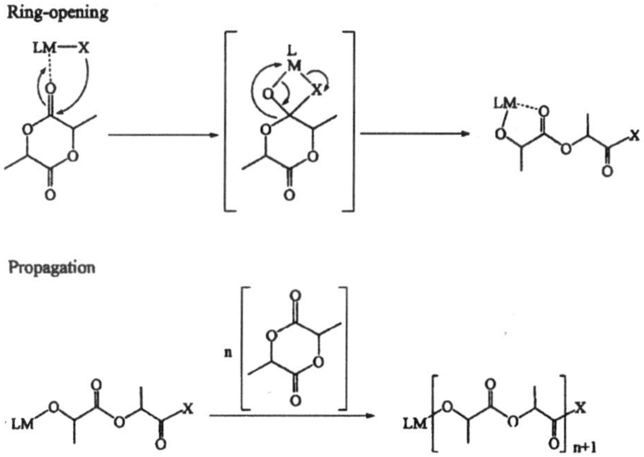

Scheme 8

and that of Darensbourg and others that poly(propylene oxide) and poly(propylene carbonate) production can occur at a single metal center (81). Moreover, there has been kinetic evidence that implicates a bimetallic mechanism in the ROP of lactones by porphyrin aluminum alkoxide initiators, but these are kinetically slow reactions. Such are the complexities of catalysis by metal complexes that we may expect to see one mechanism operate for one system and yet another for a different metal complex.

Because of the different parameters required for the polymerization of lactides/lactones and epoxides, no single-site catalysts are known that will lead to alternating copolymers, although block copolymers can be made by sequential

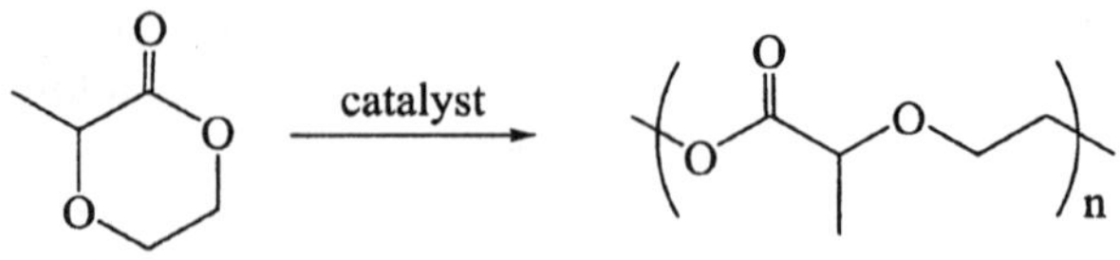

Scheme 9

ROP (83). A clever route to the regular alternating polymer of ethylene oxide and a half-lactide unit was reported by the Minnesota group who used ROP of the cyclic ether—ester as shown in Scheme 9.

The Dupont Story of 1,3-Propanediol

DuPont's engineered *Escherichia coli* strain efficiently converts corn-derived glucose into 1,3-propanediol (3G), a monomer, which on combination with terephthalic acid, produces polypropylene terephthalate (3GT) (see Scheme 10). The

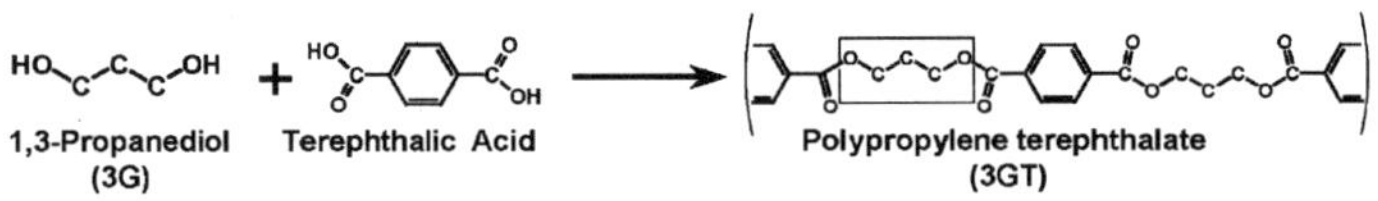

Scheme 10

traditional chemical routes that have been employed for the production of 1,3-propanediol are shown in Scheme 11. In nature, the conversion of corn sugar

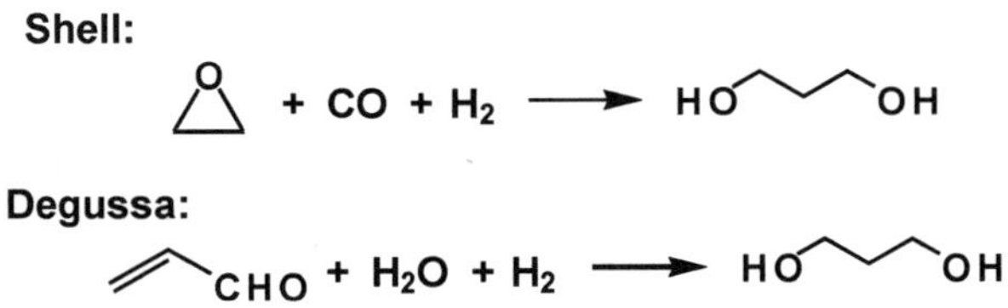

Scheme 11

(glucose) to glycerol and its further conversion to 1,3-propanediol is carried out by two separate microorganisms. DuPont have invented (83) a single bacterium (microorganism) to catalyze both steps by a fermentation process. Polypropylene terephthalate is used mainly in the manufacture of fibers for clothing. It is an advanced polymer/fiber, which is known for its vibrant color, easy care, and softness. In addition, it also possesses superior stretch and recovery properties and is UV, chlorine, and stain resistant.

Developments in the Catalytic Conversions of Carbohydrates, Fatty Compounds, Glycerol, and Terpenes

The special issue of *Topics in Catalysis*, edited by van Bekkum & Gallezot and devoted entirely to the "Catalytic Conversion of Renewables" (17) is a valuable source of information. Its relevance to the present survey prompts us to enumerate some of the key subjects discussed in this source. They are

- Gallezot et al.'s work on the conversion of native starch to hydrophilic and hydrophobic products (84)
- van Bekkum et al.'s use of TEMPO (2,2,6,6,-tetramethyl piperidine-1-oxyl), a so-called nitroxyl radical, as an agent in the oxidation of polysaccharides (85)
- Moreau et al.'s work on substituted furans from carbohydrates (86)
- Schoevaart & Kieboom's application of galactose oxidase in combined bio- and chemocatalytic transformations of polysaccharides (87)
- Jacob et al.'s esterification studies of fatty compounds (88)
- Bondioli's method of preparation of fatty acid esters by catalytic means (89)
- Perez-Pariente et al.'s use of solid catalysts for the synthesis of fatty acid esters of glycerol and sorbitol (90)
- Mol's work on catalytic metathesis of unsaturated fatty acid esters and oils (91)
- Carrettin et al.'s use of supported gold catalysts for the oxidation of glycerol and propan-1,2-diol (92)
- Barrault et al.'s development of the use of agricultural products for non-food applications, particularly in selective oligomerization of glycerol using cerium-containing mesoporous solids (93)
- Ravasio et al.'s isomerization studies of α-pinene and limonene oxides to carbonyl compounds and the selective epoxidation of terpenic alcohols using mono- and bifunctional heterogeneous catalysts (71, 72)
- Monteiro & Veloso's review on the transformation of terpenes, from renewable resources, for the production of pharmaceuticals, agrochemicals, flavors and fragrances (94), and
- Swifts's work on some key catalytic transformations (hydrogenation/hydrogenolysis, isomerization, aromatization, dehydrogenation, epoxidation, etc.) of the major monoterpene feedstocks (95).

The *Annual Review of Materials Research* is online at http://matsci.annualreviews.org

LITERATURE CITED

1. Raja R, Sankar G, Thomas JM. 2001. Bifunctional molecular sieve catalysts for the benign ammoximation of cyclohexanone: one-step, solvent-free production of oxime and epsilon-caprolactam with a mixture of air and ammonia. *J. Am. Chem. Soc.* 123:8153–54
2. Thomas JM, Raja R, Sankar G, Bell RG. 1999. Molecular-sieve catalysts for the selective oxidation of linear alkanes by molecular oxygen. *Nature* 398:227–30
3. Arnold FH. 1998. Design by directed evolution. *Acc. Chem. Res.* 31:125–31
4. Joo H, Lin ZL, Arnold FH. 1999. Laboratory evolution of peroxide-mediated cytochrome P450 hydroxylation. *Nature* 399:670–73
5. Cirino PC, Arnold FH. 2003. A self-sufficient peroxide-driven hydroxylation biocatalyst. *Angew. Chem. Int. Ed. Engl.* 42:3299–301
6. Farinas ET, Alcalde M, Arnold F. 2004. Alkene epoxidation catalyzed by cytochrome P450BM-3 139–3. *Tetrahedron* 60:525–28
7. Jaeger KE, Reetz MT. 1998. Microbial lipases form versatile tools for biotechnology. *Trends Biotechnol.* 16:396–403
8. Jaeger KE, Reetz MT. 2000. Directed evolution of enantioselective enzymes for organic chemistry. *Curr. Opin. Chem. Biol.* 4:68–73
9. Fleming BD, Tian Y, Bell SG, Wong LL, Urlacher V, Hill HAO. 2003. Redox properties of cytochrome P450(BM3) measured by direct methods. *Eur. J. Biochem.* 270:4082–88
10. Raja R, Ratnasamy P. 1996. Selective oxidation with copper complexes incorporated in molecular sieves. *Stud. Surf. Sci. Catal.* 101:181–90
11. Raja R, Ratnasamy P. 1997. Direct conversion of methane to methanol. *Appl. Catal. A* 158:L7–15
12. Ratnasamy P, Raja R, Srinivas D. 2005. Oxidation and hydroxylation of aromatic compounds over solid catalysts. *Philos. Trans. R. Soc. London Ser. A* 363: In press
13. Parton RF, Vankelecom IFJ, Casselman MJA, Bezoukhanova CP, Uytterhoeven JB, Jacobs PA. 1994. An efficient mimic of cytochrome-p-450 from a zeolite encaged iron complex in a polymer membrane. *Nature* 370:541–44
14. Thomas JM, Raja R, Sankar G, Johnson BFG, Lewis DW. 2001. Solvent-free routes to clean technology. *Chem. Eur. J.* 7:2973–78
15. Thomas JM, Raja R, Sankar G, Bell RG, Lewis DW. 2001. Benign by design. New catalysts for an environmentally conscious age. *Pure Appl. Chem.* 73:1087–101
16. Drumright RE, Gruber PR, Henton DE. 2000. Polylactic acid technology. *Adv. Mater.* 12:1841–46
17. van Bekkum H, Gallezot P, eds. 2004. *Catalytic Conversion of Renewables*. Vol. 27
18. Road map for Biomass Technologies in the United States, Biomass R&D Technical Advisory Committee, Dec. 2002. http://www.bioproducts-bioenergy.gov
19. Thomas JM, Raja R, Lewis DW. 2005. Single-site heterogeneous catalysts. *Angew. Chem. Int. Ed. Engl.* In press
20. Kresge CT, Leonowicz ME, Roth WJ, Vartuli JC, Beck JS. 1992. Ordered mesoporous molecular-sieves synthesized by a liquid-crystal template mechanism. *Nature* 359:710–12
21. Inagaki S, Fukushima Y, Kuroda K. 1993. Synthesis of highly ordered mesoporous materials from a layered polysilicate.

J. Chem. Soc. Chem. Commun. 680–82
22. Raja R, Thomas JM, Jones MD, Johnson BFG, Vaughan DEW. 2003. Constraining asymmetric organometallic catalysts within mesoporous supports boosts their enantioselectivity. *J. Am. Chem. Soc.* 125:14982–83
23. Johnson BFG, Raynor SA, Shephard DS, Mashmeyer T, Thomas JM, et al. 1999. Superior performance of a chiral catalyst confined within mesoporous silica. *Chem. Commun.* 1925–26
24. Jones MD, Raja R, Thomas JM, Johnson BFG. 2003. A new approach to the design of heterogeneous single-site enantioselective catalysts. *Top. Catal.* 25:71–79
25. Rouzaud J, Jones MD, Raja R, Johnson BFG, Thomas JM, Duer MJ. 2003. Potent new heterogeneous asymmetric catalysts. *Helv. Chim. Acta* 86:1753–59
26. Thomas JM, Raja R, Johnson BFG, Hermans S, Jones MD, Khimyak T. 2003. Bimetallic catalysts and their relevance to the hydrogen economy. *Ind. Eng. Chem. Res.* 42:1563–70
27. Thomas JM, Johnson BFG, Raja R, Sankar G, Midgley PA. 2003. High-performance nanocatalysts for single-step hydrogenations. *Acc. Chem. Res.* 36:20–30
28. Thomas JM, Raja R. 2001. Catalytically active centres in porous oxides: design and performance of highly selective new catalysts. *Chem. Commun.* 675–87
29. Thomas JM. 1999. Design, synthesis, and in situ characterization of new solid catalysts. *Angew. Chem. Int. Ed. Engl.* 38: 3589–628
30. Thomas JM, Raja R. 2004. Catalytic significance of organometallic compounds immobilized on mesoporous silica: economically and environmentally important examples. *J. Organomet. Chem.* 689: 4110–24
31. Anastas PT, Williamson TC. 1998. Frontiers in green chemistry. *Green. Chem.* 1–26
32. Bommarius AS, Riebel BR. 2004. In *Biocatalysis*. pp. 1–18. Weinheim: Wiley-VCH
33. Klibanov AM. 1990. Asymmetric transformations catalyzed by enzymes in organic-solvents. *Acc. Chem. Res.* 23: 114–20
34. Koeller KM, Wong CH. 2001. Enzymes for chemical synthesis. *Nature* 409:232–40
35. Winter G, Fersht AR, Wilkinson AJ, Zoller M, Smith M. 1982. Redesigning enzyme structure by site-directed mutagenesis—Tyrosyl transfer-RNA synthetase and ATP binding. *Nature* 299: 756–58
36. Arnold FH. 2004. *The Rosalind Franklin Lecture*. Presented at Cambridge, UK
37. Reetz MT. 2003. *Evolutionary method creates enantioselective enzymes*. Presented at Achema 2003, Frankfurt, Germany
38. Glieder A, Farinas ET, Arnold FH. 2002. Laboratory evolution of a soluble, self-sufficient, highly active alkane hydroxylase. *Nat. Biotechnol.* 20:1135–39
39. Cheetham AK, Ferey G, Loiseau T. 1999. Open-framework inorganic materials. *Angew. Chem. Int. Ed. Engl.* 38:3269–92
40. Ray S, Vasudevan S. 2003. Encapsulation of cobalt phthalocyanine in zeolite-Y: evidence for nonplanar geometry. *Inorg. Chem.* 42:1711–19
41. Thomas JM, Greaves GN. 1994. Probing solid catalyst under operating-conditions. *Science* 265:1675–76
42. Herron N, Tolman CA. 1987. A highly selective zeolite catalyst for hydrocarbon oxidation—a completely inorganic mimic of the alkane omega-hydroxylases. *J. Am. Chem. Soc.* 109:2837–39
43. DeVos DE, Meinershagen JL, Bein T. 1996. Highly selective epoxidation catalysts derived from intrazeolite trimethyltriazacyclononane-manganese complexes. *Angew. Chem. Int. Ed. Engl.* 35: 2211–13

44. Zsigmond A, Notheisz F, Csjernyik G, Bäckvall JE. 2002. Ruthenium-catalyzed aerobic oxidation of alcohols on zeolite-encapsulated cobalt salophen catalyst. *Top. Catal.* 19:119–24
45. Raja R, Jacob CR, Ratnasamy P. 1999. Direct oxidation of propane to isopropanol. *Catal. Today* 49:171–75
46. KnopsGerrits PP, Toufar H, Jacobs PA. 1997. Catalytic tuning of the olefin epoxidation with hydrogen peroxide and Faujasite Y occluded Mn bipyridine complexes. *Stud. Surf. Sci. Catal.* 105:1109–16
47. Raja R, Ratnasamy P. 1995. Activation of dioxygen by copper complexes incorporated in molecular sieves. *J. Mol. Catal.* 100:93–102
48. Raja R, Ratnasamy P. 1997. Oxyhalogenation of aromatics over copper phthalocyanines encapsulated in zeolites. *J. Catal.* 170:244–53
49. Raja R, Thomas JM. 1998. A manganese-containing molecular sieve catalyst designed for the terminal oxidation of dodecane in air. *Chem. Commun.* 1841–42
50. Sato K, Aoki M, Noyori R. 1998. A "green" route to adipic acid: direct oxidation of cyclohexenes with 30 percent hydrogen peroxide. *Science* 281:1646–47
51. Lee SO, Raja R, Harris KDM, Thomas JM, Johnson BFG, Sankar G. 2003. Mechanistic insights into the conversion of cyclohexene to adipic acid by H_2O_2 in the presence of a TAPO-5 catalyst. *Angew. Chem. Int. Ed. Engl.* 42:1520–23
52. Dugal M, Sankar G, Raja R, Thomas JM. 2000. Designing a heterogeneous catalyst for the production of adipic acid by aerial oxidation of cyclohexane. *Angew. Chem. Int. Ed. Engl.* 39:2310–13
53. Raja R, Sankar G, Thomas JM. 2000. Designing a molecular sieve catalyst for the aerial oxidation of n-hexane to adipic acid. *Angew. Chem. Int. Ed. Engl.* 39:2313–16
54. Thomas JM, Raja R, Johnson BFG, O'Connell TJ, Sankar G, Khimyak T. 2003. Bimetallic nanocatalysts for the conversion of muconic acid to adipic acid. *Chem. Commun.* 1126–27
55. Draths KM, Frost JW. 1994. Environmentally compatible synthesis of adipic acid from D-glucose. *J. Am. Chem. Soc.* 116: 399–400
56. Fahrenkamp-Uppenbrink J. 2003. Renewable nylons. *Science* 300:867
57. Taramasso M, Perego G, Notari B. 1983. *U.S. Patent No. 4410501*
58. Armor JN. 1981. Ammoximation, a direct route to cyclohexanone oxime and caprolactam from NH_3, O_2, and cyclohexanone. *J. Catal.* 70:72–83
59. Oevering H, Raja R, Wolters H, Buijs W, Thomas JM. 2003. *E.P. Patent Application. No. 03078997*
60. Hill CL. 1988. *Advances in Oxygenated Processes*, ed. AL Baumstark, London: JAI
61. Murugavel R, Roesky HW. 1997. Titanosilicates: recent developments in synthesis and use as oxidation catalysts. *Angew. Chem. Int. Ed. Engl.* 36:477–79
62. Kamata K, Yonehara K, Sumida Y, Yamaguchi K, Hikichi S, Mizuno N. 2003. Efficient epoxidation of olefins with $>=$ 99% selectivity and use of hydrogen peroxide. *Science* 300:964–66
63. Notari B. 1996. Microporous crystalline titanium silicates. *Adv. Catal.* 41:253–334
64. Ratnasamy P, Srinivas D, Knözinger. 2004. Active sites and reactive intermediates in titanium silicate molecular sieves. *Adv. Catal.* 48:1–169
65. Maschmeyer T, Rey F, Sankar G, Thomas JM. 1995. Heterogeneous catalysts obtained by grafting metallocene complexes onto mesoporous silica. *Nature* 378:159–62
66. Couves JW, Thomas JM, Waller D, Jones RH, Dent AJ, et al. 1991. Tracing the conversion of aurichalcite to a copper catalyst by combined X-ray absorption and diffraction. *Nature* 354:465–68
67. Sheldon RA. 1983. New catalytic

methods for selective oxidation. *J. Mol. Catal.* 20:1–26
68. Jarupatrakorn J, Tilley JD. 2002. Silica-supported, single-site titanium catalysts for olefin epoxidation. A molecular precursor strategy for control of catalyst structure. *J. Am. Chem. Soc.* 124:8380–88
69. Fujdala KL, Tilley TD. 2003. Design and synthesis of heterogeneous catalysts: the thermolytic molecular precursor approach. *J. Catal.* 216:265–75
70. Thomas JM, Catlow CRA, Sankar G. 2002. Determining the structure of active sites, transition states and intermediates in heterogeneously catalysed reactions. *Chem. Commun.* 2921–25
71. Guidotti M, Ravasio N, Psaro R, Gianotti E, Marchese L, Coluccia S. 2003. Heterogeneous catalytic epoxidation of fatty acid methyl esters on titanium-grafted silicas. *Green Chem.* 5:421–24
72. Ravasio N, Zaccheria F, Guidotti M, Psaro R. 2004. Mono- and bifunctional heterogeneous catalytic transformation of terpenes and terpenoids. *Top. Catal.* 27:157–68
73. He J, Ma H, Guo ZY, Evans DG, Duan X. 2003. Chemical stability of Ti-modified MCM-41 catalysts in the hydroxylation of benzene in the liquid phase. *Top. Catal.* 22:41–51
74. Thomas JM, Lewis DW. 1996. Towards rational design of solid acid catalysts. *Z. Phys. Chem.* 197:37–48
75. Haag WO, Lago RM, Weisz PB. 1984. The active-site of acidic aluminosilicate catalysts. *Nature* 309:589–91
76. Tullo A. 2003. Industry eyes growth. *Chem. Eng. News* 81:18
77. Chen JS, Thomas JM. 1994. MAPO-18 (M-Mg, Zn, Co)—a new family of catalysts for the conversion of methanol to light olefins. *J. Chem. Soc. Chem. Commun.* 603–4
78. Barrett PA, Jones RH. 1995. The pore-size modification of a layered aluminophosphate $[Al_3P_4O_{16}C_9H_{24}N_3]$ by rational selection of the intercalated template cation. *J. Chem. Soc. Chem. Commun.* 1979–81
79. Chisholm MH, Zhou ZP. 2004. New generation polymers: the role of metal alkoxides as catalysts in the production of polyoxygenates. *J. Mater. Chem.* 14:3081–92
79a. Kricheldorf HR, Boettcher C, Tonnes KU. 1992. Polylactones. 23. Polymerization of racemic and meso D,L-lactide with various organotin catalysts stereochemical aspects. *Polymer* 33:2817–24
80. von Schenck H, Ryner M, Albertsson AC, Svensson M. 2002. Ring-opening polymerization of lactones and lactides with Sn(IV) and Al(III) initiators. *Macromolecules* 35:1556–62
81. Darensbourg DJ, Yarbrough JC. 2002. Mechanistic aspects of the copolymerization reaction of carbon dioxide and epoxides, using a chiral salen chromium chloride catalyst. *J. Am. Chem. Soc.* 124: 6335–42
82. Bechtold K, Hillmyer MA, Tolman WB. 2001. Perfectly alternating copolymer of lactic acid and ethylene oxide as a plasticizing agent for polylactide. *Macromolecules* 34:8641–48
83. Laffend LM, Nagarajan V, Nakamura CE. *U.S. Patent No. 5686276*
84. Sorokin AB, Kachkarova-Sorokina SL, Donze C, Pinel C, Gallezot P. 2004. From native starch to hydrophilic and hydrophobic products: a catalytic approach. *Top. Catal.* 27:67–76
85. Bragd PL, van Bekkum H, Besemer AC. 2004. TEMPO-mediated oxidation of polysaccharides: survey of methods and applications. *Top. Catal.* 27:49–66
86. Moreau C, Belgacem MN, Gandini A. 2004. Recent catalytic advances in the chemistry of substituted furans from carbohydrates and in the ensuing polymers. *Top. Catal.* 27:11–30
87. Schoevaart R, Kieboom T. 2004. Application of galactose oxidase in chemoenzymatic one-pot cascade reactions without

intermediate recovery steps. *Top. Catal.* 27:3–9
88. Hoydonckx HE, De Vos DE, Chavan SA, Jacobs PA. 2004. Esterification and transesterification of renewable chemicals. *Top. Catal.* 27:83–96
89. Bondioli P. 2004. The preparation of fatty acid esters by means of catalytic reactions. *Top. Catal.* 27:77–82
90. Marquez-Alvarez C, Sastre E, Perez-Pariente J. 2004. Solid catalysts for the synthesis of fatty esters of glycerol, polyglycerols and sorbitol from renewable resources. *Top. Catal.* 27:105–17
91. Mol JC. 2004. Catalytic metathesis of unsaturated fatty acid esters and oils. *Top. Catal.* 27:97–104
92. Carrettin S, McMorn P, Johnston P, Griffin K, Kiely CJ, et al. 2004. Oxidation of glycerol using supported gold catalysts. *Top. Catal.* 27:131–36
93. Barrault J, Clacens JM, Pouilloux Y. 2004. Selective oligomerization of glycerol over mesoporous catalysts. *Top. Catal.* 27:137–42
94. Monteiro JLF, Veloso CO. 2004. Catalytic conversion of terpenes into fine chemicals. *Top. Catal.* 27:169–80
95. Swift KAD. 2004. Catalytic transformations of the major terpene feedstocks. *Top. Catal.* 27:143–55

Annu. Rev. Mater. Res. 2005. 35:351–95
doi: 10.1146/annurev.matsci.35.103103.120732

First published online as a Review in Advance on April 4, 2005

POROUS ALUMINOPHOSPHATES: From Molecular Sieves to Designed Acid Catalysts

H.O. Pastore,[1] S. Coluccia,[2] and L. Marchese[3]
[1]Instituto de Química, Universidade Estadual de Campinas, CP6154, CEP 13083-970, Campinas, SP, Brasil; email: gpmmm@iqm.unicamp.br
[2]Dipartimento di Chimica IFM, Universitá di Torino, I-10125, Turin, Italy; email: salvatore.coluccia@unito.it
[3]Dipartimento di Scienze e Tecnologie Avanzate, Universitá del Piemonte Orientale, I-15100, Alessandria, Italy; email: leonardo.marchese@mfn.unipmn.it

Key Words silicoaluminophosphates, materials precursors, acidity, properties design

■ **Abstract** This review covers the synthesis, characterization, and physico-chemical properties of microporous and mesoporous aluminophosphates and silicoaluminophosphates molecular sieves. Particular emphasis is given to the materials that have found applications as acid catalysts. We consider the evolution of the synthesis procedures from the first discoveries to the current methodologies and give perspectives for new possible synthesis strategies. Emphasis is given to the use of specially prepared precursors/reactants designed for the use as molecular sieves. Experimental (especially MAS-NMR and FTIR spectroscopy) and theoretical approaches to the description of the Si insertion into the ALPO framework and to the acidic properties of SAPOs and MeAPSOs materials are discussed.

INTRODUCTION

More than 20 years have passed since the seminal report on the synthesis of aluminophosphates (1). New discoveries and insights into the chemistry and structures of these materials and their derivatives continue to appear.

In fact, the late 1970s saw the search for new compositions of molecular sieves, different from silicates and aluminosilicates, and the appearance of the aluminophosphate (ALPO) and silicoaluminophosphate (2) families. The first reports indicated the possibility of 13 different elements in the $ALPO_4$ framework (Li, Be, B, Mg, Ti, Mn, Fe, Co, Zn, Ga, As) and noted that this framework could have 24 structures, 7 of which were zeolite analogs and 15 new structures (1).[4]

Nowadays aluminophosphates assume 53 structures (3), with 18 analogs of natural or synthetic zeolites. The majority is composed of microporous molecular sieves with small, medium, and large pores, but some of these structures are extra-large pore molecular sieves, with entrances limited by rings of more than 12 T atoms. The number of elements possibly included in the framework is now 17, i.e., the original elements noted above and in addition V, Cr, Ni, and Cu.

We focus here mainly on new preparation methods and properties that derive from these synthesis methods, especially the acidic properties created from Si insertion at the P sites (SAPO). Several examples of Me-containing silicoaluminophosphates (MeAPSO) are also illustrated. These underline the modification of the acidic properties of the material induced by the presence of a second heteroatom in the structure (4). For a critical overview of the redox properties of Me-substituted ALPO and SAPO, the reviews from Hartmann & Kevan (5) and from Frache et al. (6) are recommended.

The great interest in SAPO molecular sieves is noted by numerous publications; in the past five years more than 350 reports have appeared in the literature, a large part concerning the application of SAPO in catalysis. Among these, more than 100 articles regard studies on SAPO-34, an acid molecular sieve with chabazite-type structure, and derivatives that are the best candidates for commercial applications. An example is the UOP/Hydro MTO process (7) for the transformation of methanol to light olefins (MTO); plant construction is underway in Lagos, Nigeria by UOP/Eurochem Technologies and should be completed in 2006. Ethylene and propylene obtained from this plant should result in the production of 40,0000 metric tons/year of high-density polypropylene and polyethylene.

---

[4]List of acronyms: AEL, aluminophosphate structure eleven; AFI, aluminophosphate structure five; ALPO, aluminophosphate; CAL, campinas-alessandria; CHA, chabazite; CrAPSO, cromosilicoaluminophosphate; DAF, Davy Faraday; DGC, dry gel conversion; DnR, double n-ring; EO-PO-EO, copolymer ethylene oxide-propylene oxide; EXAFS, extended X-ray absorption fine structure; FSM, folded sheet material; HF, high frequency; HMI, hexamethyleneimine; LEV, levyne structure; LF, low frequency; MCM, Mobil composition of matter; MeAPSO, metalsilicoaluminophosphate; MeAPO, metaloaluminophosphate; nMR, n-membered ring; MTO, methanol-to-olefins; NU-87, new (ICI) structure 87; SAPO, silicoaluminophosphate; SAS, Saint Andrews structure; SDA, structure directing agent; SM 1, 2, or 3, substitution mechanisms 1; 2 or 3; SSZ, Standard Oil synthetic zeolite; SWNT, single wall nanotube; T atoms, structure atoms, Si, P, Al, etc.; TAPO, titanoaluminophosphate; TAPSO (or TiAPSO), titanosilicoaluminophosphate; TEAOH, tetraethylammonium hydroxide; TMAOH, tetramethylammonium hydroxide; TPAOH, tetrapropylammonium hydroxide; TPD, temperature programmed desorption; VFI, Virginia structure five; VPI, Virginia Politechnique Institute; VPT, vapor phase transport; XANES, X-rays absorption near edge spectroscopy; ZON, ZAPO-M one (zincoaluminophosphate-Mulhouse-one); ZSM-5, Zeolite Socony Mobil structure 5.

Acid and structural properties of SAPO-34 and of chabazite-related materials are of special interest and are compared with those of other largely studied SAPOs. It is our hope that readers will appreciate the beginnings of this technology and be aware of the new aspects of the chemistry of aluminophosphates molecular sieves.

CHEMISTRY AND BONDING IN PHOSPHATES

Several observations led the way to the discovery of aluminophosphates. The incorporation of phosphorus atoms into a zeolite was reported as early as 1971 (8). The material was obtained from an aluminosilicate gel containing phosphate groups and, as a result of phosphorus/phosphate incorporation, the ion exchange capacity diminished as the phosphorus concentration in the material increased. $AlPO_4$-5 was synthesized in 1978 (9) from a gel containing tetrapropylammonium hydroxide at low pH. These conditions were markedly different from the ones known to generate zeolites.

The change in properties upon inclusion of phosphate/phosphorous in the gel, especially the ion exchange capacity, could be explained not only by the simultaneous presence of the original SiO_2 and $[AlO_2]^-$ but also by $[PO_2]^+$ in the framework, in such a way that each $[PO_2]^+$ moiety cancelled the charge of a $[AlO_2]^-$. Crystal chemistry concepts confirmed that phosphorous was a good candidate for the construction of tetrahedral nets and that its use, together with aluminum, could probably render a neutral network similar to silica. Several elements can be incorporated, individually or collectively, into the framework of ALPO molecular sieves; however, not all the possible known structures can be obtained with these elements (Figure 1). The possibility of having oxide networks with tetrahedral elements that were not expected to be in that type of site, using Pauling's radius ratio (10), T-O distance, and electronegativity, was assigned to the larger spatial accommodation provided by the microporous structure (9). Nonetheless, larger spatial accommodation should also be expected in microporous silicates in which some of the elements incorporated in aluminophosphates were never seen sitting in such a framework. It appears that crystal chemical concepts may be flexible in cases where ionicity/covalence balance on the bonds may be less straightforward, as in the case of phosphates, probably because *d* orbitals in phosphorous atoms influence the nature of the network-bonding character (11, 12), making it more covalent and allowing tetrahedral inclusion of otherwise forbidden elements.

Although zeolites and aluminophosphates are somewhat alike, the chemistry behind the formation of these materials is very different. Zeolites are formed mostly in alkaline medium, where silanol groups may be deprotonated. Low pH values in the synthesis gels may lead to the formation of dense phases of aluminosilicates.

Low to neutral pH values are ideal for aluminophosphate synthesis, although pH lower than 3 causes the appearance of dense phases and one higher than 10 provokes a reduced yield owing to solubilization or a lower crystallization degree.

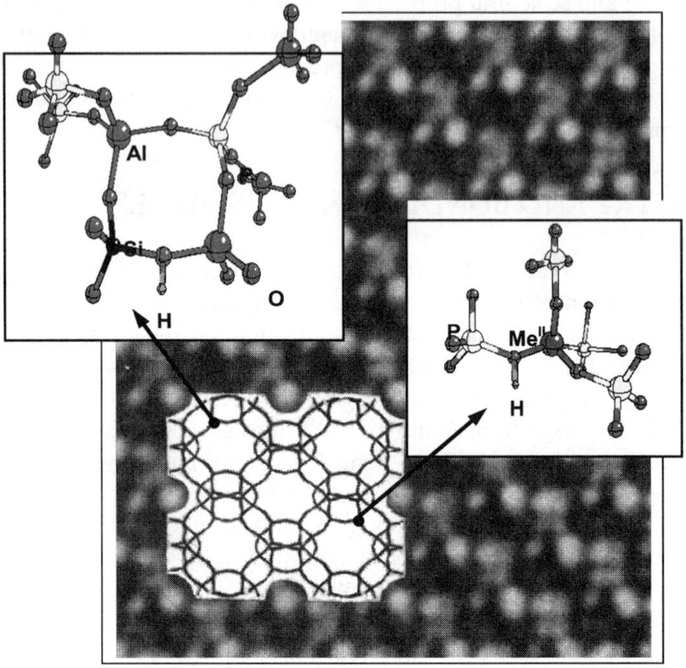

Figure 1 The sites of silicon and/or other elements in the tetravalent state and of metal ions in the divalent state (Mg, Zn, Ni, Co, . . .) in the aluminophosphate 18.

Within the above mentioned pH range, aluminum ions may expand the coordination sphere to accommodate five or six ligands (13). In the aluminophosphates, four of the six positions may be occupied by oxygen atoms from phosphate groups and the remaining two by water, hydroxyl groups, or fluoride ions. The effect of these extra ligands is to cause structure deformation, phase change, and/or diminished thermal stability (1, 14).

Deriving Acidity in Aluminophosphates: The Insertion of Silicon

The introduction of silicon atoms in ALPO frameworks leads to the appearance of Brönsted acidity relevant for acid-catalyzed reactions such as methanol-to-olefins (MTO) (15–17), *n*-alkane cracking and hydrocracking (18–20), oxidative dehydrogenation of alkanes (21), and even more complex hydrocarbon transformations (22). The origin of the Brönsted acidity in SAPOs is the insertion of Si at the phosphorous site, which leads to the formation of negatively charged frameworks that are balanced by protons attached to Si-O-Al bridges; the mechanism is denoted SM2. The substitution of silicon for aluminum, SM1, is another possible mechanism of Si insertion; however, this would lead to the formation of Si-O-P bridges, which are energetically unfavorable as suggested by a computational study on SAPO-5 (23) and have never been observed experimentally. The double substitution of neighboring Al and P by two Si atoms has been often proposed for the Si insertion; this third mechanism, SM3, is responsible for the formation of extended Si islands, e.g., silicate-type patches, within the ALPO frameworks. The three mechanisms are represented in Scheme 1.

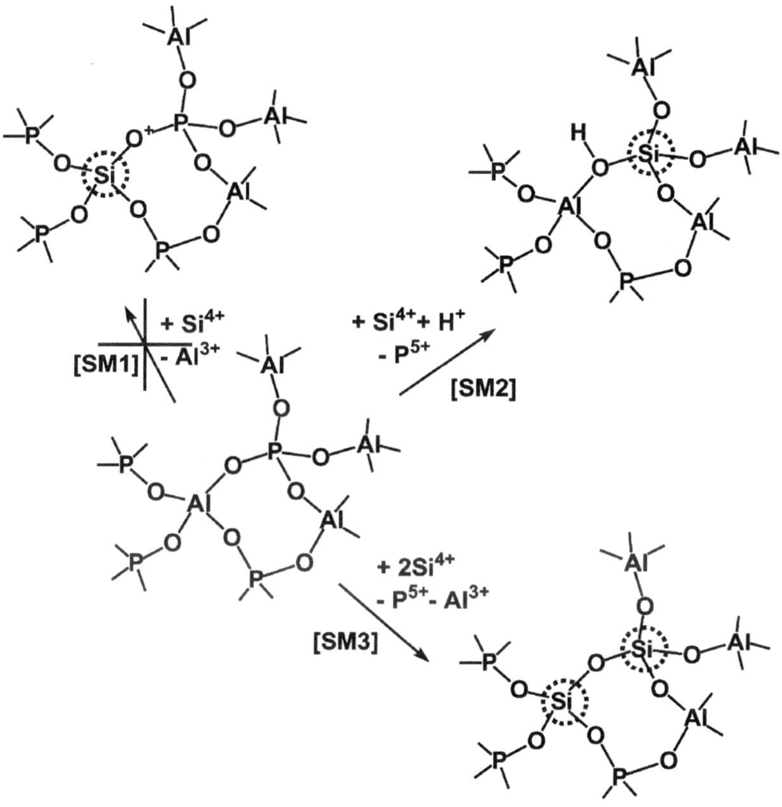

Scheme 1

It is obvious that the local structure of Brönsted acid sites in zeolites and SAPOs is similar, constituted by Si-O(H)-Al units; however, the acidity of these materials may vary greatly, even in the case of structurally similar systems. The acidity depends on several factors, for example bond angles, bond lengths, and the electrostatic potential around the acid centers and within the zeolite cages. Among these factors, the atomic environment around the acid centers plays a relevant role. The first coordination sphere of T atoms around the aluminum is formed exclusively of Si atoms in zeolites, whereas Si and P can be present in SAPOs. In addition, the formation of Si islands in SAPOs leads to an even larger number of acid site

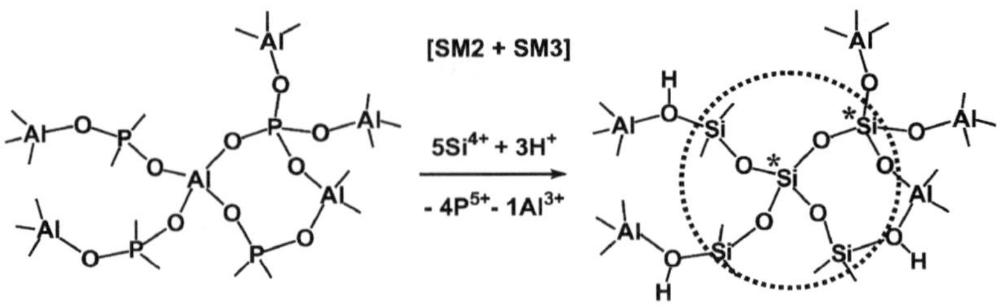

Scheme 2

environments; this makes the acidity of SAPOs more tunable than in zeolites, providing that the Si insertion can be controlled. Scheme 2 shows a representation of the formation of silica patches by substitution of 5Si for 4P and 1Al through a cooperative [SM2 + SM3] mechanism. The asterisks mark the pair of Si atoms that underwent [SM3] substitution.

One fundamental step toward the understanding of the acidity of SAPOs and zeolites came from the elegant work of Barthomeuf, who rationalized the mechanism of silicon insertion in ALPO frameworks and the trends governing number and strength of Brönsted acid sites produced in this way (24). A topological approach, which took into account both the structure topology and the Al content, was first developed for several zeolites (25) and subsequently applied to SAPOs.

With the aid of planar schemes for representing the distribution of Si and Al in zeolites and Si, Al, and P in SAPOs networks (Figure 2), Barthomeuf (24, 25) showed that the occupation by Al, Si, and P atoms of the first and second shell of T atoms around a central Si governs the acid strength through both the

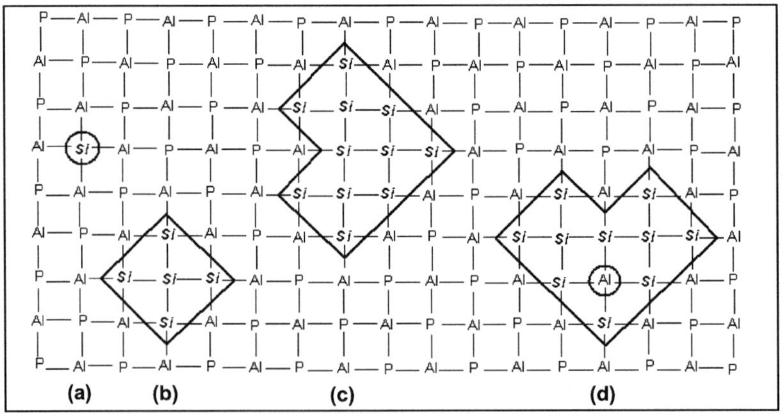

Figure 2 Planar scheme for the Si, Al, and P distribution in a SAPO network: (*a*) isolated Si, (*b*) 5Si island, (*c*) 11Si island, (*d*) Si-Al phase with 1 Al in the siliceous island (adapted from Reference 24).

electronegativity of the atoms and the connectivity of tetrahedral units. The acid strength increases as the first shell contains more Si and as the second shell contains more P. The presence of Si islands leads to the formation of stronger acid sites by forming Si(*n*Al) species ($n < 4$) at the border of the islands.

The numbers of these acid sites and of the *n* value depend on the dimension of the islands, which is related to the Si concentration and to the structure topology (Figure 3). A correlation between acid strength and ^{29}Si MAS NMR chemical shift, i.e., the Si environment in the first and second shell of the tetrahedral is richer in silicon, was finally established.

The results of the topological approach proposed by Barthomeuf (24, 25) were later supported by a theoretical study of Sastre et al. (26), who investigated the Si island formation and confirmed that siliceous patches within SAPO-5 and SAPO-34

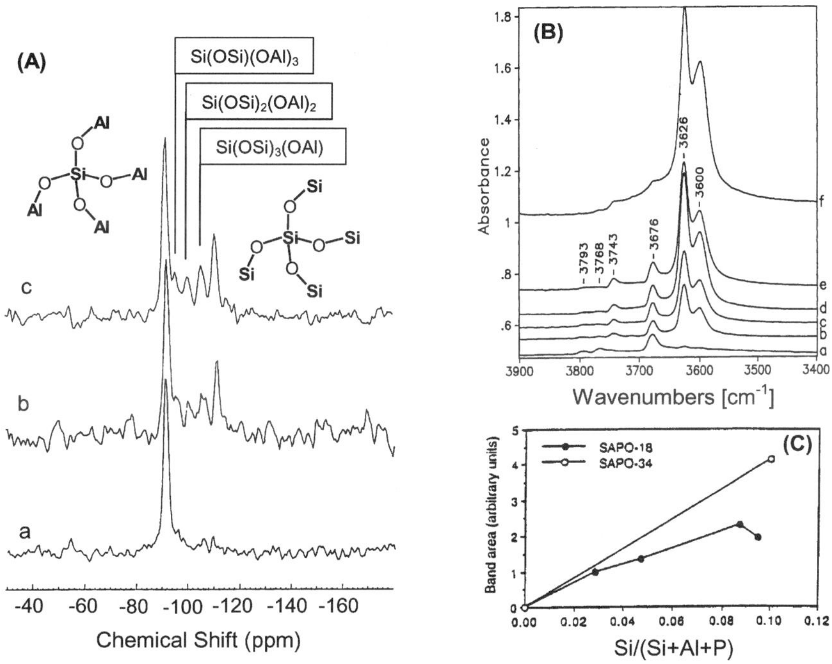

Figure 3 Combined ^{29}Si MAS NMR (*A*) and FTIR (*B*,*C*) study showing that SAPO-34(x_{Si} = 0.10) displays essentially isolated Si centers (with ^{29}Si chemical shift at –92 ppm), Brönsted acidity bridged (Si-O(H)-Al), hydroxyls at 3626, and 3600 cm^{-1}. (*B*) Si islands in SAPO-18 materials with even lower silicon fraction (x_{Si} = 0.047, *curve b* and x_{Si} = 0.087, *curve c*); $\delta = -111$ ppm for $Si(OSi)_4$, and −96, −100, and −105 ppm for $Si(OSi)_n(OAl)_{4-n}$ species. Accordingly, the concentration of Brönsted acid sites in the SAPO-18 is much lower than SAPO-34 (*C*). Panel *B* shows the FTIR spectra of (*a*) AlPO-18, (*b–e*) SAPO-18 with different silicon concentration, and (*f*) SAPO-34 (adapted from Reference 25).

are energetically stable and are more favorable in SAPO-5 where they form at lower silicon concentration. The literature concerning both experimental and theoretical results on the mechanisms of Si substitutions in SAPOs has been carefully reviewed and is recommended for the reader interested in this topic (26).

However, it is important to remember that the role of the structure-directing agent (SDA) in governing the Si distribution in SAPOs is a relevant aspect. Again the work of Barthomeuf and her group (27) is exemplary; using X-ray diffraction (XRD) and ^{27}Al and ^{29}Si MAS NMR spectroscopy, they described the Si distribution in SAPO-34 prepared in the presence of morpholine and tetraethylammonium hydroxide (TEAOH) as SDA (Figure 4). The model proposed for SAPO-34 was subsequently generalized to other SAPOs (27).

The Si distribution depends on the number of framework charges, which is influenced by both the number of SDA molecules embedded within the cages and the Si atoms introduced within the framework. The chabazite cages may host up to two molecules of morpholine, whereas only one molecule of TEAOH was found in the cavities. As the maximum number of isolated tetrahedral [SiO_4] groups is limited by the number of SDA molecules, it is predicted that in the presence of morpholine the amount of isolated Si is higher and may reach the upper limit determined by the structure topology ($x_{\mathrm{si}} \approx 0.11$ in the case of chabazite).

One of the main results of the theoretical work by Sastre et al. (26) is the estimation of the deprotonation energies of different acid sites, which fully supported the model indicating that the acidity is greater at the edges of Si islands. A correlation between acidity and island dimension was established by considering 5Si and 8Si islands in SAPO-5 and SAPO-34 (Figure 5, see color insert). In addition, a structure dependence on the formation energy of 8Si islands was found, which showed that larger islands were more favorable in SAPO-5 because of its more flexible framework structure. Subsequently, these studies were extended to SAPO-11, SAPO-18, and SAPO-31, showing that small differences were found on passing from isolated Si to Si islands, which suggested that both types of distribution are possible in these SAPOs (28).

The validity of this theoretical approach is proven by the fact that several experimental results could be theoretically simulated. However, when very similar structures are considered, such as ALPO-18 and ALPO-34, for instance, their different behavior upon Si insertion (Figure 3) remains to be rationalized, and the computational strategy needs to be more finely tuned.

The acidity of SAPOs, similar to that of zeolites, is commonly studied by NH_3 and pyridine adsorption and/or their temperature-programmed desorption (TPD). Although reliable quantitative data on the overall number of acid sites are obtained, especially by pyridine adsorption, these techniques are unable to determine the presence of sites with slightly different acidity. For monitoring the acidity of catalysts, the use of small and weakly interacting probe molecules, such as CO, H_2, N_2, etc., which induce only a mild perturbation of the catalytic sites, has been recommended (29–31). Nowadays, notwithstanding the above guidelines for experimental determination of relative acidities, the use of computational methods

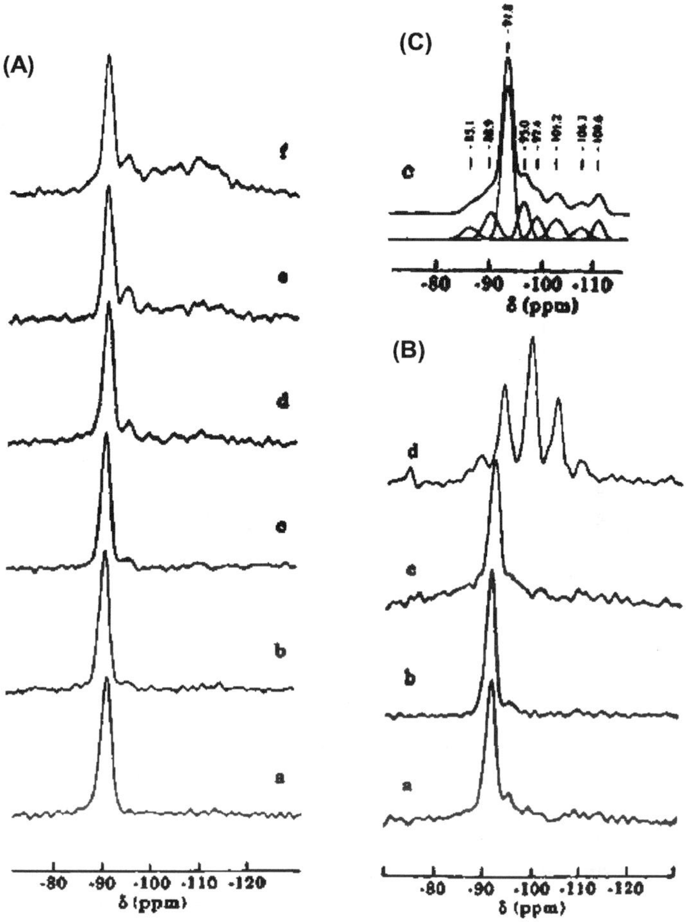

Figure 4 ^{29}Si MAS NMR of SAPO-34 prepared in the presence of morpholine (*A*) and TEAOH (*B*), at increasing Si concentration (x_{Si}). (*A*): x_{Si} = 0.10 (*a*), 0.14 (*b*), 0.16 (*c*), 0.18 (d), 0.23 (*e*) and 0.33 (*f*). (*B*): x_{Si} = 0.10a (*a*), 0.10b (*b*), 0.11 (*c*); a chabazite with Si/Al = 2.34 (*d*) is also reported for comparison. (*C*) Simulated spectra of the SAPO-34 prepared with TEAOH at x_{Si} = 0.11. It is evident that silica islands are formed at lower x_{Si} when TEAOH is used as SDA (adapted from Reference 27).

for the determination of acidity differences is becoming increasingly popular, and thus the modeling of Brönsted acidity in SAPOs by computational tools is becoming more and more effective (32–34). Many energy contributions, such as the electrostatic potential and electric field, can now be taken into account in the calculations for determining the vibrational frequency of Brönsted sites and the energy required to elongate the OH bond (35). The reliability of the computational tools is confirmed by results that describe the correlation between the OH stretching

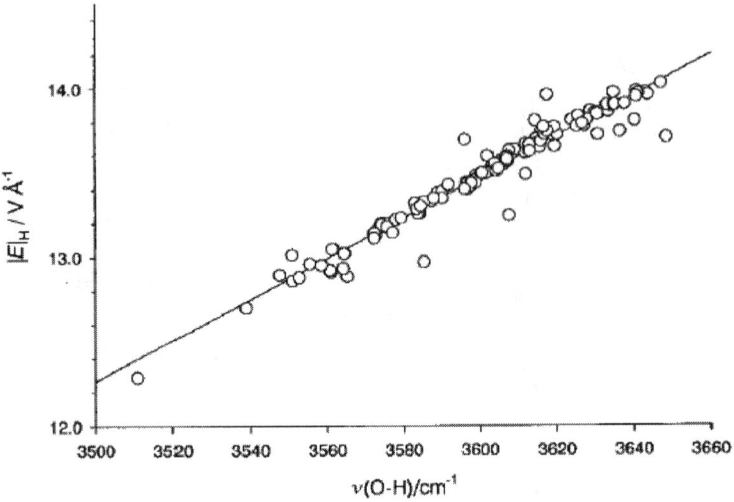

Figure 6 Correlation between OH stretching frequency (ν_{OH}) of zeolites and SAPOs and the gradient norm of the electrostatic potential at the proton ($|E|_H$). (Reproduced from Reference 36, with the permission of the R. Soc. Chem., © 1999.)

frequency of Brönsted acid centers in zeolites and SAPOs and the gradient norm of the electrostatic potential (Figure 6), which describes the vibrational properties of microporous solids having a large variety of structures and compositions (36).

Despite the effectiveness of computational methods, the acidity of Brönsted acid centers is experimentally monitored by the changes induced by the OH stretching frequency upon adsorption of molecular probes (4, 29, 37). Makarova et al., for instance, provided an elegant example of the use of Fourier transform infrared (FTIR) spectroscopy of molecular probes in combination with ^{1}H MAS NMR spectroscopy for determining the distribution of Brönsted hydroxyls in SAPO-37. With this study it was possible to evaluate the extinction coefficients of OH stretching vibrations in HZSM-5 and faujasite-type zeolites (38). The sensitivity of FTIR spectroscopy, with the use of CO as molecular probe, for describing acidity strength and OH location in SAPO-40 (Figure 7) and SAPO-34 (Figure 8, see color insert) was described by Onida et al. (39) and by Smith et al. (40), respectively. SAPO-40 displays five main OH species, described as high-frequency (HF) (HF, at 3737, 3730, and 3724 cm^{-1}) and low-frequency (LF) (LF$_1$ and LF$_2$, at 3559 and 3528 cm^{-1}). The former, probably hydroxyl groups located in 12-member-ring channels, are sterically free, whereas the latter ones, sitting either in side pockets or in 8-member-ring channels of the AFR structure, are sterically hindered and interact poorly with probe molecules. SAPO-34 shows two main OH groups at 3601 cm^{-1} (LF mode) and 3630 cm^{-1} (HF mode) located in the chabazite cage and H-bonded to framework oxygen atoms, respectively (a third component at 3625 cm^{-1} could be determined only through curve fitting procedures). The acidity

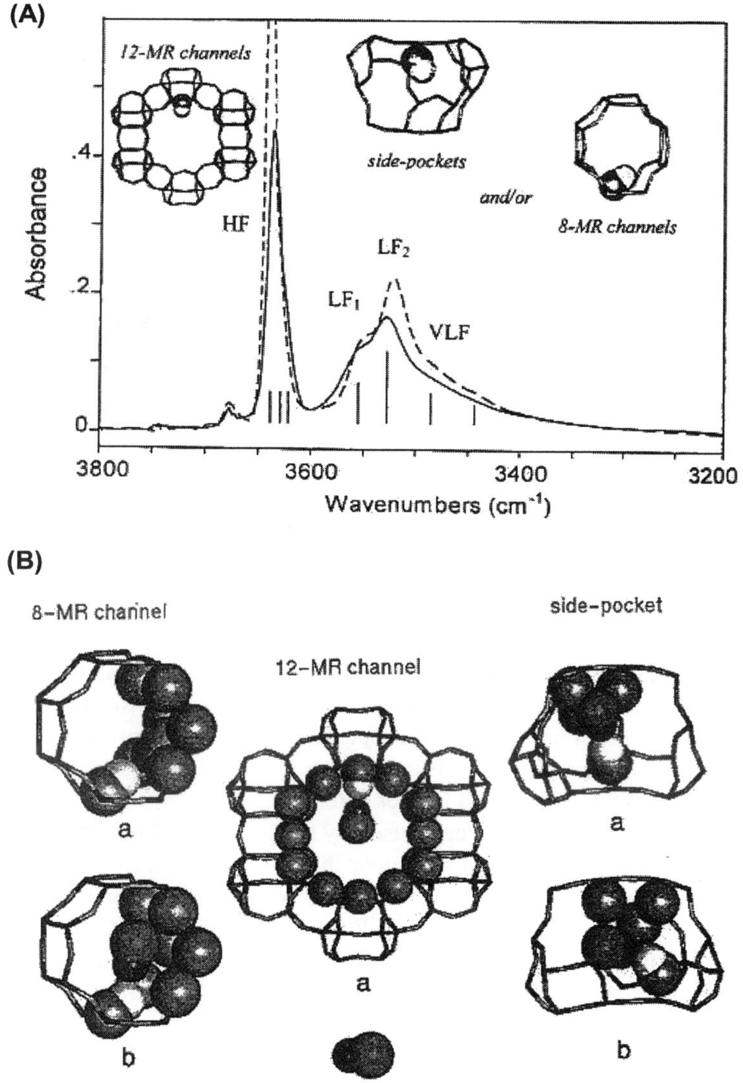

Figure 7 (*A*) FTIR spectra of SAPO-40 and structures corresponding to different OH species: solid curve, room temperature; broken curve, nominal 77 K. (*B*) Molecular graphics description of the CO/OH interaction in the 12-member-ring channel, side pockets in the 12-member-ring channel, and in 8-member-ring channels: (*a*) structures, OH · · · CO linear arrangement; (*b*) structures, bent OH · · · CO arrangement avoiding overlap. (Reproduced from Reference 34 with the permission of the Am. Chem. Soc., © 1997.)

of HF species in SAPO-40 and SAPO-34 was determined by CO adsorption and the results were similar to those of HY and mordenite zeolites, producing stretching frequencies similarly red-shifted ($\Delta\nu_{OH}$ = 285–290 cm^{-1}). A minor fraction of hydroxyl groups absorbing at 3625 cm^{-1} in the case of SAPO-34 showed a very large shift ($\Delta\nu_{OH} \approx 350\ cm^{-1}$), signifying that these protons should be strongly acidic, comparable to the acidity of HZSM-5 zeolites. This result is quite unusual for SAPOs and needs further experimental and theoretical evidence to be accepted. The acidity of LF groups determined by CO adsorption could not be compared with that of the HF groups in that complexes with different geometries (non-linear configuration in the case of LF species) were formed.

The theoretical and experimental results reported in this section elucidate the role of both approaches in describing the properties derived upon Si insertion within ALPO frameworks. The combination of FTIR of molecular probes with MAS NMR spectroscopy appears to provide invaluable results in the description of the silicon distribution and site location of hydroxyls of SAPOs. Computational studies effectively support the interpretation of the spectroscopic results and, more generally, the description of the physico-chemical properties of these acid microporous solids.

The Insertion of Metal Ions in Microporous $ALPO_4$ and SAPOs

The inclusion of transition metal ions in the framework sites of aluminophosphates and silicoaluminophosphates of several structure types affords materials with interesting adsorption and catalytic properties. These transformations and their already known consequences have been recently reviewed by Hartmann & Kevan (5) and are not discussed in depth further here.

In this review the effects caused by the presence of a second heteroelement in SAPO frameworks are of particular interest. The nature of the second heteroelement largely defines the sites where it is going to substitute for, either Al^{III}, P^{V} or Si^{IV} and in siliceous or aluminosilicate domains. Alternatively, depending on its nature, it may even form domains of heteroelement in the framework. One recently reported material where the occurrence of both types of substitution was observed is the chabazite-like structure, TAPSO-34 (41). Figure 9, see color insert, shows the structures of both sites, as well as the UV-Vis spectra for each of the situations.

Other factors influencing the type/site of substitution are the reaction conditions and the relative concentrations of Si and heteroelement. In particular, substitution of metal, for instance, for Si^{IV} in silica domains, is most probable under high Si and low metal concentration conditions. In this case, one may expect that the properties of the MeAPSO are similar to those of the Me-substituted silicalite.

The substitution of Ti^{IV} for Si^{IV} ions in SAPO-37, for example, may provide materials with interesting properties in oxidation reactions (42). NMR studies showed that the incorporation of Ti into a SAPO-37 framework (SiO_2/TiO_2 = 17) did not modify the distribution of the framework species, meaning that it has no preference for substitution in the ALPO or in the silica domains. Titanium

such as cyclohexylamine. Therefore, SDA interaction with the growing framework also plays an important role in the synthesis of molecular sieves. An elegant example of the dual function of SDAs is the production of porous aluminophosphates using fine-tuned combinations of different molecules, such as TPAOH/TMAOH for SAPO-37 (59) and SAPO-40 (60), TEAOH/Pr_3N for ALPO-52 (61), or TMAOH/diethanolamine/F for SAPO-LTA (62), where the different molecules are found in different cavities of the material.

It is important to point out that molecular sieves with large cages may be templated by multiple small SDAs that are occluded in the cages. The best example of this is cloverite, which contains eight quinuclidinium cations per large cage (63), and exemplifies the idea of a supramolecular templating process used for the preparation of mesoporous molecular sieves.

The aggregation of SDA molecules was probed by FT-Raman spectroscopy, which showed that small cages may also be created by aggregates. Using this technique, the ring-breathing vibration of nonprotonated and protonated morpholine molecules was shown to be approximately 10 cm^{-1} apart (64). This observation was further extended by the use of FTIR and thermogravimetry with the aim of understanding the SDA framework interactions. It was discovered that two molecules of morpholine are retained in the cages of ALPO-34, whereas for SAPO-34, water molecules are also present in the cages. The vibrational features of protonated morpholine within the intrazeolite spaces are highly sensitive to interaction with other molecules in the same space and with the framework itself (65).

A combined single-crystal XRD and FTIR study (66) showed that both neutral and protonated organic molecules can be found in three different positions inside the chabazite cage of SAPO-34. The presence of water close to morpholinium ions suggests that HM^+-H_2O-HM^+ (or M) complexes are located inside each chabazite cage. H-bonds between H_2O molecules located within 6-ring windows may connect these complexes and form chains that may extend through several cages, which may be the building units of the supramolecular architecture (Figure 11, see color insert).

In contrast to the chaining of molecules in the chabazite structure of SAPO-34/ALPO-34, in mono-dimensional ALPO and SAPO-5 (at 90 K to room temperature), the triethylamine molecules are free to reorient (67). The only restriction observed by ^1H-NMR is an interaction of the Brönsted acid sites of SAPO-5 with the N-lone pair of electrons in the SDA. Such results imply that molecules in these types of molecular sieves generally assume distributed orientations in micropores, which are even more marked in systems with a homogeneous pore structure such as ALPO-5, which has many possible orientations with different potential energies.

In both porous and nonporous aluminophosphates, water molecules are highly organized, as indicated by Raman, thermogravimetric analysis (TGA), X-rays diffraction, and NMR data (68). In fact, water interaction with lattice aluminum through its oxygen atoms causes the formation of $AlO_4(OH_2)_2$. These water molecules interacting with the lattice may bind to other water molecules through H-bond; the last ones may use both oxygen and hydrogen atoms for that purpose. In some cases, lattice oxygen within the structure can take over the hydrogen-bonding acceptor function. This specific pattern of bonding results in the formation of water clusters well structured within the aluminophosphate network.

These recent examples show that the SDA and solvent molecules may form complexes capable of specific functions in directing structures, in addition to individual templating and space-filling roles.

SYNTHESIS METHODS AND THEIR INFLUENCE ON MOLECULAR SIEVES PROPERTIES

The Benefits of Fluoride in the Synthesis

Guth (69) was the first to report the benefits brought by the use of fluoride ions as mineralizers in the synthesis of zeolites. As a matter of fact, larger and more crystalline particles are produced because they are formed at a smaller rate in the presence of this ion. Later on it was found that $ALPO_4$-34 could be prepared only in the presence of fluoride (65). In this case, it served to hold morpholinium ions in the cavities; otherwise the neutral ALPO framework would not be formed. Recently it was discovered (70) that fluoride ions are not only a mineralizing agent but also act as specific templates of D4R units in gallophosphates. Since this discovery, it has been possible to synthesize a series of structures by using the latter mineralizing agent. In the majority of the as-synthesized samples, fluoride ions act as bridges between aluminum ions that become 5- or 6-coordinated depending on the particular structure. One of the most recent examples of new structures possible in the presence of F^- is the ZON structure type (71). $ALPO_4$-ZON materials were prepared at pH between 3.5 and 7.5, with TMA^+ as a templating agent only if enough F^- anions were present. Diethanolamine is used to control the pH in this range and is not found in the final product. Silicon may be incorporated in the framework if the concentration of F^- ions is diminished; otherwise, SAPO-SOD is formed, confirming the co-templating role of F^-.

SSZ-51, an aluminophosphate whose structure is closely-related to that of SAPO-40, was prepared in the presence of 4-dimethylaminopyridine as SDA and, in addition, in the presence of F^- as a mineralizing agent. The solid presents aluminum that is probably coordinated to F^- ions (as shown by ^{27}Al-NMR), which becomes 4-coordinated upon calcination. Despite the high thermal stability (the material preserves its structure after heated at 800°C), it loses a large part of its crystallinity upon hydration. This is a reversible behavior because an XRD profile corresponding to a crystalline material reappears upon heating to 100°C (72).

Aluminophosphates with structures related to the chabazite and to SAS (Saint Andrews Structure) were prepared by the simultaneous use of cyclam (1, 4, 8, 11-tetraazacyclotetradecane) and fluorine, and cyclam, fluoride, and Cu^{+2} ions, respectively (73). In that last case, careful choice of synthesis conditions, such as the use of F^- ions in combination with macrocycles containing secondary amines, minimizes the chances of metal ions being incorporated into the framework; furthermore, the choice of a metal cation with a large preference for octahedral coordination and a high-binding constant with cyclam helped in reducing the chances of Cu being found in the framework (73).

F^- ions are instrumental in producing specific structures, such as the neutral ALPO ones; F^- serves as a binding site for the cationic SDA. Indeed, the incorporation of silicon into the framework of these aluminophosphate-fluorides is possible only if the fluoride concentration diminishes reasonably; otherwise framework-incorporated silicon is not going to be compensated by the SDA cation and will not be stable in the structure. In addition, in the presence of transition metals in the reaction mixture, their location in the final structure may be finely controlled by the possibility of forming stable complexes with F^- ions and/or the SDA.

The Dry-Gel Conversion Method

Dry-gel conversion is a relatively new synthesis method that allows an almost complete conversion of gel to a molecular sieve in high yield while enabling minimization of waste and reduction in reactor volume. It favors the production of uniform crystals with small particle sizes. This method was originally introduced as vapor phase transport (VPT) in the 1990s (74), when it was used extensively for the production of zeolites. It took almost ten more years before this method was used in the synthesis of aluminophosphates (75).

In the synthesis of ALPO-5 and ALPO-11 (76), the amount of water needed to perform the gel transformation into the molecular sieves structures is very small as the reaction gel is heated at 80°C for 1–2 h; it forms a white solid that crystallizes into a molecular sieve structure. The physico-chemical characteristics of samples prepared by DGC are very much the same as the ones prepared by conventional hydrothermal crystallization, including the crystal sizes and morphologies.

Influence of Synthesis Conditions on Particle Size and Morphology

The advantages of using microwave irradiation for zeolite synthesis are mostly due to (*a*) the very fast crystallization (77, 78), (*b*) increased phase purity (79), (*c*) the narrow particle size distribution (80), and (*d*) easy preparation of unstable microporous materials (81). Caro et al. (82) proposed that several mechanisms for fast crystallization of a structure could be acting simultaneously, some of them based on the following assumptions: (*a*) Microwave irradiation could destroy H-bonds causing a fast gel dissolution and formation of Al–O–P bondings; (*b*) the gel contains self-assembled microarrays, which transform directly into the AFI structure; and (*c*) heating is not disturbed by convection.

ALPO-AFI particle morphology was shown to change supposedly under microwave irradiation. Aggregated spheres, plates, rods, prisms, and barrels can be synthesized by varying the reaction conditions. The fact that silica, HF, and other acids were also added to the reaction mixtures made it difficult to ascertain the influence of microwave irradiation alone on the particles morphologies observed (83, 84).

The real growth behavior of AFI-type crystals in the aluminophosphates and silicoaluminophosphates was studied by the use of a direct observation cell. It revealed a difference in the growth pattern for ALPO-5 compared with that of SAPO-5 crystals mostly due to the influence of silica over the growth in the *c*-axis direction. Dilution was also an important variable; however, the duration of the synthesis was the most influential parameter owing to a secondary nucleation process and the formation of a growth ring pattern (85). In fact, Williams et al. (86) showed that there was a preference of ALPO-5 to crystallize as thin hexagonal platelets in concentrated reaction mixtures, whereas with more diluted gels, spherical particles were obtained.

As for the sizes of AFI crystals, the more diluted gels, H_2O/P_2O_5 of 450, produced SAPO-5 crystals as large as 500 μm (87).

Fast crystallization of more crystalline $ALPO_4$ materials was also the first outcome of gradual and controlled heating of the precursor gel. This method allowed the preparation of particles of uniform size and morphology, with silicon incorporation preferably in framework phosphorous sites (88). If ethylene glycol was used as solvent and with the same gradual heating procedure, highly crystalline SAPO-5 samples with very high silicon contents were prepared. ^{29}Si-NMR showed that the samples contained silica islands but displayed very high acidity (89) (see above).

The synthesis with microwave irradiation no doubt promotes the rapidly desired crystallization and thus favors the kinetic products. Nonetheless, the above mentioned studies reinforce the idea that the morphology of aluminophosphate and silicoaluminophosphate crystals is largely controlled by the relative composition of the synthesis gel. A possible role for microwave heating on the general morphology of the products obtained is still open to investigation.

The Influence of Surfactants in the Gel Synthesis

VPI-5 (VFI-type code) is a member of the ALPO family; its framework is formed by a system of 18-membered ring-lined channels, with 1.21 nm diameter, which gives VPI-5 the potential capacity of processing large organic molecules (Figure 12). VFI topology crystals may be prepared with framework silicon; however, only a

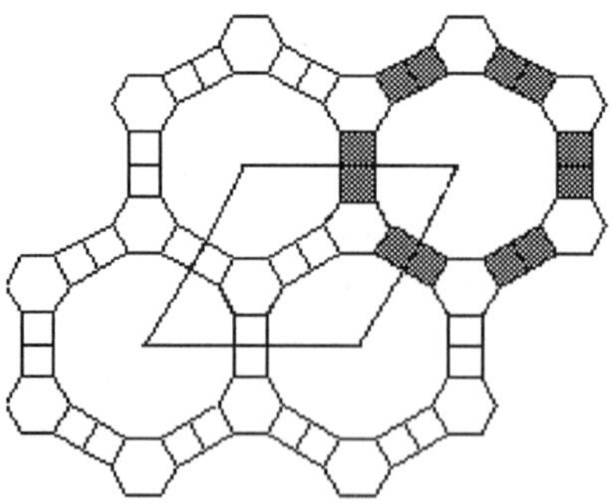

Figure 12 Framework [001] projection of VPI-5. The 1.21-nm 18-membered ring channels run perpendicular to the sheet. The structure is the result of inserting one 4-membered ring adjacent to the ones previously existing in $ALPO_4$-5 (adapted from Reference 90).

very small amount of silicon substitutes phosphorous and generates acid sites in Si-VPI-5. In fact, the largest part substitutes the couple Al + P, thus creating silica islands (90).

It was found that the larger the molecular volume of the amines used to prepare VPI-5 samples the more favorable the Si to P substitution mechanism. Therefore, in 1995 (91) the use of surfactants, e.g., hexadecyltrimethylammonium bromide, was proposed. In this case, ~15 mol% of *n*-hexanol in relation to water as solvents was used to produce higher acidity from the insertion of silicon at the framework P-sites. After calcination, the SAPO-8 produced contained many fewer silica islands and its acidity was approximately half of that in a SAPO-37 sample, with 33 Si atoms/unit cell, which was taken as model of acidity.

Similarily, the preparation of SAPO-5 with surfactant in water/hexanol solvent system caused the modification of Si distribution within the ALPO framework. ^{29}Si-NMR showed that this new strategy afforded a diminished number of silica islands and the appearance of the intermediate sites—Si(3OAl,1OSi), Si(2OAl,2OSi), and Si(1OAl,3OSi)—instead of only Si(4OAl) and Si(4OSi), as had appeared in the sample prepared in the absence of surfactant (92). The two-phase solvent system favors the slow release of silicon species into the aqueous counterpart, where it hydrolyzes. The fact that neutral and anionic surfactants play no role on silicon incorporation led to the conclusion that cationic surfactants were used to stabilize the small silicon species. More recently, however, sodium 3-bromopropanesulfonate, sodium 1-butanesulfonate, sodium naphthalene-1-sulfonate, and sodium *n*-decylsulfate, although anionic, were proven to be good additives for the two-phase synthesis and allowed the incorporation of larger amounts of silicon into the framework, presumably because of the emulsion abilities of the sulfate/sulfonate additives (93).

COMPUTATIONAL AID TO THE SYNTHESIS: FACTORIAL EXPERIMENTAL DESIGN, COMPUTATIONAL METHODS FOR THE DESIGN OF FRAMEWORKS, AND DE NOVO DESIGN

Factorial experimental design is a powerful method for designing the smallest number of individual experiments to map the important variables in the synthesis of a molecular sieve. Optimization of SAPO-47 synthesis using *sec*-butylamine was studied by a 2^k factorial experimental design (94). A series of 32 experiments, $k = 5$, was performed by varying the temperature, crystallization times, and SiO_2/Al_2O_3, P_2O_5/Al_2O_3, and *sec*-butylamine/Al_2O_3 molar ratios in the synthesis gel. The results of interest were the yield of solid powder, crystallinity degree, and phase purity. The effects of the parameters were somewhat predictable.

The time of crystallization (2 or 10 days) did not influence the product yield and crystallinity degree and had a weak influence on the product purity. The influence of temperature (170 or 200°C) was also weak, especially for crystallinity and phase purity. The molar ratios of P_2O_5/Al_2O_3 (0.4 or 1.0) and SiO_2/Al_2O_3 (0.6 or 1.2) were important for the product purity. The template concentration had a strong influence on the responses because it showed an important combined binary effect upon the product purity with the phosphorous content in the gel. This reinforces the importance of pH in the synthesis. The individual importance of the phosphorous content in the gel, which is a more important factor than the silicon content, also supports the importance of pH because the source of P is phosphoric acid.

There are some drawbacks associated with this method when applied to studies on synthesis. Because the model depends heavily on statistics, the first assumptions of the important variables are made by the operator him/herself and are generally the ones important according to his/her experience. The second problem is that these variables in a regular synthesis of molecular sieves cannot be varied in a large range; temperatures, for instance, cannot assume values too high or too low. The same is true for reactants that control pH and other factors. Lastly, but no less important, is the fact that when a large number of variables need to be examined, the number of experiments increases exponentially up to unreasonable values; in this case one's own bench experience is a better advisor.

The computational methods for design of frameworks might be divided into two types: the design of frameworks using methods such as constrained assembly of atoms (95) or automated assembly of secondary building units (96) and the approaches to rational syntheses of target materials, which deal with designing the template itself or predicting the ability of pre-existing molecules to template a desired framework. Among these, the de novo design is the process of greater interest from the point of view of templates and their variants (97).

The de novo computer-modeling ligand design has been applied in the search for biologically active materials. Broadly speaking, the technique allows a molecule to be grown into a targeted space, filling it accurately, i.e., favorable interactions are maximal between the molecule and the cavity. In this process, approximations such as restricted growth or annealing of large groups of atoms into molecules are possible.

For the cavities found in microporous materials, the process is simplified because the cavity may be crystallographically described; the cavities are reasonably rigid and therefore do not suffer changes of conformation as easily as do biological cavities, and van der Waals forces mainly determine the templating process. In addition, the SDAs for microporous cages are usually smaller in size and simpler in nature than the ones found in biological systems.

This procedure was used initially in two ways: the first was a type of control procedure, where existing molecular sieves were tailored by the molecules already known as SDAs. One study involved the levyne structure. In this case, quinuclidine and adamantane were used as fragments and modified during the process into forming the analogs of the templates known for NU-3 (LEV), *N*-methylquinuclidium, and 1-amino-adamantane. In the process other templates were generated, e.g., 1,2-dimethylcyclohexane, which was considered an excellent candidate for templating of LEV structure. In fact, the 2-methylcyclohexylamine formed the CoAPO,

DAF-4 with the desired levyne structure. The growth process with its different stages has the acronym ZEBEDDE (zeolites by evolutionary de novo design). Figure 13 (see color insert) shows the template-building process in detailed steps.

The other direction was the preparation of specific structures with templates designed for them. Using this approach, another CoAPO with chabazite structure, DAF-5, also grown using the ZEBEDDE code was prepared from 4-piperidinopiperidine (Figure 14, see color insert) (97).

COMBINATORIAL APPROACH TO THE SYNTHESIS OF ALUMINOPHOSPHATES AND SILICOALUMINOPHOSPHATES

The combinatorial approach was first reported in 1998 for the production of zeolites (98). The combination of a block of very small autoclaves, commercial pipette robots, and automated powder XRD analysis allowed the study of a large number of formulations in a single experiment. The ingenious construction of a combinatorial material library using hydrothermal synthesis as an example was then reported (99).

The procedure was made even more automated (100) by a methodology that efficiently dispensed reagent into the autoclave block. The product washing and recovery was made in the autoclave block using a custom-designed centrifuge apparatus. The products were identified by XRD in transmission mode with the aid of an automated XY stage for sample translation. In this case, the synthesis of possible aluminophosphate structures was studied using a $[Co(Cp)_2]OH/Pr_2NH$ combination, as well as each compound alone as a SDA (Figure 15). Pure AFI phase was obtained at $[Co(Cp)_2]OH/Pr_2NH$ molar ratios of 0.2:0.8 or 0.1:0.9. Tridimite was observed when $[Co(Cp)_2]OH$ or Pr_2NH were used alone at that concentration. The AEL phase was obtained from a single SDA when the concentration achieved 1.0 mol per mol Al_2O_3. This methodology demonstrated that by using a direct-dosing system for reagents at the micromolar scale, one need not manipulate individual samples in the construction of libraries of molecular sieves. Efficient sample recovery and low reagent consumption allow exploring many different reaction conditions for the fastest discovery of new phases.

A slight modification of the multiautoclave stage permitted the use of the vapor-phase transport method (101).

With the application of the above strategies, new open framework zinc phosphates were prepared. The influence of reagent contents was examined, and two phases were isolated: (*a*) $[Zn_6P_5O_{20}(H_2O)] \cdot 0.5C_6H_{16}N_2 \cdot C_5H_{14}N_2 \cdot 3H_2O$ and (*b*) $Zn_3(PO_4)_2H_2O$. Their synthesis parameters were systematically evaluated in a fast and efficient way by the use of the combinatorial methodology (102).

When using combinatorial strategies, care should be exercised in upgrading the synthesis procedures. Products obtained from autoclaves with volumes in the

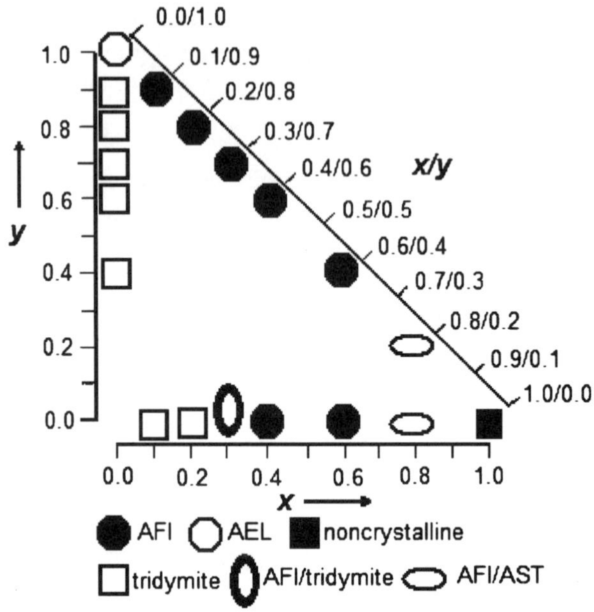

Figure 15 The effect of single and mixed templates on the resulting aluminophosphate phases. The chemical composition of a starting gel was $1Al_2O_3{:}1P_2O_5$: $x[Co(Cp)_2]OH$: $yPr_2NH{:}190H_2O$ (x and y represent the molar amounts of each of the components in the synthesis) (adapted from Reference 100).

order of μL may not be the same when larger autoclaves are used. Effects such as convection might vary greatly in autoclaves with different volumes.

With the precautions noted above in mind, the de novo approach to crystal engineering, the factorial experimental design, and the combinatorial approach can be powerful procedures in the development of new structures and modifications not only of aluminophosphates and derivatives but also of molecular sieves in general.

INTERMEDIATE PHASES AND USE OF ISOLATED PRECURSORS IN THE SYNTHESIS

In some special cases, intermediate phases are observed during the synthesis of aluminophosphates. Some studies involved the isolation of these intermediate phases and their synthesis to use them as precursors. In general, these studies aimed at a higher control of the synthesis process, which may lead to an improvement of the yield and quality of the products and a deeper knowledge of the assembling process that forms these solids. In other studies, ALPO and SAPO molecular sieves are prepared directly from an isolated precursor that already displays Al–O–P or Al–O–Si–O–Al–O–P bonds.

Intermediate Phases

Intermediate phases were isolated during the preparation of ALPO-18 and studied with the aid of NMR techniques in combination with powder XRD (103). Four different stages were identified:

1. Mixing the Al and the P source, in the presence of the template, produces an ordered aluminophosphate material at ambient conditions.
2. One hour at 150°C causes the dissolution of this initial ordered material to produce an aluminophosphate amorphous phase containing both tetrahedral and octahedral aluminum sites connected to phosphorous ions,
3. After 2 h at 150°C, the previously amorphous material transforms to ALPO-5, which becomes the only phase after 24 h heating.
4. ALPO-18 is the product obtained if the former ALPO-5 is heated at 150°C for more than 24 h. A period of 72 h was enough to obtain the complete conversion of ALPO-5, the intermediate phase, into ALPO-18.

Although a very thorough NMR study was made of the materials obtained for each step of the preparation, it was not possible to identify the first ordered aluminophosphate, which was formed before heating by simply mixing of the reactants.

Intermediate phases were also detected in the development of an aluminophosphate solid in studies that involved the isolation, characterization, and separate synthesis to use these phases as precursors.

Several aluminophosphate molecular sieves, ALPO-5, ALPO-22, ALPO-16, and SAPO-35, are formed via a common lamellar aluminophosphate material, $ALPO_4$-L, during hydrothermal synthesis from a gel containing Al_2O_3, P_2O_5, SiO_2 and hexamethyleneimine (HMI), in aqueous media or in ethylene glycol (104). The intermediate, $ALPO_4$-L or $SAPO_4$-L, is a lamellar aluminophosphate intercalating HMI and water and is stable up to 250°C. $SAPO_4$-L, the precursor of SAPO-35 in ethylene glycol, is stable up to 350°C. Both are believed to be formed by the rearrangement of $3HMIH^+Al_3P_4O_{16}^{-3}$ sheets formed in the reaction medium, as in Figure 16.

Monitoring of SAPO-34 crystallization led to the proposal of a two-stage crystallization process (105). During the time of the first stage, frequently less than 2.5 h, most of the crystallization takes place; as a matter of fact, 80% of the final crystallinity is achieved at this point of the synthesis. Preliminary structure units similar to the framework of SAPO-34 are already formed before crystallization begins. This work provided evidence that the formation of SAPO-34 may occur via a type of gel conversion mechanism. NMR investigations showed that the major part of silicon (~80%) is incorporated in the beginning of the crystallization as well-distributed isolated Si(4OAl) sites. The progress of the reaction at times longer than 2.5 h favored the formation of silica islands by the substitution of 2Si atoms per P + Al atoms (see above). At this stage, apart from the formation of silica patches indicated by NMR, very little happens in terms of crystallization.

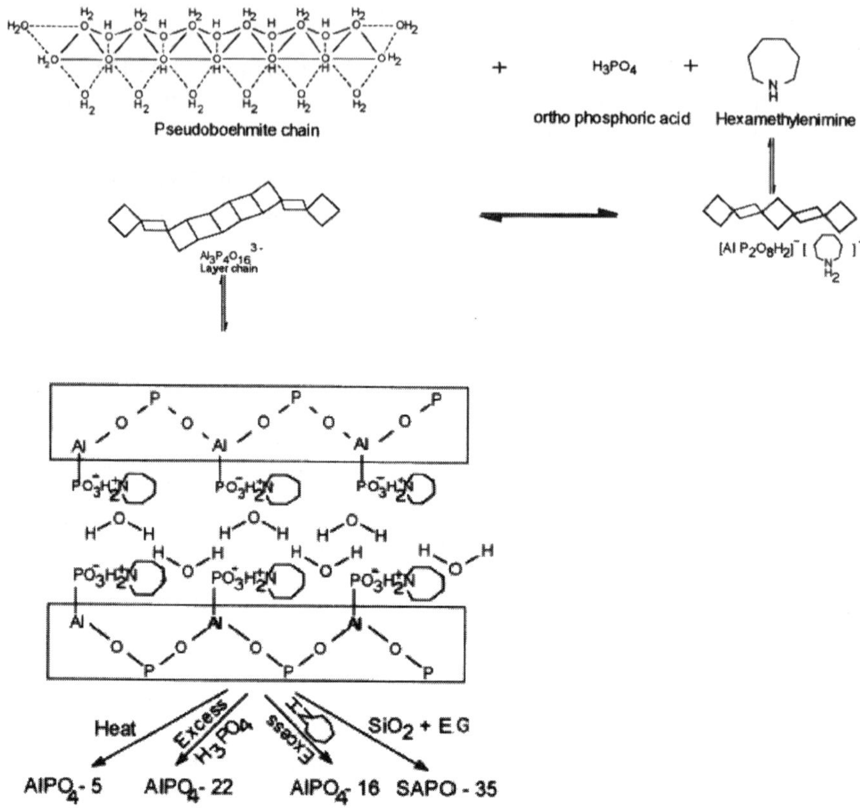

Figure 16 A schematic route for the formation of $AlPO_4$ molecular sieves from the layer $AlPO_4$-L material. *Top*: The formation of $[AlP_2O_8H_2]^-$ chains and of $[Al_3P_4O_{16}]^{-3}$ layer chains by the reaction of alumina with phosphoric acid in the presence of a template; *middle*: assembling of charged layer chains generating $AlPO_4$-L; *bottom*: representation of the several transformations that $AlPO_4$-L undergoes to produce molecular sieves (reproduced from Reference 104 with the permission of Elsevier Sci., © 1998).

A thorough examination of the SAPO-34 crystallization was made by Vistad et al. (from 1999 to 2003). The first study (106) demonstrated that the presence or absence of HF causes significant spectral changes, even at room temperature. The addition of H_3PO_4 and/or HF to the source of aluminum resulted in no direct interaction between fluoride and phosphate although both reacted with the aluminum in acidic environments.

In the presence of fluoride ions, only octahedral Al was observed; the addition of the silicon source provoked the formation of SiF_6^{-2}. Different P-containing species were found in phosphate-based mixtures; addition of HF or morpholine caused the disappearance of these species, some by the formation of P-Al-F complexes of the

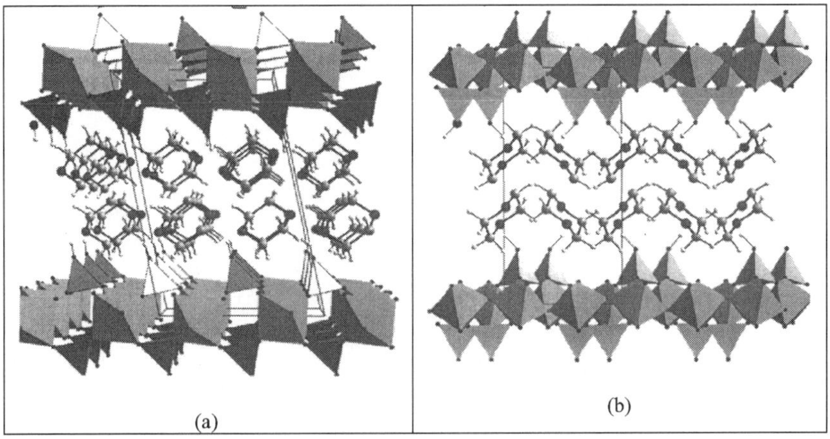

Figure 17 Proposed structure of the prephase viewed along the (*a*) *b*-axis and (*b*) *c*-axis. The octahedral Al-coordination is the result of fluorine bridges (reproduced from Reference 107 with the permission of Am. Chem. Soc., © 2001).

form $[Al(H_2O)_{5-x}(H_2PO_4)(F,OH)_x]^{+2-x}$. At very low pH, AlF_x^{+3-x} (≤ 3 x ≤ 6) is formed and transforms to tetrahedral Al species by the addition of morpholine.

The same gel-containing fluoride, studied at room temperature, undewent hydrothermal treatment at 185°C (107). A layered crystalline precursor phase, the prephase, was obtained after just 1 h, still during the heating of the autoclave, at 170°C. After 4 h, only pure triclinic SAPO-34 phase was present with no important changes occurring at later times. The prephase could be synthesized separately. Its structure consists of $AlPO_4F$ layers separated by a double layer of morpholine molecules, as shown in Figure 17.

Starting from the prephase, the addition of water resulted in the formation of triclinic $ALPO_4$-34 at 180°C after 24 h, which indicated that the prephase contained the elements necessary for the preparation of the molecular sieve.

If the formation of the prephase was experimentally avoided, for instance, by rapid heating to the final temperature for the synthesis of SAPO-34, the formation of SAPO-5 was observed.

Well-designed heating profiles and careful control of the synthesis time resulted in the appearance of the prephase and its transformation into SAPO-34 with time (Figure 18).

The final study in the series (108) dealt with the in situ NMR examination of the SAPO-34 crystallization in the presence and absence of fluoride ions. Figure 19 (see color insert) shows the formation of the different phases, prephase, and SAPO-34 observed in this series of studies. The proposed mechanisms involve only 4-ring-type structures because they display the charge density that matches that of the morpholinium cation; larger rings or double rings would have lower charge density.

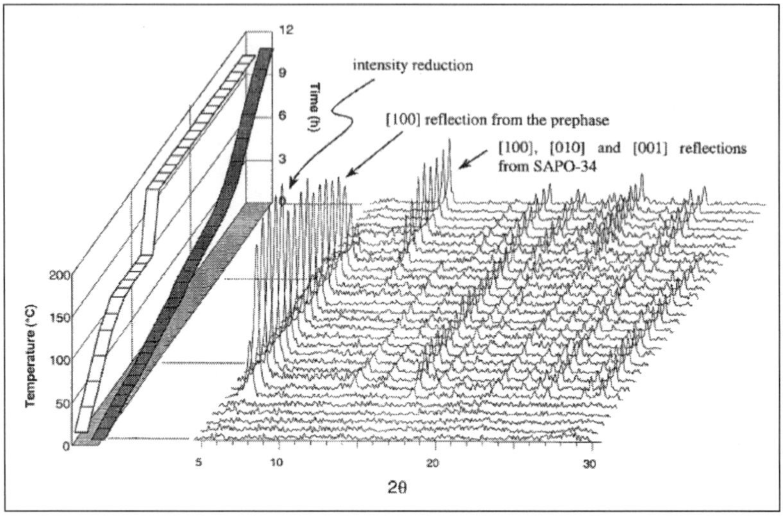

Figure 18 In situ XRD patterns (background removed) obtained during the synthesis of SAPO-34 in the presence of HF and final set temperature of 177°C. The dark and light line graphs on the left-hand side represent the synthesis time and temperature, respectively (reproduced from Reference 107 with the permission of the Am. Chem. Soc., © 2001).

The authors proposed the formation of 4-ring-type I from the dissolution of the gel. This structure assembles to form the layered $AlPO_4F$ prephase that dissolves upon increase in the temperature to form 4-ring-type II. Defluorination processes allow the formation of 4-ring-type III units with pentacoordinated aluminum. If the defluorination process continues, then 4-ring-type V is formed which, by condensing with 4-ring-type III, forms the triclinic chabazite structure.

Silicon incorporation is proposed to occur on the 4-ring-type I unit by substitution of aluminum or phosphorous, i.e., a layered silicoaluminophosphate phase that reorganizes and condenses.

Isolated Precursors

Single-source precursors are useful for understanding the local structure of aluminophosphates and are soluble model compounds for secondary building units (SBU) found in molecular sieves.

Soluble molecular phosphonates are found as (*a*) D4R, $M_4P_4O_{12}$ (M = Al, Ga); (*b*) 8-ring stable intermediates in the formation of D4R; and (*c*) $Al_6P_6O_{18}$, D6R. The skeleton of the molecular structure resembles the D6R building units found in $ALPO_4$-5, $SAPO_4$-5, and others (Scheme 5) (109).

The formation of D6R was possible by tailoring the nature of the alkyl groups on Al and on P; the same strategy may allow the formation of even larger double rings.

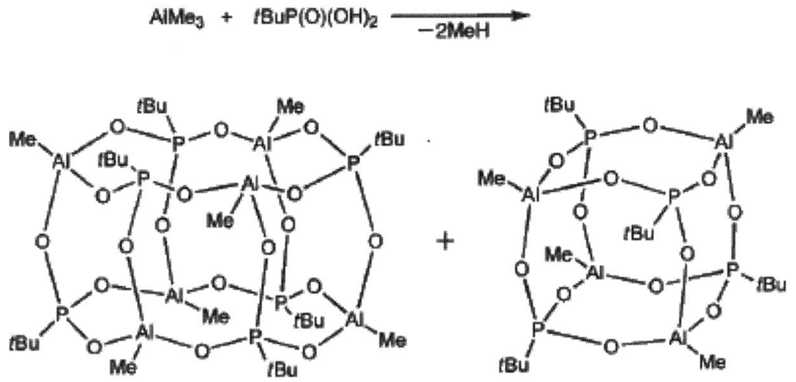

Scheme 5

Single-source precursors are also useful in the sense that a simple thermolytic route is often enough to produce porous aluminophosphates and derivatives.

$L_nM[OSi(O^tBu)_3]_m$ and $L_nM[OP(O)(O^tBu)_2]_m$ were used as single sources of porous metalosilicate and metalophosphate with designed properties (110). The heteroelement alcoxide, $[(O^tBu)_3SiO]_2Al(\mu\text{-}O)_2P(O^tBu)_2]_2Al(Me)OSi(O^tBu)_3$, contains the three elements found in the silicoaluminophosphates (Figure 20) (111); its thermolysis in the solid state at 150°C produced a SAPO material with 32.2% yield. The stability of the Si–O–Al–O–P bonds was confirmed by alcoxide hydrolysis in benzene solution.

Another single-source precursor was reported by Tiemann & Fröba (112) with an empirical formula $[Al(PO_4)(HCl)(C_2H_5OH)_4]_4$ and a structure (Figure 21) composed of a core unit with a cube-like arrangement of alternating aluminum and phosphorous atoms connected to each other by oxygen atoms. This precursor is not stable in aqueous conditions; therefore, the synthesis that employs this precursor ought to be performed in a nonaqueous medium.

These two last precursors were used in the preparation of aluminophosphate molecular sieves and are discussed further below.

ALPO-*ntu* is a lamellar aluminophosphate whose structure mimics that of the naturally occurring silicate mineral kanemite (113); it is also called ALPO-kanemite. The chemical formulas of these materials are $AlPO_3(OH)_2[NH_2(CH_2)_xCH_3]$, with $x = 3$, 5, or 7; the alkylamines act as templates. In the as-prepared state, for either aluminum or phosphorous, three of the four oxygen atoms are used in bonding within the layer, whereas the fourth oxygen points to the interlamellar space (Figure 22). These anionic oxygen atoms bear a proton when bound to aluminum or are compensated by the ammonium ion formed in the reaction when the anionic oxygen is bound to the phosphorous atom. These ammonium cations are exchangeable, although the proton bound to aluminum is not. The structure is best described as a thin layer, composed of 6-membered rings that share all corners (Figure 22*a*).

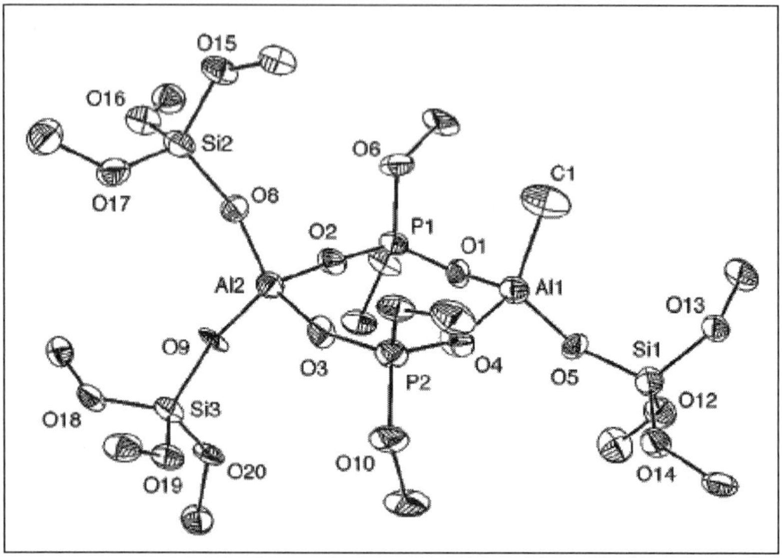

Figure 20 Thermal ellipsoid plot of three heteroelement alcoxide. The *t*Bu methyl groups and all hydrogen atoms have been omitted for clarity (reproduced from Reference 111 with the permission of the Am. Chem. Soc., © 2001).

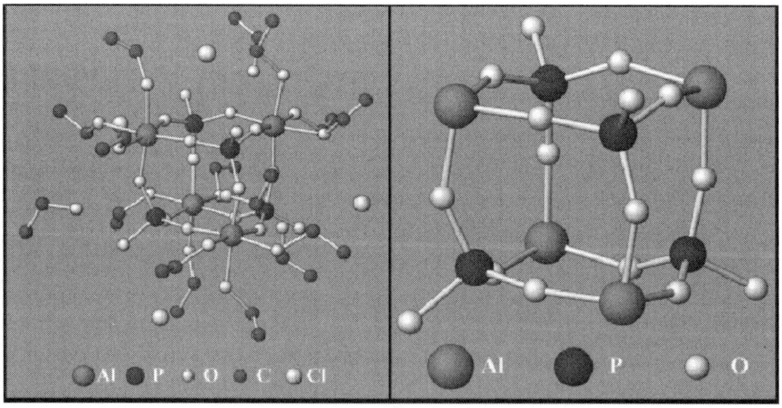

Figure 21 *Left*: schematic representation of the molecular structure of the precursor $[Al(PO_4)(HCl)(C_2H_5OH)_4]_4$; *Right*: the same view without the ethanol and Cl ligands clarifies the cube-like structure of the core unit (H atoms not shown) (reproduced from Reference 112 with the permission of the R. Soc. Chem., © 2002).

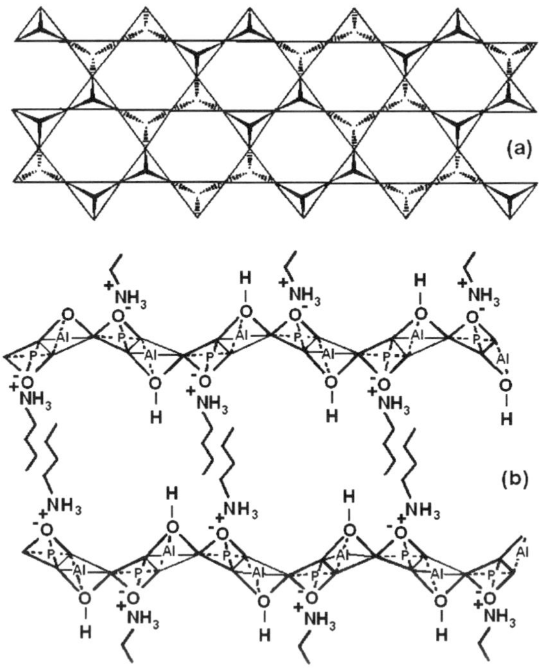

Figure 22 Schematic representation of (*a*) the layer projection of 6-T-membered ring and (*b*) the layer stacking in ALPO-*ntu* or ALPO-kanemite with *n*-butylamine in the interlamellar space (reproduced from Reference 113 with the permission of the Am. Chem. Soc., © 1997).

From *n*-butylamine-ALPO-*ntu*, SiO_2, and HMI, a silicoaluminophosphate, CAL-1, was prepared (114). SAPO-44 was prepared when cyclohexylamine substituted HMI in the synthesis gel (115). Both present a chabazite-type structure and convert to SAPO-34 on calcination. In the absence of SiO_2 and with the addition of morpholine and HF, a phase indistinguishable from ALPO-34 was formed (E.C. Oliveira, L. Marchese & H.O. Pastore, manuscript in preparation). If HMI, SiO_2, and small amounts of water were added to ALPO-kanemite, SAPO-35 was obtained (L.G. Pedroni & H.O. Pastore, manuscript in preparation). These reactions are displayed in Scheme 6.

Although the molecular level properties of CAL-1, SAPO-44, ALPO-34, and SAPO-35 prepared from ALPO-kanemite are virtually the same as the samples produced with the conventional methodologies (starting from soluble sources of T atoms), their morphology can be very different (Figure 23).

ALPO-kanemite is a material composed of disordered, thin blades of various shapes and dimensions. The particles of molecular sieves produced from it have two morphologies that can be identified in Figure 23. The most evident is the presence of more regular layers, either large and separated or small and aggregated. These

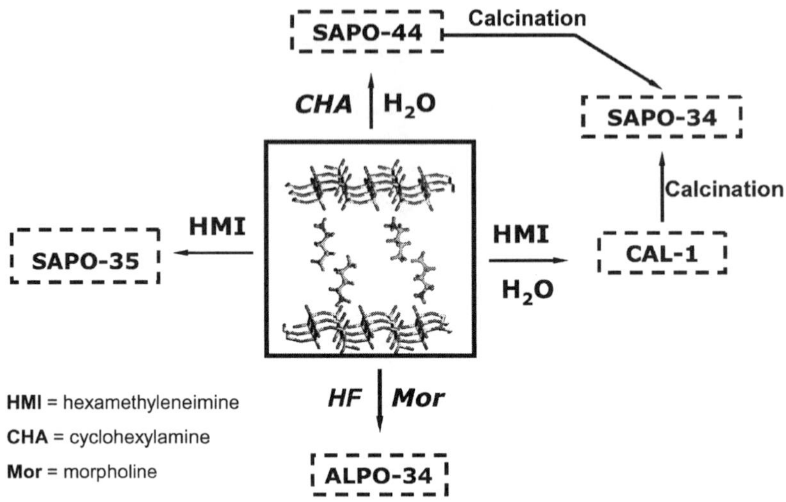

Scheme 6

are most clearly seen in ALPO-34 and SAPO-35. The larger layers are so thin that they are even transparent under the scanning electron microscopy beam.

The second most evident object is the large rhombohedral particle; it is present in all the molecular sieves produced from the ALPO-kanemite and seems to be produced from the sintering and packing of smaller layers, which might be typical of this specific synthesis procedure.

One first observation to make when considering these morphologies is their effect upon the materials properties. In this particular case, diffusion and mass transport properties should be particularly affected because acid sites are more easily accessed in materials composed of thin blades than they are in larger, compact particles. Thus reactions are expected to be facilitated and possibly lead to higher conversions when using materials prepared in this way.

In the reaction of ALPO-kanemite to each of these structures, a material composed of only 6-membered rings is transformed into another one with 6- and 4-membered rings. This process ought to involve condensation of layers to some extent, but also opening of 6R. The process represented in Scheme 7 is a proposal for the initial steps of this reaction.

The high pH caused by HMI provokes the formation of silicates from the dissolution of silica. These groups may attack Al sites in (I) because these sites can sustain an expansion of the coordination sphere. This attack may provoke the formation of a 5-coordinated aluminum (II) or the opening of one 6-membered ring (III). Another oxygen atom in the structure may then bind to the silicon atom, causing the formation of a 4-membered ring of the CHA-like structure of CAL-1.

Figure 23 ALPO-kanemite and the molecular sieves prepared from it. See text for details.

If this proposal is valid, then it may lead to materials with more homogeneous silicon distributions and, as a consequence, to more homogeneous acid site distribution. As a matter of fact, the FTIR spectra in the region of hydroxyl groups on SAPO-34 (kan) (from the calcinations of CAL-1) compared with SAPO-34 (gel) (prepared from the conventional method starting from aluminum hydroxide and phosphoric acid) show slightly narrower bands corresponding to bridging hydroxyl groups, thus signaling a probably more homogeneous distribution of acid sites (Figure 24).

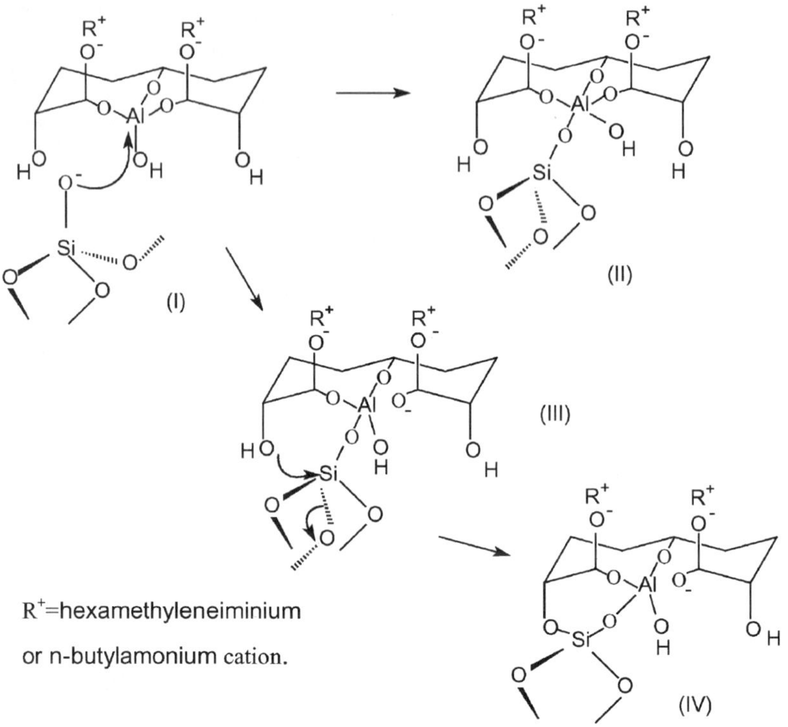

Scheme 7

MESOPOROUS ALUMINOPHOSPHATES AND SILICOALUMINOPHOSPHATES

Since the discovery of mesoporous silicate and aluminosilicate in 1992 (118), the idea of performing the reactions already known for microporous aluminophosphates and silicoaluminophosphates in larger environments became extremely attractive. In fact, this was a search started long before the discovery of mesoporous molecular sieves and resulted in the preparation of VPI-5 and cloverite, which are microporous materials with extra-large pores. The use of surfactants as structure-directing agents led first to the description of mesolamellar aluminophosphates by Sayari & coworkers (119).

It was not until 1997 that the synthesis of mesoporous aluminophosphate and silicoaluminophosphate was reported (120, 121).

From these reports, syntheses of mesostructured aluminophosphate molecular sieves have been reported by several groups over the years, using different types of surfactants (120, 122–129) in aqueous or nonaqueous solvents (121, 122, 130). The materials obtained were either hexagonal or tubular mesoporous aluminophosphate molecular sieves. They may have heteroelements such as Ti (121, 131, 132), Mg

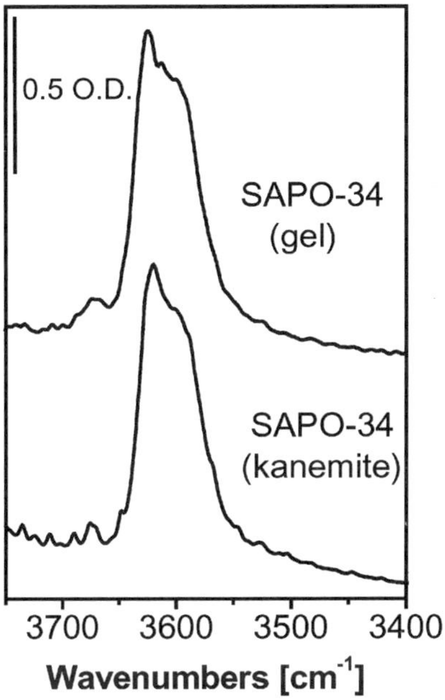

Figure 24 FTIR spectra in the hydroxyl region of SAPO-34 prepared from layered precursors [SAPO-34(kan)] and from conventional gel procedure [SAPO-34(gel)].

(133, 134), Si (127, 134, 135), Fe (137), B (138), Mn, and V (127). Grafted Ti on aluminophosphate was also reported (139). A good review on the subject was published in 2001 (140).

For *meso*-aluminophosphate and *meso*-silicoaluminophosphate, an accurate acidity analysis based on the FT-IR spectra of the adsorption of CO and NH_3 probe molecules is available (141–143).

Figure 25 shows FTIR spectra in the OH stretching region of siliceous MCM-41, *meso*-ALPO and *meso*-SAPO, after calcination at 600°C in O_2.

In the *meso*-ALPO spectrum, the band at 3675 cm^{-1} is assigned to the stretching mode of isolated P-OH (15, 144), whereas the weak band at 3789 cm^{-1} is assigned to the stretching vibration mode of free Al-OH groups of Al ions in tetrahedral coordination; this is evidence that Al ions are present in framework positions (133).

In contrast, the band between ~3743 and 3729 cm^{-1} is caused by the stretching vibration of free Al-OH groups of Al ions in octahedral coordination, as found in transition aluminas (37) and in agreement with the ^{27}Al-MAS-NMR signals of *meso*-ALPO (121, 132, 139).

In the spectrum of *meso*-SAPO, in addition to the bands from the Al-OH and P-OH groups, a new narrow band at 3745 cm^{-1} appears. This signal is due to the stretching mode of isolated Si-OH groups (145, 146). The Al-OH band has its relative intensity diminished with the introduction of silicon into the framework at Si/Al = 0.35 and disappears from the spectra of the *meso*-SAPO prepared at Si/Al = 0.70. The CO adsorption at 100 K on both mesoporous ALPO and SAPO led to the formation of Si-OH · · · CO, Al-OH · · · CO and P-OH · · · CO complexes.

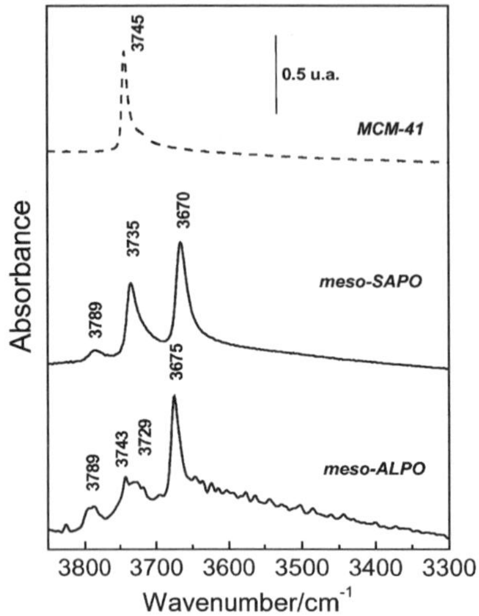

Figure 25 FTIR spectra of calcined siliceous MCM-41, *meso*-ALPO and *meso*-SAPO.

Table 1 reports on the shift of the OH stretching ($\Delta\nu_{OH}$) upon CO adsorption on *meso*-ALPO, *meso*-SAPO, mesoporous MCM-41, and microporous SAPO-34; the larger the shift, the higher the acidity of the sites (29). This study showed that the acidity of P-OH is comparable to that of Al-OH in *meso*-ALPO. The acidity of these groups, however, is definitely larger than that of Si-OH groups present both in *meso*-SAPO and pure siliceous MCM-41. Conversely, the acidity of both P-OH and Al-OH is lower than that of the protons located at the bridged hydroxyls [Si-O(H)-Al] in zeolites or zeotypes. As an example, the shift of bridged OH upon CO adsorption is reported (Table 1) in the case of microporous SAPO-34, (15).

At lower CO pressures, a new band was visible. This was assigned to an interaction of CO with Al^{3+} ions. The band is similar to Lewis sites observed in an Al-containing zeolite, which are additional sites of partially extra framework Al^{3+} (147, 148). AlO-H in partially extra framework groups is not observed at higher

TABLE 1 Values of ν_{OH} and $\Delta\nu_{OH}$ for surface groups on MCM-41, SAPO-34, *meso*-ALPO, and *meso*-SAPO

Sample	Groups	ν_{OH}/cm^{-1}	$\Delta\nu_{OH}$/cm^{-1}
meso-ALPO	P-OH	3675	175
	Al-OH	3730	150–200
meso-SAPO	P-OH	3675	185
	Si-OH	3745	90
MCM-41	Si-OH	3745	70
SAPO-34	Si(OH)Al	3630	287

Si loadings in the ALPO framework. This means that these groups are probably bound by silicon in the formation of *meso*-SAPO.

NH_3 adsorption on *meso*-ALPO (142) showed that Al-OH and P-OH are sufficiently acidic to promote H^+ transfer and produce NH_4^+ ions. For this work, ammonia adsorption was also performed on *meso*-SAPO, where Si-OH is not acidic enough to create NH_4^+ ions but only creates weak hydrogen-bonded Si-OH $\cdots$ NH_3. In highly siliceous mesoporous materials such as MCM-41 or MCM-48, in which only Si-OH groups are present, intact NH_4^+ species were not formed (146). This confirmed that P-OH and Al-OH groups in mesoporous aluminophosphates and silicoaluminophosphates are more acidic than Si-OH. Moreover, the presence of Al^{3+} Lewis acid sites was inferred by a strong band at 1620 cm^{-1}, caused by the asymmetric bending mode of coordinated NH_3.

Aluminophosphate and silicoaluminophosphate with pores in the range of mesoporosity were the paradigm for many reactions that would have been made possible by the use of larger intrazeolitic spaces. However, as shown above (see the comparison of *meso*-SAPO and SAPO-34 in Table 1, for instance) and as seen for the aluminosilicate analogs, the acidity of these materials is not comparable to the acidities of microporous molecular sieves and zeolites, respectively. Nevertheless, the potential of these molecular sieves for organic reactions, where a mild acidity is necessary, is large and will be ready to be explored if the hydrothermal stability of these structures is improved.

Like the microporous counterparts VPI-5 (90) and cloverite (63), the main problem of the mesostructures mentioned is their low structural stability on calcination or when submitted to conventional neutral or acid solvent extraction for the removal of the SDA. An alkaline extraction was proposed for *meso*-ALPO and *meso*-MAPO, the magnesium-containing aluminophosphates, which prevents collapse of the mesostructure while promoting simultaneous ion exchange in Mg-aluminophophates (133). The combined procedure, alkaline ion exchange and calcination to eliminate residues, is effective in preserving the structure after calcination, but when these samples are hydrated, the high-temperature dehydration procedure may cause structure collapse.

One possible solution would be to use a single-source precursor, that is, a reactant that presents Al and P already bound through an oxygen atom. The above-mentioned ALPO-kanemite (113) was the first option, mostly because the mesoporous aluminosilicate, FSM-16 (149), was prepared via the so-called folded sheet mechanism using hydrated Na-silicate kanemite as reactant. The mesoporous aluminophosphate from ALPO-kanemite was prepared at room temperature, upon ion-exchange of the cetyltrimethylamonium cation. However, the resulting porous aluminophosphate compound was thermally unstable, and calcination at 540°C led to the collapse of the ordered structure and formation of an X-ray amorphous phase.

A toluene solution of the compound displayed in Figure 21, a single-source precursor of three structure T atoms (111), with a structure-directing polymer, $EO_{20}PO_{70}EO_{20}$ or $PO_{19}EO_{33}PO_{19}$, produced a mesoporous silicoaluminophosphate with disordered hexagonal or wormhole-type pore structure. Different from

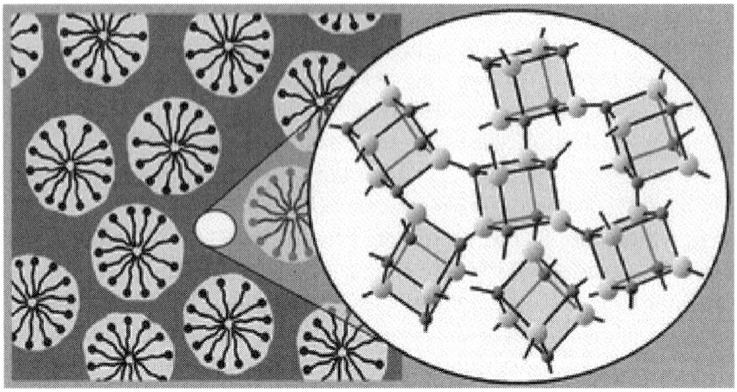

Figure 26 Representation of the proposed structure for *meso*-ALPO prepared from a single-source precursor (reproduced from Reference 112 with the permission of R. Soc. Chem., © 2002).

silicoaluminophosphates in all the previous studies, this was the first mesoporous silicoaluminophosphate where only tetrahedral aluminum was observed by ^{27}Al-NMR. It showed structure stability upon removal of the SDA, similar to the materials discussed above, but its hydrothermal stability was not examined. Also from a single source and in a nonaqueous synthesis, a *meso*-aluminophosphate (112) was prepared from a precursor displayed in Figure 21. The authors proposed the structure (Figure 26) for the mesostructured aluminophophate but, again, no comments were made on the hydrothermal stability.

Thus the field of mesostructured ALPO and SAPO is still open and needs procedures to produce mesostructures stable not only to the elimination of the SDA but also to hydration-dehydration cycles and hydrothermal treatments. The procedures already reported for *meso*-aluminosilicates, which involve partial crystallization of pore walls, may also be a good solution for the *meso*-aluminophosphates and derivatives (150).

PERSPECTIVES

This review deals mainly with synthesis strategies for the production of Si-substituted aluminophosphates used as acid catalysts. However, some of these materials (and their parent ALPOs) have found applications in other fields, such as molecular sieving and supports for optical devices. The use of molecular sieves as advanced functional materials was reviewed by Caro (151). We report on a few examples of recent applications of ALPO-5: (*a*) dye composites with laser activity (152), obtained by in situ inclusion of several dyes (coumarin 466, pyridine 2, etc.); (*b*) Cr-containing SAPO-5 materials with activity in the photoinduced charge separation of pyrene under UV irradiation, which may open the road for photochemical conversion and storage devices (153); and (*c*) SAPO-5 used as a template for the fabrication of single-walled carbon nanotubes (SWCNs). The presence of Brönsted acidity in this material led to the catalytic formation of SWCNs with better quality compared with those SWCNs obtained previously on ALPO-5 (154).

Studies on precursors to molecular sieves seem to be of particular interest. The examples discussed in this review are but a small part of what can be achieved by careful study and synthesis design.

Layered aluminophosphates are suitable precursors not only because of the flexibility of the interlayer space, whose distance can be adjusted by design, but also because interesting changes can be made to the layers themselves. Due to the wealth of chemical transformations, such as ion exchange and grafting, for example, that can be made on the interlamellar functional groups, even a change on the hydrophobic/hydrophilic character is possible. Acidic groups can be created according to necessity by changing the lamella composition. Lamella capable of performing redox reactions can be prepared by inserting transition metal atoms into its structure. All these properties are then transported to the final molecular sieve prepared using these specially engineered layered materials as precursors.

Therefore, in terms of layered precursors, the ALPO-kanemite is an example of the sort of elegant processes that can be operated in the direction of making them into molecular sieves.

Precursors such as the silicoaluminophosphate and aluminophosphate octamers reported by Fujdala & Tilley (111) and Tiemann & Fröba (112) hold very attractive possibilities. One is the increase in the organization of pores walls of mesoporous molecular sieves. This will certainly bring the thermal and hydrothermal stabilities of these materials to levels suitable for industry needs, without mentioning the possible changes of chemical composition to tune the acidity of materials or to finely design redox structures.

ACKNOWLEDGMENTS

The authors are deeply indebted to Drs. D.W. Lewis and G. Sastre for allowing the use of original Figures (13 and 14). Eraldo Luiz Lehmann, Alberto Frache, and Enrica Gianotti are acknowledged for helping with references and figures.

LITERATURE CITED

1. Wilson ST, Lok BM, Messina CA, Cannan TR, Flanigen EM. 1982. Aluminophosphate molecular sieves: a new class of microporous crystalline inorganic solids. *J. Am. Chem. Soc.* 104: 1146–47
2. Lok BM, Messina CA, Patton RL, Gajek RT, Cannan TR, Flanigen EM. 1984.

Silicoaluminophosphate molecular sieves: another new class of microporous crystalline inorganic solids. *J. Am. Chem. Soc.* 106:6092–93
3. International Zeolite Association Website, www.iza-online.org (October 2004)
4. Zecchina A, Lamberti C, Bordiga S. 1998. Surface acidity and basicity: general concepts. *Catal. Today* 41:169–77
5. Hartmann M, Kevan L. 2002. Substitution of transition metal ions into aluminophosphates and silicoaluminophosphates: characterization and relation to catalysis. *Res. Chem. Intermed.* 28:625–95
6. Frache A, Gianotti E, Marchese L. 2003. Spectroscopic characterisation of microporous aluminophosphate materials with potential application in environmental catalysis. *Catal. Today* 77:371–84
7. Wilson ST, Barger PT, Reynolds TM. 1999. *U.S. Patent No. 5912393*
8. Flanigen EM, Grosse RW. 1971. Molecular sieve zeolites I. *Adv. Chem.* 101:76
9. Wilson ST. 2001. Phosphate-based molecular sieves: novel synthetic approaches to new structures and compositions. *Stud. Surf. Sci. Catal.* 37:229–60
10. Pauling L. 1960. The sizes of ions and the structure of ionic crystals. In *The Nature of the Chemical Bond*, pp. 332–400, Ithaca, CA: Cornell Univ. Press. 2nd ed.
11. Pastore HO. 1996. A lógica da substituição isomórfica em peneiras moleculares. *Quím. Nova.* 19:372–76
12. Corà F, Catlow CRA. 2001. Ionicity and framework stability of crystalline aluminophosphates. *J. Phys. Chem. B* 105:10278–81
13. Greenwood NN, Earnshaw A. 1989. *Chemistry of the Elements*. Singapore: Maxwell Macmillan Int. Ed.
14. Chen J, Thomas JM, Wright PA. 1994. Silicoaluminophosphate mumber-18 (SAPO-18)—a new microporous solid acid catalyst. *Catal. Lett.* 28:241–48
15. Chen J, Wright PA, Thomas JM, Natarajan S, Marchese L, et al. 1994. SAPO-18 catalysts and their Brönsted acid sites. *J. Phys. Chem.* 98:10216–24
16. Stöcker M. 1999. Methanol-to-hydrocarbons: catalytic materials and their behavior. *Micropor. Mesopor. Mater.* 29: 3–48
17. Zhu ZD, Hartmann M, Kevan L. 2000. Catalytic conversion of methanol-to-olefins on SAPO-n (n = 11, 34, and 35), CrAPSO-n, and Cr-SAPO-n molecular sieves. *Chem. Mater.* 12:2781–87
18. Martens JA, Grobet PJ, Jacobs PA. 1990. Catalytic activity and Si, Al, P ordering in microporous silicoaluminophosphates of the SAPO-5, SAPO-11, and SAPO-37 type. *J. Catal.* 126:299–305
19. Su BL, Lamy A, Dzwigaj S, Barthomeuf D. 1991. Oxidizing and reducing properties of SAPO-37 molecular-sieve—comparison with acidity and catalysis. *Appl. Catal.* 75:311–20
20. Wang LJ, Guo CW, Yan SR, Huang XD, Li QZ. 2003. High-silica SAPO-5 with preferred orientation: synthesis, characterization and catalytic applications. *Micropor. Mesopor. Mater.* 64:63
21. Marchese L, Frache A, Gatti G, Coluccia S, Lisi L, et al. 2002. Acid SAPO-34 catalysts for oxidative dehydrogenation of ethane. *J. Catal.* 208:479–84
22. Lopez CM, Rodriguez K, Mendez B, Montesa A, Machado FJ. 2000. Influence of the silicon content upon the acidity and catalytic properties of AFI-like catalysts. *Appl. Catal. A* 197:131–37
23. Sastre G, Lewis DW, Catlow CRA. 1996. Structure and stability of silica species in SAPO molecular sieves. *J. Phys. Chem.* 100:6722–30
24. Barthomeuf D. 1994. Topological model for the compared acidity of SAPOs and SiAl zeolites. *Zeolites* 14:394–401
25. Barthomeuf D. 1993. Topology and maximum content of isolated species (Al, Ga, Fe, B, Si, . . .) in a zeolitic framework. An approach to acid catalysis. *J. Phys. Chem.* 97:10092–96

26. Sastre G, Lewis DW, Catlow CRA. 1997. Modeling of silicon substitution in SAPO-5 and SAPO-34 molecular sieves. *J. Phys. Chem.* 101:5249–62
27. Vomscheid R, Briend M, Peltre MJ, Man PP, Barthomeuf D. 1994. The role of the template in directing the Si distibution in SAPO zeolites. *J. Phys. Chem.* 98:9614–18
28. Sastre G, Lewis DW, Catlow CRA. 1997. Mechanism of silicon incorporation in aluminophosphate molecular sieves. *J. Mol. Catal. A* 119:349–56
29. Coluccia S, Marchese L, Martra G. 1999. Characterization of microporous and mesoporous materials by the adsorption of molecular probes: FTIR and UV-Vis studies. *Micropor. Mesopor. Mater.* 30:43–56
30. Knözinger H, Huber S. 1998. IR spectroscopy of small and weakly interacting molecular probes for acidic and basic zeolites. *J. Chem. Soc. Faraday Trans.* 94:2047–59
31. Hadjiivanov KI, Vayssilov GN. 2002. Characterization of oxide surfaces and zeolites by carbon monoxide as IR probe molecule. *Adv. Catal.* 47:307–511
32. Sastre G, Lewis DW. 1998. Modelling of Brönsted acidity in AFI and CHA zeotypes. *J. Chem. Soc. Faraday Trans.* 94:3049–58
33. Sauer J, Schröder KP, Termath V. 1998. Comparing the acidities of microporous aluminosilicate and silicoaluminophosphate catalysts: a combined quantum mechanics-interatomic potential function study. *Collect. Czech. Chem. Commun.* 63:1394–408
34. Elanany M, Koyama M, Kubo M, Selvam P, Miyamoto A. 2004. Periodic density functional investigation of Brønsted acidity in isomorphously substituted chabazite and ALPO-34 molecular sieves. *Micropor. Mesopor. Mater.* 71:51–56
35. Sastre G, Lewis DW, Corma A. 2000. The role of the electrostatic potential, electric field and electric field gradient on the acidity of AFI and CHA zeotype. *Phys. Chem. Chem. Phys.* 2:177–85
36. Lewis DW, Sastre G. 1999. Rationalization of the IR stretching frequencies of Brønsted acid centres in microporous solids. *Chem. Commun.* 349–50
37. Morterra C, Magnacca G. 1996. A case study: surface chemistry and surface structure of catalytic aluminas, as studied by vibrational spectroscopy of adsorbed species. *Catal. Today* 27:497–532
38. Makarova MA, Ojo AF, Karim K, Hunger M, Dwyer J. 1994. FTIR study of weak hydrogen bonding of Brönsted hydroxyls in zeolites and aluminophosphates. *J. Phys. Chem.* 98:3619–23
39. Onida B, Gabelica Z, Lourenço JP, Ribeiro MF, Garrone E. 1997. Spectroscopic characterization of the hydroxyl groups in SAPO-40. 2. Interaction with CO and N_2. *J. Phys. Chem. B* 101:9244–49
40. Smith L, Cheetham AK, Marchese L, Gianotti E, Thomas JM, et al. 1996. A quantitative description of the active sites in the dehydrated acid catalyst HSAPO34 for the conversion of methanol to olefins. *Catal. Lett.* 41:13–16
41. Gianotti E, Frache A, Coluccia S, Thomas JM, Maschmeyer T, Marchese L. 2003. The identity of titanium centres in microporous aluminophosphates compared with Ti-MCM-41 mesoporous catalyst and titanosilsesquioxane dimer molecular complex: a spectroscopic study. *J. Mol. Catal. A* 204–205:483–89
42. Jappar N, Tanaka Y, Nakata S, Tatsumi T. 1998. Synthesis and characterization of a new titanium silicoaluminophosphate: TAPSO-37. *Micropor. Mesopor. Mater.* 23:169–78
43. Barker CM, Gleeson D, Kaltsoyaqnnis N, Catlow CRA, Sankar G, Thomas JM. 2002. On the structure and coordination of the oxygen-donating species in

Ti-MCM-41/TBHP oxidation catalysts: a density functional theory and EXAFS study. *Phys. Chem. Chem. Phys.* 4:1228–40
44. Shetti VN, Manikandan P, Srinivas D, Ratnasamy P. 2003. Reactive oxygen species in epoxidation reactions over titanosilicate molecular sieves. *J. Catal.* 216:461–67
45. Escalante D, Giraldo L, Pinto M, Pfaff C, Sazo V, et al. 1997. A study of the feasibility of incorporation of chromium into the molecular sieve framework: the transformation of 1-butene over Cr-silicoaluminophosphate molecular sieves. *J. Catal.* 169:176–87
46. Chen JD, Sheldon RA. 1995. Selective oxidation of hydrocarbons with O_2 over chromium aluminophosphate-5 molecular-sieve. *J. Catal.* 153:1–8
47. Thomas JM, Raja R, Sankar G, Bell RG. 2001. Molecular sieve catalysts for the regioselective and shape-selective oxyfunctionalization of alkanes in air. *Acc. Chem. Res.* 34:191–200
48. Modén B, Oliviero L, Dakka J, Santiesteban JG, Iglesia E. 2004. Structural and functional characterization of redox Mn and Co sites in AlPO materials and their role in alkane oxidation catalysis. *J. Phys. Chem. B* 108:5552–63
49. Höchtl M, Jentys A, Vinek H. 2001. Isomerization of 1-pentene over SAPO, CoAPO (AEL, AFI) molecular sieves and HZSM-5. *Appl. Catal. A* 207:397–405
50. Chen JS, Thomas JM. 1994. MAPO-18 (M-Mg,Zn,Co)—a new family of catalysts for the conversion of methanol to light olefins. *J. Chem. Soc. Chem. Commun.* 603–4
51. Franklin IL, Beale AM, Sankar G. 2003. On the activity, longevity and recyclability of Mn(II) and Co(II) substituted AlPO-18 catalysts for the conversion of methanol to light olefins. *Catal. Today* 81:623–29
52. Thomas JM, Greaves GN, Sankar G, Wright PA, Chen J, et al. 1994. On the nature of the active site in a CoAPO-18 solid acid catalyst. *Angew. Chem. Int. Ed. Engl.* 33:1871–73
53. Marchese L, Chen JS, Thomas JM, Coluccia S, Zecchina A. 1994. Brönsted, Lewis and redox centres on CoAPO-18 catalysts. Vibrational modes of adsorbed water. *J. Phys. Chem.* 98:13350–56
54. Marchese L, Martra G, Damilano N, Coluccia S, Thomas JM. 1996. Elucidating the nature of the cobalt centres in CoAPO-18 acid catalysts. *Stud. Surf. Sci. Catal.* 101:861–70
55. Gianotti E, Marchese L, Martra G, Coluccia S. 1999. The interaction of NO with Co^{2+}/Co^{3+} redox centres in CoAPOs catalysts: FTIR and UV-Vis investigations. *Catal. Today* 54:547–52
56. Lourenço JP, Ribeiro MF, Ribeiro FR, Rocha J, Gabelica Z, et al. 1997. Generation of acid sites by incorporation of cobalt in the AFR structure. *Stud. Surf. Sci. Catal.* 105:1973–80
57. Kang M. 2000. Methanol conversion on metal-incorporated SAPO-34s (MeAPSO-34s). *J. Mol. Catal. A* 160:437–44
58. Kang M. 2000. Effect of cobalt incorporated into the framework of SAPO-34 (CoAPSO-34s) on NO removal. *J. Mol. Catal. A* 161:115–23
59. Briend M, Lamy A, Peltre MJ, Man PP, Barthomeuf D. 1993. Thermal-stability of tetrapropylammonium (TPA) and tetramethylammonium (TMA) cations occluded in SAPO-37 molecular-sieves. *Zeolites* 13:201–11
60. Lourenço JP, Ribeiro MF, Ribeiro FR, Rocha J, Gabelica Z, et al. 1994. Study of catalytic properties of SAPO-40. *Stud. Surf. Sci. Catal.* 84:867–74
61. Wilson ST, King L. 1988. *European Patent No. 293938*
62. Sierra L, Patarin J, Derouche C, Gies H, Guth JL. 1994. Novel molecular-sieves of the aluminophosphate family—$ALPO_4$ and substituted derivatives with

the LTA, FAU and AFR structure-types. *Stud. Surf. Sci. Catal.* 84C:2237–44
63. Merrouche A, Patarin J, Kessler H, Soulard M, Delmotte L, et al. 1992. Synthesis and characterization of cloverite—a novel gallophosphate molecular-sieve with 3-dimensional 20-membered ring channels. *Zeolites* 12:226–32
64. Ashtekar S, Barrie PJ, Heargreaves M, Gladden LF. 1997. An FT-Raman study of the template-framework interaction in $ALPO_4$-based molecular sieves. *Angew. Chem. Int. Ed. Engl.* 36:876–78
65. Marchese L, Frache A, Gianotti E, Martra G, Causà M, Coluccia S. 1999. ALPO-34 and SAPO-34 synthesized by using morpholine as templating agent. FTIR and FT Raman studies of the host-guest and guest-guest interactions within the zeolitic framework. *Micropor. Mesopor. Mater.* 30:145–53
66. Martucci A, Alberti A, Cruciani G, Frache A, Coluccia S, Marchese L. 2003. Combined single crystal X-ray diffraction and FTIR study of morpholinium-water molecular complexes embedded in a chabasite network. *J. Phys. Chem. B.* 107:9655–61
67. Gotoh K, Ishimaru S, Ikeda R. 2000. Dynamic behaviour of triethylamine molecules adsorbed in aluminophosphate ($ALPO_4$-5) and silicoaluminophosphate ($SAPO_4$-5) molecular sieves. *Phys. Chem. Chem. Phys.* 2: 1865–69
68. Knop-Gerrits P-P, Toufar H, Li X-Y, Grobet P, Schoonheydt RA, et al. 2000. The structure of water in crystalline aluminophosphates: isolated water and intermolecular clusters probed by Raman spectroscopy, NMR and structural modeling. *J. Phys. Chem. A* 104:2410–22
69. Guth JL, Kessler H, Higel JM, Lamblin JM, Patarin J, et al. 1989. Zeolite synthesis in the presence of fluoride ions—a comparison with conventional synthesis methods. *ACS Symp. Ser.* 398:176–95
70. Matijasic A, Paillaud J-L, Patarin J. 2000. Synthesis, characterization and structure determination of Mu-15: a new fluorogallophosphate with fluoride as a specific template of the D4R units. *J. Mater. Chem.* 10:1345–51
71. Sierra L, Patarin J, Guth JL. 2000. Synthesis of aluminophosphates and derived materials with the ZON structure type. *Micropor. Mesopor. Mater.* 38:123–33
72. Morris RE, Burton A, Bull LM, Zones SI. 2004. SSZ-51—a new aluminophosphate zeotype: synthesis, crystal structure, NMR and dehydration properties. *Chem. Mater.* 16:2844–51
72a. Briend M, Vomscheid R, Peltre MJ, Man PP, Barthomeuf D. 1995. Influence of the choice of the template on the short- and long-term stability of SAPO-34. *J. Phys. Chem.* 99:8270–76
73. Wheatley PS, Morris RE. 2002. Cyclam as a structure directing agent in the crystallization of aluminophosphate open framework materials from fluoride media. *J. Solid. State Chem.* 167:267–73
74. Xu WY, Dong JX, Li JP, Li JQ, Wu F. 1990. A novel method for the preparation of zeolite ZSM-5. *J. Chem. Soc. Chem. Commun.* 755–56
75. Zhang L, Gavalas GR. 1999. Vapor-phase transport synthesis of ZnAPO-34 molecular sieve. *Chem. Commun.* 97–98
76. Bandyopadhyay R, Bandyopadhyay M, Kubota Y, Sugi Y, 2002. Synthesis of $ALPO_4$ molecular sieves with AFI and AEL structures by dry gel conversion method and catalytic application of their SAPO counterparts on isopropylation of biphenyl. *J. Porous Mater.* 9:83–95
77. Arafat A, Jansen JC, Ebaid AR, van Bekkum H. 1990. Microwave preparation of zeolite-Y and ZSM-5. *Zeolites* 13:162–65
78. Park M, Komarneni S. 1998. Rapid synthesis of $AlPO_4$-11 and cloverite by microwave-hydrothermal processing. *Micropor. Mesopor. Mater.* 20:39–44
79. Zhao JP, Cundy C, Dwyer J. 1997. Synthesis of zeolites in a microwave

heating environment. *Stud. Surf. Sci. Catal.* 105:181–88
80. Xu XC, Yang WS, Liu J, Lin LW. 2000. Synthesis of a high-permeance NaA zeolite membrane by microwave heating. *Adv. Mater.* 12:195
81. Jhung SH, Chang J-S, Hwang JS, Park S-E. 2003. Selective formation of SAPO-5 and SAPO-34 molecular sieves with microwave irradiation and hydrothermal heating. *Micropor. Mesopor. Mater.* 64:33–39
82. Girnus I, Jancke K, Vetter R, Richter-Mendau J, Caro J. 1995. Large $ALPO_4$-5 crystals by microwave heating. *Zeolites* 15:33–39
83. Jhung SH, Chang J-S Hwang, YK, Park S-E. 2004. Crystal morphology control of AFI molecular sieves with microwave irradiation. *J. Mater. Chem.* 14:280–85
84. Jhung SH, Chang J-S, Kim DS, Park S-E. 2004. Effects of silica on the synthesis of AFI molecular sieve in acid and base conditions under microwave irradiation. *Micropor. Mesopor. Mater.* 71:135–42
85. Iwasaki A, Sano T, Kodana T, Kyozumi Y. 2003. Growth behavior of AFI-type crystals. *Micropor. Mesopor. Mater.* 64:145–53
86. Wan Y, Williams CD, Duke CVA, Cox JJ. 2000. Systematic studies on the effect of water content on the synthesis, crystallization, conversion and morphology of $ALPO_4$-5 molecular sieve. *J. Mater. Chem.* 10:2857–62
87. Rakoczy RA, Ernest S, Hartmann M, Traa Y, Weitkamp J. 1999. Synthesis of large molecular sieve crystals with the AFI ($ALPO_4$-5) topology. *Catal. Today* 49:261–66
88. Sinha AK, Sainkar S, Sivasanker S. 1999. An improved method for the synthesis of the silicoaluminophosphate molecular sieves, SAPO-5, SAPO-11 and SAPO-31. *Micropor. Mesopor. Mater.* 31:321–31
89. Seelan S, Sinha AK. 2004. Crystallization and characterization of high-silica silicoaluminophosphate SAPO-5. *J. Mol. Catal. A* 215:149–52
90. Davis ME, Montes C, Hathaway PE, Garcés JM. 1989. Synthesis of aluminophosphates and element-substituted aluminophosphate VPI-5. *Stud. Surf. Sci. Catal.* 49A:199–206
91. del Val S, Blasco T, Sastre E, Pérez-Pariente J. 1995. Synthesis of SiVPI-5 with enhanced activity in acid-catalysed reactions. *J. Chem. Soc. Chem. Commun.* 731–32
92. Franco MJ, Mifsud A, Pérez-Pariente J. 1995. Study of SAPO-5 obtained from surfactant-containing gels: Part 1. Crystallization parameters and mechanisms of Si substitution. *Zeolites* 15:117–23
93. Han XS, Williams CD, Lee DF, Round CI. 2002. Synthesis of high silicon content SAPO-5 using anionic surfactants in a hexagonal/aqueous two-phase media. *Chem. Commun.* 166–67
94. Dimitriu E, Lutic D, Hulea V, Dorohoi D, Azzouz A, et al. 1999. Synthesis optimization of Chabasite-like SAPO-47 in the presence of sec-butylamine. *Micropor. Mesopor. Mater.* 31:187–93
95. Li Y, Yu J, Liu D, Yan W, Xu R, Xu Y. 2003. Design of zeolite frameworks with defined pore geometry through constrained assembly of atoms. *Chem. Mater.* 15:2780–85
96. Mellot-Draznieks C, Girard S, Férey G. 2002. Novel inorganic frameworks constructed from double-four-ring (D4R) units: computational design, structures, and lattice energies of silicate, aluminophosphate, and gallophosphate candidates. *J. Am. Chem. Soc.* 124: 15326–35
97. Lewis DW, Willock DJ, Catlow C Richard A, Thomas JM, Hutchings GJ. 1996. De novo design of structure-directing agents for the synthesis of microporous solids. *Nature* 382:604–6
97a. Thomas JM. 1999. Design, synthesis, and in situ characterization of new solid

catalysts. *Angew. Chem. Int. Ed.* 38: 3588–628
98. Akporiaye DE, Dahl JM, Karlsson A, Wendelbo P. 1998. Combinatorial approach to the hydrothermal synthesis of zeolites. *Angew. Chem. Int. Ed. Engl.* 37: 609–11
99. Klein J, Lehmann CW, Schmidt H-W, Maier WF. 1998. Combinatorial material libraries on the microgram scale with an example of hydrothermal synthesis. *Angew. Chem. Int. Ed. Engl.* 37:3369–72
100. Deleted in proof
101. Zhang L, Yao J, Zeng C, Xu N. 2003. Combinatorial synthesis of SAPO-34 via vapor phase transport. *Chem. Commun.* 2232–33
102. Song Y, Yu J, Li G, Li Y, Wang Y, Xu R. 2002. Combinatorial approach for the hydrothermal synthesis of open-framework zinc phosphates. *Chem. Commun.* 1720–21
103. Huang Y, Demko BA, Kirby CW. 2003. Investigation of the evolution of intermediate phases of $ALPO_4$-18 molecular sieve synthesis. *Chem. Mater.* 15:2437–44
104. Venkatathri N, Hedge SG, Ramaswamy V, Sivasanker S. 1998. Isolation and characterization of a novel lamellar-type aluminophosphate, ALPO-L, a common precursor for $ALPO_4$ molecular sieves. *Micropor. Mesopor. Mater.* 23:277–85
105. Tan J, Liu Z, Bao X, Liu X, Han X, et al. 2002. Crystallization and Si incorporation mechanisms of SAPO-34. *Micropor. Mesopor. Mater.* 53:97–108
106. Vistad ØB, Hansen EW, Akporiaye DE, Lillerud KP. 1999. Multinuclear NMR analysis of SAPO-34 gels in the presence and absence of HF: the initial gel. *J. Phys. Chem. A* 103:2540–52
107. Vistad ØB, Akporiaye DE, Lillerud KP. 2001. Identification of a key precursor phase for synthesis of SAPO-34 and kinetics of formation investigated by in situ X-ray diffraction. *J. Phys. Chem.* 105:12437–47
108. Vistad ØB, Akporiaye DE, Taulelle F, Lillerud KP. 2003. In situ NMR of SAPO-34 crystallization. *Chem. Mater.* 15:1639–49
109. Yang YM, Walawalkar G, Pinkas J, Roesky HW, Schmidt H-G. 1998. Molecular aluminophosphonate: model compound for the isoelectronic double-six-ring (D6R) secondary building unit of zeolites. *Angew. Chem. Int. Ed.* 37:96–98
110. Fujdala KL, Tilley TD. 2001. Thermolytic transformation of tris(alkoxy) siloxychromium(IV) single-source molecular precursors to catalytic chromia-silica materials. *Chem. Mater.* 13: 1817–27
110a. Lugmair CG, Tilley TD, Rheingold AL. 1999. Di(*tert*-butyl)phosphate complexes of aluminum: precursors to aluminum phosphate xerogels and thin films. *Chem. Mater.* 11:1615–20
110b. Terry KW, Lugmair CG, Tilley TD. 1997. Tris(*tert*-butoxy)siloxy complexes as single-source precursors to homogeneous zirconia- and hafnia-silica materials. An alternative to the sol-gel method. *J. Am. Chem. Soc.* 119:9745–56
111. Fujdala KL, Tilley TD. 2001. An efficient, single-source molecular precursor to silicoaluminophosphates. *J. Am. Chem. Soc.* 123:10133–34
112. Tiemann M, Fröba M. 2002. Mesoporous aluminophosphates from a single source pecursor. *Chem. Commun.* 406–7
113. Cheng S, Tzeng J-N, Hsu B-Y. 1997. Synthesis and caracterization of a novel layered aluminophosphate of kanemite-like structure. *Chem. Mater.* 9:1788–96
114. Albuquerque A, Coluccia S, Marchese L, Pastore HO. 2004. Synthesis of SAPO-34 from the lamellar ALPO-kanemite. *Stud. Surf. Sci. Catal.* 154: 966–71
115. Martins GAV, Pastore HO. 2005. The synthesis of SAPO-44 from lamellar

ALPO-kanemite. *Stud. Surf. Sci. Catal.* In press
116. Deleted in proof
117. Deleted in proof
118. Beck JS, Vartuli JC, Roth WJ, Leonowicz ME, Kresge CT, et al. 1992. A new family of mesoporous molecular sieves prepared with liquid crystal templates. *J. Am. Chem. Soc.* 114:10834–43
119. Sayari A, Karra VR, Reddy JS, Moudrakovski IL. 1996. Synthesis of mesostructured lamellar aluminophosphates. *Chem. Commun.* 411–12
120. Chakraborty B, Pulikottil AC, Das S, Viswanathan B. 1997. Synthesis and characterization of mesoporous SAPO. *Chem. Commun.* 911–12
121. Kimura T, Sugahara Y, Kuroda K. 1997. Synthesis of a hexagonal mesostructured aluminophosphate. *Chem. Lett.* 26:983–84
122. Feng P, Xia Y, Feng J, Bu X, Stucky GD. 1997. Synthesis and characterization of mesostructured aluminophosphates using the fluoride route. *Chem. Commun.* 949–50
123. Cabrera S, Haskouri JE, Guillen C, Beltrán-Porter A, Beltrán-Porter D, et al. 1999. Tuning the pore size from micro- to mesoporous in thermally stable aluminophosphates. *Chem. Commun.* 333–34
124. Chakraborty B, Pulikottil AC, Viswanathan B. 1998. Physico-chemical and MAS NMR characterization of mesoporous SAPOs. *Appl. Catal. A* 167: 173–81
125. Perez OJO, Borade BB, Clearfield A. 1998. Synthesis of a mesoporous aluminophosphate. *J. Mol. Struct.* 470:221–28
126. Zhao D, Luan Z, Kevan L. 1997. Synthesis of thermally stable mesoporous hexagonal aluminophosphate molecular sieves. *Chem. Commun.* 1009–10
127. Luan Z, Zhao D, He H, Klinowski J. 1998. Tubular aluminophosphate mesoporous material containing framework silicon, vanadium and manganese. *Stud. Surf. Sci. Catal.* 117:103–10
128. Khimiyak YZ, Klinowski J. 1998. Synthesis of new mesostructured aluminophosphates. *J. Chem. Soc. Faraday Trans.* 94:2241–47
129. Khimiyak YZ, Klinowski J. 2002. Synthesis of mesostructured aluminophosphate using cationic templating. *Phys. Chem. Chem. Phys.* 2:5275–85
130. Kapoor MP, Raj A. 2000. Synthesis of mesoporous hexagonal titanium aluminophosphate molecular sieves and their catalytic applications. *Appl. Catal. A* 203:311–19
131. Zhao XS, Lu GQ. 2001. Aluminophosphate-based mesoporous molecular sieves: synthesis and characterization of TAPOs. *Micropor. Mesopor. Mater.* 44–45:185–94
132. Mohapatra SK, Hussain F, Selvam P. 2001. Titanium substituted hexagonal mesoporous catalysts for the oxidation of phenols at room remperature. *Catal. Commun.* 4:57–62
133. Masson NC, Pastore HO. 2001. Synthesis and characterization of aluminophosphate mesoporous material containing framework magnesium. *Micropor. Mesopor. Mater.* 44–45:173–81
134. Khimiyak YZ, Klinowski J. 2001. Incorporation of magnesium in mesostructured and mesoporous aluminophosphate. *Phys. Chem. Chem. Phys.* 3: 1544–51
135. Oliveira EC, Pastore HO. 2002. Synthesis of mesoporous silicoaluminophosphates (SAPO). *Stud. Surf. Sci. Catal.* 141:299–304
136. Deleted in proof
137. Wang L, Tian B, Fan J, Liu X, Yang H, et al. 2004. Block copolymer templating synthesis of ordered large-pore stable mesoporous aluminophosphates and Fe-aluminophosphate based on an "acid-base air" route. *Micropor. Mesopor. Mater.* 67:123–33
138. Khimiyak YZ, Klinowski J. 2002.

Synthesis and characterization of mesoporous aluminophosphates containing boron. *J. Mater. Chem.* 12:1079–85
139. Gianotti E, Oliveira EC, Coluccia S, Pastore HO, Marchese L. 2003. Synthesis and surface properties of Ti-containing mesoporous aluminophosphates. A comparison with Ti-grafted mesoporous Silica Ti-MCM-41. *Inorg. Chim. Acta* 349:259–64
140. Tiemann M, Fröba M. 2001. Mesostructured aluminophosphates synthesized with supramolecular structure directors. *Chem. Mater.* 13:3211–17
141. Gianotti E, Oliveira EC, Dellarocca V, Coluccia S, Pastore HO, Marchese L. 2002. Meso-ALPO prepared by thermal decomposition of the organic-inorganic composite. A FTIR spectroscopic study. *Stud. Surf. Sci. Catal.* 141:417–24
142. Gianotti E, Oliveira EC, Dellarocca V, Coluccia S, Pastore HO, Marchese L. 2002. Acidity of mesoporous aluminophosphates and Silicas MCM-41. A combined FTIR and UV-Vis-NIR study. *Stud. Surf. Sci. Catal.* 142B:1419–26
143. Gianotti E, Oliveira EC, Coluccia S, Pastore HO, Marchese L. 2004. The surface acidity of mesoporous silicoaluminophosphates. A FTIR study. *Stud. Surf. Sci. Catal.* 154:1498–504
144. Marchese L, Chen JS, Wright PA, Thomas JM. 1993. Formation of hydronium at the Broensted site in SAPO-34 catalysts. *J. Phys. Chem.* 97:8109–12
145. Marchese L, Gianotti E, Dellarocca V, Maschmeyer T, Rey F, et al. 1999. Structure-functionality relationships of grafted Ti-MCM41 silicas. Spectroscopic and catalytic studies. *Phys. Chem. Chem. Phys.* 1:585–92
146. Gianotti E, Dellarocca V, Marchese L, Martra G, Coluccia S, Maschmeyer T. 2002. NH_3 adsorption on MCM-41 and Ti-grafted MCM-41. FTIR, DR UV–Vis–NIR and photoluminescence studies. *Phys. Chem. Chem. Phys.* 4:6109–15
147. Katovic A, Giordano G, Bonelli B, Onida B, Garrone G, et al. 2001. Preparation and characterization of mesoporous molecular sieves containing Al, Fe or Zn. *Micropor. Mesopor. Mater.* 44–45:275–81
148. Onida B, Borello L, Bonelli B, Geobaldo F, Garrone E. 2003. IR study of the acidity of ITQ-2, an "all-surface" zeolitic system. *J. Catal.* 214:191–99
149. Yanagisawa T, Shimizu T, Kuroda K, Kato C. 1990. Trimethylsilyl derivatives of alkyltrimethylammonium-kanemite complexes and their conversion to microporous SiO_2 materials. *Bull. Chem. Soc. Jpn.* 63:1535–37
150. Pinnavaia TJ, Kim S-S, Zhang Z, Liu Y. 2004. Mesoporous metal oxides with improved atomic ordering in the pore walls. *Stud. Surf. Sci. Catal.* 154:14–19
151. Caro J. 2004. Zeolites and mesoporous materials as advanced functional material. *Stud. Surf. Sci. Catal.* 154:82–97
152. Weitß Ö, Loerke J, Wüstefeld U, Marlow F, Schüth F. 2002. Host-guest interactions and laser activity in ALPO-5/laser dye composites. *J. Solid State Chem.* 167:302–9
153. Ranjit KT, Kevan L. 2003. Photoinduced charge separation of pyrene in chromium containing silicoaluminophosphate (SAPO-5) microporous materials at room temperature. *J. Phys. Chem.* B 107:2610–17
154. Li ZM, Zhai JP, Liu HJ, Li IL, Chang CT, et al. 2004. Synthesis of 4 Å single-walled carbon nanotubes in catalytic Si-Substituted ALPO-5 molecular sives. *Appl. Phys. Lett.* 85:1253–55

Annu. Rev. Mater. Res. 2005. 35:397–426
doi: 10.1146/annurev.matsci.35.102003.142526

First published online as a Review in Advance on April 1, 2005

CHEMICAL DESIGN AND IN SITU CHARACTERIZATION OF ACTIVE SURFACES FOR SELECTIVE CATALYSIS

Mizuki Tada and Yasuhiro Iwasawa
Department of Chemistry, Graduate School of Science, The University of Tokyo, Hongo, Bunkyo-ku, Tokyo 113-0033, JAPAN; email: iwasawa@chem.s.u-tokyo.ac.jp

Key Words design of catalyst surfaces, active structures, molecular imprinting, supported metal complexes, time-resolved XAFS

■ **Abstract** This review illustrates how catalytically active molecular structures are created on oxide surfaces by attachment of metal complexes with subsequent structural transformation and molecular imprinting on the surfaces. Also discussed is how the designed surfaces are characterized in situ by physical techniques including conventional and time-resolved X-ray absorption fine structure. The structural transformation and molecular imprinting for attached metal complexes can provide a new class of catalytic materials with a high complexity for selective catalysis including shape-selective and asymmetric catalysis.

INTRODUCTION

Chemical design of artificial enzyme catalysts with high activity and molecular recognition has been a long-term challenge in catalytic materials research. Ligands coordinated to metal center significantly modify and regulate not only reactivity of the metal electronically but also reaction space around the metal geometrically. As a result, sharp selectivity can be achieved on metal-complex catalysts that may be difficult to obtain on metal particles and metal single-crystal surfaces. Application of the regulation by ligands in metal complexes allows selective catalysis and asymmetric synthesis.

However, homogeneous metal complexes in solutions are generally easy to gather and can cause undesired decomposition during catalytic cycles, resulting in loss of the catalytic activities. Transformation of homogeneous catalysts to heterogeneous catalysts with molecular-level active structures has been accomplished by attaching metal complexes on oxide or polymer supports to develop new, environmentally friendly chemical processes (1–8). The attached metal complexes may be further treated chemically at the surface, which provides unique structures different from their homogeneous analogues and often having remarkable catalytic properties. Understanding and controlling catalyst surfaces are the key issues for development of new catalysts, but conventional supported metal catalysts are generally heterogeneous and complicated. Their characteristics are difficult to ascertain, especially under catalytic reaction conditions, which results in the restriction of the development of new catalytic materials. Molecular-level control of catalyst surfaces is necessary for design of multi-functionalized catalytic sites.

New and distinct materials and chemistry prepared stepwise in a controllable manner by using organometallic and inorganic complexes as precursors provide an opportunity for the development of efficient, catalytic, molecularly organized surfaces. The key factors of chemical design of supported catalyst surfaces include composition, structure, oxidation state, distribution, morphology, and polarity, which should be organized at the surface. In the development of novel catalysts, new chemical concepts and strategies regarding composition or structure are conceived that can be associated with in situ characterization.

Whereas it has been recognized on the basis of in situ characterization that design of active structures and ensembles and control of their reactivity are critical in achieving high catalytic performance with good selectivity, these issues have not been adequately addressed and are still a serious challenge in catalysis and materials research. In this review, we summarize recent examples of the chemical design of catalytically active structures by attaching metal complexes on oxide surfaces and the subsequent structural transformation and molecular imprinting on the attached metal complexes. We also document how the designed surfaces are characterized by physical techniques, including conventional and time-resolved X-ray absorption fine structure (XAFS).

DESIGN AND IN SITU CHARACTERIZATION OF THE REACTION INTERMEDIATE FOR ALKENE HYDROFORMYLATION ON A Rh-DIMER/SiO_2 CATALYST

Design and Characterization of Each Step for the Catalysis

Molecular design of the active structure regarded as a reaction intermediate for a target catalytic reaction provides efficient catalysis. In order to achieve the catalyst design, the reaction mechanism, including the structural and electronic change of active metal sites, must be known on a molecular level.

Attached Rh particle catalysts show a maximum activity for ethene hydroformylation with an optimum particle size, suggesting the contribution of metal ensembles with Rh—Rh bonding to the catalytic performance; in homogeneous systems, mononuclear Rh complexes catalyze the reaction. There is a clear contrast between both systems. Demonstration of the role of metal-metal bonding for catalysis is one of the crucial issues for design of new catalytically active sites. In order to prepare supported metal species with metal-metal bonds, there have been many attempts to support di- and multi-nuclear metal complexes on oxide surfaces (1–3). Alkene hydroformylation on Rh complexes proceeds via an Rh-acyl intermediate. An example of the Rh-acyl intermediate is described below.

Attached Rh dimers were prepared by reaction of a *trans*-[$(RhCp^*CH_3)_2(\mu$-$CH_2)_2$] (Cp^* = pentamethylcyclopentadienyl) dimer with surface OH groups of SiO_2, TiO_2, Al_2O_3, and MgO (9). Each step for the transformation of the incipient-attached Rh species at the surfaces was characterized by Fourier transform infrared (FT-IR) spectroscopy and extended X-ray absorption fine structure (EXAFS) spectroscopy, as shown in Figure 1 (10). A CH_3 group and a μ-CH_2 group were reacted to form C_2H_5 on a Rh atom (species **1** in Figure 1). The Rh dimer (**1**) was exposed to CO to form *gem*-carbonyls (**2**), which appeared at 2032 and 1969 cm^{-1}. One of the carbonyls inserted into ethyl ligand to form an acyl ligand (1710 and 1394 cm^{-1}) (species **3**) by heating at 423 K, as illustrated in Figure 1. Note that the obtained Rh dimer (**3**) is regarded as a catalytic intermediate for ethene hydroformylation. Rh-dimer species with a Rh—Rh bond was also produced on a TiO_2 surface by using the *trans*-[$(RhCp^*CH_3)_2(\mu$-$CH_2)_2$] dimer precursor, as shown in Figure 2. On the other hand, mononuclear Rh species was formed on Al_2O_3, and aggregated Rh species was formed on MgO, irrespective of the use of the same precursor (Figure 2).

Catalytic activities and selectivities for ethene hydroformylation on the catalysts Rh_2/SiO_2, Rh_2/TiO_2, Rh_2/Al_2O_3, and Rh_2/MgO are summarized in Table 1 (10). The Rh_2/SiO_2 catalyst showed high activity and selectivity compared with that of a conventional impregnated Rh catalyst. On the other hand, the other catalysts, including Rh dimer on TiO_2, were inactive.

The transformation of Rh species during the ethene hydroformylation on Rh_2/SiO_2 was characterized by FT-IR and EXAFS, as shown in Figure 1. The acyl ligand is hydrogenated to the corresponding aldehyde product (propanal). CO molecules coordinate to Rh again, whereas Rh–Rh bonding is broken, leading to two mononuclear sites (Figure 1). The mononuclear species with dicarbonyl groups and ethyl group (species **4**) is transformed to the acyl species (species **3**) again at 423 K, accompanied with rebonding of Rh–Rh. The transformation between the structures (species **3** and **4**) occurs reversibly. Note that the catalytic ethene hydroformylation on Rh_2/SiO_2 proceeds in conjunction with breaking and rebonding of the Rh–Rh bond. This is the first example that clearly illustrates how the metal-metal bonding plays an essential role in solid catalysis. In the Rh_2/SiO_2 catalyst, one Rh atom with the acyl ligand works as the reaction site, whereas another Rh atom promotes the Rh reaction site for the CO insertion (10).

Rh_2/SiO_2 and Rh_2/TiO_2 have dimer structures, but their reactivities differ greatly. The dimer structure of Rh_2/TiO_2 did not act as a catalytic site for ethene hydroformylation, whereas a CO-bridged Rh dimer structure was formed, as shown in Figure 2. Further CO treatment brought about dissociation of the Rh–Rh bond to mononuclear species similar to the case of Rh_2/SiO_2, but acyl species as a key intermediate for the hydroformylation was not formed at the reaction temperature.

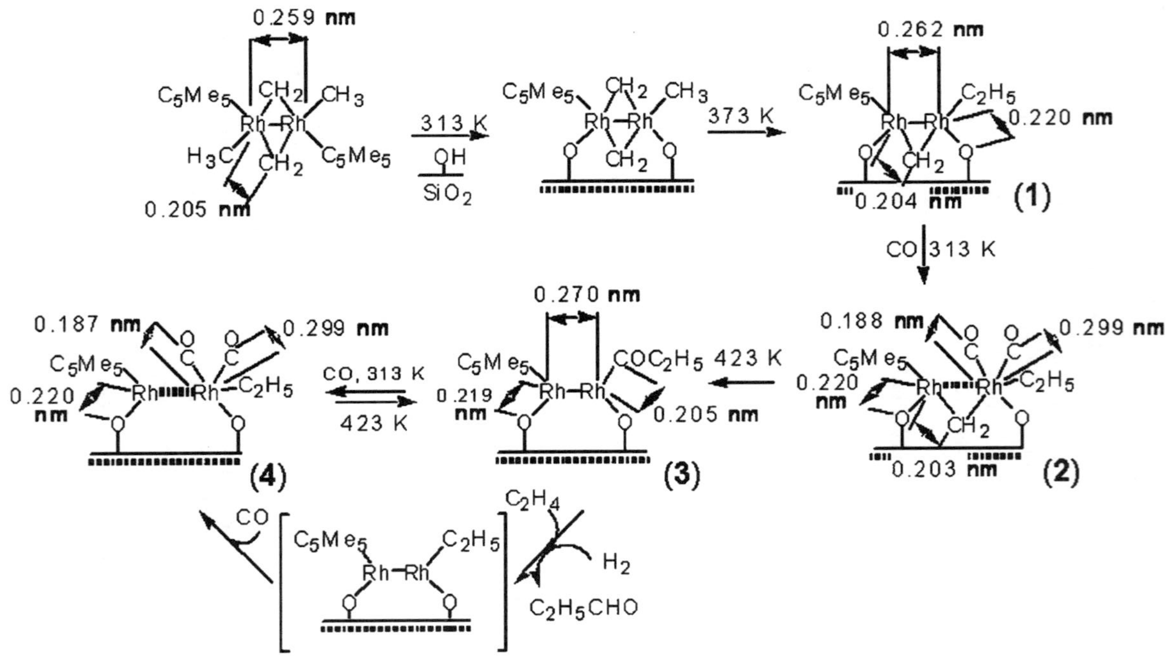

Figure 1 Structural transformations of Rh_2 species on SiO_2 and the structural change during ethene hydroformylation.

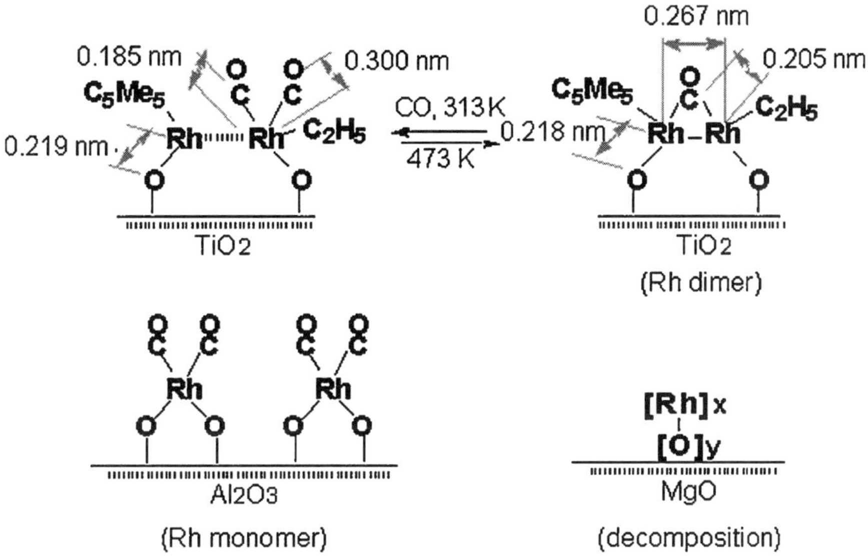

Figure 2 Structures of Rh_2/TiO_2, Rh_2/Al_2O_3, and Rh_2/MgO.

The difference may be caused by electronic and geometric states of the Rh sites. The flexibility of Rh–Rh bond in Rh_2/SiO_2 is such that it makes CO insertion promoted by Rh–Rh bond formation possible. On Al_2O_3, metal-support interaction becomes stronger, the Rh dimer is supported as monomers, and the formation of Rh–Rh bond is prohibited. Again, metal-metal-assisted CO insertion does not occur on Al_2O_3 support. On MgO, the Rh dimer is converted to clusters during heating the surface of the Rh dimers. No hydroformylation reaction occurs. On Rh_2/TiO_2, Rh_2/Al_2O_3, and Rh_2/MgO, promotion of CO insertion by the adjacent Rh atoms is not observed, possibly because the interaction between Rh dimers and supports are too strong (9).

TABLE 1 Catalytic performances (TOF: turnover frequency; selectivity) of ethene hydroformylation on the Rh-dimer catalysts at 413 K

Catalyst	TOF (total)	TOF (ethane)	TOF (propanal)	Selectivity/%
Impreg. Rh/SiO_2	22.8	21.5	1.3	5.6
Rh_2/SiO_2	36.9	4.1	32.8	88.9
Rh_2/TiO_2	6.3	6.3	0	0
Rh/Al_2O_3	6.8	6.8	0	0
Rh_x/MgO	56.0	56.0	0	0

TOF: 10^{-4} min^{-1}.

$CO:H_2:C_2H_4$ = 1:1:1 (total pressure = 40.0 kPa).

Hydroformylation reaction of alkenes is known to proceed on mono-metal sites such as mononuclear metal complexes in homogeneous systems. However, mono-metal sites are generally disadvantageous because of the necessity to eliminate some of the intimate ligands to form an unsaturated site on the metal atom. On the other hand, on the attached Rh dimer, it is not necessary to dissociate any ligands during the catalytic cycle, and hence the Rh dimer acts as an effective catalytic site. Chemistry of the Rh dimer on the surface is different from chemistry of the homogeneous Rh monomer in a homogeneous system. CO insertion into alkyl on the Rh monomer proceeds only in the presence of excess CO, and acyl decomposes to CO and alkyl under vacuum. In contrast, CO insertion proceeds under vacuum without excess CO, and acyl decomposes in the presence of CO. The catalytic ethene hydroformylation on Rh_2/SiO_2 is referred to as metal-assisted catalysis by the two adjacent Rh atoms.

Characterization of the Reaction Steps by Time-Resolved DXAFS

SETUP OF A DXAFS SYSTEM AND ANALYSIS OF DXAFS DATA Local structures of noncrystalline dispersed metal species have been extensively studied by XAFS because it does not require long-range order of the metal species (2, 11–13). XAFS can directly provide the interatomic distances and coordination numbers between the particular metal and the surrounding atoms. XAFS is a suitable technique for the structural analysis of active sites under reaction conditions because X rays are not absorbed by the reaction gases present in the catalytic systems.

However, it takes tens of minutes to measure an EXAFS spectrum for dispersed species by conventional XAFS technique. Thus the conventional XAFS technique cannot monitor time-dependent structures during dynamic chemical processes. Time-resolved energy-dispersive XAFS (DXAFS) (14–27) can be applied to studies on dynamic structural changes of metal sites during catalytic reactions. DXAFS is a technique to measure the whole range of X-ray absorption spectra using a bent crystal and a position-sensitive detector (1024 ch photodiode arrays), as shown in Figure 2. Because no mechanical motion for monochromatizing X rays is required in this technique, EXAFS spectra for catalysts can be measured in less than one second. The DXAFS technique has been used in making clear structural changes during catalyst preparation, chemical treatment, and catalytic reaction, which cannot be elucidated by other time-resolved methods such as time-resolved IR, Raman, XRD, and luminescence techniques. Another technique for time-resolved XAFS, called quick-scanning XAFS, is achieved by using an electrical gear, which results in a smooth quasi-continuous operation of the monochromator stepping motors. A DXAFS measurement setup and detector system may be complex relative to quick-scanning XAFS, but DXAFS has a better time-resolution capability. Because a DXAFS spectrum is measured without scanning the X-ray energies, it may suitable for study of transient and short-lived catalytic species. We have succeeded in observing and analyzing the structural changes in the

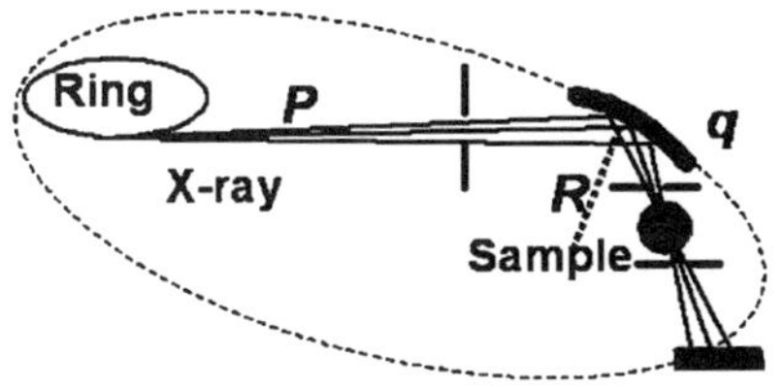

Figure 3 Diagram of energy dispersive XAFS (DXAFS) technique (Bragg type). The X-rays stemming from synchrotron radiation are reflected by an elliptically bent crystal to focus at a particular point where the sample is placed.

temperature-programmed reduction of Cu/ZSM-5 (1 s time resolution for Cu K-edge EXAFS), the temperature-programmed decarbonylation of $Mo(CO)_6$/YZ (1 s time resolution for Mo K-edge EXAFS), the kinetics of carbonylation and decarbonylation of [Ru_6C] clusters on MgO (1 s time resolution for Ru K-edge EXAFS), and the disintegration of Rh clusters on Al_2O_3 by CO adsorption (100 ms time resolution for Rh K-edge EXAFS) by DXAFS (21–23, 25, 27–29).

DXAFS measurements were carried out on BL-9C at KEK-PF with a Si(311) polychromator. The Si(311) bent-crystal polychromator (Bragg-type) was used to achieve elliptical optics for focusing incident X rays at the sample. The relationship for the curved polychromator in Figure 3 is given by Equation 1 (27).

$$\frac{1}{p} + \frac{1}{q} = \frac{2}{R \sin \theta_B}, \qquad 1.$$

where p, q, R, and θ_B are the source-to-crystal distance, the crystal-to-focus distance, the radius of curvature of the Si crystal, and the Bragg angle at the center of the crystal, respectively. The energy range (ΔE) is given by Equation 2.

$$\Delta E/E = L \sin \theta_B \cot \theta_B / p, \qquad 2.$$

where E and L are the energy at the center of the crystal and the length of the irradiated crystal, respectively. A self-scanning photodiode array (PDA: 1024 sensing elements: 25 μm width and 2.5 mm height for each element) manufactured by Hamamatsu Photonics (S3904-1024FX SPL3402) was used as a position-sensitive linear detector. A fiber optical plate embrocated by a CsI (Tl) fluorescent (150 μm thickness) was directly coupled at the active area of PDA to convert X ray to visible light. The energy calibration at each sensing element of the PDA was performed by comparison with an XAFS spectrum of a Rh foil recorded using an Si(311) channel-cut monochromator. The X-ray absorption spectra are calculated by ln (I_0/I), where I and I_0 are X-ray intensities with and without a sample, respectively.

The in situ DXAFS cell was made of stainless steel and had two slit windows with capton films to measure the incident and transmitted X rays. EXAFS data were analyzed by the UWXAFS package (30). After background subtraction using AUTOBK (31), the k^3-weighted EXAFS functions were Fourier-transformed into R-space, and the structural parameters were determined by a curve-fitting procedure in the R-space using the FEFFIT program involving multiple scattering effects. The back-scattering amplitude and phase shift functions and the mean free paths were generated by using the EXCHANGE flag of an input file in the FEFF 8.0 code taking account of an energy resolution of the DXAFS spectrometer (4.0 eV) (32).

The time-resolved DXAFS spectra were fitted using a model function composed of four shells, i.e., Rh–C–(O) and Rh–(C)–O for carbonyl ligands, Rh–COC_2H_5 for acyl ligand, and Rh–Rh for Rh dimer. The coordination numbers (CNs) of Rh–C–(O) and Rh–(C)–O bonds were fixed to be the same. Rh–O bonds at the interface between the Rh dimer and the SiO_2 surface were assumed to remain unchanged during the chemical processes because the process is completely reversible, and the Rh–O bond remains unchanged in the two stable species (acyl and dicarbonyl species). The Debye-Waller factors of all the contributions were fixed at the values obtained by conventional XAFS to reduce the number of independent parameters in the fitting procedure (33). The validity of this procedure was confirmed for stable Rh species by comparing the Debye-Waller factors and the Rh–O bond distances obtained by the DXAFS and conventional XAFS. The residual factors for the fitting were typically 4%.

DYNAMIC MECHANISM FOR CO INSERTION AND CARBONYLATION IN Rh_2/SiO_2 BY TIME-RESOLVED DXAFS Figure 4*a* (see color insert) shows the dynamic changes in the CNs of Rh–Rh, Rh–COC_2H_5, and Rh–CO during the carbonylation of the acyl species (species **3** in Figure 1) to form the dicarbonyl species (species **4** in Figure 1), which were analyzed for the DXAFS data recorded every 100 ms at 313 K and CO = 12.8 kPa (A. Suzuki, M. Tada, M. Nomura & Y. Iwasawa, unpublished data). Immediately after CO exposure, the CN of the Rh–CO bond appeared and increased with time, and reached a value that corresponds to half of the saturated value at a final stage after 9 s. The CN of the V–O bond at 0.270 nm began to decrease with a 0.5 s delay after the CO exposure, and eventually the Rh–Rh bond was broken completely at about 5 s. In contrast to these behaviors, Rh–COC_2H_5 ligand remained unchanged until 5 s, which indicates that the acyl CO is not converted to carbonyl directly to Rh at this stage. Pressure change of CO measured in the same in situ DXAFS cell obviously supports the DXAFS analysis, where one CO molecule per Rh dimer is adsorbed for the first 5 s. These results demonstrate that adsorption of one CO molecule per Rh dimer breaks the Rh–Rh bond to form mononuclear species.

After the formation of Rh(CO)(COC_2H_5) species, the acyl ligand began to decompose, whereas the CN of Rh–CO increased to a saturated value. During this process, the CO pressure did not change further. Thus it is obvious that the Rh(CO)(COC_2H_5) decomposed to form Rh-dicarbonyl species. In conclusion, the carbonylation of species **3** proceeds by two steps via an intermediate species Rh(CO)(COC_2H_5), as shown in Figure 5.

In the process of CO insertion under vacuum at 463 K, the Rh-dicarbonyl species (**4**) decreased with time, whereas Rh–COC_2H_5 increased, as shown in Figure 4*b* (A. Suzuki, M. Tada, M. Nomura & Y. Iwasawa, unpublished data). The decrease in the CN of Rh–CO in the Rh dicarbonyl (**4**) is equivalent to the increase in the

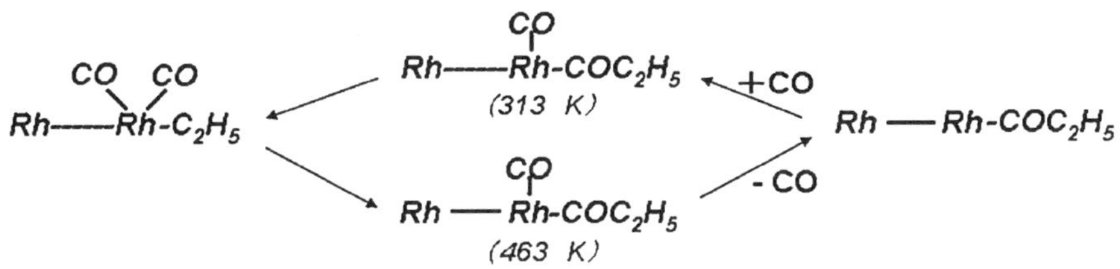

Figure 5 Structural transformation with hysteresis between species **3** and species **4** (see Figure 1) determined by time-resolved DXAFS.

CN of Rh–COC_2H_5, which indicates the direct transformation of $Rh(CO)_2$ (**4**) to Rh(CO)(COC_2H_5). The adjacent two Rh atoms were bonded again, and the CN of the Rh–Rh bond increased with time from the beginning of CO insertion. Thus Rh–Rh bond formation occurred simultaneously with the insertion of CO. It is known that CO insertion to alkyl on Rh monomers proceeds only in the presence of excess CO, and the acyl decomposes to CO and alkyl under vacuum. Accordingly, the results suggest that the CO insertion on the Rh_2/SiO_2 catalyst is assisted by the Rh–Rh bond formation.

There was a break in the change of the CNs of Rh–CO and also in the change of the CNs of Rh–COC_2H_5, as shown in Figure 4*b*. The result suggests the presence of an intermediate species, Rh(CO)(COC_2H_5). This intermediate species holds the Rh–Rh bonding, whereas the intermediate species Rh(CO)(COC_2H_5) in the carbonylation has no Rh–Rh bonding. The CN of Rh–COC_2H_5 did not change after 10 s, whereas the CN of Rh–Rh bond continued to increase until it reached a CN of unity. The difference between the intermediates in the carbonylation of species **3** and the CO insertion of species **4** may be due to the different temperatures at 313 K and 463 K for the carbonylation and the CO insertion, respectively. Thus the reversible process between species **3** and **4** involves a structural hysteresis of the intermediate. It should be noted that the time-resolved DXAFS can provide the details of the reaction mechanism involving the sequence and time scale of bond rearrangements on the active metal center during the catalytic reaction.

SELF-DIMERIZATION OF V COMPLEXES ON A SiO_2 SURFACE AND THEIR ASYMMETRIC CATALYSIS FOR OXIDATIVE COUPLING OF 2-NAPHTHOL TO BINOL

Catalytic asymmetric synthesis, which requires the fine regulation of catalytic sites, is one of the most difficult problems in catalytic chemistry. In the past decade, many chiral ligands were synthesized and applied to organic reactions such as hydrogenation, hydroformylation, epoxidation, etc. However, such chiral metal-complex catalysts are generally employed in homogeneous systems, where they often decompose or aggregate to inactive metal ensembles in an undesired manner. The fine design of chirality on heterogeneous catalyst surfaces is still a serious challenge. Heterogenization by attaching such metal complexes on oxide supports is one of the effective ways to solve these problems. However, the application of asymmetric metal complexes with chiral ligands to heterogeneous catalysts is neither easy nor simple. A new strategy to construct novel chirality on surfaces is indispensable for heterogeneous asymmetric catalysis. It is also difficult to realize both high activity and high enantioselectivity because restriction of reaction space around the metal center to increase enantioselectivity results in a decrease of the catalytic activity. There are few reports of asymmetric metal-complexes on oxide surfaces that are stable for organic reagents and under severe catalytic conditions. We have found novel chiral self-dimerization of V monomers on a SiO_2 surface (35). The obtained V dimer is highly enantioselective for the asymmetric oxidative coupling of 2-naphthol, 96% conversion, 100% selectivity to 1,1′-binaphthol (BINOL), and 90% enantioselectivity, although the V monomer is inactive for the coupling (35). The oxidative coupling of 2-naphthol is a representative reaction mode for a direct synthesis of BINOL, whose optically pure derivatives are regarded as versatile chiral auxiliaries and ligands in asymmetric syntheses (36, 37). Nevertheless, only a few studies have been reported for the asymmetric coupling of 2-naphthols in homogeneous systems (38–45). To our knowledge, there is no report on heterogeneous asymmetric catalysts for the catalytic coupling reaction to date.

Chiral Self-Dimerization of V Complexes on SiO_2

The V-monomer complexes coordinated with chiral, α-amino-acid-based Schiff bases [α-amino acid (L-valine, L-isoleucine, L-leucine, L-*tert*-leucine, and L-phenylalanine)] were dissolved in dehydrated ethanol and impregnated under N_2 atmosphere with SiO_2 (Aerosil 200), Al_2O_3 (Allon C), and TiO_2 (P-25) precalcined at 773 K for 2 h. After stirring for 1 h at ambient temperature, the solvent was evaporated, and the sample was dried under vacuum for 7 h. V loadings on SiO_2 were changed in the range of 0.3–3.4 wt%, which were determined by XRF (35).

Assembled structures often provide amazing events in catalysis, different from monomer catalysis, by creating new reaction sites/spaces and controlling dynamic behavior of metal sites. Thus, ultimately, it may be important to explore a new way to design active assembly of metal complexes on oxide surfaces in a controllable manner.

The V precursors coordinated with Schiff-base ligands are monomeric in alcohol solutions, and two diastereomers exist owing to two possible directions of V=O to the face of the tridentate ligands. Energy difference between the two diastereomers was calculated to be less than 1 kJ mol^{-1}, indicating that the two diastereomers coexist in similar concentrations in solutions. In crystal, no direct V–V interaction is observed; the shortest interatomic length is 0.6816 nm (46). There is no attractive force producing V–V interaction in the structure of the V precursor itself.

The pre-edge position of the V K-edge XANES spectra and the XPS V 2p binding energies reveal that V valence in the supported V complexes stays at four, the same as that of the V precursor. On the support, the V=O bond is preserved, but the symmetry of the V complexes becomes distorted, as seen by the appearance of a d-d transition in the DR-UV/VIS spectra. These behaviors were observed irrelevant to the amount of V loading. Thus an increase in V loading on SiO_2 is not a reason for the distortion of V species.

The curve-fitting analysis of V K-edge EXAFS spectra was performed to examine local structures around V center. The curve-fitting results are presented in Table 2. The XANES spectra for both the precursor and the supported V complex demonstrated the existence of a V=O double bond, thus the CN of the V=O bond was fixed to be 1.0 for the fitting, to minimize fitting errors. The EXAFS curve-fitting analysis for the V precursor (L-leucine-based Schiff base) agreed with XRD data. EXAFS data for the supported V complexes were analyzed in a similar way. A V=O bond at 0.157 ± 0.001 nm is similar to that of the precursor, whereas the CN of a V–O bonds at 0.199 ± 0.002 nm decreased from 3.8 ± 0.4 to 2.8 ± 0.5, as shown in Table 2. The decrease in the CN of V–O upon supporting was also observed for 0.8, 1.6, and 3.4 V wt% samples independent of V loading. Longer-distance bonding was not observed with all the catalysts, indicating that there was no direct V–V bonding in the supported V catalysts. The treatment of the supported V catalyst with 2-naphthol caused recovery of the CN of V–O bonds to 4.0 ± 0.6 (Table 2). The CN of Rh–Rh did not increase after the addition of excess 2-naphthol (the molar ratio of 2-naphthol/V = 200). No more 2-naphthol molecules coordinated to the supported V complex. The V=O double bond was at

TABLE 2 Curve-fitting results of V K-edge Fourier-transformed EXAFS spectra for a V precursor (L-leucine): its fresh supported V complex (V: 3.4 wt%); treated with 2-naphthol and measured at 16 K

Shell	Coordination number	Distance/nm	σ^2/nm^2
Precursor[a]			
V=O	1.0	0.158 ± 0.001	$3\,(\pm 2) \times 10^{-5}$
V–O	3.8 ± 0.4	0.198 ± 0.001	$1\,(\pm 1) \times 10^{-5}$
(b) Supported V complex[b]			
V=O	1.0	0.157 ± 0.001	$1\,(\pm 1) \times 10^{-5}$
V–O	2.8 ± 0.5	0.199 ± 0.002	$3\,(\pm 2) \times 10^{-5}$
Treated with 2-naphthol[c]			
V=O	1.0	0.157 ± 0.002	$1\,(\pm 2) \times 10^{-5}$
V–O	4.0 ± 0.6	0.199 ± 0.002	$2\,(\pm 2) \times 10^{-5}$

[a] $k = 30$–$120\ nm^{-1}$, $R = 0.10$–0.20 nm, $R_f = 0.34\%$.
[b] $k = 30$–$120\ nm^{-1}$, $R = 0.10$–0.20 nm, $R_f = 0.87\%$.
[c] $k = 30$–$120\ nm^{-1}$, $R = 0.10$–0.20 nm, $R_f = 1.70\%$.
The coordination number of V=O was fixed to be unity.

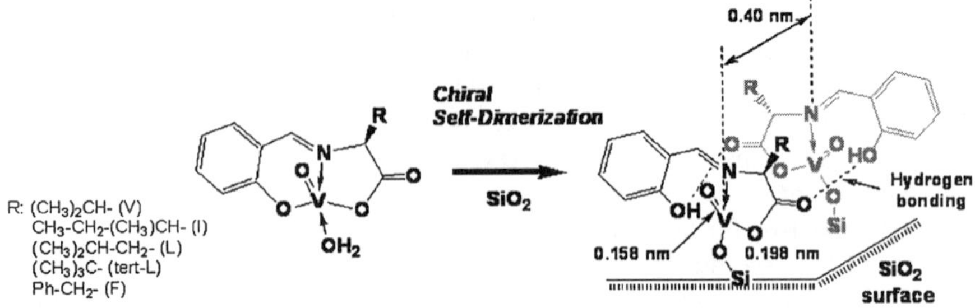

Scheme 1 Preparation of the chiral self-dimerized V complexes on SiO_2. Two V monomers are dimerized with each other through two sets of hydrogen bonds between Ph–OH and COO. The V–V distance was evaluated in the presence of O_2 by electron spin resonance (ESR).

0.157 ± 0.002 nm, similar to that of the fresh sample before the 2-naphthol coordination. There was no direct V–V bonding after the coordination of the reactant.

The EXAFS and FT-IR spectra of the supported V complexes reveal the unique surface-attaching reaction promoted by surface Si–OH groups. The decrease in the CN of V–O(N) bonds indicates that the V center becomes unsaturated on the surface. The large shift of $\nu_{(Ph-O)}$ and the change in the intensity ratio of $\nu_{(Ph)}$, together with the comparison of FT-IR spectra between Ph–OH and Ph–ONa, demonstrate that the Ph-O coordination is dispatched from the V center by the reaction of the V precursor with the Si–OH to form Ph–OH group as illustrated in Scheme 1 (35).

The unsaturated V centers are stabilized by the two sets of hydrogen bonds between the two V complexes, which are the driving force for the self-dimerization of the V complexes at the surface, as shown by DFT calculations (see below). The X-band ESR spectra at 6 K for the supported V complexes (V: 3.4 wt%) in Figure 6 demonstrate the existence of two V complexes near by. Hyperfine signals for the V monomer precursor in methanol were observed owing to d^1 configuration of the V^{4+} ions. Relationships $g_{\parallel} < g_{\perp}$ and $A_{\parallel} > A_{\perp}$ demonstrate an axially compressed ${d_{xy}}^1$ configuration of the V=O complex. For the V complexes supported on SiO_2, two different signals were observed in Figure 6. A broad signal was superposed with the hyperfine feature representing a ${d_{xy}}^1$ configuration, and the hyperfine coupling constants were $g_{\parallel} = 1.948$, $A_{\parallel} = 17.01\ cm^{-1}$ and $g_{\perp} = 1.991$, $A_{\perp} = 6.49\ cm^{-1}$. Its half-field signal was negligible, as shown in Figure 6*B*. After the coordination of 2-naphthol, oxygen was admitted on the supported V complex and the ESR spectra were measured at 6 K, where the complexes are saturated with adsorbed O_2. The adsorption of O_2 on the supported V complex dramatically increased the broad signal as shown in Figure 6*b1*. A new peak also appeared at half field of the main signal, which is attributed to a forbidden transition, the half band of $|\Delta Ms| = 2$ (Figure 6*B*). The transition ($|\Delta Ms| = 2$) is normally forbidden, but the triplet state of two d^1 electrons can be observed as a weak transition around half-field

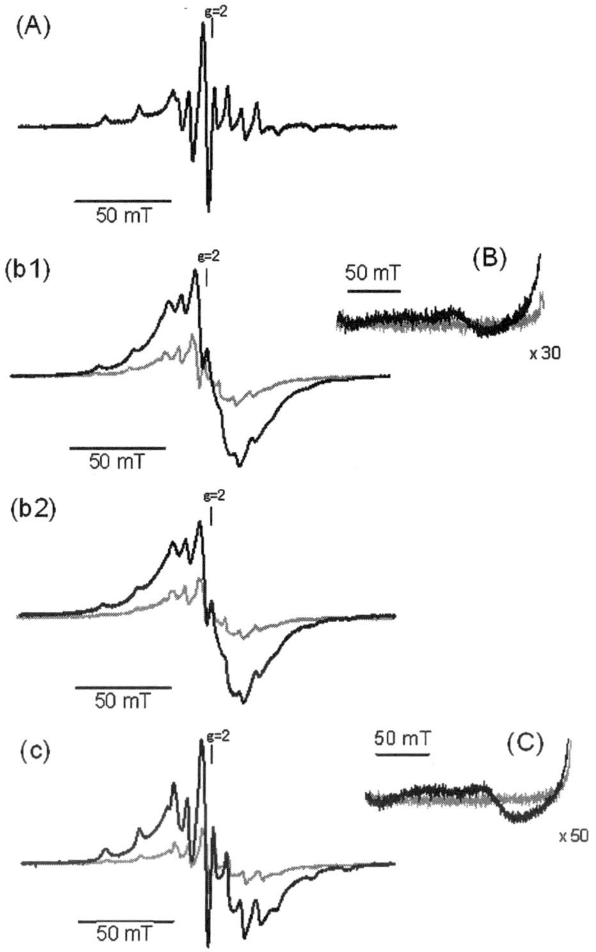

Figure 6 *A*: ESR spectra at 6 K for a V precursor (L-leucine) in methanol. The SiO_2-supported V complex (V: 3.4 wt%) in the absence (*gray line*) and presence (*black line*) of O_2 (*b*: main signal, *B*: half band), and that treated with 2-naphthol in the absence (*gray line*) and presence (*black line*) of O_2 (*c*: main signal, *C*: half band). ESR measurements after the adsorption and desorption of O_2 were performed twice on the supported V complex: the first run (*b1*) and the second run (*b2*) are shown.

position when two metal species are located nearby. Thus these signals prove the existence of V–V interaction stimulated by the O_2 adsorption. The ratio of the intensity of $|\Delta Ms| = 2$ to that of $|\Delta Ms| = 1$ is related to the sixth power of the interatomic distance of V–V by Equation 3 (47),

$$\text{intensity}\,(|\Delta Ms| = 2)/\text{intensity}\,(|\Delta Ms| = 1) = \mathrm{C}r^{-6}, \qquad 3.$$

where r is the V–V distance and C is a constant. On the basis of two V d^1 dimers with O=V–V=O coupled structures, whose V=O are directed outside of V–V, and by using reference compounds, the V–V distance of the supported V complex (3.4 wt%) was evaluated to be 0.40 ± 0.05 nm by Equation 3 (35, 48). After the evacuation of oxygen at 293 K, the ESR signal returned to the original intensity, i.e., before the adsorption, as shown in Figure 6*b2*, and the half-band signal disappeared. The behavior in the ESR spectra was reversible for the evacuation at 293 K and the readsorption of oxygen at 6 K, as shown in Figure 6*b2*.

The coordination of 2-naphthol maintained the axially compressed d^1 configuration of the V complex. A broad signal was observed, similar to that seen before the 2-naphthol coordination (Figure 6*c*). The effect of O_2 adsorption on the ESR signal was also observed after the coordination of 2-naphthol as a large increase in the broad main signal and the appearance of a weak half band (Figure 6*C* and *c*). The distance of V–V interaction was evaluated to be 0.40 ± 0.05 nm, identical to the case of the supported V complex without 2-naphthol in the presence of O_2 (M. Tada, N. Kojima, Y. Izumi, T. Taniike & Y. Iwasawa, unpublished data). The bond length of O_2 and the average distance of V–O coordination are 0.111 nm and 0.20–0.21 nm, respectively; therefore, the two V atoms should be located at a distance greater than 0.40 nm and less than 0.53 nm. The V–V distance determined by the ESR analysis agrees well with these values. Adsorbed oxygen bridges the two V atoms.

On the basis of the characterizations by ESR, XPS, XANES, EXAFS, DR-UV/VIS, and FT-IR, we investigated the following two issues: the chemical attachment of the V precursor on SiO_2 and the assembly of V complexes on the surface by DFT calculations. The coupling of two V monomers on a silica cluster was successfully modeled by DFT calculations (35; M. Tada, N. Kojima, Y. Izumi, T. Taniike & Y. Iwasawa, unpublished data). Figure 7 (see color insert) shows the assembled dimer structure of the L-leucine-based V complexes, where two monomers are bonded to each other with two sets of hydrogen bonding between the Ph–OH and the carbonyl oxygen of COO. We have examined all possible hydrogen bondings of the Ph–OH with functional groups of the adjacent V complexes by DFT calculations; for example, Ph–OH · · · HO–Ph, Ph–OH · · · O=V, etc. The other combinations for hydrogen bonding did not result in reasonably assembled structures. The conformation with two V=O bonds directed toward the outside of these complexes in Figure 7 was the only model to totally agree with the EXAFS, FT-IR, and ESR data. The hydrogen bonding driving the dimerization unbinds the free Ph–OH, which easily returns to V and allows the unsaturated structure of V center to be active for the catalytic coupling reaction. Furthermore, the fixing of V=O direction by the tridentate ligand in the V dimeric structure brings about fastening chirality on the V center. Two chiral isobutyl groups derived from L-leucine overhang outside the assembly, and a chiral reaction space is constructed inside the space surrounded by the two V complexes, as shown in Figure 7.

V loading on SiO_2, including an L-leucine segment, was varied in the V loading range 0.3–3.4 wt% in order to maximize enantioselectivity for the coupling reaction (Table 3). No leaching of V complexes was observed for any of the supported catalysts in toluene. With an increase in V loading (0.3, 0.8, 1.6, and 3.4 wt%), the enantioselectivity dramatically increased (32, 39, 48, and 90% ee, respectively), as shown in Table 3. The 3.4 wt% V catalyst corresponding to full coverage of the V complex exhibited 90% ee, which was comparable to the best performance thus far reported for the coupling of 2-naphthol on a homogeneous catalyst. Furthermore, the supported V catalysts can be reused after filtration and exhibited similar catalytic performances, as shown in Table 3 (35; M. Tada, N. Kojima, Y. Izumi, T. Taniike & Y. Iwasawa, unpublished data).

It is noteworthy that the enantioselectivity of the coupling reaction is determined by the chiral conformation of the V center rather than the chirality of Schiff-base ligands. The chirality of the ligands sterically affects the chiral self-dimerization of V complexes on the surface, and the chiral ligands themselves do not determine the enantioselectivity for the 2-naphthol coupling. The increase in V loading on SiO_2 achieved a higher regulation of mobility of the assembled V species on the surfaces, which resulted in the high enantioselectivity of 90% at 93% conversion on the V 3.4 wt% catalyst compared with the case of the V 0.3 wt% catalyst with the same V structure (Table 3).

Mechanism for the Asymmetric Oxidative Coupling of 2-Naphthol

The unsaturated V dimer is easily coordinated with 2-naphthol, accompanied with an increase in CN of V–O from 3 to 4, as characterized by EXAFS in Table 2. In spite of the addition of excess amounts of 2-naphthol, a V site accommodates only one 2-naphthol molecule. The coupling reaction for synthesis of one BINOL molecule requires two 2-naphthol molecules, and the V dimer structure can react two 2-naphthol molecules activated on two adjacent V sites in the V dimer to form BINOL. Monomer structures are entropically disadvantageous for the coupling (M. Tada, N. Kojima, Y. Izumi, T. Taniike & Y. Iwasawa, unpublished data).

The dimer structure can also efficiently activate oxygen molecules. The dissociation of a strong O=O bond requires extremely unsaturated metal species in low valence states, but the following reactions often stop because the metal-oxygen bonds produced after the bond breaking of O=O are too stable. Similar amounts of oxygen are adsorbed before and after the coordination of 2-naphthol. Hence, an oxygen molecule weakly adsorbs on the V dimer without losing its reactivity. The catalytic reaction did not proceed in the absence of oxygen. Thus weakly adsorbed oxygen molecules are probably active species for the hydrogen extraction of 2-naphthol.

The asymmetric oxidative coupling of 2-naphthol is as follows (M. Tada, N. Kojima, Y. Izumi, T. Taniike & Y. Iwasawa, unpublished data). At first, 2-naphthol coordinates to an unsaturated V center, then an O_2 molecule adsorbs on the V dimer in a bridging fashion, as indicated from the V–V separation determined by ESR. In the third step, the activated oxygen extracts the hydrogen atom of the hydroxyl group of the coordinated 2-naphthol. Finally, two dehydrogenated 2-naphthols couple together to form BINOL in the hollowed slit of the V dimer structure (Figure 7). The coordination and oxygen adsorption of 2-naphthol were too fast to monitor, and thus these two steps are not rate determining for the coupling reaction. The reaction rate was of first order to 2-naphthol concentration. Hence, it is suggested that the third step, H transfer from 2-naphthol to oxygen molecule, is a rate-determining step and the subsequent coupling reaction to form BINOL should be faster. This also implies that the H abstraction of the second 2-naphthol by the half-hydrogenated oxygen (V-OOH) is much faster than that of the first 2-naphthol in the third step.

DESIGN OF A MOLECULAR IMPRINTING Rh-DIMER CATALYST

Principle of Molecular Imprinting for Metal Complexes on Surfaces

To prepare artificial enzymatic systems possessing recognition ability for particular substrate molecules, molecular imprinting methods that create template-shape cavities (with memory of the template molecules in polymer matrices) have been developed and established; e.g., in receptor, chromatographical separations, and fine chemical sensing. They are regulated mainly by simple adsorption on the imprinted sites (49–56). However, regulation of catalytic reactions is still difficult because catalysis is a dynamic process involving rearrangement of chemical bonds in the substrate molecules. Nevertheless, artificial enzymatic materials synthesized by molecular imprinting techniques using a variety of template molecules provide promising molecular recognition catalysis with 100% selectivity for a variety of catalytic reactions where natural enzymes can not be employed (55, 56).

Recently, in addition to imprinted acid-base catalysts (57–63), attempts to imprint metal complexes, have been reported, which represent the current state of the art (55, 56). In most cases of metal-complex imprinting, ligands of the complexes are used as template molecules, the aim of which is to create a cavity near the metal site. Several noteworthy examples of molecular imprinting of metal complexes allow (*a*) attachment of metal complex on robust supports, (*b*) enclosure of the metal complex by polymer matrix, and (*c*) production of a shape-selective cavity on the metal site. Metal complexes thus imprinted have been applied to molecular recognition (64, 65), reactive complex stabilization (66, 67), ligand exchange reaction (68), and catalysis (69–78).

Most of the imprinted metal-complex catalysts have been prepared by imprinting in bulk polymer. However, active sites prepared by bulk imprinting may often be located inside the bulk polymer matrices, where access of substrate molecules to the sites is difficult. Furthermore, such organic polymers tend to be unstable in organic solvents or under severe catalytic conditions, such as in the presence of oxidants or at high temperatures. As described above, however, oxide-supported metal complexes often exhibit unique properties and remarkable reactivity so that molecular imprinting of metal complexes at oxide surfaces has many possibilities to produce new excellent catalysts similar to enzymatic systems. However, molecular imprinting of metal complexes at oxide surfaces had not been reported before our study. Very recently, we have succeeded in preparing imprinted rhodium complexes at an Ox.50 (SiO_2) surface for catalytic shape-selective hydrogenation of simple alkenes without any functional group (75–78). Recognition of simple alkenes without any functional group is generally difficult. Here, we summarize our recent works on molecular imprinting of metal-complex catalysts, which show remarkable catalytic performances beyond simply supported metal-complex catalysts.

Molecular Imprinting of Rh Dimers at a SiO_2 Surface

A template with the same shape as the transition state of a rate-determining step is most appropriate for catalyst design, but it is hard to estimate its accurate structure. Reaction intermediates before and after the rate-determining step may have shapes similar to the early and late transition states, respectively, and they are generally used as templates for molecular imprinting catalysts. Our strategy is to use a ligand as template. Ligands of a metal complex not only influence its catalytic activity but also provide an unsaturated, reactive metal site with a ligand-shaped space on removal of a ligand. We chose a ligand of the attached metal complex with a shape similar to a reaction intermediate of alkene hydrogenation as a template molecule. $P(OCH_3)_3$ ligand has a shape similar to one of the half-hydrogenated species of 3-ethyl-2-pentene, which can produce the intermediate-shape cavity by extraction of the ligand.

Our strategy to design active and selective catalysts was based on the following five factors for regulation: (*a*) conformation of ligands coordinated to Rh atom, (*b*) orientation of vacant site on Rh, (*c*) cavity with the template molecular shape for reaction space produced behind template removal, (*d*) architecture of the cavity wall, and (*e*) micropore in inorganic polymer-matrix overlayers stabilizing the active species at the surface.

We prepared a new Rh-dimer molecular imprinting catalyst (Rh_{2imp}) at a SiO_2 (Ox.50) surface for hydrogenation of alkenes. The preparation steps are illustrated in Scheme 2 (see color insert) (76, 77). In heterogeneous systems, the metal-metal interaction is in general crucial to dissociate hydrogen molecules, which is most relevant to hydrogenation. The metal dimer is regarded as a minimum active structure for heterogeneous metal catalysis involving metal-metal bonding. In order to create a Rh-dimer structure with a template cavity on the Ox.50 surface, a $Rh_2Cl_2(CO)_4$ and a $P(OCH_3)_3$ were used as a Rh-dimer precursor and a template ligand, respectively. The template $P(OCH_3)_3$ is regarded as an analogue to a half-hydrogenated alkyl intermediate of hydrogenation of 3-ethyl-2-pentene. $Rh_2Cl_2(CO)_4$ was attached on the Ox.50 surface, and retention of its dimer structure was monitored by FT-IR (76). By exposure to $P(OCH_3)_3$, the surface-attached Rh carbonyl dimer was converted to a Rh monomer pair (Rh_{2sup}), with two $P(OCH_3)_3$ ligands on Rh accompanied with Rh–Rh bond breaking (Rh–P: 0.224 nm; no Rh–Rh bond by EXAFS), which was attached on the surface via Rh–O bonding at 0.203 nm in a bidentate form as characterized by EXAFS (Figure 8 and Table 4). The EXAFS spectra were measured at 15 K. After the $P(OCH_3)_3$ coordination, the first step of molecular imprinting for the attached rhodium complexes is a hydrolysis-polymerization of $Si(OCH_3)_4$, which possesses methoxy groups with positive interaction with the template. The second step is removal of a template by evacuation at 363 K to produce the imprinted Rh_{2imp} catalyst. By the imprinting procedure, Rh 3d XPS intensity reduced remarkably, indicating that the Rh species were embedded in the polymerized silica matrix overlayers that were characterized by ^{29}Si solid-state magic angle spinning (MAS) NMR (76).

The structures around Rh atoms in the Rh_{2imp} catalyst at each step of the alkene hydrogenation were investigated by EXAFS (Figure 8). EXAFS spectra for three samples—a fresh Rh_{2imp} catalyst, a Rh-dimer hydride catalyst formed by adsorption of 101.3 kPa of H_2 at room temperature, and a catalyst after reaction of the hydride species with 3-methyl-2-pentene at 348 K for 5 min—were measured at 15 K. The EXAFS Fourier transforms and the curve fitting are shown in Figure 8. Note that the surface imprinting causes dimerization of the Rh monomers to produce highly active Rh dimers with a direct Rh–Rh bond at 0.268 nm in the pores of 0.74 nm dimension in the SiO_2-matrix overlayers on the Ox.50 surface. The EXAFS analysis also indicates the removal of a single phosphite ligand per Rh by the molecular-imprinting procedure.The CN of the Rh–P bond at 0.221 nm reduced from 2.3 to 1.1, as shown in Table 4.

The change in Rh structures in the imprinting processes is explained by DFT calculation. Five structural models on a silica cluster substrate—$[Rh(P(OCH_3)_3)_2]_2$, $Rh(P(OCH_3)_3)$-$Rh(P(OCH_3)_3)_2$, $Rh_2(P(OCH_3)_3)_3$, P-bridged $Rh_2(P(OCH_3)_3)_3$, and $Rh_2(P(OCH_3)_3)_2$—were calculated, which are regarded as models for the Rh-monomer pair catalyst (Rh_{2sup}), possible intermediate species with three $P(OCH_3)_3$ ligands, and the imprinting Rh-dimer catalyst (Rh_{2imp}), respectively. The optimized structures and energy profiles for those model species are illustrated in Figure 9 (76). Initial structures of each Rh complex, including the SiO_2 substrate, were obtained by the molecular mechanics (MM) method with the universal force field 1.02. $[Rh(P(OCH_3)_3)_2]_2$ with four $P(OCH_3)_3$ ligands per two Rh atoms is the most stable complex with a Rh–Rh distance of 0.306 nm, which is indicative of no direct bonding between Rh atoms. $Rh_2(P(OCH_3)_3)_3$ with three $P(OCH_3)_3$ ligands per two Rh atoms is the second most stable complex with a Rh–Rh distance of 0.271 nm. This result implies that the removal of one phosphite ligand is compensated by the formation of Rh–Rh bond. In order to gain insight into this process, one of four $P(OCH_3)_3$ ligands in $[Rh(P(OCH_3)_3)_2]_2$ was removed, while the Rh–Rh bond was fixed at 0.306 nm. An optimized structure for $Rh_2(P(OCH_3)_3)_3$, which becomes unstable, is depicted in Figure 9. The energy difference (197 kJ mol^{-1}), roughly estimated, indicates the presence of activation energy for the ligand

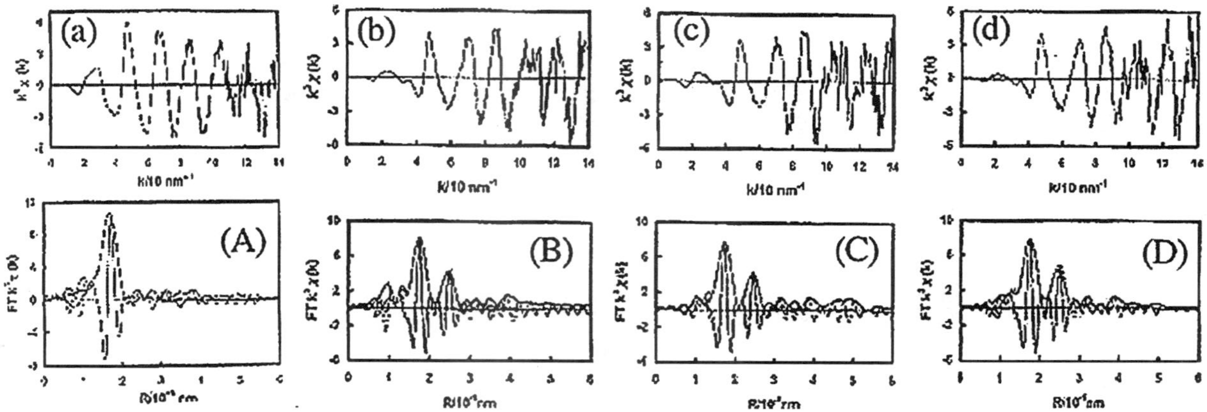

Figure 8 k^3-weighted EXAFS oscillations (*a–d*) and their associated Fourier transforms (*A–D*) at Rh K-edge measured at 15 K for the supported Rh_{2sup} catalyst (*a*, *A*), and the imprinted Rh_{2imp} catalysts (*b–d*, *B–D*). *b* and *B*, *c* and *C*, and *d* and *D* are the spectra for the fresh Rh_{2imp} catalyst, the Rh_{2imp} catalyst exposed to 101.3 kPa of H_2 at 293 K, and the Rh_{2imp} catalyst after reaction of the H_2-exposed sample with 3-methyl-2-pentene at 348 K for 5 min, respectively. Dotted and solid lines represent the absolute values and imaginary parts of the observed data and the fitted spectra, respectively.

TABLE 4 Structural parameters determined by EXAFS curve-fitting analysis for the supported and imprinted Rh catalysts measured at 15 K

Shell	CN	R/nm	σ^2/nm^2	
Fresh Rh_{2sup} catalyst				$\Delta E_0 = 6 \pm 4$ eV
Rh–O	1.6 ± 1.0	0.203 ± 0.003	$(6 \pm 3) \times 10^{-5}$	$R_f = 0.2\%$
Rh–P	2.3 ± 1.4	0.224 ± 0.005	$(2 \pm 2) \times 10^{-5}$	
Fresh Rh_{2imp} catalyst				$\Delta E_0 = 11 \pm 2$ eV
Rh–O	2.0 ± 0.5	0.211 ± 0.002	$(4 \pm 3) \times 10^{-5}$	$R_f = 0.3\%$
Rh–P	1.1 ± 0.2	0.221 ± 0.001	$(1 \pm 2) \times 10^{-5}$	
Rh–Rh	1.3 ± 0.4	0.265 ± 0.001	$(7 \pm 2) \times 10^{-5}$	
After H_2 adsorption				$\Delta E_0 = 9 \pm 4$ eV
Rh–O	1.7 ± 0.5	0.208 ± 0.003	$(6 \pm 3) \times 10^{-5}$	$R_f = 1.6\%$
Rh–P	1.2 ± 0.2	0.220 ± 0.001	$(2 \pm 4) \times 10^{-5}$	
Rh–Rh	1.3 ± 0.4	0.265 ± 0.001	$(7 \pm 2) \times 10^{-5}$	
After reaction of the H_2-adsorbed sample with 3-methyl-2-pentene				$\Delta E_0 = 12 \pm 3$ eV
Rh–O	2.2 ± 0.6	0.212 ± 0.003	$(5 \pm 4) \times 10^{-5}$	$R_f = 1.4\%$
Rh–P	1.1 ± 0.2	0.221 ± 0.001	$(1 \pm 4) \times 10^{-5}$	
Rh–Rh	1.2 ± 0.4	0.270 ± 0.001	$(6 \pm 2) \times 10^{-5}$	

removal, but the real transition state may involve Rh–Rh bonding at a shorter distance than 0.306 nm, resulting in a decrease in the activation energy. It should be noted that during the transformation of the supported Rh catalyst (Rh_{2sup}) to the imprinted Rh catalyst (Rh_{2imp}), there is a stable structure of $Rh_2(P(OCH_3)_3)_3$ with one phosphite ligand on a Rh atom and two phosphite ligands on another Rh atom. A bridge-bonded structure with a bridge $P(OCH_3)_3$, which might be possible for $Rh_2(P(OCH_3)_3)_3$, is unstable by 59 kJ mol^{-1} compared with that of the nonbridged $Rh_2(P(OCH_3)_3)_3$ according to the DFT calculation. Further loss of one of the two $P(OCH_3)_3$ ligands on a Rh atom leads to the formation of $Rh_2(P(OCH_3)_3)_2$ (Rh_{2imp}) with expense of the energy (Figure 9). Thus the removal of the template ligand was slow and required overnight evacuation (a procedure in which a vaccuum is pulled over the Rh_2 species to enhance the removal of the template ligand) at 363 K.

We assume that the elimination of the two phosphite ligands from the monomer pair to produce the $Rh_2(P(OCH_3)_3)_2$ structure (Rh_{2imp} catalyst) occurs step by step via the stable intermediate $Rh_2(P(OCH_3)_3)_3$ with one $P(OCH_3)_3$ on a Rh atom and two $P(OCH_3)_3$ on another Rh atom in the dimer at the Rh–Rh bond of 0.271 nm. Therefore, the stable intermediate $Rh_2(P(OCH_3)_3)_3$ is the imprinted Rh species

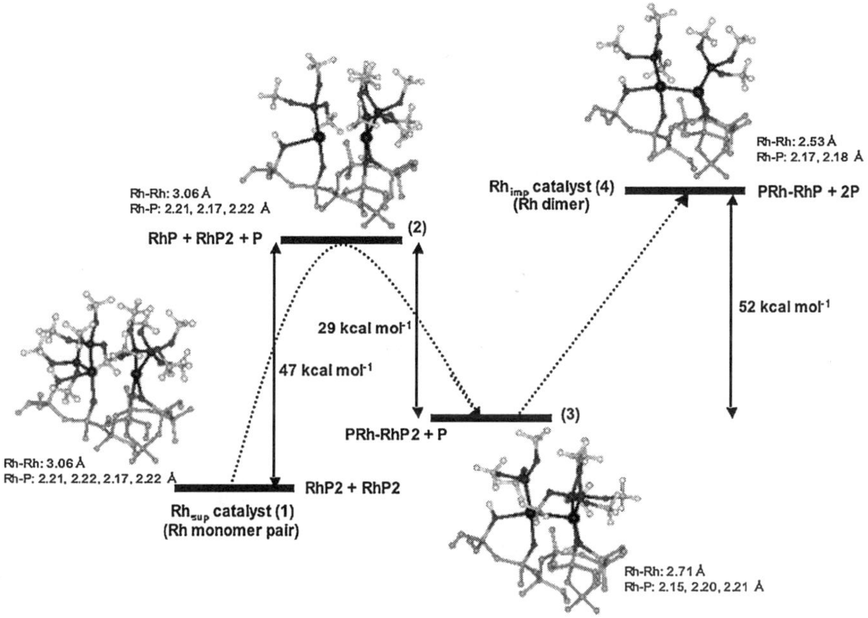

Figure 9 Relative energies and structural parameters for the models of the Rh_{2sup} catalyst: (*1*) $Rh(P(OCH_3)_3)$-$Rh(P(OCH_3)_3)_2$; (*2*) $Rh_2(P(OCH_3)_3)_3$; (*3*) and the Rh_{2imp} catalyst (*4*) by the DFT calculation.

under the chemical vapor deposition (CVD) process at 348 K. The elimination of a template $P(OCH_3)_3$ ligand from the intermediate at 363 K under vacuum generates the template-sized cavity in the SiO_2 matrix overlayers. Based on the MM method, modeling of the imprinted Rh-dimer catalyst was performed using a commercially available modeling tool, Cerius 2.4.2 (Accelrys), with the universal force field 1.02. The modeling of the Rh_{2imp} dimer coordinated with 3-ethyl-2-pentene in the cavity combined in the 0.74 nm micropore is shown in Figure 10 (see color insert).

Shape- and Size-Selective Hydrogenation of Alkenes on the Imprinted Rh-Dimer Catalyst

Table 5 shows steady-state reaction rates [turnover frequencies (TOF)] of the hydrogenation of eight alkenes at 348 K on the supported Rh-complex catalyst (Rh_{2sup}) and the molecular-imprinting catalyst (Rh_{2imp}). The homogeneous complexes, $Rh_2Cl_2(CO)_4$ and $RhCl(P(OCH_3)_3)_3$, and the supported species, $Rh_2Cl_2(CO)_4/SiO_2$, show no activities for the reaction. On the other hand, the Rh_{2sup} catalyst exhibits significant catalytic activities under the similar reaction conditions. It is to be noted that the hydrogenation reactions are greatly enhanced

TABLE 5 Degrees of enhancement of the reaction rates by molecular imprinting (ratio of TOFs), activation energies (E_a), activation enthalpies ($\Delta^{\ddagger}H$), and activation entropies ($\Delta^{\ddagger}S$) for the catalytic hydrogenation of alkenes at 348 K

Reactant	TOF/s^{-1} (Rh_{2imp} catalyst)	Ratio of TOFs[a]	Supported catalyst (Rh_{2sup}) E_a[b] $\Delta^{\ddagger}H$[c] $\Delta^{\ddagger}S$[d]	Imprinted catalyst (Rh_{2imp}) E_a[b] $\Delta^{\ddagger}H$[c] $\Delta^{\ddagger}S$[d]
2-pentene	6.6×10^{-2}	51	34, 31, −205	26, 23, −195
3-methyl-2-pentene	3.6×10^{-3}	51	44, 41, −200	43, 40, −170
4-methyl-2-pentene	5.9×10^{-3}	45	40, 37, −207	40, 37, −175
3-ethyl-2-pentene	1.5×10^{-3}	35	42, 39, −210	39, 36, −189
4-methyl-2-hexene	1.3×10^{-4}	14	40, 37, −212	10, 7, −276
2-octene	9.6×10^{-4}	10	28, 25, −215	7, 4, −257
1-phenylpropene	3.0×10^{-2}	7	29, 26, −213	8, 5, −256

[a]ratio of TOFs: TOF of the Rh_{2imp} catalyst/TOF of the Rh_{2sup} catalyst.
[b]E_a: kJ mol^{-1}, [c]$\Delta^{\ddagger}H$: kJ mol^{-1}, [d]$\Delta^{\ddagger}S$: J K^{-1} mol^{-1}.

by the surface imprinting. The imprinted Rh_{2imp} catalyst is extremely active for the alkene hydrogenation. For example, hydrogenation of 2-pentene is 51 times greater than that on the supported Rh_{2sup} catalyst. The metal-metal bonding and coordinative unsaturation of the Rh dimer are key factors for the remarkable activity of the Rh_{2imp} catalyst. The Rh_{2imp} catalyst is highly durable and surprisingly air-stable in spite of its unsaturated structure, which is advantageous in practical handling of the system. Furthermore, it can be reused without any loss of the catalytic activity.

The Rh dimer after H_2 adsorption exhibited EXAFS oscillation and Fourier transform similar to those for the freshly imprinted catalyst (Figure 8 and Table 4). The detailed analysis of the EXAFS data confirmed retention of the local conformation of the Rh dimer with a Rh–Rh bond (CN: 1.3 ± 0.4), two Rh–O bonds (CN: 1.7 ± 0.5), and a Rh–P bond (CN: 1.2 ± 0.2) (Table 4). Contraction of the Rh–Rh bond from 0.268 nm (±0.001 nm) to 0.265 nm (±0.001 nm) was found with the hydride dimer, which indicates stabilization of the dimer structure by electronic rearrangement owing to monohydride coordination on both Rh atoms in the dimer. After reaction of the Rh-dimer hydride species with 3-methyl-2-pentene, the shrunk Rh–Rh bond of the monohydride species expanded again to recover the initial bond length. During the hydrogenation process, the Rh–Rh bond is retained without any breaking of the bond. Furthermore, the bond distances and CNs for Rh–O and Rh–P did not change significantly, demonstrating a high stability of the Rh dimers under the catalytic reaction conditions at 293–353 K in toluene (Table 4). The EXAFS analysis demonstrates that the alkene hydrogenation on the molecular imprinting Rh_{2imp} catalyst proceeds on the Rh-dimer unit in each step of the catalytic cycle. It appears that the chemical attachment in a tetradentate form (Rh–O: 0.211 nm) and the location with stable fitting in the micropore (1.9 nm thick) prevented the Rh dimers from leaching to the reaction solution, as well as from decomposing and gathering of the Rh dimers.

Ratios of TOFs corresponding to the degree of enhancement of the reaction rates by the imprinting revealed that the Rh_{2imp} catalyst showed size- and shape-selectivities for the alkenes, as shown in Table 5. Selectivity for the alkene hydrogenation on the Rh_{2imp} catalyst depends on the size and shape of the template cavity as a reaction site in the micropores of the SiO_2-matrix overlayers on the Ox.50 surface, in addition to the electronic and geometric effects of the ligands. The TOF ratios reduced as the gain of alkene size increased. It should be noted that there is a large decrease between 3-ethyl-2-pentene and 4-methyl-2-hexene (4-ethyl-2-pentene) due to the difference in the shape of the alkenes. The difference in the TOF ratios between 4-methyl-2-pentene and 4-methyl-2-hexene is also large, whereas the difference in the size of a methyl group is distinguished by the Rh_{2imp} catalyst. The TOF ratio for 4-methyl-2-hexene is much smaller than that for 2-pentene, where the difference is probably from the ethyl group. Thus the molecular imprinting catalyst discerns the size and shape of the alkenes. However, the molecular imprinting catalyst can not differentiate between methyl and ethyl groups among the smaller alkenes other than by the template size, as shown in Table 5. It is noteworthy that the reaction rates of 2-pentene, 2-octene, and 1-phenylpropene on the supported Rh_{2sup} catalyst are similar to each other, whereas the rate enhancements (TOF ratios) for 2-octene and 1-phenylpropene (10 and 7 times, respectively) are much smaller than that for 2-pentene (51 times). Because the length of the linear alkene chains can not be discerned by the ligand-coordinated metal site, the difference may be caused by the size of the wall of the template cavity around the Rh-dimer site.

Activation energies for the hydrogenation on the Rh_{2sup} catalyst are divided into two values, about 30 and 42 kJ mol^{-1}, for the linear alkenes and the branch alkenes, respectively (Table 5). Such a large difference in the activation energy has also been observed with the Wilkinson complex. In contrast, activation entropies for the hydrogenation on the Rh_{2sup} catalyst are similar to each other, where values are in the range -200 to -215 J mol^{-1} K^{-1} for all the alkenes.

After imprinting, significant differences between small and large alkenes were observed in both activation energies and activation entropies. The activation energies for the hydrogenation of 3-ethyl-2-pentene and smaller alkenes on the Rh_{2imp} catalyst were 26–43 kJ mol^{-1}, which are similar to the values observed on the Rh_{2sup} catalyst. The activation entropies for the four alkenes were -170 and -195 J mol^{-1} K^{-1}, which are larger than those (-200 and -210 J mol^{-1} K^{-1}) obtained for the Rh_{2sup} catalyst. In contrast, for the larger alkenes, such as 4-methyl-2-hexene, 2-octene, and 1-phenylpropene, the activation energies were 10, 7, and 8 kJ mol^{-1}, respectively, which are small compared with those for the Rh_{2sup} catalyst and other metal-complex catalysts. Furthermore, the activation entropies reduced prominently from about -210 J mol^{-1} K^{-1} to about -260 J mol^{-1} K^{-1}.

The conspicuous change in the kinetic parameters for the larger alkenes is parallel to the change in enhancement of the reaction rates. These dramatic decreases in the activation energy and TOF ratio can be explained by the shift of the rate-determining step from the alkyl formation to the coordination of alkene to the Rh site. Attendant to this shift, the activation entropies for the larger alkene molecules also decreased, mostly to about $-260\ \mathrm{J\ mol^{-1}\ K^{-1}}$ from about $-180\ \mathrm{J\ mol^{-1}\ K^{-1}}$ for the other small molecules. This suggests that conformation of the coordinated alkene in the template cavity is regulated by the wall of the cavity and the remaining $P(OCH_3)_3$ ligands. For the alkenes with the larger sizes and different shapes, which differ from those of the template, the coordination to the Rh site through the cavity space becomes the slowest step in the reaction sequences. The location of Rh center to which the alkenes coordinate, the conformation of remaining $P(OCH_3)_3$ ligand, the orientation of a template-vacant site on Rh, the template cavity shape, the architecture of the cavity wall, and the micropore surrounding the Rh dimer in the SiO_2-matrix overlayers all provide the active Rh_{2imp} catalyst for size- and shape-selective hydrogenation of the alkenes.

Regulation of simple alkenes without any functional groups is generally difficult. The method that combines metal-complex attachment and molecular imprinting on the surface demonstrates that the strategy can regulate and design chemical reactions at a molecular level similar to those of artificial enzyme catalysts. However, achievement of a more strict recognition of molecules needs further investigation, and molecular recognition of reactants with functional groups is the next stage of the study.

SUMMARY

The unique catalytic properties of metal complexes, which differ from those of metal and metal oxides, have been used exclusively for various chemical processes including selective catalysis, shape-selective catalysis, and asymmetric catalysis. Attaching metal complexes to oxide surfaces provides new catalytic systems that have the advantages of both heterogeneous and homogeneous catalysts. Advanced catalytic materials can be designed through the combination of metal-complex attachment and molecular imprinting on oxide surfaces. The imprinting of metal complexes may be a promising way to design artificial enzymatic catalysts with 100% selectivity for target reactions. In situ characterizations, including time-resolved DXAFS, are inevitably important for understanding of the origin and mechanism of the tremendous catalysis promoted by the designed surfaces and also for development of a new class of efficient catalysts.

ACKNOWLEDGMENT

This study was supported by a Grant-in-Aid for the 21st Century COE Program for Frontiers in Fundamental Chemistry from the Ministry of Education, Culture,

Sports, Science, and Technology. The XAFS measurements have been done with the approval of the Photon Factory advisory committee (PAC) (proposal No. 2001G315; proposal No. 2004G080). The authors thank Prof. M. Nomura and Dr. A. Suzuki at PF for measurements of time-resolved DXAFS. The authors also thank Dr. T. Sasaki and Mr. T. Taniike for the DFT and MM calculations.

LITERATURE CITED

1. Iwasawa Y, ed. 1986. *Tailored Metal Cataysts*. Dordrecht: Reidel
2. Iwasawa Y. 1987. Chemical design surfaces for active solid catalysts. *Adv. Catal.* 35:187–264
3. Iwasawa Y. 1993. Chemical design and characterization of catalysts and catalysis: an approach to dynamic catalyst design. *Catal. Today* 18:21–72
4. Iwasawa Y. 1996. Characterization and chemical design of oxide surfaces. *Stud. Surf. Sci. Catal.* 101:21–34
5. Iwasawa Y. 1997. Surface catalytic reactions assisted by gas phase molecules. *Acc. Chem. Res.* 30:103–9
6. Psaro R, Recchia S. 1998. Supported metals derived from organometallics. *Catal. Today* 41:139–47
7. Gates BC. 2001. Models of metal catalysts: beyond single crystals. *Topics Catal.* 14:173–80
8. Zecchina A, Arean CO. 1993. Structure and reactivity of surface species obtained by interaction of organometallic compounds with oxidic surfaces—IR studies. *Catal. Rev. Sci. Eng.* 35:261–317
9. Bando KK, Asakura K, Arakawa H, Isobe K, Iwasawa Y. 1996. Surface structures and catalytic hydroformylation activities of Rh dimers attached on various inorganic oxide supports. *J. Phys. Chem.* 100:13636–45
10. Asakura K, Bando KK, Iwasawa Y, Arakawa H, Isobe K. 1990. Metal-assisted hydroformylation on a SiO_2-attached rhodium dimer. In situ EXAFS and FT-IR observations of the dynamic behaviors of the dimer site. *J. Am. Chem. Soc.* 112:9096–103
11. Koningsberger DC, Prins R, eds. 1988. *X-Ray Absorption: Principles, Applications, Techniques of EXAFS, SEXAFS and XANES*. New York: Wiley
12. Bart JCJ, Vlaic G. 1987. Extended X-ray absorption fine structure studies in catalysis. *Adv. Catal.* 35:1–138
13. Iwasawa Y, ed. 1996. *X-Ray Absorption Fine Structure for Catalysts and Surfaces*. Singapore: World Scientific
14. Matsushita T, Phizackerley RP. 1981. A fast X-ray absorption spectrometer for use with synchrotron radiation. *Jpn. J. Appl. Phys.* 20:2223–28
15. Flank AM, Fontaine A, Jucha A, Lemonnier M, Williams C. 1982. Extended X-ray absorption fine-structure in dispersive mode. *J. Phys. Lett.* 43:L315–19
16. Dartyge E, Fontaine A, Jucha A, Sayers D. 1984. *EXAFS and Near Edge Structure III*. Berlin: Springer-Verlag
17. Couves JW, Thomas JM, Waller D, Jones RH, Dent J, et al. 1991. Tracing the conversion of aurichalcite to a copper catalyst by combined X-ray absorption and diffraction. *Nature* 354:465–68
18. Sankar G, Thomas JM, Waller D, Couves JW, Catlow CRA, Greaves GN. 1992. Time-resolved energy-dispersive and conventional EXAFS studies of the interactions of nitrous-oxide with supported copper catalyst. *J. Phys. Chem.* 96:7485–89
19. Bazin D, Dexpert H, Lagarde P,

Bournonville P. 1988. EXAFS studies of bimetallic platinum-rhenium/alumina and platinum-rhodium/alumina catalysts in the first stages of preparation, drying, and calcinations. *J. Catal.* 110:209–15
20. Fiddy SG, Newton MA, Dent AJ, Salvini G, Corker M, et al. 1999. In situ energy dispersive EXAFS (EDE) of low loaded Pt_{acac2}/H-I SiO_2 catalyst precursors on a timescale of seconds and below. *J. Chem. Soc. Chem. Commun.* 851–52
21. Yamaguchi A, Shido T, Inada Y, Kogure T, Asakura K, et al. 2000. Time-resolved DXAFS study on the reduction processes of Cu cations in ZSM-5. *Catal. Lett.* 68:139–45
22. Yamaguchi A, Suzuki A, Shido T, Inada Y, Asakura K, et al. 2001. Energy-dispersive XAFS study on the decarbonylation process of $Mo(CO)_6$ in NaY zeolite. *Catal. Lett.* 71:203–8
23. Yamaguchi A, Shido T, Inada Y, Kogure T, Asakura K, et al. 2001. In situ time-resolved energy-dispersive XAFS study on the reduction processes of Cu-ZSM-5 catalysts. *Bull. Chem. Soc. Jpn.* 74:801–8
24. Fiddy SG, Newton MA, Campbell T, Corker JM, Turin S, et al. 2001. In situ, time resolved, and simultaneous multi-edge determination of local order change during reduction of supported bimetallic (Pt-Ge) catalyst precursors using energy dispersive EXAFS. *Chem. Commun.* 445–46
25. Yamaguchi A, Suzuki A, Shido T, Inada Y, Asakura K, et al. 2002. In situ time-resolved energy-dispersive X-ray absorption fine structure study on the decarbonylation processes of $Mo(CO)_6$ entrapped in NaY and HY zeolites. *J. Phys. Chem. B* 106:2415–22
26. Newton MA, Dent J, Diaz-Moreno S, Fiddy SG, Evans J. 2002. Rapid phase fluxionality as the determining factor in activity and selectivity of highly dispersed, Rh/Al_2O_3 in deNO(x) catalysis. *Angew. Chem. Int. Ed.* 41:2587–89
27. Iwasawa Y. 2003. In situ characterization of supported metal catalysts and model surfaces by time-resolved and three-dimensional XAFS techniques. *J. Catal.* 216:165–77
28. Suzuki A, Inada Y, Yamaguchi A, Chihara T, Yuasa M, et al. 2004. Time-scale and sequence of dynamic structural changes in a MgO-attached ruthenium cluster catalyst observed by in situ time-resolved DXAFS. *J. Phys. Chem. B* 108:5609–16
29. Suzuki A, Inada Y, Yamaguchi A, Chihara T, Yuasa M, Iwasawa Y. 2003. Time scale and elementary steps of CO-induced disintegration of surface rhodium clusters. *Angew. Chem. Int. Ed.* 42:4795–99
30. Stern EA, Newville M, Ravel B, Yacoby Y, Haskel D. 1995. The UWXAFS analysis package—philosophy and details. *Phys. B* 209:117–20
31. Newville M, Livins P, Yacoby Y, Stern EA, Rehr JJ. 1993. Near-edge X-ray-absorption fine-structure of Pb—a comparison of theory and experiment. *Phys. Rev. B* 47:14126–31
32. Ankudinov AL, Ravel B, Rehr JJ, Conradson SD. 1998. Real-space multiple-scattering calculation and interpretation of X-ray-absorption near-edge structure. *Phys. Rev. B* 58:7565–76
33. Stern EA. 1993. Number of relevant independent points in X-ray-absorption fine-structure spectra. *Phys. Rev. B* 48: 9825–27
34. Deleted in proof
35. Tada M, Taniike T, Kantam LM, Iwasawa Y. 2004. Chiral self-dimerization of vanadium complexes on a SiO_2 surface: the first heterogeneous catalyst for asymmetric 2-naphthol coupling. *Chem. Commun.* 2542–43
36. Pu L. 1998. 1,1′-binaphthyl dimers, oligomers, and polymers: molecular recognition, asymmetric catalysis, and new materials. *Chem. Rev.* 98:2405–94
37. Noyori R. 1994. *Asymmetric Catalysis in the Organic Synthesis.* New York: Wiley
38. Nakajima M, Miyoshi I, Kanayama K, Hashimoto S. 1999. Enantioselective synthesis of binaphthol derivatives by oxidative coupling of naphthol derivatives

catalyzed by chiral diamine copper complexes. *J. Org. Chem.* 64:2264–71
39. Lin X, Yang J, Kozlowski MC. 2001. Enantioselective oxidative biaryl coupling reactions catalyzed by 1,5-diazadecalin metal complexes. *Org. Lett.* 3: 1137–40
40. Gao J, Reibenspies JH, Martell AE. 2003. Structurally defined catalysts for enantioselective oxidative coupling reactions. *Angew. Chem. Int. Ed.* 42:6008–12
41. Irie R, Masutani K, Katsuki T. 2000. Asymmetric aerobic oxidative coupling of 2-naphthol derivatives catalyzed by photoactivated chiral (NO)Ru(II)-salen complex. *Synlett.* 1433–36
42. Chu CYY, Hwang DR, Wang SK, Uang BJ. 2001. Chiral oxovanadium complex catalyzed enantioselective oxidative coupling of 2-naphthols. *Chem. Commun.* 980–81
43. Hon SW, Li CH, Kuo JH, Barhate NB, Liu H, et al. 2001. Catalytic asymmetric coupling of 2-naphthols by chiral tridentate oxovanadium(IV) complexes. *Org. Lett.* 3:869–72
44. Luo Z, Liu Q, Gong L, Cui X, Mi A, Jiang Y. 2002. Novel achiral biphenol-derived diastereomeric oxovanadium(IV) complexes for highly enantioselective oxidative coupling of 2-naphthols. *Angew. Chem. Int. Ed.* 41:4532–35
45. Somei H, Asano Y, Yoshida T, Takizawa S, Yamataka H, Sasaki H. 2004. Dual activation in a homolytic coupling reaction promoted by an enantioselective dinuclear vanadium(IV) catalyst. *Tetrahedron Lett.* 45:1841–44
46. Rehder D, Schulzke C, Dan H, Meinke C, Hanss J, Epple M. 2000. Water and bromide in the active center of vanadate-dependent haloperoxidases. *J. Inorg. Biochem.* 80:115–21
47. Eaton SS, More KM, Sawant BM, Eaton GR. 1983. Use of the electron-paramagnetic-res half-field transition to determine the interspin distance and the orientation of the interspin vector in systems with 2 unpaired electrons. *J. Am. Chem. Soc.* 105:6560–67
48. Deleted in proof
49. Brady PA, Sanders JKM. 1997. Selection approaches to catalytic systems. *Chem. Soc. Rev.* 26:327–36
50. Sherrington DC. 1998. Preparation, structure and morphology of polymer supports. *Chem. Commun.* 21:2275–86
51. Davis ME, Katz A, Ahmad WR. 1996. Rational catalyst design via imprinted nanostructured materials. *Chem. Mater.* 8:1820–39
52. Sellergren B. 2000. Imprinted polymers with memory for small molecules, proteins, or crystals. *Angew. Chem. Int. Ed.* 39:1031–37
53. Whitcombe MJ, Alexander C, Vulfson EN. 2000. Imprinted polymers: versatile new tools in synthesis. *Synlett.* 6:911–23
54. Mosbach K. 2000. Molecularly imprinted polymers and their use in biomimetic sensors. *Chem. Rev.* 100:2495–504
55. Tada M, Iwasawa Y. 2003. Design of molecular-imprinting metal-complex catalysts. *J. Mol. Catal. A* 199:115–37
56. Tada M, Iwasawa Y. 2003. Approaches to design of active structures by attaching and molecular imprinting of metal complexes on oxide surfaces. *J. Mol. Catal. A* 204–205:27–53
57. Morihara K, Kurokawa M, Kamata Y, Shimada T. 1992. Enzyme-like enantioselective catalysis over chiral 'molecular footprint' cavities on a silica (alumina) gel surface. *J. Chem. Soc. Chem. Commun.* 358–60
58. Heilmann J, Maier WF. 1994. Selective catalysis on silicon dioxide with substrate-specific cavities. *Angew. Chem. Int. Ed.* 33:471–73
59. Katz A, Davis ME. 2000. Molecular imprinting of bulk, microporous silica. *Nature* 403:286–89
60. Strikovsky AG, Kasper D, Crun M, Green BS, Hradil J, Wulff G. 2000. Catalytic molecularly imprinted polymers using conventional bulk polymerization or

suspension polymerization: selective hydrolysis of diphenyl carbonate and diphenyl carbamate. *J. Am. Chem. Soc.* 122:6295–96

61. Markowitz MA, Kust PR, Deng G, Schoen PE, Dordick JS, et al. 2000. Catalytic silica particles via template-directed molecular imprinting. *Langmuir* 16:1759–65
62. Tanimura T, Katada N, Niwa M. 2000. Molecular shape recognition by a tin oxide chemical sensor coated with a silica overlayer precisely designed using an organic molecule as the template. *Langmuir* 16:3858–65
63. Suzuki A, Tada M, Sasaki T, Shido T, Iwasawa Y. 2002. Design of catalytic sites at oxide surfaces by metal-complex attaching and molecular imprinting techniques. *J. Mol. Catal. A* 182–83:125–36
64. Fujii Y, Matsutani K, Kikuchi K. 1985. Formation of a specific coordination cavity for a chiral amino-acid by template synthesis of a polymer Schiff-base cobalt(III) complex. *J. Chem. Soc. Chem. Commun.* 415–17
65. Suh J, Cho Y, Lee KJ. 1991. Macrocyclic metal-complexes built on polyethylenimine. *J. Am. Chem. Soc.* 113:4198–202
66. Krebs JF, Borovik AS. 1995. Metallonetwork polymers—reversible Co binding to an immobilized copper(I) complex. *J. Am. Chem. Soc.* 117:10593–94
67. Krebs JF, Borovik AS. 1998. Dioxygen binding to immobilized Co-II complexes in porous organic hosts: evidence for site isolation. *Chem. Commun.* 553–54
68. Brunkan NM. Gagne MR. 2000. Effect of chiral cavities associated with molecularly imprinted platinum centers on the selectivity of ligand-exchange reactions at platinum. *J. Am. Chem. Soc.* 122:6217–25
69. Matsui J, Nicholls IA, Karube I, Mosbach K. 1996. Carbon-carbon bond formation using substrate selective catalytic polymers prepared by molecular imprinting: an artificial class II aldolase. *J. Org. Chem.* 61:5414–17
70. Locatelli F, Gamez P, Lemaire M. 1998. Molecular imprinting of polymerised catalytic complexes in asymmetric catalysis. *J. Mol. Catal. A* 135:89–98
71. Santora BP, Larsen AO, Gagne MR. 1998. Toward the molecular imprinting of titanium Lewis acids: demonstration of Diels-Alder catalysis. *Organometallics* 17:3138–40
72. Polborn K, Severin K. 2000. Biomimetic catalysis with immobilised organometallic ruthenium complexes: substrate- and regioselective transfer hydrogenation of ketones. *Chem. Eur. J.* 6:4604–11
73. Polborn K, Severin K. 2000. Biomimetic catalysis with an immobilised chiral rhodium(III) complex. *Eur. J. Inorg. Chem.* 8:1687–92
74. Cammidge AN, Baines NJ, Bellingham RK. 2001. Synthesis of heterogeneous palladium catalyst assemblies by molecular imprinting. *Chem. Commun.* 2588–89
75. Tada M, Sasaki T, Iwasawa Y. 2002. Novel SiO_2-attached molecular-imprinting Rh-monomer catalysts for shape-selective hydrogenation of alkenes; preparation, characterization and performance. *Phys. Chem. Chem. Phys.* 4:4561–74
76. Tada M, Sasaki T, Shido T, Iwasawa Y. 2002. Design, characterization and performance of a molecular imprinting Rh-dimer hydrogenation catalyst on a SiO_2 surface. *Phys. Chem. Chem. Phys.* 4:5899–909
77. Tada M, Sasaki T, Iwasawa Y. 2002. Performance and kinetic behavior of a new SiO_2-attached molecular-imprinting Rh-dimer catalyst in size- and shape-selective hydrogenation of alkenes. *J. Catal.* 211:496–510
78. Tada M, Sasaki T, Iwasawa Y. 2004. Design of a novel molecular-imprinted Rh-amine complex on SiO_2 and its shape-selective catalysis for alpha-methylstyrene hydrogenation. *J. Phys. Chem. B* 108:2918–30

Annu. Rev. Mater. Res. 2005. 35:427–64
doi: 10.1146/annurev.matsci.35.100303.120733

First published online as a Review in Advance on April 1, 2005

METAL CATALYST DESIGN AND PREPARATION IN CONTROL OF DEACTIVATION

Michael S. Spencer
School of Chemistry, Cardiff University, Wales CF10 3TB, United Kingdom; email: spencerms@cardiff.ac.uk

Martyn V. Twigg
Johnson Matthey, Orchard Road, Royston, Herts SG8 5HE, United Kingdom; email: martyn.twigg@matthey.com

Key Words catalyst life, sintering, poisoning, copper catalyst, iron catalysts

■ **Abstract** Catalyst life (with activity and selectivity) is one of the three essential properties of a practical catalyst, both in laboratory and industrial usage. For most catalysts in which the active phase is a metal, high activity requires a high metal surface area. In all cases, for long life, the metal surface area has to be maintained. The various modes of catalyst deactivation leading to loss of metal surface area are described (sintering, poisoning, coking, etc.). The choice of catalyst components for both active metal phases and for support phases is shown to be subject to thermodynamic and kinetic limits. Routes for the preparation of catalysts need to be and can be designed to give not only active, but also stable catalysts. Examples described include copper/zinc catalysts for hydrogenation/dehydrogenation reactions, iron catalysts for ammonia synthesis, nickel catalysts for hydrogenation reactions, automobile exhaust catalysts, and Raney catalysts.

INTRODUCTION

In general, three catalyst properties are essential: activity, selectivity, and life. They can be assessed in this order. Clearly a catalyst of very low or negligible activity is of no use, whatever its other properties. This requirement depends on and is defined by the proposed use, whether for a single laboratory experiment or continuous use in an industrial plant.

Second, in most applications, selectivity to the desired product is also essential. Catalyst selectivity is irrelevant in only the few cases, such as ammonia synthesis from nitrogen and hydrogen, where in practice there is but a single product that can be formed (other nitrogen/hydrogen compounds, such as hydrazine and hydrazoic acid, are thermodynamically unstable under synthesis conditions). Selectivity is usually important, again in both laboratory and industrial use, to minimize problems of product separation, excess consumption of raw materials, environmental hazards of unwanted by-products, etc., and for economic considerations.

Suitable catalyst life is the third requirement, which can be defined as the time over which catalyst activity and selectivity must be maintained (1). This can vary from a period of minutes for a laboratory catalyst to 150,000 miles duration for an auto-exhaust catalyst or to some years for a catalyst in an industrial plant. Catalyst deactivation, which can include loss of selectivity as well as loss of activity, is inevitable because catalysts themselves are always thermodynamically unstable. Although loss of activity is frequently mentioned in papers on catalysis, the published literature on the topic of deactivation is in general inadequate, both on fundamental aspects and on practical application. In this review we examine the ways in which the principles of catalyst design can give long catalyst life, followed by several examples of how this is achieved in practice. In the limited space available we deal solely with metal catalysts, both supported and unsupported, where the main or sole catalytic function takes place on the metal surface. Catalysts in which the active metal catalyst is in the form of a metal compound, either surface or bulk (e.g., sulfide catalysts used for hydrodesulfurization), are omitted. In bifunctional catalysts, both the metal and the support phase have different catalytic functions. These can be of comparable importance, as in the catalytic reforming process, or catalysis on the support phase can be the more critical function, and examples of those processes are hydrocracking and hydroisomerization. Catalyst design for these processes is complex and not considered explicitly in this review, but nevertheless the principles of the design of simple metal catalysts are generally applicable to complex, bifunctional catalysts.

Loss of Activity

Catalyst life is determined by the loss of the two essential properties, activity and selectivity (2). These two properties do not necessarily decrease in parallel. The consequences of loss in activity and in selectivity need to be considered separately. Large losses in activity, e.g., down to only a few percent of the original design activity, obviously lead to a failure of the catalytic process. This is clearly so for both industrial processes and laboratory procedures. Catalyst design and preparation has to be such as to avoid catastrophic failure of the catalyst. On the other hand, smaller losses in catalyst activity usually do not have such serious consequences. In practice, loss in activity is defined as activity per catalyst pellet, particle, or unit weight, not as the specific activity of the metal surface or the activity per catalyst site. Loss in overall catalyst activity can arise from either a loss of active sites or loss of activity of the sites, or indeed various combinations of both. In terms of assessing the consequences of deactivation, the origins are irrelevant, although an understanding of the causes of deactivation is essential for its control. Catalyst decay is inevitable, at least over some timescale, and it is common industrial practice (1) to compensate for decreases in catalyst activity by increases in reactor temperature. As most industrial catalytic processes have significant positive activation energies, comparatively small and acceptable increases in reactor temperature are sufficient. For example, it is possible to calculate the required increases in temperature to maintain an overall reaction rate equal to the initial rate after a fall in catalyst activity to 50% of the original activity. Let the rate of the overall catalytic reaction be given by the simple Arrhenius equation 1.

$$r = A \exp(-E_A/RT), \qquad 1.$$

where r is the overall rate of catalytic reaction; A is the constant containing all the nontemperature variable parameters, including reactant and product partial pressures; and E_A is the overall activation energy of the catalytic reaction, including chemical reaction and heat/mass transfer terms.

Then the required increase in temperature can be shown (M.S. Spencer, unpublished data) to be given by Equation 2 derived from Equation 1.

$$\Delta T = T_1 \left\{ \left[1 + \frac{RT_1 \ln(0.5)}{E_A} \right]^{-1} - 1 \right\}, \qquad 2.$$

where ΔT is the required increase in temperature, and T_1 is the original reactor temperature at full catalyst activity.

Values of ΔT calculated from Equation 2 for various typical temperatures and activation energies are given in Table 1. As expected, these values of ΔT are smallest for reactions of higher activation energy proceeding at the lower temperature. What is acceptable in industrial practice is, of course, strongly dependent on the process and operating conditions, but most of the temperature increases in Table 1 are possible in many processes. Similarly comparable increases in reaction temperature are practical in many laboratory experiments.

An alternative to increasing reactor temperature to compensate for loss of catalyst activity is the use of a larger reactor. Here the reactor is designed for use with catalyst activity at the end of life rather than with initial activity so that any premature shutdown owing to catalyst deactivation can be avoided. Catalyst life is

TABLE 1 Values of ΔT, the increase in reactor temperature required to compensate for a loss in catalyst activity to 50% of its original value

T/K	473	673	873
E_A/kJ mol^{-1}			
50	27.3	56.6	97.3
100	13.3	27.2	46.2
150	8.6	17.9	30.3

then defined in terms of plant shutdown for other reasons (e.g., statutory inspection of boilers) (1). This approach involves extra cost in two ways. If, to follow the example above, a fall in catalyst activity to 50% can be expected over the normal time between plant shutdowns, then the catalyst reactor has to be larger to accommodate a double-size catalyst bed, and also twice as much catalyst has to be supplied for the catalyst reactor. However, the extra costs of both capital and revenue are relatively minor in most process cases. Similarly, the costs of extra catalyst is rarely a problem in laboratory or pilot plant experiments.

It follows that loss of catalyst activity, unless it is catastrophic, is not usually a serious problem in practice, and thus catalyst design is primarily concerned with other factors. This is not to deny that in most cases it is better to have more activity than less in a given catalyst (especially when the plant is already constructed). This is exemplified by processes with active catalysts operating at high temperature, e.g., methanol conversion to formaldehyde over a silver catalyst or ammonia oxidation over a platinum catalyst. Only small catalyst beds are required, and the limits to plant output are set by engineering considerations.

Loss of Selectivity

In general, the loss of catalyst selectivity during catalyst life has more severe consequences than the loss of activity considered alone. In industrial practice, loss of catalyst selectivity has two main consequences. First, a fall in the yield of the desired product means that an increased amount of raw material has to be used to give the same amount of product or, alternatively, a lower rate of production from a fixed rate of feedstock has to be accepted. Either of these obviously can have severe deleterious effects on the profitability of the plant. Second, if less of the desired product is formed from the feedstock, it follows that increased amounts of by products are formed. This will also usually increase the costs of an industrial process in terms of the necessary separation of the reaction products and of the disposal of unwanted by-products, especially those that have little value and are environmentally undesirable.

A simple example of the effects and consequences of deactivation on activity and selectivity is the high-temperature stage of the water-gas shift Reaction 3.

$$CO + H_2O = CO_2 + H_2. \qquad 3.$$

The process, usually followed by a second stage of low-temperature shift Reaction 3, is widely (see below) used (1) after methane-steam reforming, coal-gasification, and other processes for hydrogen generation. The catalyst consists of iron oxide/chromia, and the active phase in the working catalyst is magnetite, Fe_3O_4. This catalyst has a good activity (although less than that of the copper/zinc oxide/alumina catalyst used in the low-temperature water-gas shift process) and a very high selectivity for Reaction 3. Under normal operating conditions, negligible amounts of methane, which can be formed by Reaction 4,

$$CO + 3H_2 = CH_4 + H_2O, \qquad 4.$$

or of any other organic compounds are produced. Although the catalyst is robust under most operating conditions, various modes of deactivation can occur but only two are considered here: sulfur poisoning from hydrogen sulfide and over-reduction to metallic iron. Sulfur poisoning is not usually a problem in methane-steam reforming plants in which almost all of the sulfur is removed from the feed gas, but it can occur in coal-based plants. Hydrogen sulfide can convert magnetite to iron sulfide by Reaction 5.

$$Fe_3O_4 + 3H_2S + H_2 = 3FeS + 4H_2O. \qquad 5.$$

Carbonyl sulfide can effect the same conversion as it reacts first with steam to give hydrogen sulfide, Reaction 6.

$$COS + H_2O = CO_2 + H_2S. \qquad 6.$$

Only low levels of hydrogen sulfide are needed to give substantial conversions of magnetite. Values of the equilibrium constant of Reaction 5 are given in Table 2. It follows from the large values of K_5 that only traces of hydrogen sulfide can bring about Reaction 5, i.e., substantial conversion to iron sulfide. With conversion to a sulfided catalyst, the activity for Reaction 3 is decreased to about half the activity of the unsulfided catalyst, and the selectivity is unchanged. From the discussion above, it is seen that this mode of deactivation, although significant, does not cause serious problems in practical operations. The copper-based catalyst used in the low-temperature water-gas shift process is, in contrast, much more susceptible to sulfur poisoning, and large losses in activity can occur (see below).

The iron in the water-gas shift reaction catalyst is maintained as Fe_3O_4 by the hydrogen/steam ratio used in the process (3). Normally there is no difficulty in keeping a stable catalyst. However, a plant upset or incorrect reduction giving loss of steam can lead to further reduction of the magnetite to metallic iron by Reaction 7.

$$Fe_3O_4 + 4H_2 = 3Fe + 4H_2O. \qquad 7.$$

TABLE 2 Variation of equilibrium constant for Reaction 5 with temperature (adapted from Reference 3)

Temperature/°C	Equilibrium constant[a]
300	3.11×10^9
350	1.15×10^9
400	4.74×10^8
450	2.18×10^8
500	1.07×10^8
550	5.49×10^7

[a]Equilibrium constant defined by $K_5 = p^4_{H_2O}/p_{H_2}\,p^3_{H_2S}$.

TABLE 3 Standard enthalpy changes at 298 K of the water-gas shift reaction and of two undesirable side reactions in the process; data from Reference 3

Reaction		Standard enthalpy change/kJ mol^{-1}
$CO + H_2O = CO_2 + H_2$	(3)	−41.1
$CO + 3H_2 = CH_4 + H_2O$	(4)	−206.2
$2CO = C + CO_2$	(8)	−172.5

This change in the catalyst, effectively a deactivation, causes a change in catalyst selectivity, which has serious consequences. Once metallic iron is formed, further catalytic reactions become possible, principally Reactions 4 and 8.

$$2CO = C + CO_2. \qquad 8.$$

This change in selectivity means that carbon monoxide, which would have been converted to the desired product, hydrogen, is converted instead into other unwanted products. If Reaction 8 proceeds to any significant extent, there is the further problem that the deposited carbon will build up and eventually envelop the catalyst, killing all activity. There is another effect of the loss of selectivity with the onset of Reactions 4 and 8. In Table 3 the enthalpy changes of these reactions are given. These are much greater than the standard enthalpy change of the water-gas reaction itself, so quite small losses in selectivity to Reactions 4 and 8 can cause considerable increases in the heat evolved in the reactor. These can lead to temperature "runaway" and catastrophic failure of the plant. Similar problems of temperature runaway occur in many selective organic oxidation reactions, where the (negative) enthalpy change to the desired products is much smaller than the enthalpy change of complete combustion to carbon dioxide and water.

Practical Considerations

Catalyst deactivation can give varied effects and the costs involved can be much more than just that of catalyst replacement, although that is itself rarely negligible. The costs of premature plant shutdown, loss of product, use of excess raw material and, in worst case, plant failure can all be much higher than the cost of a catalyst change. Catalyst deactivation has to be taken very seriously in relation to both the design and preparation of catalysts. There is an abundant literature on catalyst deactivation, including the proceedings of a series of conferences on the subject (4). Further, as inevitably all catalysts die off over some timescale, considerations of deactivation occur in numerous papers and books that are primarily concerned with other aspects of catalysis. In this review we look at some fundamental and practical aspects of the contributions that catalyst design and preparation can make to minimize deactivation.

BASIC PRINCIPLES

Modes of Deactivation

Loss of catalyst activity can occur broadly in three ways. First there can be a loss of the catalytically active surface without any change in the specific reaction rate, i.e., the rate of reaction at a single catalytic site. Second, some change in the metal surface can occur that alters the specific reaction rate without necessarily involving any change in metal surface area. Third, changes in the catalyst structure can occur which, although not affecting the metal crystallite or its surface area, can limit the access of reactants to the active site (or the removal of products away from the active site). Causes for these different modes of deactivation are listed in Table 4. It should be noted that any of these modes can give changes in selectivity (which may be beneficial but are more often harmful) as well as loss of activity.

Metal Surface Area

For metals, the predominant sintering mechanism in the bulk is vacancy diffusion, which suggests a relationship with cohesive energy. Hughes (6) gave the following increasing order of stability for metals:

$$Ag < Cu < Au < Pd < Fe < Ni < Co < Pt < Rh < Ru < Ir < Os < Re.$$

Loss of activity in metal catalysts via loss of surface area owing to sintering is a universal phenomenon. Clearly, though, the lower the metal is in this series the more troublesome deactivation by sintering will be and the more care has to be taken to minimize the effect. For example, it is not surprising that copper-based catalysts are more susceptible to sintering than the nickel and iron catalysts used in ammonia and hydrogen plants. This is also shown by copper's low Hüttig

TABLE 4 Causes of catalyst deactivation (adapted from Reference 5)

Type	Cause	Results
Mechanical	Particle failure	Bed channelling, plugging
	Fouling	Loss of surface
Thermal	Component volatization	Loss of component
	Phase changes	Loss of surfaces
	Compound formation	Loss of component and surface
	Sintering	Loss of surface
Chemical	Poison adsorption	Loss of active sites
	Coking	Loss of surface, plugging

temperature (7), which reflects a relatively low melting point (1083°C), compared with that of iron (1535°C) and nickel (1455°C). Therefore, copper-based catalysts have to be operated at relatively low temperatures, usually no higher than 300°C. This is well below the maximum operating temperatures of iron and nickel catalysts.

In most metal catalysts there is a need to prepare and maintain a high metal surface area. In all finely divided solids there is a thermodynamic driving force to increase particle size by the reduction of surface energy. Thermal sintering, the more significant the smaller the metal particles, is almost always a factor in catalyst deactivation at the temperatures required for catalytic reaction. Catalyst design has to ensure that this inevitable process is minimal, which is achieved largely by the use of one or more other phases, commonly an oxide. This phase is termed as a support if present as a substantial fraction of the catalyst or as a structural promoter if only a minor component. In some cases, the metal may appear to be without any structural promoter, but usually this is misleading. Raney nickel catalysts (see below) are prepared by the extraction by alkali of the aluminum from a Ni/Al alloy to leave a highly porous nickel metal catalyst (8). However, the extraction of aluminum is not complete nor is it desirable for it to be so. The remaining aluminum is oxidized/hydrated to an amorphous aluminum oxide/hydroxide, which acts as a structural promoter. With higher melting point for metals such as platinum (e.g., platinum black), this appears to be relatively less important. There are a few metal catalysts that do not incorporate other phases to maintain catalyst structure. Two examples are ammonia oxidation to nitric oxide (9, 10), Reaction 9, and methanol conversion to formaldehyde (9, 11), Reaction 10.

$$NH_3 + {}^5\!/_4O_2 = NO + {}^3\!/_2H_2O. \qquad 9.$$

$$CH_3OH + {}^1\!/_4O_2 = HCHO + {}^1\!/_2H_2O + {}^1\!/_2H_2. \qquad 10.$$

The catalysts used in those processes are unsupported metals. Platinum (the presence of alloyed rhodium is not significant for the argument here) in the form of platinum wires in a woven mesh or, more recently, a knitted mesh for high gross stability, is used for ammonia oxidation, Reaction 9. Granules of unsupported silver metal in a shallow bed form the catalyst for methanol oxidation/dehydrogenation, Reaction 10, although silver gauzes are sometimes used. The diameter of the platinum wires for ammonia oxidation is typically 75 μm (9). The granular silver catalyst typically consists of crystals in the range 5 mm to less than 0.5 mm.

Selectivity is a vital requirement in both processes because complete combustion to more stable products, Reactions 11 and 12, has to be avoided.

$$2NH_3 + {}^3\!/_2O_2 = N_2 + 3H_2O. \qquad 11.$$

$$CH_3OH + {}^3\!/_2O_2 = CO_2 + 2H_2O. \qquad 12.$$

Other by products, e.g., N_2O by low-temperature oxidation of ammonia, are also possible, but they are less significant. High-temperature operation, around 900°C for ammonia oxidation and 600°C for methanol conversion, provides the key to selectivity. Typical selectivities are 95–98% in ammonia oxidation and over 90% in formaldehyde production. At these temperatures, metal sintering is fast and would prevent the use of catalysts in the form of microcrystallites even if extensively supported. However, in both cases the specific rates of reaction are very high (part of the way in which high selectivity is achieved), so only very low metal surface areas are needed to give acceptable overall process rates. Thus the very large (relatively) sizes of the particles of these metal catalysts, subject to low (relatively) sintering rates at the operating temperature, provide practical catalysts.

In most other cases specific reaction rates are much lower, so much higher metal surface areas, i.e., much smaller metal crystallites, are needed. Stable catalysts with fine metal crystallites, to operate at realistic temperatures, require other phases to keep the crystallites separate and minimize sintering. This principle is essentially the same for, e.g., a catalytic reforming catalyst of <1% Pt (+other promoters) on an alumina support (12) and an ammonia synthesis catalyst of >98% Fe (+other promoters) with ~1% alumina structural promoter (13). The optimum amount of the second phase to give maximum metal surface area is primarily a function of the relative sizes of metal and support particles. In catalytic reforming catalysts, the platinum crystallites, which are much smaller than the alumina particles, are distributed over the surface of the alumina particles with sufficient distance between the platinum crystallites to keep sintering within acceptable limits. In principle, an alternative approach would be to use catalysts with larger platinum crystallites, which sinter less readily, but then much larger amounts of platinum would be needed to provide the required metal surface area. With expensive metals such as platinum, this is not a practical approach. On the other hand, the cost of the iron metal in ammonia synthesis catalysts is not significant in plant operation. For these catalysts the maximum stable metal surface area is obtained with relatively large metal crystallites, where sintering is minimized by particles of alumina, which are much smaller than the iron crystallites and interposed between them.

These two examples represent the extremes of disparity between the sizes of metal crystallites and particles of support phases. In many other catalysts these sizes are of the same order of magnitude, and consequently the relative amounts of metal and support in the catalyst are also of the same order of magnitude.

Activity and Deactivation

The discussion in the previous section on the ways of maximizing of metal surface areas applies both to the initial activity of the catalyst and to the limiting of catalyst deactivation. Both forming and keeping activities require stable metal surface areas, but sometimes the requirements are in conflict. A high initial activity may be associated with a high rate of decay. This is less desirable than a lower initial

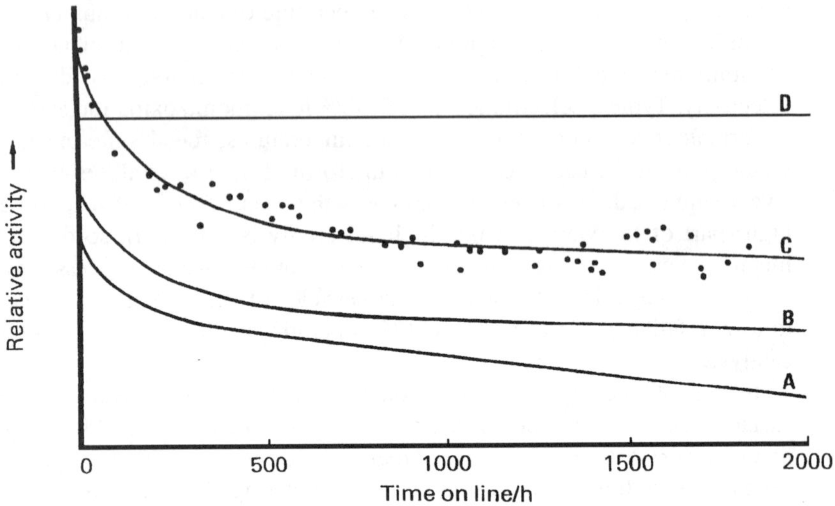

Figure 1 Relative activities of typical copper/zinc oxide low-temperature water-gas shift catalysts under poison-free conditions. (*A*) Cr_2O_3-based catalysts; (*B*) conventional Al_2O_3-based catalyst; (*C*) high copper content Al_2O_3-based catalyst; (*D*) optimized Al_2O_3-based catalyst. For clarity, experimental points are shown on only one of the curves (adapted from Reference 14).

activity, which is more stable over the period of use. The integrated activity over catalyst life represents the total output from a plant over that period.

An example is shown in Figure 1. Supported copper catalysts are used for the low-temperature process stage of the water-gas shift process, Reaction 3. The relative activities of several copper catalysts are shown for times that are a significant part of expected plant life. In particular, a comparison between catalysts C and D shows that the higher initial activity from a higher copper content is associated with a higher rate of decay of activity. Clearly catalyst D is preferred for plant operation. If the copper crystallites in both catalysts were of similar size ranges, then the higher initial activity of catalyst C would arise from the higher total copper metal surface area. The origin of the faster decay was probably also the higher copper content. Not only was the amount of refractory support inevitably less in this catalyst, but ratio support/copper decreased more. Thus there was an insufficient amount of support to hinder the sintering of the copper crystallites.

The support/physical promoter in a catalyst is required both to give high metal surface area in the initial catalyst activation by reduction and to help maintain this high metal surface area during the whole period of catalyst use.

Catalyst Preparation

Almost all techniques of the preparation of supported metal catalysts fall into three categories. The preparation of unsupported metal catalysts is not considered here.

1. Impregnation. A preformed support phase is impregnated with a solution of a compound of the catalytic metal (together with any promoters), and subsequent processing and reduction yield metal crystallites on the support.
2. Precipitation. Typically compounds of both metal and support are precipitated, either coincidentally or consecutively, from aqueous solutions of precursor salts. Subsequent processing and reduction yields metal crystallites separated by support phases.
3. Extraction. This is of more limited use than routes 1 or 2. A solid nonporous phase is made, usually by fusion. Activation and development of the porous structure, and hence of metal surface area, consists of the removal of one component of the nonporous precursor without loss of structure. Examples are (*a*) the reductive removal of oxygen from magnetite, Fe_3O_4, to give an iron catalyst; and (*b*) the removal by alkali of aluminum from a nickel/aluminum alloy to give a Raney nickel catalyst (see below). As described above, both catalysts contain physical promoters.

The choice of preparative technique obviously depends primarily on the chemistry of the metal and support components, but it is also very much a function of the necessary metal/support ratio in the final catalyst. Catalysts of low metal content, i.e., those where the metal crystallite size is much less than the size of support particles, are usually made by impregnation. This category also includes catalysts whose metals are supported on the internal pore structure of zeolites, when the metal precursor compounds are distributed in the zeolite by ion exchange. These catalysts are usually examples of bifunctional catalysts, where the zeolite support exercises a catalytic function in addition to the catalytic function of the metal. Catalysts of Pd metal in HY zeolite are used in hydrocracking processes. A nonzeolite example is the Pt/chlorided-alumina (+promoters) catalyst for catalytic reforming processes. The design of bifunctional catalysts is complex and is not considered further here. Catalysts containing broadly equivalent amounts of metal and support are most frequently made by precipitation, although sometimes repeated impregnation can be used. Extraction depends not only on the chemistry of the components but also on the avoidance of collapse of the pore structure. It is used, as shown in the two examples cited above, for catalysts of very high metal content with only small amounts of physical promoter.

Catalysts prepared by extraction have a further advantage over precipitated or impregnated catalysts in terms of lower rates of deactivation by metal sintering. This is rarely recognized. With the high metal content of those catalysts, the metal crystallites do not have to be so small to give a desired metal area per gram of catalyst. But further, as the catalysts are of higher overall density (often by a factor of several times) than catalysts consisting largely of oxide supports, the metal area per unit volume (the important criterion in industrial use) is correspondingly greater. This means that to obtain a given reaction rate, i.e., a given metal surface area, the metal crystallites are typically much larger in extraction catalysts than in impregnated or precipitated catalysts. This in turn leads to lower rates of sintering, for the driving force for sintering decreases with increasing particle radius. An example of this is the conventional ammonia synthesis catalyst (see below), prepared by an extraction technique to give iron metal crystallites of about 30 μm size, where catalyst lives of 5–10 years are not uncommon.

There is a further practical note for the support in supported-metal catalysts. Except for the few catalysts, such as the ammonia synthesis catalyst and Raney nickel, where the catalysts on activation keep their bulk structure, activated catalysts without a support consist of fine metal powders. Although catalysts in this form can be used in laboratory experiments and commercially in liquid systems [e.g., powdered nickel metal catalysts in the hydrogenation of vegetable oils to margarine (15)], processes with gaseous reactants and products require catalyst beds through which a process stream can pass. Thus the catalyst, no matter what its microcomposition is, must have a macrocomposition of pieces that are large enough to allow a gas flow. The gas flow through a bed of metal particles of 10 μm diameter would clearly be negligible, even at the laboratory. This is one reason (others will be considered below) for the prevalence of alumina, silica, and their compounds as supports for metal catalysts. Suitable-sized gross particles that would allow gas flow through a catalyst bed, typically 2–10 mm, can readily be made with these supports by standard ceramic manufacturing techniques such as pelletting and extrusion.

The failure to make suitable catalyst particles for use in industrial reactors is one reason for the lack of commercial exploitation of catalysts based on Cu/Ce intermetallic compounds for methanol synthesis (16). Nonporous Cu/Ce intermetallic compounds react with a gaseous CO/H_2 mixture to give a catalyst consisting of very small copper-containing active sites in a partially reduced CeO_2 matrix. These catalysts have an initial activity for methanol synthesis that is several orders of magnitude greater than that of the standard $Cu/ZuO/Al_2O_3$ catalyst. Unfortunately, the stability of these catalysts is poor, even in laboratory experiments, and this has prevented their industrial use (16). But there is another difficulty. The partial oxidation of the solid pieces of the Cu/Ce intermetallic compound gave a fine powder of the active catalyst, totally unsuitable for an industrial catalytic reactor, and all attempts to incorporate this material in some support failed because of coincident loss of activity.

Model of Supported Metal Catalysts

Intuitively, it might seem that, regardless of problems of catalyst decay, maximum metal surface areas would be obtained in catalysts of maximum metal content and maximum porosity. However, the situation is often more complex, as shown in Figure 2. Maximum nickel surface area of the catalyst is reached only at about 50% nickel content.

A simple model has been developed (M.S. Spencer, unpublished data) that shows why maximum metal areas are reached in catalysts with metal contents of 50% or less. Most real catalysts consist of multi-phase solids of irregular shapes, which are difficult to analyze. Maximum metal surface areas, and even more

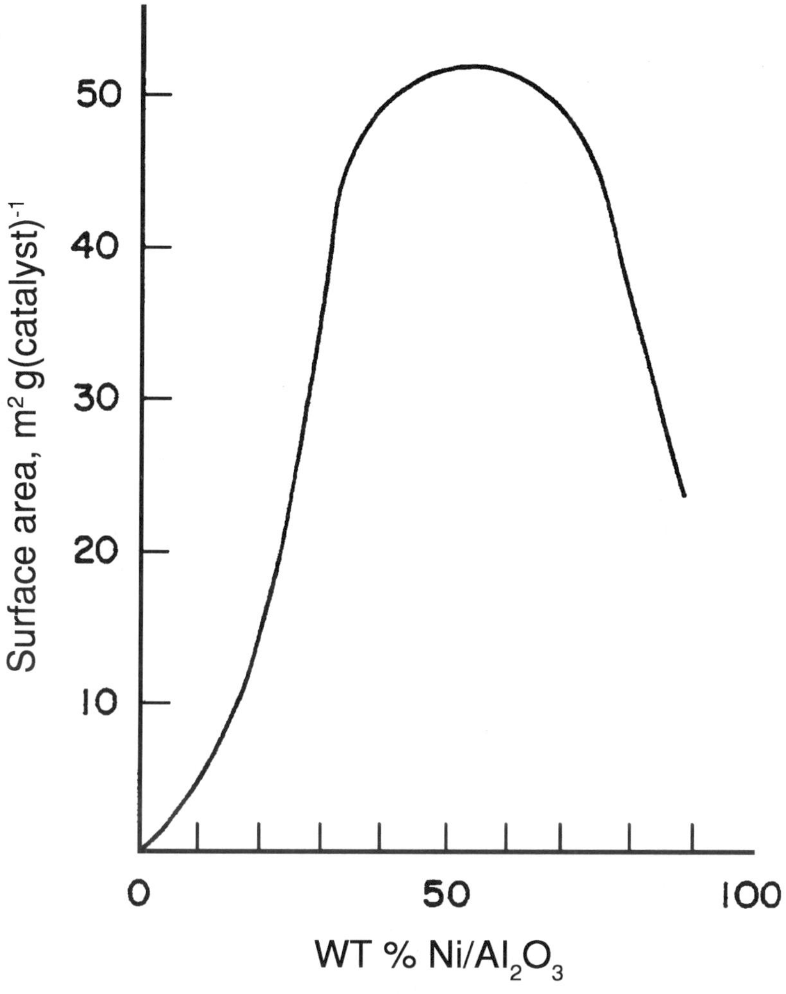

Figure 2 Surface area of metallic nickel in alumina-supported nickel catalysts as a function of nickel content. (Adapted from Reference 5.)

important, maximum sustainable surface areas, have been determined by empirical experiment. If we take first the idealized case of a catalyst consisting of unsupported, spherical metal particles, in a body-centred cubic array, then both the porosity and metal surface area of the array can readily be calculated. The porosity of the array is 0.320, independent of the sphere radius, r, and the surface area is $\sqrt{3}\pi/8r$ for the spheres in a cube of unit side. It is clear that, as expected, the maximum surface area is obtained at the minimum sphere radius. This is shown more clearly in Figure 3 where the surface area, expressed in the customary units of m^2 (g metal)$^{-1}$, is shown as a function of sphere radius for three widely used

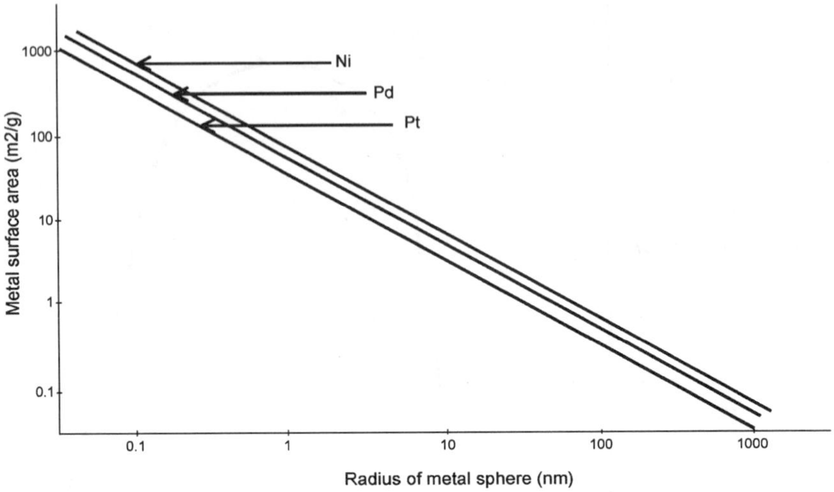

Figure 3 Surface areas of close-packed arrays of spherical metal particles of nickel, palladium, and platinum as a function of particle radius (M.S. Spencer, unpublished data).

catalytic metals, nickel, palladium, and platinum. The calculated values are in broad agreement with maximum experimental values (17).

In practice, however, as described above, catalysts consisting solely of very small metal particles are not stable. At the temperatures needed for useful catalytic reactions, sintering of the metal particles almost always leads to growth in particle size and consequent loss in surface area. A different approach is needed to obtain better models of real catalysts that include particles of support or physical promoter. It is then possible to look at the relations between surface area, porosity, and catalyst geometry. The results have been calculated for several different, simple models. None of these corresponds closely with real catalysts, but it is found that the results do not depend strongly on the geometries of the models chosen.

In the first model, there is an array of straight parallel tubes through a solid cube of unit side. The ends of the tubes, on the two appropriate faces, are in a regular, face-centered square array. To take account of stabilizers to prevent sintering, the thickness of the walls between two pores is kept constant at 2a, whereas the porosity and surface area are calculated for variable r, the pore radius. Then the calculation gives

$$\theta = \frac{\pi r^2}{2(r + \mathrm{a})^2}, \tag{13.}$$

$$\phi = \frac{\pi r}{6(r + \mathrm{a})^2}, \tag{14.}$$

where θ = porosity, and
ϕ = surface area of tubes.

TABLE 5 Relation between surface area and porosity for model catalysts

	Porosity	
Catalyst model	**For maximum surface area**	**Maximum possible**
Face centered square array of parallel hollow tubes	0.39	0.78
Body-centered cubic array of spherical holes	0.31	0.68
Body-centered array of solid cubes	0.41	1.00

In this model the maximum value of ϕ is found, not as r approaches zero but at $r = \text{a}$. At this value the porosity is 0.393. Similar calculations have been carried out for two other different models:

1) a body-centered cubic array of spherical holes in a solid matrix, where, as above, the thickness of the walls between the holes is constant;
2) a body-centered cubic array of solid cubes in what is otherwise free space. The distance between the cubes was kept constant while the size of the cubes was varied. The supports between the solid cubes were ignored in the calculations.

The results are shown in Table 5. Two features are notable. First, the values of porosity at which maximum surface area is obtained do not vary widely despite the very different geometries of the models used. This indicates that the results should be broadly applicable to the complex geometries of real catalysts. Second, and this is perhaps in contrast to intuitive suggestion, maximum surface areas are not obtained at the highest porosities but at porosities in the region of 0.3 to 0.4. This is in broad agreement with experimental optimizations of many supported metal catalysts, including those made by impregnation, precipitation, and extraction from a solid phase. Although a direct comparison is not possible, the calculated results in Table 5 are in qualitative agreement with the experimental curve for Ni/Al_2O_3 catalysts in Figure 2.

Catalyst Activation

The final stage of catalyst preparation is the activation of the catalyst before use in the reactor. With metal catalysts, the activation stage is usually the reduction of the oxide (or sometimes another compound) of the metal by hydrogen or carbon monoxide (18). This is carried out either in the reactor itself before use or separately (preactivation) in a purpose-built reactor by the catalyst manufacturer. In the case of preactivation, the catalyst is subsequently partially oxidized under carefully controlled conditions, followed by activation using rereduction in the plant reactor. There are two advantages of preactivation of the catalyst. First the prereduction reactor can be designed specifically for catalyst reduction, and this can lead to a better product, i.e., a more active or more stable catalyst. Second, as the major part of the catalyst reduction has been carried out before the catalyst is installed in the plant, the subsequent rereduction takes much less time than the full reduction of the original catalyst precursor. Thus the use of prereduced catalysts can make it possible after a shutdown to bring a plant on stream more quickly, so increasing total time on stream.

Almost all metal catalysts are activated by carefully controlled reduction with gases containing hydrogen and/or carbon monoxide. An exception, described below, is the three-way auto-exhaust catalyst, where reduction is dependent on the vagaries of the engine-exhaust gases. The need for reductive activation for a wide range of metal catalysts leads to restrictions on the metals to be used as the active phase, which needs to be reduced to the metal in an acceptable time, and on the support oxides, etc., which obviously should not undergo reductive change.

BULK REDUCTION MODEL An analysis has been made (M.S. Spencer, unpublished data) of the limits to the reductions of some oxides widely used in catalysis. The limits arise from the thermodynamics of the reduction reaction and the kinetics of the removal of the H_2O or CO_2 by-product. Anderson (19) examined some equilibrium limits to several metal catalysts but under rather arbitrary conditions. No other analysis of this aspect of choice of active metals and inactive supports appears to have been made.

In this review only reduction with hydrogen is considered, but the equivalent reductions with carbon monoxide lead to very similar conclusions. This is in accord with industrial experience that shows in practice catalyst reductions with hydrogen, carbon monoxide, or hydrogen/carbon monoxide mixtures do not usually give widely different results. This follows from the near equivalence of the two reducing agents via the water-gas shift Reaction 3. Values of the equilibrium constant (20) for this reaction at typical catalyst reduction temperature are given in Table 6. Sometimes reductions with carbon monoxide or hydrogen/carbon monoxide mixtures can give faster or more extensive reduction of metal catalysts than hydrogen alone (21, 22). This can be related to the values of the equilibrium constant for Reaction 3 being above unity in the temperature range of interest.

In the model analyzed, the catalyst precursor oxide in a catalytic reactor is in a flow system in which hydrogen is passed over the precursor oxide. The exit gas from the reactor contains bulk steam and unconverted hydrogen. The maximum possible steam content of this exit gas (i.e., the maximum possible P_{H_2O}/P_{H_2} ratio)

TABLE 6 Values of the equilibrium constant, *K*, for the water-gas shift reaction (adapted from Reference 20)

Temperature, °C	200	250	300	350	400	450
K	211	84.0	38.8	20.3	11.7	7.34

is set by the equilibrium of the metal oxide reduction. If the rate-controlling step in the catalyst reduction is, for example, the rate of the solid-state reaction or mass transfer in the catalyst bed, then the steam content of the exit gas will be below the level determined by the thermodynamics of the catalyst reduction. At a given reducing gas flow rate, the lower the exit steam content, the lower the limiting rate of reduction will be and hence the longer the time of reduction. In principle, every metal oxide can be reduced by hydrogen when the steam reduction by-product is continuously removed in a flow system, so strictly speaking no metal oxide can be described as irreducible under typical catalyst reduction procedures. The key factor is the time taken for the reduction: With many oxides the limiting, i.e., minimum, times are so long that the oxides are in practice irreducible.

The calculations were based on the following assumptions:

1) The catalyst metal/metal oxide is in equilibrium with the H_2/H_2O gas mixture.
2) The reduction is not limited by the rates of any reaction steps in the solid phase reactions.
3) The exit gas from the reactor contains the same H_2/H_2O ratio as the gas at the catalyst surface (i.e., there is no mass transfer control either in the catalyst pellets or the catalytic reactor).

The equilibrium constants were calculated from standard data for a number of metals of catalytic interest, all based on the general reduction Reaction 15.

$$^{1}/_{x}MO_{x} + H_2 = {}^{1}/_{x}M + H_2O. \qquad 15.$$

This form of reaction equation facilitates the comparison of different metals in the same way as is achieved with Ellingham diagrams (21). Values are given in Table 7.

Then by relating the oxygen content of the catalyst precursor to the hydrogen and steam contents of reduction gas after passage through the reactor, the theoretical minimum reduction time can be calculated for 1 liter of catalyst, Equation 16.

$$t_m = \frac{22.4\ aT}{213\ K\phi sp_o}\ h, \qquad 16.$$

where ϕ is the mol fraction of H_2 in exit gas; s is the space velocity through bed, h^{-1}; and p_o is the total gas pressure, bar.

Substitution of typical catalyst reduction parameters into Equation 16 gives Equation 17 for minimum reduction time:

$$t_m = \frac{1.64 \times 10^{-4}\ T}{K}\ h. \qquad 17.$$

Variations in the reduction parameters, corresponding to different reductions, do not make substantial differences to the general conclusions reached. Values of t_m for a series of metals of catalytic interest are given in Table 7.

TABLE 7 Minimum times of reduction of metal oxides with hydrogen

Metal Oxide	T/K	K_1	Time[a]
Al_2O_3	800	1.88×10^{-18}	8.0×10^{12} y
ZnO	500	3.2×10^{-9}	2900 y
MnO	800	9.49×10^{-9}	1580 y
Cr_2O_3	500	7.5×10^{-13}	12.5×10^{6} y
	800	1.23×10^{-7}	121 y
	1100	2.52×10^{-5}	0.82 y
WO_2	500	0.90×10^{-3}	91 h
	800	0.069	1.90 h
	1100	0.41	0.44 h
MoO_2	500	1.01×10^{-3}	81 h
	800	0.075	1.75 h
	1100	0.43	0.42 h
FeO	500	10.0×10^{-3}	82 h
	800	0.094	84 min
	1100	0.232	47 min
NiO	800	257	1.84 s
Cu_2O	800	1.24×10^{6}	3.8×10^{-4} s
PdO	800	6.7×10^{6}	7.0×10^{-5} s

[a]Units of time given: y, year; h, hour; min, minute; s, second.

The values of t_m are the minimum possible values for the reduction of the oxides under the given conditions. Practical reduction times are likely to be longer, in some cases much longer. The wide range of times is in good general agreement with experimental observation. The first four metal oxides (Al_2O_3, ZnO, MnO, and Cr_2O_3) are all essentially irreducible in practice, agreeing with calculated minimum times of reduction in the range 1 to 10^{12} years. It should be noted that the calculations exclude any partial reduction to defective oxides or reduction to alloys [e.g., α-brass formation in the reduction of copper/zinc oxide catalysts (23–25)]. The use of these metal oxides in catalysis is as oxide catalysts per se or as supports to active metallic phases.

In marked contrast are the three oxides at the bottom of Table 7 (NiO, Cu_2O, and PdO). The calculated minimum reduction times are in the range of about 10^{-4} to 2 s. In practice it is not possible to achieve the reduction of even small catalyst beds in such short times: Apart from kinetic considerations, the limits set by mass and heat transfer would prevent such rapid reduction. Nevertheless, all these oxides are in practice readily reduced by conventional means. The calculations show that thermodynamic/kinetic limits of the type considered play no part in the reduction of these and other similar metal oxide catalyst precursors.

The remaining three metal oxides in Table 7 (WO_2, MoO_2, and FeO) all have limiting reduction times, which are accessible in both laboratory and industrial practice, i.e., times of a few hours, so all are potentially capable of reduction to the bulk metal by conventional means. This is commonplace for iron catalysts (see discussion above). In contrast, both molybdenum and tungsten oxides are not in practice capable of conventional reduction, but it is clear that this failure of reduction is not due to thermodynamic/kinetic factors. It seems unlikely that mass or heat transfer limits are responsible, as these limits would be of little importance in small-scale laboratory reductions. It was shown (see below) that the removal of adsorbed oxygen from metallic molybdenum and tungsten surfaces cannot be achieved in practical times by conventional reductions with hydrogen. However, many metal catalysts operate, even under reducing conditions, with some surface oxygen coverage (26, 27). These calculations suggest that if MoO_2 or WO_2 can be reduced to the metallic state, e.g., by reducing agents more powerful than hydrogen to overcome chemical kinetic limitations, their use as metal catalysts under normal reducing conditions may be possible.

SURFACE REDUCTION MODEL The formulation of a metal catalyst by the activation of an oxide precursor by reduction can be divided somewhat arbitrarily into two stages. The first stage, considered in the previous section, is the reduction of the bulk oxide precursor to the bulk catalytic metal. The surface of this metallic phase can be expected, as the bulk reduction is completed, to be covered with adsorbed oxygen (or with a few atomic layers of a metal oxide). In the second stage of reduction, all or a substantial part of this adsorbed oxygen is removed reductively. An analysis of a model for surface reduction comparable with the bulk model (see above) has been carried out (M.S. Spencer, unpublished data). A major difference is that a series of well-defined compounds cannot be used. As the adsorption energies of $O_{(a)}$ on a given metal surface vary with $O_{(a)}$ coverage, crystal face, surface defect, etc., no single limiting reduction time can be calculated for a given metal covered with adsorbed oxygen. Consequently, the minimum reduction times were calculated for a series of examples defined by $O_{(a)}$ adsorption energies.

There is a further problem in this model. The adsorption energy of adsorbed oxygen on a surface is defined (28) as the enthalpy change of Reaction 18:

$$2O_{(a)} \longrightarrow O_{2(g)}. \qquad 18.$$

It follows that the bond energy of the metal oxygen bond is given by Equation 19:

$$\Delta H_{bond}(M-O) = \frac{(\Delta H_{ads} + D_{O_2})}{2}, \qquad 19.$$

where $\Delta H_{bond}(M–O)$ is the bond energy of the metal-oxygen bond; ΔH_{ads} is the adsorption enthalpy of $O_{(a)}$; and D_{O_2} is the dissociation enthalpy of O_2 molecule.

Many values of ΔH_{ads} have been tabulated by Somorjai (29) for numerous metal surfaces, so representative values of ΔH_{bond} (M–O) can be calculated. However, there remains a further difficulty. The analysis above is based on the equilibration of Reaction 18. Thus it assumes that in the adsorption of oxygen, the reverse of Reaction 18, two identical adsorbed oxygen atoms are formed and no other energetic terms need to be included in Equation 19. Recent surface science, especially with the use of scanning tunnelling microscopy, has shown that in some, possibly many, cases this assumption is not valid. For example (M.W. Roberts, personal communication), the dissociation adsorption of molecular oxygen on a metal surface can give rise to two different $O_{(a)}$ species, one of which is bonded at the adsorption site, and the other, with some of the energy of reaction converted to kinetic energy, moves across the metal surface with this energy to give a different $O_{(a)}$ species on a different site. The M–O bond energy calculated from Equation 19 would not be valid for either $O_{(a)}$ species in this example.

In the analysis of the model, therefore, a series of M–O bond energies, typical of various catalytic metals, were chosen. Minimum reduction times were calculated as for the bulk reductions, but here the assignment of times to adsorbed oxygen on given metals remains imprecise. The values of enthalpies of adsorption used 450, 600, and 750 kJ mol^{-1}, covering the range of values for the transition metals found capable of bulk reduction in practical times (Table 7). The reduction reaction for the removal of adsorbed oxygen is Reaction 20.

$$O_{(a)} + H_{2(g)} = H_2O_{(g)} \qquad 20.$$

Values of the equilibrium constant, K_{20}, were calculated from standard data, and then, as for bulk reduction (see above), minimum times were evaluated for the flow of hydrogen through a catalyst bed, using typical values of the experimental parameters. The metal surface area was taken as 50 m^{2}g^{-1}, a representative value for an active catalyst. The reduction time is given by Equation 21.

$$t = -1.368 \times 10^{-5} \frac{T}{K_{20}} \ln(\theta), \qquad 21.$$

where t is the reduction time in hours, and θ is the adsorbed oxygen coverage after time t.

Values of the minimum times for three reduction temperatures are given in Table 8. Perhaps the most striking feature, as with Table 7, is the very wide range of minimum times, over 16 orders of magnitude. The adsorption energy of $O_{(a)}$ is the dominant factor. The extra time needed to decrease $O_{(a)}$ coverage from 0.1 to 10^{-6} is, as expected, greater than the time to fall from full coverage to $\theta = 0.1$, but not as large as might be expected from the experimental difficulties of removing small coverages of adsorbed oxygen from metal surfaces. The explanation is likely to be that the heterogeneity of the adsorbed species is reflected in a range of adsorption energies, with, in accord with Table 8, small coverages of strongly adsorbed oxygen being difficult to remove.

As mentioned above, no exact coordination of the calculated times in Table 8 can be made with specific metals. However, comparison of the values with the data given by Somorjai (28) shows that, broadly, the metals that can be readily reduced from the oxide (e.g., Ni, Cu, Pd) are also those from which adsorbed oxygen is also

TABLE 8 Minimum times for the reductive removal of adsorbed oxygen from catalytic metal surfaces

			Minimum time at coverage[a]	
T/K	**ΔH_{ads}/kJ mol^{-1}**	K_{20}	$\theta = 0.1$	$\theta = 10^{-6}$
500	450	1.39×10^5	4.07×10^{-4} s	0.00245 s
	600	2.03×10^{-3}	7.76 h	46.55 h
	750	2.96×10^{-11}	6.07×10^4 y	3.64×10^5 y
800	450	5.72×10^4	1.60×10^{-3} s	0.010 s
	600	0.725	125 s	750 s
	750	9.19×10^{-6}	0.31 y	1.88 y
1100	450	4.43×10^4	2.8×10^{-3} s	1.68×10^{-2} s
	600	12.15	10.22 s	61.4 s
	750	3.36×10^{-3}	10.3 h	61.7 h

[a]Units of time given: y, year; h, hour; s, second.

readily removed. The exceptions are metallic tungsten and molybdenum, which are theoretically capable of being reduced from the oxides WO_2 and MoO_2. As the adsorption energies on these metals are of the order of 750 kJ mol^{-1}, the reduction removal of adsorbed oxygen would require temperatures of at least 1100 K. This in agreement with experimental observations of the difficulty of reducing Mo and W oxides to the metals and of removing adsorbed oxygen from Wo and W metal surfaces.

EXAMPLES OF CATALYST SYSTEMS

Copper/Zinc Oxide Catalysts for Methanol Synthesis

The problems of the deactivation of the copper/zinc oxide catalysts used in methanol synthesis have been analyzed in detail (14, 31–34). In normal operation of methanol synthesis plants, the only significant mode of deactivation is sintering of the copper crystallites and consequent loss of copper metal surface area. Poisoning by sulfur, chlorine, or other compounds, potentially powerful poisons, is negligible in normal plant operation because these are removed by the earlier stages of generation of synthesis gas. Furthermore, poisoning as a result of plant mal-operation has been described before (14, 31, 32) and is not treated here. Catalyst activity for methanol synthesis under industrial conditions is proportional to the Cu surface area (35). An earlier proposal (36) that methanol synthesis occurred on defective zinc oxide has been discredited (22, 35). Thus loss of Cu surface area results in direct equivalent loss of activity. The design of the methanol synthesis catalysis has therefore as its aim the initial generation and the subsequent maintenance of the maximum copper metal surface area.

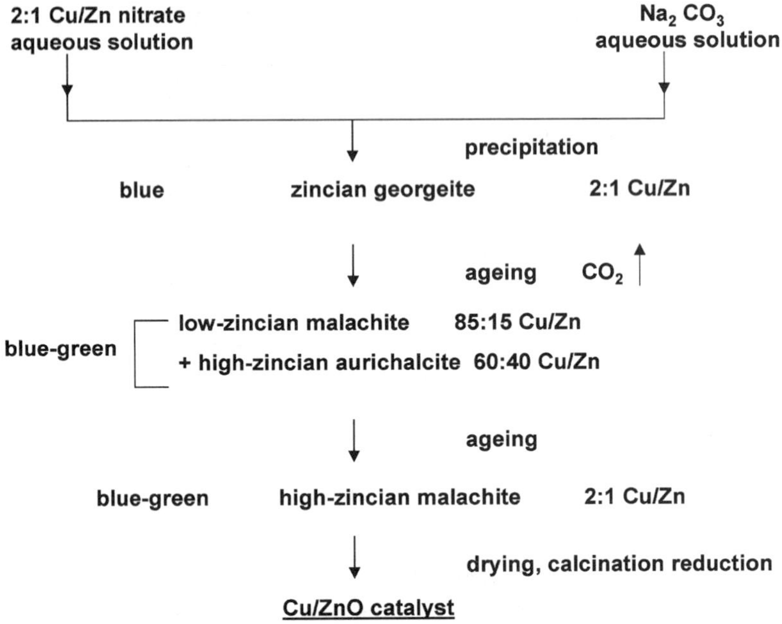

Figure 4 Sequence of precursor phases formed in the coprecipitation of hydroxy carbonates for copper/zinc oxide catalysts for methanol synthesis (adapted from Reference 37).

This aim is achieved by the continuous coprecipitation of mixed carbonates (copper + stabilizer), so as to give a uniform distribution of copper crystallites and stabilizer particles in the final catalyst. Hydroxy carbonates of zinc and copper are especially effective here (37–39). The sequence observed in the coprecipitation procedure is shown in Figure 4. Zinc oxide is peculiarly effective in giving an initial high Cu metal surface area. However, zinc oxide alone is not a sufficiently refractory support to give acceptable catalyst lives, so for the first low-pressure synthesis plants a $Cu/ZnO/Al_2O_3$ catalyst was developed, in which the alumina component hindered sintering of the zinc oxide and copper components. Continued development resulted in catalysts suitable for operation at 100 bar, around the optimum operating pressure for high-capacity plants producing more than 1000 tonnes day^{-1} of methanol. Initial tests showed that the rate of catalyst die-off was higher at the higher pressure. ZnO/Al_2O_3 is not sufficiently refractory as a support at 100 bar, but a more refractory support is provided by the introduction of some of the zinc component as zinc spinel ($ZnAl_2O_4$), which is produced in a finely divided state. In more recent work (40, 41), claims are made for increased stability in $Cu/ZuO/Al_2O_3$ catalysts containing further refractory oxides such as MgO. It is understood that these are now in commercial use. The addition of alumina and zinc spinel to ZnO gives a more refractory support for the copper metal crystallites than zinc oxide alone. The addition of even more refractory oxides, such as MgO, probably contributes more to stabilizing the mixed oxide support than to stabilizing the copper crystallites directly. Clearly the optimization of these multi-component catalysts for both maximum initial activity and maximum stability is difficult.

Raney copper catalysts, prepared by the alkaline extraction of aluminum from copper/aluminum alloys, are effective catalysts for methanol synthesis. Wainwright and his colleagues (42, 43 and references therein) have carried out an extensive and detailed study of these catalysts, especially those incorporating zinc oxide and other oxides, which are also used in precipitated catalysts for methanol synthesis. The optimum catalysts (43) are significantly more active than commercial precipitated catalysts, which is not altogether surprising in view of the intrinsic advantages of extraction catalysts (see above). In particular, the advantages of slower rates of deactivation from the larger copper metal crystallites should lead to longer catalyst lives in industrial use. However, to the authors' knowledge, Raney copper catalysts have not yet achieved commercial use for methanol synthesis. In addition to natural inertia inhibiting a complete change of catalyst manufacturing technique, there appear to be three major problems:

1) As described above, the development of the optimum precipitated catalyst formulation has taken several decades since the success of the first $Cu/ZnO/Al_2O_3$ catalysts for methanol synthesis (44). A similar program of optimization would probably be needed for Raney copper catalysts.
2) Realistic testing under industrial conditions (e.g., with a side stream reactor) (45) would be needed to obtain realistic estimates of catalyst life.
3) Catalyst activation would need to be considered in detail. Clearly it would not be possible to extract the Al component of the alloy precursor in the plant reactor (a contrast with the procedure for ammonia synthesis catalysts) (see above), as it requires the use of alkali solutions. Some form of preactivation of the catalyst by the catalyst manufacturer would be needed. For example, the appropriate and controlled extraction by NaOH solution could be followed by water washes (to remove soluble salts) and then by a limited and controlled oxidation of the copper metal surface (stopping well short of bulk oxidation). This passivated catalyst could then be transported to the plant, installed in the synthesis reactor and rereduced with a H_2/CO mixture. While this is all plausible, it is not obvious that, unlike the equivalent case of ammonia synthesis catalysts, there is necessarily any cost advantage over the precipitated catalyst.

Iron Catalysts for Ammonia Synthesis

The catalysts used industrially for ammonia synthesis are probably the most long lived (46, 47). There is rarely any observable decline in activity during catalyst life, which is frequently 10 years or more. There are two reasons for this. First, the process stages in a modern ammonia plant for preparing the reactant gases from natural gas, steam, and air are highly effective in the removal of catalyst poisons. Second, the design and manufacture of the catalyst, metallic iron with promoters, gives high catalyst stability, especially a low rate of sintering of the iron crystallites.

The catalyst is potentially sensitive to failure by poisoning. Not only common catalyst poisons such as sulfur, chlorine, and arsenic, but also compounds containing oxygen (water, CO and CO_2) can prevent nitrogen adsorption by forming strongly bound layers of adsorbed O, S, Cl, As, etc. In early plants (48), the reactant gases were made from coal, which was the source of various catalyst poisons, and the intermediate process stages were much less effective at removing, e.g., H_2S and CO, than the equivalent processes in modern plants. Thus the ammonia synthesis catalysts in the old plants were subject to extensive poisoning. As this was unavoidable with the technology of the time (there is not any possible catalyst modification that could counter the poisoning), plants had to be designed to allow for catalyst poisoning. This was done by increasing the amount of catalyst in the reactor, so that in effect the first bed of catalyst functioned much more for poison removal than for ammonia synthesis [the "guard bed" technique (49)].

Catalyst poisoning is no longer a problem in modern ammonia plants based on natural gas feedstocks. The levels of sulfur and other catalyst poisons are much lower in natural gas than in coal. Furthermore, the development of the process steps of low-temperature water-gas shift, methanation, and zeolite-based gas adsorption (50) and improved engineering now ensure that essentially no poisons reach the synthesis catalyst bed. Only plant mishap or mal-operation can lead to loss of catalyst activity by poisoning.

There is, of course, no requirement for selectivity in ammonia synthesis. Loss of activity is almost entirely due to loss of iron metal surface area through sintering of the iron metal crystallites. The catalyst is made by an extraction technique (see above), which gives relatively large, very stable metal crystallites, resistant to sintering.

The dominance of the conventional catalyst, refined but essentially unchanged since its invention about a century ago, can be explained by a comparison with a typical supported Fe catalyst prepared by a precipitation route. There have been many attempts (51, 52; S.P.S. Andrew, personal communication) over the years to prepare more active catalysts in this way, but none has been commercially successful. Although a catalyst with smaller Fe metal crystallites, and hence higher specific Fe metal area, appears attractive, this gain turns out to be illusory. The key parameter for plant activity is the Fe metal area available in the plant reactor, not specific Fe metal area.

For the catalyst comparison, the Fe metal surface is calculated (M.S. Spencer, unpublished data) for the same volume of each catalyst. For the conventional catalyst, typical properties (13) are Fe content of 97.5%, Fe crystallite size of 30 nm (giving a Fe metal surface area of 10–20 $m^2\ g^{-1}$) and porosity of 45%. If the promoters are ignored and the Fe crystallites are taken as cubes, then the number of Fe crystallites in a 1 cm^3 catalyst can be calculated to be 2×10^{16}. The surface area of each Fe crystallite is $5.4 \times 10^{-15} m^2$, so the total Fe surface area in 1 cm^3 is 110 m^2. The calculated specific Fe surface area is 25.4 m^2g^{-1}, in sufficient agreement with experimental values 10–20 m^2g^{-1} (especially when the effect of promoters is included) for validity of this simple model.

For a good, hypothetical precipitated Fe catalyst, the following values were taken: 50% wt Fe metal, 50% wt support oxide (e.g., alumina) of specific gravity 2.0; porosity 50%, Fe metal crystallite size 5 nm. Then calculation as before gives the number of Fe metal crystallites in 1 cm^3 catalyst particle as 8.1×10^{17}. In this catalyst, the surface area of each Fe metal crystallite is 1.5×10^{-16} m^2, so the total Fe metal surface area in 1 cm^3 is 122 m^2. This value is, within the limits of calculation, the same as for the conventional catalyst. The specific surface area would be 76.5 m^2 (g catalyst)$^{-1}$, which would be regarded as acceptable for this type of supported metal catalyst. These results are summarized in Table 9.

Thus although the precipitated catalyst has a much higher specific Fe metal surface area, this is not translated into a practical advantage in a plant. The reason for this is that the conventional catalyst, being almost all Fe metal, has a much higher density than the precipitated catalyst (50% oxide support) and thus carries more Fe surface area into the reactor.

There are two other factors that weigh heavily in favor of the conventional catalyst. First, no matter how effective the oxide support is in hindering Fe metal sintering in a precipitated catalyst, loss of Fe metal surface area by sintering is certain to be greater than in the conventional catalyst. The driving force for sintering is the reduction of surface energy by a reduction in surface area, and consequently the driving force is greater for small crystallites than for large crystallites (6,

TABLE 9 Calculated Fe metal surface areas for models of iron catalysts made by extraction and precipitation routes

Property	Catalyst by extraction route	Catalyst by precipitation route
Fe metal content/wt%	97.5	50.0
Oxide support + promoter content/wt%	2.5	50.0
Fe metal crystallite size/nm	30	5
Porosity/%	45	50
Specific Fe metal surface area/m^2 (g catalyst)$^{-1}$	25	76
Fe metal surface area in 1 cm^3 catalyst/m^2	110	122

54). The size of the crystallite in conventional catalysts, about 30 nm, is much greater than that in the model precipitated catalyst, 5 nm, so sintering would be much more severe in the precipitated catalyst. Secondly, the cost of manufacturing the conventional catalyst is much lower than that of a typical precipitated catalyst. Consequently, any precipitated catalyst would have to possess considerable performance advantages over the conventional catalyst for it to be accepted commercially.

In recent years a new ammonia synthesis catalyst, a potassium-promoted ruthenium on a carbon support (55), has been used commercially. It has a higher activity than the conventional iron catalysts, but it is more expensive. Therefore, ruthenium catalysts are used only in the inlet beds of reactors, where there is a premium on high activity. Nothing appears to have been published on the deactivation of these ruthenium catalysts. Poisoning would not be a problem, as discussed above, and loss of Ru surface area by sintering is presumably controlled by the use of relatively low Ru metal contents. It should also be noted that melting point of ruthenium, 2523 K, is much higher than that of iron, 1808 K, so loss of surface area by sintering of ruthenium would be much slower than that of iron.

Catalysts essentially the same as the iron catalysts for ammonia synthesis are also used industrially (sometimes without the potassium promoter) for other hydrogenations (56), such as adiponitrile hydrogenation to hexamethylenediamine, Reaction 22.

$$NC(CH_2)_4CN + 4H_2 = H_2N(CH_2)_6NH_2. \qquad 22.$$

Little has been published about catalyst deactivation in these processes. As they are operated at lower temperatures than ammonia synthesis, loss of catalyst structure should not be a problem, nor, as long as purified feedstocks are used, should catalyst poisoning arise. However, a possible cause of deactivation could be the formation of high molecular weight by-products that would block catalyst pores and surface.

Copper/Zinc Oxide Catalysts for Water-Gas Shift Reaction

The water-gas shift process (Reaction 3) is used for the conversion of carbon monoxide to hydrogen in ammonia plants and in other plants where reducing gas is needed. To get near-complete conversion, two process stages are employed. In the first, the high-temperature stage (HTWGS), the major part of the conversion of CO occurs over a magnetite/chromia catalyst. The catalyst for the second, low-temperature stage (LTWGS) is copper-based, usually copper/zinc oxide/alumina, broadly similar to the methanol synthesis catalyst discussed above. Deactivation of these catalysts (14, 57) occurs by loss of active copper metal surface area by sintering of copper metal crystallites and by poisoning of the copper metal surface.

The design of the catalysts to minimize sintering of the copper metal crystallites follows that for methanol synthesis catalysts, with the same catalyst precursors.

However, there are some differences, arising from difference between the two processes. The LTWGS process is operated typically at 200–230°C, less than the 250–300°C used for methanol synthesis. Thus the intrinsic driving force for copper metal sintering is less in LTWGS, so longer catalyst lives could be expected on sintering alone. However, the situation is more complex. Methanol synthesis is usually operated at 100 bar pressure, but the steam content of the reacting gases is low, only a few bar. The LTWGS, when part of an ammonia synthesis train, is at lower total pressure, e.g., 30 bar, but the partial pressure of steam can be 10–20 bar, i.e., higher than in methanol synthesis. Steam is a potent agent in promoting the sintering of oxides (probably via surface hydroxylation), so it is more difficult to maintain the integrity of the alumina and zinc oxide phases, and hence limit copper sintering, in the LTWGS catalyst. Lower copper contents are generally used and overall thermal stability depends strongly on manufacturing procedures, not only on composition (14).

In practice, with well-made LTWGS catalysts, excellent catalyst lives can be obtained (Figure 1), and deactivation from poisoning normally sets the limit on useful catalyst life. Sulfur and chlorine are two principal poisons for LTWGS catalysts in ammonia plants, although sometimes other poisons can be significant, some of which, e.g., silica, can exacerbate the effects of sulfur and chlorine (14). Both sulfur and chlorine poison the LTWGS catalyst by being strongly adsorbed on the copper metal surface and preventing the desired catalytic reaction. Poisons can come from various sources, e.g., sulfur from natural gas feedstock, sulfates in boiler feed water, iron sulfate in the HTWGS catalyst, chloride from sodium chloride in boiler feed water, and salt aerosols in process air. In good plant practice all these sources are kept to a minimum.

Zinc oxide is incorporated in LTWGS catalysts primarily as a structural component to minimize copper sintering, but it is also highly effective in limiting sulfur poisoning by use of Reaction 23.

$$ZnO_{(s)} + H_2S_{(g)} = ZnS_{(s)} + H_2O_{(g)}. \qquad 23.$$

The thermodynamics of Reaction 23 have been analyzed (14) for relevant process conditions. For typical values of total pressure of 20 bar and steam pressure of 10 bar, the equilibrium concentration of H_2S over ZnO is only 0.007 ppm at 227°C. Thus, at equilibrium, the ZnO in LTWGS catalysts gives excellent protection for H_2S to the copper crystallites. The small crystallite size of ZnO, coming from the method of manufacture, ensures fast reactive absorption.

The situation is different for chloride poisoning. This can operate by several parallel mechanisms (14):

1) Reaction to give adsorbed chlorine atoms that can block or modify catalytic sites.
2) The low melting points and high surface mobilities of Cu (I) halides (Table 10) mean that even extremely small amounts of copper halide are

TABLE 10 Melting points (°C) of some copper (I) compounds (adapted from Reference 14)

Metal	Chloride	Bromide	Sulfide
1083	430	492	1100

sufficient to provide mobile species that accelerate the sintering of copper crystallites.

3) Poisoning of copper catalysts by reduced sulfur components (e.g., H_2S) is exacerbated by traces of mobile Cu (I) chloride.
4) Furthermore, the zinc oxide present in LTWGS catalysts reacts to form zinc halides, which also have low melting points ($ZnCl_2$, m.p. 283°C). This causes further poisoning and sintering problems.

In practice, the only halide encountered in hydrogen and ammonia plants is chloride in the form of HCl. The formation of zinc chloride is given by Reaction 24.

$$ZnO_{(s)} + 2HCl_{(g)} = ZnCl_{2(s)} + H_2O_{(g)}. \qquad 24.$$

In contrast to the case of sulfide poisoning, zinc oxide gives no protection against chloride poisoning. The limits on HCl content to avoid catalyst poisoning are more severe than for H_2S poisoning, of the order of 1 ppb. As before with H_2S, the thermodynamics of Reaction 24 can be used (14) to show that the equilibrium concentration of HCl in the gas phase is about 80 ppm if $ZnCl_2$ is formed (in practice, depending on the amount of steam present, hydoxychloride species may also be formed). The large difference in concentration demonstrates that zinc oxide can give no protection to the LTWGS catalyst against HCl poisoning. It has been shown by similar calculations that other oxides that could be present for structural reasons, e.g., Al_2O_3, MgO, are equally ineffective in preventing chloride poisoning (32).

Chloride poisoning is a particularly difficult problem for water-gas shift plant operators because it is usually present in process gas in amounts below the normal levels of detection. However, chloride is usually retained at the top (inlet region) of the catalyst bed, typically in the form of zinc hydroxy chloride, with the result that the overall catalyst change is partially self-guarding. This technique is frequently extended to the provision of a separate guard reactor (57). In this case, replacement of the absorbent (usually the LTWGS catalyst itself) in the guard bed can be carried out while main reactor is kept on line, achieving catalyst lives of several years.

It can be seen that deactivation of LTWGS catalysts by sintering of the copper crystallites or by poisoning with H_2S can be prevented by appropriate catalyst design and careful manufacture. In contrast, deactivation by chloride poisoning can be controlled only by minimizing the trace sources of chloride and by appropriate plant design.

Platinum Group Metal Catalysts for Car Emissions Control

The internal combustion engine is not completely efficient and tail-pipe emissions from cars include hydrocarbons with partially oxidized derivatives (HCs), CO, as well as NO formed at high temperature during the explosive combustion process. In the presence of bright sunlight, HCs and NO undergo a series of photochemical free radical reactions that produce O_3 and other undesirable species such as peroxyacetylnitrate (PAN) (59). By the 1970s, air quality became a serious environmental problem in some cities, so it was necessary to significantly reduce car emissions. This was done by engine improvements and by fitting catalysts in the exhaust systems. A few years after the initial implementation of autocatalysts, urban air quality improved, and ever since, at regular intervals new increasingly demanding emissions standards have been introduced. This ensured continual air quality improvement and forced development of more advanced catalyst technologies, and the autocatalyst area grew to become one of the most important in the catalyst industry (60). It has been estimated that some 1000 million autocatalysts have been manufactured, and the production rate is now some 120 million/year.

EARLY OXIDATION CATALYSTS At first HCs and CO levels in car-exhaust gas were lowered by injecting air upstream of an oxidation catalyst, thus converting them to CO_2 and H_2O. In the early development phase, base-metal formulations, such as those containing copper and/or nickel, were found to be very susceptible to poisoning, particularly by lead, halide, and sulfur compounds, and they were unable to withstand exposure to high temperatures. Temperatures of around 1000°C are encountered when HCs and CO burn over a hot oxidation catalyst (61, 62, 62a–c). Oxidation catalysts based on platinum formulations performed much better and had acceptable lives. These catalysts were made by impregnating a fairly high surface area alumina (e.g., 120 m^2/g) in the form of pellets, or better (and now exclusively) coated on the interior channel walls of a monolithic honeycomb. This was followed by calcination to decompose the soluble platinum metal salts and thereby form highly dispersed platinum metal. The monoliths were, and continue to be, extruded cordierite that has a very low coefficient of thermal expansion (63, 63a). In some situations, monoliths made by rolling alternative flat and crimped metal foils are also used (64). Some early commercial catalyst coatings were made using high surface area platinum black, which when properly incorporated into the coating formulations performed adequately, but sintering of the active platinum phase was a problem. It was found the useful life of these early oxidation catalysts could be improved by including a second platinum group metal (as a result of alloy formation, and perhaps a mechanism involving retention and release of poison species). Platinum/palladium and later platinum/rhodium formulations were in common use.

NITRIC OXIDE REDUCTION Although nitric oxide is an endothermic compound and a free radical, in practice it cannot be readily dissociated into nitrogen and oxygen; it has to be reduced to water and nitrogen (65). When it became necessary to lower tail-pipe nitric NO_x levels (NO_x is used to signify mixtures of NO and NO_2), as well oxidize HCs and CO, a two-stage process was first used in which NO_x was reduced to nitrogen over a platinum/rhodium catalyst. Then air was injected and excess reductant oxidized over a platinum/palladium catalyst. Rhodium is active in NO_x reduction when it is in the metallic state. Prolonged exposure to oxygen at high temperature causes it to oxidize and the oxide can then react, with for example alumina, to form inactive stable species. It can take a significant time to reduce these back to the metallic state, so it is important not to expose a rhodium-containing catalyst to prolonged conditions that result in this form of catalyst deactivation. Formulations containing small amounts of platinum associated with the rhodium reactivate under reducing conditions more readily than those with no platinum because of spill-over effects from the platinum that accelerate the rhodium rereduction. There were also potential selectivity problems with the platinum/rhodium upstream catalysts because of over-reduction of nitric oxide to ammonia and of sulfur dioxide to hydrogen sulfide (Reactions 25 and 26). Both of these reactions are undesirable because of the odor of the products, and relatively small formulation changes, such as the platinum/rhodium ratio, the presence of base-metal oxides, and manufacturing procedures helped minimize ammonia and hydrogen sulfide formation. In spite of chemical complexities the two-stage emissions control arrangement worked well (66).

$$2NO + 5H_2 = 2NH_3 + 2H_2O. \tag{25}$$

$$SO_2 + 3H_2 = H_2S + 2H_2O. \tag{26}$$

THREE-WAY CATALYSTS By the early 1980s, it had been shown that oxidation of HCs and CO, and also the reduction of NO_x to nitrogen, could be achieved over a single, so-called three-way catalyst (TWC), if the air/fuel ratio was maintained around the stoichiometric point. It was possible to do this via a feedback control loop made up of an oxygen sensor in the exhaust gas, electronic fuel injection, and a control microprocessor. The composition of the exhaust gas oscillated about each side of the stoichiometric point as the feedback loop hunted around the stoichiometric point. The first TWCs were based on platinum/rhodium formulations, and it was found advantageous to have an active redox component to buffer the effects of the changing gas phase oxygen content. Perovskites (especially cobalt perovskite) were used in this role with limited success, but the introduction of ceria brought about significant improvements. This involves the reversible Ce(III)/Ce(IV) couple that is active at the operating temperatures. Maintaining the surface composition at the stoichiometric point in this way particularly enhanced the NO_x control efficiency. Since the introduction of these oxygen storage components into TWCs, there has been a move to using increasingly more stable components and to minimizing negative interactions between them. For example, it is possible to optimize the environment around platinum, and if this is different from that which is optimal for rhodium, it is advantageous to physically divide the catalyst into two (or more) layers containing well-separated, different active metal dispersions with their specific promoter packages. Usually platinum and palladium function best in oxidation roles, and they are often in the bottom part of a two-layer TWC. The rhodium present in the top layer is then exposed to all of the reductant species that reduce NO_x on rhodium active centers before they diffuse to the strongly oxidizing environment in the lower layer. Here physical separation enhances overall catalyst performance and life by preventing alloy formation, separating otherwise incompatible promoters, and encouraging desired reactivity by matching catalytic functionality with appropriate reactant diffusion.

By the correct use of promoters, particularly alkaline earths and lanthanide oxides, it was possible to modify the catalytic properties of palladium so it can function as a TWC. This entails interplay between catalysis by the metal and its oxide, the presence of which can be controlled by close contact with cations that stabilize surface oxygen. Again, separating the catalyst coating into two layers can minimize cross-contamination and help obtain long-lasting high activity. When first introduced, TWCs operated with fuel containing high levels of sulfur, and the then rudimentary engine control often resulted in overly rich operation. This in turn favored formation of undesirable hydrogen sulfide, but this was minimized by incorporating base-metal oxides so that the catalyst retains reduced sulfur species, which are later converted to sulfates under oxidizing conditions. Ultimately the sulfates decompose to the less noxious sulfur dioxide. The overall performance of modern TWCs is astonishing in relation to their counterparts of some three decades ago—today it is possible to have extended operation at extremely high temperatures while maintaining high activity with ultra-high conversions of all three pollutants. Very early catalyst light-off is now achieved by having a TWC actually mounted on the exhaust manifold so heat is not lost, but to be mounted in this location, exceptional thermal durability is required (68).

CATALYSTS FOR DIESEL CARS During recent years in Europe there has been a major increase in the number of diesel-powered cars, brought about by their better fuel economy and improved driving characteristics following significant engineering developments. At the time of writing, some 50% of all new European cars have diesel engines, and these have very lean (oxygen-rich) exhaust gas. Under these conditions, platinum catalysts should operate well, but as a consequence of their excellent fuel economy, exhaust gas temperatures are low, and during normal town driving, the temperature may well be below 130°C, and at higher speeds, a typical "high temperature" might be no more than 300°C. Under these conditions, and in the presence of the poison SO_2, providing high activity is a challenge. This is made even more difficult by inhibition of CO oxidation by the presence relatively heavy HCs that adsorb strongly at low temperature on the platinum active sites and inhibit oxidation activity. This problem was largely overcome (69, 70) by incorporating selected zeolites into the oxidation catalyst formulations: these adsorb HCs at low temperature, thereby freeing the platinum sites for CO oxidation. When higher temperatures are reached, the HCs are released and undergo oxidation on the platinum sites, and the overall oxidation performance is markedly improved. Long-term durability became a problem as emissions limits were reduced further and sulfur fuel levels remained fairly high. Over time the poisoning effects of SO_2 became increasingly more significant as sulfate accumulated within the catalyst—even at low temperatures a small fraction of adsorbed SO_2 is oxidized to SO_3 that combines with support components, converting them to sulfates that seriously degrade catalyst performance. Figure 5, see color insert, schematically illustrates some of the reactions involved in the deactivation by SO_2. These effects have been largely overcome by incorporating "sulfur getters" and employing support material with lower sulfating tendencies.

NO_x control under lean conditions is chemically difficult. So far diesel passenger car legislation has not required additional NO_x emissions control measures, and what was necessary has been achieved via engine modifications. This may not be true in the future, and two approaches are being developed to provide lean NO_x control. The first is so-called NO_x trapping, and the second is selective catalytic reduction (SCR) with ammonia, which has been used for many years in the chemical industry. Here only the former will be considered. During normal lean-running, NO_x in diesel exhaust gas is oxidized to NO_2 over a platinum component in the NO_x-absorbing catalyst (NAC). The NO_x then reacts with a basic component to form a nitrate species. Typically, the absorbing component is an alkaline metal compound that under normal operating conditions is in the form of a carbonate, and the absorption reactions concerned are shown in Equations 27 and 28.

$$2NO + O_2 = 2NO_2. \quad 27.$$

$$MCO_3 + NO_2 = MNO_3 + CO_2. \quad 28.$$

When the trapping capacity of the NAC is almost exhausted, it is regenerated by exposure to a pulse of rich gas; during this process the stored nitrate becomes thermodynamically unstable owing to the reverse reaction becoming favored, and thus NO is released. This is then reduced to nitrogen over a rhodium component in the NAC, as shown in Reactions 29 and 30; they also show reformation of the original carbonate phase. This overall process has been used successfully on lean-burn gasoline engines and is being applied to diesel engines (71).

$$MNO_3 = MO + NO + O_2. \quad 29.$$

$$NO + CO = N_2 + CO_2. \quad 30.$$

$$MO + CO_2 = MCO_3. \quad 31.$$

However, there is a gradual deactivation process involving conversion of the nitrate-forming phase into thermodynamically very stable sulfate that causes the NO_x storing capacity of a NAC to decrease during real-world use. The reactions concerned are given in Equations 31–33. Like the process that regenerates a NAC when it is in the nitrate form, so also the sulfate can be decomposed to restore the full NO_x capacity. As this requires temperatures in the region of 600°C,

also can be used. Nickel catalysts in general are active for the hydrogenolysis of carbonyl and carbon-carbon bonds, but here the combination of low-process temperatures and some degree of poisoning from sulfur compounds in the vegetable oils limits the catalytic activity to the required hydrogenations. This catalyst deactivation is essential for the process and in turn this leads to a short catalyst life, from sulfur poisoning rather than metal sintering or carbon formation (both slow at the low-process temperature). The use of batch processes allows frequent catalyst replacement (75).

This is an example where the design of the process and the catalyst is governed by the need for process selectivity rather than any need for long catalyst life.

CONCLUSIONS

With industrial metallic catalysts, obtaining high activity and appropriate selectivity is very important, but it is essential that catastrophic deactivation does not take place during use and that any loss of activity is gradual and predictable. Therefore, it is important that catalytic processes are operated in well-controlled ways so as not to induce deactivation. Moreover, it is clear from the examples discussed in this review that a great deal of effort goes into the design and preparation of commercial heterogeneous catalyst systems to prevent deactivation as much as possible and control deactivation processes during their actual use.

LITERATURE CITED

1. Twigg MV, ed. 1996. *Catalyst Handbook*. London: Manson. 608 pp. 2nd ed.
2. Spencer MS. 1996. Fundamental principles. See Ref. 1, pp. 73–84
3. Lloyd L, Ridler DE, Twigg MV. 1996. The water-gas shift reaction. See Ref. 1, pp. 283–339
4. Delmon B, Froment GF, eds. 1999. *8th International Symposium on Catalyst Deactivation*. Amsterdam: Elsevier. 435 pp.
5. Richardson JT. 1989. Catalyst deactivation: how they fail. In *Principles of Catalyst Development*, p. 191. New York: Plenum. 288 pp.
6. Hughes R. 1989. *Deactivation of Catalysts*. New York: Academic. 279 pp.
7. Spencer MS. 1985. Stable and metastable metal surfaces in heterogeneous catalysis. *Nature* 323:685–87
8. Lieber E, Morritz FL. 1953. The uses of Raney Metal. *Adv. Catal.* 5:417–55
9. Davies P, Donald RT, Harbord NH. 1996. Catalytic oxidations. See Ref. 1, pp. 469–518
10. Satterfield CN. 1996. Catalytic oxidation. In *Heterogeneous Catalysis in Industrial Practice*, pp. 312–20. Florida: Krieger. 2nd ed.
11. Satterfield CN. 1996. Methanol to formaldehyde. See Ref. 10, pp. 285–90
12. Satterfield CN. 1996. Processing of petroleum and hydrocarbons. See Ref. 10, pp. 358–70
13. Jennings JR, ed. 1991. *Catalytic Ammonia*

Synthesis Fundamentals and Practice. New York: Plenum. 451 pp.
14. Twigg MV, Spencer MS. 2001. Deactivation of supported copper metal catalysts for hydrogenation reactions. *Appl. Catal. A*. 212:161–74
15. Satterfield CN. 1996. Supported metal catalysts. See Ref. 10, pp. 202–3
16. Nix RM, Judd RW, Lambert RM, Jennings JR, Owen G. 1989. High-activity methanol synthesis catalysts derived from rare-earth/copper precursors: genesis and deactivation of the catalytic system. *J. Catal*. 118:175–91
17. Anderson JR. 1975. *Structure of Metallic Catalysts*. London: Chapman & Hall. 469 pp.
18. Goodman DR. 1996. Handling and using catalysts in the plant. See Ref. 1, pp. 140–90
19. Anderson JR. 1975. Dispersed metal catalysts. See Ref. 17, p. 167
20. Twigg MV. 1996. Equilibrium constants: for the CO conversion reaction (shift) at various temperatures. See Ref. 1, pp. 543–48
21. Warn JRW, Peters APH. 1996. Free energy in industrial processes. In *Concise Chemical Thermodynamics*. p. 151. London: Chapman & Hall. 188 pp. 2nd ed.
22. Chinchen GC, Short GD, Spencer MS, Waugh KC. 1987. Promotion of methanol synthesis and the water gas shift reactions by adsorbed oxygen on supported copper catalysts. *J. Chem. Soc. Faraday Trans I*. 83:2193–12
23. Spencer MS. 1987. Brass formation in copper-zinc catalysts. I. *Surf. Sci*. 192: 323–28
24. Spencer MS. 1987. Brass formation in copper-zinc catalysts. II. *Surf. Sci*. 192: 329–35
25. Spencer MS. 1987. Brass formation in copper-zinc catalysts. III. *Surf. Sci*. 192:336–43
26. Freund JH, Roberts MW. 1996. Surface chemistry of carbon dioxide. *Surf. Sci. Rep*. 25:227–73
27. Spencer MS. 2000. Precursors of copper/zinc oxide catalysts. *Catal. Lett*. 66: 255–57
28. Somorjai GA. 1981. The surface chemical bond. In *Chemistry in Two Dimension Surfaces*, p. 285. Ithaca, NY: Cornell Univ. Press. 575 pp.
29. Somorjai GA. 1981. The surface chemical bond. See Ref. 28, pp. 314, 315
30. Deleted in proof
31. Bridger GW, Spencer MS. 1996. Methanol synthesis. See Ref. 1, pp. 460–62
32. Twigg MV, Spencer MS. 2003. Deactivation of copper metal catalysts for methanol decomposition, methanol steam reforming and methanol synthesis. *Topics Catal*. 22:191–203
33. Chinchen GC, Denny PJ, Jennings JR, Spencer MS, Waugh KC. 1988. Synthesis of methanol: I. Catalysts and kinetics. *Appl. Catal*. 36:1–65
34. Kung HH. 1992. Deactivation of methanol synthesis catalysts. *Catal. Today*. 11:443–53
35. Chinchen GC, Spencer MS. 1991. Sensitive and insensitive reactions on copper catalysts. *Catal. Today*. 10:293–301
36. Klier K. 1982. Methanol synthesis. *Adv. Catal*. 31:243–313
37. Pollard AM, Spencer MS, Thomas RG, Williams PA, Holt E, Jennings JR. 1992 Georgeite and azurite as precursors in the preparation of co-precipitated copper/zinc oxide catalysts. *Appl. Catal A* 85:1–11
38. Waller D, Stirling D, Stone FS, Spencer MS. 1989. Copper-zinc oxide catalysts: activity in relation to precursor structure and morphology. *Faraday Disc. Chem. Soc*. 87:107–20
39. Stone FS, Waller D. 2003. Cu-ZnO and Cu-ZnO/Al_2O_3 catalysts for the reverse water-gas shift reaction. The effect of the Cu/Zn ratio on precursor characteristics and on the activity of the derived catalysts. *Topics Catal*. 22:305–18
40. Short GD, Chinchen GC, Williamson JG. 1985. *U.S. Patent No. 4788175*

41. Chinchen GC, Jennings JR. 1989. *U.S. Patent No. 4863894*
42. Ma L, Tran T, Wainwright MS. 2003. Hydrogenation of CO_2 over skeletal copper surfaces containing precipitated ZnO. *Topics Catal.* 22:287–93
43. Ma L, Tran T, Wainwright MS. 2003. Methanol synthesis from CO_2 using skeletal copper catalysts containing coprecipitated Cr_2O_3 and ZnO. *Topics Catal.* 22:295–304
44. Chinchen GC, Mansfield K, Spencer MS. 1990. How does the methanol synthesis catalyst work? *CHEMTECH* 20(11):692–99
45. Satterfield CN. 1996. Experimental laboratory reactors. See Ref. 10, pp. 523–32
46. Højlund Nielsen PE. 1991. Deactivation of synthesis catalyst. See Ref. 13, pp. 285–302
47. Jennings JR, Ward SA. 1996. Ammonia synthesis. See Ref. 1, pp. 384–440
48. Tamaru K, 1991. The history of the development of ammonia synthesis. See Ref. 13, pp. 1–18
49. Richardson JT. 1989. Catalytic deactivation: how they fail. See Ref. 5, p. 189
50. Hooper CW. 1991. Ammonia synthesis commercial practice. See Ref. 13, pp. 253–84
51. Leigh GJ. 1991. What does the future hold? A survey of possible technical developments. See Ref. 13, pp. 365–87
52. van der Wal WJJ, Geus JW. 1984. *U.S. Patent No. 4459370*
53. Deleted in proof
54. Kuczynski GC, Millar AE, Sargent GA, eds. 1984. *Materials Science Research, Vol. 16: Sintering and Heterogeneous Catalysis*. New York: Plenum. 349 pp.
55. Tennison SR. 1991. Alternative non iron catalysts. See Ref. 13, pp. 303–64
56. Thomas CL. 1970. *Catalytic Processes and Proven Catalysts*. p. 152. New York: Academic
57. Lloyd L, Ridler DE, Twigg MV. 1996. The water-gas shift reaction. See Ref. 1, pp. 331, 324
58. Deleted in proof
59. Finlayson-Pitts BJ, Pitts JN. 2000. The atmospheric system. In *The Chemistry of the Upper and Lower Atmosphere*, e.d. BJ Finlayson-Pitts, JN Finlayson Jr, pp. 15–24. New York: Academic
59a. Finlayson-Pitts BJ, Pitts JN. 2000. Analytical methods and typical concentrations for gases and particles. See Ref. 59, pp. 594–95
60. Twigg MV. 1999. Twenty-five years of autocatalysts. *Platinum Metals Rev.* 43:168–71
61. Barnes GJ. 1975. *Catalysis for the control of Automotive Pollutants, American Chemical Society, Advances in Chemistry Series 143*
62. Yao Y. 1975. Oxidation of hydrocarbons and carbon monoxide over metal oxides IV. Perovskites. *J. Catalysis* 36:226–75
62a. Tejuca LG, Fierro JLG, eds. 1992. *Properties and Applications of Perovskite-Type Oxides*. New York: Dekker
62b. Seiyama T. 1992. Total oxidation of hydrocarbons on perovskite oxides. See Ref. 62a, pp. 215–34
62c. Viswanathan B. 1992. CO oxidation and NO reduction on perovskite oxides. See Ref. 62a, pp. 271–88
63. Cybulski A, Moulijn JA, eds. 1998. *Structured Catalysts and Reactors*. New York: Dekker
63a. Gulati ST. 1998. Ceramic catalyst supports for gasoline fuel. See Ref. 63, pp. 15–58
64. Twigg MV, Webster DE. 1999. Metal and coated-metal catalysts. See Ref. 63, pp. 59–90
65. Ragsdale RO. 1973. Reactions of nitrogen (II) oxide. In *Developments in Inorganic Nitrogen Chemistry*, ed. CB Colburn, 2:1–26. Amsterdam: Elsevier
66. Twigg MV, Wilkins AJJ. 1998. Autocatalysts—past, present, and future. See Ref. 63, pp. 91–120
67. Deleted in proof
68. Twigg MV, Collins NR, Morris D, Cooper JA, Marvell DR, et al. 2002. High

temperature durable three-way catalysts to meet European stage IV emission requirements. *Soc. Auto. Eng. Tech. Pap.* 2002-01-0351
69. Fisher JM, Gray PG, Rajaram RR, Hamilton HGC, Ansell PG. 1996. *Worldwide Patent 96/39244:1–9*
70. Phillips PR, Chandler GR, Jollie DM, Wilkins AJJ, Twigg MV. 1999. Development of advanced diesel oxidation catalysts *Soc. Auto. Eng. Tech. Pap.* 1999-01-3075:1-8
71. Gill LJ, Blakeman PG, Twigg MV, Walker AP. 2004. The use of NO_x absorbers on diesel engines. *Top. Catal.* 28:157–64
72. Schwenk E, Papa D, Hankin H, Girsberg H. 1955. Propylbutyrolactone and (tetrahydrofuryl) propionic acid. In *Organic Syntheses Collective*, ed. EC Horning, 3:742–44. New York: Wiley
73. Wainwright MS. 1999. Skeletal metal catalysts in preparation of solid catalysts. In *Preparation of Solid Catalysts*, ed. G Ertl, H Knozinger, J Weitkamp, pp. 28–43. Weinheim: Wiley-VCH
74. Raney M. 1927. *U.S. Patent No. 162810*
75. Patterson HBW. 1994. *Hydrogenation of Oils and Fats: Theory and Practice.* Champaign IL: Am. Oil Chem Soc. 267 pp.

Annu. Rev. Mater. Res. 2005. 35:465–504
doi: 10.1146/annurev.matsci.35.100303.122141

First published online as a Review in Advance on April 1, 2005

Electron Microscopy in the Catalysis of Alkane Oxidation, Environmental Control, and Alternative Energy Sources

Pratibha L. Gai
DuPont, Central Research and Development Laboratories, Experimental Station, Wilmington, Delaware 19880-0356, and Department of Materials Science and Engineering, University of Delaware, Newark, Delaware 19716; email: pratibha.l.gai@usa.dupont.com

Jose J. Calvino
Departamento de Ciencia de los Materiales e Ing. Met. y Química Inorgánica, Facultad de Ciencias de la Universidad de Cádiz, Spain; email: jose.calvino@uca.es

Key Words hydrocarbon technology, environmental catalysts, in situ ETEM, TEM, nanostructures, hydrogen energy

■ **Abstract** The key role of electron microscopy in understanding and creating advanced catalyst materials and processes in selective alkane oxidation, environmental control, and alternative energy sources is reviewed. In many technological processes, catalysts are increasingly nanoscale heterogeneous materials. With growing regulatory guidelines requiring efficient and environmentally compatible catalytic processes, it is crucial to have a fundamental understanding of the catalyst nanostructure and modes of operation under reaction conditions to design novel catalysts and processes. The review highlights the pioneering development and applications of atomic resolution in situ-environmental transmission electron microscopy (ETEM) for probing dynamic catalysis directly at the atomic level, high-resolution electron microscopy, and analytical spectroscopic methods in the development of alkane catalyzation, environmental protection, and new energy sources.

INTRODUCTION

The selective oxidation of hydrocarbons to a variety of industrial polymer products and novel routes for environmental catalysis have received considerable interest (1). The selective oxidation of n-butane to maleic anhydride (MA) is an important commercial process. Maleic anhydride is used as a raw material for a variety of products ranging from tetrahydrfurans (THF) for industrial fibers and polyurethane technologies to agricultural chemicals. Several international groups are working to improve the process. The oxidation of n-butane to MA is a remarkably selective 14-electron oxidation. To date, vanadium phosphorus oxide–based catalysts are still

0084-6600/05/0804-0465$20.00

the most active and selective systems for this chemistry. It is likely, however, that other highly active and selective catalysts exist, with possibly different properties (stability and lattice oxygen capacity). However, to date, none has been discovered. Significant progress in the area is being made by studying catalysts under in situ conditions. By understanding the nature of active species, more stable and active catalysts (e.g., those having higher surface areas, or placed on high surface area supports) could be developed.

Also, environmental concern is undoubtedly a powerful driving force, particularly in catalysis. Thus the topic of environmental catalysis has established itself as an area of increasing interest that encompasses a variety of technical approaches, all of which have the common goal of improving the quality of the natural environment and contributing to so-called sustainable development (2). Generally speaking, environmental catalysis is not only related to technologies aimed at correcting or controlling emissions to the environment of noxious pollutant substances but also to areas such as green chemistry or alternative energy sources. The search for catalytic processes that allow higher selectivity (i.e., less undesired by-products); lower energy consumptions; diversification of fuels; improvements in the quality of fuels; or for alternative, more environmentally benign energy sources are also the scope of this exciting, ever-growing field. In this review we highlight the key role of electron microscopy (EM) methods in recent advances in promoted catalysts for butane oxidation as well as environmental and energy catalysis.

EXPERIMENTAL PROCEDURES

Electron Microscopy Techniques

Electron microscopy techniques are powerful and versatile research tools for investigating the local structure and chemistry of complex heterogeneous catalysts directly, from the macroscopic to the atomic scale (3, 4). Several monographic books, issues and articles on EM of catalysis clearly demonstrate the tremendous impact of these techniques in the characterization of catalytic solids (5–17).

Briefly, EM provides structural information of samples in both the real space and reciprocal space. It provides local structural information of the surface and the bulk of the sample routinely at the atomic level and also provides chemical, electronic, and three-dimensional structural information. EM is a diffraction technique in which crystals diffract electrons in accordance with Bragg's Law, $n\lambda = 2d_{hkl}\sin\theta$, where λ is the wavelength of electrons, d is the lattice planar spacing of hkl reflection, and θ is the scattering angle. The diffraction pattern thus formed may be regarded as a Fourier transform (FT) of the crystal. An apertured inverse FT in the objective lens of the EM forms the image. The interactions of electrons with matter generates elastic scattering and a variety of other signals that can be used to obtain the structural and chemical information of a sample. Elastic scattering occurs when incident electrons interact with the potential field of nuclei of matter with no energy loss in momentum transfer, whereas inelastic scattering occurs when interactions between incident electrons with electrons of matter occur and scattered electrons lose energy (3).

In a conventional transmission electron microscope (TEM) and high-resolution TEM (HRTEM), electrons transmit through the sample. The TEM has an electron gun and electromagnetic lenses, which include condenser and objective lenses, and operates at ambient temperature in high vacuum. TEM diffraction contrast is the conventional TEM imaging technique where the Bragg condition is satisfied for a single diffracted beam (3). If only the diffracted beam corresponding to the incident electron beam direction (primary beam containing the direct transmitted electrons) is used for imaging, a bright-field (bf) image is obtained; if only scattered electrons are used, a dark-field (df) image is formed. Defects such as point and extended defects control properties of catalysts, and understanding them is critical for controlling and optimizing catalytic properties. The diffraction contrast technique is particularly effective for determining the nature of defects (i.e., lattice imperfections), generally without detailed simulations of defect models. One of the most powerful methods of direct structural analysis of catalysts is HRTEM, where two or more Bragg reflections are used for imaging. The power of HRTEM in determining real-space structures of complex inorganic solids that were beyond the reach of conventional structural techniques (e.g., X-ray, neutron crystallography) has been elucidated by several workers (1, 5). Unlike conventional diffraction techniques [powder X-ray diffraction (XRD), neutron], HRTEM provides localized real-space information, now routinely at the atomic level, about the bulk and surface properties of solids, the corresponding chemical information, as well as electron diffraction information in reciprocal space.

For EM of catalysts, as-synthesized powders may be used. They are dispersed in alcohol and deposited on generally 3-mm-carbon-filmed beryllium, copper, or aluminum grids. If carbon or other support films are not desirable, finely meshed metal grids can be used. High purity grids should be used for experiments. The powders are thin enough for the electron beam to penetrate, and no further sample preparations are usually necessary.

Chemical composition analysis in the EM is carried out using electron-stimulated energy dispersive X-ray spectroscopy (EDX or EDS) or electron energy-loss spectroscopy (EELS). In EDX, X-rays with energies characteristic of elements are measured in the probed region of the sample. In EELS, energy losses produced by the interaction between incident electrons and the sample are measured. The energy losses result from inelastic scattering effects. EELS has normally been used to detect light elements in materials. The measurements via EELS spectra can provide quantitative information about chemical or oxidation states and elemental or chemical compositions and can be particularly effective in catalyst studies. Rich spectroscopic and structural information can be retrieved at atomic scale resolution, and with high energy resolution in the case of spectroscopic signals, which allows us to scrutinize the bulk, surface, and interface details of all types of materials. Thus TEM provides a set of highly complementary tools that allow a very precise examination of materials in general (3) and of catalytic materials in particular (5–17).

Recent Advances in Atomic Resolution In Situ ETEM for Probing Catalysis Directly at the Atomic Level

Heterogeneous catalysis is a dynamic process. In gas-catalyst reactions, the dynamic atomic structure of catalysts under operating conditions plays a pivotal role in governing catalytic properties. Therefore, there is a strong need for in situ EM methods for direct probing of live catalyst-gas molecule reactions, reaction mechanisms, transient species, and the chemistry of catalysts under reaction conditions. Furthermore, post-reaction examination of a static catalyst, which is taken out of the reaction environment and cooled to room temperature (RT), does not often represent the dynamic state of the catalyst and there may well be atmospheric contamination, which can cause errors in interpretation.

In ETEM, a gas reaction cell (or environmental cell, ECELL or microreactor) is integrated inside an electron microscope column, which is differentially pumped as described by Gai (7). Recent developments include atomic resolution in situ ETEM (also referred to as environmental-HRTEM), pioneered by Gai & Boyes (18, 19). It enables probing of dynamic chemical reaction processes directly at the atomic level. In this development, a new approach has been taken to design the ETEM, with radial holes in the objective lens pole pieces for gas feed, and the ECELL is integral to the ETEM (Figure 1*a*). The ETEM design of Gai & Boyes has been adopted by commercial TEM manufacturers (FEI) and copied globally.

In in situ ETEM studies, appropriate gas compositions, using a conventional reactor-type gas-manifold system, enable the inlet of flowing gases into the ETEM microreactor. A mass spectrometer is added for gas analysis. A sample stage with a furnace allows samples to be heated in the in situ ETEM (20–22) (Figure 1*b*). Lower gas pressures (a few mbar) are used for high-resolution imaging. Higher gas pressures are possible with some loss of resolution owing to multiple scattering of electrons through gas layers. Experiments on a large number of catalysts (5) have shown that the gas layer in immediate contact with the catalyst surface is the most critical to the catalytic reaction as compared with the total amount of gas present. Careful experimental conditions are used in in situ ETEM to ensure that the material is not damaged in the electron beam (12, 19, 20, 22a,b). These include the use of very-low-dose electron beam currents for imaging materials. The image signal is then amplified by a low-light television camera connected to a video monitor, or recorded on a charge-coupled device (CCD) camera. Calibration blank experiments (without the electron beam) are carried out with the beam switched on for a few seconds only to record the final state of the materials and to check with in situ data. Under these careful conditions, no invasive beam damage is observed. The pioneering atomic resolution ETEM development of Gai & Boyes has been highlighted by the American Chemical Society's *C&E News* (23, 24). Parallel chemical

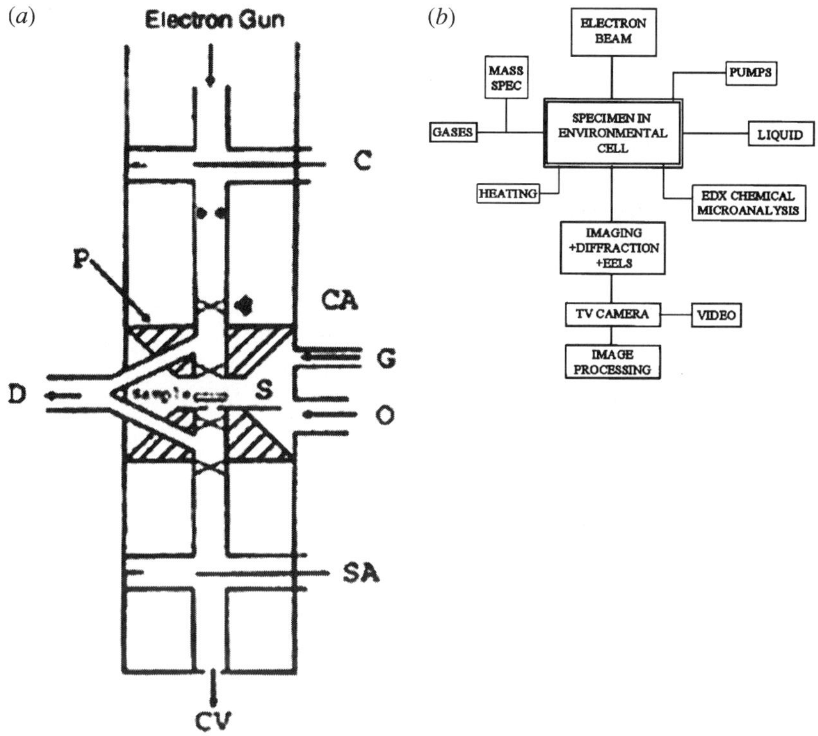

Figure 1 (*a*) Schematic of atomic resolution in situ environmental transmission electron microscope (ETEM) with environmental cell (ECELL or microreactor), showing radial holes through objective lens pole pieces for gas lines (*D*). Differential pumping is used. P, pole pieces; CA, cell aperture; O, objective aperture; SA, selected area aperture; S, sample; G, gas; C, condenser aperture (18, 19). (*b*) Schematic of various components of ETEM.

reaction experiments have been performed on larger amounts (~1 g) of the sample in a microreactor to correlate microstructural observations with catalytic reactivity properties. Complementary techniques in catalytic chemistry involve temperature programmed (TP), oxidation state, and BET (Brunauer-Emmett-Teller) surface area measurements and powder XRD of samples.

Recent in situ developments also include of wet-environmental transmission electron microscopy (hereafter referred to as Wet-ETEM) to directly probe controlled liquid-catalyst reactions in liquid phase hydrogenation and polymerization at operating temperatures, at the nanomolecular scale (25, 26). This method opens up new opportunities for high-resolution studies of a wide range of solution-solid and solution-gas-solid reactions in the chemical, polymer, and biological sciences. Other EM methods generally used in catalysis are scanning electron microscopy (SEM) and scanning TEM (STEM) (27). SEM utilizes a focused electron beam that systematically scans across the surface of a sample and is used to characterize surface topography of bulk powders or pellets (28, 29). STEM, which is essentially a combination of scanning EM and TEM, is powerful as a structural and an analytical tool and uses a finely focused probe (<1 nm) over a thin sample (30). An important method in high resolution STEM is high angle annular dark field (HAADF) for Z (or atomic number)-contrast imaging (27). It utilizes the fact that high-angle scattering (angles >30 mrad) obeys the Rutherford law, i.e., the scattering cross section is proportional to Z^2. The use of HAADF-STEM signal removes the complexity of conventional bright field scattering and the associated diffraction complications. The use of HAADF-STEM to determine the three-dimensional structure of supported metal nanocatalysts at a very high spatial resolution of <1 nm has been demonstrated recently for Pd-Ru nanocatalysts supported on mesoporous silica (31).

CATALYST SYNTHESIS AND IN SITU CHARACTERIZATION IN N-BUTANE CATALYSIS

Control (Base) Catalysts

The most successful industrial process to manufacture MA is the vapor phase oxidation of n-butane over active vanadyl pyrophosphate catalysts (32). In the following, we review highlights of the recent work and the crucial role of EM, especially atomic resolution in situ ETEM, dynamic electron diffraction, and correlations of the catalyst microstructure with its reactivity. We review the effect of steam and cation promoters on catalyst performance. In the selective oxidation of the alkane to MA, the most active and selective phase is believed to be vanadyl pyrophosphate [$(VO)_2P_2O_7$] catalysts (hereafter referred to as VPO), with active centers located at the exposed (010) basal planes (20, 32). There are also some claims that the active phase is a combination of VPO and $VOPO_4$ phases, or a P-rich phase, or a mixture of VPO with unknown or amorphous phases (32, 33). VPO is orthorhombic with a = 1.659 nm, b = 0.776 nm, and c = 0.9588 nm (32, 34). VPO is a layered structure and the basal (010) plane, considered to be the active plane, consists of edge-sharing pairs of distorted vanadium octahedra shared at corners by the phosphate tetrahedral groups (Figure 2*a*, see color insert). An organic method is generally employed to synthesize the base catalyst precursors (20, 32, 35, 36). Vanadium pentoxide and anhydrous phosphoric acid, prepared by mixing P_2O_5 with aqueous phosphoric acid, are refluxed in 11:1 isobutyl alcohol/benzyl alcohol solvent mixtures under an inert atmosphere. Enough anhydrous phosphoric acid to satisfy a 1.1/1, P/V atomic ratio is slowly added to the alcohol mixture over a period of a couple of hours. The solution is allowed to reflux for several hours. Soluble pentavalent vandyl species are reduced predominantly by benzyl alcohol to precipate the blue tetravalent precursor (which is the case of pure VPO catalysts), which crystallizes as vanadium hydrogen phosphate hemihydrate, $VO(HPO_4)\,0.5\,H_2O$. The precursors are subsequently dried in vacuum at

150°C for 18 h. Bartley (37) described vanadium (V) phosphate prepared using a solvent-free (in the absence of water) leading to an anhydrous $VOPO_4$ that can be readily hydrated to the precursor.

Promoted Catalysts

In the case of promoted VPO catalysts, promoter cations are added as the soluble chlorides and alkoxides, which subsequently coprecipitate. A number of single metal cations (e.g., Fe, Cr, Ce, Li, Zn) have been added (38–41), and the literature connected to the promotion of VPO catalysts has been reviewed recently by Hutchings and coworkers (40). ETEM, SEM, and diffraction methods play a key role in elucidating the surface and bulk microstructural changes, which are due to the catalyst promotion, and the interplay between the promoted structures and reactivity to obtain a deeper understanding of the role of promoted microstructures in the alkane oxidation catalysis.

Gai and coworkers (35, 36) have synthesized promoted catalysts with cations of two metals substituting for V. Their approach has been to replace V 4+ with an equimolar combination of cations of different valencies such as Fe 3+ and Sb 5+ or bismuth and molbydenum (36). They have compared catalyst performance from this approach with that involving the substitution of a lower valent cation such as Fe 3+ or Cr 3+ for V 4+, as well as cerium oxide additions as an oxygen reservoir to the VPO catalyst. In addition they have prepared novel, grafted-promoted VPO catalysts as alternative synthetic pathways.

To create the novel grafted-promoted vanadium phosphate catalyst, a mixture of bismuth and molybdenum alkoxides is used with the hemihydrate precursor. The alkoxides, prepared using bismuth and molybdenum chloride, form a soluble, mixed alkoxide complex when reacted with ethanol. The mixed alkoxide, dissolved in alcohol, is reactively grafted onto the preformed precursor. The catalysts in these experiments contain only 1 mol% bismuth and molybdenum supported on $VOHPO_4\ {}^1/_2H_2O$. This process is a grafting procedure, and the precursor complex is then heated to transform into the final, promoted catalyst. Calcinations are performed in a small, 3.5-cm fluidized bed on 60 g catalyst samples. Samples are heated in air at 390°C for 1 h, followed by 460°C for 18 h in 1.5% butane/air. The relatively high temperature of 460°C used for the calcination is beneficial for the rapid equilibration of the catalysts.

Microreactor Evaluations

A fixed bed microreactor is used for catalytic evaluations. Analyses of the reactor feed gases prior to and after the reaction (including for butane and reaction products such as MA, acetic acid, and acrylic acid) are carried out by gas-liquid chromatography (GLC).

The pseudo first order rate constant, k, given for the disappearance of butane, is obtained by fitting the reactor data to a classical first order rate expression:

$$d[\text{butane}]/dt = -k\,[\text{butane}],$$

$$d(x_0 - x)/dt = -k(x_0 - x),$$

where x_0 is the initial concentration of butane and x is the portion of butane reacted.

Integrating this expression gives the concentration of butane exiting the reactor as a function of contact time, t, in the reactor:

$$[\text{exit butane}] = x_0 - x = x_0 \exp(-kt).$$

In addition to describing the reaction rate of butane with a catalyst, the rate constant, k, includes several other factors including the dependence of the reaction rate on the oxygen concentration (which remains relatively constant under our conditions) and the concentration of catalyst active sites (also assumed constant).

In Situ ETEM Observation of Butane Oxidation

Direct in situ ETEM studies of dynamic butane oxidation over (010) VPO catalyst (Figure 2*b*) at operating temperatures (~360–390°C) are reported in *Science* (20).

Figure 2 (*b*) Atomic resolution ETEM image of (010) VPO in butane. The associated electron diffraction (ED) is inset. The 002 and 400 Bragg reflections are shown along with the corner positions of 201 reflections. The structural model is approximately superimposed to illustrate vanadyl and P atom columns. The image and ED show a well-ordered structure. (*c*) In situ ETEM studies of VPO reaction in butane recorded at ~390°C, showing two sets of extended defects along the symmetry related [201] and [20 −1] directions, indicated by arrows. Inset (C2) shows one set of dynamic extended defects formed close to the catalyst surface, imaged in the 201 reflection in diffraction contrast. Inset (C1) shows the corresponding ED with diffuse streaks along ⟨201⟩ indicating that (201) planar oxygen atoms are partaking in the catalysis. (*d*) In situ ETEM of one set of the extended defects at high resolution showing a stacking fault close to the catalyst surface, bounded by partial screw dislocations P1 and P2 and lattice displacement along ⟨201⟩. (*e*) (*top*) A three-dimensional projection of a partial layer of the idealized (010) VPO structure showing oxygen atoms (along ⟨201⟩). O and O″ are in front and back layers. Some V and P atoms are denoted. (*f*) (*Bottom*) Generation of glide defects and model for novel glide shear mechanism along ⟨201⟩ in butane oxidation catalysis. The atom shown by arrow (e.g., top of front layer) glides to the site created by loss of O and so on. Analysis of the extended defects, using electron microscopy, shows them to be formed by novel glide shear (18, 20). (*g*) A two-dimensional structural model for the observed extended glide shear plane defects that represent the diffusion front of anion vacancies: projection of one layer of the idealized (010) VPO, showing the novel glide shear mechanism. The glide defects accommodate misfit between the surface-containing anion vacancies (*filled circles*) and the underlying matrix and preserve anion vacancies associated with Lewis acid sites for catalysis. (*h*) Schematic of glide shear: glide plane defects at the interface between the catalyst surface-containing anion vacancies and the underlying matrix.

These studies have shown evidence for catalyst surface structure modifications accompanied by two sets of symmetry-related extended defects along ⟨201⟩ directions (shown by arrows in Figure 2*c* in diffraction contrast). One set of the defects in the HRTEM image is shown in Figure 2*d*. Defect analysis reveals that the defects are formed by the glide shear mechanism (18, 20) shown in the structural model in Figure 2*e*–*g*. The defects are partial glide (screw) dislocations separating a stacking fault in {201} planes, and the stacking fault is terminated by the dislocation. In the defect mechanism, anion vacancies are formed in the basal plane along ⟨201⟩, between corner-sharing phosphate tetrahedra and vanadyl octahedra following the catalyst oxygen loss in the butane oxidation. This important glide shear process, in which oxide surface layers undergo a structural transformation by glide shear to accommodate the surface misfit owing to anion vacancies formed during the butane

(*b*) (*c*)

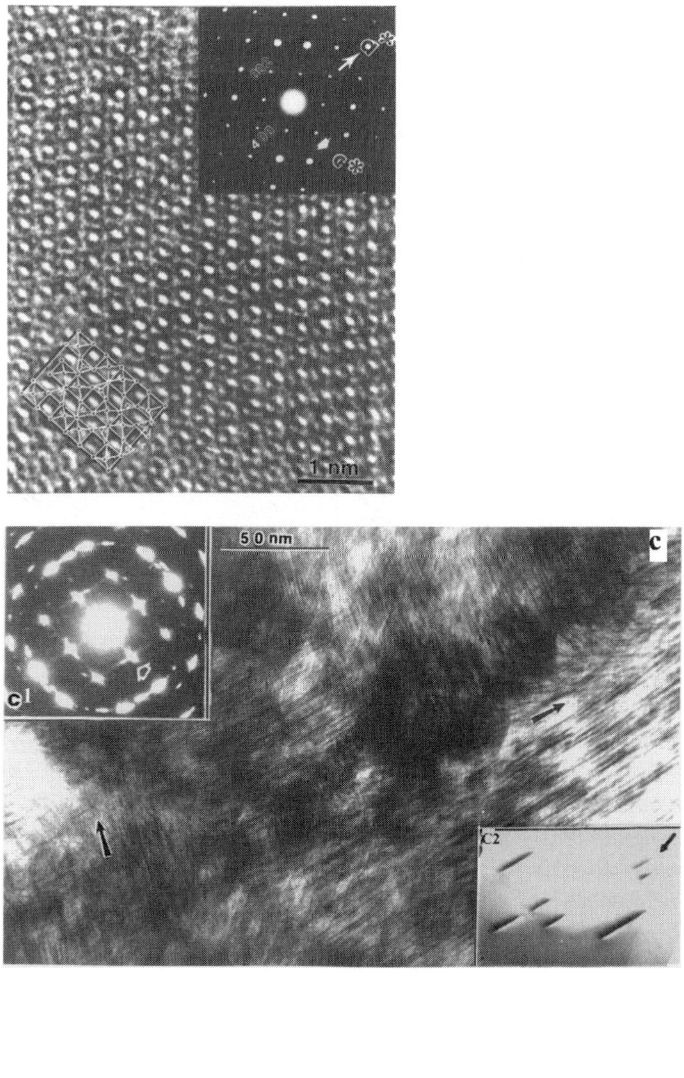

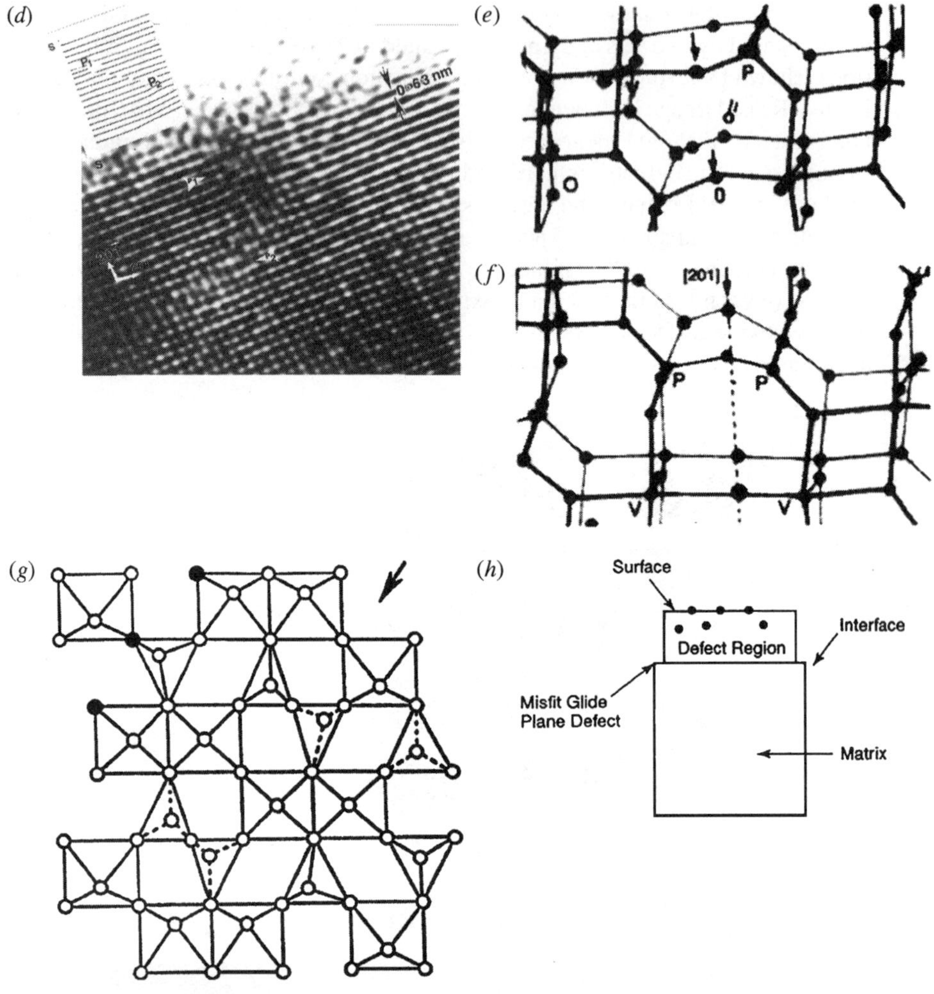

Figure 2 (*Continued*)

reaction, essentially preserves anion vacancies associated with Lewis acid sites at the catalyst surface, as shown in the schematic in Figure 2*h*. This novel mechanism explains the release of structural oxygen and preservation of active Lewis acid sites without the catalyst structural collapse. The doubly positively charged anion vacancy sites are important in the hydrocarbon activation and oxygen exchange in the catalyst regeneration. The glide shear defect regions correspond to a slightly anion-deficient defect pyrophosphate phase, with pyrophosphate groups sharing corners similar to metaphosphate $(PO_3)_n$ groups. The local glide defects regions, which maintain the anion vacancies (V^{3+} phases), do not change the overall VPO structure under the catalyst-operating conditions and as such are not visible in X-ray diffraction (XRD). In situ ETEM and electron diffraction have thus played a key role in unraveling the important defect structures in butane catalysis. The correlation of the glide defect microstructure with catalytic reactivity in hydrocarbon oxidation catalysis has shown uniformly high selectivity for MA in 1.5% butane/air in VPO for prolonged periods with maintenance of high reactivity in other catalytic oxides. These data indicate that glide shear is the most effective defect mechanism by which oxide catalysts accommodate nonstoichiometry and continue to operate in selective hydrocarbon oxidation reactions (18).

Effect of Steam Exposure on the Catalyst Performance

In any hydrocarbon oxidation, steam/water vapor is a by-product. The effect of catalyst exposure to water vapor, especially on the catalyst microstructure and performance, remains a relatively unexplored area. The complex characteristics of long-term effects of water vapor and the evolution of defect structures have been described in the literature (35). In these studies, VPO catalysts are exposed to nitrogen/steam environments for periods ranging from 24 h to 312 h. Powder XRD has shown a modest decrease in diffraction line widths in samples treated for longer periods (312 h). ETEM shows that steam plays an important role in the development of glide defect structures in the catalyst. The key observation from in situ ETEM is the gradual evolution of the extended glide defects along ⟨201⟩ orientations, with increased concentration compared with the concentration of defects in catalysts treated in nitrogen only. Prolonged steam exposure leads to a long-range order of the glide shear defects (18). In situ ETEM studies of reduction of VPO and other oxide systems have revealed the formation of extended glide shear defects, with the defect ordering leading to new homologous series of phases based on the crystal glide shear mechanism (18). The studies have important implications in oxide catalysis and, more generally, in oxide crystallography.

EFFECT OF CATION PROMOTERS

In in situ studies of promoted catalysts, compositions of gases (e.g., 1.5% butane/air) in the ETEM microreactor are comparable to those in a practical reactor. In situ data are complemented by SEM, STEM with selected area electron diffraction, and EDX. In a recent study (36), both as-prepared powders and cross-sectioned samples (for compositional depth profile) were used to understand the differences in the surface and bulk composition of a number of promoted vanadyl pyrophosphate catalysts. The catalyst particles are cross sectioned to allow access to the surface and to the interior (core) to obtain insights into relationships between the promoted catalyst microstructure and the reactor performance. Vanadium oxidation state measurements, as determined by redox titrimetry, show small changes in vanadium oxidation state after activation. Likewise, large changes in N_2/BET surface area are not observed for these systems, suggesting that the increased n-butane oxidation activity is through modification of active sites or an increase in the site density (sites/m^2) of the catalyst surface.

TABLE 1 Fixed bed microreactor data in 2% butane/air, 380°C, of grafted catalyst systems compared to control catalysts

Catalyst composition	k (s^{-1})	Selectivity at 40% conversion	Vox	N_2/BET SA (m^2/g)
(A) 1 mol% Bi, Mo $VO(HPO_4) \cdot 0.5\ H_2O$	1.02	86	4.07/4.06	34
(B) 1 mol% Bi, Mo $VO(HPO_4) \cdot 0.5\ H_2O$	0.88	83		
(C) EtOH on $VO(HPO_4) \cdot 0.5\ H_2O$, precursor before activation	0.56	88	4.07/4.05	26
(D) HCI + EtOH on $VO(HPO_4) \cdot 0.5\ H_2O$, precursor before activation	0.49	83		
(E) $VO(HPO_4) \cdot 0.5\ H_2O$, precursor before activation	0.56	86	4.00/4.01	30

Powder XRD data of Bi, Mo-promoted catalysts, catalysts with ethanol alone and control catalysts, show single-phase crystalline vanadium pyrophosphate. Surface area and redox titrimetry data are listed in Table 1. Table 2 shows data for higher butane concentrations.

The tables show that improved reactor performance is observed for the grafted promoted VPO systems. The results of grafted catalysts have been compared with data from the coprecipitated promoter catalyst.

TABLE 2 Fixed bed microreactor reactor, 2% butane/air, 9% butane/air at 380°C

	2% Butane/air		9% Butane, 10% O_2	
Catalyst	k (s^{-1})	Selectivity at 40% conversion	Yield	Selectivity
(1) Control catalyst	0.50	80	7.2	80
(2) 5BiMo	0.89	66	6.2	60
(3) 1BiMo	1.12	87	10	70
(4) Higher Bi	1.84	62	12.8	75
(5) Higher Bi, highest Bi/Mo ratio possible using anhydrous chloride alkoxides in EtOH	1.22	71	7.6	71
(6) Higher Mo	0.92	63	7.4	65
(7) Mo only	0.91	90	9.2	75
(8) 5 mol% Mo on VPO precursor	1.22	71	7.6	60

The catalyst systems (e.g., catalyst A) in Table 1 are prepared by reactive grafting of the mixed alkoxides of bismuth on molybdenum on preformed vanadium hydrogen phosphate precursor followed by calcination and activation at 460°C in 1.5% butane/air. In Table 1, fixed bed microreactor data obtained using a 2% butane/air feed mixture are shown. The catalyst activity of the catalysts grafted with 1 mol% of Bi, Mo after the calcination and activation heat treatments in 1.5% butane/air is shown for catalysts (A) and (B). Remarkably, the pseudo first order rate constant is approximately 60–80% greater than that of the untreated, control catalyst (catalyst E). N_2/BET surface area measurements show nearly negligible changes in surface area, indicating that the performance improvements of the grafted systems are most likely related to an increase in the density of active sites on the catalyst. In either case, bismuth is a significantly beneficial cocatalyst and together with molybdenum greatly enhances catalyst activity while maintaining a high selectivity to MA. Bismuth may function to stabilize molybdenum (or vanadium) in a more stable and selective oxidation state, and this role may become more important under the more reducing conditions of the 9% butane/10% O_2 reactant stream (Table 2).

Microchemical depth profiling studies show that the catalyst surface layers are molybdenum rich, and the core of the catalyst particles are enriched in bismuth, in the 1 mol% Bi, Mo VPO catalyst. Figure 3 shows the compositional depth profile of the Mo/Bi ratio, plotted (with a scaling factor) as a function of depth in the

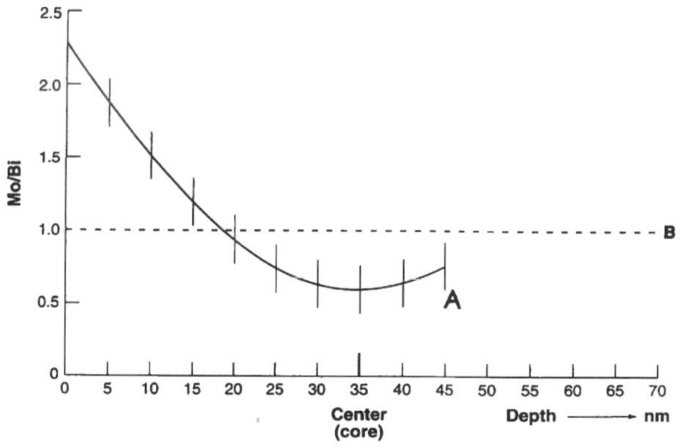

Figure 3 Compositional depth profile of Mo/Bi ratios (scaled) plotted as a function of depth in the catalyst from the surface to the interior (core) of cross-sectioned samples. Plot A is for grafted 1 mol% (Bi, Mo) VPO catalyst and plot B is for catalyst prepared by coprecipitation of promoter cations with the precursor (coprecipitated catalyst) for coprecipitated catalyst. The grafted catalysts show surface enrichment of Mo and the core enrichment of Bi and higher reactivity and selectivity compared with coprecipitated catalysts. Statistical variations are shown. The coprecipitated catalyst (B) does not show this behavior.

catalyst crystal, from the surface to the center (interior) of the particles. Plot A shows the composition depth profile for the grafted catalyst 1 mol% Bi, Mo VPO, and plot B is the composition depth profile for coprecipitated ($Bi_{0.05}$ $Mo_{0.05}$) VPO.

The presence of the Mo-rich phase may therefore be responsible for the observed enhanced catalytic activity, by altering the nature or introducing new active sites on the catalyst surface. As described previously (20), anion vacancies and glide defects are introduced in VPO following catalyst oxygen loss in butane oxidation, leading to V^{3+} phases. It is believed that anion defect concentrations from the substitution of Bi^{3+} are optimized by Mo^{6+}, thus maintaining the optimal concentration of defects and surface vanadium oxidation state (36).

FUTURE CHALLENGES IN ALKANE OXIDATION CATALYSIS

For VPO chemistry, a major challenge is to understand the nature of the equilibrated catalyst, the activation conditions for generating the equilibrated catalyst, and the effect of the catalyst compositional change on the catalyst activation. Large, parallel microreactor systems are probably best suited to investigating this large parameter space. Here too, in situ ETEM studies are set to play an important role to advance the field. In this section, the nature of anion vacancy defects, especially near the catalyst surface, and their marked effect on catalyst performance are reviewed. These vacancy defect structures may also be induced and perhaps stabilized by investigating catalyst/support interactions. These approaches may be crucial to creating advanced catalysts with higher productivities for this chemistry, as well as for other selective oxidations. Finally, it can be speculated that the stabilization of furan intermediate in the butane oxidation and its subsequent hydrogenation to THF (or THF in a single step) by overcoming the thermodynamically favorable MA production can have a huge economic impact.

ELECTRON MICROSCOPY IN ENVIRONMENTAL CONTROL CATALYSIS

In this section we review the contributions of TEM techniques in three areas at the core of environmental catalysis: three-way catalysis (TWC), De-NO_x catalysis, and catalysis for alternative, cleaner energy sources. These areas are, to a large extent, related, either through some of the reactions involved or through the formulation of the materials that are currently being employed or essayed in these catalytic processes. Thus although a TWC is actually the essential component of the emission control system of all the spark-ignited motor vehicles (42, 43), and De-NOx technology is well established for the after-treatment of nitrogen oxide emissions from stationary sources, future trends in engine concepts envisage these in combination. For instance, improved fuel economy and zero-CO_2 emissions will require the development of after-treatment catalysts for automobiles, which should guarantee the efficient elimination of nitrogen oxides under lean-burn conditions (44–46). Likewise, in a recent prospective review on automotive catalysis by Shelef & McCabe (47), the replacement of actual fossil fuels by hydrogen has been outlined as one of the ways to mitigate the contribution of vehicular transport to the accumulation of CO_2. Fuel cells and the combustion of methane from natural gas are pointed as the basis of these reduced CO_2-emission alternatives (46, 47). A very recent publication (48) also stresses the close connection between the areas mentioned previously.

Although TWC and De-NO_x processes for stationary sources are nowadays well-established technologies (2), further developments are still necessary to fulfill the more stringent emission limits of future legislation (i.e., EURO IV in 2005 and in the United States). For the TWC, extending the lifetime (mileage) of the catalytic converters up to 100,000 km is still a challenge that requires improvements in their deactivation behavior. Likewise, overcoming the problem of toxic emissions (mostly hydrocarbons) during the 30 s or so period of cold-start, in which the TWC catalyst temperature is not high enough to guarantee an efficient conversion of the pollutants in the exhaust gases, or the development of efficient catalysts for the reduction of nitrogen oxides under oxygen-rich (lean) conditions are some challenges that will likely require parallel developments from the materials side (47). Of course, this is also true for application in the field of catalysis for energy, in which interest in catalytic processes for in situ generation of H_2 for fuel cell applications has steeply risen recently.

As outlined above, the involvement of TEM in catalyst characterization and development can be categorized into three levels: (*a*) TEM is used as a routine tool to determine basic catalyst parameters, for example, TEM determination of particle size; (*b*) well established TEM techniques are exploited intensively to understand particular aspects of material chemistry problems in catalysis; and (*c*) novel EM techniques are applied to develop catalysts. These approaches are valuable in environmental catalysis, and below several examples are described. We comment on not only the most relevant results but also on suggestions about additional aspects of interest that could be potentially investigated by TEM.

It should be remembered that catalysts, owing to their usually complex formulation and their micro- or nanocrystalline nature, represent a bench-mark for most TEM techniques. In addition, TEM provides a unique view of these peculiar materials that most often is out of reach by any other macroscopic technique.

TEM in Three-Way Catalysis

According to Shelef & McCabe (47) the union of catalysts and the automobile has been one of the greatest successes of heterogeneous catalysis over the past 25 years. TWC allows the oxidation of CO and unburned hydrocarbons into H_2O and CO_2 and, simultaneously, the reduction of nitrogen oxides (NO_x) into N_2

(1) under air-to-fuel ratios (A/F) close to the stoichiometric point, A/F = 14.5 (17). The actual formulation of these catalysts include (47) noble metals (Pt, Pd, Rh), redox (oxygen storage), textural promoters ($CeO_2/Ce_xZr_{1-x}O_2$), an alumina support (γ-Al_2O_3), and some alumina stabilizers such as lanthana, silica, or baria (2, 44, 49).

The contributions of EM to basic and applied knowledge on these types of catalysts have been focused on the following topics: (*a*) influence of synthesis procedures on the nanostructure of the catalysts; (*b*) understanding redox behavior; (*c*) metal-support interaction effects in noble metal/cerium-based oxides; and (*d*) understanding deactivation phenomena. Although they have been cited separately, these topics are, to an extent, related to each other. It is also important to state that most of the TEM work has been performed on TWC model systems consisting of noble metals (NM = Pt, Rh, Pd) supported on a variety of cerium oxides including pure ceria, Ce-Zr mixed oxides, Ce-Ln (Ln = Ce,Pr) mixed oxides, or more complex ternary or quaternary mixed oxides.

NANOSTRUCTURE-SYNTHESIS CORRELATIONS AS SEEN BY TEM

In relation to synthesis, HRTEM has provided key information about the strong perturbation of the redox behavior of NM/Ceria systems associated with the use of chlorine-containing precursors (50–56). It has been suggested that the large amounts of chlorine incorporated into ceria or ceria-zirconia supports during impregnation and further activation in hydrogen result in the formation of a tetragonal-CeOCl phase, which grows in topotaxy with the ceria matrix. The stabilization of Ce^{3+} species in this crystalline phase perturbs and limits the extent of the Ce^{4+}/Ce^{3+} exchange process thus modifying the redox response of these materials (Figure 4).

Systematic approaches using TEM to establish the influence of synthesis procedures on the nanostructure of TWC or TWC model catalysts are lacking. In isolated contributions, TEM techniques are used to determine basic structural parameters such as the size of metal particles, the size of the oxygen storage or redox promoter crystallites, or some crystallographic features of the redox promoter component in Ce-Zr- and Ce-Tb-mixed oxides prepared using the microemulsion method (57–60).

Basic applications of TEM focused on the metal component can also be found in a number of papers. Bera et al. (61, 62) used TEM to investigate the effect of CeO_2 in promoting CO oxidation and NO reduction reactions on Pt/CeO_2 and Pd/CeO_2 catalysts and found an atomically dispersed metallic phase in close contact with the support. HRTEM has been applied (63) to study the effect of the support in the dispersion of the metal phase in Rh catalysts. From comparison of particle-size histograms established from HRTEM images of an $Rh/Ce_{0.5}Zr_{0.5}O_2$ and an Rh/Al_2O_3 catalyst, the authors suggest that the mixed ceria-zirconia supports

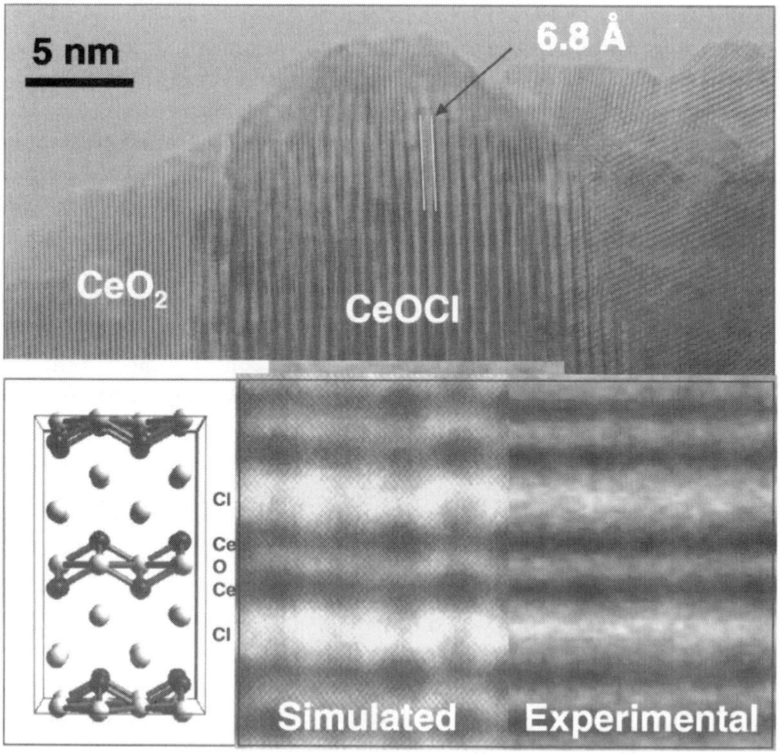

Figure 4 (*top*) HRTEM image recorded on a Rh/CeO_2 catalyst prepared from a chloride Rh precursor. A t-CeOCl crystal imaged down [010] is observed. Notice the parallel aligment with respect to the CeO_2 matrix. (*bottom*) [010]-CeOCl structure, [010]-CeOCl-simulated image and enlargement of the experimental image.

stabilize a highly dispersed state of the metallic component of the catalyst, one of the roles commonly attributed to this component of TWC catalysts.

TEM Studies of TWC Deactivation

One of the most valuable contributions of TEM in TWC has been in the deactivation of TWC catalysts. TWC catalysts work under harsh temperature and chemical environment conditions. A complex mixture of gaseous components whose redox character cyclically changes (at a frequency of about 10 Hz) between net oxidizing and net reducing, flows at a very high velocity over the catalyst at temperatures that can reach peaks of over 1000°C (2). Such redox and thermal stresses lead to severe modifications in the TWC converter.

The contribution of TEM to understanding the deactivation phenomenon refers to real as well as model TWC catalysts. Concerning the former, there is an interesting analytical EM study (BF-STEM and EDS analysis) on samples obtained from an aged catalyst removed from a 34,800 Km-driven car (64). Using TEM techniques the authors conclude that there were many different deactivation factors. According to their results, loss of the metal components (Pt, Pd), changes of the alumina carrier, and Pt-Pd alloying effects were negligible, whereas metal particle sintering and poisoning by P, Pb, and S were the most important deactivation factors. These authors highlighted the role of TEM as an indispensable tool for the diagnosis of deactivation of TWC and point to the importance of TEM for determining metal particle size in real TWC catalysts deactivated by poisoning. In these samples, conventional adsorption measurements failed to establish this parameter due to complications in the chemisorption behavior of common probe molecules.

In relation to deactivation by poisoning, the contribution of cerium sulfate (64), cerium phosphate (65), and Pb-Al_2O_3 compounds (42) formation have been studied using HRTEM and EDS analysis. Blockage of Ce^{4+}/Ce^{3+} redox exchange reaction by formation of stable Ce^{3+} compounds and modification of the texture and surface acid-base properties of alumina are the likely origins of such deactivation. From TEM observations, Graham et al. (66, 67) noted the occurrence of a peculiar deactivation effect: encapsulation of Pd particles within the matrix of the redox promoter (a $Ce_{0.5}Zr_{0.5}O_2$ mixed oxide with a surface area of 88 m^2/g in their studies). Aging at 1150°C in an atmosphere that mimics the exhaust gas conditions results in severe sintering of the Ce-Zr mixed oxide support (surface area drops down to 1 m^2/g). TEM, ED, and elemental EDS analyses, performed on cut samples, convincingly prove the presence of large (50 nm) Pd-containing particles buried within the structure of the collapsed support (67). About 85% of the metal loading becomes encapsulated. According to the authors (66), the encapsulated particles suffer a compressive lattice strain (in the range of 1–3%).

Sintering of the metal particles is undoubtedly one of the key factors that contributes to TWC deactivation (64, 68–72). Systematic HRTEM studies of sintering of Pt and Rh catalysts supported on ceria-based oxides, in a variety of oxidizing and reducing atmospheres, have been performed by Bernal et al. (73–79). These studies, shown in Figure 5*a*, reveal that in general Pt is much more prone to suffer sintering at high temperature under reducing environments than does Rh. Sintering phenomena becomes especially significant for both metals at temperatures above 700°C. Also interesting, in the case of Rh catalysts sintered at high reduction temperatures (900°C), is that oxidation treatments at temperatures above 800°C produce the redispersion of the large Rh particles (80, 81). According to these authors, Pt behaves quite differently in this respect. In contrast to Rh data, HRTEM, EDS, and SEM data indicate that after high-temperature oxidation treatments, Pt particles supported on ceria sinter heavily into micron-sized metallic aggregates (76, 81), which coexist with a small fraction of 2–3-nm-sized Pt particles. Regarding the behavior of Pd, redispersion effects have been also invoked (82) to explain the reactivation of a Pd/$Ce_{0.68}Zr_{0.32}O_2$ catalyst, aged under real conditions, after oxidation at high temperature (900°C). Others conclude that metal loading does not have any influence on the sintering behavior. However, the transition of

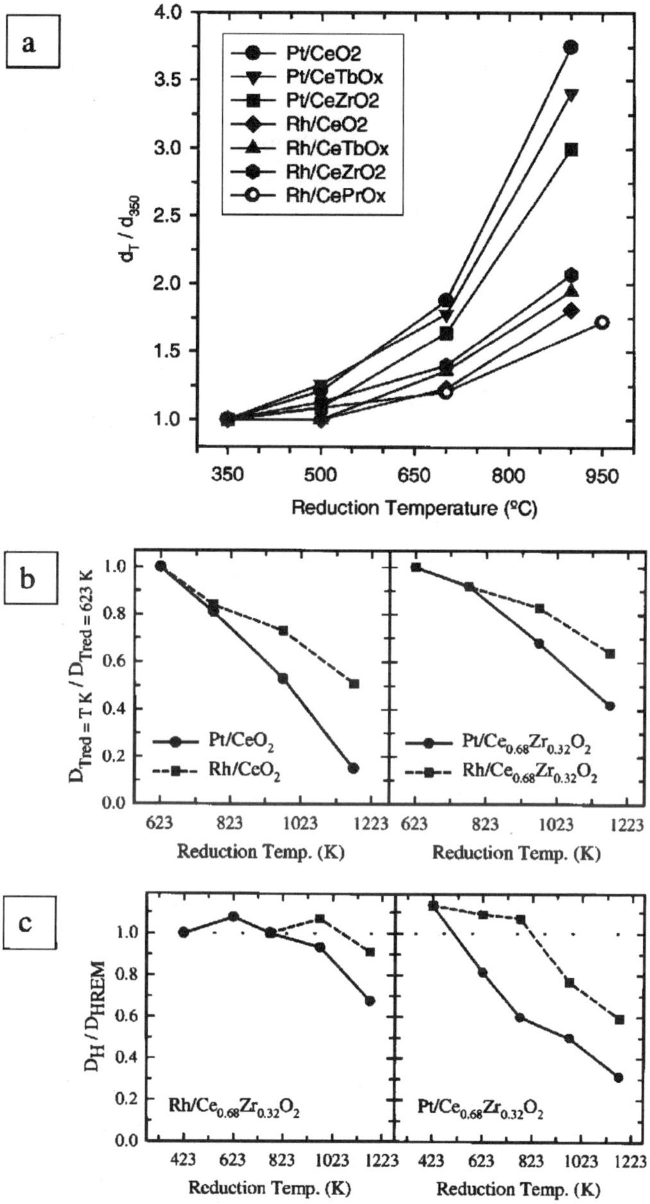

Figure 5 (*a*) Evolution of average particle size as a function of reduction temperature (T_{red}) for a series of Rh and Pt ceria-based catalysts. Data are referred to the average value at the lowest T_{red}. (*b*) Evolution of dispersion, as determined from HRTEM, as a function of T_{red} for two Rh and Pt catalysts. (*c*) Evolution with T_{red} of dispersion values determined from HRTEM (D_{HREM}) and from volumetric H_2-chemisorption experiments (D_H) for an Rh and a Pt catalyst supported on a ceriz-zirconia-mixed oxide.

alumina from the θ to the α polymorph seems to enhance the growth of the metal particles.

In a recent paper (83), where aging under real driving conditions and laboratory-scale aging tests under hydrothermal (reducing and oxidizing) conditions up to 1200°C were compared, a good correlation was found. This study also revealed that sintering is the primary deactivation factor before poisoning starts to be important. HRTEM has also become an indispensable tool to reliably estimate the dispersion of the metallic phase in TWC model catalysts, a key parameter to understand their performance (74, 84), as well as particle size distribution histograms.

Figure 5*b* plots a comparative HRTEM study of deactivation (loss of dispersion) behavior of Pt and Rh catalysts supported on CeO_2 and $Ce_{0.68}Zr_{0.32}O_2$ as a function of reduction temperature. Note how Pt deactivates more severely than Rh on both supports in terms of dispersion (54). Likewise, Figure 5*c* illustrates the effect of using (simultaneously) dispersion values determined from HRTEM (D_{HREM}) and chemisorption (D_H). Data plotted in this figure, which represent the variation of the ratio of these dispersion values as a function of reduction temperature, have been used to detect the on-set temperatures of metal-support interaction effects in Pt and Rh catalysts supported on ceria-zirconia-mixed oxides. According to these data, this temperature is much lower in the Pt catalyst. Moreover, these data also suggest that reversibility of these interaction effects is higher in the case of the Rh catalysts.

Understanding Metal-Support Interaction Effects

Metal-support interaction effects are a crucial factor for understanding the chemical and catalytic behavior of metal catalysts supported on ceria-based oxides because of the reducibility of these oxides. Metal-support interaction effects in the catalytic performance of noble metals (Pt, Rh, Pd) with ceria and ceria-mixed oxides in a wide range of reduction temperatures (200–900°C) have been the focus of a number of TEM studies (10, 54, 56, 57, 78–80, 85, 86–105). The epitaxial growth of the supported metal nanoparticles on the surface of the oxide supports is generally observed. Experimental and simulated HRTEM profile and plan-view images indicate that this growth takes place under two different kinds of orientation relationships: parallel and twined, Figure 6*a*. This is the most remarkable nanostructural feature of these systems after reduction at temperatures ranging from 200°C up to 700°C. Between 700–900°, reduced support moieties mobilize and migrate over the surface of the supported nanoparticles, which become "decorated" by crystalline support overlayers, Figure 6*b*. This decoration phenomenon has also been observed by HRTEM in other systems (12, 21, 22, 106). Recently, the role of these ceria overlayers in modifying CO adsorption, as well as in reactions involving CO or small hydrocarbons, has been investigated using model inverse systems (97, 104, 107). In the very-high-reduction temperature regime, 900°C or above, interaction between the metals and the support introduces more severe modifications (Figure 6*c*) and redispersion effects in supported Rh (Figure 6*d*), under oxidizing conditions. Thus alloying of the metals with the lanthanide

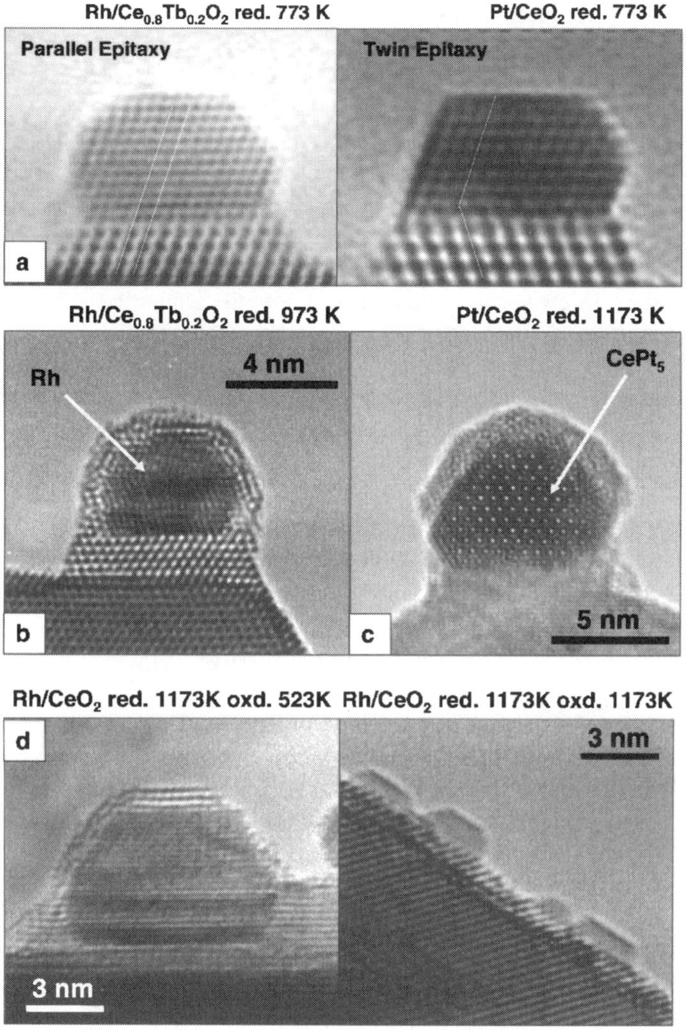

Figure 6 Evolution of metal-support interaction effects, as revealed by HRTEM: (*a*) metal-support epitaxy; (*b*) decoration of metal particles; (*c*) Pt-Ce alloying; (*d*) reversibility of metal-support decoration effects (*left*). Redispersion of Rh into small, surface-clean, particles after oxidation at a high temperature of 1173 K (*right*).

elements present in the support (Ce in the case of pure ceria supports or Ce and Tb in the case of mixed Ce-Tb oxides) has been unambiguously detected by HRTEM and EELS. In the case of Pt, an $LnPt_5$ intermetallic (Ln = Ce or Ce,Tb) phase has been identified on the basis of ED and experimental and simulated HRTEM images, as well as by measuring directly the composition of supported nanoparticles in EELS spectra.

In the case of NM supported on CeO_2-ZrO_2 mixed-oxides, appreciable differences are observed after reduction at the highest temperatures. Thus the extent of the decoration of Rh, Pt and Pd nanoparticles in samples reduced at 900°C is negligible compared with that observed on the catalysts based either on pure ceria or in mixed oxides containing two reducible lanthanides (Ce-Tb-mixed oxides). Likewise, the formation of intermetallic compounds has not been observed in Pt catalysts supported on Ce-Zr mixed oxides, whereas they have been clearly detected in Pt/$CeTbO_x$ systems. Therefore, the incorporation of zirconium into the ceria lattice seems to modulate the mobility of ceria, allowing ceria both to migrate over the particle surface and to diffuse into the fcc lattice of Pt.

In summary, the degree of metal-support interaction in NM/ceria catalysts increases with reduction temperature, going from purely structural interactions, through electronic interactions in the case of Pt, to metal-support interface rearrangements, and finally up to interactions that involve solid-state reactions between the supported phase and the support. TEM is a unique tool for unveiling the nanostructural features of metal-support interaction effects in TWC model catalysts, following their evolution in reducing and oxidizing environments as a function of temperature and, finally, indicating the precise temperature windows at which each type of manifestation appears or is removed.

Contribution of TEM to the Understanding of Redox Behavior of TWC

One of the major roles of the ceria-based oxides added as redox promoters in TWC formulations is to dump the fluctuations of A/F ratios during the catalyst operation. By keeping this parameter as close as possible to the stoichiometric value, 14.5, the simultaneous elimination of the different pollutants present in the exhaust gases becomes optimum (2). Buffering the oscillations of the A/F is possible through the Ce^{4+}/Ce^{3+} redox exchange. This change in the oxidation state of cerium is accompanied by the creation/refilling of oxygen vacancies in the structure of the oxide. Among the roles attributed to the ceria oxides, we should also mention the promotion of CO conversion into CO_2 through the so-called water gas shift reaction (WGSR): $CO + H_2O \rightarrow CO_2 + H_2$. Such a promotional effect is also linked to the redox properties of these oxides.

The incorporation of zirconia or other reducible lanthanide elements into the ceria lattice significantly improves its redox behavior, which is one reason for substituting pure ceria with ceria-zirconia mixed oxides in the modern formulations of TWCs. Redox aging at high temperatures strongly modifies the redox properties of ceria-zirconia oxides. In the same way, the presence of a supported metal phase modifies the redox response of these oxides in a synergistic way, especially at low temperatures.

Structural and spectroscopic information available at atomic scale using TEM techniques provides key information for investigating this facet of TWC materials. Moreover, the recent achievement of atomic-resolution in situ TEM has opened the extraordinary possibility of investigating the state of catalytic materials under conditions that avoid any alteration owing to their interaction with the oxidants present in atmospheric air, as well as following the dynamics of redox processes (5, 12, 48, 109, 110). This is an indispensable requirement for characterizing air-sensitive materials, as is the case for all NM/ceria catalysts.

Published TEM investigations of redox properties of TWC model catalysts encompass several topics: (*a*) monitoring changes in the oxidation state of the elements as a function of the chemical environment (110, 111); (*b*) understanding the way in which the oxygen vacancies are assimilated by the oxide (111); (*c*) elucidating the role of doping with other elements or the supported phase on the enhancement of the redox properties (112, 113); (*d*) unveiling mechanisms of modification of the redox response after high temperature redox aging (114–118); and (*e*) understanding the oxidation of the supported metal phases (80, 119–125).

The combined use of in situ EELS and in situ atomic resolution imaging of a $Rh/Ce_{0.8}Pr_{0.2}O_2$ catalyst under 5 torr H_2 pressure has been demonstrated as a function of temperature (25–900°C) (111). In situ HRTEM images obtained simultaneously with these spectra indicate that vacancies formed at low temperatures, associated to the reduction of Pr^{4+}, did not order into superstructures, and ordering was observed once only when Ce^{4+} began to become reduced. A π type phase, one of the members in the homologous series of reduced phases of the higher rare-earth oxides, has been detected in the mixed Ce-Pr oxide. Detailed analysis of HRTEM and in situ EELS spectra has allowed Sharma et al. to make semiquantitative estimations of the reduction degree in Ce-Zr mixed oxides as a function of reduction temperature and structural changes in Pr-Tb mixed oxides (110, 126, 127). Concerning the redox behavior of Ce-Zr-mixed oxides, HRTEM evidence has contributed to prove the formation of a twofold superstructure in these oxides after redox aging at high temperature (116–119). Such a superstructure could be on the basis of the enhanced oxygen storage and handling capabilities observed after such aging treatments (128, 129).

Detailed HRTEM studies are currently aimed at investigating the nanostructural changes of the supported metal phase upon treatments in redox cycles. The presence of a supported phase greatly modifies the redox response of the support. Hence, understanding the nanostructure of the metal phase becomes a requisite to characterize, in detail, the redox response of the whole system.

For Rh particles supported on silica, HRTEM evidence of the restructuring of large metallic particles, in the range of 5–10 nm in size, have been reported. Concerning the oxidation of Pd, complex nanostructural changes during the PdO $\leftrightarrow$ Pd transformation have been revealed by the HRTEM images, which are used to rationalize the activity change pattern characteristic of Pd in complete oxidation reactions (124, 125). It is also worth highlighting, from the possibilities offered by TEM techniques in catalysts characterization, that the electron holography results obtained by Allard et al. (130) show evidence of the formation of internal voids at the core of Pd particles submitted to oxidation-reduction cycles.

TEM of $DeNO_x$ Catalysts

Nowadays $DeNO_x$ catalysis encompasses the abatement of nitrogen oxide present in burnt gases emitted from stationary sources (power plants, chemical plants) and the control of NO_x emission from automobiles working under oxygen-rich (lean) conditions. The catalytic technology in the former case, which is currently well established, is based on V_2O_5 (MoO_3-WO_3)/TiO_2 formulations (2, 131). On these catalysts, the selective catalytic reduction (SCR) of nitrogen oxides using ammonia as reductant allows their total elimination in the presence of a large excess of oxygen.

In the case of automobiles, to date a large number (literally hundreds) of catalysts has been studied, but the ideal solution has not yet been found. SCR of NO_x using different types of reductants (ammonia, urea, hydrocarbons) have been assayed on two kinds of materials: metal/zeolites and metal/alumina. A suitable solution of using a NO_x storage system (NO_x-trap) combined with a TWC catalyst has also been reported. In this case, during the NO_x storage lean cycle, NO is oxidized into NO_2 on the TWC catalyst at the same time that residual CO and HC are oxidized and stored as a nitrate on a barium-containing compound. Then, for a few seconds, the engine is operated under rich conditions, which allows the reduction of NO_x into N_2 and the regeneration of the trap.

Although TEM has not been used as intensively on these materials as has TWC, relevant contributions to understanding their performance and deactivation can be found in recent literature dealing with V_2O_5/TiO_2 systems (132–136), M/Al_2O_3 [M = Ag (137–139), Pd (140, 141), Cu (142), Pt (143), Re (144, 145)], and M/zeolite (146–149) catalysts.

TEM of V_2O_5/TiO_2 Catalysts and Related Systems

The influence of the synthesis procedure on the nanostructure of V_2O_5/TiO_2-type catalysts has been investigated (134, 136). Reddy et al. (134), using HRTEM and EDS, studied the promoting effect of ZrO_2 in a V_2O_5/ZrO_2-TiO_2 catalyst and the role of calcination temperature on the structure of this catalyst. The transition from an amorphous TiO_2-ZrO_2 support to a crystalline ZrV_2O_7 compound, as the V-containing phase, dispersed on the surface of the TiO_2-ZrO_2 support, is detected as the calcination temperature is raised from 500°C up to 800°C. A critical 4% V_2O_5 loading is reported to establish the nature of vanadia-titania interaction in catalysts, which were prepared using the sol-gel route (136). For loading below this limit, HRTEM images suggest that vanadia exists in the form of titania-coordinated surface species, whereas for loading higher than 4%, three-dimensional hexagonal V_2O_5 crystallites, which are less active, are detected. The authors also claim that important changes in selectivity to N_2 in the SCR of NO by propane are linked to such nanostructural changes.

Deactivation phenomena have been reported (132, 133). EDS-TEM studies point out the interest in these techniques as deactivation diagnosis tools in the

V_2O_5-WO_3-TiO_2 SCR European standard catalyst (EUROCAT) (132). After aging in real conditions, important changes in the distribution of V and W were detected that led to the formation of particles. The structure of the vanadia-tungsta surface coating suffers an amorphization process, and sulfur is detected as the major poison. EDS studies, performed on commercial SCR catalysts and reported in Reference (133), indicate that HCl also has some detrimental impact on V, facilitating its loss by formation of volatile species.

Dynamic redox phenomena and evolution of shear structures in vanadium pentoxide catalysts using in situ ETEM have been reported by Gai (150). EELS studies of the oxidation states of V in its oxide compounds have been reported that help to explain the catalytic performance and deactivation routes (151–154). The exceptional increase in energy resolution facilitated by the use of gun monochromators and improved, aberration-corrected-EELS spectrometers enables the EELS spectra to be brought down to about 0.1 eV, thus allowing fruitful comparative studies between experimental EELS spectra and theoretical electronic states obtained from band-structure calculation. Therefore, high resolution EELS will play an important role in the characterization of this family of catalysts in the near future.

TEM Characterization of M/Al_2O_3 De-NO_x Catalysts

Concerning the nature of the active phase in M/Al_2O_3 catalysts, Reference (142) provides an example of the simultaneous application of HAADF and EELS to characterize the spatial distribution of oxidation states of the metallic component. The presence of Cu in different oxidation states on the surface and bulk of the Cu particles, even after reduction at 800°C, is evidenced in a Cu/γ-Al_2O_3 catalyst and attributed to the metal-support interaction effect. HRTEM is used to elucidate the formation of a new La-Al_2O_3 crystalline phase in Pd/Al_2O_3-La_2O_3 catalysts (140) and the redispersion role of Re in high-temperature oxidation treatments of Re/γ-Al_2O_3 catalysts (144, 145).

TEM Characterization of M/Zeolite De-NO_x Catalysts

Both the size and shape of the metallic component and its precise spatial distribution are essential parameters in the case of metal/zeolite systems. The location of the supported metal phase on the outer surface of the zeolite, or within its pore structure, can dramatically influence the intrinsic properties of the metal particles, due to a confinement effect, and also the course of the reaction within the pores. This is, therefore, an important parameter to be disclosed in this particular kind of material. For such a purpose, electron tomography becomes an indispensable tool (156). The first demonstration of three-dimensional characterization of catalysts, using bright-field (BF) tomography, is from Koster et al. (157). From a three-dimensional TEM reconstruction of an Ag/NaY catalyst, they conclude convincingly the presence of larger silver particles outside the channel structure and small ones (about 18 nm) inside the pores. With the same approach, they obtain precise information on the characteristics of pore structure and connectivity of the pore system in zeolite Y (158). Also in an inspiring work, a detailed localization of silver particles and ZrO_2 in SBA-15 zeolite is carried out (159). From the BF-tomography data, an interpretation of the hysteresis effects observed in the N_2 desorption isotherms of this material is proposed.

More recently, HAADF STEM tomography has been introduced (160) and applied to catalyst characterization (161, 162). The direct Z dependence of this imaging mode allows fulfillment of the projection requirement of tomography. Using this new approach, partitioning of Pd-Ru bimetallic particles between the outer and inner surfaces of a MCM-41 zeolite has been nicely determined. Accordingly, the presence of nanoparticles at the zeptogram level of quantity can be imaged within mesoporous hosts (162). Backscattered imaging in a low-voltage SEM is shown to yield images of catalyst nanoparticles similar to STEM-HAADF images, which has economical advantages (5, 14, 163). HAADF imaging, in combination with EDS elemental mapping, has also proved of interest in unveiling deactivation mechanisms of a Pt/mordenite catalyst, under real operating conditions, in the SCR of NO_x-using ammonia (135). Analytical data evidence in this case is the preferential association of S with the Pt particles.

Finally, we will mention the results presented in Reference (146). There, HRTEM is used to characterize the geometrical arrangement of pores in a mesoporous Tin(IV) phosphate. The phosphate, with a hexagonal arrangement of the pores, appears more active than that which has a square distribution of the pores. HRTEM also allows the authors to draw conclusions about the amorphous nature of the pore walls, whose composition and Sn/P ratio were established from EDS measurements.

In summary, as shown with different examples, the array of TEM techniques has provided indispensable tools to understand synthesis, activation, performance, and deactivation of the different families of materials nowadays involved in De-NO_x catalysis.

THE ROLE OF TEM IN THE CHARACTERIZATION OF CATALYSTS FOR ALTERNATIVE ENERGY SOURCES

Reducing dependence on fossil fuels, diversifying energy sources, and reducing the environmental impact of current energy-producing processes are major targets of research in the field of catalysis for energy. A variety of processes are currently being investigated, among which those related to the production of H_2 to feed fuel cells are most in focus. Reforming and partial oxidation of different raw chemicals as CH_4, methanol, or ethanol provide suitable ways to obtain H_2 (164). Combustion of this H_2 in a fuel cell would allow the transformation of chemical energy into electrical energy (165). Likewise, gasification of biomass into syngas ($CO + H_2$ mixture) seems a viable alternative (166, 167).

Most of the TEM work related to catalysis for energy deals with the nanoscale characterization and development of materials (*a*) as catalysts in the electrodes

HRTEM has revealed (183) the occurrence of surface reconstruction phenomena during CO oxidation.

Especially remarkable are the TEM studies conducted in a series of papers by Akita et al. (189, 200–203) on Au/TiO_2 catalysts. HRTEM performed by these authors has revealed the presence of epitaxial growth of the gold particles on top of the titania polymorphs (189, 202). ADF-STEM images and EELS data also reveal the preferential deposition of gold on the surface of the rutile poloymorph (189). The systematic combined application of HRTEM and electron holography used by these authors to understand the influence of particle size on the morphology of the gold particles and on the features of the Au-titania interface is inspiring. Changes in the contact angle between the metal and the support are determined to take place at particle sizes about 2 nm. The mean inner potential of the small Au particles, which is another parameter determined by these authors from the electron holograms, shows significant deviations from the values corresponding to bulk gold, a fact that could really be at the origin of the peculiar catalytic behavior of the small Au particles. Moreover, changes with size of this mean inner potential are measured, which can be taken as an indication of the occurrence of charge transfer effects, i.e., of metal-support interaction effects (200, 201, 203). These papers highlight the tremendous potential of the HRTEM-electron holography in tandem in catalysts characterization.

CONCLUSION

Herein we have commented on a significant numbers of papers dealing with the characterization and development of a large variety of catalytic materials using electron microscopy methods. A fully comprehensive survey of the use of EM techniques in such a widespread collection of catalysts has not been our main target. We have tried, instead, to present a collection of papers that adequately illustrate and represent the full potential of this large family of techniques in understanding and developing the catalysis of hydrocarbon oxidation, environmental protection, and energy.

A wide range of information is available because of electron microscopy, which opens a unique window to the intricate field of the structure and its relationship to reactivity of complex catalytic solids. Such information has established a guiding light in the synthesis of new catalysts with improved performances and the development of new processes and has provided a key to understanding the function and deactivation of catalysts.

ACKNOWLEDGMENTS

We thank L.G. Hanna, F. Gooding, and L. Wang of DuPont (CRD) for technical support, and K. Kourtakis (DuPont), S. Bernal (Cadiz), and A.J. Pérez-Omil (Cadiz) for helpful discussions.

The *Annual Review of Materials Research* is online at http://matsci.annualreviews.org

LITERATURE CITED

1. Thomas JM, Thomas WJ. 1997. *Principles and Practice of Heterogeneous Catalysis*. Weinheim, Ger.:VCH
2. Ertl G, Knözinger H, Weitkamp J. 2004. *Handbook of Heterogeneous Catalysis*. 4:559–696. Weinheim, Ger.: Wiley-VCH
3. Hirsch PB, Howie A, Nicholson RB, Pashley DW, Whelan MJ. 1977. *Electron Microscopy of Thin Crystals*. New York: Krieger
4. Williams DB, Carter CB. 1996. *Principles of Electron Microscopy*. New York: Plenum
5. Gai PL, Boyes ED. 2003. *Electron Microscopy Heterogeneous Catalysis*. Bristol/Philadelphia: Inst. Phys. Publ.
6. Zeitler E, ed. 1990. Monographic issue: characterization of catalysts. *Ultramicroscopy* 34:1–128
7. Gai PL. 1992. Defects in oxide catalysts: fundamental studies of catalysis in action. *Catal. Rev. Sci. Eng.* 34:1–54
8. Datye AK, Smith DJ. 1992. The study of heterogeneous catalysis by HRTEM. *Catal. Rev. Sci. Eng.* 34:129–78
9. Ledoux MJ, ed. 1995. Monographic issue: HREM for catalysis. *Catal. Today* 23:161–298
10. Bernal S, Baker RT, Burrows A, Calvino JJ, Kiely CJ, et al. 2000. Structure of highly dispersed metals and oxides: exploring the capabilities of high-resolution electron microscopy. *Surf. Interface Anal.* 29:411–21
11. Thomas JM, Terasaki O, Gai PL, Zhou W, Gonzalez-Colbet J. 2001. Structural elucidation of mesoporous and microporoues solids using HRTEM. *Acc. Chem. Res.* 34:583–93
12. Gai PL. 2002. Development of environmental electron microscopy and applications to catalysis. *Top. Catal.* 21:161–74
13. Thomas JM, Terasaki O. 2002. The electron microscope is an indispensable instrument for the characterization of catalysts. *Top. Catal.* 21:155–59
14. Midgley PA, Weyland M, Thomas JM, Boyes ED, Gai PL. 2002. Electron tomography. *Angew. Chem. Int. Ed.* 41:20–25
15. Datye AK. 2003. Electron microscopy of catalysts. *J. Catal.* 216:144–54
16. Thomas JM, Gai PL. 2004. Electron microscopy and the materials chemistry of catalysts. *Adv. Catal.* 48:171–227
17. Thomas JM, Midgley PA. 2004. High-resolution transmission electron microscopy: the ultimate nanoanalytical technique. *Chem. Commun.* 253–67
18. Gai PL. 1997. A new structural transformation mechanism in catalytic oxides. *Acta Cryst. B* 53:346
19. Boyes ED, Gai PL. 1997. Environmental high resolution microscopy and applications to chemical science. *Ultramicroscopy* 67:219–32
20. Gai PL, Kourtakis K. 1995. Solid state defect mechanism in vanadyl pyrophosphate catalysts: implications for selective oxidation. *Science* 267:661–63
21. Gai PL. 1998. Direct probing of gas molecules-solid catalyst reactions at the atomic scale. *Adv. Mater.* 10:1259–63
22. Gai PL. 1999. Environmental high resolution electron microscopy of gas-catalyst reactions.. *Top. Catal.* 8:97–114

22a. Gai PL. 1999. Probing selective oxidation catalysis under reaction conditions by atomic scale environmental high resolution electron microscopy. *Curr. Opin. Solid State Mater. Sci.* 4:63–73

22b. Gai PL. 2001. Development of electron microscopy methods in the study of

catalysts. *Curr. Opin. Solid State Mater. Sci.* 5:371–80
23. Haggin J. 1995. Catalysis at the atomic level. *C&E News* 73:39–41
24. Jacoby M. 2002. EM in catalysis. *C&E News* 80:26
25. Gai PL. 2002. Development of wet-ETEM in catalysis. *Microsc. Microanal.* 8:21–28
26. Gai PL, Kourtakis K, Boyes E. 2005. In situ wet nanoscale imaging of nitriles in a solution phase: novel hydrogenation chemistry through nanocatalysts on nano-supports. *Catal. Lett.* 102:(1–2). In press
27. Crewe A. 2000. STEM. *Rev. Sci. Instr.* 42:411
28. Boyes ED. 1998. High resolution, low voltage SEM. *Adv. Mater.* 10:1277–80
29. Boyes ED. 2001. High resolution low voltage SEM and analysis. *Inst. of Phys. Conf. Ser.* 168:115–17
30. Boyes ED, Ringnalda J, van der Stam MA, Fliervoet TF, Van Cappellen E. 2001. A 2-2-2 200 kV field emission STEM/TEM system. *Microsc. Microanal.* 7:232–33
31. Midgley PA, Thomas JM, Laffont L, Weyland M, Raja R, et al. 2004. Electron tomography. *J. Phys. Chem. B* 108:4590–92
32. Centi G, ed. 1993. Monographic issue: butane catalysis. *Catal. Today* 16:1–193
33. Sananes MT, Hutchings GJ, Volta J. 1995. VPO catalysts. *J. Catal.* 154:253–60
34. Bordes E. 1993. V-P-O systems. *Catal. Today* 16:27–42
35. Gai PL, Kourtakis K, Coulson R, Jacobsen I. 1997. HRTEM of butane oxidation catalysis and correlation with reaction chemistry. *J. Phys. Chem.* 101:9916–25
36. Kourtakis K, Gai PL. 2004. Novel microstructures and reactivity in n-butane oxidation: advances and challenges. *J. Mol. Catal.* 220:93
37. Bartley JK. 2001. Vanadium phosphate prepared using solvent-free method. *Catal. Lett.* 3:99–105
38. Haber J, Zazhigalov V, Stoch J, Bogutskaya L, Batcherikova I. 1997. Mechanochemistry: the activation method of VPO catalysts for n-butane partial oxidation. *Catal. Today* 33:39–49
39. Hutchings GJ, Sananes MT, Sajip A, Kiely CJ, Burrows A, et al. 1997. Improved methods of preparation of vanadium phosphate catalysts. *Catal. Today* 33:161–71
40. Hutchings GJ, Higgins R. 1996. Promoted VPO catalysis. *J. Catal.* 162:153–60
41. Irusta S, Boix A, Pierini B, Caspani C, Petunchi J. 1999. Catalysis. *J. Catal.* 187:298–302
42. Magnacca G, Morterra C. 2000. Surface characterization of modified aluminas. *Phys. Chem. Chem. Phys.* 2:3903–8
43. Sung S, Smaling RM, Brungard NL. 1998. Pb poisoning on Pd-only TWC catalysts. *Stud. Surf. Sci. Catal.* 116:165–74
44. Farrauto RJ, Heck RM. 1999. Catalytic converters: state of the art and perspectives. *Catal. Today* 51:351–60
45. Gandhi HS, Graham GW, McCabe RW. 2003. Automotive exhaust catalysis. *J. Catal.* 216:433–42
46. Pishinger S. 2004. The future of vehicle propulsion—combustion engines and alternatives. *Top. Catal.* 30/31:5–16
47. Shelef M, McCabe RW. 2000. Twenty-five years after introduction of automotive catalysts: what next? *Catal. Today* 62:35–50
48. Wu JCS, Wan B-ZE. 2004. Catalysis for new energy resources and environmental protection. *Catal. Today* 97:93–215
49. Kaspar J, Fornasiero P, Hickey N. 2003. Automotive catalytic converters: current status and some perspectives. *Catal. Today* 77:419–28
50. Fornasiero P, Hickey N, Kaspar J, Dossi C, Gava D, Graziani M. 2000. Redox and chemisorptive properties of ex-chloride and ex-nitrate $Rh/Ce_{0.6}Zr_{0.4}O_2$ catalysts: 1. Effect of low-temperature redox cycling. *J. Catal.* 189:326–38
51. Fornasiero P, Hickey N, Kaspar J, Montini T, Graziani M. 2000. Redox and chemisorptive properties of ex-chloride and ex-nitrate $Rh/Ce_{0.6}Zr_{0.4}O_2$ catalysts:

2. Effect of high-temperature redox cycling. *J. Catal.* 189:339–48
52. Trovarelli A, Dolcetti G, Leitenburg C, Kaspar J, Finetti P, Santoni A. 1992. Rhodium-ceria interaction induced by high-temperature reduction: characterization and catalytic behavior in transient and continuous conditions. *J. Chem. Soc. Faraday Trans.* 88:1311–14
53. Le Normand F, Hilarie L, Kili K, Krill G, Maire G. 1988. Catalytic reactions. *J. Phys. Chem.* 92:2561–68
54. Bernal S, Botana FJ, Calvino JJ, Cauqui MA, Cifredo GA, et al. 1993. Microstructural and chemical-properties of ceria-supported rhodium catalysts reduced at 773 K. *J. Phys. Chem.* 97:4118–23
55. Bernal S, Botana FJ, Calvino JJ, Cifredo GA, Omil JAP. 1993. HREM study of the phases resulting from the preparation of ceria supported rhodium catalysts. *Electron Microsc. Anal.* 1993:485–88
56. Kepinski L, Okal J. 2000. Occurrence and mechanism of formation of CeOCl in Pd/CeO_2 catalysts. *J. Catal.* 192:48–53
57. Hungria AB, Martinez-Arias A, Fernandez-Garcia M, Iglesias-Juez A, Guerrero-Ruiz A, et al. 2003. Structural, morphological, and oxygen handling properties of nanosized cerium-terbium mixed oxides prepared by microemulsion. *Chem. Mater.* 15:4309–16
58. Martinez-Arias A, Fernandez-Garcia M, Iglesias-Juez A, Hungria AB, Anderson JA, et al. 2001. New Pd/Ce_xZr_{1-x} O_2/Al_2O_3 three-way catalysts prepared by microemulsion: Part 2. In situ analysis of CO oxidation and NO reduction under stoichiometric CO+NO+O_2. *Appl. Catal. B* 31:51–60
59. Hungria AB, Iglesias-Juez A, Martinez-Arias A, Fernandez-Garcia M, Anderson JA, et al. 2002. Effects of copper on the catalytic properties of bimetallic Pd-Cu/(Ce,Zr)Ox/Al_2O_3 and Pd-Cu/(Ce,Zr)O_x catalysts for CO and NO elimination. *J. Catal.* 206:281–94
60. Martinez-Arias A, Fernandez-Garcia M, Salamanca LN, Valenzuela RX. 2000. Structural and redox properties of ceria in alumina-supported ceria catalyst supports. *J. Phys. Chem. B* 104:4038–46
61. Bera P, Patil KC, Jayaram V, Subbanna GN, Hegde MS. 2000. Ionic dispersion of Pt and Pd on CeO_2 by combustion method: effect of metal-ceria interaction on catalytic activities for NO reduction and CO and hydrocarbon oxidation. *J. Catal.* 196:293–301
62. Bera P, Gayen A, Hegde MS, Lalla NP, Spadaro L, et al. 2003. Promo effect of CeO_2 in combustion synthesis Pt/CeO_2 catalyst for CO oxidation. *J. Phys. Chem. B* 107:6122–30
63. Vlaic G, Fornasiero P, Martra G, Fonda E, Kaspar J, et al. 2000. Morphology of rhodium particles in ex-chloride Rh/$Ce_{0.5}Zr_{0.5}O_2$ catalyst. *J. Catal.* 190: 182–90
64. Herz RK, Shlnouskls EJ. 1985. Application of high-resolution analytical electron microscopy to the analysis of automotive catalysts. *Ind. Eng. Chem. Prod. Res.* 24:6–10
65. Larese C, Galisteo FC, Granados ML, Mariscal R, Fierro JLG, et al. 2003. Deactivation of real three way catalysts by $CePO_4$ formation. *Appl. Catal. B* 40:305–17
66. Graham GW, O'Neill AE, Uy D, Weber WH, Sun H, Pan XQ. 2002. Observation of strained PdO in an aged Pd/ceria-zirconia catalyst. *Catal. Lett.* 79:99–105
67. Jiang JC, Pan XQ, Graham GW, McCabe RW, Shwank J. 2004. Microstructure of a Pd/ceria-zirconia catalyst after high-temperature aging. *Catal. Lett.* 53:37–42
68. Burch R. 1997. Low NOx options in catalytic combustion and emission control. *Catal. Today* 35:27
69. Graham GW, Jen HW, Chun W, McCabe RW. 1997. Encapsulation of Pd particles by ceria-zirconia mixed oxides. *Catal. Lett.* 44:185
70. Neyestanaki AK, Klingstedt F, Salmi T, Murzin DY. 2004. Deactivation of

postcombustion catalysts, a review. *Fuel* 83:395–408

71. Skinjoh H, Muraki H, Fujitani Y. 1991. Effect of severe thermal aging on noble metal catalysts. In *Catalysis and Automotive Control II*, ed. A Cruq, p. 617. Amsterdam: Elsevier
72. Mabilon G, Durand D, Prigent M. 1991. Kinetics of the physico chemical and catalytic activity evolution of a platinum-rhodium catalyst in automotive exhaust gas. In *Catalysis and Automotive Control II*, ed. A Cruq, p. 569. Amsterdam: Elsevier
73. Bernal S, Calvino JJ, Cauqui MA, Gatica JM, Larese C, et al. 1999. Some recent results on metal/support interaction effects in NM/CeO_2 (NM: noble metal) catalysts. *Catal. Today* 50:175–206
74. Bernal S, Calvino JJ, Cauqui MA, Omil JAP, Pintado JM, Rodriguez-Izquierdo JM. 1998. Image simulation and experimental HREM study of the metal dispersion in Rh/Ce O_2 catalysts. Influence of the reduction/reoxidation conditions. *Appl. Catal. B* 16:127–38
75. Bernal S, Blanco G, Calvino JJ, Gatica JM, Omil JAP, Pintado JM. 2004. Characterisation of three-way automotive aftertreatment catalysts and related model systems. *Top. Catal.* 28:31–45
76. Bernal S, Cauqui MA, Cifredo GA, Gatica JM, Larese C, Omil JAP. 1996. Chemical and microstructural investigation of Pt/Ce O_2 catalysts reduced at temperatures ranging from 473 to 973 K. *Catal. Today* 29:77–81
77. Bernal S, Calvino JJ, Gatica JM, Lopez-Cartes C, Pintado JM. 2002. Chemical and nano-structural aspects of the preparation and characterisation of ceria and ceria-based mixed oxide-supported metal catalysts. In *Catalysis by Ceria and Related Materials*, ed. A Trovarelli, pp. 85–168. London: Imperial College Press
78. Gatica JM, Baker RT, Fornasiero P, Bernal S, Blanco G, Kaspar J. 2000. Rhodium dispersion in a Rh/$Ce_{0.68}Zr_{0.32}O_2$ catalyst investigated by HRTEM and H-2 chemisorption. *J. Phys. Chem. B* 104:4667–72
79. Gatica JM, Baker RT, Fornasiero P, Bernal S, Kaspar J. 2001. Characterization of the metal phase in NM/$Ce_{0.68}Zr_{0.32}O_2$ (NM: Pt and Pd) catalysts by hydrogen chemisorption and HRTEM microscopy: a comparative study. *J. Phys. Chem. B* 105:1191–99
80. Bernal S, Botana FJ, Calvino JJ, Cifredo GA, Perez Omil JA. 1995. HREM study of the behavior of a Rh/CeO_2 catalyst under high-temperature reducing and oxidizing conditions. *Catal. Today* 23:219–50
81. Bernal S, Blanco G, Calvino JJ, Lopez-Cartes C, Perez-Omil JA, et al. 2001. Electron microscopy (HREM, EELS) study of the reoxidation conditions for recovery of NM/CeO_2 (NM: Rh, Pt) catalysts from decoration or alloying phenomena. *Catal. Lett.* 76:131–37
82. Hickey N, Fornasiero P, Di Monte R, Kaspar J, Gonzalez-Velasco JR, et al. 2004. Reactivation of aged model Pd/$Ce_{0.68}$-$Zr_{0.32}O_2$ three-way catalyst by high temperature oxidising treatment. *Chem. Commun.* 196–97
83. Lassi U, Hietikko M, Rahkamaa-Tolomen K, Kallinen K, Savimäki A, et al. 2004. Deactivation correlations over Pd/Rh monoliths: the role of gas phase composition. *Top. Catal.* 30/31:457–61
84. Bernal S, Botana FJ, Calvino JJ, Lopez-Cartes C, Perez-Omil JA, Rodriguez-Izquierdo JM. 1998. The interpretation of HREM images of supported metal catalysts using image simulation: profile view images. *Ultramicroscopy* 72:135–64
85. Concepción P, Corma A, Silvestre-Alberto J, Franco V, Chane-Ching JY. 2004. Chemoselective hydrogenation catalysts: Pt on mesostructured CeO_2 nanoparticles embedded within ultrathin layers of SiO_2. *J. Am. Chem. Soc.* 126: 5523–32
86. Bernal S, Blanco G, Calvino JJ, Lopez-Cartes C, Perez-Omil JA, et al. 2001.

Electron microscopy (HREM, EELS) study of the reoxidation conditions for recovery of NM/CeO_2 (NM: Rh, Pt) catalysts from decoration or alloying phenomena. *Catal. Lett.* 76:131–37
87. Bernal S, Blanco G, Gatica JM, Larese C, Vidal H. 2001. Effect of mild re-oxidation treatments with CO_2 on the chemisorption capability of a Pt/CeO_2 catalyst reduced at 500 degrees C. *J. Catal.* 200:411–15
88. Deleted in proof
89. Bernal S, Botana FJ, Calvino JJ, Lopez-Cartes C, Perez-Omil JA, Rodriguez-Izquierdo JM. 1998. The interpretation of HREM images of supported metal catalysts using image simulation: profile view images. *Ultramicroscopy* 72:135–64
90. Rodriguez-Izquierdo JM, Bernal S, Cauqui MA, Finol D, Perez-Omil JA, et al. 1998. Characterization of La_2O_3/SiO_2 mixed oxide catalyst supports. *Am. Chem. Soc. Abstr.* 215:U484
91. Bernal S, Calvino JJ, Cauqui MA, Cifredo GA, Jobacho A, Rodriguez-Izquierdo JM. 1993. Metal-support interaction phenomena in rhodium ceria and rhodium titania catalysts—comparative study by high-resolution transmission electron spectroscopy. *Appl. Catal. A* 99:1–8
92. Bernal S, Calvino JJ, Cauqui MA, Gatica JM, Cartes CL, et al. 2003. Some contributions of electron microscopy to the characterisation of the strong metal-support interaction effect. *Catal. Today* 77:385–406
93. Baker RT, Calvino JJ, Lopez-Cartes C, Mira C, Perez-Omil JA, Rodriguez-Izquierdo JM. 1999. Detection and quantification of small misalignments in nanometer-sized particles on oxide support systems by the analysis of plan view HREM images. *Inst. Phys. Conf. Ser.* 161:537–40
94. Blanco G, Calvino JJ, Cauqui MA, Corchado P, Lopez-Cartes C, et al. 1999. Nanostructural evolution under reducing conditions of a Pt/CeTbOx catalyst: new alternative system as a TWC component. *Chem. Mater.* 11:3610–19
95. Bernal S, Calvino JJ, Gatica JM, Larese C, Lopez-Cartes C, Perez-Omil JA. 1997. Nanostructural evolution of a Pt/CeO_2 catalyst reduced at increasing temperatures (473–1223 k): a HREM study. *J. Catal.* 169:510–15
96. Penner S, Rupprechter G, Sauer H, Su DS, Tessadri R, et al. 2003. Pt/ceria thin film model catalysts after high-temperature reduction: a (HR)TEM study. *Vacuum* 71:71–76
97. Jenewein B, Fuchs M, Hayek K. 2003. The CO methanation on Rh/CeO_2 and CeO_2/Rh model catalysts: a comparative study. *Surf. Sci.* 532–535:364–69
98. Penner S, Wang D, Su DS, Rupprechter G, Podloucky R, et al. 2003. Platinum nanocrystals supported by silica, alumina and ceria: metal-support interaction due to high-temperature reduction in hydrogen. *Surf. Sci.* 532–35:276–80
99. Kepinski L, Wolcyrz M. 1997. Microstructure of Pd/CeO_2 catalyst: effect of high temperature reduction in hydrogen. *Appl. Catal. A* 150:197–220
100. Datye AK, Kalakkad D, Yao MH, Smith DJ. 1995. Comparison of metal-support interactions in Pt/TiO_2 and Pt/CeO_2. *J. Catal.* 155:148–53
101. Datye AK, Kalakkad D, Robota HJ. 1994. Metal support interactions. *J. Catal.* 148:729–36
102. Kalakkad D, Datye A, Robota H. 1992. Interaction of Pt and ceria probed by transmission electron microscopy and catalytic reactivity. *Appl. Catal. B* 1:191–219
103. Cochrane HD, Hutchinson JL, White D, Parkinson GM, Dupas C, Scott AJ. 1990. High resolution electron microscopy of ceria-supported catalysts. *Ultramicroscopy* 34:10–16
104. Hayek K, Fuchs M, Klötzer B, Reichl W, Rupprechter G. 2000. Studies of metal-support interactions with "real" and "inverted" model systems: reactions of CO and small hydrocarbons with hydrogen on noble metals in contact with oxides. *Top. Catal.* 13:55–66

105. Bitter H, Cauqui MA, Gatica JM, Bernal S, Ramaker DE, Koningsberger DC. 2000. Changes in the electronic structure of Pt/CeO_2 based $DeNO_x$ catalysts as a function of the reduction treatment. Detection of a Pt-H anti-bonding shape resonance and Pt-H EXAFS. *Stud. Surf. Sci. Catal.* 130D:3183–88
106. Bernal S, Botana FJ, Calvino JJ, Lopez C, Perez-Omil JA, Rodriguez-Izquierdo JM. 1996. High-resolution electron microscopy investigation of metal-support interactions in Rh/TiO_2. *J. Chem. Soc. Faraday Trans.* 92:2799–809
107. Hayek K, Bernd J, Klotzer B, Reichl W. 2001. Surface reactions on inverse model catalysts: CO adsorption and CO hydrogenation on vanadia- and ceria-modified surfaces of rhodium and palladium. *Top. Catal.* 14:25–33
108. Wang D, Penner S, Su DS, Rupprechter G, Hayek K, Schlogl R. 2003. Silicide formation on a Pt/SiO_2 model catalyst studied by TEM, EELS, and EDXS. *J. Catal.* 219:434–41
109. Gai PL, Boyes ED. 1997. Environmental HREM in materials science. In *In situ Microscopy in Materials Research*, ed. PL Gai, pp. 123–40. Dordrecht: Kluwer
110. Sharma R, Weiss K. 1998. Development of a TEM to study in situ structural and chemical changes during gas-solid interactions at elevated temperature. *Microsc. Res. Tech.* 42:270–80
111. Lopez-Cartes C, Bernal S, Calvino JJ, Cauqui MA, Blanco G, et al. 2003. In situ transmission electron microscopy investigation of Ce(IV) and Pr(IV) reducibility in a Rh (1%)/$Ce_{0.8}Pr_{0.2}O_{2-x}$ catalyst. *Chem. Commun.* 644–45
112. Bernal S, Calvino JJ, Cauqui MA, Lopez-Cartes C, Perez-Omil JA. 2001. The role of metal-support interaction on the oxidation of noble metals supported on cerium based oxides. *Inst. Phys. Conf. Ser.* 168:389–92
113. Rocchini E, Vicario M, Llorca J, de Leitenburg C, Dolcetti G, Trovarelli A. 2002. Reduction and oxygen storage behavior of noble metals supported on silica-doped ceria. *J. Catal.* 211:407–21
114. Bernal S, Calvino JJ, Cifredo GA, Finol D, Gatica JM, et al. 2002. Study of the structural modifications induced by reducing treatments on a $Pd/Ce_{0.8}Tb_{0.2}O_{2-x}/La_2O_3\text{-}Al_2O_3$ catalyst by means of X-ray diffraction and electron microscopy techniques. *Chem. Mater.* 14:1405–10
115. Kepinski L, Wolcyrz M, Marchewka M. 2002. Structure evolution of nanocrystalline CeO_2 supported on silica: effect of temperature and atmosphere. *J. Solid State Chem.* 168:110–18
116. Kang ZC, Eyring L. 2000. Lattice oxygen transfer in fluorite-type oxides containing Ce, Pr, and/or Tb. *J. Solid State Chem.* 155:129–37
117. Sasaki T, Ukyo Y, Suda A, Sugiura M, Kuroda K, et al. 2003. Oxygen absorption behavior of $Ce_2Zr_2O_{7+x}$ and formation of $Ce_2Zr_2O_{7.5}$. *J. Ceram. Soc. Jpn.* 111:382–85
118. Baker RT, Bernal S, Lopez-Cartes C, Perez-Omil JA, Montardi Y. 1999. Investigation of redox-cycled Ce-Zr mixed oxides for automotive catalysts by electron microscopy. *Electron Microsc. Anal.* 3: 521–24
119. Masui T, Ozaki T, Adachi G, Kang ZC, Eyring L. 2000. A new ceria-zirconia mixed oxide phase based on pyrochlore. *Chem. Lett.* 7:240–41
120. Rupprechter G, Hayek K, Hofmeister H. 1998. Electron microscopy of thin-film model catalysts: activation of alumina-supported rhodium nanoparticles. *J. Catal.* 173:409–22
121. Cunningham J, Cullinane D, Farrell F, Morris MA, Datye AK, Kalakkad D. 1995. Changes in microstructure and catalytic activity effected by redox cycling of rhodium upon CeO_2 and Al_2O_3. *Stud. Surf. Sci. Catal.* 96:237–48
122. Deleted in proof
123. Bernal S, Blanco G, Cauqui MA, Corchado P, Pintado JM, Rodriguez Izquierdo

JM. 1997. Oxygen buffering capacity of mixed cerium terbium oxide: a new material with potential applications in three-way catalysts. *Chem. Commun.* 1545–46

124. Lyubovsky M, Pfefferle L, Datye A, Bravo J, Nelson T. 1999. TEM study of the microstructural modifications of an alumina-supported palladium combustion catalyst. *J. Catal.* 187:275–84
125. Datye AK, Bravo J, Nelson T, Atanasova P, Lyubovsky M, Pfefferle L. 2000. Catalyst microstructure and methane oxidation reactivity during the Pd-PdO transformation on alumina supports. *Appl. Catal. A* 198:179–96
126. Sharma R, Crozier PA. 1999. In situ electron microscopy of CeO_2 and CeO_2-ZrO_2 reduction. *Inst. Phys. Conf. Ser.* 161:569–72
127. Sharma R, Crozier PA, Kang ZC, Eyring L. 2004. Observation of dynamic nanostructural and nanochemical changes in ceria-based catalysts during in-situ reduction. *Philos. Mag.* 84:2731–47
128. Baker RT, Bernal S, Blanco G, Gatica JM, Pintado JM, et al. 1999. Reversible changes in the redox behaviour of $Ce_{0.68}$-$Zr_{0.320}$ mixed oxide: effect of alternating the re-oxidation temperature after reduction at 1223 K. *Chem. Commun.* 149–50
129. Izu N, Kishimoto HOT, Ono K, Ozuka-Yao-Matsuo S. 2001. *Adv. Mater.* 2:397–440
130. Allard LF, Völkl A, Carim A, Datye AK, Rouff R. 1996. Morphology and crystallography of nanoparticulates revealed by electron holography. *Nanostruct. Mater.* 7:137–46
131. Shelef M. 1995. Selective catalytic reduction of NO_x with N-free reductants. *Chem. Rev.* 95:209–25
132. Coudurier G, Vedrine JC. 2000. EUROCAT oxide: a European V_2O_5-WO_3/TiO_2 SCR standard catalyst study: characterisation by electron microscopies (SEM, HRTEM, EDX) and by atomic force microscopy. *Catal. Today* 56:415–30
133. Lisi L, Lasorella G, Malloggi S, Russo G. 2004. Single and combined deactivating effect of alkali metals and HCl on commercial SCR catalysts. *Appl. Catal. B* 50:251–58
134. Reddy EP, Rojas TC, Fernández A. 2000. Transmission electron microscopy and energy-dispersive X-ray spectroscopy study of V_2O_5/TiO_2-ZrO_2. *Langmuir* 16: 4217–21
135. Herman RG, Sale JW, Harvey G, Stenger J, Lyman CE, et al. 2002. Monitoring aging and deactivation of emission abatement catalysts for selective catalytic reduction of NO_x. *Top. Catal.* 18:251–57
136. Hoang-Van C, Zeogaoui O, Pichat P. 1998. Vanadian-titania aerogel deNO_x catalysts. *J. Non-Cryst. Solids* 225:157–62
137. Bogdanchikova N, Meunier FC, Avalos-Borja M, Breen JP, Pestryakov A. 2002. On the nature of the silver phases of Ag/Al_2O_3 catalysts for reactions involving nitric oxide. *Appl. Catal. B* 36:287–97
138. Hickey N, Fornasiero P, Kaspar J, Graziani M, Martra G, et al. 2002. Improvement of SO_x-resistance of silver lean-DeNO_x catalysts by supporting on CeO_2-containing zirconia. *J. Catal.* 209:271–74
139. Richter M, Langpape M, Kolf S, Grubert G, Eckelt R, et al. 2002. Combinatorial preparation and high-throughput catalytic tests of multi-component deNO_x catalysts. *Appl. Catal. B* 36:261–77
140. Fuentes S, Bogdanchikova N, Avalos-Borja M, Boronin A, Farias MH, et al. 2000. Structural and catalytic properties of Pd/Al_2O_3-La_2O_3 catalysts. *Catal. Today* 55:301–9
141. Macleod N, Cropley R, Keel JM, Lambert RM. 2004. Exploiting the synergy of titania and alumina in lean NO_x reduction: in situ ammonia generation during the $Pd/TiO_2/Al_2O_3$-catalysed $H_2/CO/NO/O_2$ reaction. *J. Catal.* 221:20–31
142. Sun K, Liu J, Browning ND. 2002. Direct atomic scale analysis of the distribution

of Cu valence states in Cu/γ-Al_2O_3 catalysts. *Appl. Catal. B* 38:271–81
143. Balint I, Miyazaki A. 2004. NO/CH_4 reaction over nanodispersed Pt particles. *Top. Catal.* 30/31:123–26
144. Okal J, Kepinski L, Krajczyk L, Tylus W. 2003. Oxidation and redispersion of a low-loaded Re/γ-Al_2O_3 catalyst. *J. Catal.* 219:362–71
145. Okal J, Kepinski L, Krajczyk L, Drozd M. 1999. Oxidation and redispersion of a Re/γ-Al_2O_3 catalyst. *J. Catal.* 188:140–45
146. Serre C, Auroux A, Gervasini A, Hervieu M, Férey G. 2002. Hexagonal and cubic thermally stable mesoporous tin(IV) phosphates with acidic and catalytic properties. *Angew Chem. Int. Ed.* 41:1594–97
147. Perez-Ramirez J, Garcia-Cortes JM, Kapteijn F, Mul G, Moulijn JA, et al. 2001. Characterization and performance of Pt-USY in the SCR of NOx with hydrocarbons under lean-burn conditions. *Appl. Catal. B* 29:285–98
148. Garcia-Cortes JM, Perez-Ramirez J, Rouzaud JN, Vaccaro AR, Illan-Gomez MJ, et al. 2003. On the structure sensitivity of deNO_x HC-SCR over Pt-beta catalysts. *J. Catal.* 218:111–22
149. Wögerbauer C, Maciejewski M, Baiker A, Göbel U. 2001. Ir/H-ZSM-5 catalysts in the selective reduction of NO_x with hydrocarbons. *Top. Catal.* 16/17:181–86
150. Gai PL. 1983. In situ ETEM of redox studies of V_2O_5 catalysts. *Philos. Mag.* 48: 359–65
151. Kryukova GN, Zenkovets GA, Pfander N, Su D-S, Schlogl R. 2003. Synthesis and characterization of the titanium doped nanostructural V_2O_5. *Mater. Sci. Eng. A* 343:8–12
152. Su DS, Zandbergen HW, Tiemeijer PC, Kothleitner G, Hävecker M, et al. 2003. High resolution EELS using monochromator and high performance spectrometer: comparison of $<V_2O_5$ ELNES with NEXAFS and band structure calculations. *Micron* 34:235–38
153. Su DS, Hébert C, Willinger M, Schlogl R. 2003. Anisotropy and collection angle dependence of the oxygen K ELNES in V_2O_5: a band-structure calculation study. *Micron* 34:227–33
154. Hébert C, Willinger M, Pongratz P, Shattschneider P, Schlogl R. 2002. Oxygen K-edge in vanadium oxides: simulations and experiments. *Eur. Phys.* 28:407–14
155. Baudin F, Da Costa P, Thomas C, Calvo S, Lendresse Y, et al. 2004. No_x reduction over CeO_2-ZrO_2 supported iridium catalyst in the presence of propanol. *Top. Catal.* 30/31:97–101
156. Ziese U, de Jong KP, Koster AJ. 2004. Electron tomography: a tool for 3D structural probing of heterogeneous catalysts at the nanometer scale. *Appl. Catal. A* 260:71
157. Koster AJ, Ziese U, Verkleij AJ, Jannsen AH, Jong KP. 2000. Three-dimensional transmission electron microscopy: a novel imaging and characterization technique with nanometer scale resolution for materials science. *J. Phys. Chem. B* 104:9368–70
158. Jannsen AH, Koster AJ, Jong KP. 2002. On the shape of the mesopores in zeolite Y: a three-dimensional transmission electron microscopy study combined with texture analysis. *J. Phys. Chem. B* 106:11905–9
159. Jannsen AH, Yang CM, Wang Y, Schüth F, Koster AJ, Jong KP. 2003. Localization of small metal (oxide) particles in SBA-15 using bright-field electron tomography. *J. Phys. Chem. B* 107:10552–56
160. Weyland M, Midgley PA. 2001. High spatial resolution tomographic reconstruction from STEM high angle annular dark field (HAADF) images. *Inst. Phys. Conf. Ser.* 168:107
161. Midgley PA, Weyland M, Thomas JM, Johnson FG. 2001. Z-contrast tomography: a technique in three-dimensional nanostructural analysis based on Rutherford scattering. *Chem. Commun.* 907–8

162. Weyland M. 2002. Electron tomography of catalyst. *Top. Catal.* 21:175–83
163. Gai PL, Kourtakis K, Ziemecki S. 2000. In situ real-time environmental-TEM of nanometer size xerogel catalysts. *Microsc. Microanal.* 6:335–42
164. Avci AK, Ilsen OZ, Trimm DL. 2001. On-board fuel conversion for hydrogen fuel cells: comparison of different fuels by computer simulations. *Appl. Catal. A* 216(1–2):243
165. Carrette L, Friedrich KA, Stimming U. 2000. Fuel cells: principles, types, fuels, and applications. *Chem. Phys. Chem.* 1:162
166. Tomishige K, Asadullah M, Kunimori K. 2004. Syngas production by biomass gasification using $Rh/CeO_2/SiO_2$ catalysts and fluidized bed reactor. *Catal. Today* 89:389–403
167. Asadullah M, Ito S, Kunimori K, Yamada M, Tomishinge K. 2002. Biomass gasification to hydrogen and syngas at low temperature: novel catalytic system using fluidized-bed reactor. *J. Catal.* 208:255–59
168. Kiros Y, Schwartz S. 2000. Long-term hydrogen oxidation catalysts in alkaline fuel cells. *J. Power Sources* 87:101–5
169. Löffer MS, Grob B, Natter H, Hempelmann R, Krrajewski T, Divisek J. 2001. Synthesis and characterization of catalyst layers for direct methanol fuel cell applications. *Phys. Chem. Chem. Phys.* 3:333–36
170. Zhang YJ, Maroto-Valiente A, Rodriguez-Ramos I, Xin Q, Guerrero-Ruiz A. 2004. Synthesis and characterization of carbon black supported Pt-Ru alloy as a model catalyst for fuel cells. *Catal. Today* 93–95:619–26
171. Li W, Zhou W, Li H, Zhou Z, Zhou B, et al. 2004. Nano-stuctured Pt-Fe/C as cathode catalyst in direct methanol fuel cell. *Electrochim. Acta* 49:1045–55
172. Staiti P, Arico AS, Antonucci V, Hocevar S. 1998. Morphological variation of platinum catalysts in phosphotungstic acid fuel cell. *J. Power Sources* 70:91–101
173. Aupretre F, Descorme C, Duprez D. 2004. Hydrogen production for fuel cells from the catalytic ethanol steam reforming. *Top. Catal.* 30/31:487–91
174. Llorca J, Homs N, Sales J, de la Piscina PR. 2002. Efficient production of hydrogen over supported cobalt catalysts from ethanol steam reforming. *J. Catal.* 209:306–17
175. Verykios XE. 2003. Catalytic dry reforming of natural gas for the production of chemicals and hydrogen. *Int. J. Hydrogen Energy* 28:1045–63
176. Liguras DK, Goundani K, Verykios XE. 2004. Production of hydrogen for fuel cells by catalytic partial oxidation of ethanol over structured Ni catalysts. *J. Power Sources* 130:30–37
177. Kang ZC, Eyring L. 2001. Hydrogen production from methane and water by lattice oxygen transfer with $Ce_{0.70}$-$Zr_{0.25}Tb_{0.05}O_{2-x}$. *J. Alloys Compounds* 323–324:97–101
178. Masui T, Nakano K, Ozaki T, Adachi G, Kang ZC, Eyring L. 2001. Redox behavior of ceria-zirconia solid solution modified by the chemical filing process. *Chem. Mater.* 13:1834–40
179. Kang ZC, Wang ZL. 2003. Novel oxides for cycled hydrogen production from methane and water using a temperature swing. *Adv. Mater.* 15:521–26
180. Basinska A, Kepinski L, Domka F. 1999. The effect of support on WGSR activity of ruthenium catalysts. *Appl. Catal. A* 183:143–53
181. Hartmann M, Bischof C, Luan Z, Kevan L. 2001. Preparation and characterization of ruthenium clusters on mesoporous support. *Microporous Mesoporous Mater.* 44–45:385–94
182. Fu Q, Weber A, Flytzany-Stephanopoulos M. 2001. Nanostructured Au-CeO_2 catalysts for low-temperature water-gas shift. *Catal. Lett.* 77:87–95
183. Panzera G, Modafferi V, Candamano S,

Donato A, Frusteri F, Antonucci PL. 2004. CO selective oxidation on ceria-supported Au catalysts for fuel cell application. *J. Power Sources* 135:177–83
184. Bulushev DA, Kiwi-Minsker L, Yuranov I, Suvorova EI, Buffat PA, Renken A. 2002. Structured Au/FeO_x/C catalysts for low-temperature CO oxidation. *J. Catal.* 210:149–59
185. Choudhary TV, Sivadinarayana C, Chusuei CC, Datye AK, Fackler J, Goodman DW. 2002. CO oxidation on supported nano-Au catalysts synthesized from a $[Au_6(PPh_3)_6](BF_4)_2$ complex. *J. Catal.* 207:247–55
186. Bera P, Hegde MS. 2004. Characterization and catalytic properties of combustion synthesized Au/CeO_2 catalyst. *Catal. Lett.* 79:75–81
187. Lopez N, Norskov JK, Janssens TVW, Carlsson A, Puig-Molina A, et al. 2004. The adhesion and shape of nanosized Au particles in a Au/TiO_2 catalyst. *J. Catal.* 225:86–94
188. Liu X, Korotkikh O, Farrauto R. 2002. Selective catalytic oxidation of CO in H2: structural study of Fe oxide-promoted Pt/alumina catalyst. *Appl. Catal. A* 226: 293–303
189. Akita T, Ichikawa S, Tanaka K, Haruta M. 2001. Analytical TEM study on the dispersion of Au nanoparticles in Au/TiO_2 catalyst prepared under various temperatures. *Surf. Interface Anal.* 31:73–78
190. Chen YJ, Yeh CT. 2001. Deposition of highly dispersed gold on alumina support. *J. Catal.* 200:59–68
191. Deluga GA, Salge JR, Verykios XE. 2004. Renewable hydrogen from ethanol by autothermal reforming. *Science* 303:993–97
192. Eyring L, Kang ZC. 2000. *Patent No. WO2000025898*
193. Fatsikostas AN, Verykios XE. 2004. Reaction network of steam reforming of ethanol over Ni-based catalysts. *J. Catal.* 225:439–52
194. Luo JZ, Yu ZL, Ng CF, Au CT. 2000. CO_2/CH_4 reforming over Ni-La_2O_3/5A: an investigation on carbon deposition and reaction steps. *J. Catal.* 194:198–210
195. Bernal S, Botana FJ, Calvino JJ, Cifredo GA, Garcia R, Rodriguez-Izquierdo JM. 1990. HREM characterization of metal-catalysts supported on rare-earth-oxides—samarium oxide as support. *Ultramicroscopy* 34:60–65
196. Bernal S, Calvino JJ, Lopez-Cartes C, Pintado JM, Perez-Omil JA, et al. 1999. Nanostructural evolution of high loading Rh/lanthana catalysts through the preparation and reduction steps. *Catal. Today* 52:29–43
197. Bernal S, Calvino JJ, Cauqui MA, Gatica JM, Cartes CL, et al. 2003. Some contributions of electron microscopy to the characterisation of the strong metal-support interaction effect. *Catal. Today* 77:385–406
198. Haruta M, Tsubota S, Kobayashi T, Kageyama H, Genet M, Delmon B. 1993. Low-temperature oxidation of CO over gold supported on TiO_2, Fe_2O_3, and Co_3O_4. *J. Catal.* 144:175–79
199. Haruta M. 1997. Size- and support-dependency in the catalysis of gold. *Catal. Today* 36:153–66
200. Ichikawa S, Akita T, Okumura M, Kohyama M, Tanaka K. 2003. Electron holographic analysis of nanostructured gold catalyst. *JEOL News* 38:6–9
201. Ichikawa S, Akita T, Okumura M, Haruta M, Tanaka K, Kohyama M. 2003. Electron holographic 3D nanoanalysis of Au/TiO2 catalyst at interface. *J. Electron Microsc.* 52:21–26
202. Akita T, Tanaka K, Tsubota S, Haruta M. 2000. Analytical high resolution TEM study of supported gold catalysts: orientation relationship between Au particles and TiO_2 supports. *J. Electron Microsc.* 49:657–62
203. Ichikawa S, Akita T, Okumura M, Haruta

M, Tanaka K. 2002. Electron holographic nano-characterization of gold catalyst at interface. *Mater. Res. Soc. Symp. Proc.* 727:11–16
204. Centeno MA, Paulis M, Montes M, Odriozola JA. 2002. Catalytic combustion of volatile organic compounds on $Au/CeO_2/Al_2O_3$ and Au/Al_2O_3 catalysts. *Appl. Catal. A* 234:65–78
205. Besson M, Kallel A, Gallezot P, Zanella R, Louis C. 2003. Gold catalysts supported on titanium oxide for catalytic wet air oxidation of succinic acid. *Catal. Commun.* 4:471–76

Annu. Rev. Mater. Res. 2005. 35:505–38
doi: 10.1146/annurev.matsci.35.100303.110641

First published online as a Review in Advance on April 4, 2005

COMPOSITE MATERIALS FOR WIND POWER TURBINE BLADES

Povl Brøndsted, Hans Lilholt, and Aage Lystrup
Materials Research Department, Risoe National Laboratory, DK 4000 Roskilde, Denmark; email: povl.brondsted@risoe.dk, hans.lilholt@risoe.dk, aage.lystrup@risoe.dk

Key Words composites, properties, processing, damage, fatigue

■ **Abstract** Renewable energy resources, of which wind energy is prominent, are part of the solution to the global energy problem. Wind turbine and the rotorblade concepts are reviewed, and loadings by wind and gravity as important factors for the fatigue performance of the materials are considered. Wood and composites are discussed as candidates for rotorblades. The fibers and matrices for composites are described, and their high stiffness, low density, and good fatigue performance are emphasized. Manufacturing technologies for composites are presented and evaluated with respect to advantages, problems, and industrial potential. The important technologies of today are prepreg (pre-impregnated) technology and resin infusion technology. The mechanical properties of fiber composite materials are discussed, with a focus on fatigue performance. Damage and materials degradation during fatigue are described. Testing procedures for documentation of properties are reviewed, and fatigue loading histories are discussed, together with methods for data handling and statistical analysis of (large) amounts of test data. Future challenges for materials in the field of wind turbines are presented, with a focus on thermoplastic composites, new structural materials concepts, new structural design aspects, structural health monitoring, and the coming trends and markets for wind energy.

INTRODUCTION

The international wind energy market showed a new record in 2003 with a growth rate of 15%. Globally, a total power of 8.3 GW was installed. The total installed wind energy power has now reached more than 40 GW, and the average growth in the market during the past five years has been 26% per year. These figures were reported in the annual report on the status for wind energy in March 2004 from the consulting company BTM (1). This illustrates how during the past 25 to 30 years the use of wind turbines for electricity generation has grown from a grass-root initiative to an efficient alternative energy resource. The capacity of a commercial wind turbine has today reached the range of 2–5 MW, with large plants being built world-wide on land and offshore.

This status illustrates the increasing concern in the world, and perhaps in Europe in particular, about the supply and consumption of energy in a modern and civilized society. Over the centuries, energy has been supplied by wood, coke, coal, oil, and natural gas, as well as by uranium (nuclear energy). The early sources were natural in the sense that they renewed themselves over a short time period (years), while the later energy supplies were tapped from sources established millions of years ago. The increasing energy demands in the world today, owing to improved and expanded civilizations and to increasing populations, have led to concerns over the limited energy resources in storage on the earth.

At the same time there is an increasing concern about the pollution of the world/environment (generation of waste). This has led to the focus on a sustainable energy supply, which (probably) implies optimized use of energy, minimized pollution and, implicitly, reduction in energy consumption. These aspects have led to an increasing focus on the short-time stored energy resources; among these the most developed types today are wind energy and biomass (in various forms).

WIND ENERGY

For wind energy a converter is needed to turn the kinetic wind energy into operational energy, e.g., electricity and/or heat. The converter is based on a rotor driven by the wind, thereby extracting a power of

$$P = \alpha \rho A v^3, \qquad 1.$$

where α is an aerodynamic efficiency constant, ρ the density of air, A the area of rotor-plane, and v the wind velocity. The rotor needs some sort of aerodynamic device, e.g., a wing or rotorblade with an aerodynamic shape, to be able to rotate. The rotor is typically placed on a tower, and this converter is usually called a wind turbine (in the past, a wind mill). In the early years the United States used the designation wind energy conversion system (WECS). Today nearly all wind turbines have rotors with three blades, where the rotor is mounted in an approximately vertical plane, on a horizontal axis, facing into the wind (Figure 1, see color insert).

ROTORBLADES

The present review is focused on the rotorblades, which probably present the most challenging materials, design, and engineering problems. The rotor and its three rotorblades constitute a rather flimsy structure, consisting of cantilever-mounted blades on a central hub. The basic design aspects for a rotorblade are the selection of material and shape. The material should be stiff, strong, and light. The shape should be aerodynamic, similar to that of an airplane wing.

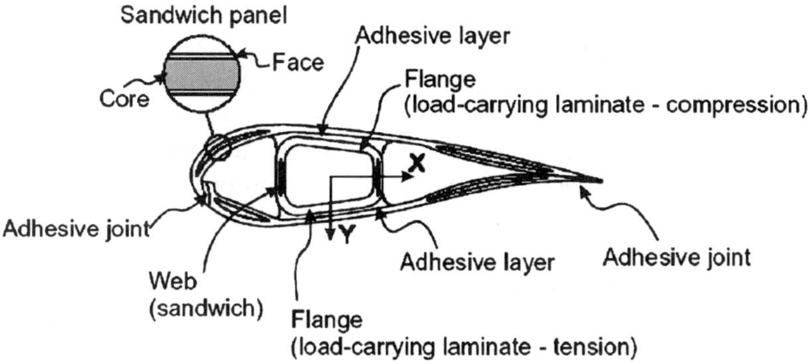

Figure 2 Cross-section principle of a rotorblade giving the nomenclature of the different blade construction elements.

The shape of a rotorblade in cross section is shown in Figure 2. The aerodynamic contours are formed by the (outer) relatively thin shells. They are supported structurally by a longitudinal beam or by webs, which carry a substantial part of the load on the blade. In the longitudinal direction, the rotorblades are tapered and twisted. The tapering is needed to economize with weight of the material because of the increasing loads from tip to root of a cantilever structure. The tapering, both in external shape and in thickness of the shells/beams/webs, is usually designed to ensure the same materials loadings, e.g., a maximum strain as design allowable.

It is interesting to note, as the pioneering work in the United States in 1970 (and earlier) observed, that a simple plank of constant cross-sectional area, rounded along the forward edge and sharpened along the rear edge, and with no twist of the (rectangular) profile, would have an aerodynamic efficiency of 75–80%, where 100% accounts for a rotorblade optimized according to the principles described above. The challenge for the designers is thus to go beyond the simple plank and integrate the aerodynamic shape, the tapering and the twist, into a design of the blade structure that is optimized with respect to materials selection and cost-effective production.

LOADS ON ROTORBLADES

The rotor and the rotorblades are exposed to external loads. These originate from the wind and from gravity. The general rotorblade geometry shown in Figure 2 has the blades arranged with their flat dimension in the plane of the rotor. This is so because the linear velocity of the outer part of the blade is high with blade tip speeds of 75–85 m/s, which are much higher than the wind speeds, even at storm conditions (~25 m/s). Therefore the relative wind direction, as seen by the blade, is nearly in the plane of the rotor, although the real wind direction is at a right angle to the rotor plane.

The blades are exposed to the wind that through the lift on the aerodynamic profile causes loads at right angle to the blades, which therefore react by bending flap-wise. These loads are both static, causing a permanent bending of the blades, and dynamic, causing a fatigue flap-wise bending because of the natural variations in wind speed. In addition, these static and fatigue load spectra vary during rotation, as seen by a given blade, when the blade points upward and downward, respectively; this is caused by the natural wind shear, which is the increase of average wind speed with increasing height over the terrain.

Normally, the maximum wind speed for operation of the wind turbine is about 25 m/s, i.e., storm conditions, beyond which the rotor is brought to a standstill by turning the rotorblades out of the wind by rotation about the longitudinal axis of the blade. This position exposes the rotorblades to the wind, coming in a right angle to blade planar surface, and causes the blade to bend as under a steady load on its surface. The natural variations in wind speed will cause dynamic flap-wise fatigue load spectra, as seen by the blade. The absolute magnitudes of these loads are comparable to the aerodynamic lift on the blades, as described above. Therefore, the rotorblades are exposed to flap-wise bending fatigue loads in varying spectra distributions under all conditions during service.

The blades are also exposed to gravity, and this is most pronounced when they are in their horizontal position. These loads cause bending in an edge-wise mode, and a given blade bends one way on the right-hand side and the opposite way on the left-hand side of rotor plane. This explains why the edge-wise bending also causes fatigue of the blade material and structure during rotation. Furthermore, the blades are exposed to centrifugal forces during the rotation of the rotor. Due to the limits of the linear blade velocity, the rotational speed is relatively low, typically from 20 rpm down to 10 rpm for large rotors/long rotorblades. Therefore, the longitudinal tensile loads in the blades are relatively low and normally are not taken into account as a design parameter. The design lifetime of modern wind turbines is normally thought to be 20 years, and the corresponding number of rotations is of the order 10^8 to 10^9.

MATERIALS REQUIREMENTS

These operational parameters and conditions lead to the following requirements focused on stiffness, density, and long-time fatigue:

- high material stiffness is needed to maintain optimal aerodynamic performance,
- low density is needed to reduce gravity forces,
- long-fatigue life is needed to reduce material degradation.

The optimal design of the rotorblades is today a complex and multifaceted task and requires optimization of properties, performance, and economy. It is not the objective of this review to discuss the design process in detail, but rather to focus

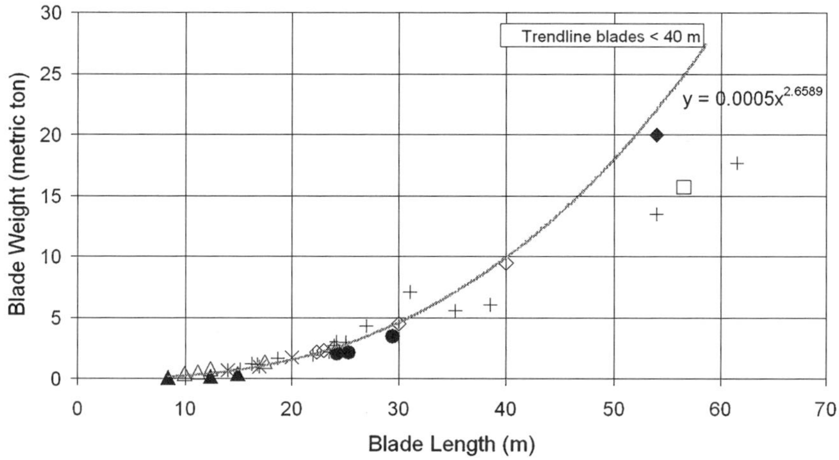

Figure 3 Development in rotorblade weight versus length. Symbols indicate different manufacturers and processing technologies.

on the materials and how they meet the requirements of wind turbine performance, as discussed above.

The combined result of the design process and the materials can be illustrated in the form of rotorblade weight as a function of rotorblade length, as presented in Figure 3. The lower end of the curve represents the relatively short blades of length 12–15 m, as were common in the early years of wind turbine design (1980s). The points represent rotorblades of increasing length, as developed until the present. An empirical curve is shown for the points representing blades with lengths below 40 m, giving a power law with an exponent of about 2.6; this is lower than expected for a simple up-scaling of the design on a volume basis (exponent 3), but it corresponds more closely to up-scaling of only two dimensions (length and thickness, exponent 2). This rather low exponent is a good indication of the high quality of the design process. Three recent and very long rotorblades of 54 to 61.5 m (2), plotted in Figure 3, show the further improvement in the design process by the points being below the extrapolated empirical curve, although it should be noted that they might refer to different wind load classifications. An additional discussion of these recent design improvements is given below.

The materials properties requirements of high stiffness, low weight, and long-fatigue life can be used to perform a materials selection, initially looking at all materials. The property combinations presented in Reference (3) are illustrative for a first selection of potentially usable materials classes. In a simplified form, the diagram of stiffness versus density in Figure 4 shows the procedure to be used (4). For details of materials selection, see Reference (3). The mechanical design of a rotorblade corresponds nominally to a beam, and the merit index is for this case

$$M_b = E^{1/2}/\rho, \qquad 2.$$

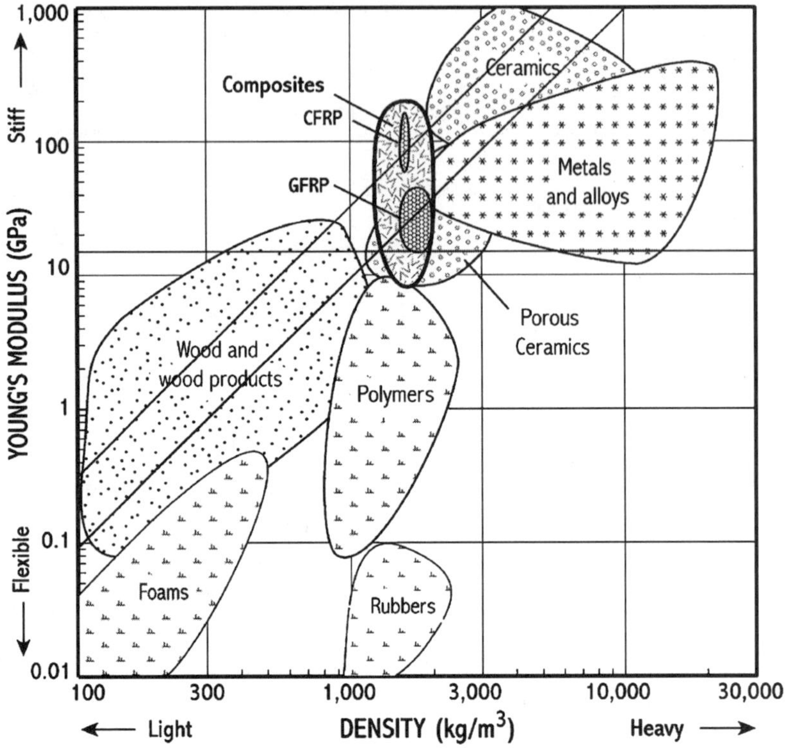

Figure 4 Diagram showing stiffness versus density for all materials. The merit index for a beam $M_b = E^{1/2}/\rho$ is represented by sloping lines with M_b equal to 0.003 (*lower line*) and 0.006 (*upper line*). The criterion for absolute stiffness E = 15 GPa is indicated by the horizontal line (4).

where E is the material stiffness and ρ is the material density. Lines of constant M_b are superimposed on the diagram, and materials that fulfill the criterion (partially or completely) are on the line and to the upper-left of the line.

The two lines shown in Figure 4 are arbitrary and illustrate lines of materials that are equally good in terms of stiffness and density for a cantilever beam. The lower of the two lines indicates that potential candidates materials are wood, composites, porous ceramics, metals, and ceramics. The lower line has a merit index of $M_b = 0.003$ with units of E in GPa and ρ in kg/m^3. If the merit index is doubled to $M_b = 0.006$, the upper line is valid. This line indicates that candidate materials are woods, composites, and ceramics.

A second criterion is stiffness on an absolute scale; a stiff material causes less deflection of a cantilever beam than does a flexible material. In the diagram of Figure 4, a stiffness criterion corresponds to a horizontal line. Deflection considerations are related to rotorblade geometry and dimensions, as well as to the overall design of tower and rotor, in particular the blade deflection when the rotorblade passes the tower. A sensible deflection requires a material stiffness of 10–20 GPa; for illustration, the line for $E = 15$ GPa is drawn in Figure 4. Materials on or above this line satisfy the criterion, and it is seen that most woods, some composites, and some porous ceramics are excluded.

In terms of materials strength, i.e., resistance against long-time fatigue loads, and extra high loads, it is also important to consider materials fracture toughness. The toughness in relation to density is collected in a similar diagram, which can be found in Reference (3). The merit index for high fracture toughness and low density shows that candidate materials are woods and composites.

MATERIALS FOR ROTORBLADES

The combined materials performance criteria identify the candidates for rotorblades as woods and composites. Woods are potentially interesting because of their low density, but their rather low stiffness makes it difficult to limit the (elastic) deflections for very large rotorblades. Even wood materials with cellulosic fibers all aligned in the major load-bearing directions are close to the maximum performance possible for wood. Furthermore, wood is a natural material and thus environmentally attractive, but at the same time difficult to obtain in reproducible and high quality, which is a requirement for stable and economical manufacturing of rotorblades and thus economically attractive wind energy.

Partly for these reasons, composites have until now been most extensively used. Below we focus on these materials.

Fibers

The materials science basis for these composites and their potential use is the existence of strong and stiff fibers (5–10) and compliant and easily processable polymers, such as thermosets and thermoplastics (11).

The strong and stiff fibers are themselves not usable for structural purposes; their good properties can be exploited only as an important component of composites. Nonetheless, it is important to describe the fibers even if the composites are the real structural materials for rotorblades.

By far the most widely used fibers are glass fibers. In recent years carbon fibers have become of increasing interest because of the requirements presented by the ever-larger rotorblades and the decreasing price of carbon fibers. Other potentially interesting fibers are aramid, polyethylene, and cellulose, all of which have moderate mechanical properties, and low or very low densities. Key data for these fibers and their composites are collected in Table 1, and the information is discussed in the following section with a focus on glass and carbon fibers.

TABLE 1 Composite materials based on the fibers listed and a polymer matrix with properties $E_m = 3$ GPa, $\sigma_m = 100$ MPa, and $\rho_m = 1.2$ g/cm^3. The composite properties are calculated from the simple composite theory (law of mixtures); the orientation factor is 1 for aligned composites and 1/3 for random composites

	Fibers			Composites					
Type	Stiffness E_f GPa	Tensile strength σ_f MPa	Density ρ_f g/cm^3	Volume fraction V_f	Orientation θ	Stiffness E_c GPa	Tensile strength σ_c MPa	Density ρ_c g/cm^3	Merit $E_c^{1/2}/\rho_c$
Glass-E	72	3500	2.54	0.5	0°	38	1800	1.87	3.3
				0.3	Random	9.3	420	1.60	1.9
Carbon	350	4000	1.77	0.5	0°	176	2050	1.49	8.9
				0.3	Random	37	470	1.37	4.4
Aramid	120	3600	1.45	0.5	0°	61	1850	1.33	5.9
				0.3	Random	14.1	430	1.27	2.9
Polyethylene	117	2600	0.97	0.5	0°	60	1350	1.09	7.1
				0.3	Random	13.8	330	1.13	3.3
Cellulose	80	1000	1.50	0.5	0°	41	550	1.35	4.7
				0.3	Random	10.1	170	1.29	2.5

GLASS FIBERS Glass is generally composed of SiO_2, Al_2O_3, and smaller amounts of other oxides (5–7). The atoms of Si and O form a lattice with no crystallographic order, and glass fibers are therefore amorphous with isotropic properties, e.g., stiffness and thermal expansion. Glass fibers are produced in several chemical compositions, for specific purposes. The glass fiber called type E (electrical) is the most widely used for composites. Glass fibers have diameters normally in the range of 10 to 20 μm and are produced from molten glass by pulling fibers from spinnerets into bundles of hundreds to thousands of individual fibers. Their surfaces are normally coated immediately with a polymer sizing, typically a silane compound, to protect the fiber against cracks and adhered water; the sizing, which is a multicomponent compound, is also designed to improve the bonding of the glass fiber surface to the polymer matrix, typically a thermoset, and thereby enhance the properties of the composite.

Glass fibers for composites have a good combination of properties: moderate stiffness, high strength, and moderate density, as listed in Table 1.

CARBON FIBERS Carbon fibers (5–7) are composed of nearly pure carbon, which forms a crystallographic lattice with a hexagonal shape called graphite. The atoms are held together by strong covalent forces inside the hexagon plane, where the atom spacing is 1.4 Å, the bonds between the hexagon planes are weak, and the spacing is large, 3.4 Å. This crystallography means that the mechanical properties have very high values in the plane of the hexagonal planes and rather low values at right angle to the planes. This also implies a high degree of anisotropy, both in mechanical properties and in thermal expansion. The crystallographic structure and anisotropy of graphite dictates the arrangement of the hexagonal planes, such that the planes must be oriented parallel to the fiber axis in order to achieve useful properties of the fibers. In all carbon fibers there is a high degree of alignment of the graphite planes, which varies somewhat depending on the type of fiber and its production method.

The carbon fibers are produced by two different methods. The first and most widely used method starts with polyacrylonitrile (PAN) fibers. These textile fibers are oxidized, stretched, and finally heat-treated at temperatures of 1500 to 2500°C. These process steps align and couple the original C-C backbone of PAN into the graphite hexagon-planes required for the fibers. The second method starts from natural tar, which contains the graphite units in a random mixture. Various processing steps lead to the production of fibers through spinnerets, which ensure alignment of the graphite planes and thus again the required properties of the fibers. The first method achieves the necessary graphite plane alignment through coupling of C-C backbones, whereas the second method aligns pre-existing graphite planes. Both methods include rather expensive raw materials and numerous and expensive processing steps. Recent efforts (in the United States) have been initiated to find cheaper raw materials, e.g., lignin from biomass, and to develop the processing economy, e.g., by reducing the number of processing steps.

Carbon fibers for composites have an excellent combination of very high stiffness, high strength, and low density, as listed in Table 1. However, the use of carbon fibers in a hybrid combination with glass fibers has today placed focus on carbon fibers with moderate to low stiffness and relatively high failure strain, so that the carbon fibers can share the loads and deform in concert with the glass fibers of moderately high failure strain.

Matrix Materials

The polymeric composites with the above mentioned fibers have matrices of polymers, typically thermosets or thermoplastics (11). Both are rather soft and flexible (low stiffness of less than 4 GPa), and their main purpose is to bind the fibers together so that they can act in concert and give a functional composite for structural purposes. The toughness and especially failure strain is moderate for thermosets, 5–8%, and large for thermoplastics, 50–100%, and the matrices thus induce toughness in the composites, in particular via energy absorbing mechanisms related to the interface. The early composite materials for rotorblades were glass fibers combined with polyester, a material, as well as a processing method, taken from the boat industry. The easy availability of materials and processing was part of the reason for the rather fast development of rotorblades and wind energy in the early years.

The thermosets most used are polyesters, vinylesters, and epoxies. All have stiffness values of 3–4 GPa and densities of 1.1–1.3 g/cm^3 and thus match the fibers reasonably well. The thermosets go through an irreversible curing reaction, which implies a contraction giving internal stresses in the composites.

The thermoplastics for composites for rotorblades have been under development over the past 10–15 years. The properties have also rather low values: stiffness values of 1–3 GPa, and densities of 0.9–1.4 g/cm^3. The interest is centered on the potential recycling of thermoplastic polymers, and this has been investigated, although no clear procedure has been established for recycling of thermoplastic polymer composites. The thermoplastics go through a melting and solidification process during processing, which is potentially fast. The solidification also introduces a thermal contraction leading to internal stresses; these could be high because of the rather high melting point/processing temperature for thermoplastics, and better thermoplastics typically have higher melting points. The internal stresses, of whatever origin, affect the mechanical properties often detrimentally.

Composite Materials

The fibers and the matrix are combined into the composite (9, 10, 12–15). Many combinations and mixing ratios are possible, and the composite properties, most generally, are governed by the fibers, the matrix, and the interface established between these (normally) two components. Properties of fibers and matrices are, of course, important, and so are the ways in which they are arranged in the composite. The important parameters are their relative amounts, often described by the fiber volume fraction, and the spatial orientation of the fibers. As a simple illustration, the stiffness of the composite E_c is controlled and calculated according to

$$E_c = \eta \cdot V_f \cdot E_f + V_m \cdot E_m, \qquad 3.$$

where E is the stiffness (elastic modulus), V is the volume fraction, η is an orientation factor for the fibers, index f is fiber, and m is the matrix. For a perfect composite with no porosity, $V_f + V_m = 1$. The orientation factor is equal to 1 for aligned parallel fibers loaded along the fiber direction; for a randomly oriented fiber assembly in two dimensions (in a plane; often called a fiber-mat) the orientation factor is 1/3.

A range of values, listed in Table 1, illustrates the simple, basic mechanical properties of composites based on the various fibers mentioned above. The fibers are normally the dominant contributor to the composite properties, and for ease of comparison, all composite properties are calculated for a matrix of thermoset with parameters $E_m = 3$ GPa, $\sigma_m = 100$ MPa and $\rho_m = 1.2$ g/cm^3. For each composite based on a given fiber, the first line describes a composite with 50 volume% fibers, fully aligned $\eta = 1$, whereas the second line describes a composite with 30 volume% fibers arranged in a random planar mat.

In Table 1 the first column lists the typical fiber properties and the following columns list the calculated composite properties. The final column shows the elastic merit index for a cantilever beam, as discussed above. It is clear from the composite stiffness that all aligned fiber composites have rather high values, whereas nearly all random fiber composites have stiffnesses below the limit of 15 GPa (discussed above). Glass fiber composites have moderate properties and were selected in the early days because of availability and known processing technology. The other potential composites are slightly better, and only carbon fiber composites show significant improvements over the well-established glass fiber composites. This is the reason for the increasing use of carbon fiber composites in large and advanced rotorblades. The modern materials are in most cases a combination of glass fibers and carbon fibers in a hybrid construction. The hybrid aspects can be used on a macro scale, with selected parts of the rotorblade, typically the outer weight critical elements, made in carbon fiber composites, or the hybrid aspects can be on the material structure level where glass and carbon fiber are intermixed on a layer basis or perhaps even on the basis of individual fiber bundles or fibers. This hybrid concept is often a compromise between the improved performance of carbon fibers and the high cost of carbon fibers. Few rotorblades have been made completely of carbon fiber composites.

MANUFACTURING TECHNOLOGIES

During and after the energy crises in the early 1970s, many smaller companies and grass-root organizations started to build wind turbines for energy production, and shortly after, in the mid-1970s, utility companies took up the challenge and started building larger wind turbines, often with push and financial support from the government. New manufacturing technologies were gradually developed and introduced by the blade manufacturers as the blades grew in size and in number produced (up through the 1990s), and an increasing concern about the working environment and more strict legislation pushed the technologies away from wet and open processes toward prepreg technology and closed mold infusion techniques.

Wet Hand-Lay-Up

In the early days, smaller glass fiber reinforced polyester blades were manufactured using the traditional wet hand-lay-up technique in open molds, which has been used for decades for building boats. The reinforcements were all glass fibers and mainly chopped strand mats (CSM) with random fiber orientation; in some cases woven fabrics were added to increase stiffness and strength. The upper and lower shells were adhesively bonded together to form the airfoil-shaped blade, see Figure 2. As the blades became longer (approaching 8 m), webs were inserted to support the airfoil and to take up both bending and shear loads, and a demand for higher stiffness and strength introduced a more dedicated fiber orientation with more fibers in the longitudinal direction of the blade. This was achieved either by using unidirectional woven fabrics (unbalanced fabrics with more fibers in one of the directions, normally the warp direction) or by laying down many parallel rovings in the length direction in between the CSMs. In some designs, these rovings served another purpose: they were used to form the root-end attachment as well. At the root-end the rovings were wound around steel bushes (small tubes), and continued back into the blade. The steel bushes formed the holes in what became the flange of the blade to be mounted to the hub. This principle is illustrated in Figure 5 and is known as a Hütter flange after the inventor Hütter, who used this technique for a wind turbine blade as early as the 1950s (16).

In the mid-1970s, at Tvind in Denmark, a group of pioneers built a three-bladed wind turbine with blade length of 27 m (17). The blades were manufactured from glass fiber and epoxy in a one-sided open mold. The lower side of the airfoil was laid up in a mold using a combination of CSM and rovings. A double Hütter flange,

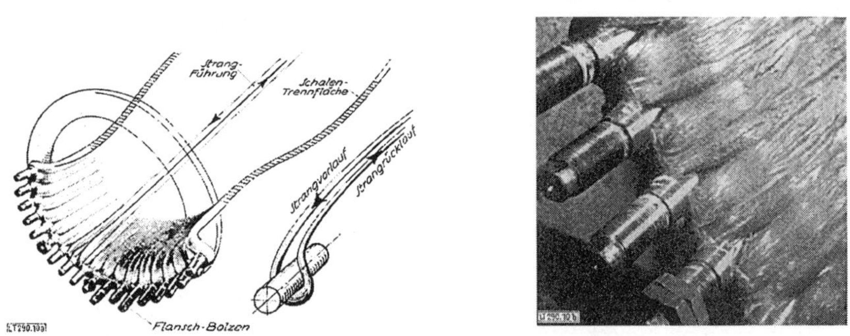

Figure 5 Principle and photo of a Hütter flange, illustrating how the fiber rovings are wound around steel bushes to form the flange (16).

Figure 6 The Tvind blades were built in a one-sided mold. Sandwich spars and webs form the upper contour of the blade (17).

with both an external and an internal flange with bolt holes, was constructed at the same time. After curing of the lower airfoil laminate, many transverse foam spars, with the shape of the airfoil profile, were placed along the entire length of the blade at equidistant distance, and laminated onto the lower airfoil, as shown in Figure 6. Longitudinal webs were integrated, as well. The skin laminates of the webs also formed a flange toward the upper airfoil, which was built as the last part of the blade. The Tvind turbine is still in operation, but the blades were replaced in 1993 (18).

Filament Winding

When government-supported development programs for larger wind turbines started in the mid-1970s, the companies involved looked for more rational manufacturing techniques than the labor intensive hand-lay-up technique, and filament winding was investigated by several nations. Filament winding is a rational way of placing a huge amount of roving in a controlled manner around a rotating mandrel. The shape of a wind turbine blade is not cylindrical, and the majority of the fibers have to be placed along the blade length; therefore, the filament winding technique had to be developed further for this specific application.

In the United States, a technique for winding the entire blade was developed by Kaman Aerospace Corporation and Structural Composites Industries, for blades up to 45 m in length in glass fibers (19–21). A set of three mandrels was used to gradually build up the airfoil with integrated webs. First, the leading edge part and the forward shear web were wound around the first mandrel, then a second mandrel was attached to the already wound leading edge structure, and more fibers were wound onto the two combined mandrels adding more materials to the first

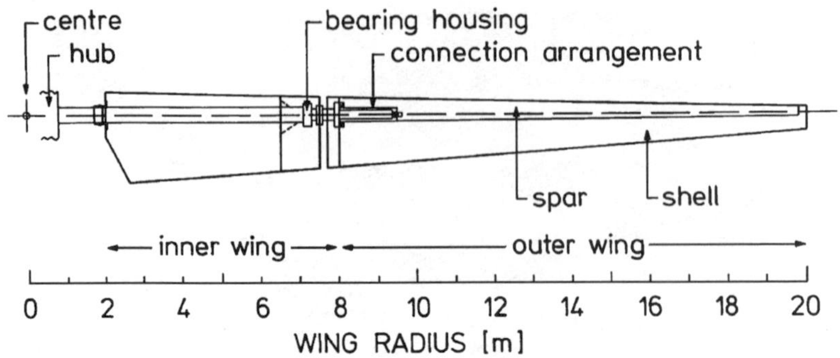

Figure 7 Principle of the Nibe blades. The inner 8 m has a steel spar and the outer 12 m is an all-composite blade (22).

structure and forming a second shear web. Finally, a third mandrel was attached, and more winding created the after body with the trailing edge. The winding was done with a combination of glass fiber roving and a glass fiber tape with most of the fibers in the transverse direction of the tape. This special tape was necessary to achieve sufficient bending stiffness and strength of the completed blade, as it is difficult and inefficient to place rovings along the axis of the mandrel with a filament-winding machine.

A tape-winding technique for winding the spar, including the leading edge and the shear web, was also developed in Denmark by Risoe National Laboratory and the Vølund company in the late 1970s (22). The blades for the two Nibe wind turbines were 20 m long and were a hybrid construction consisting of a steel spar for the inner 8 m and a glass fiber/polyester spar for the outer 12 m of the blade, Figure 7. The airfoil was hand-laid-up glass/polyester with a balsa core sandwich in the after-body, and the airfoil was adhesively bonded to the spars. Because most of the fibers must be placed along the length of the spar, a special 20-cm-wide woven glass fiber tape, with 90% of the fibers oriented transverse to the tape direction, was wound around the spar mandrel with an overlap of half the width of the tape, as illustrated in Figure 8 (see color insert). The thickness of the spar was easily varied along the length, from 24 mm at the root-end to 6 mm at the tip, by changing the length of travel of the tape placement apparatus along the length of the spar. Because the spar was tapered toward the tip-end and the warp glass fibers in the tape could not stretch, it was necessary to insert and withdraw a wedge underneath the tape, at the shear web position of the spar, for each revolution of the mandrel, in order to control and maintain straight fibers with the desired orientation in the upper and lower part of the spar. All the slack at the one side of the tape was taken up by the wedge, and the surplus of tape and mismatch of fiber orientation was concentrated at the web, where it actually improved the shear properties of the cured laminate. After the final layer of glass fiber tape was wound, a peel ply was applied, and the spar was parked, lined up, and cured.

Prepreg Technology

The prepreg (pre-impregnated) technology is adapted from the aerospace and aircraft industry, and it is based on the use of a semi-raw product where the fiber fabrics are pre-impregnated with resin, which is not yet cured. At room temperature the resin is like a tacky solid, and the tacky prepregs can be stacked on top of each other to build the desired laminate. By increasing the temperature, the resin becomes liquid/viscous, and the laminate can be consolidated under pressure and cured into the final component. Prepregs are available in many varieties and combinations of types of fibers, style of fabrics, and resin systems, all having different process and curing temperatures ranging from about 70 to 225°C. For large wind turbine blades, a process temperature around 80°C is most common. It results in sufficient temperature resistance of the cured laminate, and it keeps down the process and equipment costs. The required pressure to consolidate the stacked layers of prepregs is achieved by vacuum. The whole lay-up of prepreg is covered by a polymer film, which is sealed to the mold along its edge. A vacuum is pulled underneath the polymer film, and the atmospheric pressure outside the film presses the film, and thus the prepreg layers, toward the mold surface. The shelf life of the prepreg at room temperature normally ranges from days to a few weeks depending on the resin system. Therefore, the prepreg is typically stored at −18°C, resulting in a shelf life from 6 to 12 months.

The prepreg technology offers some advantages: it is easier to control and obtain constant materials properties, and a higher fiber content gives higher specific stiffness and strength of the material, which leads to lighter blades. Also, a clean process is obtained, which leads to a better working environment, with fewer requirements for the workshop ventilation systems (resulting in cost savings on smaller ventilation systems and energy needed for heating).

Vestas Wind Systems, one of the wind turbine manufacturers, uses the prepreg technology for their blade production. Vestas has used glass/epoxy prepreg technology for many years, and they have now introduced carbon fibers in their 45-m-long blades (23, 24).

Resin Infusion Technology

There are many names and variations of the resin infusion technology, which in principle consists of placing dry fibers in a mold, encapsulating and sealing off the fiber package, injecting the liquid resin into the fiber package, and curing the component. This technology was developed in the 1950s, but was used only rarely by the industry until the 1990s. Since then, the fibers, the resins, the accessories for the process, and the process equipment have been intensively developed, and resin infusion technology is now a widely used industrial process. The most important issue for the process is to ensure that all fibers are thoroughly wetted by the resin, in other words, there should be no areas with dry fibers in the final product. Most developments have concentrated on this single issue, and this problem has been tackled from many sides:

- fiber sizing with improved wettability to facilitate complete wetting without manipulating the rovings,
- fiber fabrics with special architecture to control flow pattern of the resin,
- resin with lower viscosity (at room or moderate temperature) to improve wettability and lower the process time for large components,
- resin that does not release volatiles under vacuum,
- accessories that help control resin flow pattern over large areas and ensure complete wetting of thick laminates (resin distribution mesh, special sandwich core, etc.),
- equipment for continuous mixing of resin without introducing air,
- design of molds (placement of inlet and outlet for the resin, sealing around the edge) to control resin flow to prevent entrapped area with dry fibers,
- sensors for monitoring the flow front,
- computer models for predicting and optimizing the flow pattern for a given component.

LM Glasfiber in Denmark is one of the turbine blade manufacturers that uses resin infusion technology, and it has proven to be successful in the company's most recently developed 61.5-m-long blade (25). Generally, it is normal to use a vacuum infusion technique, where all the dry fibers are covered with a polymer film sealed to the edge of the mold and held in place against the mold by a vacuum pulled underneath the polymer film, Figure 9 (see color insert). Thereafter, the resin line is opened, and the resin is pulled into the fiber package by the vacuum. When all the fibers are wetted, the resin line is closed, and the vacuum sucks out surplus resin into a resin trap. The atmospheric pressure outside the polymer film consolidates the laminate toward the mold surface, and the resin is allowed to cure. The two airfoils and the webs are manufactured separately, and all parts are adhesively bonded to complete the blade in a subsequent process.

Bonus Energy (now Siemens Wind Power) in Denmark has developed a special fiber placement and resin infusion technology for the company's 40-m-long blade, where the fibers for the entire blade, including airfoil, spars and webs, are placed inside a mold cavity, and the resin is infused to complete the blade in one pass (26).

The advantages of resin infusion technologies, compared with those of hand-lay-up, are the same as mentioned for the prepreg technology, except for the easy process control. As mentioned above, the challenge is to control the resin flow to avoid spots with dry fibers in the finished product.

MATERIAL MECHANICS

The mechanical behavior of composite materials is influenced by a variety of parameters. The homogeneity and quality of the materials manufactured are essential. Defects such as wrinkles, misalignment, and porosities formed during the manufacturing process can lead to failure mechanisms such as fiber fracture, interface cracking, and matrix cracking, which are sources for microbuckling, translaminar crack growth, and delaminations. These failure mechanisms have essential influence on the resulting properties, and it is the obtainable material quality that sets the limits for the mechanical properties and their variations. From a basic materials approach the mechanical properties are controlled by the damage initiation and the damage evolution.

Damage

Talreja (27) described the mechanisms of damage in fiber, matrix, and interface in order to understand the first damage evolution in composite materials. The basic understanding of damage forms the basis for the damage continuum mechanics that has rapidly developed during the past two decades. With a focus on tension loading, Talreja (28) describes the fatigue failure mechanisms in three regions: at high loads, fiber failure; at medium loads, combined matrix and interface failure; and at low loads, matrix cracking. The damage mechanisms are different in the different regimes of the S-N curve, and accelerating the fatigue tests into another regime and extrapolating between the different regimes might lead to wrong results.

The degree of damage in a polymer matrix composite can be followed by measuring the decrease in elastic stiffness from the beginning to the end of the lifetime of the material (29). The stiffness versus number of cycles relation shows a progress that can be divided into stages (30–32). The first stage is a short initial phase with a rapid decrease in stiffness. The second stage is a steady-state phase where the rate of decrease in modulus is constant, and the third stage is a failure stage where the stiffness falls drastically because of fatal failures. Independent of damage stage, the rates of change in stiffness and the lifetime depend on the failure mode region where the material is loaded. The rates of change in stiffness of the material are defined as the damage growth rate. In order to generalize these considerations, this rate is normalized, and a linear relation between the normalized stiffness and the corresponding number of cycles is likely to be valid. Using this relation, the stiffness degradation at a given loading level for a prescribed number of cycles can be predicted. Under the assumption that the stiffness change is history independent, a model was suggested (30). On the basis of the constant amplitude S-N-curves, this model can predict the lifetime of the material when subjected to variable loading conditions, for example, stochastically defined or block-defined loading histories (31).

Recently, continuum damage mechanics have also been used to model the influence of environment, frequency and other state variables on the mechanical properties. This modeling, which is a part of the OPTIMAT (33, 34) project, was initially presented by Megnis, Brøndsted, and others (35, 36). In these papers the use of internal state variables describing the degradation of the material caused by external conditions is introduced. Each variable has an effect on the damage evolution, exemplified by the change in stiffness of the material. In its final form, it is expected that this material model can be used in the design considerations.

MECHANICAL PROPERTIES

The wind turbine blades are designed toward stiffness and fatigue. Hence, the materials required for the blades are materials with a very high performance. Their exceptionally high stiffness to weight and high strength to weight ratios, together with an excellent fatigue performance, are the reasons for composite materials being the dominant choice. Basically, the elastic properties can be calculated on the basis of the properties of the individual constituents (fiber and matrix) and the knowledge of the lay-up (i.e., fiber orientation, fiber volume fraction), Equation 3. In these cases, either the simple rule of mixture can be used or more advanced laminate calculation procedures are available. The calculations are experimentally verified by static testing measuring tension, compression, and shear properties. The demands on higher stiffness affect the fatigue behavior because materials with a higher stiffness are generally more fatigue sensitive. Consequently, the fatigue strength is approaching the performance limits, and the fatigue limits of the materials become the main limiting factor.

Stiffness, static strength, and fatigue performance are measured by testing test coupons or components. The properties measured are used to qualify the materials and to ascertain that the materials fulfill the design demands originated in the aero-elastic modeling of the blades under expected wind loads. Uniform static tests are carried out under quasistatic loading, and fatigue tests are carried out under varying loading conditions. Properties are measured under tensile loads, compressive loads, shear loads, or combinations of these in bi- and multiaxial loading modes.

During the past two decades valuable experience has been obtained in different national and international research programs. In three European projects, from 1986 to 1996 (37–39), research was focused on the mechanical behavior of the wind turbine blades and the materials. In these programs static properties, as well as the high cycle fatigue behavior of glass-polyester and glass-epoxy materials, were measured. Test techniques were developed and design curves based on constant load amplitude tests and on variable amplitude tests using stochastic load histories were established. The results from these projects were collected in two books edited by Kensche (40) and Mayer (41). De Smet & Bach collected the data into a database FACT (42).

Supported by Sandia National Laboratories, Montana State University has worked intensively since 1989 on characterizing composite materials for wind turbines. The results from this nationally supported program are collected in a large database, the DOE/MSU database, first presented in 1997 by Mandell & Samborsky (43). Since 2001 the database has been updated annually (44).

The European activities have continued in a currently running European project (2001–2006), OPTIMAT (33, 34), under the 5th EU framework program. This project aims to provide accurate design recommendations for the optimized use of materials within wind turbine rotorblades and to achieve improved reliability. The project is investigating the structural behavior of the composite material (glass-epoxy) under the unique combination of conditions experienced by rotor-blades such as variable amplitude loading, complex multiaxial stress states (45, 46), extreme environmental conditions, thick laminates, and their possible interactions. Techniques will be developed for damage mechanics (35), life extension, condition assessment, and repair. The increasing demands for design toward stiffness and fatigue strength have required development of more accurate and reliable testing techniques.

Static Tests

TENSILE TESTING Tensile testing of composite materials (48) is performed using well-established techniques and is governed by standards, where the ASTM D3039 (49) and ISO 527 (50) are the most well documented and widely used. Making between five and ten tests on small specimens usually generates the static strength data of the materials, e.g., modulus, Poisson ratio, strength, and strain properties. The standards deal with tensile testing of both unidirectional and multidirectional fiber composites. Figure 10 (see color insert) shows a tensile set-up with non-contact extensometry.

COMPRESSION TESTING Compression strength is the most critical static property of many composite materials. The compression strength of unidirectional carbon fiber epoxy is about 60% of the static tensile strength. Basically, the failures in these cases come from microbuckling of the fibers, and several papers have treated this problem on a micromechanical approach (48). Fleck (51) showed in a review on the main mechanics of compressive failure that from a stability point of view the fiber alignment must be within $\pm 1^\circ$ in order to avoid microbuckling and instability. Figure 11 shows a kink-band formed through two porosities.

Because the main fiber direction in the flanges of the wind turbine blades is unidirectional, it has become the main topic for optimizing the manufacture of unidirectional materials. In order to document these properties, many attempts have been made to develop the most reliable test methods for documentation. Compression test of unidirectional composites is probably the most contentious type of testing. In a compression test, the critical factor appears to be from instability. This sets the requirements for the test set-up and alignment of test fixture. Early suggestions were to use supported test fixtures, such as the Celanese fixture and the IITRI test fixture (Lamothe & Nunes) (52), adopted in the ASTM D3410 (53), and the Boeing fixture defined by ASTM D695 (54), which suggested a supporting jig for thick specimens. Several other techniques of antibuckling support have been suggested, but all have given contradicting results. Adams & Welsh (55) suggested a combined loading compression test method where the load transfer to the test specimen are controlled end loading and shear loading introduced from the sides of the specimens. Also Adams (56) discussed the use of tabbed and untabbed compression specimens, but neither of these test techniques has resulted in convincing

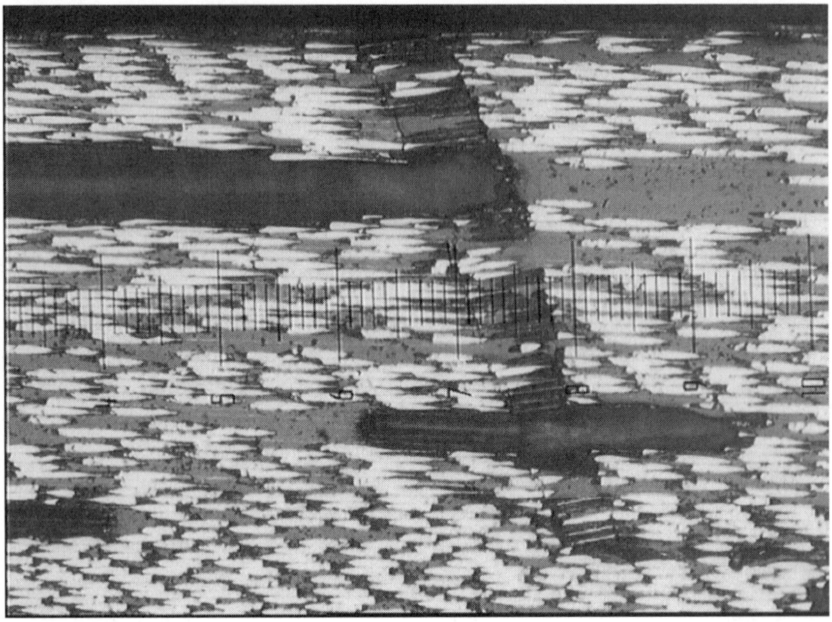

Figure 11 Microbuckling initiated in compression between two porosities. (Scale unit: 10 μm.)

results, especially when focusing on unidirectional carbon fiber composites. In 1999 an ISO standard for compression testing (ISO 14126) (57) was released. This standard takes the bending during test into account and gives a maximum bending ratio for accepting the test results.

In order to further reduce the bending and to gain a closer control of the end to shear load ratio, a new concept of the combined compression test fixture was developed at Risoe National Laboratory (Figure 12, see color insert).

This concept of performing compression testing appears to be very successful. Bending is almost avoided, and it has been possible to obtain compression strengths of carbon-epoxy composites that are 30–40% higher than those measured earlier. In the OPTIMAT project it has been shown that even for glass/epoxy, the compression strengths measured are 20% higher when using a combined loading fixture compared with those used testing a short-gauge-length free specimen.

SHEAR TESTING The knowledge of the shear properties of the composite is also vital when modeling structural behavior (48). The shear modulus is most successfully measured by the V-notch beam method [Iosipescu 1967 (58)] standardized in ASTM D 5379 (59). However, when measuring laminate shear properties, the ±45° tensile test can be used [ISO 14129 (60) or ASTM D 3518 (61)]. Interlaminar shear is measured by the rail shear test [ASTM D3846 (62)] or by the short

Figure 1 The world's largest wind turbine (summer 2004) at Risoe National Laboratory test site, Høvsøre, Denmark. Tower height is 120 m, rotor diameter is 110 m, and generator power is 3.6 MW.

Figure 8 Tape winding of a spar for the Nibe wind turbines. The thickness of the spar was varied along the length, by varying the length of travel of the tape placement apparatus along the length of the spar. A wedge was manually inserted and withdrawn underneath the tape, at the shear web position of the spar, for each revolution of the mandrel, in order to control and maintain straight fibers with the desired orientation in the upper and lower part of the spar.

Figure 9 Set-up for resin infusion test of a root-end section of a large rotorblade at LM Glasfiber, Lunderskov, Denmark (25).

beam bending test ILSS [ASTM D2344 (63)], which, however, is not suitable for measuring design data and is mainly used in production control and data comparison.

Fatigue

On the basis of the load histories predicted for the blades, strength requirements are calculated for the materials. It must be verified that the fatigue strength for a given lifetime meets the requirements. These requirements are established by different national (64) and international recommendations, i.e., the DNV-Risoe guidelines (65) or the Germanischer Lloyd Wind Energy guidelines (66).

The results from the large research programs (33, 37–39, 42, 44) have contributed considerably to the general knowledge of the fatigue behavior of the composite materials. A variety of models have been used for describing behavior and degradation, and damage evolution in the materials has been suggested (67). Some models are constitutive ones based on physical and mechanical observations, and other models are empirical mathematical descriptions of experimental data gained. Using an extensive collection of papers and reports, Sutherland (68) published an extensive review paper on the fatigue properties of materials for wind turbines. In this review the behavior of metallic materials and woods are included.

In order to verify the models and approve the materials used to provide a given fatigue lifetime, extensive testing has focused on design considerations. Only a few standards exist for measuring the fatigue performance of the composites materials. For tension-tension fatigue, ASTM D3479 (69) gives very brief guidelines for a tensile test specimen, with an axial tension-tension cyclic loading. This standard does not prescribe clamping procedures, frequencies, or data reduction, and thus it has little value in practical applications. Late in December 2003 another international standard, ISO 13003 (70), was published defining general procedures for fatigue testing. This standard addresses all modes of testing and test machines. However, suggestions for procedures and data evaluations are still very vague and do not consider any newer findings and research results.

LOADING HISTORIES The wind speeds and the wind resources on the turbines have been extensively studied and are described in the design manuals (65, 66). On the basis of the measurements on nine different wind turbines in Europe, and representing approximately two months in service of a virtual wind turbine, a model simulating the load conditions in a blade (based on the loads in the flapwise direction measured at a point close to the blade root) has been introduced in the WISPER load sequence. WISPER (WInd SPEctrum Reference) has been used as a standardized load spectrum for horizontal axis wind turbine blades. The spectrum was developed in an IEA working group in 1987 (71). A truncated spectrum WISPERX, containing only 10% of the cycles, was later developed in order to reduce test time. The WISPERX was applied as testing sequences in the beginning of the 1990s.

However, to a considerable degree fatigue testing has been carried out as constant amplitude tests. The constant amplitude tests form the basis for understanding the fatigue damage models and are the basis for formulating models predicting the behavior under stochastic loadings. The actual loading on the blades brings the material into tension-tension fatigue and compression-compression fatigue and in some parts of the blades into tension-compression fatigue. The loading type is described by the parameter R = minimum load/maximum load. Because of the typical loads on the blades, three loading modes are normally tested. They are tension-tension fatigue (R = 0.1), tension-compression fatigue (R = −1), and compression-compression fatigue (R = 10). Most tests are performed at relatively low frequency (5–10 Hz) in order to keep down the temperature increase from hysteresis loss heating of the test specimen. In order to gain very high cycle fatigue data, Mandell et al. (72) have developed a test set-up in which they tension-tension loaded few-fiber specimens. The test equipment is based on a set-up with a loudspeaker as loading medium. At frequencies up to 200 Hz, they report data into the 10^9–10^{10} cycles regime. Although showing a large scatter, the data are consistent with extrapolated low-frequency data. From the University of Latvia, Apinis (73) has recently published a description of a high-frequency fatigue test machine based on an electrodynamic vibration exciter. Using this machine, a series of high-frequency tests on standard test coupons from the European Joule projects have been tested into the ultra-high cycle regime at 300 to 500 Hz. Results were compared with the low-frequency results reported in the FACT database (42). In that paper the authors deal with the results as if they represented a single population. However, it is our opinion that the results from high-frequency tests are grouped in two populations of which the data from the low-frequency tests show the highest performance, which indicates that a frequency effect exists. It is still not clear if viscoelastic effects and temperature effects can explain the frequency effect.

Data Handling

Making between 5 and 20 tests on small specimens usually generates the fatigue strength data of the materials. The Wöhler diagram (or S-N curve) describes the relation between the loading parameter, which can be maximum values, amplitudes, or ranges (two × amplitude) and the number of cycles to failure with a constant mean value or R-value. The Wöhler diagram is illustrated in Figure 13 using selected data from the OPTIMAT database.

A number of experimental investigations (59, 60, 74–78) on fiber-reinforced materials for rotorblades have shown that the fatigue strength, S, at a constant R-ratio can be described by the analytical relation suggested by Basquin in 1910.

$$N_f \cdot S^m = C, \qquad 4.$$

where C is a constant, N_f is the number of cycles to failure, and m is a shape parameter (Wöhler exponent) reflecting the slope of the log-log line in the Wöhler

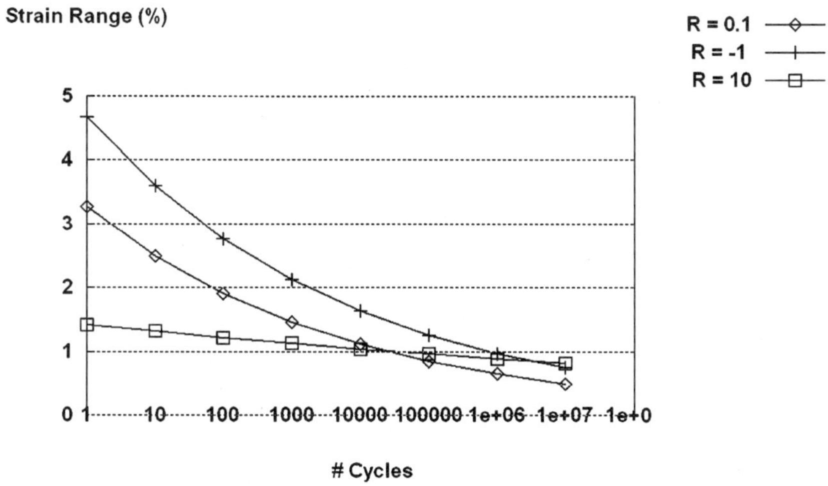

Figure 13 Wöhler diagram showing 50% regression lines based on data from the OPTMAT database.

diagram. A high m-value (>10) reflects a flat curve and a moderate m-value (1–10) reflects a steeper line. (The fatigue strength, S, can be either on a strain, stress or load basis.) Continuous fiber-reinforced polymers cannot be assumed to have a fatigue limit but will continue to degrade according to the equation above.

Alternatively, as mentioned by Sutherland, the S-N presentation can be fitted using an exponential function

$$N \cdot 10^{m_1 S} = C_1, \quad 5.$$

which results in a straight line relationship in a log-linear diagram. This empirical model has mainly been applied to present the data from the DOE/MSU database (43, 44).

However, the dependent parameter N is a two-parameter function depending on both the applied amplitude and the applied mean value. The third dimension in the fatigue parameter space can be illustrated in the Haigh (or Goodman) diagrams (79). These diagrams can be plotted showing the relationship between mean value and amplitude or range at a particular fatigue life. This is illustrated in Figure 14. A point in this type of diagram represents a given number of cycles to failure, i.e., one point marks at which range and mean level the lifetime will be the preset number of cycles. Straight lines through the origo in this diagram show "iso-R" lines and test results from tests run at constant R-values lie on these lines. If the test results are positioned on vertical lines, it means that the tests were run at constant mean values. Combining the Haigh diagram and the Wöhler diagram gives a three-dimensional plotting of the fatigue results, as shown in Figure 15. The relationships between mean and amplitude or range for metals were suggested by Goodman (80) and Gerber (81) in the 19th century, long before the synthetic

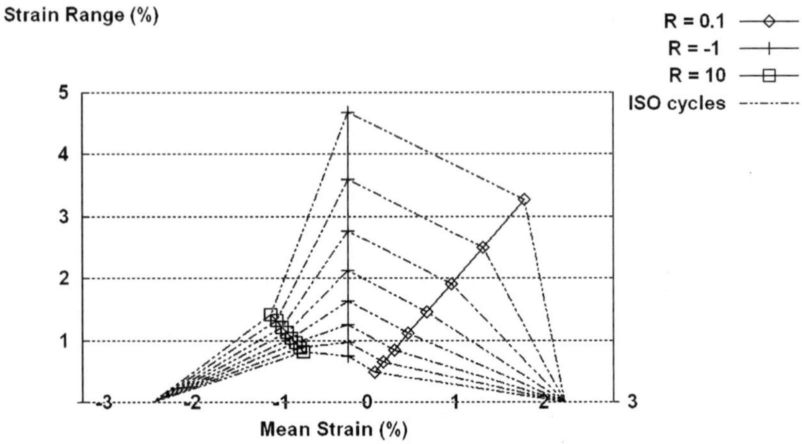

Figure 14 Haigh diagram based on the fitted data in Figure 13.

polymers were invented and nearly a century before continuous fiber composites were introduced. The Gerber and Goodman relationships can be formulated in the following expression:

$$S_a = S_c \cdot \left(1 - \frac{S_m}{S_u}\right)^c, \qquad 6.$$

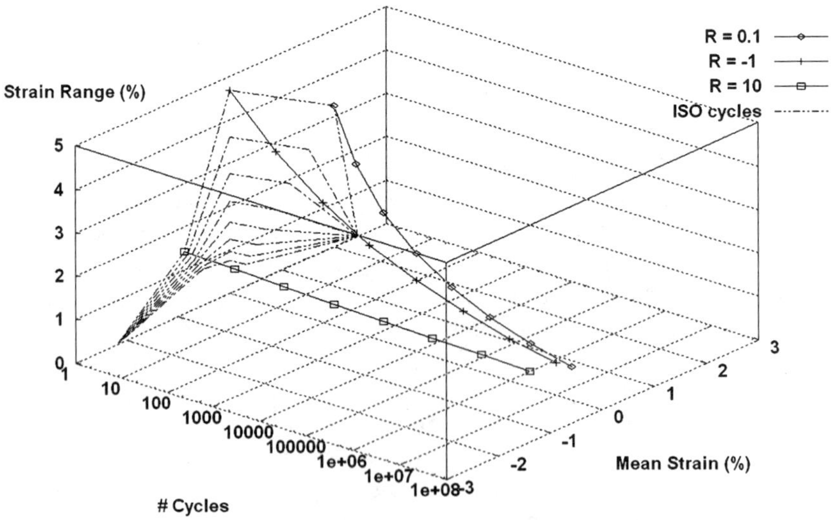

Figure 15 Combining Figure 13 and Figure 14 gives a three-dimensional illustration of the relationship between Haigh diagrams and Wöhler diagrams.

where S_a and S_m are the amplitude and mean values, respectively, S_c is the fatigue limit, and S_u is the ultimate value of the strength parameter, and c is a constant equal to 1 for the Goodman relationship and equal to 2 for the Gerber suggestion.

Sutherland (67, 68) has reported several efforts to plot data mainly from the DOE/MSU database into Haigh diagrams. An attempt has been made to use Gerber or Goodman relationships, but for composites no empirical or analytical relationship between mean values and amplitudes at fixed numbers of cycles could be established.

It is advantageous to be able to compare variable amplitude (VA) fatigue test results to constant amplitude (CA) fatigue test results in order to present them on a simple parametric form.

Palmgren (82) and Miner (83) suggested that cumulative damage in fatigue during a VA test can be reflected in one parameter, the Palmgren-Miner sum, calculated as

$$M = \sum_{i=1}^{j} \frac{n_i}{N_i}, \qquad 7.$$

where i is the level number in the VA test, n_i is the number of cycles at level i in the VA test, N_i is the cycles to failure in the CA test at level i, and j is the number of different load levels. This sum forms the basis for the Palmgren-Miner rule, which states that failure takes place when M equals 1. This relationship is based on fatigue of metals, and for this group of materials the exponent m of the S-N curve is of the order of 3. The S-N line for metals is much steeper than the S-N lines for glass and carbon fiber composites where the exponent m is 10–25.

A method of predicting VA lifetimes based on CA findings was suggested by Dover (83a) and descriptions of methods for comparing fatigue lives for spectrum loading was reported by Solin (83b), which together have been successfully used in evaluating stochastic load data on steel structures. The most successful methods are based on the calculation of an equivalent loading value. An equivalent fatigue stress or strain, S_{eq}, of a VA fatigue test is defined as the CA fatigue level at which failure takes place after the same number of cyclic loadings. Brøndsted et al. adapted this concept (84) for analyzing fatigue tests using block loading and WISPERX spectrum loading.

Combining the Basquin equation and Palmgren-Miner rule with n_i as the number of cycles at different load levels, S_i, the equivalent stress or strain, S_{eq}, is calculated from

$$S_{eq} = C^{\frac{1}{m}} \cdot \left(\sum n_i\right)^{-\frac{1}{m}}, \qquad 8.$$

and an expression for the normalized equivalent stress is given as

$$S_{eq} = \left(\frac{\sum (n_i \cdot S_i^m)}{\sum n_i}\right)^{\frac{1}{m}} \cdot \left(\frac{1}{M}\right)^{\frac{1}{m}}. \qquad 9.$$

M is the Miner sum. It is seen that the equivalent stress value depends on $M^{\frac{1}{m}}$. Knowing the Palmgren-Miner sum, VA fatigue lifetimes can now be predicted based on the CA fatigue test results and the knowledge of the variable load history.

Hence, a comparison between CA test results and VA test results is possible by assuming that $M^{\frac{1}{m}} = 1$, but the validity of the Palmgren-Miner rule, $M = 1$, can be discussed. A typical m-value of metallic materials is 3. Allowing S_{eq} to vary within values of 10% means that $M^{\frac{1}{m}}$ can vary between 0.9 and 1.1. Now taking the power into account, this means that an acceptable variation in M is in the range from 0.73 to 1.33. For glass fiber polymer composites the exponent m is of the order of 10, and the acceptable variation in M ranges from 0.3 to 2.6. Furthermore, for carbon fiber composites, ceramics, and other very brittle materials the exponent m is found to be up to 20 or 25, which leads to a very large acceptable variation in the Miner sum. Therefore, using the Palmgren-Miner sum when comparing VA and CA fatigue test results, the slope of the constant amplitude S-N curve is a key parameter. The Miner sum M is hereby shown to be too dependent on the slope m and is therefore not a suitable parameter to use because high exponent (m) values give a large scatter in the Miner sum (M).

By testing this rule it has been found that failure can take place in cases where the sum is significantly lower than 1 (unsafe as a design rule) and where it is considerably higher than 1 (safe as a design rule); therefore, lifetime predictions alone cannot be based on this rule. A large scatter in the Palmgren-Miner sum is often explained by the degree of sequence dependence, which can be beneficial ($M > 1$) or detrimental ($M < 1$).

Environmental conditions, which affect the materials fatigue behavior, should be considered and taken into account. Such conditions in particular include humidity and temperature, both of which may lead to degradation of strength and stiffness. Their design effects calculated by introducing appropriate partial coefficients as safety factors should be applied in the strength analysis. Not much has been published thus far on environmental effects, but such investigation is a part of the OPTIMAT project.

STATISTICAL DATA ANALYSIS

The S-N curve is often analyzed by linear regression analysis of the S-N data pairs in log-log or in log-linear mode depending on the expression adopted. However, because the statistical distribution is unknown, the ASTM E739 (85) does not recommend either extrapolating the lines outside the interval tested or using the common practice of introducing a 95% confidence line as a design line. Furthermore, it cannot be recommended to use another common practice, which is to define the upper design bounds by subtracting two or three times the standard deviation. Because the design lifetime of a wind turbine blade goes into the very high cycle range, one could either adopt dubious accelerated high cycle fatigue test data or turn to conservative statistically calculated upper bounds for the design lines. A statistical method for predicting lifetime was suggested by Echtermeyer et al. (86), and Ronold & Echtermeyer (87) presented the theoretical basis. They showed the possibilities for establishing characteristic values for unconditioned and conditioned variables and introduced a mathematical concept of calculating 95% tolerance bounds with a 95% confidence interval.

CHALLENGES

Thermoplastic Composites

Since the 1980s different types of thermoplastic composites with continuous fibers have been on the market, ranging from carbon fiber/PEEK to glass fiber/PP. The drive for using thermoplastic composites is higher toughness, faster processing, unlimited shelf life of the semi-raw materials, a clean working environment, and potential for much easier recycling of the material. Factors that have held back the thermoplastic composites from being used for wind turbine blades are high process temperature (requires high temperature resistant tooling and accessories), introduction of new process technologies, and more difficult adhesive joining.

In principle, the manufacturing technology for traditional thermoplastic composites is simple. It consists of (*a*) heating the material to melt the thermoplastic polymer, (*b*) consolidating the material under pressure to form the composite material and the component, and (*c*) cooling the material to return the polymer to a solid state. For those thermoplastic matrix materials that can be used continuously at elevated temperatures, such as 80–250°C, the processing temperature is in the range of 180–390°C, and special care is required in designing tools and selecting accessories regarding thermal expansion and thermal resistance (88).

Recently, a new type of cyclic thermoplastic polybutylene terephthalate (PBT) resin has been developed. It has a very low viscosity at the processing temperature and is suitable for the resin infusion technology (89). At the process temperature, which is around 180°C, a reaction transforms the liquid polymer into a solid thermoplastic polymer.

Several wind turbine blade manufacturers have been or are involved in programs where the use of thermoplastic composites for blades has been or is being investigated. In 1996, LM Glasfiber, Comfil, and Risoe National Laboratory developed a concept for manufacturing airfoils for blades in glass fiber/L-PET (a modified PET with process temperature of 220°C) (88, 90, 91). The concept was demonstrated on a 3.2-m-long airfoil. In 1999 Bonus Energy and Risoe National Laboratory developed a concept for manufacturing complete blades in one pass in a glass fiber/PP composite material at a process temperature of 180°C. That concept was demonstrated on a short blade section of 0.5 m, as illustrated in Figure 16 (see color insert) (92). In 2004, Gaoth Tec, Cyclics, and Mitsubishi initiated a development program for a 12.6-m-long blade manufactured by a resin infusion technology using a thermoplastic resin from Cyclics (93).

Structural Materials

New materials and materials concepts are illustrated in Table 1, where the fibers of polymer origin have potential, in particular due to their low density. The merit index $E^{1/2}/\rho$ is more affected by (reduced) density than (increased) stiffness, so (very) light materials are of particular interest for weight critical components such as large rotorblades. At present none of the composites based on aramid, polyethylene, or cellulose fibers has been developed to an industrial technical level, which could make them practically interesting for large rotorblades. Two of the polymer fibers, aramid and polyethylene, are based on crude oil, i.e., a non-renewable resource, whereas the cellulose fibers are based on biomass, which is a renewable resource. The increasing focus on environment and ecology, as well as on limited resources, may make the cellulose fibers particularly interesting for demanding structural applications, such as large rotorblades.

New Structural Design

An important goal for rotorblade manufacturers is the reduction of blade weight relative to the length. This is both structurally necessary, as discussed above, and economically attractive because it will lower the cost of rotorblades and thus eventually reduce the cost per energy unit produced (cost/kWh). In addition to the new materials concept, the structural design has also been improved over the years. The three new data points in Figure 3 show the weight reduction achieved for essentially glass fiber composite blades, by improved design ideas (as well as some use of carbon fiber composites). The structural design of the heavy root section of rotorblades has been optimized (although details are not available), leading to thinner root diameters for longer blades and thus smaller, lighter, and less expensive hubs.

The special behavior of anisotropic non-symmetrical laminates under combined bending and twisting has been exploited to design composite laminates and blade structures where the blade bends and twists simultaneously under heavy loads. Thus the aerodynamic profile of the rotorblade can be made to adjust itself optimally at low wind speeds for maximum aerodynamic performance, and to twist itself out of the wind at high wind speeds so that the loads on the blade are reduced. An additional advantage under heavy winds is that the normal pitch control, which physically rotates the blades around the blade axis, is slow to respond to gusts, so here the active blade twisting helps in controlling the blade and rotor performance under the always varying wind conditions, i.e., wind speed and wind direction.

New Structural Health

With still larger rotorblades and the associated risks, and with the constant striving to improving the economy of wind energy, the rotorblades need to be designed both safely and to the limit, to avoid over-design and over-weight. In this context, materials research needs to better understand the continuous degradation of fiber composites under (long-time) fatigue loading and to develop models for prediction of lifetimes.

Structural health monitoring has been under development recently to allow continuous recording of important parameters for the state of degradation/damage of the materials. The state of degradation is a measure of the remaining lifetime and when reliable models exist accurate prediction of lifetime can be made. Any extraordinary change in the state of degradation can be used to initiate inspection, repair, or replacement of rotorblades. Recently, large rotorblades have been fitted with monitoring systems having embedded optical fibers and related software, which record loads, temperature, damage, and lightning strikes (25).

Trends and Markets

The trends for rotorblades are for increasing size and, most critical, reduced weight. This couples with the ever-increasing size of turbines, which are supposed to deliver reduced costs for wind energy (cost/kWh). The materials and, in particular, composites are important elements in this development because they deliver low density, good stiffness, and potentially long lifetimes. New trends in the materials occur with respect to properties, design, and monitoring, as well as manufacturing technology. Today's markets are world wide, with suppliers of wind turbines mainly located in Europe (72%), with activities in United States (18%) and Asia (10%) (1). The wind turbines are generally being placed off-shore in parks or farms, both to exploit the more free and generally stronger winds and also to reduce the environmental effects on land.

In order to meet the challenge as wind turbine blades grow longer and to allow for the next generation of larger wind turbines with intelligent blades, higher demands are placed on materials and structures. This requires more thorough knowledge of materials and safety factors, as well as further investigation into new materials. Furthermore, a change in the whole concept of structural safety of the blade is required. Kensche (94) reviewed the trends in the fatigue investigations of materials for wind turbines. He points out that many topics remain to be explored. In particular, there should be a focus on the basic understanding of damage and failure mechanism, effects and interpretation of the stochastic loadings, multiple stress states, environmental effects, size effects, and thickness effects. New initiatives are in progress in the field of condition monitoring (structural health monitoring), and several manufacturers of wind turbine blades are already fitting the blades with sensors. The challenge is to be better able to interpret the signals and to know the actions to take. This requires an even better basic understanding of the life cycle behavior and the degradation mechanisms.

The *Annual Review of Materials Research* is online at http://matsci.annualreviews.org

LITERATURE CITED

1. BTM Consult ApS, World Market Update 2003—March 2004. 90 pp.
2. NewsLetter, May 2004. LM Glasfiber A/S. www.lm.dk
3. Ashby MF. 1992. *Materials Selection in Mechanical Design*. Oxford, UK: Pergamon. 312 pp.
4. Lovatt A, Shercliff H. 2002. Material selection and processing. http://www-materials.eng.cam.ac.uk/mpsite/interactive_charts/
5. Watt W, Perov PV. 1985. Strong fibres. In *Handbook of Composites*, ed. A Kelly, YN Rabotnov. Amsterdam: North-Holland/Elsevier. 755 pp.
6. Bunsell AR. 1988. *Fibre Reinforcements for Composite Materials*. Amsterdam: Elsevier. 537 pp.
7. Chou TW, ed. 2000. Fibre reinforcements and general theory of composites. See Ref. 95, 1:802
8. Lilholt H, Lawther JM. 2000. Natural organic fibers. See Ref. 95, 1:303–325
9. Lilholt H, Toftegaard H, Thomsen AB, Schmidt AS. 1999. Natural composites based on cellulose fibres and polypropylene matrix. *Proc. Int. Conf. Compos. Mater.,12th, Paris, Pap. 1115*. 9 pp.
10. Lilholt H, Madsen B, Toftegaard H, Cendre E, Megnis M, et al., eds. 2002. Sustainable natural and polymeric composites—science and technology. *Proc. 22nd Risoe Int. Symp. Mater. Sci. Risoe Natl. Lab., Roskilde, Denmark.* 371 pp.
11. Talreja R, Månson J-AE. 2000. Polymer matrix composites. See Ref. 95, 2:1129
12. Jones RM. 1975. *Mechanics of Composite Materials*. New York: McGraw-Hill. 355 pp.
13. Hancox NL, Mayer RM. 1994. *Design Data for Reinforced Plastics, A Guide for Engineers and Designers*. London: Chapman & Hall. 326 pp.
14. Lilholt H, Talreja R, eds. 1982. Fatigue and creep of composite materials. *Proc. 3rd Risoe Int. Symp. Metall. Mater. Sci. Risoe Natl. Lab., Roskilde, Denmark.* 341 pp.
15. Andersen SI, Brøndsted P, Lilholt H, Lystrup Aa, Rheinlaender JT, et al., eds. 1997. Polymeric composites—expanding the limits. *Proc. 18th Risoe Int. Symp. Mater. Sci. Risoe Natl. Lab., Roskilde, Denmark.* 525 pp.
16. Hütter U. 1960. Tragende Flugzeugteile aus glasfaserverstärkten Kunststoffen. *Luftfahrttechnik* 6:2:34–44
17. Tvindskolerne. 1977. Lad 100 møller blomstre. 6 ed. *Forlaget Skipper Klement. Special-Trykkeriet A/S. Viborg, Denmark* (In Danish)
18. Fakta om Vindenergi. 2003. *Faktablad M5*. Danish Wind Turbine Owners' Assoc. http://www.dkvind.dk/fakta/fakta_pdf/M5.pdf. (In Danish)
19. 1976. Design study of wind turbines 50 kW to 3000 kW for electric utility application. *Kaman Rep. No. R-1382, ERDA/NASA—19404—76/2*. pp. 4–5, 4–51, 4–79. Kaman Aerospace Corp., Bloomfield, CT
20. White ML, Weigel WD. 1979. A low cost composite blade for a 300 foot diameter wind turbine. *34th Annu. Tech. Conf., Reinforced Plastics/Compos. Inst.*, Sec. 15-C: 1–6. Soc. Plastic Ind., Washington, DC
21. Weingart O. 1979. Fabrication of large composite spars and blades. See Ref. 20, Sec. 15-B:1–4
22. Johansen BS, Lilholt H, Lystrup Aa. 1980. Wingblades of glass fibre reinforced polyester for a 630kW wind turbine—design, fabrication and materials testing.

In *Advances in Composite Materials,* 2:1355–67. Oxford, UK: Pergamon
23. http://www.vestas.com/produkter/pdf/updates_020804/V90_3_UK.pdf
24. Griffin DA. 2004. Growing opportunities and challenges in wind turbine blade manufacturing. *High-Performance Composites.* http://www.compositesworld.com/hpc/issues/204/May/450
25. http://www.lm.dk
26. Griffin DA. 2004. Cost and performance tradeoffs for carbon fibres in wind turbine blades. *SAMPE J.* June/July, pp. 20–28
27. Talreja R. 1982. Damage models for fatigue of composite materials. See Ref. 14, pp. 137–53
28. Talreja R. 2000. Fatigue of polymer matrix composites. See Ref. 95, 2:529–52
29. Stinchcomb WW, Bakis CE. 1990. Fatigue behavior of composite laminates. *Fatigue of Composite Materials,* ed. KL Riefsnider, pp. 105–80. Amsterdam: Elsevier
30. Andersen SI, Brøndsted P, Lilholt H. 1996. Fatigue of polymeric composites for wingblades and the establishment of stiffness-controlled fatigue diagrams. *1996 Eur. Union Wind Energy Conf. EWEC '96, Göteborg,* pp. 954–59. Bedford: HS Stephens & Assoc.
31. Brøndsted P, Lilholt H, Andersen SI. 1997. Fatigue damage prediction by measurements of the stiffness degradation in polymer matrix composites. In *Int. Conf. Fatigue Compos.,1st, Paris, 8. International spring meeting of ICFC, Paris,* pp. 370–77. Paris: ICFC
32. Brøndsted P, Johansen BS. 1999. Measurement of damage progress in FRP materials. *Plast. Rubber Compos.* 28:9:458–62
33. EU FP5. 2001–2006. Reliable optimal use of materials for wind turbine rotor blades (OPTIMAT BLADES). ENK6-CT-2001-00552. http://www.wmc.tudelft.nl/optimat_blades/
34. van Wingerde AM, Nijssen RPL, van Delft DRV, Janssen LGJ, Brøndsted P, et al. 2003. Introduction to the OPTIMAT BLADES project. In *Proc. CD-ROM CD 2, Eur. Wind Energy Conf. Exhib. 2003, Madrid, Spain, 16–19 June,* p. 9. Brussels: Eur. Wind Energy Assoc.
35. Megnis M, Brøndsted P, Rehman SA, Ahmad T. 2004. Life predictions of long fiber composites in extreme environmental conditions using damage evolution approach. *Proc. 11th Eur. Conf. Compos. Mater., Rhodes, Greece,* 10 pp. London: Eur. Soc. Compos. Mater.
36. Megnis M, Brøndsted P, Rehman SA, Ahmad T. 2004. Characterization of internal state variable for fiber fracture in UD composite. *Int. Conf. Advances in Computational and Experimental Engineering and Sciences, Madeira, Port,* pp. 451–56. Encino, CA: Tech Science Press
37. 1987–1989. EC Proj. No. EN3W.0024. *Design basis for wind turbines.*
38. EC-JOULE-1 programme. 1990–1993. *Fatigue Properties and Design of Wingblades for Windturbines.* Contract no. JOUR-0071-DK (MB)
39. EC-JOULE-2 programme. 1992–1995. *Development of Advanced Blades for Integration into Wind Turbine Systems.* Contract no. JOU2-CT92-085
40. Kensche CW, ed. 1996. *Fatigue of Materials and Components for Wind Turbine Rotor Blades, EUR 16684.* Eur. Comm., Luxembourg
41. Mayer RM, ed. 1996. *Design of Composite Structures against Fatigue. Mechanical Engineering.* Suffolk, UK: Publications
42. De Smet BJ, Bach PW. 1994. Database FACT: Fatigue of composites for wind turbines. *3rd IEA Symp. Wind Turbine Fatigue, ECN, Petten, The Netherlands, April 22–23*
43. Mandell JF, Samborsky DD. 1997. *DOE/MSU Composite Materials Fatigue Database. Test Methods, Materials, and Analysis,* SAND97-3002, UC-1210, Sandia Natl. Lab., Albuquerque, NM

44. Update: *DOE / MSU Composite Material Fatigue Database*. February 24, 2004. http://www.sandia.gov/Renewable_Energy/wind_energy/other/973002upd0204.pdf
45. Smits A, van Hemelrijck D, Philippidis T. 2004. The digital image correlation technique as full field strain technique on biaxial loaded composites using cruciform specimens. *ICEM12. 12th Int. Conf. Exp. Mech., Politecnico di Bari, Italy*. 8 pp.
46. Smits A, van Hemelrijck D, Philippidis T, van Wingerde AM, Cardon A. 2004. Optimisation of a cruciform test specimen for bi-axial loading of fibre reinforced material systems. *Proc. 11th Eur. Conf. Compos. Mater., Rhodes, Greece*. London: Eur. Soc. Compos. Mater. 9 pp.
47. Deleted in proof
48. Adams DF. 2000. Test methods for mechanical properties. See Ref. 95, 5:113–48
49. ASTM D3039–00. 2000. *Standard Test for Tensile Properties of Polymer Matrix Composite Materials*. W. Conshohocken, PA: Am. Soc. Test. Mater.
50. ISO 527. Plastics—Determination of tensile properties. Geneva: ISO
51. Fleck NA. 1997. Compressive failure in fibre composites. In *Advances in Applied Mechanics*, ed. JW Hutchinson, TY Wu, 33:43–117. New York: Academic
52. Lamothe RM, Nunes J. 1983. Evaluation of fixturing for compression testing of metal matrix and polymer/expoxy composites. *Compression Testing of Homogeneous Materials and Composites. ASTM STP 808*, ed. R Chait, R Papino, pp. 241–53. W. Conshohocken, PA: Am. Soc. Test. Mater.
53. ASTM D 3410–95. 1995. *Standard Test Method for Compressive Properties of Polymer Matrix Composite Materials with Unsupported Gage Section by Shear Loading*. W. Conshohocken, PA: Am. Soc. Test. Mater.
54. ASTM D695-02a. 2002. *Standard Test Method for Compressive Properties of Rigid Plastics*. W. Conshohocken, PA: Am. Soc. Test. Mater.
55. Adams DF, Welsh JS. 1997. The Wyoming combined loading compression (CLC) test method. *J. Compos. Technol. Res.* 19:3:123–33
56. Adams DF. 2002. Tabbed versus untabbed compression specimens. *Composite Materials: Testing, Design, and Acceptance Criteria, ASTM STP 1416*, ed. AT Nettles, A Zureick, pp. 1–14. W. Conshohocken, PA: Am. Soc. Test. Mater.
57. ISO 14126. 1999. *Fibre-Reinforced Plastic Composites—Determination of the Compression Properties in the In-Plane Direction*. Geneva: ISO
58. Iosipescu N. 1967. New accurate procedure for single shear testing of metals. *J. Mater.* 2(3):537–66
59. ASTM Standard D 5379. 1998. *Standard Test Method for Shear Properties of Composite Materials by the V-notched Beam*. W. Conshohocken, PA: Am. Soc. Test. Mater.
60. ISO 14129. 1998. *Fibre-reinforced Plastic Composites—Determination of the In-Plane Shear Stress/Shear Strain Response, Including the In-Plane Shear-Modulus, by the ±45° Tension Test Method*. Geneva: ISO
61. ASTM D3518. 2001. *Standard Test Method for In-Plane Shear Response of Polymer Matrix Composite Materials by Tensile Test of a ±45° Laminate*. W. Conshohocken, PA: Am. Soc. Test. Mater.
62. ASTM D3846. 1994. *Standard Test Method for In-Plane Shear Strength of Reinforced Plastics*. W. Conshohocken, PA: Am. Soc. Test. Mater.
63. ASTM D2344. 1995. *Standard Test Method for Apparent Interlaminar Shear Strength of Parallel Fiber Composites by Short-Beam Method*. W. Conshohocken, PA: Am. Soc. Test. Mater.
64. Vestergaard B, Wedel-Heinen J, Halling KM, Sønderby O, et al. 2002. In *Type Approval Scheme for Wind Turbines. Recommendation for Design*

Documentation and Test of Wind Turbine Blades, ed. C Skamris. Copenhagen: Danish Energy Agency. http://www.dawt.dk/Common/RecomBlades.pdf
65. *Guidelines for Design of Wind Turbines*. 2001. Det Norske Veritas, Copenhagen, and Risoe Natl. Lab., Roskilde, Denmark. 261 pp.
66. *Guideline for Certification of Wind Turbines*. 2003. Hamburg, Ger.: Lloyd WindEnergie
67. Sutherland HJ. 1999. *On the Fatigue Analysis of Wind Turbines. SAND99-0099*. Albuquerque, NM: Sandia Natl. Lab.
68. Sutherland HJ. 2000. A summary of the fatigue properties of wind turbine materials. *Wind Energy* 3:1–34
69. ASTM D3479. 2002. *Standard Test Method for Tension-Tension Fatigue of Polymer Matrix Composite Materials*. W. Conshohocken, PA: Am. Soc. Test. Mater.
70. ISO 13003. 2003. *Fibre-Reinforced Plastics—Determination of Fatigue Properties Under Cyclic Loading Conditions*. Geneva: ISO
71. TenHave AA. 1991. *WISPER and WISPERX Final Definition of Two Standardised Fatigue Loading Sequences for Wind Turbine Blades*. NLR Rep. NLR TP 91476
72. Mandell JF, Samborsky DD, Wang L, Wahl NK. 2003. New fatigue data for wind turbine blade materials. *J. Solar Energy Eng. Trans. ASME* 125:506–14
73. Apinis R. 2004. Acceleration of fatigue tests of polymer composite materials by using high-frequency loadings. *Mech. Compos. Mater*. 40:2
74. Brøndsted P, Andersen SI, Lilholt H. 1996. Fatigue performance of glass/polyester laminates and the monitoring of material degradation. *Mech. Compos. Mater.* 32:21–29
75. Deleted in proof
76. Echtermeyer AT, Kensche C, Bach P, Poppen CM, Lilholt H, et al. 1996. Method to predict fatigue lifetime of GRP wind turbine blades and comparison with experiments. *1996 Eur. Union Wind Energy Conf. EWEC '96, Göteborg*, pp. 907–13. Bedford: HS Stephens & Assoc.
77. Andersen SI, Brøndsted P, Lilholt H, Lystrup Aa. 1996. Risoe investigations of GI-UP materials. See Ref. 40, pp. 71–119
78. Brøndsted P, Lilholt H. 1998. *Vingedesign EFP-95 Slutrapport for Materialer*. Risoe-R-1040(DA). (In Danish)
79. Frost NE, Marsh KJ, Pook LP. 1974. *Metal Fatigue*. Oxford, UK: Clarendon
80. Goodman J. 1899. *Mechanics Applied to Engineering*. London: Longman, Green, & Co.
81. Gerber WZ. 1874. *Bayer. Archit. Ing.* 6:101
82. Palmgren A. 1924. Die Lebensdauer von Kugellagern. *Z. Ver. Dtsch. Ingenieuren* 68:14:339–41
83. Miner MA. 1945. Cumulative damage in fatigue. *J. Appl. Mech.* 12:A159–64
83a. Dover WD. 1979. Variable amplitude fatigue of welded structures. In *Fracture Mechanics: Current Status, Future Prospects*, ed. RA Smith, pp. 129–47. Toronto: Pergamon
83b. Solin J. 1990. Methods for comparing fatigue lives for spectrum loading. *Int. J. Fatigue* 12:1:35–42
84. Brøndsted P, Andersen SI, Lilholt H. 1997. Fatigue damage accumulation and lifetime prediction of GFRP materials under block loading and stochastic loading. See Ref. 15, pp. 269–78
85. ASTM E739. 1998. *Standard Practice for Statistical Analysis of Linear or Linearized Stress-Life (S-N) and Strain-Life (e-N) Fatigue Data*. W. Conshohocken, PA: Am. Soc. Test. Mater.
86. Echtermeyer AA, Hayman B, Ronold KO. 1996. Comparison of fatigue curves for glass composite laminates. See Ref. 41, Ch. 14
87. Ronold KO, Echtermeyer AT. 1996. Estimation of fatigue curves for design of composite laminates. *Composite A* 27:6:485–91

88. Lystrup Aa. 1997. Processing technology for advanced fibre composites with thermoplastic matrices. See Ref. 15, pp. 69–80
89. http://www.cyclics.com
90. Lystrup Aa. 1998. Hybrid yarns for thermoplastic fibre composites. *Final Rep. MUP2 Framework Program. Risoe-R-1034(EN)*. Roskilde, Denmark: Risoe Natl. Lab.
91. http://www.comfil.biz/rapport/takwind.htm
92. 1999. *Mater. Res. Dep. Annu. Rep.* Risoe-R-1155(EN). Roskilde, Denmark: Risoe Natl. Lab., April 2000. http://www.risoe.dk/rispubl/AFM/afmpdf/ris-r-1155.pdf
93. http://www.gaoth-tec.com
94. Kensche CW. 2004. Fatigue of composites for wind turbines. *3rd Int. Conf. Fatigue Composites, Kyoto, Japan*
95. Kelly A, Zweben C, eds. 2000. *Comprehensive Composite Materials*, Vols. 1–6. Amsterdam: Elsevier/Pergamon

Annu. Rev. Mater. Res. 2005. 35:539–69
doi: 10.1146/annurev.matsci.35.102103.090513

First published online as a Review in Advance on April 1, 2005

Materials Characterization in the Aberration-Corrected Scanning Transmission Electron Microscope

M. Varela,[1] A.R. Lupini,[1] K. van Benthem,[1]
A.Y. Borisevich,[1] M.F. Chisholm,[1] N. Shibata,[2]
E. Abe,[3] and S.J. Pennycook[1]
[1]Oak Ridge National Laboratory, Condensed Matter Science Division, Oak Ridge, Tennessee 37831; email: mvarela@ornl.gov, arl1000@ornl.gov, Benthem@ornl.gov, albinab@ornl.gov, chisholmmf@ornl.gov, pennycooksj@ornl.gov
[2]Institute of Engineering Innovation. University of Tokyo, Tokyo 113-8656 Japan; email: shibata@sigma.t.u-tokyo.ac.jp
[3]Dept. of Materials Engineering, University of Tokyo, 7-3-1 Hongo, Bunkyo-ku, Tokyo 113-8656; email: abe@material.t.u-tokyo.ac.jp

Key Words *Z*-contrast microscopy, electron energy loss spectroscopy, nanotechnology, complex oxides, catalysis, semiconductors

■ **Abstract** In the nanoscience era, the properties of many exciting new materials and devices will depend on the details of their composition down to the level of single atoms. Thus the characterization of the structure and electronic properties of matter at the atomic scale is becoming ever more vital for economic and technological as well as for scientific reasons. The combination of atomic-resolution *Z*-contrast scanning transmission electron microscopy (STEM) and electron energy loss spectroscopy (EELS) represents a powerful method to link the atomic and electronic structure to macroscopic properties, allowing materials, nanoscale systems, and interfaces to be probed in unprecedented detail. *Z*-contrast STEM uses electrons that have been scattered to large angles for imaging. The relative intensity of each atomic column is roughly proportional to Z^2, where Z is the atomic number. Recent developments in correcting the aberrations of the lenses in the electron microscope have pushed the achievable spatial resolution and the sensitivity for imaging and spectroscopy in the STEM into the sub-Ångstrom (sub-Å) regime, providing a new level of insight into the structure/property relations of complex materials. Images acquired with an aberration-corrected instrument show greatly improved contrast. The signal-to-noise ratio is sufficiently high to allow sensitivity even to single atoms in both imaging and spectroscopy. This is a key achievement because the detection and measurement of the response of individual atoms has become a challenging issue to provide new insight into many fields, such as catalysis, ceramic materials, complex oxide interfaces, or grain boundaries. In this article, the state-of-the-art for the characterization of all of these different types of materials by means of aberration-corrected STEM and EELS are reviewed.

INTRODUCTION

The characterization of structure and electronic properties of materials at the atomic scale is a key issue for nanotechnology and many other fields of materials science. Transmission electron microscopy (TEM) techniques are so far one of the most useful for structure characterization at the atomic scale. At present, the TEM field is undergoing major, revolutionary changes thanks to the introduction of aberration correction. As a result, the sub-Å regime has been successfully achieved in the STEM, which as of today holds the record in spatial resolution for a direct image of a crystal lattice (1). More importantly, the combination of atomic-resolution *Z*-contrast STEM and EELS represents a powerful method to link the atomic and electronic structure of solids to their macroscopic properties (2–6, 131, 132), which allows the properties of materials, nanoscale systems, and interfaces to be probed in unprecedented detail. The detection and measurement of the response of individual atoms have become challenging issues to provide new insight into many fields, as distributions of isolated atoms deeply modify the physical properties of many of the technologically most relevant and scientifically most interesting materials. Therefore, the possibility of identifying, localizing, and even measuring the electronic properties of single atoms represents the ultimate advance for understanding the atomic origins of materials properties.

Here, we review the state-of-the-art in characterization of materials and nanostructures by means of aberration-corrected STEM and EELS. As many different systems are dealt with, we intend to show some of the most representative examples published so far, ranging from catalysts to semiconductors and several complex oxides. In the second section we discuss aberration correction, which has allowed direct imaging of a crystal lattice at sub-Å resolution and the routine imaging of single atoms. This part is technical in nature, so readers interested purely in materials characterization per se can move directly to the third section, in which the most recent advances in the application of STEM to a number of materials are reviewed. We present a few examples to show the tremendous impact of STEM developments in catalysis and also in ceramic materials science. The next subsections review the latest applications of STEM-EELS to the characterization of complex metallic alloys and oxides, enabling the possibility of single-atom spectroscopic detection and including the most recent accomplishments in light atom detection. The most recent structural and electronic characterization of several heterophase interfaces are described. Finally, we deal with the possibilities of three-dimensional depth slicing, an unexpected benefit of aberration correction that may finally overcome the projection problem from which microscopy has suffered for the past 70 years.

ABERRATION CORRECTION IN THE SCANNING TRANSMISSION ELECTRON MICROSCOPE

Achieving sub-Å resolution has been one of the main goals of advanced electron microscopy in order to allow atomic resolution characterization of a variety of materials. However, the spatial resolution is limited by the lens aberrations. In the 21st century, the promise of aberration correction has finally turned into a reality. For the first time, a direct image of a crystal lattice with sub-Å resolution was acquired by Nellist et al. in a dedicated VG Microscopes HB 603 U_STEM equipped with a Nion aberration corrector (1).

Although there are several excellent reviews on how the STEM works (7–9), we focus here on describing the basics of aberration correction in the STEM and how it has created a unique structural and electronic probe to study the most fascinating materials science problems.

A schematic of a STEM is shown in Figure 1. The essential difference between the optics of a STEM and a conventional transmission electron microscope

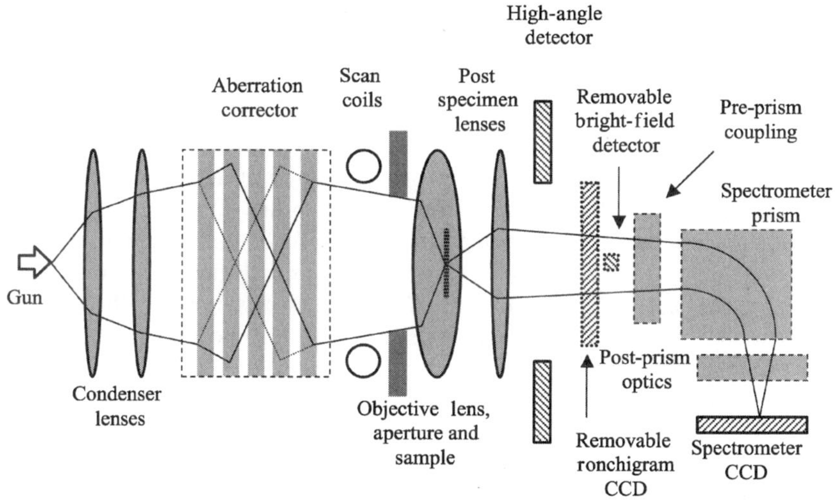

Figure 1 Simplified schematic of an aberration-corrected STEM; examples of electron trajectories are indicated. In the (quadrupole-octupole) aberration corrector, the different trajectories in the XZ and YZ planes are indicated with dotted and solid lines, respectively. The electrons are accelerated from the gun; condenser lenses are used to adjust the beam current and beam coherence and to couple to the aberration corrector. The objective lens focuses the probe, which is scanned across the sample by the scan coils. The high-angle detector collects electrons scattered to high angles. Removable bright field and Ronchigram detectors are shown as well as the important components of the spectrometer.

(CTEM) is that for a CTEM, a large area of the sample is illuminated and the significant magnification is performed by the lens system after the specimen, allowing the whole image to be recorded at once. In a STEM, the focusing is done before the electrons reach the specimen to form a very tiny probe (of the order of 1 Å), which is scanned over the sample (10). The scattered electrons are then collected in order to form the image as a function of position. In practice, a STEM will also have a series of condenser lenses that allow the beam current and coherence to be varied. Postspecimen lenses allow the detector angles to be changed.

One advantage of a STEM is that a variety of detectors can be used simultaneously. For example, it is possible to record both low- and high-angle-scattered electrons, as well as those on axis, which can be used to form a bright field (BF) image. The principle of reciprocity (10) shows that the BF STEM image should be equivalent to a CTEM image recorded under equivalent conditions. Electrons at other angles can be used to acquire several BF images at different tilts simultaneously, thus allowing schemes such as super-resolution (11). Other useful signals include back-scattered and secondary electrons, photon emission (including X-rays), and the current absorbed by the sample (12). Those electrons that lose energy as they pass through the sample can be used to record electron energy-loss spectroscopy (EELS) spectra as a function of probe position, which can provide information on the local electronic structure with atomic resolution (13, 14).

One of the main advantages of the TEM over the STEM is that the entire image is recorded simultaneously, with most of the electrons incident on the area of interest being used to form the image. The major disadvantage in TEM is that the image obtained does not necessarily relate to the structure in a simple manner. Image contrast is a complicated function of focus and specimen thickness, even in an aberration-corrected TEM, which means that there is no unique image that gives the best view of the structure.This makes it particularly difficult to investigate samples or defects with unknown structure or composition, which are of course the most interesting!

In contrast, probably the principle advantage of the STEM configuration, in addition to the variety of signals available, is that it is possible to obtain a *Z*-contrast image that relates far more simply to the sample structure. The *Z*-contrast image is formed by collecting the electrons scattered out to high angles on a high-angle annular dark field (HAADF) detector. The benefit is that this image is, to a very good approximation, a convolution of the intensity of the probe with an object function representing the sample in a simple manner. The underlying physical principle is that high *Z* atoms scatter to higher angles more strongly than lighter atoms. In the limit of Rutherford scattering, this goes as Z^2, so the contrast in the image will also vary approximately as Z^2. The exact contrast details will depend on the experimental set up and are affected by factors such as electron channeling or detector size. But over a fairly large ranges of defocus or sample thicknesses, the image will nonetheless represent the sample structure in a straightforward way. Thus one can obtain a qualitative understanding of all of the images presented in this review by considering them to be convolutions between the atomic resolution probe and a series of points at the atomic locations with a brightness of Z^2.

All of the work reported here was obtained using the two VG Microscopes' STEMs at Oak Ridge National Laboratory (ORNL), both equipped with Nion aberration correctors. One of the key features of these instruments is the use of a cold field emitter as a gun. This has a number of advantages over a conventional Schottky emitter: It has a lower energy spread and a higher brightness, although the Schottky emitter has higher total current and better long-term current stability. The ORNL HB603U STEM, operated at 300 kV, produces a beam with a diameter of 0.6 Å, whereas the VG Microscopes HB501UX dedicated STEM operated at 100 kV STEM produces a beam with a diameter of 0.9 Å. The latest instrument also features a quadrupole coupling module from Nion Co. This allows the recording of energy loss spectra while collecting almost all of the electrons transmitted through the sample with better than 0.3 eV resolution, which is limited by the energy spread from the tip (15–17).

If the probe is positioned on a specific site of the specimen without scanning, a Ronchigram is formed (17a). The Ronchigram is used during the microscope alignment process. A Ronchigram is formed by collecting the transmitted electrons as a function of angle on a charge-coupled device (CCD) and can be thought of as a shadow image of the (partly transparent) sample. For an amorphous sample, this resembles a distorted BF image of the sample and directly reflects the symmetry of the microscope aberrations. For a crystalline sample, a convergent beam electron diffraction (CBED) pattern is formed, which is used to adjust the specimen orientation.

Aberrations in Electron Microscopy

Spherical aberration takes its name from light optics, where it was a common fault of the slightly too spherical shape of early lenses. It is a geometrical aberration, meaning that electrons traveling at different angles to the axis are focused at different distances behind the lens. A perfect point would be focused to a disc in the image, as shown in Figure 2 (see color insert). Chromatic aberration is generated if electrons with different energies (different wavelengths) are present. Then they are focused differently depending on their energy. As any electron source will emit electrons with a range of energies and some electrons lose energy as they interact with the sample, this can degrade the resolution.

The aberration function is defined as the phase change across the back focal plane of the objective lens. Another equivalent and sometimes useful definition from geometric optics is that the gradient of the aberration function gives the ray displacement at the sample plane. The aberration function is normally given in a power series with angle to axis. Off axial aberrations can normally be ignored in a STEM, although we note that they have to be considered in detail when designing aberration correctors or aiming for the highest resolution with a large field of view. After aberration correction, higher-order aberrations may be significant, by which we mean the next terms in the expansion of the aberration function; these could previously be neglected when the aberration function was dominated by spherical aberration, C_S. Therefore, in a C_S-corrected system, C_5, which is the next higher order aberration becomes the dominant aberration (18). Non-rotationally symmetric aberrations also exist, of which the most common example is astigmatism.

Aberration Correction

In a modern TEM, the resolution is ultimately limited by either the spherical or chromatic aberrations of the electron lenses. Scherzer showed in 1936 (19, 20) that spherical and chromatic aberrations were unavoidable for conventional rotationally symmetric magnetic lenses. These aberrations set a limit on the attainable resolution in a microscope for over 60 years. However, Scherzer assumed several conditions, namely that the optical system is round, produces a real image, does not vary with time, and has no charge on the optic axis. In 1947, Scherzer therefore proposed to overcome this limitation by breaking one of these conditions (20). Both of the aberration correctors that have recently demonstrated a practical resolution improvement rely on breaking the rotational symmetry of the system (21–26).

There are two main classes of aberration correctors, named after the type of multipole that they employ: quadrupole-octupole correctors and hexapole correctors. These multipoles, like the alignment coils in the TEM, are normally magnetic; although a chromatic aberration corrector would have to have both magnetic and electrostatic poles. Multipoles are named by their rotational symmetry (dipoles, quadrupoles, hexapoles, octupoles, etc.).

Quadrupole-octupole (QO) correctors work because octupoles have a field that varies with distance off the optic axis in the same way as does spherical aberration but has a fourfold rotational symmetry about this axis (with an alternating sign on rotation by 45°). Thus the QO corrector relies on using a series of quadrupoles to distort the beam in such a way that it acquires the negative spherical aberration from the octupoles and leaves the corrector as a round beam. Three octupoles are used in order to correct the round aberration in the X, Y, and 45° directions.

Hexapole correctors, stemming from an idea by Beck (27), rely on the fact that an extended hexapole imparts spherical aberration to a round beam in addition to the primary effect of threefold astigmatism. Further refinements (28, 29) allow the projection of one hexapole to another and then into the objective lens in such a manner as to eliminate the spherical aberration without increasing off axial or higher-order aberrations.

Aberration correction, like any new field, is evolving extremely rapidly. At the time of writing, aberration correctors offer only third-order aberration correction and in general are expected to make higher-order aberrations slightly worse. The next generations of aberration correctors are already being produced and are expected to offer correction of the higher-order aberrations in order to provide better resolution and larger field of view (22).

Spherical aberration correction, although initially devised for conventional TEM, may have more benefits in STEM because of the reduced sensitivity of a STEM in the *Z*-contrast mode to chromatic aberration (9, 30–32). However, just as in a CTEM, chromatic aberration correction may offer some advantages, better resolution, larger pole piece gap or more flexible designs. Kel'man & Yavor (1961) (33) showed that crossed electrostatic and magnetic quadrupole fields can produce negative chromatic aberration. Early attempts to correct the chromatic aberration of an electron microscope encountered difficulty because of the increase in spherical aberration (34), so it is likely that a combined C_s-C_c corrector would provide further benefit. The use of a monochromator would reduce chromatic abberration effects and also help improve energy resolution in EELS. The price paid for such a solution is reduced current, which then limits resolution due to an insufficient signal-to-noise ratio.

We make a distinction between aberration correction and removal of aberrations by postprocessing in the computer. Correcting aberrations through the use of hardware to alter the beam is significantly different from using postprocessing techniques to remove the effects from the image. These techniques should be regarded as quite distinct but possibly complementary.

The Corrected Electron Probe

Aberration correction has a dramatic effect on both the distribution and the spatial localization of the intensity of the electron probe. As an example, Figure 3 shows the calculated probe intensities before and after correction, with parameters based on the ORNL HB603U (these can vary depending on the microscope settings). The two probes have been normalized to the same total current and are shown on the same scale. Significant current is lost in the tails of the uncorrected probe,

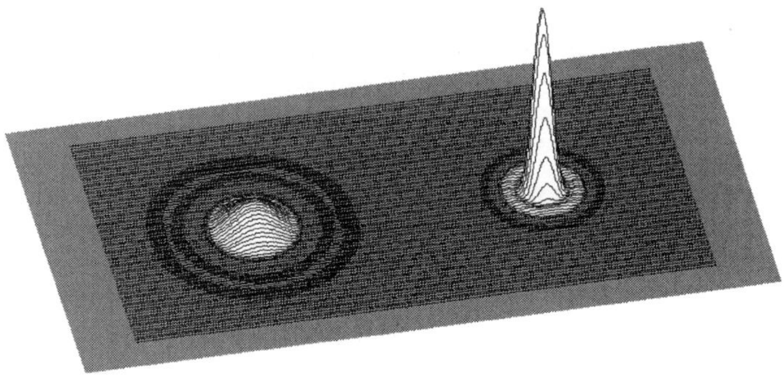

Figure 3 Calculated probe intensities before and after correction, with parameters based on the ORNL HB603U with beam energy 300 kV and energy spread 0.3 eV. (*a*) Before correction, defocus $C_1 = -63$ nm, $C_s = 1.5$ mm, $C_c = 1.3$ mm, resulting in a probe FWHM $= 0.12$ nm. (*b*) After correction, now $C_1 = 2$ nm, $C_s = -0.04$ mm, $C_c = 1.6$ mm and $C_5 = 100$ mm giving a probe FWHM $= 0.05$ nm.

but after correction the probe has a smaller full width at half maximum (FWHM), and also a far higher current density at the center. This enables a better spatial resolution, and also the higher peak intensity gives far better sensitivity to single atoms.

Sub-Å Imaging of Crystal Lattices

The highest resolution so far accomplished with aberration-corrected STEM was by Nellist et al. (1). They demonstrated direct sub-Å resolution with the aberration-corrected HB603U at ORNL. The improved resolution allows the imaging of a Si crystal observed in the [112] direction, containing pairs of Si columns that are 78 pm apart. The image of Si shown in Figure 4 (see color insert) was recorded in the annular dark field (HAADF) mode, and the pairs of atomic columns are seen directly resolved. The Fourier transform of this image shows clear lattice information down to 0.71 Å, and there is also apparent lattice information at 0.61 Å (1), although this is very close to the noise level of the data. This represents the highest resolution direct image ever obtained at the date of this writing.

CHARACTERIZATION OF MATERIALS IN THE ABERRATION CORRECTED STEM

In the following we review some recent examples where aberration-corrected STEM was used to investigate atomistic and electronic structures of grain boundaries, catalyst particles, single impurity atoms, and heterophase interfaces for correlation to the macroscopic materials properties.

Heterogeneous Catalysts

Heterogeneous catalysts, generally composed of metal nanoparticles on oxide supports, are used in a large number of processes from synthesis of pharmaceuticals to automotive emission control. *Z*-contrast STEM is uniquely suited for observations of heavier species on light substrates, and with the sub-Å resolution and single-atom sensitivity achieved via aberration correction, it can provide unprecedented information about the structure of such catalysts. Most excitingly, atomic-level information obtained by STEM can then be correlated with the results of first-principles calculations and chemical reactivity measurements, thus resulting in new insights into catalytic functionality, fueling the design of more advanced catalysts for the future. This section of the review provides several examples of such studies.

CASE 1: La DOPING OF γ-Al_2O_3 γ-alumina is one metastable polytype of Al_2O_3 that is used extensively as a catalytic support material because of its high porosity and large surface area (35). At temperatures in the range 1000–1200°C, however,

γ-alumina transforms rapidly into the thermodynamically stable α-alumina phase (corundum), drastically reducing the surface area and thus suppressing the catalytic activity of the system (36). The phase transformation can be shifted to higher temperatures by doping γ-alumina with elements such as La (37), but it was not previously possible to establish if dopants enter the bulk, adsorb on surfaces as single atoms or clusters, or form surface compounds (38–43). The lack of this information, as well as the lack of understanding of the stabilization mechanism itself, hinders performance and cost optimization of the catalytic supports. Wang and coworkers (6) used a combination of *Z*-contrast STEM and first-principle calculations to address this problem.

Figure 5*a* (see color insert) shows a *Z*-contrast image of a flake of La-doped γ-Al_2O_3 in the [100] orientation. The square arrangement of Al-O columns is clearly resolved. Single La atoms are visible in the form of brighter spots on the background of thicker but considerably lighter γ-Al_2O_3 support. Most of the La atoms are located directly over Al-O columns (site A, Figure 5*a*), but a small fraction also occupies a position shifted from the Al-O column (site B, Figure 5*a*). Note that, in this second configuration, the La atom appears much dimmer because its intensity is no longer superimposed on that of the Al-O column. Intensity variations among different La atoms on the A sites correspond to La atoms located on the entrance or the exit surfaces of the alumina flake, with the latter appearing brighter owing to electron channeling. The images reveal clearly that there is no correlation in the distribution of dopant atoms. Additional experiments utilizing a novel technique of depth slicing via focal series in STEM (see details in previous section) have demonstrated that La atoms are located on the surface of the alumina flakes, not in the bulk.

Density functional theory calculations have demonstrated that La atoms are strongly bound to the γ-Al_2O_3 surfaces (binding energy 7–8 eV), considerably stronger than to the α-alumina surface (binding energy 4.3 eV). In particular, the position over Al-O columns on the [100] surface of the γ-Al_2O_3 (Site A on Figure 5*a*) was identified as the lowest-energy position for the dopant, with La replacing surface Al and forcing it into a subsurface tetrahedral void (Figure 5*b*). Additionally, surface positions for La atoms are energetically preferred by as much as 3 to 5 eV per atom compared with bulk or subsurface locations. No tendency to form correlated arrays or clusters on the surface was found. Thus by comparing *Z*-contrast images and density functional theory calculations it became apparent that the stabilization is achieved by single La atoms adsorbed on the γ-Al_2O_3 surface. Dopants improve the stability with respect to the phase transition and make sintering highly unfavorable, as it would result in placing the surface La into the bulk (6).

CASE 2: Pt TRIMERS ON γ-Al_2O_3 The tendency of Pt to form trimer-like structures on the surface of γ-Al_2O_3 was observed by uncorrected STEM (44), but the precision of the experimental data on the interatomic distances was insufficient to choose among the competing mechanisms of the trimer formation. In a report by

Sohlberg et al. (45) aberration-corrected STEM was used to characterize Pt trimers in more detail. The images demonstrated that the Pt-Pt distances in the trimer differ significantly from those in bulk Pt (Figure 6*a*, see color insert), thus putting both substitutional and adatom models under doubt. A thorough theoretical study has shown that the experimentally observed geometry can be achieved only in an adatom cluster capped by an OH-group (Figure 6*b*). Additionally, it was demonstrated that unlike the metastable Pt_3 cluster, the Pt_3OH formation was stable and had a slightly electron-deficient character (Pt_3 was found to be slightly electron-rich). This finding relates directly to the Lewis acidity and catalytic activity of the Pt/γ-Al_2O_3 catalysts.

CASE 3: Au NANOPARTICLES ON TiO_2 It is remarkable that although Au is not a good catalyst in bulk or in the form of large particles, when prepared as nanoparticles on an oxide support, it becomes one of the most active catalysts for the oxidation of CO to CO_2 (46, 47). The cause of this activity has been a mystery for many years, with several explanations proposed. Early work showed that there is a correlation between particle size and activity (46), although one problem is that it is difficult to see the smallest nanoparticles by conventional TEM, hindering the identification of likely sites and mechanisms. More recent work on model systems has shown that the thickness may be more important than the lateral extent, with bilayer structures having the highest activity. Some of the models for the activity (48) required the presence of oxidized Au, but XANES revealed that in the most active catalysts, almost no oxidized Au is present.

Lupini et al. (49) used aberration-corrected STEM to show that for a highly active catalyst, most of the nanoparticles are 1–2 nm diameter, as shown in Figure 7*a* (see color insert). Quantifying the thicknesses by comparing to image simulations, including Au particle models on a substrate, revealed that these are indeed one or two layers thick. On the basis of the form of nanoparticles seen in the *Z*-contrast images, it was then possible for the first time to perform first-principles calculations to investigate the properties of realistic Au nanoparticles supported on TiO_2. These calculations revealed that the nanoparticles are anchored to oxygen vacancies on the surface, as depicted in Figure 7*b*, which allows the formation of a high areal density of nanoparticles and shows that such nanoparticles have the ability to bind both CO and O_2, unlike gold steps or surfaces. Thus it seems that the ability to bind both of the reactants is the key to the high activity of this unusual catalyst.

Application to Ceramic Materials: Si_3N_4

Silicon nitride ceramics are expected to be used in a wide variety of applications because of their superior high-temperature properties (50). The mechanical properties of Si_3N_4 are strongly influenced by the microstructure, and the introduction of high-aspect-ratio grains is well known to significantly improve fracture toughness (51). Such microstructures provide toughening mechanisms—for instance, crack

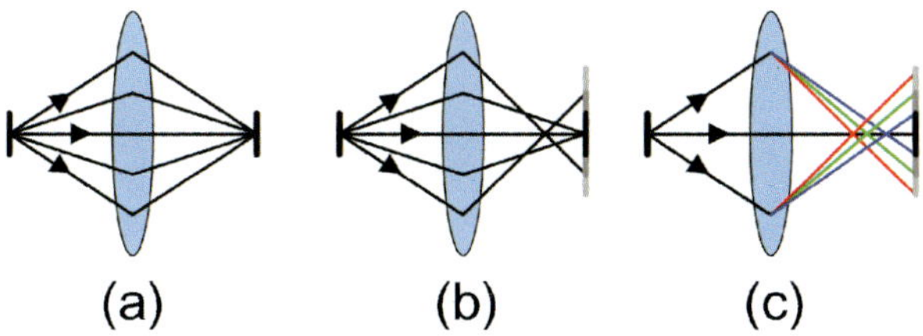

Figure 2 Illustration of the main lens aberrations. (*a*) A perfect lens focuses a point source to a single image point. (*b*) Spherical aberration causes rays at higher angles to be overfocused. (*c*) Chromatic aberration causes rays at different energies (indicated by color) to be focused differently.

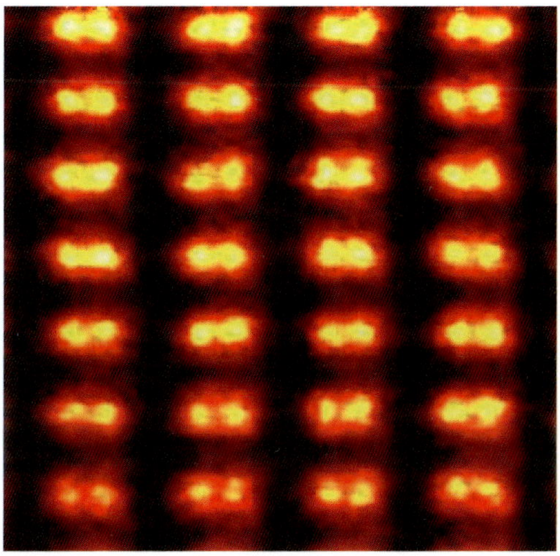

Figure 4 HAADF image of Si [112] recorded with an aberration-corrected STEM. "Dumbbells" just 78 pm apart are resolved.

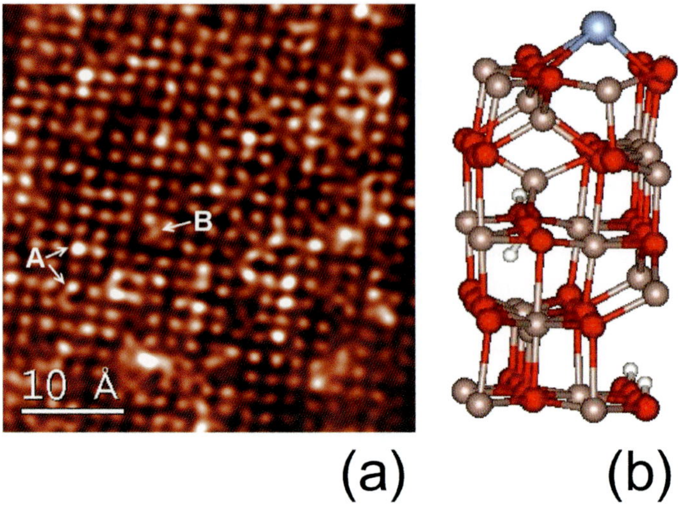

Figure 5 (*a*) *Z*-contrast image of La atoms on γ-Al_2O_3 flake in [100] orientation. Two distinct sites (A and B) for individual La atoms are seen (images are high- and low-pass filtered). (*b*) Schematics of the configuration for the La atom on the (100) surface of γ-Al_2O_3, determined by first-principles calculations.

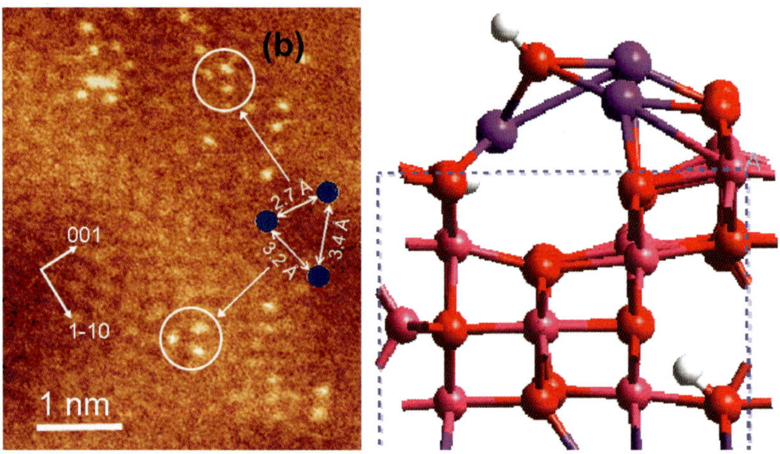

Figure 6 (*a*) *Z*-contrast STEM image of Pt on the surface of γ-Al_2O_3 close to (110) orientation. Two Pt_3 trimer structures are circled; inset, measured interatomic distances. A hint of the underlying alumina lattice is also visible. (*b*) Schematics of the configuration for the Pt_3OH unit on the (110) surface of γ-Al_2O_3, determined by first-principles calculations, with closely matching interatomic distances.

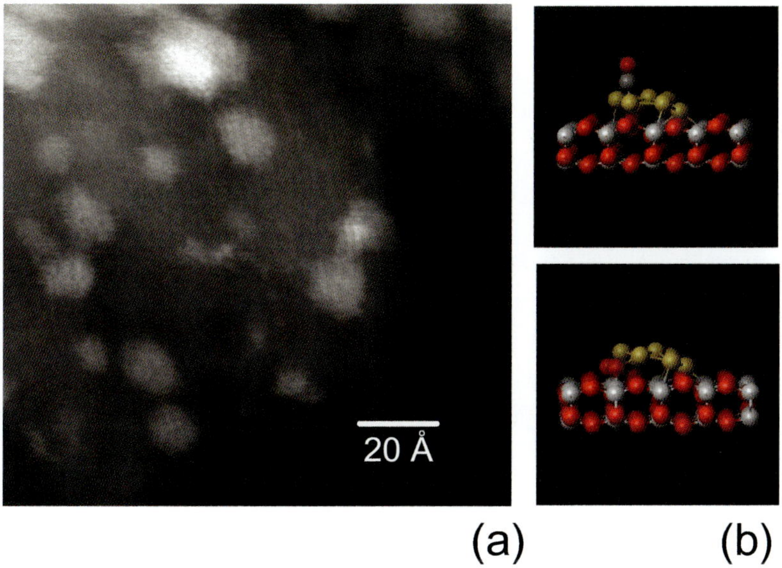

Figure 7 (*a*) *Z*-contrast STEM image showing Au nanoparticles on a titania flake. Most of the nanoparticles are between 1 and 2 nm in diameter and are one to two layers thick. (*b*) First-principles calculations for the observed nanoparticles indicate that they are capable of bonding both CO (*right top*) and O_2 (*right bottom*), shown in the lowest energy configurations, enabling the catalytic activity.

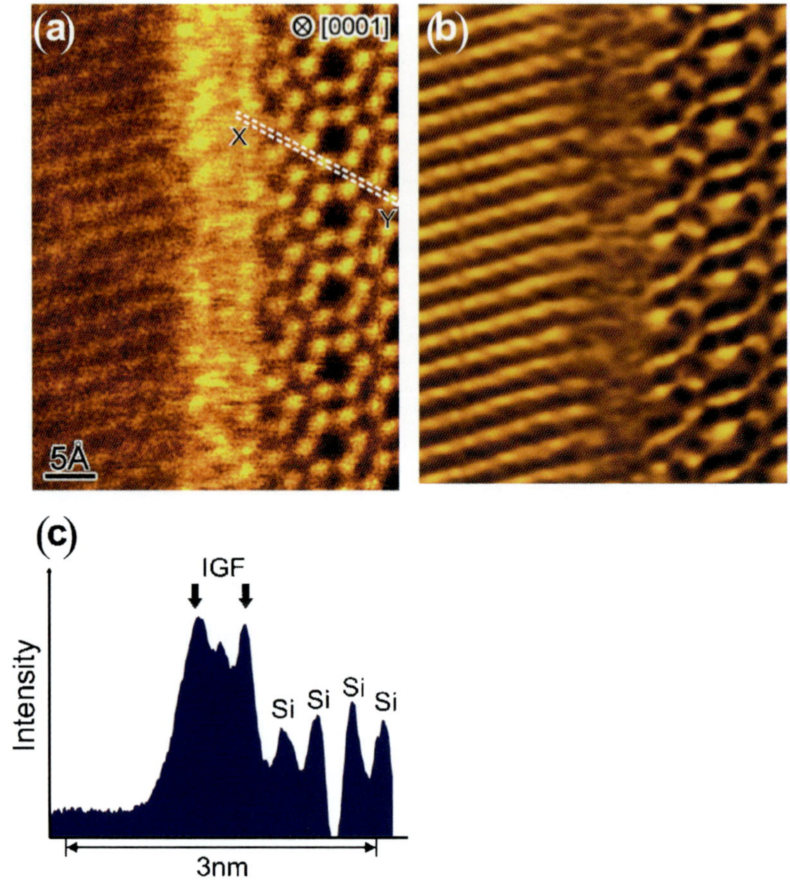

Figure 8 HAADF-STEM (*a*) and bright-field (*b*) STEM images of an intergranular film. (*c*) Intensity line profile taken from the area shown in (*a*).

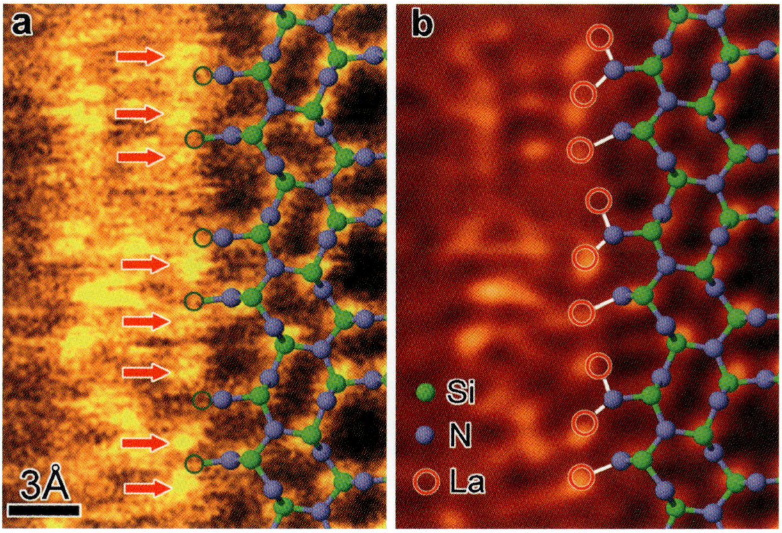

Figure 9 (*a*) Magnified HAADF-STEM image of the IGF/Si_3N_4 interface. A projection view of the β-Si_3N_4 lattice model is superimposed on the image. (*b*) Pixon reconstructed image.

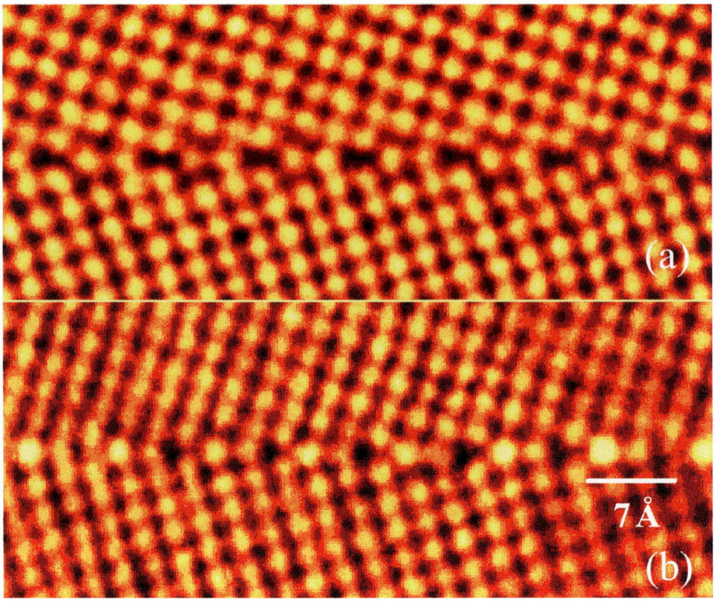

Figure 11 *Z*-contrast images of the grain boundary region of a symmetric 36.8° ⟨001⟩ tilt boundary in (*a*) pure copper and (*b*) bismuth-doped copper. The bismuth segregation site in (*b*) is seen to be distinctly bright [see Duscher et al. (62)].

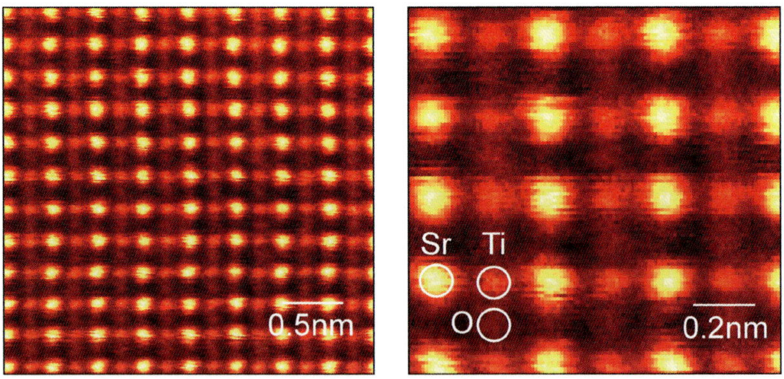

Figure 12 *Z*-contrast images of $SrTiO_3$ viewed down the [011] orientation recorded on the 300 kV aberration-corrected STEM. The oxygen columns are faint but visible between the Ti columns. No such features are seen between the Sr columns, showing that the faint features are not produced by probe tails.

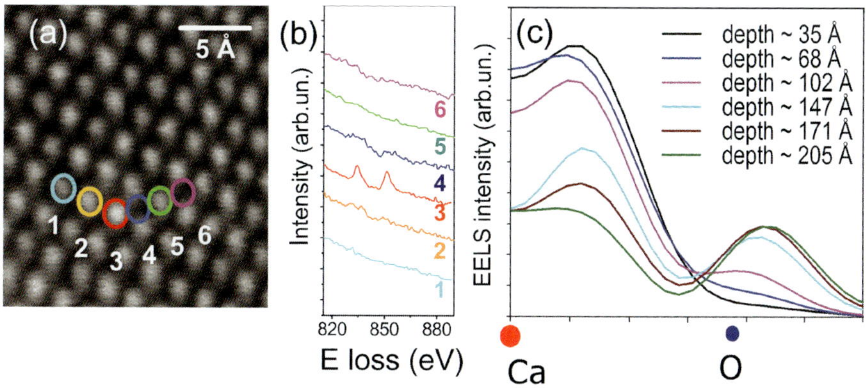

Figure 13 (*a*) HAADF image of a single La atom in a $CaTiO_3$ matrix, acquired in the 100 kV HB501UX STEM. (*b*) Electron energy loss spectra obtained from individual columns of the calcium titanate crystal. The red spectrum reveals the presence of a single La atom within the calcium column circled in red. The signal is substantially reduced from neighboring columns. (*c*) Dynamical simulation of the La spectroscopic image. The ratio of signals on adjacent columns indicates the depth of the atom in the crystal. This atom is approximately 100 Å below the surface.

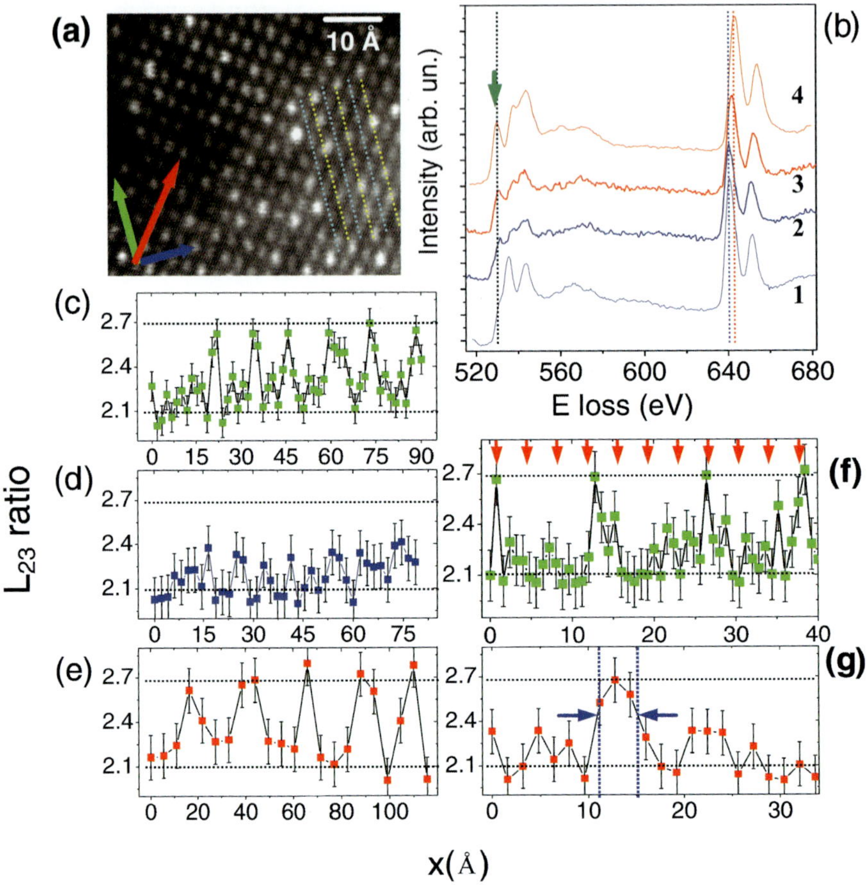

Figure 14 *Z*-contrast image (*a*) and dependence of the L_{23} ratio on position within the BCMO sample for a few specific scan directions. (*b*) Electron energy loss spectra typical of a $Mn^{+3}O$ column (*2*) and a $Mn^{+4}O$ column (*3*). Spectrum (*4*) comes from $CaMnO_3$, while (*1*) corresponds to bulk $LaMnO_3$. The oxygen K edge (at 530 eV) and the Mn L edge (around 644 eV) are shown. (*c*) L_{23} ratio along the direction marked with a green arrow in Figure 14*a*. Horizontal lines represent the L_{23} ratio expected for a pure Mn^{+3} oxidation state (*top*) and a pure Mn^{+4} oxidation state (*bottom*). (*d*) L_{23} ratio along the direction marked with a blue arrow in Figure 14*a*. (*e*) L_{23} ratio along the direction marked with a red arrow in Figure 14*a*. Also shown is the dependence of the L_{23} ratio with position in the sample obtained from high magnification EELS linescans: (*f*) scanning along the pseudocubic (100) direction. The positions of Mn planes along the scan have been marked with red arrows. (*g*) Along the pseudocubic (110). Blue arrows mark the estimated width of the stripe, about 4 Å.

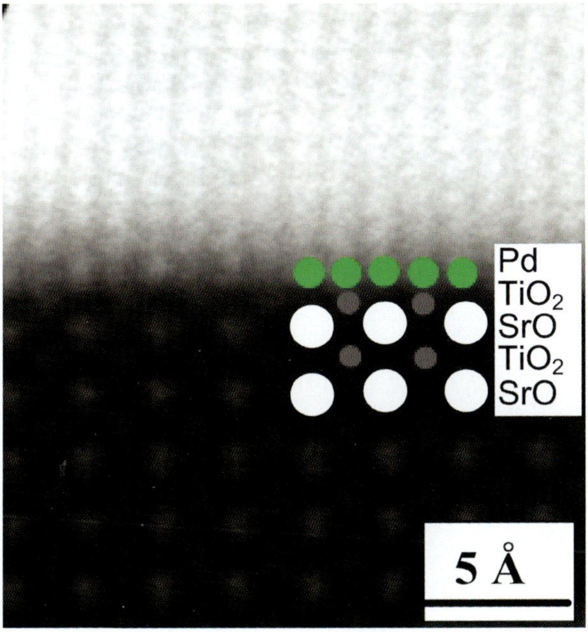

Figure 17 HAADF image of the Pd/$SrTiO_3$ interface recorded in a ⟨100⟩ zone-axis orientation. The intensity line profile clearly reveals Pd atoms sitting on top of oxygen columns of the TiO_2-terminated (100) substrate surface.

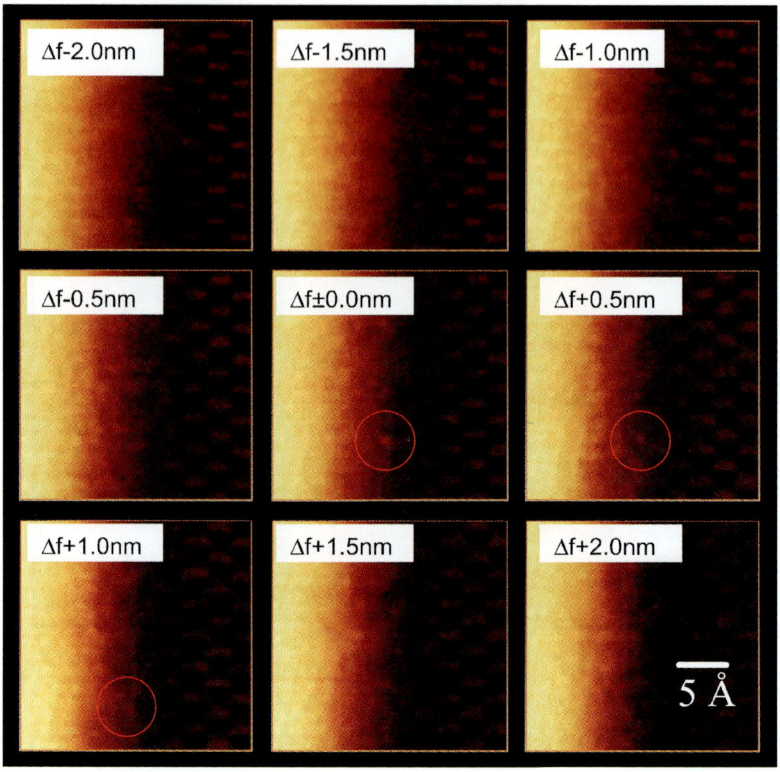

Figure 20 Through-focal-series of a $Si/SiO_2/HfO_2$ interface. The figure shows 9 consecutive micrographs extracted from a series of 21 images recorded at the focal values shown. The focus step size was 0.5 nm between frames. One single Hf atom coming into and going out of focus is marked by a red circle. The micrographs are displayed using a nonlinear contrast scale.

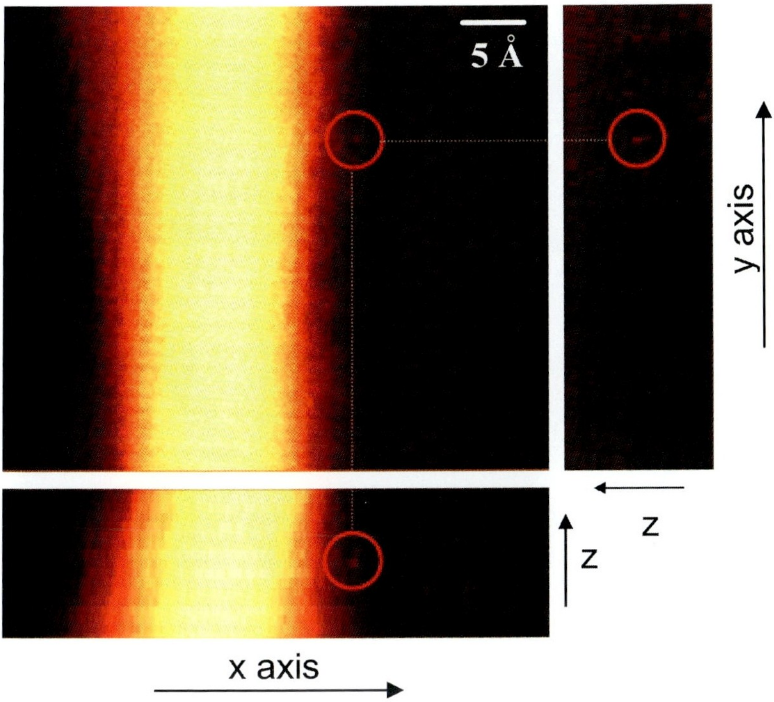

Figure 21 Slice view of the $Si/SiO_2/HfO_2$ interface. The figure shows three different views through the three-dimensional data set. Note that the pixel sizes are about 0.01 nm for the *x* and *y* directions and 0.5 nm in the *z* direction, respectively. The *x*-*y* image shows a single atom in the middle of the SiO_2 interlayer. The *y*-*z* and the *x*-*z* slices clearly show the localization of this single Hf atom in all three dimensions. The *x*-*z* slice even shows columns of the Si dumbbells from the side.

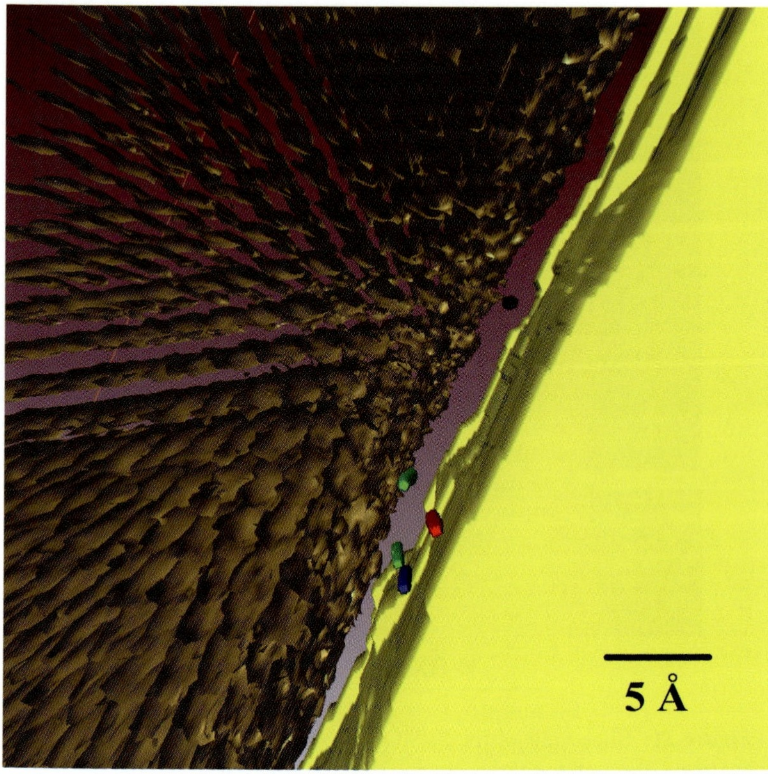

Figure 22 Snapshot of a three-dimensional representation of the $HfO_2/SiO_2/Si$ interface. A three-dimensional representation of the $HfO_2/SiO_2/Si$ interface structure was created by setting thresholds on intensity values from the through-focus-series. The Si substrate was color-coded in gold, whereas the HfO_2 film is marked solid yellow. Single Hf atoms in the interface layer are coded separately in green, black, red, and blue according to Figures 20 and 21. A movie of the rotating structure can be found in Reference 122. Note that even some surface roughness of the HfO_2 film is observable at the interface with the SiO_2 layer.

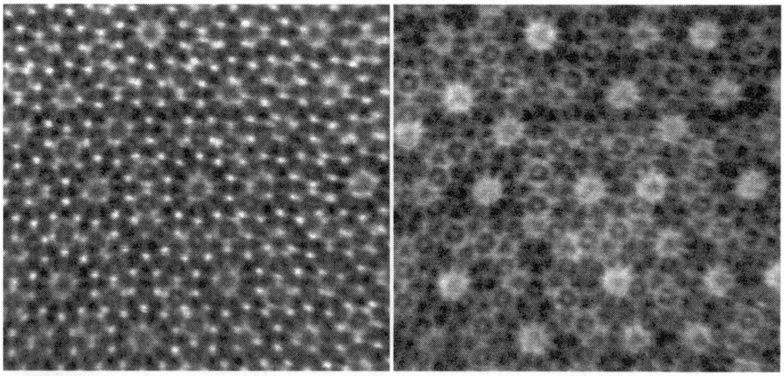

Figure 10 *Z*-contrast images of an $Al_{72}Ni_{20}Co_8$ quasicrystal viewed along the decagonal axis. (*a*) High-angle range (inner-angle ~50 mrad) showing transition metal columns bright, Al columns less bright. (*b*) Low-angle range (inner-angle ~35 mrad), showing Debye-Waller factor contrast quasiperiodically correlated (61).

Embrittlement of Structural Alloys: Bi in Cu

Embrittlement of copper by bismuth is the classic example of the phenomenon and a nearly ideal system for an atomic-scale investigation of embrittlement. Only two elements are involved, no Cu-Bi intermetallic compounds exist, and there is no evidence that hydrogen embrittles copper. Figure 11 (see color insert) (62) compares *Z*-contrast images of the nondoped (*a*) and the Bi-doped (*b*) boundaries. In this [001] projection, the atomic columns in the copper grains form a square grid of bright dots separated by 0.18 nm. The boundary in both samples is seen to consist of a periodic array of a defect structural unit (as indicated in Figure 11). One of the four undercoordinated atomic columns in each structural unit of the bismuth-doped boundary is seen to be distinctly bright. The extra intensity is consistent with the presence of an impurity with an atomic number much greater than copper, in this case Bi. The images make it clear that this grain boundary configuration is little affected by the presence of bismuth. No new phases are formed, and no drastic atomic rearrangements occur as the result of bismuth segregation to this boundary. Bismuth simply substitutes for Cu on a preferred atomic site in the boundary core. The electronic structure of the grain boundary region can be measured directly with EELS. It is found that locally the Cu-L_3 edge is affected by the presence of bismuth in the boundary. Three small peaks superimposed on the step-like Cu-L_3 edge are found to be considerably reduced in the spectrum from the doped boundary (62). This observation indicates a change in bonding of the copper atoms at the doped grain boundary. These changes are consistent with a reduction of the density of *d*-like states at the Fermi level and reduced hybridization between *d* and *s* states for the copper atoms that surround the bismuth. Both of these results also imply reduced covalent (directional) bonding between Cu 3*d* electrons in the boundary region.

The choice of a small period tilt boundary makes it possible to combine high-resolution imaging and spectroscopy with first-principles calculations. Density functional calculations of the Cu L_3 edge in the Bi-doped grain boundary show a reduction of all three maxima in near-edge structure, compared with the calculated near-edge structure for bulk Cu, just as was seen experimentally. The calculations confirmed the interpretation of the experimental EELS results, i.e., there are a reduced number of unoccupied *d* holes at or above the Fermi level and reduced hybridization between *d* and *s* states for the copper atoms that surround the bismuth impurities. The calculated charge density difference also revealed that bismuth induces a slight reduction in the directionality of bonding between Cu atoms. In addition, the calculations show enhanced electron density at the copper sites that surround the Bi impurity (62). This means that electrons further fill the usually half-filled *s*-bands of copper. This *s*-band filling can be described as a population of antibonding states, which reduces the bond strength between the copper atoms that surround the bismuth impurity.

Using a combination of *Z*-contrast imaging, EELS, and ab-initio theoretical materials simulations, the geometric and electronic structure of a grain boundary with and without the impurity can be investigated on the atomic scale. By knowing the atomic configuration, reliable electronic structure calculations can be made to determine how the impurity affects bonding and induces embrittlement. This investigation revealed that copper atoms surrounding the segregated bismuth exhibit a closed 3*d* shell, reduced *s-d* hybridization and an *s*-band that is more than half full.

The improved resolution that aberration correction offers will expand the number of boundaries that can be imaged. Atomic column positions will be revealed with more contrast and better signal-to-noise ratio with correction, thus increasing the detectability of segregants. However, the most exciting development is the possibility of imaging the atomic configurations of defects in all three dimensions that will come with further improvements in aberration correction. This development will make it possible for electron microscopy to move beyond the limits set by the projection problem and allow the determination of the atomic positions and compositions in X, Y, and Z directions of any boundary or defect.

Complex Oxide Characterization in the STEM

Many different complex oxides are of wide interest to electronics and the emerging field of spintronics. Most, such as ferroelectrics, high T_c superconductors, or colossal magnetoresistant materials, have a perovskite structure. In these materials, the aberration-corrected STEM has successfully achieved the imaging of light atoms, such as O. Seeing oxygen columns is quite a significant achievement in microscopy, which is also of enormous relevance in complex oxide science. Oxygen columns can be now resolved in the HAADF images in the STEM, even when they are surrounded by heavier columns. As an example, Figure 12 (see color insert) shows an image from $SrTiO_3$ in the [110] orientation where the light oxygen atomic columns can easily be seen.

Perovskite materials in general are highly suitable candidates for EELS analysis due to the wide availability of theoretical reports and computational techniques needed for the interpretation of the local density of states and, therefore, of the EEL spectra. Also, one of the great advantages of aberration correction in the STEM is the higher intensity and spatial localization of the electron probe. As shown earlier, not only is the probe more spatially localized, but also the intensity is higher. This is a key issue when it comes to carrying out atomic resolution EELS: the signal increases, which allows shorter acquisition times and reduction of beam damage, and the excited volume decreases, which allows a true column-by-column probe of the material's electronic properties (2, 3, 63).

Figure 13*a*, see color insert, shows a high resolution *Z*-contrast image of a La single atom from a one-unit-cell thick $La_{0.002}Ca_{0.998}TiO_3$ layer in a $CaTiO_3$ matrix (4). The higher *Z* of La as opposed to Ca, O, or Ti makes single La atoms appear as bright spots in this environment, although this brightness is dependent on atom depth, as discussed below (32, 64). By placing the STEM probe directly over the image of an isolated La atom, a clear EELS signature is obtained. In Figure 13*b*, the red spectrum shows two distinct peaks that correspond to the La $M_{4,5}$ lines at 832 and 849 eV. The intensity is low, as one would expect from the small number of counts associated with the excitations of a single atom, but clearly observable.

There has been much discussion in recent years about detection limits and the role that the delocalization of the underlying ionization process plays in limiting the resolution of the STEM (3, 14, 65–67). In order to analyze the localization of the EELS signal, Varela et al. performed line scans across the individual La cations (4). Figure 13*a* shows the directions of the scan with respect to the atom position. Given the fact that neighboring atomic columns correspond to different atomic species (O or TiO columns), the channeling and therefore the localization of the signal would be different. As depicted in Figure 13*b*, the EELS signal attributed to the La single atom is clearly localized on the atomic column containing this ion. When the electron beam is placed on any adjacent column, the signal is substantially reduced, to only $(10 \pm 5)\%$ on the neighboring TiO column, and to $(20 \pm 5)\%$ on the neighboring O site, only 0.19 nm away (shown in blue). This clearly indicates that the La can be localized to significantly better than 0.2 nm. The impact parameter for La-M excitation is $\sim$0.01 nm (3), which suggests that the residual intensity from neighboring columns is almost certainly the effect of dechanneling (68, 69).

Through dynamical simulations of probe spreading, an indication of atom depth was also obtained. EELS images were simulated using mixed dynamic form factor theory (68, 69). To estimate the depth of the impurity, Figure 13*c* shows the simulated signal strength for an aberration-free probe situated on the Ca column, O column, and TiO column as a function of the impurity depth. (A simplified cubic model for the structure was used for this simulation). The contribution on the Ca column initially increases through to the electrostatic attraction of the electron to the column (channeling), but then decreases as elastic and inelastic scattering mechanisms broaden the probe. As a function of dopant depth in the crystal, the contribution of the dopant atom to the EELS intensity on the O column rises gradually; simulations of the electron density make clear that this is due to probe spreading. From this figure it is estimated that a (20 $\pm$ 5)% signal on the O column results from an impurity depth of around 10 nm. From EELS simulations, show in Figure 13*c*, the impurity depth was estimated to be 12.2 nm. The simulations support the localization of the ionization interaction and demonstrate that the dynamical propagation, in transferring some electron intensity to adjacent columns, is responsible for the signals obtained. This level of sensitivity allows us to analyze in great detail the structure and electronic structures of complex oxides at the atomic scale.

CASE 1: CHARGE ORDERING IN MANGANITES A representative example concerns the study of charge ordering in manganese oxides exhibiting colossal magnetoresistance (CMR) (70). In some of these manganites a form of spatial modulation known as the charge-ordered state is encountered. These phases are insulating and are often antiferromagnetic so that understanding the origin of the phenomenon may provide the key to the origin of ferromagnetism and CMR in these materials. However, in spite of extensive studies dedicated to this topic, the underlying physics remains an open issue (70–74). So far, experimental techniques, such as diffraction and scanning probe microscopy, have provided only limited information, averaged over long-length scales and confined to the materials surfaces, respectively. CMR materials have a chemical formula of $A_xB_{1-x}MnO_3$, where A is a trivalent cation (La, Nd, Bi, Pr) and B a divalent cation (Sr, Ca, Ba). A HAADF image of a $Bi_{0.37}Ca_{0.63}MnO_3$ (BCMO) manganite is shown in Figure 14*a* (see color insert) along the (100) direction (75). MnO planes have been marked with blue dotted lines; BiCa planes are marked with yellow lines. The columns with a higher Bi concentration are brighter, whereas the Ca-rich columns are darker. No chemical ordering was observed from these images. But, interestingly, the resulting mixed-valence state within the Mn sublattice produces a complex electronic structure that we can image using EELS. The common view so far is that a fraction 1-x Mn ions per unit cell are in a +4 formal oxidation state, with a $3d^3$ electronic configuration ($t_{2g}{}^3\ e_g{}^0$), whereas the rest, being Mn^{+3}, have a $3d^4$ configuration $(t_{2g}{}^3 e_g{}^1)^1$. The charge-ordered state is believed to arise as a spatial stabilization of the e_g orbital distribution. For the BCMO compound the transition temperature to a disordered state is above room temperature, making it unnecessary to cool the sample to observe charge ordering.

EELS can probe the occupancy of the outer *d*-states of the Mn atoms with atomic resolution. The white lines at the onsets of the L_3 and L_2 absorption edges of transition metals are a characteristic signature of electronic transitions from the $2p_{1/2}$ and $2p_{3/2}$ Mn core states to unoccupied *d*-like states near the Fermi level (76–80). In particular, the ratio of the intensity of the L_3 peak to the L_2 peak, the L_{23} ratio, increases as the formal Mn oxidation number decreases. For example, Mn atoms in the parent compounds $LaMnO_3$ and $CaMnO_3$ have +3 and +4 formal oxidation states ($3d^4$ and $3d^3$ configurations), respectively. EEL spectra for those are shown in Figure 14*b*. The value of the L_{23} ratio in the first case is $L_{23} = 2.7 \pm 0.2$, whereas in the second, it is $L_{23} = 2.1 \pm 0.2$. In the experiment reported by Varela et al. (75), noticeable variations of this ratio can be observed as the electron beam is swept across the sample. Typical EEL spectra from two inequivalent types of Mn columns are shown in Figure 14*b*. The L_{23} ratios are again 2.7 ± 0.2 and 2.1 ± 0.2, as found in the parent compounds $LaMnO_3$ and $CaMnO_3$ respectively. These data therefore establish that the Mn atoms in BCMO coexist in two distinct formal oxidation states, +3 and +4. Further confirmation of the oxidation state is seen from the chemical shifts. In both the mixed valent manganite and the parent $LaMnO_3$ compound, a shift of approximately 1 eV is observed to lower energy losses.

Figure 14*c* shows the variation of the L_{23} ratio with position in the sample as the beam is swept parallel to the pseudocubic [100] direction (green arrow in Figure 14*a*). Clearly, the L_{23} ratio exhibits a modulation with a periodicity varying between 11 Å and 15 Å. When rotating the scanning direction by 90° (blue arrow in Figure 14*a*) *no* spatial modulation of the Mn L_{23} ratio was observed (Figure 14*d*), which points to a stripe-like geometry for the *3d* occupancy ordering. Additional scans were performed at different angles to confirm this conclusion. For example, the results obtained for a 45° angle revealed a modulation of the L_{23} ratio with a period of 22 Å (Figure 14*e*), close to the projection of the initially measured modulation length along a direction at 45°. This is consistent with the existence of a family of inequivalent planes or stripes that run along the pseudocubic (100) planes.

Higher magnification data along the [100] cubic axis are shown in Figure 14*f*, where individual Mn planes have been marked with red arrows. Jumps in the L_{23} ratio are visible only on specific individual Mn planes. From similar line scans (Figure 14*g*), the stripe width was measured to be below 3 Å, evidence that the stripes are atomically wide. Also, the uneven spacing between maxima in Figure 14*f* is noticeable, showing how the incommensurate number of doping electrons per unit cell cannot be accommodated in an equi-spaced striped pattern. Fourier transforms of the L_{23} ratio data showed clear peaks at frequencies around 0.83 Å^{-1}, denoting a spatial periodicity of about 12 Å. The stripe geometry evidenced by EELS is perfectly consistent with the macroscopically averaged periodicity observed in electron diffraction (75).

Interfaces

Most of the physical properties of composite materials are determined by low-dimensional active regions such as interfaces or grain boundaries. Hence, the understanding of the relations between their nature, electronic properties, and macroscopic behavior requires the use of probes with ultimate atomic resolution. A complete understanding of the atomistic and electronic structure of an interface can be used to predict physical properties and assist the development of new interfaces with unique properties. A few examples of atomic resolution studies of complex oxides and of the relationship between structure and electronic properties of oxide interfaces are presented below.

CASE 1: SUPERCONDUCTING/FERROMAGNETIC SUPERLATTICES The first example concerns the analysis of interfaces in superconducting/ferromagnetic $YBa_2Cu_3O_{7-x}/La_{0.67}Ca_{0.33}MnO_3$ (YBCO/LCMO) superlattices. Experimental studies on LCMO/YBCO structures suggest a strong interplay between ferromagnetism and superconductivity that results in the injection of spin-polarized carriers (81) or other proximity phenomena, which occur at nanometer distances from the interface (82–86). The critical temperatures Tc of thin high Tc superconducting films sandwiched between CMR spacers are comparatively lower than those obtained when using a non-magnetic material. Furthermore, heterostructures with ultrathin YBCO layers (one or two unit cells thick) are non-superconducting (82, 83). A deep understanding of the physics of oxide-based superconducting/ferromagnetic heterostructures needs a careful analysis of the relation between interface structure and electronic properties at the atomic scale. Recently, a detailed analysis of the structure/property relations at such interfaces has been reported by Varela et al. (87, 88). The structural quality of the samples was shown to be rather high. The YBCO/LCMO interface is coherent and free of defects, as seen in Figure 15. Occasionally, several steps one unit cell high are observed consistent with the small

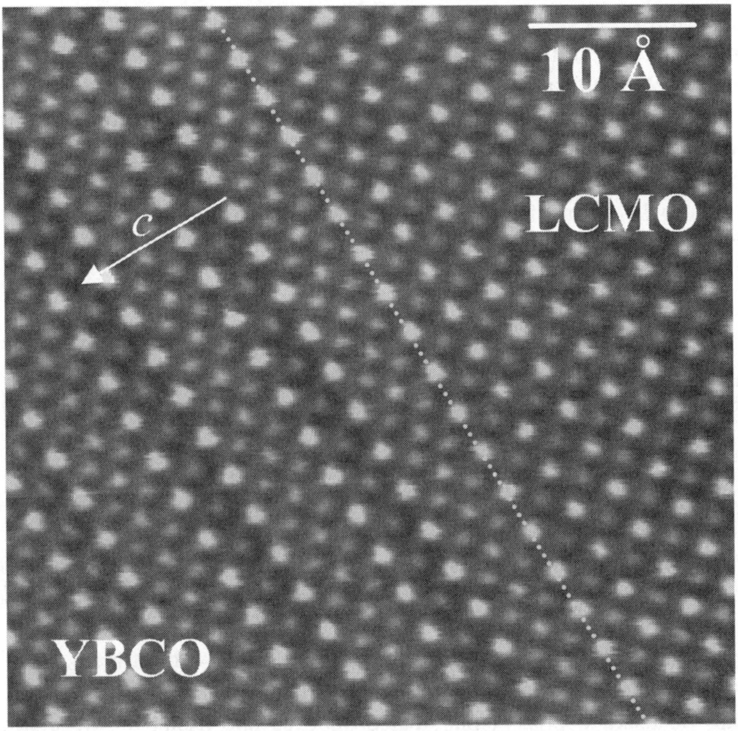

Figure 15 High-resolution *Z*-contrast image of a typical YBCO/LCMO interface. The dotted line marks the interface position.

step disorder found by X-rays (82, 83). EELS studies conclude that there is no major chemical interdiffusion (87).

Once structural effects were ruled out as a source of T_c reduction in these samples, the study of the interface electronic properties found several interesting effects. Direct evidence for charge transfer, i.e., doping of carriers from the ferromagnet into the superconductor, has been shown. Spectroscopic measurements with atomic resolution show the existence of a noticeable decrease in the density of holes in the YBCO by the interface. In HTCS materials, the O 2*p* and the Cu 3*d* bands lie very close to the Fermi energy. Therefore, the analysis of the O K edge at 530 eV and the Cu L edge at 930 eV has proven to be very useful to illuminate their electronic properties. In particular, the oxygen K edge, which results from exciting transitions from the oxygen 1*s* core level to the oxygen 2*p* bands can be used to quantitatively probe the occupancy of the oxygen 2*p* bands, i.e., the carrier density in the superconductor (89, 90). The prepeak in the O K-edge contains mostly O 2*p* band contribution. The ratio of the prepeak to the main edge provides a quantitative measure of the local density of holes. Figure 16*a* shows the prepeak intensity as a function of distance to the interface, within the YBCO, in a $[YBCO_{10\ u.c.}/LCMO_{15\ u.c.}]1000$ Å superlattice. Right on the interface plane, the hole density in the superconductor is reduced, and it then slowly increases till it achieves the bulk value at 3 nm from the interface. Also, the Mn oxidation state within the adjacent LCMO layer was found to be slightly higher (by ~5%) than the value expected for the given chemical doping. This situation is consistent with having some electron doping from the manganite into the superconductor. Very likely, this is the origin of the 3-nm-thick hole-depleted layer in the YBCO side of the interface. These results should be taken into account when trying to understand the reduced T_c values in these superlattices.

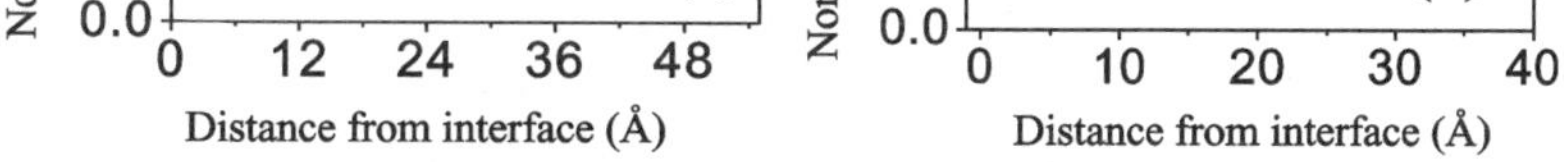

Figure 16 (*a*) Normalized prepeak intensity for the O K-edge acquired at the YBCO side of a YBCO/LCMO interface, as a function of the distance to the interface. The horizontal dotted line marks the prepeak intensity in bulk YBCO. (*b*) Normalized prepeak intensity within the YBCO in a YBCO/PBCO interface, as a function to the distance to the interface.

The situation in a superconducting/insulator $YBa_2Cu_3O_{7-x}/PrBa_2Cu_3O_7$ (YBCO/PBCO) interface is quite different, as summarized in Figure 16*b*, where again the hole density in the superconductor versus the distance to the interface is plotted. In this case, the CuO_2 planes in the first YBCO unit cell by the interface already show a doping level close to optimal. Interestingly, EELS measurements in these samples show a spatial oscillation of the density of holes with a period equal to the lattice parameter, $c = 11.7$ Å. The hole density decreases on the CuO chains, whereas it increases on the CuO_2 planes to match the bulk value. These measurements, which were also reproduced in YBCO thick films, constitute a direct proof that the holes responsible for superconductivity are localized within the CuO_2 planes, in agreement with the commonly accepted belief.

CASE 2: ELECTROCERAMICS Another very interesting example comes from electroceramic materials, also with perovskite structure. These are gaining increasing importance for technological applications as functional ceramics in electrical and microelectronic devices (91–94). By contacting these functional ceramics with typically 3*d* transition metals, the resulting metal/ceramic interfaces are known to strongly affect the electronic and mechanical properties of such devices (95, 96). From the point of view of fundamental science, metal/perovskite interfaces represent contacts between two materials with different bonding mechanisms: a metal with a purely metallic bonding characteristic and a perovskite with mixed ionic-covalent bonds (97–99).

The Pd/$SrTiO_3$ interface was chosen as a model system in which the face-centered cubic transition metal is in contact with the (100) surface of $SrTiO_3$, which crystallizes in the cubic perovskite structure at room temperature. HAADF imaging, high-spatial resolution EELS, and density functional theory in the local density approximation (100, 101) were applied to delimit how adhesion is achieved at metal/perovskite interfaces (102).

Thin Pd films with a nominal thickness of 35 nm were deposited on (100) surfaces of single crystalline $SrTiO_3$ by molecular beam epitaxy. The (100) surface of cubic $SrTiO_3$ can be terminated either by a TiO_2 or by a SrO layer. Ion etching of the substrate in a buffered HF solution and subsequent heating cycles prior to deposition resulted in a TiO_2-terminated surface (103, 104). TEM observations showed Pd islands on the substrate surface with an epitaxial orientation relationship with (100) planes and $\langle 100 \rangle$ directions in both materials parallel to each other. Figure 17 shows an HAADF image of the Pd/$SrTiO_3$ interface in $\langle 100 \rangle$ zone-axis orientation, recorded with the HB603U STEM (102). A coherent interface structure as well as the epitaxial cube-on-cube orientation relationship between the Pd film and the $SrTiO_3$ substrate is observed. An analysis of relative image intensities for the atomic columns indicates that the interface is terminated by a TiO_2 layer rather than by SrO (Figure 17, see color insert). No interfacial area with a SrO termination was observed throughout all the recorded data sets, although some areas with free SrO-terminated surfaces were observed. Thus the conclusion is that Pd prefers to nucleate on TiO_2-terminated (100) $SrTiO_3$ surfaces, which is in perfect agreement with theoretical predictions by Ochs et al. (105, 106), who found the Pd adsorption on the TiO_2-terminated substrate surface to be energetically favored compared to the nucleation on a SrO-terminated substrate surface. Ochs and coworkers also predicted an adsorption of Pd atoms on top of oxygen atoms of the (100) surface rather than above Ti atoms and the adjacent hollow sites.

To investigate this aspect in more detail, van Benthem and coworkers (102, 107) used fully relaxed atomistic structure models provided by Ochs et al. for the two different interface terminations (105) to calculate site- and symmetry-projected densities of unoccupied states (PDOS). The calculations were performed with a mixed-basis pseudopotential approach (107–109). To a good approximation the calculated O p-PDOS are proportional to electron energy-loss near-edge structures (ELNES) of O K absorption edges. van Benthem et al. (102, 110) measured the O K-, the Ti $L_{2,3}$, and the Sr $L_{2,3}$ absorption edges in electron energy-loss spectra with a high-spatial resolution to evaluate the theoretically predicted structure models. Figure 18 shows the O K-edge for bulk $SrTiO_3$ with high energy resolution (*a*) in comparison with the O K-edge measured from a 2 $\times$ 20-nm area containing the interface (*b*). The interfacial O K-edge reveals four major spectral peaks labeled A, B, C, and D characteristic of the bulk (Figure 18), along with an additional spectral feature X_3. The peak labels are chosen to coincide with a more detailed spectral analysis in references (107) and (110).

The interface spectrum (Figure 18*b*) still contains some contributions from oxygen atoms in bulk $SrTiO_3$ substrate. To extract the purely interface-related signal, the contribution of bulk spectral intensity to the interface spectrum was estimated to be 11% through the quantitative spatial difference technique (111–114). This estimate is based on geometrical considerations and beam broadening effects within the transmitted specimen. The extracted interface-specific component is plotted in Figure 18*c*. Because the signal-to-noise ratio of this spectrum is rather small, it was smoothed with a 1.0 eV-wide top-hat function and then plotted in Figure 18*d*. The interface specific line shapes in Figure 18*c*,*d* clearly show a damping of the pure bulk intensities A, B, C, and D along with the appearance of new spectral features labeled X_1, X_2 and X_3, which are absent in the bulk (see Figure 18*a*).

Comparing first the calculated line shapes with the experimental ELNES data for bulk $SrTiO_3$ (i.e., spectra a and h), it can be inferred that the calculated PDOS reproduces the experimental ELNES data (102, 107). Focusing on the interface specific components of the O K-ELNES (see Figure 18*c*,*d*), a high level of agreement is observed with the calculated PDOS for the TiO_2-termination plotted in Figure 18*e*. The calculated data not only reproduces the general ELNES line shape concerning the bulk peaks A, B, and C, but also clearly exhibits the peaks labeled X_1, X_2, and X_3 as observed by experiment. The calculated PDOS obtained from the SrO-terminated structure model (Figure 18*f*) does not show any agreement with the experimental data (Figure 18*c*,*d*). Especially in the energy regime where peak X_3 was observed, the densities of states at the SrO-terminated interface exhibits a minimum, which is contrary to the experimental findings. Therefore, it is concluded that the experimentally observed Pd/$SrTiO_3$ interface is characterized by a TiO_2 termination of the substrate surface. This is in good agreement with the direct observations by HAADF-STEM imaging (102, 110).

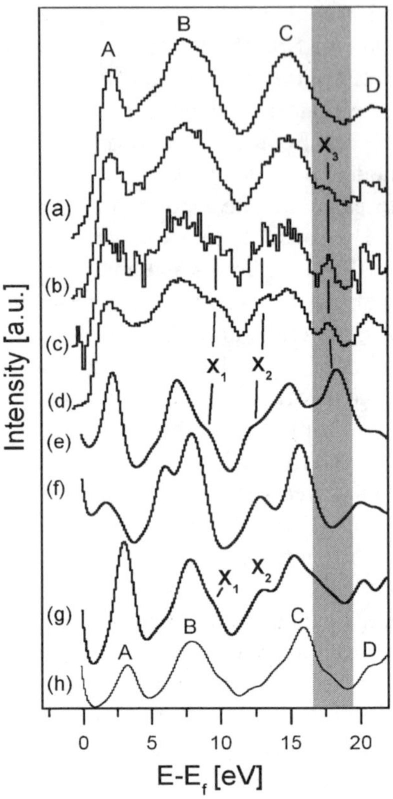

Figure 18 Experimentally and theoretically determined O K-edges recorded from bulk $SrTiO_3$ (*a*) and from a 2×20-nm^2 area containing the Pd/$SrTiO_3$ interface (*b*). The interface-specific component of (*b*) is plotted in (*c*) and (*d*), before and after smoothing of the curve, respectively. Line profiles (*e*) and (*f*) are the calculated O p-PDOS for a TiO_2- and a SrO-terminated interface structure. Spectrum (*g*) is the calculated O p-PDOS for the free TiO_2-terminated $SrTiO_3$ surface; (*h*) is the O p-PDOS in bulk $SrTiO_3$. All spectra are plotted as a function of energy relative to the Fermi level E_F.

Figure 19 is a sketch of the local oxygen atom coordination in bulk $SrTiO_3$ (*a*), at the TiO_2-terminated surface (*b*) and at the Pd/$SrTiO_3$ interface (*c*). Because oxygen atoms are coordinated by either four Sr atoms, two Sr atoms, or two Sr atoms plus one Pd atom, the crystal fields at the oxygen sites are different for the three arrangements. A detailed analysis of the corresponding p_x, p_y and p_z projected PDOS data revealed that the crystal fields at the interfacial O sites are modified compared with those of bulk $SrTiO_3$. This results in a reduced splitting of the peaks labeled C and D in Figure 18*a*, hence forming the extra peak X_3.

The agreement of the experimental electronic structure results with the calculated unoccupied densities of states positively evaluates the validity of the theoretically predicted atomistic structure model (102). Therefore, it is germane to determine the bonding characteristics across the interface in terms of the calculated occupied densities of states. The occupied PDOS shows a strong overlap of O p_z orbitals and Pd 4d orbitals across the interface. Hence, the adhesion between the (100) $SrTiO_3$ surface and the Pd film is determined by a hybridization of O p_z and Pd 4d states, establishing σ-type bonds. Parallel to this effect, an increased overlap of the O p_xPDOS with the Ti $3d_{x^2-y^2}$ PDOS can be observed, reflecting

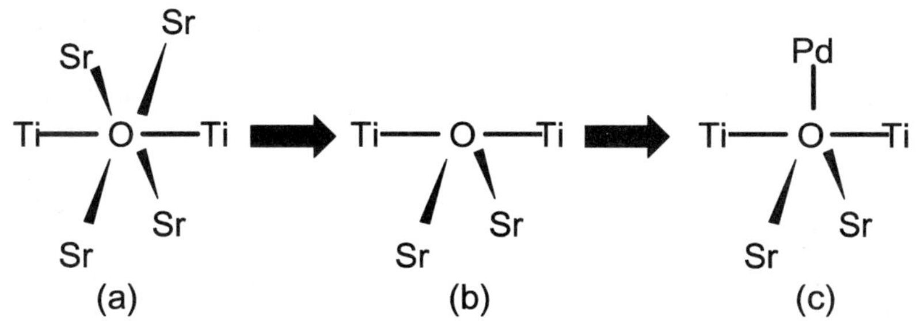

Figure 19 Sketch of (*a*) bulk, (*b*) free surface, and (*c*) interface models showing the ligands for the interfacial oxygen atoms. The modification in the coordination of interfacial oxygen columns when forming the $Pd/SrTiO_3$ interface changes the crystal field. The modified ligand field is responsible for the appearance of peak X_3 in Figure 18.

an increased bonding between O and Ti within the substrate surface (102, 107). However, Ti $3d_{x^2-y^2}$ to Pd $4d_{xy/xz}$ interactions are also observable and contribute weakly to the adhesion (107).

In summary, a combination of aberration-corrected HAADF-STEM imaging, high-spatial resolution EELS, and ab initio calculations resulted in a complete determination and understanding of the origin of adhesion forces at the $Pd/SrTiO_3$ interface.

Three-Dimensional Imaging of Semiconducting Interfaces

The dimensions of silicon-based semiconductor devices have been reducing steadily over the past few decades, as frequently described by Moore's law (115–117). Owing to the shrinking dimensions, which are close to the atomic scale, the presence of single-dopant atoms in gate dielectric films and interconnects derives ever more significance (117). Even a single impurity atom at the interface between Si and a high dielectric constant (high-*k*) film can introduce defect states into the local band structure (118) and therefore significantly alter the interface properties, such as the complex dielectric function, leakage current, and breakdown field. Similarly, it can introduce scattering centers into the Si gate, which can reduce the carrier mobility. Early work on imaging dopant atoms relied on statistical arguments to show that the imaged dopant atoms are in the bulk (119), and work done at lower resolution has used approximate matches to image simulations (120), neither of which is unequivocal proof for the location of the dopants. It has been largely overlooked that the tremendously decreased spot sizes, δ_s, due to aberration-correction of the probe-forming lenses, also result in a significantly reduced depth of focus T in the STEM (121).

$$T \leq \frac{\delta_s}{\alpha} = \frac{\lambda}{\alpha^2}, \quad 1.$$

where α is the illumination semi-angle and λ is the electron wavelength. The equation shows that the availability of larger illumination angles decreases the depth of focus much faster than it improves the lateral resolution. Hence, only very thin slices of the specimen (2–3 nm thick) come into focus at a time. By recording image stacks at different focus values it is possible to "slice" through the TEM sample in a manner comparable to that used in confocal optical microscopy (15). With the HB603U STEM, a volume resolution better than 0.08 nm $\times$ 0.08 nm $\times$ 4 nm is achievable (122), i.e., approximately a factor of 10^{12} higher than in X-ray tomography (123–125) and a factor of 100 higher than recent electron tomography techniques based on tilt-series reconstruction (126–128).

Hafnium dioxide (HfO_2) is a high-k material ($k = 25$) under consideration as a replacement for SiO_2 ($k = 3.9$) as a gate dielectric. For this study, approximately 6-nm-thick HfO_2 films were deposited on Si substrates using atomic layer deposition (129) at a temperature of 320°C. The samples were then rapid-thermal annealed at 950°C for 30s in N_2 and afterward capped with undoped polycrystalline Si by chemical vapor deposition. To avoid artifacts from ion milling, samples were prepared for cross-sectional TEM purely by mechanical polishing. Both imaging and EELS measurements revealed a thin, thermally grown SiO_2 interlayer in between the polycrystalline HfO_2 film and the silicon substrate. High-resolution ADF STEM imaging revealed single Hf atoms within the SiO_2 interlayer (122). Figure 20 (see color insert) shows 9 images extracted from a 41-frame through-focal-series recorded in $\langle 110 \rangle$ zone-axis orientation. The full series was acquired in less than 5 min. At various positions throughout the series, single Hf atoms appear and disappear over about 3 to 5 frames, each corresponding to intervals of 1.5 nm to 2.5 nm in depth (see for instance the red circle in Figure 20). To demonstrate the location of the Hf atom marked in Figure 20 more clearly, Figure 21 (see color insert) shows a projection view of the three-dimensional data cube.

The image in the x-y plane reveals the polycrystalline HfO_2 layer separated from the Si substrate by a roughly 1-nm-thick SiO_2 layer. The single Hf atom visible in the x-y plane is also observed in the x-z plane and the y-z plane as a single atom, visible about half way through the z -range of the acquired data set. The voxels in the data set are not cubic but cuboid due to the limited vertical resolution compared with the high lateral resolution. Throughout the whole data set, the two vertically outermost Hf atoms are about 6 $\pm$ 1 nm apart from each other (122). We assume that these two atoms sit at the top and bottom surfaces of the TEM specimen. As a result, the sample thickness can thus be estimated to be 6 $\pm$ 1 nm.

The through-focal-series technique to obtain three-dimensional information at atomic resolution is the first direct proof of single (dopant) atoms located within the volume of a TEM sample (122) (compare references 4, 119, 120). Volume-rendering techniques can be used to visualize these single atoms in three dimensions. By thresholding intensity values throughout the three-dimensional data set it becomes possible to visualize the location of single atoms in movies with rotating isosurfaces. A screenshot of the derived model in three dimensions, extracted from the movie, is shown in Figure 22 (see color insert). In this representation, owing to the limited vertical resolution, atoms appear cigar-shaped rather than spherical compared with that seen with the much higher lateral resolution. Note that even columns of Si dumbbells within the Si substrate are visible in the three-dimensional visualization, demonstrating the true three-dimensional character of the acquired data. As can be seen from Figure 22, even some information about interface roughness can be extracted from the recorded data set. However, it should be noted that the depth-slicing method at present does not provide good depth resolution in an aligned crystal because of strong electron channelling effects. Future generations of aberration-corrected microscopes with wider apertures may overcome this limitation (130).

CONCLUDING REMARKS

Throughout this article, we have reviewed the state-of-the-art of aberration-corrected scanning transmission microscopy and electron energy loss spectroscopy. We have demonstrated that these techniques can provide new, unique insights into the relations between the structure and the physical properties of a wide variety of heterogeneous systems. The improved sensitivity, both in imaging and spectroscopy, down to the single-atom level has greatly advanced our understanding of catalysts, semiconductors, complex oxides, and many other systems that are of enormous importance to materials science and condensed matter physics today. Furthermore, as we have shown, the basis for the microscopy of the future has been set—a three-dimensional approach to the characterization of solids, nanoscale systems, or interfaces is now possible. It is our belief that the unique role of aberration-corrected STEM (and related techniques) will be pivotal in the development of nanoscience.

ACKNOWLEDGMENTS

We express our most sincere acknowledgment to our collaborators: M. Rühle, S.T. Pantelides, A. Franceschetti, S. Rashkeev, G. Duscher, J. Santamaria, N.D. Browning, S. Overbury, O.L. Krivanek, N. Dellby, P.D. Nellist, L.J. Allen, M.P. Oxley, S.D. Findlay, K. Sohlberg, S. Wang, M. Glazov, M. Amiridis, R. Adams, C. Elsaesser, C. Scheu, T. Ochs, U. Alber, D.G. Mandrus, B. Sales, R. Jin, and H.M. Christen. We also want to acknowledge the collaboration of J.T. Luck for TEM specimen preparation and W.H. Sides, Jr. for microscope support. This research was sponsored by the Laboratory for Directed Research and Development Program of ORNL, managed by UT-Batelle, LLC, for the U.S. Department of Energy under Contract No. DE-AC05-00OR22725 and by appointments to the ORNL postdoctoral Research Program administered jointly by ORNL and ORISE and by the Alexander-von-Humboldt Foundation.

The *Annual Review of Materials Research* is online at http://matsci.annualreviews.org

LITERATURE CITED

1. Nellist PD, Chisholm MF, Dellby N, Krivanek OL, Murfitt MF, et al. 2004. Direct sub-angstrom imaging of a crystal lattice. *Science* 305:1741
2. Duscher G, Buczko R, Pennycook SJ, Pantelides ST. 2001. Core-hole effects on energy-loss near-edge structure. *Ultramicroscopy* 86:355
3. Rafferty B, Pennycook SJ. 1999. Towards atomic column-by-column spectroscopy. *Ultramicroscopy* 78:141–51
4. Varela M, Findlay SD, Lupini AR, Christen HM, Borisevich AY, et al. 2004. Spectroscopic imaging of single atoms within a bulk solid. *Phys. Rev. Lett.* 92:095502
5. Batson PE, Dellby N, Krivanek OL. 2002. Sub-angstrom resolution using aberration corrected electron optics. *Nature* 418:617–20
6. Wang SW, Borisevich AY, Rashkeev SN, Glazoff MV, Sohlberg K, et al. 2004. Dopants adsorbed as single atoms prevent degradation of catalysts. *Nat. Mater.* 3:143–46
7. Pennycook SJ. 2002. Structure determination through Z-contrast microscopy. In *Advances in Imaging and Electron Physics* 123:173–206. San Diego: Academic
8. Pennycook SJ, Nellist PD. 1999. Z-contrast scanning transmission electron microscopy. In *Impact of Electron and Scanning Probe Microscopy on Materials Research*, ed. DG Rickerby, U Valdre, G Valdre, pp. 161–207: Amsterdam: Kluwer
9. Nellist PD, Pennycook SJ. 1998. Accurate structure determination from image reconstruction in ADF STEM. *J. Microsc.* 190:159–70
10. Cowley JM. 1969. Image contrast in a transmission scanning electron microscope. *Appl. Phys. Lett.* 15:58
11. Nellist PD, Rodenburg JM. 1994. Beyond the conventional information limit—the relevant coherence function. *Ultramicroscopy* 54:61–74
12. Williams DB, Carter CB. 1996. *Transmission Electron Microscopy*. New York: Plenum
13. Lupini AR, Chisholm MF, van Benthem K, Allen LJ, Oxley MP, et al. 2005. Limitations to the measurement of oxygen concentrations by HRTEM imposed by surface roughness. *Microsc. Microanal.* 11:111–115
14. Allen LJ, Findlay SD, Lupini AR, Oxley MP, Pennycook SJ. 2003. Atomic-resolution electron energy loss spectroscopy imaging in aberration corrected scanning transmission electron microscopy. *Phys. Rev. Lett.* 91:105503
15. Lupini AR, Varela M, Borisevich AY, Travaglini SM, Pennycook SJ. 2004. Advances in aberration corrected STEM at ORNL. *Inst. Phys. Conf. Ser.* 179:211–14
16. Lupini AR, Varela M, Borisevich AY, Peng Y, Pennycook SJ. 2003. Prospects for single atom location and identification with aberration-corrected STEM. *Inst. Phys. Conf. Ser.* 180:523–32
17. Lupini AR, Krivanek OL, Dellby N, Nellist PD, Pennycook SJ. 2001. Developments in Cs-corrected STEM. *Inst. Phys. Conf. Ser.* pp. 31–34

17a. Cowley JM. 1979. Adjustment of a STEM instrument by use of shadow images. *Ultramicroscopy* 4:413–18

18. Dellby N, Krivanek OL, Nellist PD, Batson PE, Lupini AR. 2001. Progress in aberration-corrected scanning transmission electron microscop. *J. Electron Microsc.* 50:177–85
19. Scherzer O. 1936. Uber einige Fehler von

Elektronenlinsen. *Zeit. Phys.* 101:593–603
20. Scherzer O. 1947. Sparische und chromatische Korrektur con Elektronen-Linsen. *Optik* 2:114–32
21. Krivanek OL, Dellby N, Lupini AR. 1999. Towards sub-angstrom electron beams. *Ultramicroscopy* 78:1–11
22. Krivanek OL, Nellist PD, Dellby N, et al. 2003. Towards sub-0.5 angstrom electron beams. *Ultramicroscopy* 96:229–37
23. Rose H. 2004. Outline of an ultracorrector compensating for all primary chromatic and geometrical aberrations of charged-particle lenses. *Nuclear Instr. Method Phys. Res. A* 519:12–27
24. Weissbacker C, Rose H. 2002. Electrostatic correction of the chromatic and of the spherical aberration of charged-particle lenses (Part II). *J. Electron Microsc.* 51:45–51
25. Weissbacker C, Rose H. 2001. Electrostatic correction of the chromatic and of the spherical aberration of charged-particle lenses (Part I). *J. Electron Microsc.* 50:383–90
26. Urban K, Kabius B, Haider M, Rose H. 1999. A way to higher resolution: spherical-aberration correction in a 200 kV transmission electron microscope. *J. Electron Microsc.* 48:821–26
27. Beck V. 1979. A hexapole spherical aberration corrector. *Optik* 53:241–55
28. Crewe AV, Kopf D. 1980. Limitations of sextupole correctors. *Optik* 56:391–99
29. Haider M, Uhlemann S, Schwan E, Rose H, Kabius B, Urban K. 1998. Electron microscopy image enhanced. *Nature* 392:768–69
30. Nellist PD, Pennycook SJ. 1998. Subangstrom resolution by underfocused incoherent transmission electron microscopy. *Phys. Rev. Lett.* 81:4156–59
31. Nellist PD, Pennycook SJ. 1999. Incoherent imaging using dynamically scattered coherent electrons. *Ultramicroscopy* 78:111–24
32. Nellist PD, Pennycook SJ. 2000. The principles and interpretation of annular dark-field Z-contrast imaging. *Adv. Imaging Electron Phys.* 113:147–203
33. Kel'man VM, Yavor SY. 1961. Achromatic quadrupole electron lenses. *Sov. Phys. Tech. Phys.* 6:1052–54
34. Deltrap JHM. 1964. *Correction of spherical aberration of electron lenses.* PhD thesis. Univ. Cambridge, Cambridge, UK
35. Satterfield CN. 1980. *Heterogeneous Catalysis in Practice.* New York: McGraw-Hill
36. Wefers K, Misra C. 1987. *Oxides and Hydroxides of Aluminum.* Pittsburgh: Alcoa Labs
37. Church JS, Cant NW, Trimm DL. 1994. Surface-area stability and characterization of a novel sulfate-based alumina modified by rare-earth ions. *Appl. Catal. A* 107:267–76
38. Glazov MV, Novak J, Hector LG. 1999. Stabilization of γ-alumina: how much lanthanum do we actually need? *Alcoa Tech. Pap.* 99:168
39. Kumar KNP, Tranto J, Kumar J, Engell JE. 1996. Pore-structure stability and phase transformation in pure and M-doped (M = La, Ce, Nd, Gd, Cu, Fe) alumina membranes and catalyst supports. *J. Mater. Sci. Lett.* 15:266–70
40. Oudet F, Courtine P, Vieux A. 1988. Thermal stabilization of transition alumina by structural coherence with $LaAlO_3$, $PrAlO_3$, $NdAlO_3$. *J. Catal.* 114:112–20
41. Beguin B, Garbowski E, Primet M. 1991. Stabilization of alumina by addition of lanthanum. *J. Catal.* 75:119–32
42. Vazquez A, López T, Gómez R, Morales A, Novaro O. 1997. X-ray diffraction, FTIR, and NMR characterization of sol–gel alumina doped with lanthanum and cerium. *J. Solid State Chem.* 128:161–68
43. Yamamoto T, Tanaka T, Matsuyama T, Funabiki T, Yoshida S. 2001. Structural analysis of La/Al_2O_3 catalysts by La K-edge XAFS. *J. Synchrothron Rad.* 8:634–36
44. Nellist PD, Pennycook SJ. 1996. Direct

imaging of the atomic configuration of ultradispersed catalysts. *Science* 274:413–15
45. Sohlberg K, Rashkeev S, Borisevich AY, Pennycook SJ, Pantelides ST. 2004. Origin of the anomalous Pt-Pt distances in the Pt/Alumina catalytic system. *Chem. Phys. Chem.* 5:1893–97
46. Valden M, Lai X, Goodman DW. 1998. Onset of catalytic activity of gold clusters on titania with the appearance of nonmetallic properties. *Science* 281:1647–50
47. Haruta M. 1997. Size and support dependency in the catalysis of gold. *Catalysis Today* 36:153–66
48. Bond GC, Thompson DT. 2000. Gold catalyzed oxidation of carbon monoxide. *Gold Bull.* 33:41–51
49. Lupini AR, Franceschetti AG, Pantelides ST, Dai S, Chen B, et al. 2004. Aberration corrected STEM analysis of gold nanoparticle catalytic activity. *Microsc. Microanal.* 10:462–63
50. Hoffmann MJ, Petzow G. 1994. Tailored microstructures of silicon nitride ceramics. *Pure Appl. Chem.* 66:1807–14
51. Lange FF. 1973. Relation between strength, fracture energy, and microstructure of hot-pressed Si_3N_4. *J. Am. Ceram. Soc.* 56:518–22
52. Rühle M, Evans AG. 1989. High toughness ceramics and ceramic composites. *Prog. Mater. Sci.* 33:85–167
53. Hoffmann MJ, Gu H, Cannon RM. 2000. Influence of the interfacial properties on microstructural development and properties of Si_3N_4 ceramics. In *Interfacial Engineering for Optimized Properties II*, ed. CB Carter, EL Hall, CL Briant, S Nutt, pp. 65–74. Warrendale, PA: Mater. Res. Soc.
54. Shibata N, Pennycook SJ, Gosnell TR, Painter GS, Shelton WA, Becher PF. 2004. Observation of rare-earth segregation in silicon nitride ceramics at subnanometre dimensions. *Nature* 428:730–33
55. Wang CM, Pan WQ, Hoffmann MJ, Cannon RM, Rühle M. 1996. Grain boundary films in rare-earth-glass-based silicon nitride. *J. Am. Ceram. Soc.* 79:788–92
56. Kleebe H-J, Rühle M. 1992. Analytical electron microscopy and high-resolution electron microscopy studies of grain-boundary films in silicon nitride-based ceramics. *Mater. Res. Soc. Symp. Proc.* 238:859–64
57. Averill FW, Painter GS. 1994. Symmetrized partial-wave method for density-functional cluster calculations. *Phys. Rev. B* 50:7262–67
58. For more details, visit http://www.pixon.com
59. Painter GS, Becher PF, Shelton WA, Satet RL, Hoffmann MJ. 2004. First-principles study of rare-earth effects on grain growth and microstructure in β-Si_3N_4 ceramics. *Phys. Rev. B* 70:144108
60. Abe E, Yan Y, Pennycook SJ. 2004. Quasicrystals as cluster aggregates. *Nat. Mater.* 3:759–67
61. Abe E, Pennycook SJ, Tsai AP. 2003. Direct atomic-scale observation of a local thermal vibration anomaly in a quasicrystalline $Al_{72}Ni_{20}Co_8$ compound. *Nature* 421:347–50
62. Duscher G, Chisholm MF, Alber U, Ruhle M. 2004. Bismuth-induced embrittlement of copper grain boundaries. *Nat. Mater.* 3:621–26
63. Browning ND, Chisholm MF, Pennycook SJ. 1993. Atomic-resolution chemical analysis using a scanning transmission electron microscope. *Nature* 366:143–46
64. Nakamura K, Kakibayashi H, Kanehori K, Tanaka N. 1997. Position dependence of the visibility of a single gold atom in silicon crystals in HAADF-STEM image simulation. *J. Electron Microsc.* 46:33–43
65. Kohl H, Rose H. 1985. Theory of image-formation by inelastically scattered electrons in the electron microsscope. *Adv. Imaging Electron Phys.* 65:173–227
66. Muller DA, Silcox J. 1995. Delocalization in inelastic-scattering. *Ultramicroscopy* 59:195–213

67. Holbrook OF, Bird DM. 1995. Theoretical modelling of atomic images formed with inelastically scattered electrons. In *Electron Microscopy and Analysis 1995*, pp. 175–78. Bristol, UK: IOP Publishing
68. Allen LJ, Findlay SD, Oxley MP, Rossouw CJ. 2003. Lattice-resolution contrast from a focused coherent electron probe. Part I. *Ultramicroscopy* 96:47–63
69. Findlay SD, Allen LJ, Oxley MP, Rossouw CJ. 2003. Lattice-resolution contrast from a focused coherent electron probe. Part II. *Ultramicroscopy* 96:65–81
70. Jin S, Tiefel TH, McCormack M, Fastnacht RA, Ramesh R, Chen LH. 1994. Thousandfold change in resistivity in magnetoresistive La-Ca-Mn-O films. *Science* 264:413–15
71. Von Helmolt R, Wecker J, Holzapfel B, Schultz L, Samwer K. 1993. Giant negative magnetoresistance in perovskitelike $La_{2/3}Ba_{1/3}MnO_X$ ferromagnetic films. *Phys. Rev. Lett.* 71:2331–33
72. Dagotto E, Hotta T, Moreo A. 2001. Colossal magnetoresistant materials: the key role of phase separation. *Phys. Rep.* 344:1–153
73. Salamon MB, Jaime M. 2001. The physics of manganites: structure and transport. *Rev. Mod. Phys.* 73:583–628
74. Coey JMD, Viret M, von Molnar S. 1999. Mixed-valence manganites. *Adv. Phys.* 48:167–293
75. Varela M, Franceschetti A, Lupini AR, Tian W, Jin R, et al. 2005. New insights into charge ordering phenomena in colossal magnetoresistant oxides. *Nature*. Submitted
76. Rask JH, Miner BA, Buseck PR. 1987. Determination of manganese oxidation-states in solids by electron energy loss spectroscopy. *Ultramicroscopy* 21:321–26
77. Leapman RD, Grunes LA, Fejes PL. 1982. Study of the L_{23} edges in the 3D transition-metals and their oxides by electron energy loss spectroscopy with comparisons to theory. *Phys. Rev. B* 26:614–35
78. Krivanek OL, Paterson JH. 1990. ELNES of 3D transition metal oxides. 1. Variations across the periodic table. *Ultramicroscopy* 32:313–18
79. Paterson JH, Krivanek OL. 1990. ELNES of 3D transition metal oxides. 2. Variations with oxidation state and crystal structure. *Ultramicroscopy* 32:319–25
80. Pearson DH, Ahn CC, Fultz B. 1993. White lines and *d*-electron occupancies for the 3D and 4D transition-metals. *Phys. Rev. B* 47:8471–78
81. Vasko VA, Larkin VA, Kraus PA, Nikolaev KR, Grupp DE, et al. 1997. Critical current suppression in a superconductor by injection of spin-polarized carriers from a ferromagnet. *Phys. Rev. Lett.* 78:1134–37
82. Sefrioui Z, Varela M, Peña V, Arias D, Villegas JE, et al. 2002. Superconductivity depression in ultrathin $YBa_2Cu_3O_{7-\delta}$ layers in $La_{0.7}Ca_{0.3}MnO_3$ /$YBa_2Cu_3O_{7-\delta}$ superlattices. *Appl. Phys. Lett.* 81:4568
83. Sefrioui Z, Arias D, Peña V, Villegas JE, Varela M, et al. 2003. Ferromagnetic/superconducting proximity effect in $La_{0.7}Ca_{0.3}MnO_3/YBa_2Cu_3O_7$ superlattices. *Phys. Rev. B* 67:214511
84. Peña V, Sefrioui Z, Arias D, Leon C, Santamaria J, et al. 2004. Coupling of superconductors through a half-metallic ferromagnet: evidence for a long-range proximity effect. *Phys. Rev. B* 69:224502
85. Przyslupski P, Komissarov I, Paszkowicz W, Dluzewski P, Minikayev R, Sawicki M. 2004. Magnetic properties of $La_{0.67}Sr_{0.33}MnO_3/YBa_2Cu_3O_7$ superlattices. *Phys. Rev. B* 69:134428
86. Holden T, Habermeier HU, Cristiani G, Golnik A, Boris A, et al. 2004. Proximity induced metal-insulator transition in $YBa_2Cu_3O_7/La_{2/3}Ca_{1/3}MnO_3$ superlattices. *Phys. Rev. B* 69:064505
87. Varela M, Lupini AR, Pena V, Sefrioui Z, Arslan I, et al. 2005. Direct measurement of charge transfer phenomena at ferromagnetic/superconducting oxide interfaces. *Phys. Rev. Lett.* Submitted

88. Varela M, Lupini AR, Pennycook SJ, Sefrioui Z, Santamaria J. 2003. Nanoscale analysis of $YBa_2Cu_3O_{7-x}$ /$La_{0.67}Ca_{0.33}MnO_3$ interfaces. *Solid State Electron* 47:2245–48
89. Browning ND, Yuan J, Brown LM. 1992. Determination of the local oxygen stoichiometry in $YBa_2Cu_3O_7$ by electron energy loss spectroscopy in the scanning transmission electron microscope. *Physica C* 202:12–18
90. Browning ND, Chisholm MF, Pennycook SJ, Norton DP, Lowndes DH. 1993. Correlation between hole depletion and atomic-structure at high-angle grain-boundaries in $YBa_2CuO_{7-\delta}$. *Physica C* 212:185–90
91. Henrich VE, Dresselhaus G, Zeiger HJ. 1978. Energy-dependent electron energy loss spectroscopy-application to the surface and bulk electronic structure of MgO. *Phys. Rev. B* 17:4908–21
92. Uchino K. 1994. Ferroelectric ceramics. In *Materials Science and Technology A Comprehensive Treatment—Structure and Properties of Ceramics*, ed. M Swain, RW Cahn, P Haasen, EJ Kramer, 11:635–77. Wenheim, GR: VCH
93. Meixner H, Gerblinger J, Lampe U, Fleischer M. 1995. Thin film gas sensors based on semiconducting metal oxides. *Sensors Actuators B* 23:119–25
94. Setter N, Waser R. 2000. Electroceramic materials. *Acta Mater.* 48:151–78
95. Howe JM. 1993. Bonding, structure and properties of metal-ceramic interfaces. 1. Chemical bonding, chemical reaction and interfacial structure. *Int. Mater. Rev.* 38:233–56
96. Howe JM. 1993. Bonding, structure and properties of metal-ceramic interfaces. 2. Interface fracture-behavior and property measurement. *Int. Mater. Rev.* 38:257–71
97. Finnis MW. 1996. The theory of metal-ceramic interfaces. *J. Phys. Condens. Matt.* 8:5811–36
98. van Benthem K, Elsasser C, Rühle M. 2003. Core-hole effects on the ELNES of absorption edges in $SrTiO_3$. *Ultramicroscopy* 96:509–22
99. van Benthem K, Scheu C, Sigle W, Rühle M. 2002. Electronic structure investigations of Ni and Cr films on (100)$SrTiO_3$ substrates using electron energy-loss spectroscopy. *Z. Metall.* 93:362–71
100. Hohenberg P, Kohn W. 1964. Inhomogeneous electron gas. *Phys. Rev.* 136:B864
101. Kohn W, Sham LJ. 1965. Self-consistent equations including exchange and correlation effects. *Phys. Rev.* 140:A1133
102. van Benthem K, Classen T, Scheu C, Elsasser C, Pennycook SJ, Rühle M. 2005. *Phys. Rev. Lett.* Submitted
103. Kawasaki M, Takahashi K, Maeda T, Tsuchiya R, Shinohara M, et al. 1994. Atomic control of the $SrTiO_3$ crystal surface. *Science* 266:1540–42
104. Polli AD, Wagner T, Rühle M. 1999. Effect of Ca impurities and wet chemical etching on the surface morphology of $SrTiO_3$ substrates. *Surf. Sci.* 429:237–45
105. Ochs T, Kostlmeier S, Elsasser C. 2001. Microscopic structure and bonding at the Pd/$SrTiO_3$ (001) interface—an ab-initio local-density-functional study. *Integr. Ferroelectr.* 32:959–70
106. Ochs T, Elsasser C. 2002. Thin Pd films on $SrTiO_3$ (001) substrates: ab initio local-density-functional theory. *Z. Metall.* 93:406–16
107. van Benthem K, Elsaesser C, Rühle M. 2005. Bonding of thin Pd films on (100)$SrTiO_3$ substrates from density functional theory. *Phys. Rev. B.* Submitted
108. Elsasser C, Takeuchi N, Ho KM, Chan CT, Braun P, Fahnle M. 1990. Relativistic effects on ground state properties of 4d transition metals and 5d transition metals. *J. Phys. Condens. Matt.* 2:4371–94
109. Meyer B, Elsaesser C, Lechermann F, Fahnle M. 2003. Fortran 90m program for mixed-basis pseudopotential calculations for crystals. Sttutgart: Max-Planck-Institut fuer Metallforschung
110. van Benthem K, Scheu C, Pennycook SJ,

Rühle M. 2005. Electronic and atomistic structure investigations of the $Pd/SrTiO_3$ interface using scanning transmission electron microscopy. *Phys. Stat. Sol.* Submitted
111. Bruley J. 1993. Spatially resolved electron energy loss near edge structure analysis of a near Σ11 tilt boundary in sapphire. *Microsc. Microanal. Microstruct.* 4:23
112. Muellejans H, Bruley J. 1995. Electron energy loss near edge structure of internal interfaces by spatial difference spectroscopy. *J. Microsc.* 180:12
113. Scheu C, Stein W, Rühle M. 2000. Electron energy-loss near-edge structure studies of a $Cu/(11_20)alpha\text{-}Al_2O_3$ interface. *Phys. Status Solidi B* 222:199–211
114. Scheu C. 2002. Electron energy-loss near-edge structure studies at the atomic level: reliability of the spatial difference technique. *J. Microsc.* 207:52–57
115. Packan PA. 1999. Perspectives: device physics—pushing the limits. *Science* 285:2079
116. Moore G. 1965. Cramming more components onto integrated circuits. *Electronics* 38:38
117. Tuomi I. 2002. The lives and death of Moore's law. *First Monday*, p. 7
118. van Benthem K. 2005. *Science. Submitted*
119. Lupini AR, Pennycook SJ. 2003. Localization in elastic and inelastic scattering. *Ultramicroscopy* 96:313–22
120. Voyles PM, Muller DA, Grazul JL, Citrin PH, Gossmann HJL. 2002. Atomic-scale imaging of individual dopant atoms and clusters in highly n-type bulk Si. *Nature* 416:826–29
121. Reimer L. 1997. *Transmission Electron Microscopy*. Berlin: Springer
122. van Benthem K. 2005. 3D imaging of individual Hf atoms at a $Si/SiO_2/HfO_2$ dielectric. *Appl. Phys. Lett.* Submitted
123. Flannery BP, Deckman HW, Roberge WG, Damico KL. 1987. 3-Dimensional X-ray microtomography. *Science* 237: 1439–44
124. Williams GJ, Pfeifer MA, Vartanyants IA, Robinson IK. 2003. Three-dimensional imaging of microstructure in Au nanocrystals. *Phys. Rev. Lett.* 90:175501
125. Miao J, Ohsuna T, Terasaki O, Hodgson KO, O'Keefe MA. 2002. Atomic resolution three-dimensional electron diffraction microscopy. *Phys. Rev. Lett.* 89:155502
126. de Rosier DJ, Klug A. 1968. Reconstruction of 3D structures from electron micrographs. *Nature* 217:130
127. Midgley PA, Weyland M. 2003. 3D electron microscopy in the physical sciences: the development of Z-contrast and EFTEM tomography. *Ultramicroscopy* 96:413–31
128. Koster AJ, Ziese U, Verkleij AJ, Janssen AH, de Jong KP. 2000. Three-dimensional transmission electron microscopy: a novel imaging and characterization technique with nanometer scale resolution for materials science. *J. Phys. Chem. B* 104:9368–70
129. Ritala M, Kukli K, Rahtu A, Raisanen PI, Leskela M, et al. 2000. Atomic layer deposition of oxide thin films with metal alkoxides as oxygen sources. *Science* 288:319–21
130. Peng YP, Nellist PD, Pennycook SJ. 2004. HAADF-STEM imaging with sub-angstrom probes: a full Bloch wave analysis. *J. Electron Microsc.* 53:257–66
131. Falke U, Bleloch A, Falke M, Teichert S. 2004. Atomic structure of a (2×1) reconstructed $NiSi_2/Si(001)$ interface. *Phys. Rev. Lett.* 92:116103
132. Arslan I, Bleloch A, Stach EA, Browning ND. 2005. Atomic and electronic structure of mixed and partial dislocations in GaN. *Phys. Rev. Lett.* 94:025504

effect as the electric field enhancement due to collective electron resonances in the islands of the thin metal films.

The properties of this new phenomenon were very similar to SERS (3, 4), which was detected in 1977 (5, 6) after the first observation of a Raman signal (7); thus this new phenomenon was designated surface-enhanced infrared absorption spectroscopy (SEIRAS) (8) by Osawa & Ikeda.

Hartstein et al. had also detected the dependence of the enhancement on the angle of incidence and on the thickness of the metal layer (2). The enhancement factor increased at increasing metal layer thickness until no infrared emission was transmitted by the increasing absorption. Moreover, the enhancement factor increased if the angle of incidence was close to the critical angle for total internal reflection. The polarization of light did not seem to have any influence on the enhancement mechanism. Furthermore, both metals (gold and silver) were observed to show the same enhancement, but overlayers showed larger enhancement than underlayers. In 1984, Hatta et al. (9) examined the enhancement of the infrared absorption of *m*- and *p*-nitrobenzoic acid (MNBA and PNBA) deposited on vacuum-evaporated silver films with the use of the Kretschmann ATR method and proved, by means of the absorption spectra, that the observed enhancement is important only for the first molecular monolayer. All vibrational modes obeyed the usual dipole selection rule. In contrast to the study of Hartstein et al., Hatta et al. showed that there are two means of enhanced infrared absorption: The first enhancement for both p- and s-polarized light is based on the stimulation of transversal collective electron resonance; the second enhancement is limited to p-polarized light and is stimulated by delocalized plasmons (9).

The latter explanation of the enhancement of the infrared absorption was unconvincing, so Osawa et al. predicted that the local electromagnetic field is located in gaps between the so-called metal islands or in defects of the metal layer, such as pits or columns, that are necessary for the enhancement. Osawa computed the thickness of the electromagnetic field with the aid of the Fresnel formula (10).

Using transmission geometry, Osawa & Ikeda examined the enhancement of the infrared absorption through silver layers for the first time on calcium fluoride substrates. They presumed that the contribution of a chemical mechanism to the infrared enhancement, in addition to the well-established electromagnetic mechanism, was the result of chemical interaction with the surface metal (8).

Nishikawa et al. investigated the influence of the enhancement mechanism on morphology of the metal island films. The structure of the metal islands was analyzed with a scanning electron microscope, and it it was noted that the largest enhancement factor occurred when the metal islands were clumped densely but did not touch each other (11).

The relationship between the island structure and the infrared enhancement could be computed easily on the basis of this simple model. SEIRAS enhancement was shown to depend strongly on the size, form, and distance of the metal islands from each other (12, 13).

The magnitude of the absorbed infrared intensity also depends on the refraction index and on the surface properties of the substrate in the case of vacuum-evaporated films. Furthermore, the conditions under which the metal islands are vacuum evaporated, such as thickness, evaporating rate, and temperature, are important (14–17).

Osawa et al. detected a surface selection rule for the SEIRAS effect, so that only those vibrational modes whose dipole changes are perpendicular to the metal surface are infrared active. The origin of this effect could be identified as an electrical field perpendicular to the metal surface that excites the surface molecules and is stronger than the incident photon field. Osawa et al. assumed that the vibrations of the adsorbed molecules are excited by the polarization of the metal islands. The observed spectra could be explained by means of the effective medium model of Brüggemann (18).

Merklin & Griffith presumed that the surface selection rule is not applicable under all circumstances with regard to the surface roughness. A model for the absorption must consider the structural elements that cause the exceptions in the surface selection rule (19).

The enhancement factor in the SEIRAS is between 10 and 1000 according to the adsorbed molecule. It is now commonly accepted that the dominant role in the enhancement mechanism is the electromagnetic interaction between the incident photon field with the metal and the adsorbed molecules. The chemical interaction of the molecules with the surface gives additional enhancement (20). The authors observed that the enhancement of absorption decreases strongly within 5 nm above the surface (20). A detailed description of the enhancement mechanisms and some applications are given by Osawa (21, 22).

SURFACE ENHANCED RAMAN SPECTROSCOPY The application of SERS was started in the mid-1970s. In 1974, Fleischmann et al. first observed the phenomenon for pyridine adsorbed on an electrochemically roughened silver electrode (7). The vibrational bands of adsorbed species appeared different from those of pyridine in aqueous electrolyte. In 1977, Jeanmaire & Van Duyne (5), as well as Albrecht & Creighton (6), concluded that the rough silver electrode produced a Raman spectrum that was a million-fold more intense than what was expected. This enormously strong surface signal had to be caused by a true enhancement of the Raman scattering efficiency itself. SERS can improve the sensitivity significantly by several orders of magnitude for the roughened surface of many metals, including noble and transition metals. Typical enhancements are $10^2 \sim 10^6$ (23, 24). Recently, there were some reports that single-molecule SERS could show enhancement up to 10^{14} (25–27). SERS is currently one of the most sensitive techniques available to surface science. Its ability of delivering specific chemical identification and to couple with a wide range of instruments has led to its continuing use in both new and traditional areas of surface science. These attributes are of prime importance to both fundamental research and to a range of industrial fields related to surface oxidation, adhesion, corrosion, and catalytic processes, and in advanced materials, biology, and sensor technology.

This review is focused on the application of SERS in adhesion science. The following explains some SERS applications and achievements in this field.

The application of SERS for the study of the adsorption and adhesion of organic molecules on engineering metals under control of the electrode potential (28) is of special interest because it aids in understanding the adhesion of functional groups in polymers in corrosive environments.

The SERS mechanism has attracted much interest from various communities of surface science, spectroscopy, condense-phase physics, and nanoscience (29). SERS has two mechanisms that are widely used: The first is the classical electromagnetic (EM) enhancement mechanism. This mechanism involves the creation of a localized surface plasmons (LSP) on the substrate surface, which transfers energy through an electric field to the target molecules, thus allowing otherwise inaccessible vibrational structures to be determined.

The second mechanism is the charge transfer (CT) or chemical enhancement mechanism. This mechanism involves the incident radiation striking the roughened metallic surface, which results in a photon being excited within the metal to higher energy levels. From this excited state, a charge transfer process to a vibrational level of the same energy within the target analysis takes place. Vibrations in vibrational energy states occur that result in the transfer of a photon of different frequency being passed back to the metallic energy levels and returned to the ground state of the metal.

However, the origin of the SERS still remains a matter of controversy. Nonetheless, scientists now hold that the EM is the most important enhancement contribution, it can enhance the Raman intensity up to $10^3 \sim 10^5$ (30). The CT has a minor enhancement, it can give an enhancement of about $10 \sim 10^2$ (31, 32). The enhancement is a combination of these two effects.

The key experimental challenge of SERS is to produce a roughened surface that can activate the SERS phenomenon. In the past, SERS-active surfaces could be obtained only on some coinage metals, namely, copper, silver, and gold (33). Recently, it became possible to perform SERS on some transition metals, for example, Fe, Ni, Pt, Co, Ru, and their alloys (23). During the past three decades, many methods have been proposed to achieve SERS-active surfaces: for example, electrochemical oxidation-reduction cycle (ORC) (34), current-controlled ORC, film deposition, template synthesis of metal nanowire arrays (25), deposition of colloid Au on aminosilane-modified surfaces (35), electrodeposition of silver on silver films (36), filtration (37), metallic double layers (38), electron-beam lithography (EBL) (39, 40), photochemical deposition of SERS on active silver nanoparticles (41), self-assembled metal colloid monolayers (42), and chemical etching (43).

In Situ Study of the Adsorption of Organic Molecules on Metals from Water Solutions

ADSORPTION OF ORGANIC MOLECULES ON METAL SUBSTRATES FROM LIQUID ENVIRONMENTS SERS has been successfully used to investigate the geometry, conformation, and chemical reaction of molecules adsorbed on metal surfaces. The most prominent systems studied by SERS are pyridine and its derivatives adsorbed on silver electrodes (28–32), the surface reaction of thiols adsorbed on silver surfaces

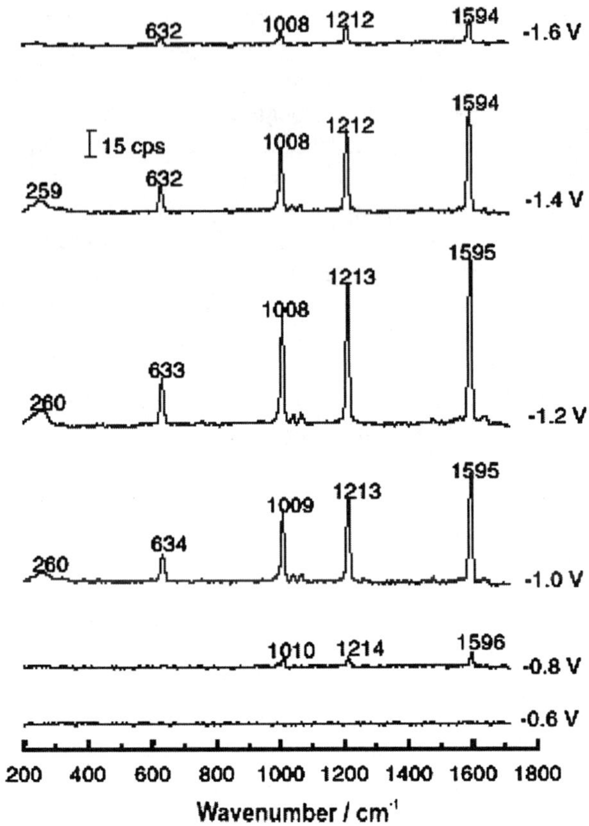

Figure 1 Potential dependent SERS spectra of pyridine adsorbed on the Ni electrode in a solution containing 0.01 M pyridine and 0.1 M KCl. The excitation line is 514.5 nm. Reprinted from (28), with permission from Elsevier.

(33, 34), the orientation of organic molecules adsorbed on metal surfaces (36–40), and protein adhesion (44).

In Figure 1 [figure 6 of Huang et al. 2004 (28)] SERS spectra of pyridine on a roughened Ni electrode in a solution containing 0.01 M pyridine and 0.1 M KCl are shown as a function of the electrode potential. No adsorption of the molecule is observed at −0.6 V (SCE). However, if the potential is shifted more negatively, four peaks arise and grow in intensity until at −1.6 V (SCE) the peak intensity decreases again due to hydrogen evolution. The slight shift in the peak positions were assigned to the chemical interaction between the adsorbate and the Ni substrate. The broad peak at 260 cm^{-1} was interpreted as a Ni-N stretching vibration.

Królikowska et al. (34) studied the structure of thioglycolic acid (TGA) monolayers on silver and gold. These molecules bond via the thiol group to the metal

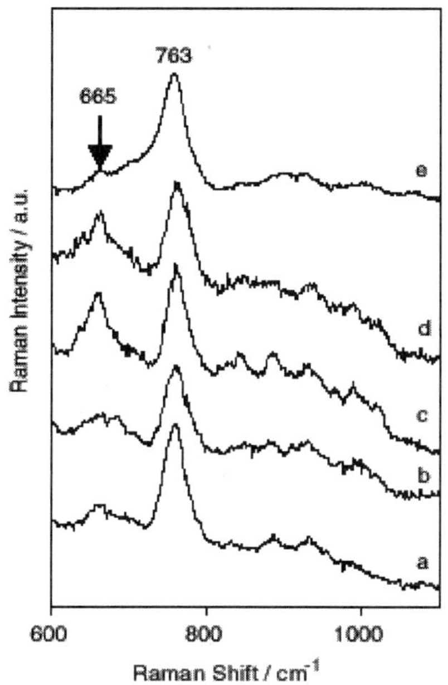

Figure 2 Evoution of SERS spectrum during formation of TGA monolayer on Ag from a 10-mM TGA aqueous solution. Layer grown for (*a*) 3 min, (*b*) 7 min, (*c*) 1 h, (*d*) 2 h, and (*e*) 24 h. Spectra were scaled and shifted. Reprinted from (34), with permission from Elsevier.

and thereby provide a carboxylic acid group as a surface functionality that can be used, for example, for improved adhesion at metal/polymer interfaces. From the in situ measurement of the adsorbed functional thiol, the authors concluded that the conformation of adsorbed TGA monolayers depends on the type of substrate and is influenced by the pH value of the solution.

Figure 2 shows an in situ measurement of the monolayer formation on Ag from a 10-mM TGA aqueous solution at different times of adsorption. In the first stage of incubation, the relative concentration of the gauche conformer increases (peak at 665 cm^{-1}), but it disappears at longer times. The results show that SERS is a versatile technique for the in situ study of model adhesion promoter systems in aqueous solutions.

Self-assembled monolayers (SAMs) have been investigated excessively in recent years as model systems (45, 46). SAMs are formed when functional molecules spontaneously adsorb as a monomolecular layer on a solid surface. Monofunctional SAMs, such as organothiols, organosilanes, or organophosphonates, adsorb with their functional group on the substrate surface. An ordering process follows that leads to two-dimensional crystalline layers. Whereas thiol groups bind only to oxide-free surfaces such as gold, silver, or copper, organosilanes and organophosphonates can strongly bond to oxide- and hydroxide-covered metals and are thus applicable for engineering metals. Most studies, however, deal with the adsorption of organothiols on gold. Bifunctional organic molecules are used as adhesion-promoting SAMs. One end of the organic molecule is a functional group that can bond to the metal surface; the other end, which consists of hydroxyl groups, epoxy, amino, or carboxyl groups, can then react with the polymer. For the study of the adsorption, scanning tunnelling microscopy (STM) has been applied extensively to organothiols on gold. However, the application of STM is limited mainly to this system because in the case of engineering materials the native oxides on the metal surface disable the application of STM.

As an optical spectroscopy, SERS is applied in the field of SAMs. For example, the self-assembly of monodisperse gold and silver particles, which forms monolayers on polymer-coated substrates, yields macroscopic surfaces that are highly active for surface-enhanced Raman scattering. Particles are bound to the substrate through multiple bonds between the colloid and functional groups on the polymer such as cyanide (CN), amine (NH_2), and thiol (SH) (47).

Monolayer structure on the different substrates (48, 49) and molecule orientation (50, 51) can be investigated in real time by SERS.

Zhang et al. (52) examined SAMs deposited on metal substrates by the use of surface plasmon resonance, UV-VIS spectroscopy, and SEIRAS. The multi-COOH terminal groups of the self-assembled porphyrin film were chemisorbed onto the gold surface of the substrate. The results gained by SEIRAS led to the conclusion that the porphyrin macrocycle is oriented flatly with regard to the metal surface of the substrate. Moreover, it was shown that that one of the four acid groups was not bonded to the gold surface.

STUDIES OF POLYMER FILM FORMATION ON METAL SUBSTRATES In situ studies of polymer film formation on metal substrates were mainly limited to the deposition of ultrathin films from the gas phase by means of reflection absorption spectroscopy (53–55). Extensive reviews on this well established spectroscopic technique have been published by Chabal (56) and Golden (57).

Infrared reflection-adsorption spectroscopy (IRRAS) is generally used for characterization of thin films or monolayer on metal substrates. The advantages are the high surface sensitivity and the surface selection rule. Polarization modulation is most often done to distinguish between the surrounding bulk medium and adsorbate on the metal surface. The modulated reflectivity is independent of the isotropic adsorption from gas or bulk water. Consequently, the interfering effect of water vapor and carbon oxide can be eliminated. IRRAS is dependent upon the optical constants of the thin film and substrate, the angle of incidence, as well as the polarization of the incident IR radiation. Only p-component radiation interacts with the sample surface. Consequently, the active vibrations that can be detected in IRRAS must have a component of the dynamic dipole polarized in the direction normal to the sample surface. These results give the so-called surface selection rule for PM-IRRAS. The surface selection rule is used to study the molecular orientation of thin films or monolayers deposited on metal or dielectric substrates. Grundmeier & Stratmann developed a spectroscopic cell that enabled the combination of quartz crystal microbalance (QCM) with in situ IRRAS for the investigation of the growth of chemical vapor deposition (CVD) and plasma CVD films and their stability in corrosive environments (53, 54). The advantage of the combination of QCM and IRRAS is the simultaneous measurement of mass change and chemical structure changes on the surface.

Deposition of alkoxyorganosilanes showed that a minimum water adsorption layer is necessary on the metal surface before hydrolysis of the alkoxy to silanols starts in the ultrathin adsorbed water film, which then leads to the film formation the surface (53). However, in this case no special spectral features in the infrared spectra could be assigned to interfacial bonds between the organosiloxane structures and the iron oxide.

In contrast, in cases where ultrathin Si-organic films were deposited from water-free, low-pressure gas phases by means of plasma polymerization, Grundmeier and Stratmann observed spectral features in the first 1 to 2 nm of the film adjacent to the oxide surface that could be assigned to the interaction between the oxide and the plasma polymer (see Figure 3) (54).

Turner & Boerio made similar measurements for pure SiO_2 films deposited from oxygen/organosilane mixtures on aluminum and titanium substrates (55). Relatively thick films (~8.0 nm) had infrared spectra that were typical of bulk amorphous silicon dioxide (a-SiO_2). When thinner films were analyzed (~0.6 nm), other infrared bands emerged that were from the formation of silicon suboxide at the interface. Infrared bands resulting from oxidation of the substrate during deposition were also observed. It was determined that during the initial stages of plasma deposition, metal atoms from the substrate migrated to the metal-oxide surface. This resulted in preferential oxidation of metal atoms with the formation of silicon suboxide at the film/metal interface. In addition, interfacial suboxide formation was shown to depend on the diffusivity of the metal substrate atoms through the surface

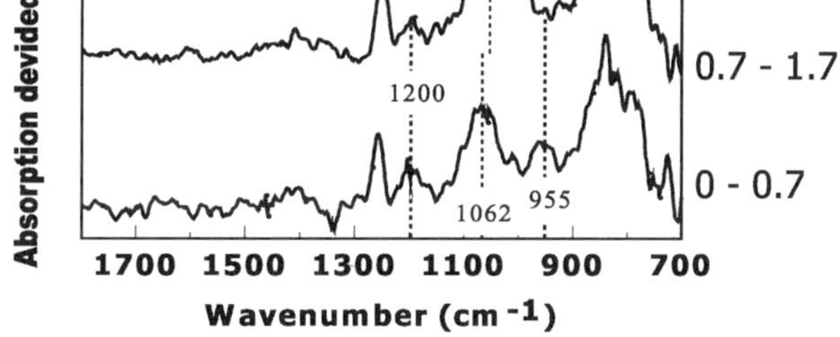

Figure 3 Fourier transform-IRRAS (FT-IRRAS) difference spectra revealing the structure of the subsequently deposited three layers of hexamethyldisilane plasma polymer on plasma oxidized iron. The polymer was deposited at a partial pressure of $p_{HMDS} = 5$ Pa and $p_{Ar} = 25$ Pa. Reprinted from (54), with permission from Elsevier.

oxide of the metal. As a result, more interfacial suboxide formed for depositions on titanium substrates compared with depositions on aluminum substrates.

ADSORPTION OF WATER AT BURIED METAL SURFACES The adsorption of water at polymer/metal interfaces is of great interest for de-adhesion phenomena. In particular, spectroscopic techniques can reveal information about the replacement of adsorbed organic molecules by water.

In the past, no surface-enhanced ATR-IR spectroscopy was done to observe the incorporation of water in polymer/metal interfaces (58, 59). However, because the electric field probes not only the interface region but also a few hundred-nanometer -thick interphase region, the assignment of spectral feature to the chemical changes at the interface is difficult.

In contrast, based on the strong enhancement of the absorption of molecules that are directly bonded to the metal surface, SEIRAS is applicable for the characterization of very thin films with thicknesses of less than 1 nm on top of the ATR crystal and thus is powerful for the in situ characterizations of molecules that are adsorbed on electrode surfaces from water-based solutions (60, 61).

The in situ observations of electrode/electrolyte interfaces with ATR-SEIRAS offer some comparisons with IRRAS measurements, especially during the adsorption from the solution: The sensitivity is 10 times higher, adsorbed molecules can be detected selectively with less interference origin from the adjacent medium, and the dynamic electrochemical process can be observed because the diffusion of the electro active substances is not limited (62).

It is possible to observe electrode/electrolyte interfaces with high sensitivities with SEIRAS; thus Ataka et al. were able to examine the potential dependent reorientation of water molecules at gold electrodes successfully in perchloric acid. The interference from the bulk solution could be avoided. Four different models could be derived from the spectra from water molecules with different orientations whose existence is dependent on the potential. Direct statements about the interface could be found within 10 nm above the surface (63).

Osawa described SEIRAS as a sensitive technique to make time-resolved monitoring possible for reactions and other time-controlled phenomena at electrochemical interfaces. He noted that being able to observe the potential-dependent reorientation of water molecules at interfaces is one of the biggest achievements of the SEIRAS. Simultaneous IR and electrochemical measurements offer interesting insights into electrochemical reactions. Kuhne et al. proved that SEIRAS is a necessary procedure in order to examine the interaction of proteins at the interface. With the aid of SEIRAS, it is possible to examine structure differences within a functional membrane. The enhancement factor depends on the orientation of the structure groups to the silver films (64). By means of SEIRAS, Jensen et al. investigated the influence of the sandwich architecture on the SEIRAS enhancement from PNBA on thermally evaporated silver films (66). Zhang et al. examined self-assembled molecular monolayers on gold substrates by means of SEIRAS and recognized the orientation of components on top of the gold film (67).

The structure and the orientation of water at gold electrodes in contact with aqueous sulfuric acid with ATR-SEIRAS was examined by Wandlowski et al. in a spectro-electrochemical cell (68).

Surface-enhanced Raman scattering of water molecules has been a great concern of many SERS studies since 1981 (69–77). Generally, the frequency of the ν(O-H) band decreases as the extent of intermolecular interaction (hydrogen bonding) between H_2O and other species increases. Moreover, it has been found that the number and bandwidth of the ν(O-H) bands are dependent on the nature of the cation present in aqueous solutions. Thus SERS provides interesting information about the state of adsorbed water at metal/electrolyte interfaces.

Shen & Pemberton studied the trace behavior of trace interfacial water and its interaction with Li^+ and Br^- ions at Ag electrodes in a series of alcohols (70) (see Figures 4 and 5).

It is obvious that SERS spectra in the ν(O-H) region exhibit four bands from the interfacial water species that is a function of the electrode potential. These bands are assigned to water species interacting with specially adsorbed Br^- and Li^+ in

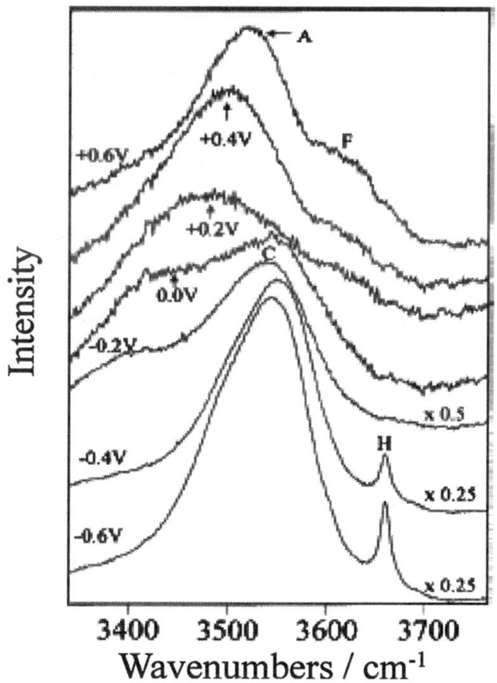

Figure 4 SERS spectra at Ag electrode in ν(O-H) region for ethanol containing 0.4-M LiBr (integration time 60 s). Reprinted from (70), with permission from Elsevier.

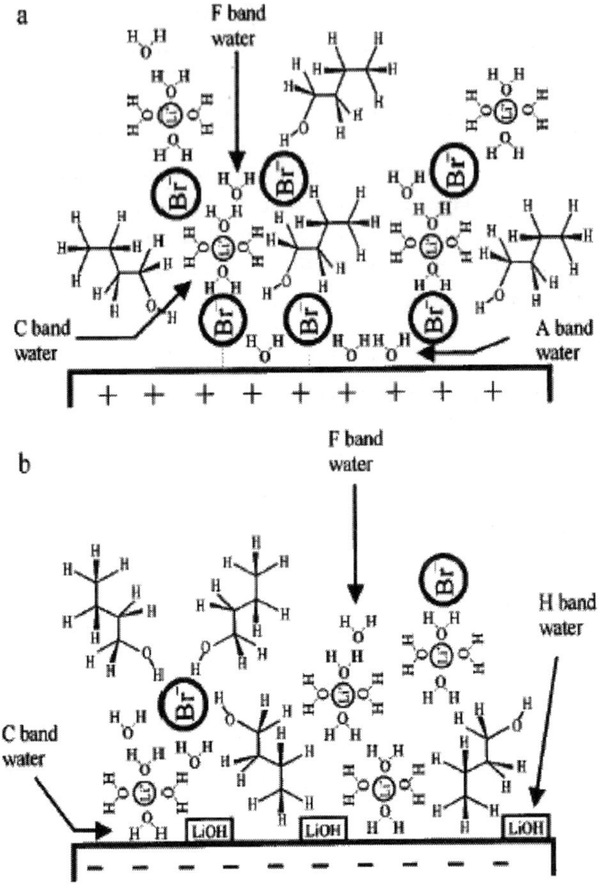

Figure 5 Schematic of proposed interfacial structure at Ag electrodes for potentials (*a*) positive and (*b*) negative of the point of zero charge (PZC). Reprinted from (70), with permission from Elsevier.

the outer Helmholtz plane (A and C bands), free water species in the interface (F bands), or hydroxide in LiOH microcrystallites precipitated onto the electrode surface (H bands). Interfacial water molecules cluster around Li^+ and Br^- ions. The spectra data suggest pronounced restructuring of the water in the interface in the vicinity of the potential of zero charge.

Using the function of the applied potential, Cao et al. studied SERS of water on silver electrodes in a solution of acetonitrile with perchlorates as anions (74). Two SERS bands were observed at 3487 and 3586 cm^{-1} and assigned to the interfacial H_2O molecules. The authors could distinguish between water molecules that interact via the oxygen atom (3487 cm^{-1}) and those that interact via the hydrogen atoms (3586 cm^{-1}).

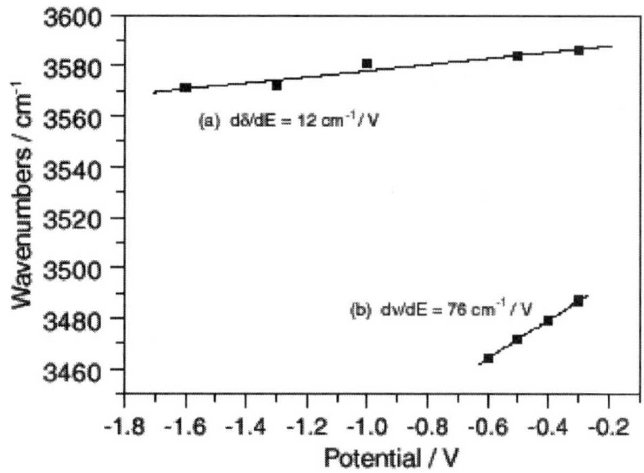

Figure 6 Dependence of the band frequency versus applied potential for (*a*) 3586 cm^{-1} and (*b*) 3487 cm^{-1}. Reprinted from (74), with permission from Elsevier.

In Figure 6 (see Figure 3 in Reference 74), the dependence of the band frequency versus the applied potential for the two peaks is shown.

DESORPTION OF ADSORBED ORGANIC MOLECULES Desorption of molecules or monolayers from metal surfaces can be studied effectively by SERS. Usually, desorption is dependent on the temperature, electrode potentials, or concentrations in solution (78–86).

Bandyopadhyay & Vijaymohanan (79) studied the self-assembled monolayer desorption as a function of temperature by in situ SERS. Litorja and colleagues carried out detailed research on thermal desorption of silver film on SiO_2 nanosphere surfaces (82). Otto et al. used desorption of molecules to study the mechanism of SERS (83, 84). Brolo (85) and Tian & Ren (69) combined SERS and electrochemistry to study the potential-dependent adsorption/desorption of molecules. Figure 7 (80) and Figure 8 (85) present two examples to show the ability of SERS for the investigation of the desorption phenomena on metal surfaces.

Figure 7 indicates that the thermal stability of the monolayer is significantly affected upon exposure to the $CuCl_2$ solution. In the case of Cu^{2+} ion-modified BDMT monolayers, it appears that as the temperature is increased, the copper ions become mobile, and it is likely that they reach the Au-BDMT interface and participate in bonding (80). This makes the Au-BDMT binding weaker, and the monolayer desorbs at a lower temperature than in the unmodified case.

Figure 8 shows the L-Cys and Cl^- coadsorption/desorption is dependent on the applied potentials (85). When the potential is above –650 mV, the Cl^- is likely to interact with the amino group, and it will repel the carboxylate of L-Cys; if the potential is more negative than – 650 mV, Cl^- will desorb from the silver electrode, and then the carboxylate will reorient and adsorb on the electrode.

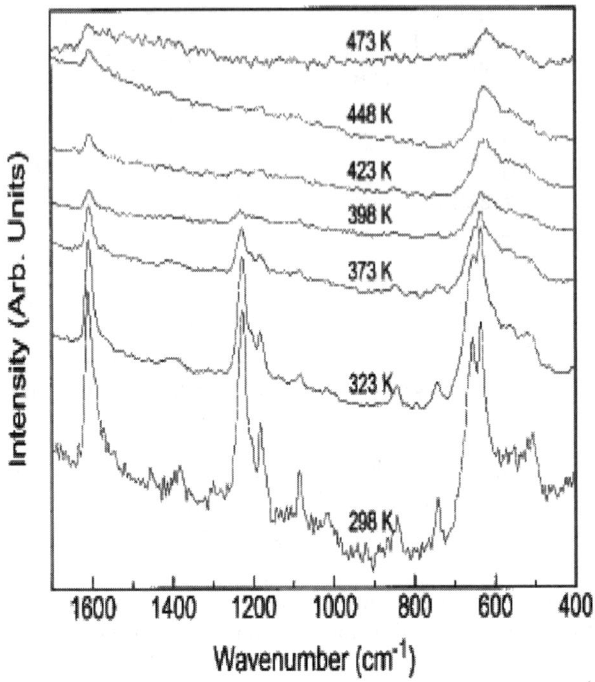

Figure 7 Surface-enhanced Raman spectra Cu^{2+} exposed to a BDMT (1,4-benzenedimethanethiol) monolayer on Au as a function of temperature. Corresponding temperatures are indicated. Reprinted from (80), with permission from ACS.

The above applications of Raman spectroscopy have demonstrated that SERS is a powerful tool for the investigation of adhesion of organic molecules on metal electrodes. In recent years, SERS has been successfully used to investigate the adhesion of biomolecules (87), DNA molecules (85, 88), and to measure the adhesion strength in composites (89).

ESTIMATION OF ADHESION FORCES AT METAL/POLYMER INTERFACES

Atomic Force Microscopy Force-Distance Curves of Electrolyte Covered Metals

Since 1989, atomic force microscopy (AFM) has emerged as a useful tool for studying adhesion by means of force-distance curves (90). For typical force-distance curve values no interaction between tip and substrate can be observed at large Z displacement. During the approach of the tip to the surface, van der Waals forces occur and the tip is abruptly pulled into contact with the sample (jump to contact

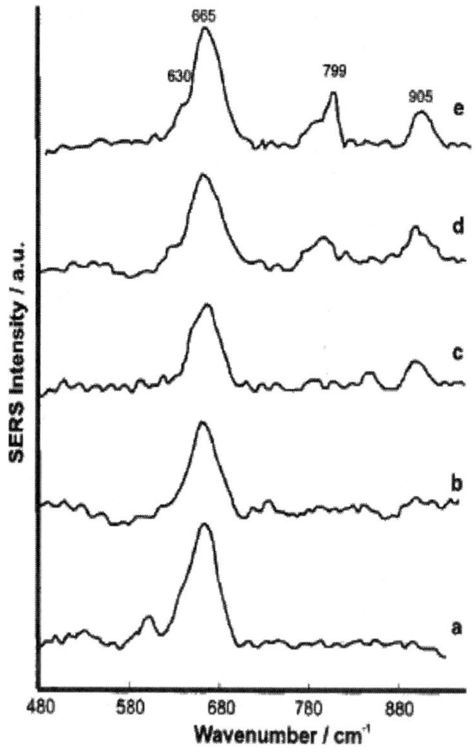

Figure 8 SERS spectra of L-Cys on Ag at different applied potentials (Ag/KCl = 0.2 M, [L-Cys] $= 5 \times 10^{-2}$ M). (*a*) E(mV_{SHE})$= -300$, (*b*)$= -450$, (*c*)$= -600$, (*d*) $= -700$, (*e*) $= -850$. Reprinted from (85), with permission from ACS.

point). A deflection of the cantilever can be observed during further approach. On the way back to the start position of the tip, the force-distance (FD) curve has the same shape but a hysteresis is observed owing to adhesion to the substrate surface. At the jump-off contact point, the adhesion force can be analyzed (91). AFM is able to acquire force-distance curves on various surfaces in gaseous and liquid environments, with high lateral (25 nm) (92), vertical (0.1 Å), and force (1 pN) (93) resolution. Force-distance curves, measured in air, are dominated by meniscus forces of condensed water (the amount of hysteresis in air is much higher than in water) (94). To avoid these forces, measurements should be realized in a controlled atmosphere, vacuum, or in liquids.

Originally, force-distance curves were performed only to reduce the necessary force between tip and sample during topographic scans to ensure the AFM to be a nondestructive method for sensitive systems such as biological films (94).

More recently, this method was chosen to focus on the analysis of surfaces and especially adhesion forces and on material properties such as elasticity. In contrast to other methods (e.g., the surface force apparatus), the force balance or the osmotic stress method (see review in Reference 95), can be applied to more systems because only smoothness on a nanometer scale is necessary to measure reliable data. The reader is referred to an excellent review on the application of AFM for chemical force microscopy (91).

Within this review, the authors have focused on the recent applications of AFM force distance curve measurements for

- measuring force-distance curves of organically functionalized tips or polymer nanospheres on metals in electrolytes, and
- the recent developments in single-molecule adhesion measurements on metals.

Moreover, we have highlighted the emerging field of simulation of the force-distance curves for an improved understanding of the measured data.

At early stages the comparison of theoretically calculated and experimental results was difficult because in theory a spherical tip was estimated, whereas in reality the geometry of the tip was undefined. Hüttl et al. tried to avoid this problem by preparing tips with a defined spherical contact area. One way was the oxidation of a commercial Si tip under defined conditions (1000°C under normal atmosphere for 150 min) to get a homogeneous SiO_2 layer of some 100 nm (96, 97). It was also possible to get a TiO_2 layer by deposition of Ti on a Si tip and oxidizing it in an oxygen atmosphere at 500°C (96). Both oxide layers afford better mechanical stability than that of commercial tips, and the tip geometry is better defined.

A more universal solution is replacing the tip by a colloidal particle of well-defined spherical shape and a size of 1–50 μm. This so-called colloidal probe technique was first applied by Ducker et al. (98, 99), and since then it has become a well-established and powerful tool for the study of surface forces (100).

In 2002, Cho & Sigmund (101) began making surface force measurements with a nanosized colloidal probe technique using a multiwalled carbon nanotube (<10 nm).

These possible variations in tip preparation also give room for chemical modifications of the tip surface, for example by changing the composition of the colloidal particle. Thus probing of molecular interactions between well-defined functional groups on the tip and the sample and imaging with chemical sensitivity is possible. For a detailed review of this chemical force microscopy up to 1997 see Noy et al. (91). They focused on measurement and quantification of adhesion and friction forces between different functional groups on a chemically modified tip and the sample. The modifications were made by deposition of Au on the tip and the substrate in the first step. Then a dense self-assembled monolayer (SAM) was formed after immersion in a solution of organic thiols with different functional groups at the free SAM surface. Reactive silanes were directly deposited on clean Si or Si_3N_4 tips.

Functional groups that cause hydrophobic effects dominate both adhesion and friction forces; hence, lateral force images taken with hydrophobic tips in aqueous solutions can map hydrophobic regions on a sample.

In the fields of biology and biochemistry force-distance curve measurements have become a reliable method for detecting interactions of different molecules such as receptor-ligand pairs.

At first, measurements were carried out with functionalized AFM tips, which were covered with monolayers of biomolecules such as biotin or avidin (e.g., 102, 103). But these measurements resulted in force-distance curves between an undefined number of molecules that depend on the receptor density, the duration of contact, and the area of contact. This resulted in a "sawtooth" pattern in the retraction area of the force-distance curve and an interpretation became difficult. Thus Hinterdorfer et al. (104), Kienberger et al. (105), Riener et al. (106), and Dammer et al. (107) developed new methods of AFM tip preparation that yielded tips with 0, 1, 2, and sometimes 3 tip-bound probe molecules (for a schematic representation, see Figure 9).

Willemsen et al. (108) did a further step in the field of single-molecule adhesion analysis. They used the tip preparation of other groups (106), but with a home-built AFM, which was modified for sensitive force-distance mesurements, and

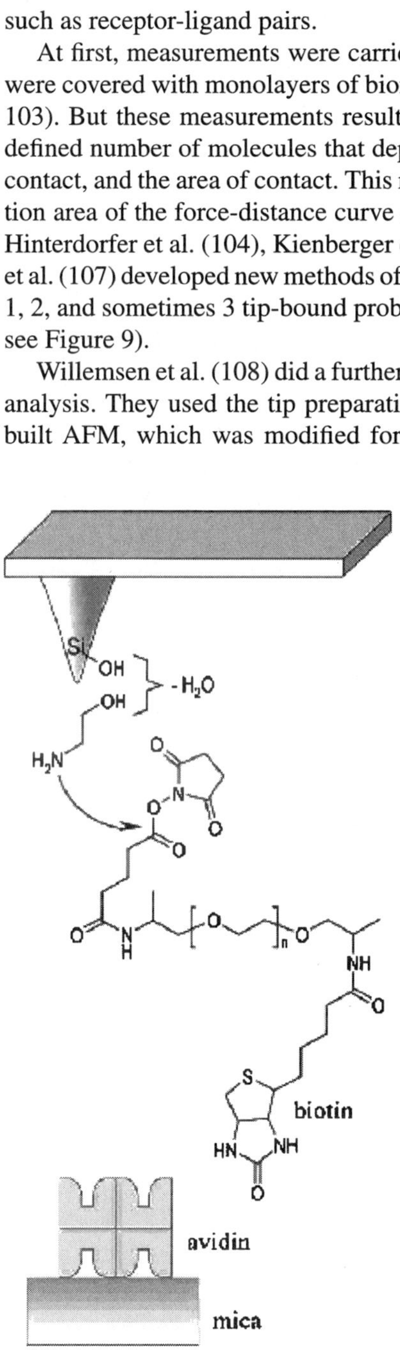

Figure 9 Schematic representation of the SMRFM (single-molecule recognition force microscopy) testing system. Avidin is electrostatically adsorbed to freshly cleaved mica. Amino groups are introduced on the AFM tip and biotin-PEG$_{800}$-NHS (n ~ 18) is attached via amide bond formation. Reprinted from (106), with permission from Elsevier.

developed a method to classify the different kinds of adhesion that occur during force-distance curve measurements. By mapping on the sample surface, they obtained images of topography and adhesion (after 100 × 100 force-distance curves, at least). With the help of simulation calculations it was possible to decrease the number of nonspecific tip-sample interactions by optimizing the tip parameters, especially the cantilever spring constant.

For measurements in liquid environments, both the functionalization of the tip and the pH were varied. On hydrophilic surfaces, pH-dependent changes in friction forces of ionizable groups were observed, and thus the hydrophilic functional groups could be mapped. Also, changes in adhesion owing to pH changes were observed. It was possible to determine local pK of acidic or basic groups. These results are of significant utility in studies of biological and polymeric systems.

In addition to variation of the pH value of the liquid, force-distance curves were performed under potential control of the substrate, which is important in obtaining information about electrostatic forces. At the beginning of force-distance measurements it was impossible to analyze the electrostatic double layer at surface potentials significantly higher than 0.1 V. The reason was that high surface potentials can be obtained only by applying an external potential to metallic or conducting samples, but all materials used at that time were not conducting. Ishino et al. (109) were one of the first groups to use an Au-coated AFM cantilever to detect electrostatic forces in aqueous solution. The surface potential of the cantilever was controlled electrochemically and force-distance curves were measured on monolayers with different kinds of functional groups.

Force variation was detected with changing voltage of the tip, and the direction of force variation was found to agree with the polarity of surface charges owing to the dissociation of the functional groups (Figure 10).

Butt and coworkers performed fundamental research on the electrostatic double layer forces on different systems. Together with Raiteri et al. (111, 112), Butt analyzed the interaction forces between two gold substrates. In contrast to Ishino et al. (109) or Yamamoto et al. (110), Butt used a gold sphere glued on an AFM cantilever instead of a gold-coated cantilever for two reasons: First, the sphere has a well-defined geometry so modeling of the probe/sample interaction is easier; second, gold cannot be removed as on coated tips. During these measurements, three potential areas were observed. For $U < -0.6$ V [versus standard hydrogen electrode (SHE)] repulsive forces were observed that decayed roughly exponentially. In the second area of $-0.6 \text{ V} < U < +0.4$ V (SHE), the force became very small and sometimes no repulsion was observed. For $U > 0.4$ V (SHE), higher repulsive forces occurred again. These experimental results were compared with predictions of the Poisson-Boltzmann theory and they agreed qualitatively. Around the point of zero charge [assumed to be at +0.1 V (SHE)], the force should be small, whereas far above and far below this point, the repulsion is expected to be strong. The current flowing through the probe and sample was measured simultaneously with the force-distance measurements and gave additional information about the interfacial processes and properties.

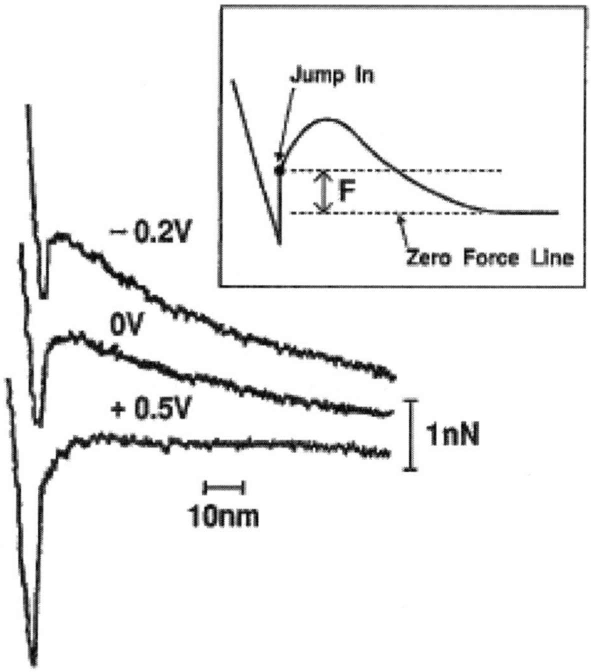

Figure 10 Force versus distance curves measured on a stearic acid monolayer while electro-chemically controlling surface potential of a Au-coated cantilever. Reprinted from (109).

Complementary analysis was performed on other substrates such as platinum and highly ordered pyrolitic graphite (HOPG) to investigate the electrostatic forces between the tip and the sample surface (113). In these experiments Butt et al. used a silicon nitride tip instead of a gold sphere. The tip was not connected to the potentiostat but it was negatively charged because the experiments were done at pH 9–10. The potential control of the force-distance measurements was ensured by a classical electrochemical three-electrode assembly (Figure 11).

With this change in experimental set up, the systems could be analyzed with conductive tips, thus providing a wider field of possible material interactions for these experiments.

In addition, more complex processes including the adsorption of charged molecules such as ionic surfactants, polyelectrolytes, and proteins from aqueous solutions can be separated into the effects of changing different parameters, e.g., pH, electrolyte concentration, and applied potential. It is possible to calculate double-layer potentials by fitting of the force-distance curves according to the DLVO (Derjaguin-Landau-Verwey-Overbeek) theory. Thus force-distance measurements within a three-electrode assembly give direct information of changes in the double layer owing to potential changes. Barten et al. (114) found only small

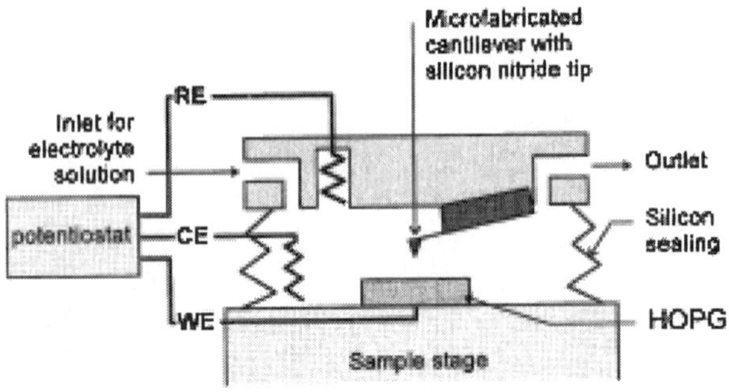

Figure 11 Schematic of the experimental set up with the sample acting as a working electrode (WE) in a three-electrode assembly (taken from Reference 113, with permission from Elsevier).

changes of the double-layer potential as result of an externally applied potential. But adsorbing molecules near the sample surface (in this case polycrystalline gold) experience the local potential shift so that this effect influences the adsorption process.

As mentioned above (see 109) it is important to perform theoretical calculations in addition to analyzing experimental results to either obtain models of the respective system or improve the experimental set up by optimizing certain parameters. Since 1999 more and more groups have started to involve molecular dynamics simulation during the investigations of adhesion phenomena of certain systems or molecules. The theories of Bradley (115), Derjaguin-Müller-Toporov (DMT; 116), and Johnson-Kendall-Roberts (JKR; 117) are often used for calculations that were developed before the force-distance curve method was invented. Other groups such as Shinto et al. (118), Gotsmann et al. (119), and Kendall et al. (120, 121) are especially active in the field of molecular dynamics simulation.

THE SCANNING KELVIN PROBE AS AN ELECTROCHEMICAL TECHNIQUE TO STUDY METAL/POLYMER INTERFACES

The stability of metal or oxide/polymer interfaces is frequently limited by electrochemical reactions that take place at the interface and lead to a chemical degradation of bonds at the interface. Electrochemical reactions dominate, in particular at room temperature and on reactive metals, which are commonly used as materials in a vast number of applications.

Electrochemical reactions require

- the establishment of an electrified interface, and
- the conservation of charge that, in particular, requires a balance between cathodic (electrons transfer the interface from the substrate) and anodic

reactions (electrons transfer the interface to the substrate or positively charged ions transfer the interface from the substrate).

Electrified interfaces are predominantly built up by mobile charged species, which, in the case of metal/polymer interfaces, may be identified with ions embedded into the polymeric matrix. Therefore, electrochemical-driven reactions prevail in environments that allow the presence of such species. Furthermore, the conservation of charge may require the presence of ion transfer reactions, and this is almost always true for reactive metals such as iron, copper, zinc, aluminum, etc.

Because the nature of the electrified interface dominates the reaction kinetics, it is most desirable to measure the potential drop across the interface, even in cases where the buried interface is not accessible for electrochemical techniques. The scanning Kelvin probe (SKP), which measures the local Fermi level, is the only technique that allows the measurement of such data, and therefore all modern models that deal with electrochemical de-adhesion reactions are based on such techniques.

In the following, the principles of the SKP are only briefly summarized, as extensive surveys on this technique may be found elsewhere.

The principal set up of a SKP for corrosion and adhesion studies is shown in Figure 12.

Principle Aspects of the Kelvin Probe

In principle the Kelvin probe measures the work-function of a sample, using the vibrating condenser method (122–125). The Kelvin probe, which may consist of a sharp wire made up of a conducting and stable phase such as graphite or gold, is electrically connected to the sample under investigation. This leads to identical Fermi levels in both sample and probe and in case of differing work functions of both materials to charging of one phase with respect to the other (Volta potential

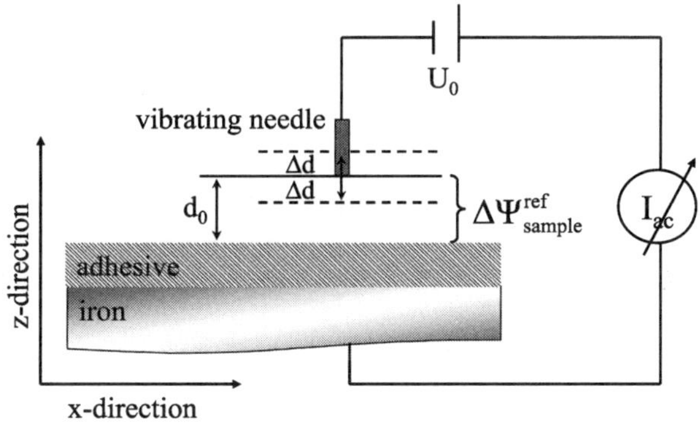

Figure 12 Schematic of the measurement of electrochemical potentials at buried interfaces with the SKP. Reprinted from (127), with permission from The Electrochemical Society, Inc.

difference). Because of the charged condenser between sample and sharp probe, an alternating current is established by a sinusoidal vibration of the probe

$$i(t) = \frac{d\,Q(t)}{dt} = (\Delta\Psi + U_0)\,\frac{d\,C(t)}{dt}. \qquad 1.$$

For a parallel plate capacitor with a sinusoidal plate, separation d(t) is given by

$$d(t) = d_0 + \Delta d \cos \omega\, t, \qquad 2.$$

with d_0 = average distance of the parallel plates, Δd = vibration amplitude, ω = angular frequency of vibration.

The time-dependent capacitance, C(t), is then given by

$$C(t) = \frac{\varepsilon \cdot A}{d(t)} = \frac{\varepsilon \cdot A}{d_0 + \Delta d \cos \omega t}. \qquad 3.$$

Thus the time-dependent Kelvin current i(t) becomes

$$i(t) = (\Delta\Psi + U_0)\,\varepsilon\,A \frac{d}{dt}\left[\frac{1}{d_0 + \Delta d \cos \omega t}\right], \qquad 4.$$

$$= (\Delta\Psi + U_0)\,\frac{\varepsilon\omega A\,\Delta d \sin \omega t}{(d_0 + \Delta d \cos \omega t)^2}. \qquad 5.$$

Assuming that the average distance of the parallel plates is much larger than the vibration amplitude ($\Delta d \ll d_0$), this term can be simplified using $(d_0 + \Delta d \cos \omega t)^2 \approx d_0^2$.

The amplitude of the current I demodulated at frequency ω with a lock-in amplifier (LIA) is now

$$I = (\Delta\Psi + U_0)\,\varepsilon\,\omega\,A \frac{\Delta d}{d_0^2}. \qquad 6.$$

The measurement of the Volta potential difference is performed by zeroing the Kelvin current amplitude automatically, with a control circuit, changing the compensation voltage:

$$\text{if}\, I = 0 \Rightarrow \Delta\Psi = -U_0. \qquad 7.$$

Under those conditions, the external voltage, which is identical to the Volta potential difference, is given by the energy difference between the Fermi level of the sample with respect to the Fermi level of the probe, which is known in many cases.

Theoretical Basis of the Scanning Kelvin Probe as Used in Adhesion Science

The work function of a metal is defined as the work to remove an electron from the metal to a position far away. Thus only the well-defined surface has to be passed during this transfer and, therefore, under ultra-high vacuum conditions, the work function of the metal is a well-defined quantity.

If the metal is covered by other phases such as an aqueous electrolyte or a polymer, additional interfaces have to be passed and the measured Volta potential difference is strongly changed by additional potential differences across these interfaces. The same is true for the electrode potential, which is essentially determined by the potential difference across the electrified interface. Therefore, it is not surprising that a simple relation exists between the Volta potential difference, $\Delta\Psi^{\text{Ref}}_{\text{sample}}$, and the electrode potential, E_{corr}, of the buried interface. In case of a metal surface covered by a liquid phase, a simple linear relation exists between the two quantities:

$$E_{Corr} = \left\{ \frac{W_{\text{Ref}}}{F} - \chi_{El} - \varepsilon^{\text{Ref}}_{1/2} \right\} + \Delta\Psi^{\text{Ref}}_{Pol}, \qquad 8.$$

where W_{Ref} is the work function of the reference metal, χ_{el} the surface dipole potential of the electrolyte and $\varepsilon^{ref}_{1/2}$ ref is the half-cell potential of the reference metal.

For less highly oriented polymers with rather small dipole potentials, a similar result is obtained for the correlation of the corrosion potential E_{Corr} and the measured Volta potential difference $\Delta\Psi$:

$$E_{Corr} = \left\{ \frac{W_{\text{Ref}}}{F} - \chi_{Pol} - \varepsilon^{\text{Ref}}_{1/2} \right\} + \Delta\Psi^{\text{Ref}}_{Pol}, \qquad 9.$$

where χ_{Pol} is the surface dipole potential of the polymer surface. Here, the Kelvin probe monitors the electrode potential of the buried interface. Equation 9 demonstrates the main advantage of the SKP: It allows the measurement of electrode potentials without touching the surface under investigation across a dielectric medium of infinite resistance. For metals or metal/oxides substrates covered by organic coatings, the electrode potential of the interface is measured with high-spatial resolution.

For reactive metals, a metal/electrolyte interface usually does not exist. However, the metal surface is covered by a thin native oxide layer that influences the reactivity of the composite significantly. In this case, the Volta potential difference is given by the sum of the potential differences over all interfaces (126):

$$\Delta\Psi^{\text{Ref}}_{Pol} = \Delta\Phi^{Me}_{Ox} + \Delta\Phi_{Ox} + \Delta\Phi^{Ox}_{Pol} - \frac{1}{F}\mu^{Me}_{e} - \frac{W_{\text{Ref}}}{F} + \chi_{Pol}, \qquad 10.$$

where $\Delta\phi$ is the respective Galvani potential difference and μ_{e} is the chemical potential of the electron in the bulk metal.

If the Galvani potential across the oxide layer is substituted by the corresponding change in chemical composition (keeping the electrochemical potential constant) then, for example, for an oxide-covered iron surface the following result is obtained:

$$\Delta\Psi^{\text{Ref}}_{Pol} = -\frac{\Delta\mu^{0}_{\text{Fe}^{3+}/\text{Fe}^{3+}}}{F} + \Delta\Phi^{Ox}_{Pol} - \frac{W_{\text{Ref}}}{F} + \chi_{\text{Pol}} + \frac{\text{RT}}{F}\ln\frac{[\text{Fe}^{3+}]}{[\text{Fe}^{2+}]}. \qquad 11.$$

Thus the Volta potential difference represents the oxidation level within the oxide scale at the metal/polymer interface.

As shown below, during de-adhesion an additional liquid phase is formed between the substrate and the organic phase. Then the metal/electrolyte interface can be treated as a conventional electrochemical interface but an additional Galvani-potential difference $\Delta\Phi_D$ (Donnan potential or membrane potential) has to be taken into account at the electrolyte/polymer interface. The latter is directly correlated with the incorporation of ions into the polymeric membrane:

$$E_{Corr} = \frac{W_{\mathrm{Ref}}}{F} - \chi_{Pol} - \varepsilon_{1/2}^{\mathrm{Ref}} + \Delta\Phi_D + \Delta\Psi_{Pol}^{\mathrm{Ref}}. \tag{12.}$$

Therefore, the Volta potential difference, $\Delta\Psi_{Pol}^{\mathrm{Ref}}$, allows the measurement of the corrosion potential at the buried metal/electrolyte interface only if the Donnan potential is known or small. Usually the Donnan potential is significant for polymers with a high density of fixed charges (ion exchange membranes), because polymers with fixed cationic functional groups will exchange anions exclusively and vice versa (126). Lacquers used for corrosion protection may have some fixed ionic groups; however, their concentration is orders of magnitudes lower than that of typical ion exchange membranes.

Latest Developments of the Scanning Kelvin Probe

THE HEIGHT-REGULATED SCANNING KELVIN PROBE The conventional Kelvin probe suffers from the drawback that is does not follow the topography of the surface under investigation. Therefore it has to be lifted off from the surface by some distance, which significantly reduces the local resolution.

Recently, a height-regulated SKP with significantly improved local resolution has been developed using a double-modulation technique (127). The conventional Kelvin probe modulates only the distance between sample and probe, which leads to an alternating current between the two metals that is detected, amplified, and nulled by applying a dc voltage that is identical to the Volta potential difference. In addition to the distance modulation (some kHz) the dc compensation voltage is modulated with an additional ac voltage, $\mathrm{U_m}$, of the modulation frequency, ω_{m},

$$\mathrm{U(t)} = \mathrm{U_0} + \mathrm{U_m} \sin \omega_{\mathrm{m}} \mathrm{t}. \tag{13.}$$

The Kelvin current with the additional modulated voltage is given by

$$\mathrm{i(t)} = \frac{\mathrm{dQ(t)}}{\mathrm{dt}} = \frac{\mathrm{d}}{\mathrm{dt}}\{(\Delta\Psi + \mathrm{U(t)}) \cdot \mathrm{C(t)}\}, \tag{14.}$$

$$= \frac{\mathrm{d}}{\mathrm{dt}}\left[(\Delta\Psi + \mathrm{U_0} + \mathrm{U_m} \sin \omega_{\mathrm{m}} \mathrm{t}) \frac{\varepsilon \mathrm{A}}{\mathrm{d_0} + \Delta \mathrm{d} \cos \omega \mathrm{t}}\right], \tag{15.}$$

$$= (\Delta\Psi + \mathrm{U_0}) \frac{\varepsilon\, \omega\, \mathrm{A}\, \Delta \mathrm{d} \sin \omega \mathrm{t}}{(\mathrm{d_0} + \Delta \mathrm{d} \cos \omega \mathrm{t})^2} + \mathrm{U_m}\, \varepsilon \mathrm{A} \frac{\mathrm{d}}{\mathrm{dt}}\left[\frac{\sin \omega_{\mathrm{m}} \mathrm{t}}{\mathrm{d_0} + \Delta \mathrm{d} \cos \omega \mathrm{t}}\right]. \tag{16.}$$

The first part of this equation is now similar to the Kelvin current i(t) without the additional modulated voltage as it was given in Equation 5. From the derivation of the second part, we get two additional terms:

$$i(t) = (\Delta\Psi + U_0)\frac{\varepsilon\omega A\,\Delta d \sin\omega t}{(d_0 + \Delta d \cos\omega t)^2} + \frac{U_m\,\omega_m\,\varepsilon A \cos\omega_m t}{d_0 + \Delta d \cos\omega t}$$
$$+ \frac{\varepsilon A U_m\,\omega\,\Delta d \sin\omega_m t \sin\omega t}{(d_0 + \Delta d \cos\omega t)^2}. \qquad 17.$$

The demodulation is now done with two LIAs, one working at the frequency ω of the needle (i.e., the reference electrode) vibration and the second working at the frequency ω_m of the added modulation voltage. Again assuming $\Delta d \ll d_0$, the current amplitudes measured by the two LIAs are

$$I_\omega = (\Delta\Psi + U_0)\,\varepsilon\omega A\frac{\Delta d}{d_0^2}, \qquad 18.$$

$$I_{\omega_m} = U_m\,\varepsilon A\omega_m\frac{1}{d_0}. \qquad 19.$$

The current amplitude on the frequency of the vibrating needle in Equation 18 is equal to the case without the additional modulated voltage in Equation 5, thus the measurement of the Volta potential $\Delta\Psi$ is not affected by the use of the height regulation.

The current demodulated at the frequency ω_m of the additional voltage in Equation 19 is proportional to $1/d_0$ and can be directly used for the control circuit of the height regulation. For the SKP measurement with a constant sample/needle distance, this control circuit keeps I_{ω_m} at a constant starting value by changing the z-position of the needle.

As a consequence, the local resolution is improved by some micrometers without any expense to the operating conditions of the Kelvin probe. In order to derive valuable information on the electrochemical properties of buried interfaces, the Kelvin probe has to operate under well-controlled atmospheres, such as a relative humidity close to 100%. These conditions are guaranteed even for the height-regulated SKP. The accuracy of the height regulation is on the order of ±300 nm.

Wapner et al. introduced this technique and combined it with a blister test, so that an in situ study of de-adhesion under corrosive and mechanical load becomes possible (127).

THE SCANNING KELVIN PROBE FORCE MICROSCOPE In order to improve the local resolution of the SKP, conducting AFM tips have been used as probes for Volta potential measurements. In this case, the distance between tip and sample is no longer modulated, but instead a voltage modulation is established between the two materials. This leads to a distance fluctuation of the tip, which is registered by a reflecting laser beam (as usual in AFM set ups), and the distance fluctuation is used as a regulation tool to vary the potential difference between probe and sample such that the distance variation will eventually stop. Under those conditions the voltage difference is again identical to the Volta-potential difference (128, 129):

$$F_\omega = V_{dc} \cdot V_{ac} \cdot \mathrm{d}C/\mathrm{d}z = 0 \qquad 20.$$

$$\text{at } V_{dc} = \Delta\Psi - V_{dc,\,applied} = 0 \qquad 21.$$

(F_W, tip force; V_{dc}, dc voltage tip/sample; V_{ac}, ac bias tip/sample; C, capacitance; z, distance tip/sample; $V_{dc,applied}$, applied control bias tip/sample; $\Delta\Psi$, Volta potential difference.)

The SKP force microscopy (SKPFM) shows a significantly improved local resolution (less than 100 nm is reported for metal surfaces). However, for polymer-coated metals the resolution is limited by the polymer thickness, and to make use of the capabilities of the SKPFM technique requires the preparation of metal polymer interfaces with ultrathin organic coatings. As an example, Figure 13 shows SKPFM images of nickel markers on gold covered by polyurethane and then by a plasma polymer of different thickness (129).

The resolution of the 70-nm plasma polymer coating is the best. However, actual conflicting results seem to show that a simple relation between the thickness of the coating and the lateral resolution does not exist. Also the chemical nature of the polymer seems to influence the lateral resolution. Furthermore, SKPFM measurements under well-controlled atmospheres of high humidity are difficult to achieve.

For further information concerning this interesting technique, the reader should refer to recently published monographs (128, 129).

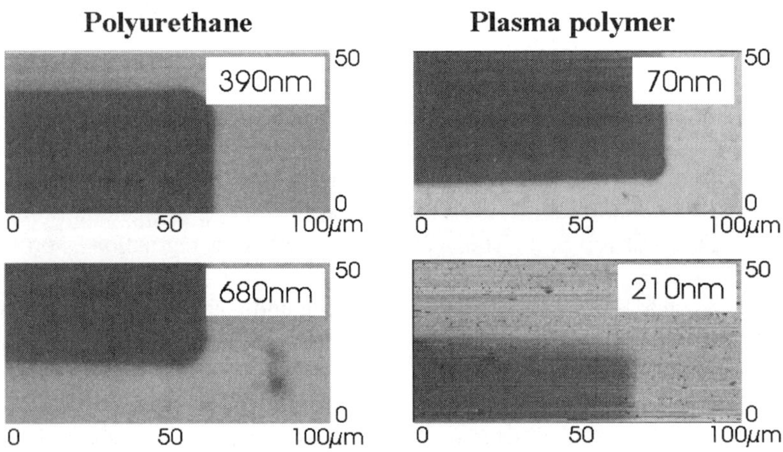

Figure 13 SKPFM images of evaporated nickel (low Volta potential) on gold, coated with polyurethane representing a plasma polymer film of different thickness. For the case of a 210-nm plasma polymer the nickel corner is significantly blurred, whereas for the 680-nm polyurethane coating it is still sharp. Reprinted from (129), with permission from CCC.

ELECTROCHEMICALLY DRIVEN DE-ADHESION REACTIONS AT METAL/POLYMER INTERFACES

De-adhesion of polymers from metallic or inorganic substrates is identical to breaking bonds at the interface. Depending on the prevailing nature of the bonds at the interface, this may occur by chemical displacement reactions such as the displacement of a hydrocarbon chain by water molecules, by mechanical forces, or by chemical and/or electrochemical reactions, which alter the nature of the chemical bonds at the interface.

The focus of the following is solely on the chemical/electrochemical attack that leads to de-adhesion. The mechanical forces are not discussed. We do discuss the transport of hydrated ions along the interface, the build-up of an electrified interface at the interface, and summarize those electrochemical reactions that lead to de-adhesion.

Incorporation of Ions into the Interface

In the case of an interfacial layer that completely inhibits electron transfer reactions at the interface or in an oxygen-free atmosphere, hydrated ions that ingress into the polymer/metal interface lead to a lowering of the measured potential in this area because the potential is polarized to the defect level. Moreover, because no oxygen reduction is involved, the potential does not shift when oxygen is added to the gas phase. Thus studies of the transport of hydrated ions and chemical de-adhesion processes can be done by means of the SKP. Figure 14 shows the potential distribution measured by means of the SKP when a thin electrolyte layer ingresses into the interface between an adhesive and an iron surface covered by a very thin, nonconducting SiO_2 layer (130, 131). The SiO_2 layer inhibits the electron transfer reaction. Consequently, no corrosive degradation of the interface takes place (see Figure 14). However, because the adhesion of the epoxy adhesive to the silica-like layer is weak, the polymer is replaced by the electrolyte starting from the defect. After

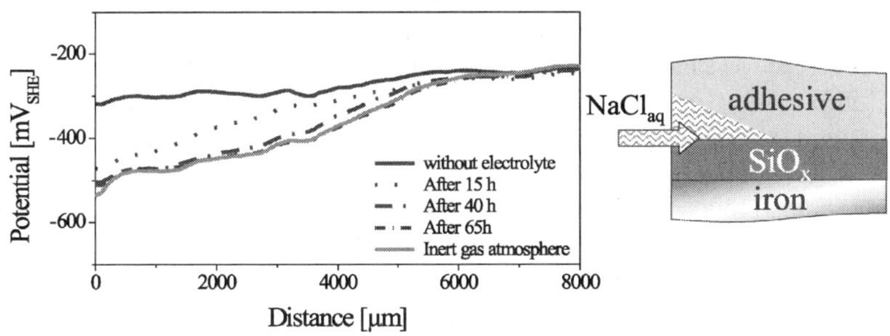

Figure 14 SKP potential profiles of the delamination of an epoxy resin adhesive from a HMDS/O_2 plasma polymer-coated iron surface, with 0.5 M NaCl as electrolyte. Reprinted from (131), with permission from Wiley-VCH.

the atmosphere changes from oxygen to nitrogen, no change is observed, which proves that only wet de-adhesion, not electrochemical de-adhesion, has occurred.

The Electrified Interface

Very little is known about the structure and electrical properties of the electrified interface. Charge separation at the interface can result from one of the following reactions:

- Electron-transfer-reactions to electro-active species such as oxygen may lead to charged species embedded in the polymeric matrix:

$$\text{metal} + z/4\ O_2 + z/2\ H_2O \rightarrow \text{metal}^{z+} + z\ OH^- \qquad 22.$$

- Chemical reactions between the organic and the inorganic phase may result in charge separation:

$$\text{R-COOH} + \text{HO-Fe}_{ox} \rightarrow \text{R-COO}^- + {}^{+}\text{Fe}_{ox} + H_2O. \qquad 23.$$

- The incorporation of ions may lead to charged interfaces.

Nazarov et al. have used again the Kelvin probe to study some of these phenomena, as the charge separation shows up immediately in a change in the Volta potential (132). If nonconducting polymers such as alkyd or epoxy resins are used, characteristic potential shifts are observed relative to the uncoated surface. These shifts are interpreted according to the reaction schemes as shown above; the alkyd resin leads to a positively charged substrate, and the epoxy resin leads to a corresponding negative charge. In the presence of water the Volta potential may change significantly. This is partly explained by a hydrolysis of bonds, e.g., by shifting the equilibrium of the second equation to the left side. Obviously, the Kelvin probe is an important tool for studying the electrified interface in the presence and absence of water, and further research will have to concentrate on these issues.

The Kelvin probe detects only the potential drop across the buried interface; it does not provide information on the potential distribution that ultimately defines the reaction kinetics of electron transfer and ion transfer reactions. However, some information has been obtained from the impedance spectra of the interface as measured by embedded microreference electrodes (133). These measurements allow differentiation between the overall impedance of the polymer film—which is determined by the dielectric properties of the film—and the interfacial impedance—which is dominated by the interfacial potential distribution. From the interfacial impedance the interfacial capacitance can be calculated. This capacitance seemed to point to the potential of zero charge (maximum capacity) but it also proved that the interfacial capacitance is significantly lower than the one in aqueous electrolytes and, therefore, the potential distribution is smeared out compared with that of the rigid electrochemical double layer. The diffuse double layer is the main reason for the slow kinetics of many electrochemical reactions at metal/polymer interfaces.

Electron Transfer Reactions at the Metal/Polymer Interface

Electrochemical reactions that lead to a degradation of the metal/polymer interface are influenced by the following properties: the electron-transfer properties at the interface, the redox properties of the oxide between the metal and the polymer, and the chemical stability of the interface with respect to those species, which are formed during the electron transfer reaction.

- The rate of electron transfer reactions (ETR) is strongly influenced by the surface composition of the metal. As most materials are covered by oxides, their electronic properties will determine the rate of ETR. Therefore, metals that are covered by electron-conducting or semiconducting oxides such as iron or zinc will show a higher ETR rate at the substrate/polymer interface compared with that of materials that form highly insulating oxides such as aluminum.
- Certain oxides are characterized by a fixed ratio of anions and cations (e.g., Al_2O_3), whereas others have a strongly potential-dependent composition such as iron-oxides owing to the flipping of valence states (Fe^{2+}, Fe^{3+}) in the cation sublattice. Any change of the electrode potential is reflected in a change of the valence states, and this will change the semiconducting properties, e.g., Fe^{2+} states can be regarded as donors of the n-type semiconductor (134, 135). Furthermore, during reduction of the oxide, the base material will be oxidized, and this may limit the adhesion of the coating.
- During the electron transfer, highly reactive intermediates and reaction products are formed that will chemically react with the interfacial layer. It is shown below that major reaction products such as OH^- ions, which are generated during the reduction of molecular oxygen, are of significance for the delamination process. Certain materials such as iron are very stable under those alkaline conditions, whereas others such as aluminum or zinc are covered with oxides that are unstable in alkaline electrolytes (136).

DE-ADHESION REACTIONS BASED ON CATHODIC DISBONDING This type of de-adhesion reaction dominates on surfaces covered by conducting oxides in the presence of defects that penetrate the coating.

Electrochemical Elements

Figure 15 (see color insert) shows as an example a typical potential map as measured by the SKP for an iron/polymer interface during cathodic undermining.

In the presence of oxygen, the electrode potential of the metal/polymer interface changes in a well-defined manner with increasing distance from the defect: Close to the defect the potential is negative, whereas at a distance from the defect, anodic potentials are observed. For most coatings, the steep increase of the electrode potential also marks the delamination front, and the region of anodic potentials represents the intact interface. The physical origin of the sudden change of the electrode potential is, however, given by the migration of ions from the defect into the interface, which results in the polarization of the highly polarizable interface to the potential of the nonpolarizable defect (137).

The intact interface is characterized by an anodic potential plateau. This plateau results from the high electronic conductivity of the oxide-covered iron surface, which allows electron transfer but no ion transfer reactions. Therefore, oxygen will be reduced at this interface, and the reaction is balanced by the anodic oxidation of the oxide. As the electrode potential of the oxide is given by the activity of Fe^{2+} and Fe^{3+} states, any oxidation results in an anodic potential shift accompanied by a steady decrease of the donor density and therefore by a decreasing rate of the electron transfer reaction. Above a certain anodic potential, the rate of the oxygen reduction is extremely small, and no further anodic potential shift is observed. This is the actual potential as measured by the Kelvin probe. The sudden potential step at the delamination front marks the most interesting position because here reactions will occur that are responsible for the loss of adhesion. As discussed above, the potential step is caused by the ingress of ions into the interface and by the galvanic coupling of the interface to the defect. The cathodic potential step also marks the partial reduction of the iron oxide and the increase in its donor density. Obviously, this must result in an increase in the rate of the electron transfer. The anodic counter reaction of the oxygen reduction is the dissolution of the base material within the defect. Between the defect and the steep potential increase, a steady potential increase to more anodic values is observed. Experiments have proven that the steady potential increase is linked to the galvanic current between defect and the frontier of incorporation of ions and is measured only if oxygen is reduced within the latter zone (137–139).

In order to compensate for the charge, soluble cations have to migrate to the zone of oxygen reduction. On the other hand, negative ions such as chlorides are repelled by the negative charge of generated hydroxyl ions from the delamination zone. This was confirmed by spatially resolved X-ray photo-electron spectroscopy (XPS) measurements that reveal the distribution of chlorides and sodium ions underneath the coating after the separation of the latter from the iron substrate (137–139). The electrochemical situation is summarized in Figure 16.

Degradation of the Interface by the Oxygen Reduction Process

During oxygen reduction, a strongly alkaline electrolyte is formed. This electrolyte may either stabilize (iron) or destabilize (zinc) the oxide on the metal. The delamination of the organic coating is caused only by bond breaking within the organic zone. It has been proven that intermediate radicals such as HO_2^-, OH, and O_2^- form during oxygen reduction (140) and may contribute to the oxidation of the organic layer (141, 142).

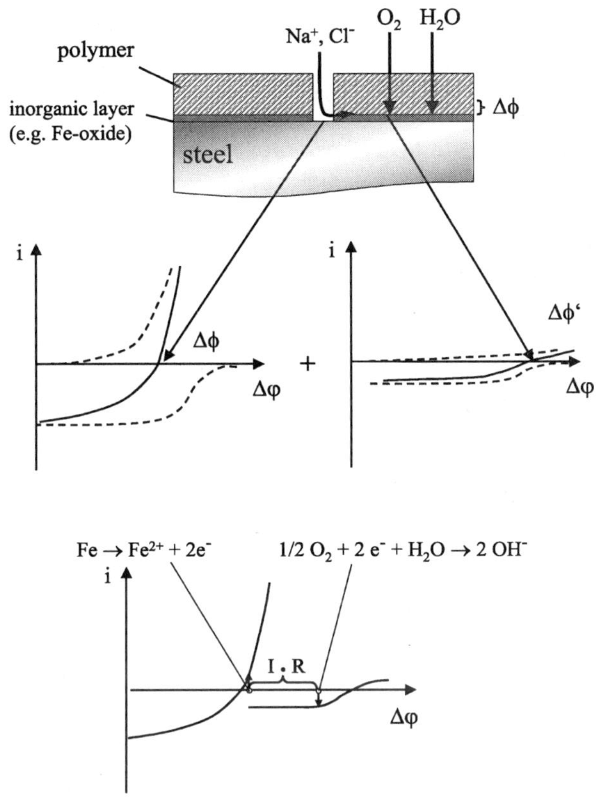

Figure 16 Schematic illustration of the electrochemical processes that led to cathodic delamination of polymer films from iron substrates. Reprinted from (123), with permission from Elsevier.

The mechanism of oxidative degradation of the polymer was postulated by Wroblowa (140), who proposed a initial step of the reaction as

$$O_2^- + R - H \rightarrow R^\bullet + HO_2^-, \tag{24.}$$

and as a subsequent reaction mechanism

$$R^\bullet + O_2 \rightarrow ROO^\bullet \tag{25.}$$

$$ROO^\bullet + R - H \rightarrow ROOH + R^\bullet \tag{26.}$$

In addition to the chemical attack of oxygen reduction products on the organic layer, several groups noted that a change in the oxide thickness owing to the high pH (~12 to 14) is possible and contributes to the delamination process (143, 144). However, in situ XPS measurements showed that, even in an alkaline media, a significant iron oxide reduction is possible only at potentials more negative than $-0.8\ V_{SHE}$ (145).

In case of zinc or galvanized steel, zinc oxides are less stable within the induced alkaline environment. Auger electron spectroscopy sputter profiles measured within the delaminated zone prove the significant growth of the oxide scale, which indicates anodic reactions (146).

MICROSCOPIC UNDERSTANDING OF ELECTROCHEMICAL DRIVEN DE-ADHESION REACTIONS Whereas the basic mechanisms of anodic and cathodic delamination, i.e., the electrochemically driven de-adhesion of polymer coatings from the metal substrate, are already well understood, little is known about the microscopic and submicroscopic aspects of delamination, which is mainly because of the lack of suitable experimental techniques and the limited local resolution of the SKP. The development of the SKPFM expanded the application of the Kelvin probe technique to the submicroscopic scale. For the first experiments, a custom-made glass cell purged with humid air to control the relative humidity of the environment was used. Flame-annealed gold was covered by an ultrathin plasma polymer film made from hexamethyldisilane and electrochemically coupled to iron to set the electrode potential of the polymer-coated gold to $-0.4\ V_{SHE}$. The electrolyte was 0.1 M NaCl in agar applied onto the surface. The gold surface, which allows electron transfer but no ion transfer reactions, behaves similarly to a passive iron electrode below the organic coating and serves as the local cathode during delamination while the iron acts as the local anode.

This special sample setup successfully initiated cathodic delamination, which is similar to the cathodic delamination on iron. Figure 17 compares the AFM

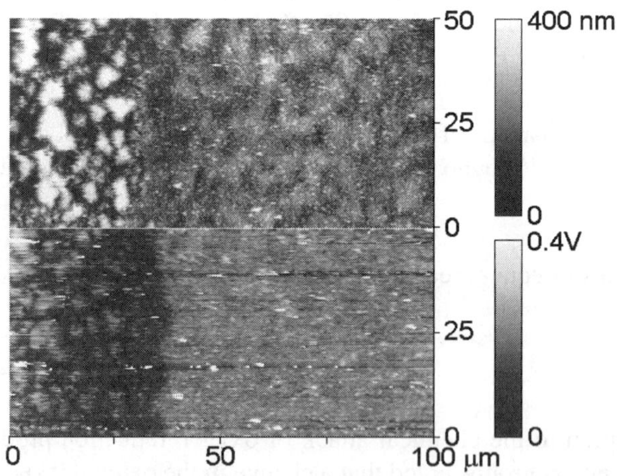

Figure 17 (*a*) Topography of a plasma polymer-coated gold surface at the delamination front (~5000 μm from the defect, 20 μm from the optically determination front). (*b*) Correlated SKPFM potential map. Reprinted from (147), with permission from Elsevier.

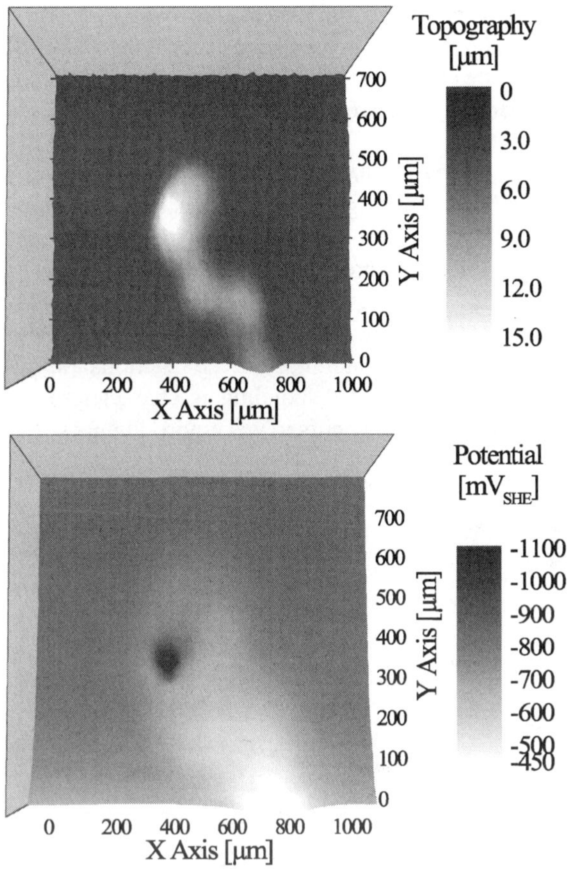

Figure 18 Height-regulated SKP (HR-SKP) measurement of a single filiform thread on an organically coated aluminum substrate under active FFC conditions: (*a*) HR-SKP topography map and (*b*) HR-SKP potential map. Reprinted from (127), with permission from The Electrochemical Society, Inc.

topography with an SKPFM potential image (147). The images show the delamination front at approximately 5000 μm from the defect. The delamination area is seen in the AFM map as an elevated topography because the ultrathin polymer film is lifted up by the undermining electrolyte. The delamination front as deduced from the height profile is marked by the white dashed line. In the Volta potential map, the delamination is also clearly visible as a region of low potential, similar to the conventional SKP results. However, the front of the negative electrode potential exceeds the delamination front by a constant distance of approximately 10 μm, indicating that an electrochemical delaminating zone advances the already mechanically delaminated area by 10 μm.

The dimension of the reaction zone was further confirmed by time-of-flight secondary ion mass spectroscopy (TOF-SIMS) measurements (147). In contrast to the continuous potential profile measured by SKPFM, the chemistry between the reaction zone and the delaminated zone according to TOF-SIMS changes abruptly for reasons not yet understood. The chemistry and electrochemistry occurring within this reaction zone are the subjects of ongoing studies.

DE-ADHESION REACTIONS BASED ON ANODIC DISBONDING Cathodic disbonding requires electron-conducting oxides and interfaces that allow electrochemical reactions to occur. The latter usually requires the presence of ions and water molecules at the interface and therefore will dominate at high relative humidity.

At low humidity and in the presence of insulating oxides such as MgO or Al_2O_3, cathodic disbonding cannot occur and anodic-driven de-adhesion reactions will prevail. In the corrosion community, these reactions are called filiform corrosion (FFC) because this type of disbonding is characterized by growing threadlike filaments (148–151). FFC occurs in wet environments with relative humidity well

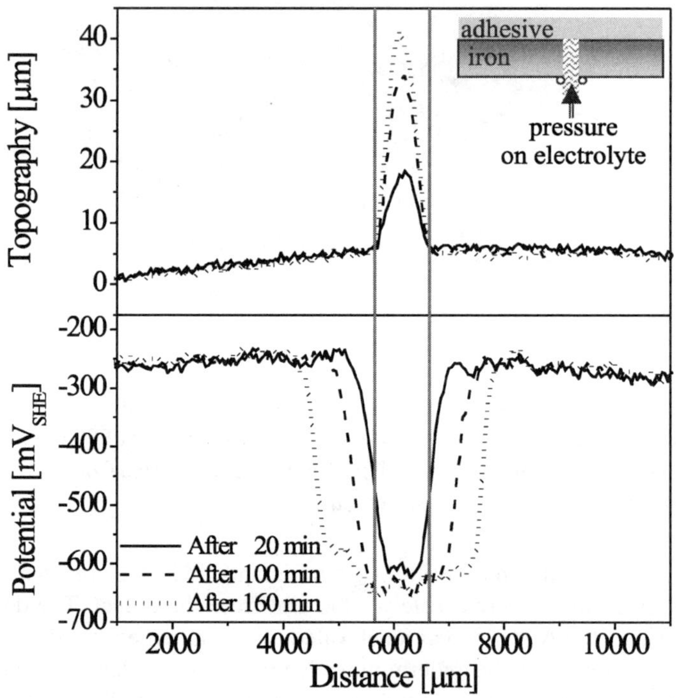

Figure 19 In situ measurement of a pressurized blister test experiment with the HR-SKP on a pure iron substrate coated with an epoxy resin adhesive and pressurized with 0.5 M NaCl at 500 mbar. HR-SKP topography and potential line-scans after different exposure times. Reprinted from (127), with permission from The Electrochemical Society, Inc.

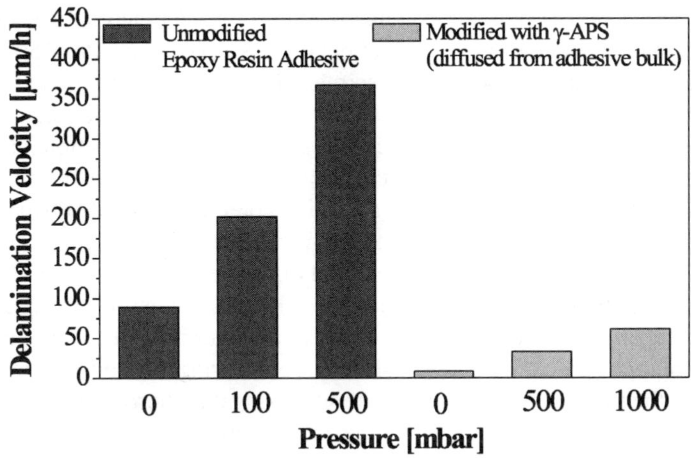

Figure 20 Increase of the delamination velocity as a function of a mechanical stress in the blister test experiment measured with the HR-SKP; comparison of an unmodified epoxy resin adhesive and of the adhesive modified with an organosilane adhesion promoter.

below 100% and usually arises from surface defects in the protective film in the presence of soluble ionic species. Various metals such as Al, Fe, and Mg show this kind of corrosion underneath a (polymer) coating. However, zinc-coated steel and even electroplated silver plate and gold plate can also be affected by FFC (152). An extended literature review of FFC investigations on aluminum is given by Bautista (153), earlier reviews have been presented by Hahin (154), Hoch (152), and Ruggeri & Beck (155). Some recent results on FFC on aluminum alloys are also available (156–163).

Although the mechanism of FFC is not totally understood, it is generally believed that the filament propagates by an anodic undermining of the surface, which arises through a differential aeration cell within the filament head. In terms of this model, the metal substrate is oxidized at the head front (anode) and oxygen is reduced in the vicinity of the head-tail junction (cathode) (164). Between the anode and cathode, a potential gradient is established that forces anions to migrate to the front and cations to the back. On aluminum substrates, an especially low pH and the oxygen deficiency at the leading edge of the filament thread can result in a secondary cathodic reaction involving hydrogen evolution in the head.

Latest in situ Volta potential profiles using a SKP give some insight into this type of de-adhesion reaction for an aluminum substrate (157, 159, 164). Because the height and the potential in this case are measured simultaneously (Figure 18), these plots allow a direct comparison between the reaction sites (potential plot) and subsequent delamination (topographic plot).

The potential plot corresponds well to the SKP potential maps that have been reported by other groups; the filament tail has a potential about 150 mV higher than the passive state of the intact specimen. This can be interpreted with regard to the formation of aluminum-oxide/-hydroxide corrosion products and, additionally, to the occurrence of reduction reactions, such as oxygen reduction, that possibly take place in the tail. The active part of the filament head, where the anodic reaction of active metal dissolution takes place, has a potential of about 500 mV lower than the intact area. If potential and topographic maps are directly compared, it is obvious that the area where the Volta potential differs from the one of the intact specimen is much broader than the area where the coating is deformed and lifted from the substrate. Assuming that the lateral resolution of the Volta potential and the topography are approximately similar (considering the simultaneous measurement with the same technique), it is quite surprising that these two profiles do not match. Additionally, it can be seen that the anodic active area is confined to a rather small region at the left border of the filament head and not in the center or across the whole area of the head as might be expected.

ELECTROCHEMICAL-DRIVEN DE-ADHESION REACTIONS BY SUPERIMPOSED MECHANICAL STRESS The pressurized blister test is an excellent method to combine electrochemical reactions at polymer/metal interfaces with a mechanical load. It allows the application of a mechanical stress from a homogeneously pressurized electrolyte on the adhesive/metal interface in a sample geometry that is accessible for the HR-SKP (127). Depending on the adjusted conditions, information on the synergy of mechanical stresses, elastic or inelastic deformations of the adhesive, transport processes, and corrosive reactions could be obtained with this method.

As an example, an iron sample was coated with a two-component epoxy resin adhesive film of 70 μm thickness. A circular hole in the iron substrate of 1 mm diameter was prepared by a milling and chemical etching process. This procedure leads to an undamaged adhesive film above the hole. The hole may now be filled by an electrolyte and pressurized to establish a mechanical stress at the interface. Figure 19 shows typical potential profiles in the vicinity of the hole compared with the topographic profile. Starting from the defect, a cathodic-driven de-adhesion process proceeds along the adhesive/metal interface with time. A corrosive delamination of the adhesive takes place over a continuous increasing area of the specimen, leading to a weakening of the adhesive/metal interface and eventually to the formation of a thin electrolyte layer in the delaminated area. As the inflection point of the potential profile marks the delamination front, the delamination rate can now be measured as a function of the pressure within the blister (Figure 20). Most interestingly, the rate strongly depends on the mechanical stress at the interface, which may be from a change of the molecular structure, the stretching of bonds, or from the pressure-induced inward migration of ions and water molecules.

The *Annual Review of Materials Research* is online at http://matsci.annualreviews.org

LITERATURE CITED

1. Faupel F, Willecke R, Thran A. 1998. Diffusion of metals in polymers. *Mater. Sci. Eng. Rep.* 22(1):1–55

1a. Tian WJ, Zhang HY, Shen JC. 1997. Some properties of interfaces between metals and polymers. *Surf. Rev. Lett.* 4(4):703–8

1b. Bohn PW. 1997. Localized optical phenomena and the characterisation of materials interfaces. *Annu. Rev. Mater. Sci.* 27:469–98

2. Hartstein A, Kirtley JR. 1980. Enhancement of infrared-absorption from molecular monolayers with thin metal overlayers. *Bull. Am. Phys. Soc.* 25:425–25

3. Wokaun A. 1984. Surface enhanced electromagnetic processes. *Solid State Phys.* 38:223

4. Otto A, Mrozek I. 1992. Surface-enhanced Raman scattering. *J. Phys. Condens Matter.* 4:1143

5. Jeanmaire DL, Van Duyne RP. 1977. Surface Raman spectroelectrochemistry. 1. Heterocyclic, aromatic, and aliphatic-amines adsorbed on anodized silver electrode. *J. Electr. Chem.* 84:1–20

6. Albrecht MG, Creighton JA. 1977. Anomalously intense Raman-spectra of pyridine at a silver electrode. *J. Am. Chem. Soc.* 99:5215–17

7. Fleischmann M, Hendra PJ. 1974. Raman-spectra of pyridine adsorbed at a silver electrode. *Chem. Phys. Lett.* 26:163–66

8. Osawa M, Ikeda M. 1991. Surface-enhanced infrared-absorption of para-nitrobenzoic acid deposited on silver island films—contributions of electromagnetic and chemical mechanisms. *J. Phys. Chem.* 95:9914–19

9. Hatta A, Suzuki Y. 1984. Infrared-absorption enhancement of monolayer species on thin evaporated Ag films by use of a Kretschmann configuration—evidence for 2 types of enhanced surface electric-fields. *Appl. Phys. Mater. Sci. Proc.* 35:135–40

10. Osawa M, Kuramitsu M. 1986. Electromagnetic effect in enhanced infrared absorption of adsorbed molecules on thin metal-films. *Surf. Sci.* 175:L787–93

11. Nishikawa Y, Fujiwara K. 1993. Trace analysis of human skin secretions by surface-enhanced infrared-spectroscopy—detection of lactate. *Anal. Sci.* 9:811–15

12. Osawa M, Ataka K. 1992. Electromagnetic mechanism of enhanced infrared absorption of molecules adsorbed on metal island films. *Surf. Sci.* 262:L118

13. Nishikawa Y, Fujiwara K, Ataka K, Osawa M. 1993. Surface-enhanced infrared external reflection spectroscopy at low reflective surfaces and its application to surface analysis of semiconductors, glasses, and polymers. *Anal. Chem.* 65:556

14. Sennet RS, Scott GD. 1950. The structure of evaporated metal films and their optical properties. *J. Opt. Soc. Am.* 40:203–11

15. Bauer E, Poppa H. 1972. Recent advances in epotaxy. *Thin. Solid Films* 12:167

16. Schlegel VL, Cotton TM. 1991. Silver-island films as substrates for enhanced Raman scattering: effect of deposition rate on intensity. *Anal. Chem.* 63:241

17. Ishida H, Fukuda H, Katagiri G, Ishitani A. 1986. An application of surface-enhanced Raman-scattering to the surface characterization of carbon materials. *Appl. Spectrosc.* 40:322–30

18. Osawa M, Ataka K. 1993. Surface-enhanced infrared-spectroscopy—the origin of the absorption enhancement and band selection rule in the infrared-spectra of molecules adsorbed on fine metal particles. *Appl. Spectrosc.* 47:1497–502
19. Merklin GT, Griffiths PR. 1997. Effect of microscopic surface roughness in surface-enhanced infrared absorption spectrometry. *J. Phys. Chem. B* 101:5810–13
20. Osawa M. 2001. Surface-enhanced infrared absorption. In *Near-Field Optics and Surface Plasmon Polaritons*, ed. S. Kawata. Berlin: Springer-Verlag, 81:163–87
21. Osawa M. 1997. Dynamic processes in electrochemical reactions studied by surface-enhanced infrared absorption spectroscopy (SEIRAS). *Bull. Chem. Soc. Jpn.* 70:2861–80
22. Osawa M. 2002. Surface-enhanced infrared absorption spectroscopy. *Handbook Vibr. Spectrosc.* 1:785
23. Tian Z, Ren B. 2003. Raman spectroscopy of electrode furfaces. In situ structural and spectroscopic probes of electrochemical aystems. In *Encyclopedia of Electrochemistry. Vol. 3: Instrumentation and Electroanalytical Chemistry*, ed. AJ Bard, M Stratmann, PR Unwin, pp. 572–656. New York: Wiley
24. Tian Z, Ren B, Wu D. 2002. Surface-enhanced Raman scattering: from noble to transition metals and from rough surfaces to ordered nanostructures. *J. Phys. Chem. B.* 106:9463–83
25. Nie S, Emory SR. 1997. Probing single molecules and single nanoparticles by surface-enhanced raman scattering. *Science* 275:1102
26. Kneipp K, Wang Y, Kneipp H, Perelman LT, Itzkan I, et al. 1997. Single molecule detection using surface-enhanced Raman scattering (SERS). *Phys. Rev. Lett.* 78:1667–70
27. Michaels AM, Nirmal M, Brus LE. 1999. Surface enhanced Raman spectroscopy of individual rhodamine 6G molecules on large Ag nanocrystals. *J. Am. Chem. Soc.* 121:9932
28. Huang QJ, Lin XF, Yang ZL, Hu JW, Tian ZQ. 2004. An investigation of the adsorption of pyrazine and pyridine on nickel electrodes by in situ surface-enhanced Raman spectroscopy. *J. Electoanal. Chem.* 563:121–31
29. Kneipp K, Kneipp H, Itzkan I, Dassri RR, Feld MS. 2002. Surface-enhanced Raman scattering and biophysics. *J. Phys. Condens. Matter* 14:599–602
30. Shibamoto K, Katayama K, Fujinami M, Tsuguo T. 2001. Detection using transient reflecting grating spectroscopy for the ultrafast interaction between photoexcited electrons and adsorbed molecules at a gold surface. *Anal. Sci.* 17:S255–58
31. Kambhampati P, Child CM, Foster MC, Campion A. 2002. On the chemical mechanism of surface enhanced Raman scattering: experiment and theory. *J. Chem. Phys.* 108:5013–16
32. Campion A, Kambhampati P. 1998. Surface enhanced Raman scattering. *Chem. Soc. Rev.* 27:241–50
33. Otto A, Mrozek I, Grabhorn H, Akemann W. 1992. Surface-enhanced Raman scattering. *J. Phys. Condens. Matter* 4: 1143
34. Królikowska A, Kudelski A, Michota A, Bukowska J. 2003. SERS studies on the structure of thioglycolic acid monolayers on silver and gold. *Surf. Sci.* 532–535:227–32
35. Seitz O, Chehimi MM, Deliry EC. 2003. Preparation and characterisation of gold nanoparticle assemblies on silanised glass plates. *Coll. Surf. A* 218:225–29
36. Sauer G, Nickel U, Schneider S. 2000. Preparation of SERS-active silver film electrodes via electrocrystallization of silver. *J. Raman Spect.* 31:359–63
37. Maurizio, Miranda M. 2004. AFM and micro-Raman investigation on filters

coated with silver colloidal nanoparticles. *Appl. Surf. Sci.* 226:125–30
38. Markwitz A, Matz W, Short K, Waldschmidt K. 2003. Surface structuring and phase formation in thin metallic layers deposited at various temperatures. *Surf. Interf. Anal.* 33:1–6
39. Felidj N, Aubard J, Lévi G. 2002. Controlling the optical response of regular arrays of gold particles for surface-enhanced Raman scattering. *Phys. Rev. B* 65:074519
40. Grand J, Kostcheev S, Bijeon JL, et al. 2003. Optimization of SERS-active substrates for near-field Raman spectroscopy. *Synt. Met.* 139:621–24
41. Kundu S, Mandal M, Ghosh SK, Pal TK. 2004. Photochemical deposition of SERS active silver nanoparticles on silica gel. *J. Photochem. Photobiol. A* 162:625–32
42. Freeman RG, Katherine C. 1995. Self-assembled metal colloid monolayers: an approach to SERS substrate. *Science* 267:1629–32
43. Pérez R, Rupérez A, Rodriguez-Castellón E, Laserna JJ. 2000. Study of experimental parameters for improved adsorbate detectability in SERS using etched silver substrates. *Surf. Interf. Anal.* 30:592–96
44. Ooka AA, Garell L. 2000. Surface-enhanced Raman spectroscopy of DOPA-containing peptides related to adhesive protein of marine mussel, *Mytilus edulis*. *Biopolymers* 57:92–102
45. Ulman A. 1996. Formation and structure of self-assembled monolayers. *Chem. Rev.* 96:1533
46. Bishop AR, Nuzzo RG. 1996. Self-assembled monolayers: recent developments and applications. *Curr. Opin. Coll. Int. Sci.* 1:127
47. Freeman RG, Grabar KC, Allison KJ, Bright RM, Davis JA, et al. 1995. Self-assembled metal colloid monolayers: an approach to SERS substrate. *Science* 267:1629–32
48. Kudelski A. 2003. Structures of monolayers formed from different HS—(CH2)2—X thiols on gold, silver and copper: comparitive studies by surface-enhanced Raman scattering. *J. Raman Spect.* 34:853–62
49. Kudelski A. 2002. Raman study on the structure of 3-mercaptopropionic acid monolayers on silver. *Surf. Sci.* 502:219–33
50. Dai QP, Xue CC, Xue G, Jiang LX. 1995. Fourier-transform surface-enhanced Raman-scattering studies of molecular self-assembly of disulfides on metals and the application in adhesion promotion. *J. Adhes. Sci. Techol.* 9: 1465–74
51. Joo SW, Han SW, Kim K. 2000. Multilayer formation of 1,2-ethanedithiol on gold: surface-enhanced Raman scattering and ellipsometry study. *Langmuir* 16:5391–96
52. Zhang ZJ, Yoshida N, Imae T, Xue QB, Bai M, et al. 2001. A self-assembled monolayer of an alkanoic acid-derivatized porphyrin on gold surface: a structural investigation by surface plasmon resonance, ultraviolet-visible, and infrared spectroscopies. *J. Coll. Int. Sci.* 243:382–87
53. Grundmeier G, Stratmann M. 1995. Nucleation and growth of plasma-polymerised hexamethyldisilazane on iron substrates. *Ber. Bunsenges. Phys. Chem.* 11:1387–92
54. Grundmeier G, Stratmann M. 1999. Interfacial processes during plasma polymer deposition on oxide covered iron. *Thin Solid Films* 352:119–27
55. Turner RH, Boerio FJ. 2002. Molecular structure of interfaces formed with plasma-polymerized silica-like primer films: Part I. Characterization of the primer/metal interface using infrared spectroscopy in situ. *J. Adhes.* 78:447–65
56. Chabal YJ. 1988. Surface infrared-spectroscopy. *Surf. Sci. Rep.* 8:211–357

57. Golden WG. 1985. Infrared-reflection-absorption-spectroscopy. In *Fourier Transform Infrared Spectroscopy*, ed. JR Ferraro, LJ Basile, pp. 315. Orlando, FL: Academic
58. Nguyen T, Byrd E, Bentz D. 1996. In situ measurement of water at the organic coating substrate interface. *Progr. Org. Coat.* 27:181–93
59. Fieldson GT, Barbari TA. 1993. The use of FTIR-ATR spectroscopy to characterize penetrant diffusion in polymers. *Polymer* 34:1146–53
60. Nishikawa Y, Nagasawa T. 1993. Silver island films for surface-enhanced infrared-absorption spectroscopy—effect of island morphology on the absorption enhancement. *Vibr. Spectr.* 6: 43–53
61. Nishikawa Y, Fujiwara K, Shima T. 1990. Qualitiative analysis of nanogram samples with Fourier-transform infrared transmission surface electromagnetic-wave spectroscopy. *Appl. Spectr.* 44: 691–94
62. Osawa M, Ataka K. 1993. Surface-enhanced infrared ATR spectroscopy for in situ studies of electrode-electrolyte interfaces. *J. Electr. Spectr. Rel. Phenom.* 64–5:371–79
63. Ataka K, Yotsuyanagi T. 1996. Potential-dependent reorientation of water molecules at an electrode/electrolyte interface studied by surface-enhanced infrared absorption spectroscopy. *J. Phys. Chem.* 100:10664–72
64. Kuhne C, Steiner G. 1998. Surface enhanced FTIR spectroscopy on membranes. Fres. *J. Anal. Chem.* 360:750–54
65. Ataka K, Hara Y. 1999. A new approach to electrode kinetics and dynamics by potential modulated Fourier transform infrared spectroscopy. *J. Electro. Chem.* 473:34–42
66. Jensen TR, Van Duyne RP. 2000. Surface-enhanced infrared spectroscopy: a comparison of metal island films with discrete and nondiscrete surface plasmons. *Appl. Spectr.* 54:371–77
67. Zhang ZJ, Imae T. 2001. Study of surface-enhanced infrared spectroscopy-1. Dependence of the enhancement on thickness of metal island films and structure of chemisorbed molecules. *J. Coll. Interf. Sci.* 233:99–106
68. Wandlowski T, Ataka K. 2004. Surface enhanced infrared spectroscopy—Au(1 1 1–20 nm)/sulphuric acid—new aspects and challenges. *Electrochim. Acta* 49:1233–47
69. Tian ZQ, Ren B. 2004. Adsorption and reaction at electrochemical interfaces as probed by surface-enhanced Raman spectroscopy. *Annu. Rev. Phys. Chem.* 55:197–229
70. Shen A, Pemberton JE. 1999. Investigation of trace interfacial water at silver electrodes in a series of normal alcohols using surface enhanced Raman scattering. *Phys. Chem. Chem. Phys.* 1:5677–84
71. Lee BW, Kwon MY, Kim JJ. 1990. Surface enhanced Raman scattering of water molecules in water-ethanol mixed solution. *Chin. J. Phys.* 28:1–8
72. Chen TT. 1985. Surface enhanced Raman scattering of H_2O, D_2O and HDO molecules on Ag electrodes *Chin. J. Phys.* 23:203–15
73. Chang YC, Chen TT. 1996. Influence of organic solvents on the intensity of surface-enhanced Raman scattering of water molecules. *Chin. J. Phys.* 34: 1352–62
74. Cao P, Gu R, Qiu L, Sun R, Ren B, et al. 2003. SERS investigation of interfacial water at a silver electrode in acetonitrile solutions. *Surf. Sci.* 531:217–25
75. Itoh T, Sasaki Y, Maeda T, Horie C. 1997. In situ SERS study of water molecules on silver electrodes in alkali-hydroxide solutions. *Surf. Sci.* 389:212–22
76. Pemberton JE, Joa SL. 1994. Water and electrolyte structure at Ag electrodes

in nonaqueous butanol solutions using surface-enhanced Raman scattering. *J. Electroanal. Chem.* 378:149–58
77. Gu R, Sun Y. 2002. Investigation on the adsorption behavior of water on Ag electrode/acetonitrile interface by using surface-enhanced Raman spectroscopy. *Chem. J. Chin. Univ.* 23:1960–64
78. Venkataramanan M, Pradeep T. 2000. A method to study the phase transition and desorption of self-assembled monolayers on planar gold surfaces. *Anal. Chem.* 72:5852–56
79. Bandyopadhyay K, Vijayamohanan K. 1999. Self-assembled monolayers of small aromatic disulfide and diselenide molecules on polycrystalline gold films: a comparative study of the geometrical constraint using temperature-dependent surface-enhanced Raman spectroscopy, X-ray photoelectron spectroscopy, and electrochemistry. *Langmuir* 15:5314–22
80. Venkataramanan M, Murty KVGK, et al. 2000. Metal ion reactivity with 1,4-benzenedimethanethiol monolayers on gold. *Langmuir* 16:7673–78
81. Deleted in proof
82. Litorja M, Haynes CL, Haes AJ, Jensen TR, Van Duyne RP. 2001. Surface-enhanced Raman scattering detected temperature programmed desorption: optical properties, nanostructure, and stability of silver film over SiO_2 nanosphere surfaces. *J. Phys. Chem B* 105:6907–15
83. Otto A, Bruckbauer A, Chen YX. 2003. On the chloride activation in SERS and single molecule SERS. *J. Mol. Struct.* 661:501–14
84. Erturk U, Pockrand I, Otto A. 1983. Surface enhanced Raman scattering from pyridine on copper in UHV. *Surf. Sci.* 131:367–84
85. Brolo AG, Germain P, Hager G. 2002. Investigation of the adsorption of L-cysteine on a polycrystalline silver electrode by surface-enhanced Raman scattering (SERS) and surface-enhanced second harmonic generation (SESHG). *J. Phys. Chem. B* 106:5982–87
86. Kudelski A. 2003. Chemisorption of 2-mercaptoethanol on silver, copper, and gold: direct Raman evidence of acid-induced changes in adsorption/desorption equilibria. *Langmuir* 19: 3805–13
87. Chen HY, Long YT. 1999. Study of biomolecules by combining electrochemistry with UV/Vis, IR and surface enhanced Raman scattering spectroscopy by a novel flow microcell. *Anal. Chem. Acta* 382:171–77
88. Maruyama Y, Ishikawa M, Futamata M. 2001. Single molecule detection with SERS. *Anal. Sci.* 17:i1181–83
89. Lange PJD, Maeder E, Mai K, Young RJ, Ahmad I. 2001. Characterization and micromechanical testing of the interphase of aramid-reinforced epoxy composites. *Comp. Part A* 32:331–42
90. Capella B, Dietler G. 1999. Force-distance surves by atomic force microscopy. *Surf. Sci. Rep.* 34:1–104
91. Noy A, Vezenov DV, Lieber CM. 1997. Chemical force microscopy. *Annu. Rev. Mater. Sci.* 27:381–421
92. Rotsch C, Radmacher M. 1997. Mapping local electrostatic forces with the atomic force microscope. *Langmuir* 13:2825
93. Aoki T, Hiroshima M, Kitamura K, Togunaga M, Yanagida T. 1997. Non-contact scanning probe microscopy with sub-piconewton force sensitivity. *Ultramicroscopy* 70:45–55
94. Weisenhorn AL, Hansma PK, Albrecht TR, Quate CF. 1989. Forces in atomic force microscopy in air and water. *Appl. Phys. Lett.* 54:2651–53
95. Butt HJ, Janschke M, Ducker W. 1995. Measuring surface forces in aqueous electrolyte solution with the atomic force microscope. *Bioelectrochem. Bioenerg.* 38:191–201
96. Hüttl G, Beyer D, Müller E. 1997. Investigation of electrical double layers on SiO_2 surfaces by means of force versus

distance measurements. *Surf. Interf. Anal.* 25:543–47
97. Hüttl G, Klemm V, Popp R, Simon F, Müller E. 2002. Tailored colloidal AFM probes and their TEM investigation. *Surf. Interf. Anal.* 33:50–53
98. Ducker WA, Senden TJ, Pashley RM. 1991. Direct measurement of colloidal forces using an atomic force microscope. *Nature* 353:239–41
99. Ducker WA, Senden TJ, Pashley RM. 1992. Measurement of forces in liquids using a force microscope. *Langmuir* 8:1831–36
100. Kappl M, Butt HJ. 2002. The colloidal probe technique and its application to adhesion force measurements. *Part. Syst. Charact.* 19:129–43
101. Cho JM, Sigmund WM. 2002. Direct surface force measurement in water using a nanosize colloidal probe technique. *J. Coll. Int. Sci.* 245:405–7
102. Florin EL, Moy VT, Gaub HE. 1994. Adhesion forces between individual ligand-receptor pairs. *Science* 264:415–17
103. Yuan C, Chen A, Kolb P, Moy VT. 2000. Energy landscape of streptavidin—biotin complexes measured by atomic force microscopy. *Biochemistry* 39:10219–23
104. Hinterdorfer P, Baumgartner W, Gruber HJ, Schilcher K, Schindler H. 1996. Detection and localization of individual antibody-antigen recognition events by atomic force microscopy. *Proc. Natl. Acad. Sci. USA* 93:3477–81
105. Kienberger F, Pastushenko VP, Kada G, Gruber HJ, Riener C, et al. 2000. Static and dynamical properties of single poly(ethylene glycol) molecules investigated by force spectroscopy. *Single Molecules* 1:123–28
106. Riener CK, Stroh CM, Ebner A, Klampfl C, Gall AA, et al. 2003. Simple test system for single molecule recognition force microscopy. *Anal. Chim. Acta* 479:59–75
107. Dammer U, Hegner M, Anselmetti D, Wagner P, Dreier M, et al. 1996. Specific antigen/antibody interactions measured by force microscopy. *Biophys. J.* 70:2437–41
108. Willemsen OH, Snel MME, Kuipers L, Figdor CG, Greve J, et al. 1999. A physical approach to reduce nonspecific adhesion in molecular recognition atomic force microscopy. *Biophys. J.* 76:716–24
109. Ishino T, Hieda H, Tanaka K, Gemma N. 1994. Electrical double-layer forces measured with an atomic force microscope while electrochemically controlling surface potential of the cantilever. *Jpn. J. Appl. Phys.* 33:L1552–54
110. Yamamoto K, Tanuma Ch, Gemma N. 1995. Competition between electrostatic and capillary forces acting on a single particle. *Jpn. J. Appl. Phys.* 34:4176–80
111. Raiteri R, Preuss M, Grattarola M, Butt HJ. 1998. Preliminary results on the electrostatic double-layer force between two surfaces with high surface potentials. *Coll. Surf. A* 136:191–97
112. Raiteri R, Grattarola M, Butt HJ. 1996. Measuring electrostatic double-layer forces at high surface potentials with the atomic force microscope. *J. Phys. Chem.* 100:16700–5
113. Döppenschmidt A, Butt HJ. 1999. Measuring electrostatic double-layer forces on HOPG at high surface potentials. *Coll. Surf. A* 149:145–50
114. Barten D, Kleijn JM, Duval J, von Leeuwen HP, Lyklema J, et al. 2003. Double layer of a gold electrode probed by AFM force measurements. *Langmuir* 19:1133–39
115. Bradley RS. 1932. The cohesive force between solid surfaces and the surface energy of solids. *Philos. Mag.* 13:853
116. Derjaguin BV, Müller VM, Toporov YP. 1975. Effect of contact deformations on adhesion of particles. *J. Coll. Interf. Sci.* 53:314
117. Johnson KL, Kendall K, Roberts AD.

1971. Surface energy and contact of elastic solids. *Proc. R. Soc. London Ser. A* 324:301
118. Shinto H, Miyahara M, Higashintani K. 1999. Evaluation of interaction forces between macroparticles in simple fluids by molecular dynamics simulation. *J. Coll. Interf. Sci.* 209:79–85
119. Gotsmann B, Anczykowski B, Seidel C, Fuchs H. 1999. Determination of tip-sample interaction forces from measured dynamic force spectroscopy curves. *Appl. Surf. Sci.* 140:314–19
120. Yong CW, Smith W, Kendall K. 2003. Molecular dynamics simulations of (001) MgO surface contacts: effects of tip structures and surface matching. *Nanotechnology* 14:829–39
121. Kendall K, Yong CW, Smith W. 2004. Particle adhesion at the nanoscale. *J. Adhes.* 80:21–36
122. Stratmann M, Streckel H. 1990. On the atmospheric corrosion of metals, which are covered with thin electrolyte layers—Part 1: verification of the experimental technique. *Corrosion Sci.* 30:681–696
123. Leng A, Streckel H, Stratmann M. 1999. The delamination of polymeric coatings from Steel—Part 1: calibration of the Kelvin probe and basic delamination mechanism. *Corrosion Sci.* 41:547–78
124. Leng A, Streckel H, Stratmann M. 1999. The delamination of polymeric coatings from steel—Part 3: effect of the oxygen partial pressure on the delamination reaction and current distribution at the meal/polymer interface. *Corrosion Sci.* 4:599–620
125. Grundmeier G, Jüttner K, Stratmann M. 2000. Novel electrochemical techniques in corrosion research. In *Corrosion and Environmental Degradation*, ed. M Schütze, pp. 285–383. Weinheim: Wiley-VCH
125a. Bard AJ, Stratmann M, Frankel GS, eds. 2003. *Encyclopedia of Electrochemistry Corrosion and Oxide Films*, Vol. 4. Weinheim: Wiley-VCH
126. Grundmeier G, Simoes A. 2003. Corrosion protection by organic coatings. See Ref. 125a, pp. 500–66
127. Wapner K, Schönberger B, Stratmann M, Grundmeier G. 2005. Development of a new height-regulating scanning Kelvin probe for simultaneous measurement of surface topology and electrode potentials at buried polymer/metal interfaces. *J. Electrochem. Soc.* 152(3):E114–22
128. Frankel GS, Rohwerder M. 2003. Electrochemical techniques for corrosion. See Ref. 125a, pp. 687–723
129. Rohwerder M, Frankel GS, Leblanc P, Stratmann M. 2005. Application of scanning Kelvin probe in corrosion science. In *Methods for Corrosion Science and Engineering*, ed. F Mansfeld, P Marcus. New York: Dekker. In press
130. Wapner K, Grundmeier G. 2004. Application of the scanning Kelvin probe for the study the corrosion resistance of interfacial thin silicon organic films at adhesive/metal interfaces. *Silicon Chem.* In press (online-prepub.)
131. Wapner K, Grundmeier G. 2004. Scanning Kelvin Probe measurements of the stability of adhesive/metal interfaces in corrosive environments. *Adv. Eng. Mater.* 6:163–67
132. Nazarov AP, Thierry D. 2004. Scanning Kelvin probe study of metal/polymer interfaces. *Electrochim. Acta* 49:2955–64
133. Feser R, Stratmann M. 1991. New model of corrosion protection of organic coatings on iron. *Werkst. Korr.-Mater. Corrosion* 42:187–95
134. Stimming U, Schultze JW. 1976. Capacity of passivated iron electrodes and band-structure of passive layer. *Ber. Bunsenges Phys. Chem.* 80:1297–302
135. Stimming U, Schultze JW. 1979. Semiconductor model of the passive layer on iron electrodes and its application to electrochemical reactions. *Electrochim. Acta* 24:859–69

136. Pourbaix M. 1974. *Atlas of Electrochemical Equilibria in Aqueous Solutions*. Houston: Natl. Assoc. Corrosion Eng.
137. Leng A, Streckel H, Stratmann M. 1999. The delamination of polymeric coatings from steel—Part 1: calibration of the Kelvin probe and basic delamination mechanism. *Corrosion Sci.* 41:547
138. Leng A, Streckel H, Hofmann K, Stratmann M. 1999. The delamination of polymeric coatings from steel—Part 2: first stage of delamination, effect of type and concentration of cations on delamination, chemical analysis of the interface. *Corrosion. Sci.* 41:579–97
139. Leng A, Streckel H, Stratmann M. 1999. The delamination of polymeric coatings from steel—Part 3: effect of the oxygen partial pressure on the delamination reaction and current distribution at the meal/polymer interface. *Corrosion Sci.* 4:599–620
140. Wroblowa HS. 1992. Intermediate products of atmospheric oxygen reduction and the integrity of metal organic coating interface. *J. Electroanal. Chem.* 339:31–40
141. Hammond JS, Holubka JW, Dickie RA. 1979. Surface-analysis of interfacial chemistry in corrosion induced paint adhesion loss. *J. Coat. Technol.* 51:45–49
142. Grundmeier G, Reinartz C, Rohwerder M, Stratmann M. 1998. Corrosion properties of chemically modified metal surfaces. *Electrochim. Acta* 43:165–74
143. Leidheiser H Jr. 1983. Towards a better understanding of corrosion beneath organic coatings. *Corrosion* 39:189–201
144. Ritter JJ, Kruger J. 1981. Studies on the subcoating environment of coated iron. In *Corrosion Control by Organic Coatings*, ed. H Leidheiser Jr, p. 28. Houston: Natl. Assoc. Corrosion Eng.
145. Kendig MW, Lumsden JB, Stocker PP. 1994. In *Surface and Interface Characterization in Corrosion*, ed. S Shah. Houston: Natl. Assoc. Corrosion Eng.
146. Fürbeth W, Stratmann M. 2001. The delamination of polymeric coatings from electrogalvanised steel—a mechanistic approach. Part 1: delamination from a defect with intact zinc layer. *Corrosion Sci.* 43:207–27
147. Rohwerder M, Hornung E, Stratmann M. 2003. Microscopic aspects of electrochemical delamination: an SKPFM study. *Electrochim. Acta* 48:1235–43
148. Van Loo M, Laiderman DD, Bruhn RR. 1953. Filiform corrosion. *Corrosion* 9:277
149. Kaesche H. 1959. Untersuchungen über die Filigrankorrosion lackierter Stahlbleche. *Werkstoffe Korrosion* 11: 668
150. Slabaugh WH, Dejager W, Hoover SE, Hutchinson LL. 1972. Filiform corrosion of aluminium *J. Paint Technol.* 44:7
151. Rudolf A, Kaiser WD. 2002. Investigation of filiform corrosion of coated aluminium. *Macromol. Symp.* 187:121–27
152. Hoch GM. 1974. A review of filiform corrosion. In *Localized Corrosion*, ed. RW Staehle, BF Brown, J Kruger, A Agarwal, p. 134. Houston: Natl. Assoc. Corrosion Eng.
153. Bautista A. 1996. Filiform corrosion in polymer-coated metals. *Prog. Org. Coat.* 28:49–58
154. Hahin C. 1987. Filiform corrosion. In *Metals Handbook*, ed. LJ Korb, DL Olson. 13:104. Ohio: ASM Int. 9th ed.
155. Ruggeri RT, Beck TR. 1983. An analysis of mass transfer in filiform corrosion. *Corrosion* 39:452–65
156. LethOlsen H, Nordlien JH, Nisancioglu K. 1997. Formation of nanocrystalline surface layers by annealing and their role in filiform corrosion of aluminum sheet. *J. Electrochem. Soc.* 144:L196–97
157. Williams G, McMurray HN. 2003. The kinetics of chloride-induced filiform corrosion on aluminium alloy AA2024-T3. *J. Electrochem. Soc.* 150:B380–88

158. McMurray HN, Williams G. 2004. Inhibition of filiform corrosion on organic-coated aluminium alloy by hydrotalcite-like anion-exchange pigments. *Corrosion* 60:219–28
159. McMurray HN, Williams G, O'Driscoll S. 2004. Chromate inhibition of filiform corrosion on organic coated AA2024-T3 studied using the scanning Kelvin probe. *J. Electrochem. Soc.* 151:B406–14
160. LeBozec N, Persson D, Thierry D, Axelsen SB. 2004. Effect of climatic parameters on filiform corrosion of coeated aluminium alloys. *Corrosion* 60:584–93
161. Lunder O, Heen KF, Nisancioglu K. 2004. Pretreatment of aluminium alloy 6060 by selective removal of surface intermetallics. *Corrosion* 60:622–31
162. LeBozec N, Persson D, Nazarov A, Thierry D. 2002. Investigation of filiform corrosion on coated aluminium alloys by FTIR microscopy and scanning Kelvin probe. *J. Electrochem. Soc.* 149:B403–8
163. Lunder O, Lapique F, Johnson B, Nisancioglu K. 2004. Effect of pre-treatment on the durability of epoxy-bonded AA6060 aluminium joints. *Int. J. Adhes. Adhes.* 24:107–17
164. Leblanc PP, Frankel GS. 2004. Investigation of filiform corrosion of epoxy-coated 1045 carbon steel by scanning Kelvin probe force microscopy. *J. Electrochem. Soc.* 151:B105–13

Subject Index

E

G

I

M

U

Z

Cumulative Indexes

CONTRIBUTING AUTHORS, VOLUMES 31–35

CHAPTER TITLES, VOLUMES 31–35

Bio and Nanobiomaterials

Catalysis

Computational Materials Research

Experimental and Theoretical Methods

Materials for Fuel Cells

Preparation, Processing, and Structural Change

Properties and Phenomena

Special Materials

Structure